New Look to Phytomedicine

New Look to Phytomedicine

Advancements in Herbal Products as Novel Drug Leads

Edited by

MOHD SAJJAD AHMAD KHAN
Department of Basic Sciences, College of Medicine, Imam Abdulrahman Bin Faisal University, Dammam, Kingdom of Saudi Arabia

IQBAL AHMAD
Department of Agricultural Microbiology, Aligarh Muslim University, Aligarh, India

DEBPRASAD CHATTOPADHYAY
Scientist G & Director, ICMR-National Institute of Traditional Medicine, Nehru Nagar, Belagavi, Karnataka, India

ELSEVIER

ACADEMIC PRESS
An imprint of Elsevier

Academic Press is an imprint of Elsevier
125 London Wall, London EC2Y 5AS, United Kingdom
525 B Street, Suite 1650, San Diego, CA 92101, United States
50 Hampshire Street, 5th Floor, Cambridge, MA 02139, United States
The Boulevard, Langford Lane, Kidlington, Oxford OX5 1GB, United Kingdom

British Library Cataloguing-in-Publication Data
A catalogue record for this book is available from the British Library

Library of Congress Cataloging-in-Publication Data
A catalog record for this book is available from the Library of Congress

ISBN: 978-0-12-814619-4

For Information on all Academic Press publications
visit our website at https://www.elsevier.com/books-and-journals

Publisher: Mica Haley
Acquisition Editor: Erin-Hill Parks
Editorial Project Manager: Megan Ashdown
Production Project Manager: Poulouse Joseph
Cover Designer: Matthew Limbert

Typeset by MPS Limited, Chennai, India

CONTENTS

LIST OF CONTRIBUTORS

Ashish K. Agrawal
James Graham Brown Cancer Center, University of Louisville, Louisville, KY, United States

Iqbal Ahmad
Department of Agricultural Microbiology, Aligarh Muslim University, Aligarh, Uttar Pradesh, India

Mohd Sajjad Ahmad Khan
Department of Basic Sciences, Biology Unit, Health Track, Imam Abdulrahman Bin Faisal University, Dammam, Kingdom of Saudi Arabia

Malika Ait-Sidi-Brahim
Laboratory of Biotechnology, Protection and Valorization of Plant Resources; Phytochemistry and Pharmacology of Medicinal Plants Unit, (URAC35 Association Unit) Faculty of Sciences Semlalia, Cadi Ayyad University, Marrakech, Morocco

Saleh Bakheet Al-Ghamdi
Biology Department, Faculty of Science, Al Baha University, Al Baha, Kingdom of Saudi Arabia

Nasser A. Al-Shabib
Department of Food Science and Nutrition, College of Food and Agriculture, King Saud University, Riyadh, Kingdom of Saudi Arabia

Mohd Musheer Altaf
Department of Agricultural Microbiology, Aligarh Muslim University, Aligarh, Uttar Pradesh, India; Faculty of Life Science, Institute of Information Management and Technology, Aligarh, India

Abdullah Safar Althubiani
Department of Biology, Faculty of Applied Science, Umm Al-Qura University, Mecca, Kingdom of Saudi Arabia

Firoz Ahmad Ansari
Department of Agricultural Microbiology, Aligarh Muslim University, Aligarh, Uttar Pradesh, India

Farrukh Aqil
Department of Medicine, University of Louisville, Louisville, KY, United States; James Graham Brown Cancer Center, University of Louisville, Louisville, KY, United States

Jamal M. Arif
Department of Biosciences, Integral University, Lucknow, Uttar Pradesh, India

Anwesha Banerjee
ICMR-Virus Unit Kolkata, ID & BG Hospital Campus, Kolkatta, West Bengal, India

Rathindranath Baral
Department of Immunoregulation and Immunodiagnostics, Chittaranjan National Cancer Institute (CNCI), Kolkata, West Bengal, India

Ushasi Bhaumik
Division of Bacteriology, National Institute of Cholera and Enteric Diseases, Kolkata, West Bengal, India

Priyanka Bhowmik
ICMR-Virus Unit Kolkata, ID & BG Hospital Campus, Kolkatta, West Bengal, India

Anamika Bose
Department of Immunoregulation and Immunodiagnostics, Chittaranjan National Cancer Institute (CNCI), Kolkata, West Bengal, India

Angela I. Calderon
Department of Drug Discovery and Development, Harrison School of Pharmacy, Auburn University, Auburn, AL, United States

Swaranjit Singh Cameotra
1103, Sector 11-C, Chandigarh, Punjab, India

Debprasad Chattopadhyay
ICMR-National Institute of Traditional Medicine, Belagavi, Karnataka, India; ICMR-Virus Unit Kolkata, ID & BG Hospital Campus, Kolkata, West Bengal, India

Puneet Singh Chauhan
Division of Plant Microbe Interactions, CSIR-National Botanical Research Institute, Lucknow, Uttar Pradesh, India

Apurba Das
ICMR-Virus Unit Kolkata, ID & BG Hospital Campus, Kolkatta, West Bengal, India

Jyoti Das
Immunology Division, National Institute of Malaria Research, New Delhi, India

Anupam Dhasmana
Himalayan School of Biosciences and Cancer Research Institute, Swami Rama Himalayan University, Dehradun, Uttarakhand, India

Deepak Dwivedi
Department of Microbiology, Barkatullah University, Bhopal, Madhya Pradesh, India

S. Farooq
The Himalaya Drug Company, Dehradun, Uttarakhand, India

Govind Gupta
Department of Microbiology, Barkatullah University, Bhopal, Madhya Pradesh, India

Ramesh Gupta
James Graham Brown Cancer Center, University of Louisville, Louisville, KY, United States; Department of Pharmacology and Toxicology, University of Louisville, Louisville, KY, United States

Debaki Ranjan Howlader
Division of Bacteriology, National Institute of Cholera and Enteric Diseases, Kolkata, West Bengal, India

Fohad Mabood Husain
Department of Food Science and Nutrition, College of Food and Agriculture, King Saud University, Riyadh, Kingdom of Saudi Arabia

Huma Jafri
Department of Agricultural Microbiology, Aligarh Muslim University, Aligarh, Uttar Pradesh, India

Roshan Jahan
Department of Biosciences, Integral University, Lucknow, Uttar Pradesh, India

Meenu Kalkal
Immunology Division, National Institute of Malaria Research, New Delhi, India

Altaf Khan
Central Laboratory Research Center, College of Pharmacy, King Saud University, Riyadh, Kingdom of Saudi Arabia

Mohammad Shavez Khan
Department of Agricultural Microbiology, Aligarh Muslim University, Aligarh, Uttar Pradesh, India

Mohd Shahnawaz Khan
Department of Biochemistry, College of Science, King Saud University, Riyadh, Kingdom of Saudi Arabia

Rais Ahmad Khan
Department of Chemistry, College of Science, King Saud University, Riyadh, Kingdom of Saudi Arabia

V.K. Khanna
System Toxicology and Health Risk Assessment Group, CSIR-Indian Institute of Toxicology Research (CSIR-IITR), Lucknow, Uttar Pradesh, India

Hemanta Koley
Division of Bacteriology, National Institute of Cholera and Enteric Diseases, Kolkata, West Bengal, India

Mustapha Larhsini
Laboratory of Biotechnology, Protection and Valorization of Plant Resources; Phytochemistry and Pharmacology of Medicinal Plants Unit, (URAC35 Association Unit) Faculty of Sciences Semlalia, Cadi Ayyad University, Marrakech, Morocco

Temitope O. Lawal
Department of Pharmaceutical Microbiology, University of Ibadan, Ibadan, Nigeria; Schlumberger Faculty for the Future Fellow, College of Pharmacy, WHO Collaborating Centre for Traditional Medicine, Department of Pharmacy Practice, University of Illinois at Chicago, Chicago, IL, United States

Mohtashim Lohani
College of Applied Medical Sciences, Jazan University, Jazan, Kingdom of Saudi Arabia

Gail B. Mahady
College of Pharmacy, WHO Collaborating Centre for Traditional Medicine, Department of Pharmacy Practice, University of Illinois at Chicago, Chicago, IL, United States

Ananya Das Mahapatra
ICMR-Virus Unit Kolkata, ID & BG Hospital Campus, Kolkatta, West Bengal, India

Hesham A. Malak
Department of Biology, Faculty of Applied Science, Umm Al-Qura University, Mecca, Kingdom of Saudi Arabia

Mohammed Markouk
Laboratory of Biotechnology, Protection and Valorization of Plant Resources; Phytochemistry and Pharmacology of Medicinal Plants Unit, (URAC35 Association Unit) Faculty of Sciences Semlalia, Cadi Ayyad University, Marrakech, Morocco

Zafar Mehmood
Department of Microbiology, The Himalaya Drug Company, Dehradun, Uttarakhand, India

Sankalp Misra
Division of Plant Microbe Interactions, CSIR-National Botanical Research Institute, Lucknow, Uttar Pradesh, India

Debashis Mitra
National Centre for Cell Science, Pune University Campus, Pune, Maharashtra, India

Radha Munagala
Department of Medicine, University of Louisville, Louisville, KY, United States; James Graham Brown Cancer Center, University of Louisville, Louisville, KY, United States

Arathi Nair
National Centre for Cell Science, Pune, Maharashtra, India

Saba Noor
Rajiv Gandhi Centre for Diabetes and Endocrinology, Jawaharlal Nehru Medical College and Hospital, Aligarh Muslim University, Aligarh, Uttar Pradesh, India

Durbadal Ojha
ICMR-Virus Unit Kolkata, ID & BG Hospital Campus, Kolkatta, West Bengal, India

A. Pandey
System Toxicology and Health Risk Assessment Group, CSIR-Indian Institute of Toxicology Research (CSIR-IITR), Lucknow, Uttar Pradesh, India

A.B. Pant
System Toxicology and Health Risk Assessment Group, CSIR-Indian Institute of Toxicology Research (CSIR-IITR), Lucknow, Uttar Pradesh, India

Faizan Abul Qais
Department of Agricultural Microbiology, Aligarh Muslim University, Aligarh, Uttar Pradesh, India

Sana Raza
Department of Biosciences, Integral University, Lucknow, Uttar Pradesh, India

Bhaskar Saha
National Centre for Cell Science, Pune, Maharashtra, India; ICMR-National Institute of Traditional Medicine, Belagavi, Karnataka, India

Mohammad Sajid
Cell Biology and Immunology Laboratory, Institute of Microbial Technology, Chandigarh, Punjab, India

Samreen
Department of Agricultural Microbiology, Aligarh Muslim University, Aligarh, Uttar Pradesh, India

Sudhanshu Sharma
Pharmacognosy & Ethnopharmacology Division, CSIR-National Botanical Research Institute, Lucknow, Uttar Pradesh, India

Vivek Kumar Sharma
Pharmacognosy & Ethnopharmacology Division, CSIR-National Botanical Research Institute, Lucknow, Uttar Pradesh, India

Brahma N. Singh
Pharmacognosy & Ethnopharmacology Division, CSIR-National Botanical Research Institute, Lucknow, Uttar Pradesh, India

A. Srivastava
System Toxicology and Health Risk Assessment Group, CSIR-Indian Institute of Toxicology Research (CSIR-IITR), Lucknow, Uttar Pradesh, India

P. Srivastava
System Toxicology and Health Risk Assessment Group, CSIR-Indian Institute of Toxicology Research (CSIR-IITR), Lucknow, Uttar Pradesh, India

Anjali Tripathi
National Centre for Cell Science, Pune University Campus, Pune, Maharashtra, India

Parul Tripathi
Department of Microbiology, Government Medical College, Ambedkar Nagar, Uttar Pradesh, India

Jay Trivedi
National Centre for Cell Science, Pune University Campus, Pune, Maharashtra, India

Sheila M. Wicks
Department of Molecular and Cellular Medicine, Rush University, Chicago, IL, United States

PREFACE

Herbal products have been in wide spread use since time immemorial and ancient texts in different cultures and civilizations contain exhaustive description of their ethnomedicinal values to manage or cure various ailments. However, their traditional uses need standardization of source materials, scientific validation using modern technologies and tools, identification of active molecules or ingredient, mode and molecular mechanism of action, toxicity profile, and efficacy in ex-vivo or in-vivo model system to create scope for delivering newer bioactive compounds to support existing chemotherapy. Discovery of new bioactive compounds from herbs/medicinal plants and their development into modern drugs chemotherapy are well recognized, but had limited success in recent past years due to the lack of novelty, drug-associated toxicity, problems in managing complex and chronic diseases, and also, age- and life style-related disorders. In addition to these, the development of multidrug resistance in microbial pathogens, emergence of new diseases, and lack of vaccines are also hindering the success of current chemotherapeutics. To combat all these, researchers have developed renewed interest in herbal or phytomedicine with improved formulation, enhanced quality, safety, and consistent performance. Most of the developing and advanced countries have now accepted herbal medicine as an alternative system of therapy in the form of Ayurvedic, Unani medicine, pharmaceuticals, botanicals, and functional foods as advocated by WHO.

In the past two decades, new dimensions in research and innovation in phytomedicine have been documented globally, mainly due to the increased understanding of the evidence-based therapy, development of new standardization techniques, improved delivery of herbal drugs, maintenance of quality of herbal formulation with little or no toxicity as well as recent progresses in nanotechnology. In this context our book provides up to date information on use of plant products as successful medicinal input in various diseases. The role of alternative medicines in terms of herbal products, their safety and cost effectiveness is being highlighted as an effective remedy complementary to the existing chemotherapy. A huge amount of literature, on medicinal plants and their therapeutic properties, is available in the form of books, reviews, and research articles contributed by scientific community. However, it is difficult to cover all the areas of research, where tremendous growth in phytomedicine is observed, in one book. Yet, we have made a sincere effort to bring together recent research works and new trends in the field of phytomedicine from different parts of the world.

"New look to phytomedicine: Advancements in herbal products as novel drug leads" is a compilation of wholesome information on phytopharmaceuticals used in modern medicine for the management and cure of several difficult-to-treat and challenging diseases such as cancer, HIV, diabetes, malaria, neurological and other metabolic disorders, and various infectious diseases. Readers will gain cutting edge knowledge on the use of plant products with scientific validation, along with advanced herbal medicine with respect to pharmacokinetics and drug delivery. The previous books written in this area dealt with plant products as a remedy for various ailments, but no systematic information was provided on their scientific validation using modern therapeutic approaches to claim their use in modern medicine. This book is destined to fill this gap with organized information on pharmacokinetics of herbal drugs, supported by mechanistic approach for combating various ailments. The drug leads developed following scientific and regulatory principles can be used to manage newer and emerging challenges as well.

This book is first of its kind in the related area that provides comprehensive collection of information on research-based, validated biological activities of plant products, from experts across the globe. A step by step information on chemistry, bioactivity, and functional aspects of bioactive compounds and their physiological roles have been highlighted. There are 24 Chapters arranged into four sections. First section describes the introduction to herbal therapeutics and its uses. Second section deals with the biological activity and discovery of new compounds from herbs, medicinal plants, and herbal medicine, and also the mechanism of action and pharmacology of plant-derived products/medicines. Third section sheds light on pharmacokinetics, interaction, and toxicity profile of phyto-compounds. Lastly, fourth section emphasizes to accommodate new dimensions in phytotherapy research and applications. We have covered the most important aspects of drug discovery of newer molecules from existing or known bioactive phytocompounds using high throughput screening, chemical modifications, use of nanoparticles for enhanced drug delivery, and combinatorial biosynthesis approaches. Pharmacognosy of plant products with mechanistic description of their action including pathogenicity is being updated with the information on use of nanotechnology and molecular tools in herbal drug research. Also, the advancements of bioinformatics tools in identifying drug leads from phyto-compounds has been incorporated in this compilation, which could be of special interest to pharmaceutical industries. These salient features of our book appeared to be quiet attractive for investigators and budding scientists and will certainly help to solve the readers' most recent queries regarding the use of herbal medicines at par with currently existing drugs. Overall, the book is useful to students, teachers, and researchers working in the universities, research organizations, and pharmaceutical/herbal industries.

With great pleasure and respect, we extend our sincere thanks to all the contributors for their timely response, excellent contribution, and consistent support. We express our deep gratitude to Prof. Faizan Ahmad (FNA, Jamia Millia Islamia, New Delhi, India), Prof. S. K. Puri (ex-director, Central Drug Research Institute, Lucknow, India), Dr. S. Farooq (director, The Himalaya Drug Co., Dehradun, India), Prof. Tariq Mansoor (vice-chancellor, Aligarh Muslim University, Aligarh, India), Prof. Gerard Bodeker (University of Oxford, Oxford, UK) and Dr. Naif Al-Qurashi (head, Department of Basic Sciences, Imam Abdulrahman Bin Faisal University, Dammam, KSA) for their encouragement and support.

The technical support and continued monitoring received from the editorial and publishing team at Elsevier, especially, Megan Ashdown, Erin Hill-Parks, and Kattie Washington, is thankfully acknowledged.

Finally we are thankful to the Almighty God, who has provided us best thoughts and strength to complete this task.

Mohd Sajjad Ahmad Khan[1], Iqbal Ahmad[2] and Debprasad Chattopadhyay[3]

[1]*Dammam, Saudi Arabia*
[2]*Aligarh, India*
[3]*Kolkata, India*

SECTION 1

Introduction to Herbal Therapeutics

CHAPTER 1

Herbal Medicine: Current Trends and Future Prospects

Mohd Sajjad Ahmad Khan[1] and Iqbal Ahmad[2]
[1]Department of Basic Sciences, Biology Unit, Health Track, Imam Abdulrahman Bin Faisal University, Dammam, Kingdom of Saudi Arabia
[2]Department of Agricultural Microbiology, Aligarh Muslim University, Aligarh, Uttar Pradesh, India

1.1 INTRODUCTION

In the 21st century, with the increased efficacy in pharmacological effects of medicinal plants, herbal medicine has been considered as a promising future medicine for the management of health care. Recently, there has been a swing in universal trend from synthetic to herbal medicine, which is claimed as "Return to Nature." Medicinal plants have been exploited since ancient times and are highly esteemed all over the world as a rich source of therapeutic agents for the prevention of diseases and ailments. Ancient Chinese and Egyptian papyrus writings describe medicinal uses of plants as early as 3000 BC. Indigenous cultures like African and Native American have used herbs in their healing rituals. Whereas, other developed traditional medical systems viz. Siddha, Ayurveda, Unani, and traditional Chinese medicine (TCM); in which herbal therapies are being used successfully (Ampofo et al., 2012). The consumption of plant-based phytomedicines and other botanicals in the West has increased multifariously in recent years. About two centuries ago, our medicinal practices were largely dominated by plant-based medicines. But, the medicinal use of herbs went into a rapid decline in the West since the introduction of more predictable synthetic drugs with their fast effects and easy availability. In contrast, many developing nations continued to benefit from the rich knowledge of herbal medicine. Siddha and Ayurveda medicines in India, Kampo Medicine in Japan, TCM, and Unani medicine in the Middle East and South Asia are still being used by a large majority of people (Mosihuzzaman and Choudhary, 2008). Overall, now a days, the demand for plant-based medicines, health products, food supplements, and cosmetics is being amassed in both developing and developed countries. The reason behind it is the growing recognition that the natural

New Look to Phytomedicine
DOI: https://doi.org/10.1016/B978-0-12-814619-4.00001-X

products are nontoxic, have less side effects, and are easily available at affordable prices (Evans, 1994).

Medicinal plants are considered as rich sources of phytochemical ingredients which play a vital role for the development of new drugs. People have been using plants as a medicine without scientific knowledge and proper guidance for thousand years ago. It has been scientifically established that every part of plants have medicinal properties including roots, stems, leafs, flowers, fruits, and seeds. However, it has also been witnessed that some plants are not safe for consumption as being toxic and show adverse effects in the body (Wink, 2010). Therefore, to develop drug from the phytocompounds, the bioactive extract should be standardized on the basis of active compound and should also undergo limited safety studies. In recent years, there has been a resurgence of interest to rediscover medicinal plants as a source of potential drug candidate. Therefore, the aim of this review is to understand the knowledge and current status of the medicinal plants and turning it to as a future source of herbal drugs.

1.2 HERBAL MEDICINE: DEFINITION AND ITS PROSPECTS

Traditional medicine refers to health practices and approaches which are based on knowledge and beliefs incorporating plants as medicines, spiritual therapies, and physical therapy; either applied singularly or in combination to treat, diagnose, and prevent illnesses or maintain well-being. In developed countries, adaptations of traditional medicine are termed complementary and alternative medicine (Gunjan et al., 2015). Whereas, herbal medicine or phytomedicine is the use of merely plants for medicinal and therapeutic purpose for curing of diseases and improve human health. World Health Organization (WHO) has defined herbal medicines as finished labeled medicinal product that contain an active ingredient, aerial, or underground parts of the plant or other plant material or combinations (WHO, 2008; Parveen et al., 2015). At pharmacodynamics scale herbal medicines are classified as (1) herbal drugs with proven efficacies with known active compounds and doses, (2) herbal drugs with expected efficacies and active compound need to be standardized, and (3) herbal drugs with uncertain efficacies but documented history of its traditional use (Parveen et al., 2015). Plants being used as food or raw material in traditional medicine are more likely to yield pharmacologically active compounds. Plants are also rich dietary sources of biomolecules, vitamins, and minerals which are crucial for maintaining the healthy body (Shakya, 2016).

Herbal medicine is widely practiced for centuries, and people have turned to natural remedies to cure common ailments such as colds, allergy, upset stomachs, and toothaches; and the inclination toward is continuously growing. However, herbal products were discarded from conventional medical use in the mid-20th century. This was not necessarily because they were ineffective but also because they were not as economically profitable as the newer synthetic drugs (Tyler, 1999). Later on, with the advancements of scientific methods, the herbal medicines could find place with research and documented for effective use as drugs. Furthermore, in the 1960s, with concerns over the toxic and iatrogenic effects of conventional medicine resulted in desire for more safe and economically cheaper drugs to promote "natural health." Thus, afterwards, there has been a shift in universal trend from synthetic to herbal medicines. Moreover, herbal medicine received a worldwide boost when the WHO exhilarated developing countries to use traditional plant medicine to accomplish needs unmet by modern systems (Miller, 1998). WHO has reported that 4 billion people (80% of the world's population) use herbal medicines for one or other aspect of primary health care (Fabricant and Farnsworth, 2001).

Indeed, the pharmacological effects of plants are indebted into the presence of metabolites, which are organic compounds and classified into primary and secondary metabolites. Primary metabolites such as glucose, starch, polysaccharide, protein, lipids, and nucleic acids are beneficial for growth and development of the human body. Whereas, plants produce secondary metabolites including alkaloids, flavonoids, saponins, terpenoids, steroids, glycosides, tannins, volatile oils, etc. to protect plants against microbial infections or invasions by pests. The therapeutic efficacy of plants is because of these secondary metabolites and these are actually termed as "phytocompounds." Which are pharmacologically active ingredients and are exploited as drugs because of their therapeutic properties (Martinez et al., 2008). The use of such compounds has reduced the risk of many human diseases including cardiovascular diseases, hepatorenal diseases, diabetes, cancers, and neurodegenerative disorders. Additionally, plants are bestowed with several other pharmacological characters such as antioxidant, antiviral, antimicrobial, and antiparasitic for human use. Especially, alkaloids reported to possess an antispasmodic, antimalarial, analgesic, and diuretic activities; terpenoids are known for their antiviral, anthelmintic, antibacterial, anticancer, antimalarial, and antiinflammatory properties; glycosides are reported for antifungal and antibacterial properties; phenols and flavonoids have an antioxidant, antiallergic, and antibacterial properties; and saponins have shown antiinflammatory and antiviral activities (Chopra and Doiphode, 2002; Maurya et al., 2008).

1.3 CURRENT STATUS OF HERBAL MEDICINE: SOURCE OF MODERN MEDICINE FROM HIGHER PLANTS

Medicinal plants play a vital role for the development of new drugs. According to WHO, nearly 25% of the modern medicines have been derived from plants being used in traditional medicine. Many others are synthetic analogs fabricated on model compounds isolated from plants. And now WHO has recognized herbal medicine as a crucial components for primary health care (Leslie, 2000). Plant-based drugs have contributed revolutionarily to modern therapeutics. Like, vinblastine from the *Catharanthus rosesus* is successfully used in treating Hodgkins, choriocarcinoma, non-Hodgkin's lymphomas, leukemia in children, testicular, and neck cancer (Farnsworth and Bingel, 1977). Phophyllotoxin, isolated from *Phodophyllum emodi*, is efficaciously used against testicular, lung cancer, and lymphomas. Taxol isolated from *Taxus brevifolius* is used for the treatment of metastatic ovarian cancer and lung cancer. Moreover, in 1953, a compound named serpentine isolated from the root of *Rauwolfia serpentina* is an noteworthy discovery in the treatment of hypertension and reducing the blood pressure (Hasan et al., 2009). It has been reported that during 1950–70 about 100 new drugs based on plants were introduced in the US pharmaceutical industry including deserpidine, reseinnamine, reserpine, vinblastine, and vincristine. From 1971 to 1990 new drugs isolated from plants such as ectoposide, eguggulsterone, teniposide, nabilone, plaunotol, Z-guggulsterone, lectinan, artemisinin, and ginkgolides. From 1991 to 1995, some more drugs of plant origin including paciltaxel, toptecan, gomishin, and irinotecan find their place in pharmaceutical industries (Vickers and Zollman, 1999). Moreover, many researchers in recent decades have recognized several other chemical compounds derived from plant sources including quinine, digoxin, aspirin, ephedrine, atropine, and colchicine (Moteriya et al., 2015; Ram et al., 2015).

1.3.1 Trends in Herbal Medicine Use: Increasing Use and Popularity

Currently herbal medicine has continued its popularity in majority of the developing countries and its use is speedily disseminating in industrialized countries as well. It has been estimated that 70% of all medical doctors in France and Germany are regularly prescribing herbal medicine. Also, the number of patients seeking herbal therapies is growing exponentially (Cragg et al., 1997). Worldwide it is expected that 80% of the population uses herbs, and in the developing countries rates could be as high as 95% (Tilburt and Kaptchuk, 2008). In China, the use of traditional herbal medicine amounts about 30%–50% of the total drug consumption. In Ghana, Mali, Nigeria, and Zambia, the herbal medicines are accounting for 60% of first line of treatment at home. It is estimated that in Europe, North America, and other developed countries, more than 50% of the population have used herbal medicinal approaches at least once in life (Gunjan et al., 2015). In San Francisco, London, and South Africa, 75% of HIV/AIDS–affected patients use herbal formulations. About 70%–90% of the population in Canada and Germany have used herbal medicines at least once in their life. In the United States, it is believed that 158 million of the adult population use herbal medicines and its use is continually being increased. In over all, the global market for herbal medicines currently stands at over US$60 billion annually and it is growing progressively (Robinson and Zhang 2011; Gunjan et al., 2015). Moreover, it is interesting to note that the adult populations are more likely to use equally both conventional and herbal medicines. Since this population has a higher incidence of chronic diseases, which more often discourage long-term use of complex conventional drug therapies due to their long-lasting side effects. On the other hand herbal medicines provide a therapy with no side effects upon long-term use. With such specifications, herbal medicine has gained interest and enjoying the worldwide faster acceptance.

1.4 FUTURE PROSPECTS OF HERBAL MEDICINE

Since decades the practitioners of traditional herbal medicine have been verbally passing on instructions as how to prepare medicine from herbs. They usually don't keep records, but now WHO has publicized in documenting the use of medicinal plants by traditional practitioners across the world. Therefore efforts have been increased in many developing countries to document ethnomedical data on herbs. It has made easier to scientifically validate their pharmacological values. Once these local ethnomedical preparations are scientifically evaluated and disseminated properly, people will be better aware and satisfied regarding efficacious drug treatment and improved health status (Kamboj, 2000; Modak et al., 2007; Gunjan et al., 2015).

Plants remain a potential source of therapeutic agents, and also serving as raw material base for the extraction of semisynthetic chemical compounds such as cosmetics, perfumes, and food industries (Modak et al., 2007; Shakya and Shukla, 2011). Moreover, the popularity of plant-derived health care products has got increasing acceptance and use in the cosmetic industry as well. Therefore, in the twofold role as a source of health care and also commercial income, medicinal plants are making an important contribution to the larger economy development process. The demand is projected to raise in the years to come in the form of sales of herbal supplements and remedies, and supplying this need by herbals will be a flourishing business (Barnes and Bloom, 2007; Gunjan et al., 2015; Kalia, 2017). This means that researchers, doctors, and pharmaceutical industries will be looking at countries like China, India, and other developing countries for their supplies. Because these countries have the most diverse number of medicinal plant species and are the top exporters of herbal raw material. With the increasing popularity as being safe and a cheaper alternative of conventional therapeutic agents, the exploitation of plants as a whole or in the form of drugs will continue in the future.

1.4.1 Development of Herbal Drugs and Its Challenges

Undoubtedly the demand for plant-derived products has increased worldwide. Keeping this in view, the efficacy of herbal drugs requires development of quality awareness. In this perspectives, main concern is limited to clinical trials to determine efficacy and safety of traditional herbal medicines. However, this shortcoming of lack of research does not obstruct most of the people from using them. As, these remedies are often have associated beliefs from long standing cultural traditions (Heinrich, 2000; Shakya et al., 2012). Moreover, when trials are conducted, due to the regulations and classifications defined by modern medical system which are suitable for conventional chemotherapeutic agents; it is not applicable to phytodrugs (Murray and Pizzorno, 2000; Rivera et al., 2007).

Therefore it becomes inappropriate to measure efficacy and safety of phytocompounds in relation to the use of modern drugs. Now, WHO has also issued operational guidelines regarding regulatory requirements needed to support clinical trials of herbal products (WHO, 2008).

There are many scientific questions that reveal the difficulties of conducting research with herbal medicines worldwide. A key challenge is to analyze and document toxicological, epidemiological, and other pharmacognosy-based data and the verification of herbal materials used. Additional challenges may include evaluation of drug interactions, constrains with clinical trials and availability of people, design of the study, and standardization. It is proclaimed that herbal medicine does not require clinical trials as it is endorsed and surviving very well at large scale at the international market alongside modern medicines. But it has become need of the hour to overcome this issue (Mills, 2003; Parveen et al., 2015).

Standardization of herbal medicines is often a very challenging due to the presence of complex and diverse secondary metabolites. Additionally, the therapeutic actions depend fundamentally on age, geographical location, and parts of the plant species used (Firenzuoli and Gori, 2007). The variability in phytochemical constituents in herbal products from the same plant species leads to intense differences in pharmacological activity. Also, the timing of harvesting process and incidents of adulterations with microorganisms affects in attaining the absolute standards of herbal medicines globally (Fong, 2002). Sometimes finding appropriate ways to address this type of challenge and conducting research is a difficult task. But high-throughput screening of phytocompounds analogs and modern biotechnological approaches is enabling herbal medicines to be assessed and recommended after clinical trials validation for human use. Furthermore, the quality of herbal medicines could be improved by implementation of good agricultural practices at the point of cultivation of medicinal plants, and also by adopting good manufacturing practices throughout the course of manufacturing and packaging of herbal products. Nevertheless, the postmarketing quality assurance should be under constant observation. Following the current improvements in quality control and regulatory measures in many countries of the world, it is envisaged that in the near future, herbal medicinal practices will be integrated into the conventional medicines (Sane, 2002; Chikezie and Ojiako, 2015).

1.4.2 Research Efforts on Herbal Medicine

As evidenced by enormous publications of scientific research papers, there is an increased interest among pharmacologist, microbiologist, biochemist, botanist, and natural product chemists, to explore medicinal plants for newer phytochemicals leading to discovery of drugs for the treatment of several ailments (Fokunang et al., 2011; Acharya and Shrivastava, 2008). Most of these research workouts cover the areas of isolation, purification, bio-analytical methodology, and characterization of the bioactive principles of phytocompounds. Furthermore, research efforts in herbal medicine are aiming to elucidate their molecular structures, and establishing their mechanism of action and probable toxicological properties (Chikezie and Ojiako, 2015). Because of the evidenced-based research conducted on herbal medicine over the years, the 21st century is witnessing a paradigm shift toward therapeutic standardization of herbal drugs. Their efficacies have been supported and confirmed through many in vivo clinical trials (Alvari et al., 2012). The safety apprehensions of consuming certain herbal medicine has also been assessed and recognized using in vitro and in vivo systems (Haq, 2004).

In this regard, it has been witnessed in last decade that phytochemical and pharmacological research activities on medicinal plants are dynamically being carried out in research institutes and universities. The scientists are making huge efforts to isolate and identify bioactive chemical constituents and to corroborate the claims of their efficacy and safety (Trusheim et al., 2007; Parekh et al., 2009). Consequently, many scientific proofs from randomized clinical trials have provided favorable outcomes toward the use of most of the herbal preparations (Bubela et al., 2008). Furthermore, the Omic techniques have helped in understanding of mechanism of action of herbal bioactive principles, which has flagged the way for the modernization and standardization of several herbal medicines (Buriani et al., 2012). It is worthwhile to note that novel approaches and current insights into herbal medicine research have made a great impact on herbal remedies to compete enough in the mainstream biomedical science.

1.5 CONCLUSION

For centuries almost every known human civilization has been using herbal medicines as effective medications for the prevention and treatment of multiple health conditions like eczema, wounds, skin infections, swelling, aging, mental illness, cancer, asthma, diabetes, jaundice, scabies, venereal diseases, snakebite, gastric ulcers, and many more. This is primarily because of the general belief that herbal drugs are without any side effects, cheap, and locally available. As a result of increasing demand for it, there are also increasing concerns about the safety, standardization, efficacy, quality, availability, and commercialization of herbal products by policy-makers, health professionals, as well as the general public. As in many countries, including the United States, herbal medicines are not regulated as extensively as conventional drug therapy. Also, globalization has greatly increased accessibility of herbal medicines from all parts of the world to any of the consumer and anywhere. Obviously, there is a urging need for coordinated efforts to conduct the necessary clinical trials to study the efficacy and safety of herbal medicines. A large number of researchers are investigating herbal medicines with reference to their therapeutic uses as documented in old books from Ayurveda, Unani, TCM, and others. Recent introduction of cutting-edge analytical techniques and research methodologies has resulted in evaluation and validation of herbal medicinal use as per standard of modern drug. Moreover, due to incorporation of improved quality control and regulatory measures, it is foreseen that in the near future, herbal medicine will be integrated into conventional medical systems, as the "herbal age is about to come."

ACKNOWLEDGMENT

We acknowledge the department of Scientific Research, Imam Abdulrahman Bin Faisal University, for financially supporting to complete this work.

REFERENCES

Acharya, D., Shrivastava, A., 2008. Indigenous Herbal Medicines: Tribal Formulations and Traditional Herbal Practices, 1st ed. Aavishkar Publishers, Jaipur, India, 440 pp.

Alvari, A., Mehrnaz, S.O., Ahmad, F.J., Abdin, M.Z., 2012. Contemporary overview on clinical trials and future prospects of hepato-protective herbal medicines. Rev. Recent Clin. Trials 7 (3), 214–223.

Ampofo, A.J., Andoh, A., Tetteh, W., Bello, M., 2012. Microbiological profile of some ghanaian herbal preparations-safety issues and implications for the health professions. Open J. Med. Microbiol. 2 (3), 121–130.

Barnes, P.M., Bloom, B., 2007. Complementary and alternative medicine use among adults and children: United States. Natl Health Stat. Rep. 12, 1–24.

Bubela, T., Boon, H., Caulfield, T., 2008. Herbal remedy clinical trials in the media: a comparison with the coverage of conventional pharmaceuticals. BMC Med. 6, 35.

Buriani, A., Garcia-Bermejo, M.L., Bosisio, E., Xu, Q., Li, H., Dong, X., et al., 2012. Omic techniques in systems biology approaches to traditional chinese medicine research: present and future. J. Ethnopharmacol. 140 (3), 535–544.
Chikezie, P.C., Ojiako, O.A., 2015. Herbal medicine: yesterday, today and tomorrow. Altern. Integr. Med. 4, 195.
Chopra, A., Doiphode, V., 2002. Ayurvedic medicine: core concept, therapeutic principles and current relevance. Med. Clin. North Am. 86 (1), 75–89.
Cragg, G.M., Newmann, D.J., Snader, K.M., 1997. Natural product in drug discovery and development. J. Nat. Prod. 60 (1), 52–60.
Evans, M., 1994. A Guide to Herbal Remedies. Orient Paperbacks, New Delhi, India.
Fabricant, D.S., Farnsworth, N.R., 2001. The value of plants used in traditional medicine for drug discovery. Environ. Health Pers. 109 (Suppl. 1), 69–75.
Farnsworth, N.R., Bingel, A.S., 1977. Problems and prospects of discovery new drugs from higher plants by pharmacological screening. In: Wagner, H., Wolff, P. (Eds.), New Natural Products and Plant Drugs with Pharmacological, Biological and Therapeutical Activity. Springer Verlag, Berlin, pp. 1–22.
Firenzuoli, F., Gori, L., 2007. Herbal medicine today: clinical and research issues. Evid. Based Complement Altern. Med. 4 (Suppl. 1), 37–40.
Fokunang, C.N., Ndikum, V., Tabi, O.Y., Jiofack, R.B., Ngameni, B., Guedje, N.M., et al., 2011. Traditional medicine: past, present and future research and development prospects and integration in the national health system of Cameroon. Afr. J. Tradit. Complement Altern. Med. 8 (3), 284–295.
Fong, H.H., 2002. Integration of herbal medicine into modern medical practices: issues and prospects. Integr. Cancer Ther. 1 (3), 287–293.
Gunjan, M., Naing, T.W., Saini, R.S., Ahmad, A., Naidu, J.R., Kumar, I., 2015. Marketing trends and future prospects of herbal medicine in the treatment of various disease. World J. Pharm. Res. 4 (9), 132–155.
Haq, J., 2004. Safety of medicinal plants. Pak. J. Med. Res. 43 (4), 1–8.
Hasan, S.Z., Misra, V., Singh, S., Arora, G., Sharma, S., Sharma, S., 2009. Current status of herbal drugs and their future perspectives. Biol. Forum Int. J. 1 (1), 12–17.
Heinrich, M., 2000. Ethnobotany and its role in drug development. Phytother. Res. 14 (7), 479–488.
Kalia, A.N., 2017. Text Book of Industrial Pharmacognosy. Oscar Publication, New Delhi, India.
Kamboj, V.P., 2000. Herbal medicine. Curr. Sci. 78, 35–39.
Leslie, T.N.D., 2000. Plant Based Drugs and Medicines. Raintree Nutrition.
Martinez, M.J.A., Lazaro, R.M., del Olmo, L.M.B., Benito, P.B., 2008. Anti-infectious activity in the anthemideae tribe. Stud. Nat. Prod. Chem. 35, 45–516.
Maurya, R., Singh, G., Yadav, P.P., 2008. Antiosteoporotic agents from natural sources. Stud. Nat. Prod. Chem. 35, 517–545.
Miller, L.G., 1998. Herbal medicinals: selected clinical considerations focusing on known or potential drug-herb interactions. Arch. Intern. Med. 158 (20), 2200–2211.
Mills, S., 2003. Herbal medicine. In: Lewith, G.T., Jonas, W.B., Walach, H. (Eds.), Clinical Research in Complementary Therapies: Principles, Problems and Solutions. Elsevier Science: Churchill Livingstone, London, UK, pp. 211–227.
Modak, M., Dixit, P., Londhe, J., Ghaskadbi, S., Paul, T., Devasagayam, A., 2007. Indian herbs and herbal drugs used for the treatment of diabetes. J. Clin. Biochem. Nut. 40 (3), 163–173.
Mosihuzzaman, M., Choudhary, M.I., 2008. Protocols on safety, efficacy, standardization, and documentation of herbal medicine. Pure Appl. Chem. 80 (10), 2195–2230.

Moteriya, P., Satasiya, R., Chanda, S., 2015. Screening of phytochemical constituents in some ornamental flowers of Saurashtra region. J. Pharmacog. Phytochem. 3 (5), 112–120.

Murray, M.T., Pizzorno Jr., J.E., 2000. Botanical medicine – a modern perspective. In: Pizzorno Jr., J.E., Murray, M.T. (Eds.), Text Book of Natural Medicine, Vol. 1. Churchill Livingstone, London, UK, pp. 267–279.

Parekh, H.S., Liu, G., Wei, M.Q., 2009. A new dawn for the use of traditional Chinese medicine in cancer therapy. Mol. Cancer 8, 21.

Parveen, A., Parveen, B., Parveen, R., Ahmad, S., 2015. Challenges and guidelines for clinical trial of herbal drugs. J. Pharm. Bioall. Sci. 7 (4), 329–333.

Ram, J., Moteriya, P., Chanda, S., 2015. Phytochemical screening and reported biological activities of some medicinal plants of Gujarat region. J. Pharmacog. Phytochem. 4 (2), 192–198.

Rivera, J.O., Ortiz, M., Gonzalez-Stuart, A., Hughes, H., 2007. Bi-national evaluation of herbal product use on the United States/Mexico border. J. Herb Pharmacother. 7 (3–4), 91–106.

Robinson, M.M., Zhang, X., 2011. Traditional Medicines: Global Situation, Issues and Challenges. The World Medicines Situation, 3rd ed. WHO, Geneva, pp. 1–14.

Sane, R.T., 2002. Standardization, quality control and GMP for herbal drugs. Indian Drug. 39 (3), 184–190.

Shakya, A.K., 2016. Medicinal plants: future source of new drugs. Int. J. Herbal Med. 4 (4), 59–64.

Shakya, A.K., Shukla, S., 2011. Evaluation of hepatoprotective efficacy of Majoon-e-Dabeed-ul-ward against acetaminophen induced liver damage: a Unani herbal formulation. Drug Develop. Res. 72 (4), 346–352.

Shakya, A.K., Sharma, N., Saxena, M., Shrivastava, S., Shukla, S., 2012. Evaluation of the antioxidant and hepatoprotective effect of Majoon-e-Dabeed-ul-ward against carbon tetrachloride induced liver injury. Exp. Toxicol. Pathol. 64 (7–8), 767–773.

Tilburt, J.C., Kaptchuk, T.J., 2008. Herbal medicine research and global health: an ethical analysis. Bull. World Health Organ. 86 (8), 594–599.

Trusheim, M.R., Berndt, E.R., Douglas, F.L., 2007. Stratified medicine: strategic and economic implications of combining drugs and clinical biomarkers. Nat. Rev. Drug Discov. 6 (4), 287–293.

Tyler, V.E., 1999. Phytomedicine: back to the future. J. Nat. Prod. 62 (11), 1589–1592.

Vickers, A., Zollman, C., 1999. ABC of complementary medicine: herbal medicine. Brit Med. J. 319 (7216), 1050–1053.

Wink, M., 2010. Introduction: Biochemistry, physiology and ecological functions of secondary metabolites, Annual Plant Reviews, 40, 2nd ed.

World Health Organisation, 2008. Media Centre, Traditional Medicine, 20.

CHAPTER 2

Diversity of Bioactive Compounds and Their Therapeutic Potential

Mohd Musheer Altaf[1,2], Mohd Sajjad Ahmad Khan[3] and Iqbal Ahmad[1]

[1]Department of Agricultural Microbiology, Aligarh Muslim University, Aligarh, Uttar Pradesh, India
[2]Faculty of Life Science, Institute of Information Management and Technology, Aligarh, India
[3]Department of Basic Sciences, Biology Unit, Health Track, Imam Abdulrahman Bin Faisal University, Dammam, Kingdom of Saudi Arabia

2.1 INTRODUCTION

For thousands of years, medicinal plants have been serving as a precious resource of therapeutics. Moreover, many of the modern drugs or therapeutics are either plant based or their products (Newmann and Cragg, 2012; Kinghorn et al., 2011). On the other hand, plant product-based drug discovery is linked with several complexities; drug manufacturers have diverted their major attention to synthetic or semisynthetic route for the development of novel agents (David et al., 2015; Beutler, 2009). However, the results achieved from plant-based drug discovery, still, did not meet up the promise as apparent in the availability of decreasing quantity of novel therapeutic compounds (Atanasov et al., 2015; David et al., 2015; Scannell et al., 2012; Kingston, 2011). This situation revived the interest in plant-based therapeutic product development, regardless of its immense complexity and intricacy, which successively requires multidisciplinary studies (Heinrich, 2010).

Medicinal plants, with healing properties and toxicology, along with additional qualities, are methodically acceptable by the public for the maintenance of normal health, depending on their availability and applicability through civilizations. In view of the fact that medicinal plants can be categorized as natural products safe to use, the rule permits them to be advertised by circular to the health-governing departments (Brazil, 2010), and they may be grown by persons who pursue high-quality agricultural procedure. Therefore the supportive self-medication is adapted under

New Look to Phytomedicine
DOI: https://doi.org/10.1016/B978-0-12-814619-4.00002-1

conditions of health problems so as to be assumed easy and more common inside a society. This practice can decrease the requirement of medical practitioners, justifying and reducing the cost of community health facility (Lorenzi and Matos, 2012).

Owing to the growth of toxicity and bacterial resistance for the synthetic therapeutic agents, humans focused on ethnopharmacognosy. They were able to discover large numbers of phyto-compounds from medicinal plants that are safe and vitally efficient substitutes with little or no toxicity. Several important medicinal properties like antioxidant, antimicrobial, anticancer, pain-relieving characteristics, etc., were accounted. Under several conditions, communities report the beneficial effects of many plant-based products. Still, scientific experiments both *in vitro* and *in vivo* are required to prove the benefits of phyto-compounds. Scientific experiments aimed in the direction of understanding pharmacokinetics, bioavailability, efficiency, security, and medicine relations of recently discovered phyto-compounds and their extracts needs a cautious assessment. Scientific experiments are carefully designed to protect the fitness of the volunteers in addition to respond exact investigation query by examining for both instant and long-standing adverse effects and their results are calculated prior to the application of plant-based therapeutic compounds to the patient. Keeping this in mind, the authors of this chapter desire to sum up here the different types of phyto-compounds, their diversity, current strategies involved in their development and their therapeutic potential.

2.2 CLASSIFICATION AND MAJOR REPRESENTATIVE OF ACTIVE COMPOUNDS

The earliest record on therapeutic uses of plants-based drugs dates back to 2600 BC and accounts for the continuation of a complicated healing arrangement in Mesopotamia consisting of about 1000 plant-based therapeutic agents. Egyptian drugs date back to about 2900 BC, but its most practical preserved evidence is the "Ebers Papyrus" from about 1550 BC, consisting nearly 1000 compounds, primarily plant-based (Cragg and Newman, 2013; Sneader, 2005; Borchardt, 2002). Conventional Chinese medicines have been widely recognized more than thousands of years (Unschuld, 1986), and the citations of the Indian Ayurveda date back to the 1st century BC (Patwardhan, 2005). The information on the therapeutic use of plants in the Western world is primarily derived from the Greek and Roman traditions. Related to this, the important manual was written by the Greek physician Dioscorides (1st century AD) and by the Romans Pliny the Elder (1st century AD) and Galen (2nd century AD) (Sneader, 2005). The Arabs preserved large quantities of the Greco-Roman information through the Dark and Middle ages (i.e., 5th to 12th centuries), and mix it together with their personal medicinal skills, herbs from Chinese and Indian Ayurveda (Cragg and Newman, 2013). Throughout this period, medicinal plants were merely used on an experimental basis, lacking mechanistic understanding of their pharmacological behavior or active ingredient. It was only in the 18th century that Anton von Störck, who examined toxic aromatic plant such as aconite and colchicum, and William Withering, who examined foxglove for the management of edema, put down the foundation for the scientific medical studies of medicinal herbs (Atanasov et al., 2015).

Plants are sunlight-based biochemical manufacturing units, which produce a large collection of bioactive compounds that are collected and extracted. These plant-based molecules are classified as primary or secondary compounds. Primary compounds are extensively distributed in the environment and are required for normal growth and development of plants (Applezweig, 1980). In contrast, secondary compounds are biosynthetic derivative of primary compounds but are present in restricted numbers and small quantities among plants. They regularly take part in ecologically important role in the interactions of plants with their environment and help in the continued existence of plants.

In general, natural compounds are obtained from plants and they serve as the favored basis for new drug development (Lahlou, 2007); it is established that primary compounds cannot be employed as intermediary in the production of high-price semisynthetic medicinal compounds, because secondary compounds typically hold extremely intricate stereo structures with many chiral centers; these bioactive molecules are being established to have medicinal function as model compounds for synthetic and semisynthetic therapeutic agents (Atanasov et al., 2015; Moses et al., 2013).

Secondary compounds are manufactured in a special type of plant cells, at a particular developmental stage (Fig. 2.1). The important secondary metabolites found in higher plants include alkaloids, flavonoids, phenols, glycosides, saponins, tannins, volatile oils, gums, and resins (Moses et al. 2013); some of the major bioactive metabolites of higher plants are mentioned as follows:

1. *Alkaloids*: they are usually found in the form of salts of inorganic and organic acids. They normally contain a nitrogen atom as part of a

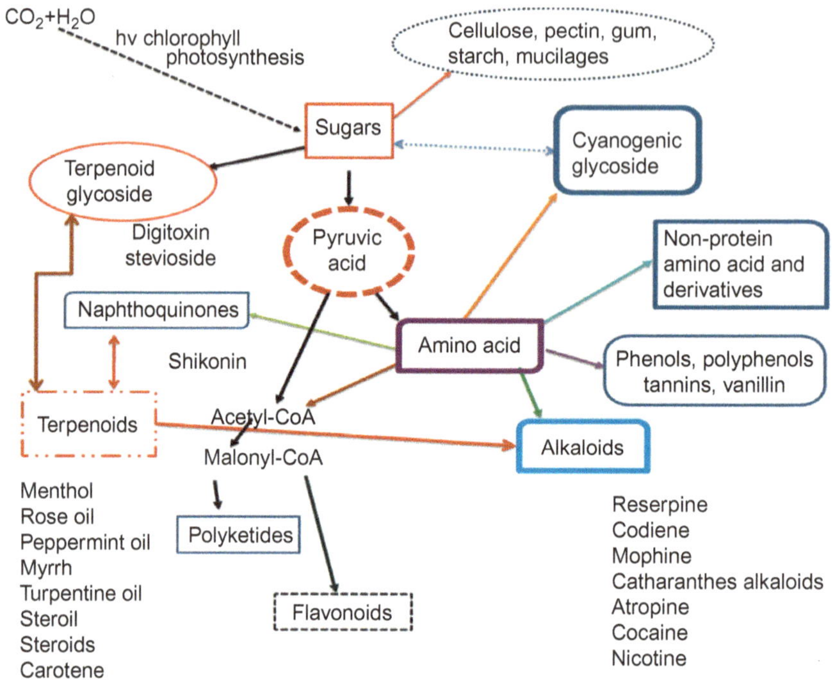

Figure 2.1 Biosynthetic pathway of important phyto-compounds. *Partly adapted from Balandrin, M.F., Klocke, J.A., Wurtele, E.S., Bollinger, W.H., 1985. Natural plant chemicals: sources of industrial and medicinal materials. Science. 228(4704), 1154–1160.*

heterocyclic ring. Classification is usually based on the type of ring system and on biosynthetic origin. Major groups include Amaryllidaceae (only found in daffodil family), betalain (yellow or purple pigments of Centrospermae), lycopodium (restricted to club mosses), indole (found in family Apocynaceae and Loganiaceae), diter-eponoid, monoterpene, sesquiterpenes (commonly found in orchids), isoquinoline (the largest group of alkaloids, extensively distributed), peptide, pyrrolidine, peperidine, quinoline (generally found in Rutaceae family), quinolizidene (generally found in Leguminosae), steroidal and tropane alkaloids (generally found in Solanaceae).

a. *Phenolics*: They are compounds containing aromatic ring bearing one or more hydroxyl groups. The majority are water soluble or hydrophilic and combined with sugars in glycosidic form and are classified according to structural complexity and biosynthetic origin as follows:

b. *Simpler phenols*:

   i. Phenolic acids and ketones

   ii. Phenyl propanoids; coumarins, benzofuran, choromones, chromenes

   iii. Phenolic quinines: benzoquinone, naphthaquinones, and anthraquinones
  c. *Polymeric phenolics*
   i. Lignins
   ii. Melanins
   iii. Tannins (polyphenols of high molecular weight found as condensed salts or gallo/epitannins)
   iv. Xanthones
   v. Stilbenoid
  d. *Flavonoids*: C15 heterocyclic nucleus of flavones varying in substitution groups.
   i. Anthocyanins and anthoclors
   ii. Flavones and flavonols
   iii. Isoflavonoids
2. *Terpenoids*: Derived from isoprene precursors which condense to form various terpenes, and are classified on the sequence of their biosynthetic complexity as follows:
  a. Monoterpenoids
  b. Iridoids
  c. Sesquiterpenoids
  d. Sesquiterpene
  e. Lactones
  f. Diterpenoid
  g. Triterpenoid saponins
  h. Steroid saponins
  i. Cardenolides and bufadienolides
  j. Phytosterols, curcum-bitacums
  k. Nortriterpenoids
  l. Carotenoids
3. *Glycosides*: Hydrolysis of these compounds yields a sugar (glycone) and a nonsugar moiety (aglycone). The linkage between the reducing sugars and the phenolic hydroxyl or alcoholic hydroxyl group of aglycone is called hemiacetal linkage. Based on the aglycone structure they are grouped as
  a. *Steroidal glycosides*: Terpenoid aglycone-
   i. 5-membered lactone ring, cardenoloids ex. digitalis
   ii. 6-membered lactone ring, bufadienolides ex. urginea
  b. *Anthracene glycosides*: Phenolic aglycone-reduced and ixodised anthraquinones

  c. *Saponin glycosides*: Terpenoid aglycone
     1. Triterpenoid
     2. Steroidal
  d. *Cyanogenic glycosides*: Hydrocyanic acid aglycone
  e. *Flavonoid glycosides*: Phenolic aglycone-2 phenylbenzopyrenes
  f. *Resinous glycosides*: Resin aglycone
4. *Volatile oils*: steam distillation of plant yields volatile compounds known as volatile oils which possess characteristics odors. The active constituents are terpenoids in association with alcohols, phenols before aldehydes or esters.
5. *Gums*: Amorphous polysaccharides that are found as plant exudates.
6. *Resins*: Naturally occurring complex organic compounds found as exudates from tree bark.

Normally, compounds extracted from plants are high-volume, low-cost bulk substances. Largely used as industrial unprocessed material, foods or food preservatives such as vegetable oils, carbohydrates (sucrose, starch, pectin, and cellulose), and proteins. The plants consist of secondary metabolites and have "medicinal" properties that have physiological consequence on mammalian structure known as the active principle. With the discovery of the physiological effect of a particular plant or its component, efforts are being made to know the exact chemical nature of these therapeutically active phyto-components and to produce these molecules by chemical production (Ramawat, 2008).

2.3 MAJOR BIOLOGICAL ACTIVITIES OF PHYTO-COMPOUNDS: OCCURRENCE AND MECHANISMS

Phyto-compounds are known to have several activities. Some might slow down microbial growth, obstruct metabolic pathways, or alter gene expression and signal transduction (Manson, 2003; Surh, 2003; Kris-Etherton et al., 2002). Phyto-compounds can also act as chemotherapeutic or chemopreventive (compounds that slow down, annul, or delay tumourigenesis), and used for cancer treatment (Sarkar and Li, 2006; D'Incalci et al., 2005). Many herbal concentrate and essential oils revealed diverse antibacterial activity, like the interaction with phospholipids bilayer of the cell membrane resulting in enhanced cell permeability and cell damage, loss of the enzymes associated with cellular respiration and manufacturing of different cellular constituents along with the damage to genetic components of the cell. Normally, these are measured by the interruption of the cytoplasmic membrane, disturbing the proton motive force, electron flow, active transport, and clotting of cell constituents (Doughari, 2012). Some specific modes of actions are discussed later.

2.4 AS ANTIOXIDANTS

Antioxidants defend cells against the harmful result of reactive oxygen species or else known as, free radicals like singlet oxygen, superoxide, peroxyl radicals, hydroxyl radicals, and peroxynite which give rise oxidative stress causing cell injury (Mattson and Cheng, 2006). Natural antioxidants participate as an important component in fitness protection and deterrence of the persistent and degenerative ailments; for example atherosclerosis, cardiac and cerebral ischemia, carcinogenesis, neurodegenerative problems, diabetic pregnancy, rheumatic disorder, injury to DNA, and ageing (Jayasri et al., 2009; Uddin et al., 2008). Antioxidants exercise their actions through hunting the "free-oxygen radicals" thus producing comparatively "stable radical." The free radicals are metastable chemical entities, which catch the electrons from the compounds from direct environments. These radicals unless shunted efficiently over time they could cause injury to vital biomolecules such as lipids, proteins as well as those exist at membranes, mitochondria, and the DNA creating aberrations eventually causing disease situation (Uddin et al., 2008). Therefore free radicals are associated with several ailments, like tumor soreness, hemorrhagic shock, atherosclerosis, diabetes, infertility, gastrointestinal ulcerogenesis, asthma, rheumatoid arthritis, cardiovascular problems, cystic fibrosis, neurodegenerative diseases (such as Parkinsonism and Alzheimer's diseases), and premature ageing. The human body manufacturesinadequate quantity of antioxidants which are vital in avoiding oxidative stress. Free radicals that produced inside human body might be eliminated by the body's individual natural antioxidant protection mechanism like glutathione or catalases). Thus this shortage had to be remunerated using external plant-based natural compounds, antioxidants, such as vitamin C, vitamin E, flavones, and carotene (Mattson and Cheng, 2006).

Plants have a broad diversity of free radicals scavenging compounds such as phenols, flavonoids, vitamins, terpenoids that have antioxidant properties (Cai and Sun, 2003; Madsen and Bertelsen, 1995). Several plants, citrus fruits, and leafy vegetables supply sufficient amount of ascorbic acid, vitamin E, carotenoids, flavonols, and phenolics which hold the capability to hunt the free radicals in human body. Important antioxidant characters have been documented in phyto-compounds that are essential for the decrease in the incidence of many ailments (Anderson and Teuber, 2001; Hertogand Feskens, 1993). Numerous nutritional polyphenolic components extracted from plants are highly efficient antioxidants *in vitro* than vitamins E or C, and therefore could participate considerably to defensive property *in vivo* (like methanol extract of *Cinnamon* include an amount of antioxidant molecules which can efficiently hunt reactive oxygen species together with superoxide anions and hydroxyl radicals as well as other free radicals *in vitro*). The fruit of *Cinnamon*, an under-utilized and alternative fraction of the plant, consist of large volume of phenolic antioxidants to neutralize the destructive activities of free radicals and can defend against mutagenesis. Antioxidants are normally mixed with several food items in order to stop the radical chain reactions of oxidation, and they work by preventing the beginning and transmission process which stops the reaction and interrupt the oxidation process. Owing to security problems associated with artificial chemical compounds, the food manufacturing business have turned to plant-based antioxidants. Also, there is rising fashion in customer inclination for natural antioxidants, all of which has given more impetus to explore natural sources of antioxidants (Doughari, 2012).

2.5 AS ANTICANCER

Polyphenols chiefly are among the different phytochemicals that have the capacity to inhibit the cancer growth (Liu, 2004). Phenolics acids generally drastically reduce the development of the exact cancer-causing nitrosamines from the nutritional nitrites and nitrates. Glucosinolates from different vegetables like broccoli, cabbage, cauliflower, and Brussel sprouts apply a significant defensive support against the colon cancer. Normal use of Brussel sprouts by volunteers (≈ 300 g/day) unbelievably causes a quick increment in the glutathione-transferase and a consequent obvious decrease in the urinary concentration of a specific purine metabolite that work as a marker of DNA-degradation in cancer. Isothiocyanates and the indole-3-carbinols may obstruct positively metabolism of carcinogens consequently inhibiting procarcinogen creation, and thus stimulating the "phase-II" enzymes, namely NADPH quinone-reductase or glutathione S-transferase that exclusively detoxify the preferred electrophilic metabolites which are able to altering the arrangement of nucleic acids. Sulforaphane (rich in broccoli) has demonstrated to be an exceptionally powerful phase-2 enzyme stimulant. It principally creates specific cell cycle stop and also the apoptosis of the neoplasm (cancer) cells. Sulforaphane firmly creates d-D-gluconolactone which has been recognized to be an important inhibitor of breast cancer. Indole-3-carbinol (very important and vital indole found in broccoli) exclusively slow down the human papilloma virus which results in uterine cancer. It inhibits the estrogen receptors particularly found in the breast cancer cells as well as down regulates CDK6, and up regulates p21 and p27 in prostate cancer cells. It can arrest G1 cell cycle and apoptosis of breast and prostate cancer cells considerably and augments the p53 expression in cells applied with benzopyrene. It also discourages Akt, NF-kappaB, MAPK, and Bel-2 signaling pathways to a large extent. Phytosterols inhibit the growth of tumors (neoplasms) in colon, breast, and prostate glands even though the exact and precise methods which are responsible for this action are poorly understood. But still they can alter tremendously the resulting cell membrane movement in the occurrence of tumor development and thus decreasing the inflammation (Doughari, 2012).

2.6 AS ANTIMICROBIAL

Phyto-compounds used by plants to defend them against phyto-pathogens have established relevance in human drugs (Nascimento et al., 2000). A number of phyto-compounds like phenolic acids perform fundamentally by playing a key role in the decrease of attachment of microorganisms to cell lining of bladder, and the teeth, which eventually decreases the occurrence of urinary-tract infections and the typical dental caries. Plants products have both bacteriostatic and bactericidal properties. The volatile gas phase of mixture of *Cinnamon* oil and clove oil demonstrated high-level capability to hinder the development of pathogenic fungi, yeast, and bacteria usually present on intermediate moisture foods when mixed with a customized environment containing an elevated concentration of CO_2 (40%) and low concentration of O_2 ($<0.05\%$) (Jakhetia et al., 2010). *Aspergillus flavus*, a toxin-producing fungi, was found to be the highly resistant microbe. It is important to note down that antimicrobial activity of identical plant fraction examined on majority of occasion was found different from one scientist to another because of the concentration of phytocompounds of the identical plant parts can differ from one ecological site to another based on the life of the plant, dissimilarity in geographical issues, the nutrient concentrations of the soil, derivation techniques, as well as process employed for antimicrobial evaluation. Therefore it is significant that systematic method be unmistakably recognized and effectively used and documented.

2.7 AS ANTIULCER

Phyto-compounds have been documented to hinder the development of *Helicobacter pylori in vitro* in addition to its urease activity. The effectiveness of several plant products in water and at low pH levels improves their effectiveness even in the human stomach. Their preventive activity on the intestinal and kidney Na^+/K^+ ATP-ase action and on alanine transport in rat jejunum has also been observed (Doughari, 2012; Jakhetia et al., 2010).

2.8 AS ANTIDIABETIC

Cinnamaldehyde, a phyto-compound, has been demonstrated to display important antihyper glycemic properties which causes the decrease in entire cholesterol and triglyceride intensity and, simultaneously, escalating high density lipoproteins-cholesterol in streptozotocin-induced diabetic rats. This study exposes the capability of cinnamaldehyde which can be used as a normal oral agent, having both hypoglycaemic and hypolipidemic properties. Current data point out that *Cinnamon* derivatives and polyphenols with procyanidin type-A polymers shows the capability to enhance the quantity of thrombotic thrombocytopenic purpura, insulin resistance, and glucose transporter-4 in 3T3-L1 adipocytes. It was recommended that the method of *Cinnamon*'s insulin-like action might be partially owing to augmentation in the quantity of thrombotic thrombocytopenic purpura, insulin resistance, and glucose transporter-4 and that *Cinnamon* polyphenols could have extra functions such as antiinflammatory and/or antiangiogenesis (Jakhetia et al., 2010).

2.9 AS ANTIINFLAMMATORY

Essential oil of *Cinnamomum osmophloeum* branches has exceptional antiinflammatory properties and cytotoxicity in opposition to HepG2 (human hepatocellular liver carcinoma cell line) cells. Earlier information as well point out that the ingredients of *C. osmophloeum* branches displayed tremendous antiinflammatory properties in repressing nitric oxide (NO) formation by LPS (lipopolysaccharide)-motivated macrophages (Jakhetia et al., 2010).

2.10 MULTIFUNCTIONAL TARGETS

Several molecular objectives of nutritional phyto-compounds have been recognized, varying from pro and antiapoptotic proteins, cell cycle proteins, cell adhesion molecules, protein kinases, transcription factors to metastasis, and cell development approaches (Aggarwal and Shishodia, 2006; Choi and Friso, 2006; Awad and Bradford, 2005). Phyto-compounds such as epigallocatechin-3-gallate (EGCG) from green tea, curcumin from turmeric, and resveratrol from red wine lean to mean at a large number of molecular targets. It is for this reason these perfect methods of action are not obtainable inspite of decades of study. The multitarget character of phyto-compounds might be useful in defeating cancer medicine resistance. This complicated style of act almost certainly delays the cancer cell's capacity to build up resistance to the phyto-compounds. It has also been established that EGCG has preventing action on the extracellular secretion of verotoxin (VT) from *E. coli* O157: H7. Ethanol pericarp derivative from *Punica granatum* was in addition accounted to suppress VT manufacturing in periplasmic space and cell supernatant. Methods accountable for this are so far poorly understood. Nevertheless, the live molecules from the plant are considered to meddle by the transcriptional and translational procedures of the bacterial cell (Voravuthikunchai and Kitpipit, 2003). Additional efforts are required to be made one by one to set up this hypothesis. Phyto-compounds can also alter transcription aspects (Andreadi et al. 2006), redox-sensitive transcription characters, redox signaling, and inflammation. For example, NO, a signaling compound having significance in inflammation, is altered by plant polyphenols and additional plant-based derivatives. Many phyto-compounds have been categorized as phytoestrogens, with health-promoting capabilities ensuing phyto-compounds to be advertised as dietary supplements (Doughari, 2012; Moutsatsou, 2007). Some of the functions of major phyto-compounds are summarized in Table 2.1.

Table 2.1 Major class of phyto-compounds and their actions

Categories	Plant source	Major compounds	Mechanism of action	References
Anticancer activities	Grapes, peanuts, mulberries, green tea, wheat, flaxseeds, *Platycodon grandiflorum*, *Draba nemorosa*, *Perilla frutescens* and *Prunus armeniaca*	Carotenoids, polyphenols, coumarins, flavonoids, terpenoids, isoflavone, quinone, lycopene, curcumin, cyclopamine	Tumor repressor, delayed growth of lung cancer, antimetastatic property, anticarcinogenic, efficient against hepatic fibrosis	Atanasov et al. (2015), Greenwell and Rahman (2015), Yin et al. (2013), Doughari (2012)
Antimicrobial activities	Ginger, turmeric, goldenseal, oregano, and knob wood	Terpenoids, alkaloids, phenolics	Antibacterial and antifungal	Doughari (2012), Jakhetia et al. (2010), Perumal Samy and Gopalakrishnakone (2010)
Detoxifying activities	*Solanum nigrum*, *Trigonella foenum-graecum* and *Nigella sativa*	Reductive acids, tocopherols, phenols, indoles, aromatic Isothiocyanates, coumarins, flavones, carotenoids, retinoids, cyanates, phytosterols	Suppress procarcinogen growth, stimulate coupling of drugs with carcinogens, repress tumorigenesis	Goyal et al. (2017), Saxena et al. (2013), Lahlou (2013), Zhou et al. (2012)
Antioxidant activities	Citrus fruits, oats, garlic, legumes, onions, walnuts and green leafy vegetables	Polyphenolic compounds, flavonoids, carotenoids, tocopherols, ascorbic acid	Oxygen free radical quenching, suppression of lipid peroxidation	Doughari (2012), Jayasri et al. (2009)
Additional activities	*Curcuma longa*, *Annona squamosa*, *Withania somnifera*	Alkaloids, terpenoids, volatiles, biogenic amines	Neuropharmacological anticancer and antioxidants properties	Ramalingum and Mahomoodally (2014), Lahlou (2013)

Source: Partly adapted from Saxena, M., Saxena, J., Nema, R., Singh, D., Gupta, A., 2013. Phytochemistry of medicinal plants. J. Pharmacogn. Phytochem. 1, 168–182.

2.11 APPROACHES FOR DRUG DISCOVERY FROM PHYTO-COMPOUNDS

Discovery and development of novel drugs is an intricate, lengthy, and costly procedure. The duration, starting from discovery of a new therapeutic agent to its completion to the hospital is about 12 years, involving a cost of over 1 billion US dollar. Presently about 80% of antimicrobial, cardiovascular, immunosuppressive, and anticancer therapeutic agents are derived from plants. Their retailing goes beyond 65 billion US dollar in 2003 (Gordaliza, 2009). It is known that over 80% of therapeutic agents' are either derivative of plants or extracted from natural products (Maridass and John de Britto, 2008). Actually, about 50% of drugs are isolated from molecules identified from plants or animals (Pan et al., 2013). Therapeutic agents' invention from plants can be separated into three stages, namely predrug stage, quasi-drug stage, and full-drug stage. They are described in detail in Fig. 2.2.

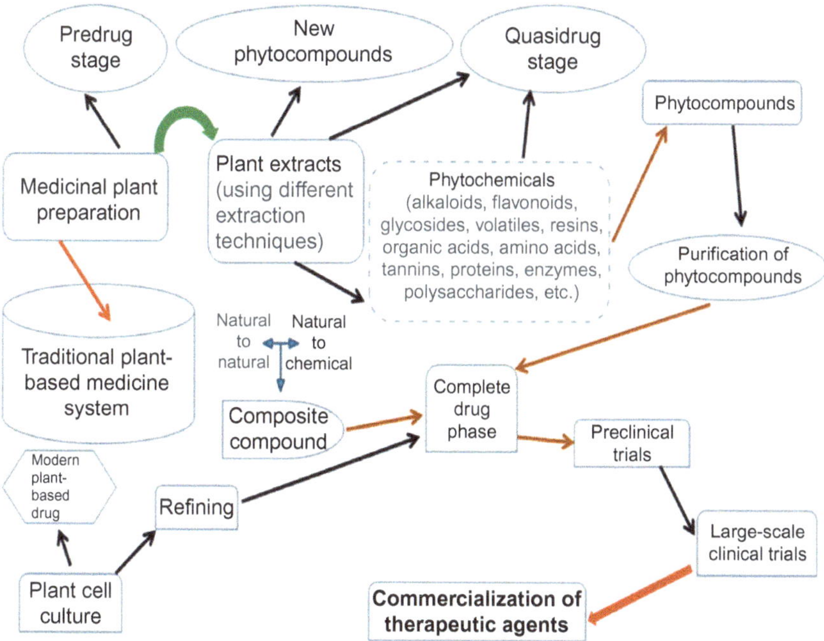

Figure 2.2 Recent methods used for therapeutic agent development from plants. *Partly adapted from Pan, S.Y., Zhou, S.F., Gao, S.H., Yu, Z.L., Zhang, S.F., Tang, M.K., et al., 2013. New perspectives on how to discover drugs from herbal medicines: CAM's outstanding contribution to modern therapeutics. Evidence Based Complement. Altern. Med. 627375.*

2.12 CHALLENGES IN THE DISCOVERY OF NEW PHYTO-COMPOUNDS

Ever since plants are brought together straightforwardly from their niche, the exact recognition and classification are necessary which serve as the basis for the discovery of new phyto-compounds. For an immediate recognition, a combination of schemes may be required, like genetic and chemical investigation additionally to morphological and anatomical description. Continuous alterations in plant taxonomy in addition to synonymy give the complexity to the demanding job of plant taxonomist. Furthermore, gathering of the plant matter and precise citations, botanical classification, with creation of the herbarium coupons are jobs that cannot be computerized and requires exceptional expertise (Bucar et al., 2013). Significant problems associated with the employment of plants as a resource for discovery of new phyto-compounds are linked with the availability of the preliminary matter. Frequently the accessible quantity of plant material is little even though several plant-based compounds have already been separated and illustrated. Moreover, available material quantities are regularly inadequate for examining different kind of bioactivities.

Furthermore, restricted accessibility creates difficulty when a bioactive plant-based product is recognized to possess an extremely capable bioactivity and develop into a pharmaceutical product. Reassemblies of natural varieties could become complicated, because plant niche can be quickly vanishes under anthropogenic factors. Moreover, the environment of plants, mainly of sheltered varieties, needs to be valued when assembling from the wild and season-dependent chemical constituents of plant matter could bind the time window for reassemblies (e.g., blossom collection demands compilation throughout the blossoming period). In cases of bring in plant matter, also a complete range of further problem may influence its availability, for instance limited wars, environmental calamity, or altering lawful system for cross-border roaming and sell to other countries. The significance of plant matter availability is demonstrated by a latest examination (Amirkia and Heinrich, 2014), studying the association among category, large quantity of alkaloids incidence and their use as medicines. Varieties allocation was judged on the beginning of Global Biodiversity Information Facility (GBIF) data. The instigators establish that 93% of all alkaloids in therapeutic use have more than 50 episodes in the GBIF database, and only two have less than 10 occurrences. For that reason, the investigators conclude that phyto-compounds present in several dissimilar species are further necessary for therapeutic application, and that delivery limitations are key hindrance to the successful study, growth, and marketing of phyto-compounds.

In several instances, when a plant is marketed into phyto-compound or when one of its components receive fame as a therapeutic agent, the plant population turn into endangered one because of its widespread untenable cultivation (Cordell, 2011; Vines, 2004). The traditional case for this was "Paclitaxel supply disaster" (Kingston, 2011). When the mixture turned out to have extraordinary medical value in ovarian cancer, abruptly the order for paclitaxel increased enormously. On the other hand, at that point, this compound was only available from the bark of the western yew (*Taxus brevifolia* L.). This was on one hand challenging, since the whole manufacturing procedure involves the tedious bark collection and aeration, removal, and cleansing. Alternatively, anxieties on the environmental effect of exhaustive bark collection were increased (Cragg et al., 2012; Kingston, 2011). Farming would be an extra feasible option to wild crafting (the practice of harvesting plants from their natural, or wild habitat, primarily for food or medicinal purposes), however, about two-third of the 50,000 medicinal plant varieties under application globally are still wild crafted (Canter et al., 2005). Thus organizations like World Health Organization (WHO) and European Medicines Agency (EMA) formulated plans on good agricultural and collection practices for therapeutic plants in turn to encourage feasible collection methods and to decrease the environmental troubles created by wild crafting of therapeutic plants.

Aside from this problem, conservation and lawful concern also have a significant pressure on availability of plants as a basis of new drug discovery, particularly regulation connected with plant access and distribution of profit, as well as patentability matters with local governments in the countries of origin. The United Nation's Convention on Biological Diversity (CBD; http://www.cbd.int/doc/legal/cbd-en.pdf), signed in 1992 by the international community in Rio de Janeiro, Brazil, aims at (1) preserving the biodiversity; (2) feasible use of its genetic assets; and (3) distributing the benefits from their use in a fair and reasonable way (Atanasov et al., 2015). Further the quality of plant material has high significance. Accessible plant matter frequently differs on value and constituents and this could slow down the evaluation for its beneficial properties. The chemical constituent is not only dependent on plant species characteristics and collection time but also on soil variety, constituent, altitude, actual climate, processing, and storage situations. An additional feature that influences the chemical constituents of the preparatory plant matter is related to endophytic microorganisms. Consequently, natural compounds found in the composed plant matter may be in some case are the compounds of the endophytic microorganism, or plant compounds produced in response to association with endophytic microorganisms (Atanasov et al., 2015).

More trouble arise due to the truth associated with the exact molecular mode of action of phyto-compounds is a daunting task (*curcumin, triptolide*) (Corson and Crews, 2007). On the other hand, complete information of the relations of a therapeutic agent with its molecular target is extremely beneficial for the growth of plant-based drug discovery, as it permits assets optimization by medicinal chemistry methods, and on several times a more suitable clinical trial plan.

The manner in which the precise clinical trials carried out on humans for the authorization of phyto-compounds as therapeutic agents creates new complications. While such research studies or clinical trials are frequently possible now by means of industrialized support owing to the elevated expenditure, simultaneously the curiosity of pharmaceutical companies in plant-based compounds that are not artificially modified is incomplete owing to the complexities associated with their patents. Apart from the problems associated with the patents of phyto-compounds, it should also be noted that there has been a general change in the therapeutic agent manufacturing, from small molecule-based drug discovery toward biologically big molecules (e.g., therapeutic proteins or nucleic acids). Nevertheless, the costs of biologically big molecules are significantly higher than the costs for small molecule therapeutic agents, and their growing clinical application is placing an increasing pressure on state-run health insurances. For that reason, to revive the increasing economic cost, reversing to a few "old" small molecule-based methods is feasible (Appendino et al., 2010). Owing to the problems explained here, the curiosity in phyto-compounds discovery has been gradually declining. Even in the last 10 years, several big- and medium-sized drug manufacturing plants, which were still active in the 1990s, leave their phyto-compounds investigation up to large extent to academic institutions and established business houses (David et al., 2015).

2.13 CONCLUSION AND FUTURE PROSPECTS

Historically, medicinal plants used as a rich source for successful drugs still represent an important pool for the identification of new pharmacological leads. Plants are producing numerous chemically diverse secondary metabolites which are optimized for bioactivities and are still far from being exhaustively investigated. Although plant-based pharmaceutical discovery and progress indicates a difficult effort involving extremely complex combined pathways; the explored systematic growth, latest scientific approaches, and scientific developments which unmistakably demonstrated that plant-based products will be among the most significant future resources for novel therapeutic compounds.

There is a need for improved technology for rapid isolation of active compounds from such preparations in large quantities for scientific evaluation and maintenance of biodiversity. The efficacy and safety of these phyto-compounds as medicine have to be supported by clinical studies. It is concluded that once the local ethnomedical preparations of traditional sources are scientifically evaluated, they should replace existing drugs commonly used for the treatment of various diseases. It is also desirable to have collaborative work with profit-sharing agreements between leading institutes, pharmaceutical companies of developed countries, and organizations in developing countries where most medicinal plants are still unexplored. This could lead a potential approach to come up with more diverse bioactive compounds for treatment of various diseases.

REFERENCES

Atanasov, A.G., Waltenberger, B., Pferschy-Wenzig, E.M., Linderd, T., Wawroscha, C., Uhrin, P., et al., 2015. Discovery and resupply of pharmacologically active plant derived natural products: a review. Biotechnol. Adv. 33 (8), 1582–1614.

Aggarwal, B.B., Shishodia, S., 2006. Molecular targets of dietary agents for prevention and therapy of cancer. Biochem. Pharmacol. 71, 1397–1421.

Amirkia, V., Heinrich, M., 2014. Alkaloids as drug leads—a predictive structural and biodiversity-based analysis. Phytochem. Lett. 10, xlviii–liii.

Anderson, K.J., Teuber, S.S., 2001. Walnut polyphenolics inhibit *in vitro* human plasma and LDL oxidation, biochemical and molecular action of nutrients. J. Nutr. 131, 2837–2842.

Andreadi, C.K., Howells, L.M., Atherfold, P.A., Manson, M.M., 2006. Involvement of Nrf2, p38, B-Raf, and nuclear factor-kappaB, but not phosphatidylinositol 3-kinase, in induction of hemeoxygenase-1 by dietary polyphenols. Mol. Pharmacol. 69, 1033–1040.

Appendino, G., Fontana, G., Pollastro, F., 2010. 3.08-Natural products drug discovery. In: Liu, H.-W., Mander, L. (Eds.), Comprehensive Natural Products II. Elsevier, Oxford, pp. 205–236.

Applezweig, N., 1980. Renewable Resources: A Systemic Approach. Academic Press, London, New York.
Awad, A.B., Bradford, P.G., 2005. Nutrition and Cancer Prevention. Taylor & Francis, Boca Raton.
Balandrin, M.F., Klocke, J.A., Wurtele, E.S., Bollinger, W.H., 1985. Natural plant chemicals: sources of industrial and medicinal materials. Science 228, 1154–1160.
Beutler, J.A., 2009. Natural products as a foundation for drug discovery. Curr. Protocol. Pharmacol. 46, 9.11.1–9.11.21.
Borchardt, J.K., 2002. The beginnings of drug therapy: ancient Mesopotamian medicine. Drug News Perspect. 15, 187–192.
Bucar, F., Wube, A., Schmid, M., 2013. Natural product isolation—how to get from biological material to pure compounds. Nat. Prod. Report. 30, 525–545.
Cai, Y.Z., Sun, M., 2003. Antioxidant activity of betalins from plants of the Amaranthacea. J. Agric. Food Chem. 51, 2288–2294.
Canter, P.H., Thomas, H., Ernst, E., 2005. Bringing medicinal plants into cultivation: opportunities and challenges for biotechnology. Trends Biotechnol. 23, 180–185.
Choi, S.W., Friso, S., 2006. Nutrient-Gene Interactions in Cancer. Taylor & Francis, Boca Raton.
Cordell, G.A., 2011. Sustainable medicines and global health care. Planta Med. 77, 1129–1138.
Corson, T.W., Crews, C.M., 2007. Molecular understanding and modern application of traditional medicines: triumphs and trials. Cell 130, 769–774.
Cragg, G.M., Katz, F., Newman, D.J., Rosenthal, J., 2012. The impact of the United Nations convention on biological diversity on natural products research. Nat. Prod. Rep. 29, 1407–1423.
Cragg, G.M., Newman, D.J., 2013. Natural products: a continuing source of novel drug leads. Biochim. Biophys. Acta 1830, 3670–3695.
D'Incalci, M., Steward, W.P., Gescher, A.J., 2005. Use of cancer chemopreventive phytochemicals as anti-neoplastic agents. Lancet Oncol. 6, 899–904.
David, B., Wolfender, J.L., Dias, D.A., 2015. The pharmaceutical industry and natural products: historical status and new trends. Phytochem. Rev. 14 (2), 299–315.
Doughari, J.S., 2012. Phytochemical: extraction methods, basic structures, and mode of action of potential chemotherapeutic agents. Phytochemicals-A Global Perspective of Their Role in Nutrition and Health. In Tech Publisher, pp. 2–15.
Gordaliza, M., 2009. Terpenyl-purines from the sea. Mar. Drugs 7, 833–849.
Goyal, S.N., Prajapati, C.P., Gore, P.R., Patil, C.R., Mahajan, U.B., Sharma, C., et al., 2017. Therapeutic potential and pharmaceutical development of thymoquinone: a multitargeted molecule of natural origin. Front. Pharmacol. 8, 656.
Greenwell, M., Rahman, P.K.S.M., 2015. Medicinal plants: their use in anticancer treatment. Int. J. Pharmaceut. Sci. Res. 6 (10), 4103–4112.
Heinrich, M., 2010. Ethnopharmacology in the 21st century-grand challenges. Front. Pharmacol. 1, 8.
Hertog, M.G.L., Feskens, E.J.M., 1993. Dietary antioxidant flavonoids and risk of coronary heart disease: the Zutphen Elderly study. Lancet 342, 1007–1011.
Jakhetia, V., Patel, R., Khatri, P., Pahuja, N., Garg, S., Pandey, A., et al., 2010. *Cinnamon*: a pharmacological review. J. Adv. Sci. Res. 1 (2), 19–23.
Jayasri, M.A., Mathew, L., Radha, A., 2009. A report on the antioxidant activities of leaves and rhizomes of *Costus pictus* D. Don. Int. J. Integret. Biol. 5 (1), 20–26.
Kinghorn, A.D.,, Pan, L., Fletcher, J.N., Chai, H., 2011. The relevance of higher plants in lead compound discovery programs. J. Nat. Prod. 74, 1539–1555.
Kingston, D.G., 2011. Modern natural products drug discovery and its relevance to biodiversity conservation. J. Nat. Prod. 74, 496–511.

Kris-Etherton, P., Hecker, M., Bonanome, K.D., Coval, A., Binkoski, S.M., Hilpert, A.E., et al., 2002. Bioactive compounds in foods: their role in the prevention of cardiovascular disease and cancer. Am. J. Med. 113, 71S–88S.

Liu, R.H., 2004. Potential synergy of phytochemicals in cancer prevention: mechanism of action. J. Nutr. 134 (12 Suppl.), 3479S–3485S.

Lorenzi, H., Matos, F.J.A., 2012. Plant as medicinais no Brasil: nativas e exóticas. Editora Instituto Plantarum. Nova Odessa, São Paulo.

Lahlou, M., 2007. Screening of natural products for drug discovery. Expert Opin. Drug Discov. 2 (7), 697–705.

Lahlou, M., 2013. The success of natural products in drug discovery. Pharmacol. Pharm. 4, 17–31.

Maridass, M., John de Britto, A., 2008. Origins of plant derived medicines. Ethnobot. Leaflet. 12, 373–387.

Madsen, H.L., Bertelsen, G., 1995. Spices as antioxidants. Trends Food Sci. Technol. 6, 271–277.

Manson, M.M., 2003. Cancer prevention—the potential for diet to modulate molecular signalling. Trends Mol. Med. 9, 11–18.

Mattson, M.P., Cheng, A., 2006. Neurohormetic phytochemicals: low-dose toxins that induce adaptive neuronal stress responses. Trends Neurosci. 29 (11), 632–639.

Moses, T., Pollier, J., Thevelein, J.M., Goossens, A., 2013. Bioengineering of plant (tri) terpenoids: from metabolic engineering of plants to synthetic biology in vivo and in vitro. New Phytologist 200, 27–43.

Moutsatsou, P., 2007. The spectrum of phytoestrogens in nature: our knowledge is expanding. Hormones (Athens) 6, 173–193.

Nascimento, G.G.F., Locatelli, J., Freitas, P.C., Silva, G.L., 2000. Antibacterial activity of plant extracts and phytochemicals on antibiotic resistant bacteria. Braz. J. Microbiol. 31, 247–256.

Newman, D.J., Cragg, G.M., 2012. Natural products as sources of new drugs over the 30 years from 1981 to 2010. J. Nat. Prod. 75, 311–335.

Pan, S.Y., Zhou, S.F., Gao, S.H., Yu, Z.L., Zhang, S.F., Tang, M.K., et al., 2013. New perspectives on how to discover drugs from herbal medicines: CAM's outstanding contribution to modern therapeutics. Evidence-Based Complement. Alternat. Med. 627375.

Patwardhan, B., 2005. Ethnopharmacology and drug discovery. J. Ethnopharmacol. 100, 50–52.

Perumal Samy, R., Gopalakrishnakone, P., 2010. Therapeutic potential of plants as anti microbials for drug discovery. Evidence-Based Complement. Alternat. Med. eCAM 7 (3), 283–294.

Ramawat, K., 2008. Herbal Drugs: Ethnomedicine to Modern Medicine. Springer, Berlin-Heidelberg.

Ramalingum, N., Mahomoodally, M.F., 2014. The therapeutic potential of medicinal foods. Adv. Pharmacol. Sci. 2014, 354264.

Sarkar, F.H., Li, Y., 2006. Using chemopreventive agents to enhance the efficacy of cancer therapy. Cancer Res. 66, 3347–3350.

Saxena, M., Saxena, J., Nema, R., Singh, D., Gupta, A., 2013. Phytochemistry of medicinal plants. J. Pharmacogn. Phytochem. 1, 168–182.

Scannell, J.W., Blanckley, A., Boldon, H., Warrington, B., 2012. Diagnosing the decline in pharmaceutical R & D efficiency. Nat. Rev. Drug Discov. 11, 191–200.

Sneader, W., 2005. Drug Discovery: A History. Wiley.

Surh, Y.J., 2003. Cancer chemoprevention with dietary phytochemicals. Nat. Rev. Cancer 3, 768–780.

Uddin, S.N., Akond, M.A., Mubassara, S., Yesmin, M.N., 2008. Antioxidant and antibacterial activities of *Trema cannabina*. Middle-East J. Sci. Res. 3, 105–108.

Unschuld, P.U., 1986. Medicine in China: A History of Pharmaceutics. University of California Press.
Vines, G., 2004. Herbal harvests with a future: towards sustainable sources for medicinal plants. Plantlife International, Salisbury, UK, <http://www.plantlife.org.uk/uploads/documents/HerbalHarvests-with-a-Future.pdf> (accessed 01.01.18).
Voravuthikunchai, S.P., Kitpipit, L., 2003. Activities of crude extracts of Thai medicinal plants on methicillin-resistant *Staphylococcus aureus*. J. Clin. Microbiol. Infect. 9, 236.
Yin, S.Y., Wei, W.C., Jian, F.Y., Yang, N.S., 2013. Terapeutic applications of herbal medicines for cancer patients. Evidence-Based Complement. Alternat. Med. 302426.
Zhou, J., Chan, L., Zhou, S., 2012. Trigonelline: a plant alkaloid with therapeutic potential for diabetes and central nervous system disease. Curr. Med. Chem. 19, 3523–3531.

CHAPTER 3

Ethnomedicinal Wisdom: An Approach for Antiviral Drug Development

Ananya Das Mahapatra[1], Priyanka Bhowmik[1], Anwesha Banerjee[1], Apurba Das[1], Durbadal Ojha[1] and Debprasad Chattopadhyay[1,2]
[1]ICMR-Virus Unit Kolkata, ID & BG Hospital Campus, Kolkatta, West Bengal, India
[2]ICMR-National Institute of Traditional Medicine, Belagavi, Karnataka, India

3.1 ETHNOMEDICINE: A BOON

A famous quote of Norwegian Artist Edvard Munch "*Nature is not all that is visible to the eye...* also includes the inner pictures of the soul" speaks many things about the mother "Nature" and her creation. For the past 100 years, mankind have been trying to unravel the secrets of nature, yet continue to be surprised by its revelations day by day. Nature is vast in its resources which not only provide food, clothes, shelter, and antiques but also provide medicine since ages. Out of these resources the greatest gift of nature is her atmosphere with food and medicine, which not only to help mankind to sustain but also to grow, develop, and survive, particularly from the diseases and sufferings. The study of natural resources traditionally used to cure or manage ailments in diverse ethnic culture is collectively termed as "Ethnomedicine" (Chattopadhyay, 2010; Chattopadhyay et al., 2009a).

Scientifically, "Ethnomedicine" is the study of "traditional medicine" of ethnic communities, their knowledge and practices that transmitted orally over centuries, and evolved over millennia of human existence (Chattopadhyay, 2009, 2010). The indigenous people of India till date used their medicaments or "so called medicines" which might be more appropriately defined as the use of plants in the treatment of diseases and should be more accurately termed as "Ethnobotanical medicine" (Fabricant and Farnsworth, 2001). For a considerable period of time, traditional medicine and ethnomedicine were ignored by the clinicians and biomedical practitioners due to a number of factors including the

New Look to Phytomedicine
DOI: https://doi.org/10.1016/B978-0-12-814619-4.00003-3

questionable purity, safety, and potency. The raw material to prepare those medicines is not standardized through modern quality control parameters; its chemical profile and their quantification are not known or maintained, and thus their purity is in question. Similarly, their toxicity profile at low doses for long term is unknown, so long-term use in tolerable dosage need to be monitored to rule out the question of long-term toxicity, if any. Although the traditional medicines are used for generations with limited or no major toxic manifestations, which can be considered as "Proof of Concept" in nature's laboratory. In 21st century the efficacy of any therapeutically useful product should be quantified in terms of modern medicine, so the efficacy of useful traditional and ethnomedicinal plants needs to be validated in modern laboratory by establishing the dosage and the exact mode and mechanism of action in one hand and *in vivo* efficacy in suitable animal models on the other. Moreover, if possible use of ex vivo model than can mimic the human body for better and global acceptability. With the upsurge research, ethnomedicinally useful plants have become one of the most acceptable resources not only for the pharmaceutical industry to develop therapeutically useful leads, or pharmaceuticals, but also to help in developing supplements or naturaceuticals, and even cosmetics. Reports from WHO-Traditional Medicinal Programme showed that a total of 122 compounds were isolated from 94 plant species, 80% of which were used for the same or related ethnomedical purposes (Fabricant and Farnsworth, 2001). Since these compounds were derived only from 94 species of plants out of an estimated 250,000 flowering plants, one might imagine the abundance of drugs remaining to be identified from plant kingdom.

3.2 ETHNOMEDICINAL WISDOM OF DIVERSE COMMUNITIES

Ethnomedicine broadly refers to the traditional medical practices concerned with the cultural interpretation of health, diseases, and illness that addresses the healthcare process and healing practices (Krippner and Staples, 2003). It is a vast interdisciplinary science that includes the knowledge about the use of natural pharmaceuticals and the ethnic group from which the same pharmacologically active ingredients belong as well. From Indian Ayurveda to Traditional Chinese Medicine (TCM) of China, from Muti of Africa to Unani medicine of Mughal India, it has been widely practiced in diverse ancient civilizations. The practitioners of traditional medicines follow their traditions, observations, and belief but unaware about the modern theory of treatment. However, their "proof of concept" was based on the end result of using such therapy for generations. The main theme of their treatment was to provide relief to the sufferer, and then find the real cause of the suffering with the belief of "healing from within." Those traditional practitioners did have a great knowledge about herbalism and ethnobotany as well as about human nature which may not be based on modern anatomy, physiology, biochemistry, and genetics. Species of *Hydnocarpus* were used by the ancient people of China for the treatment of Leprosy between 3000 and 2730 BC. Finding of opium poppy and castor bean from Egypt tombs revealed the use of phytomedicine in Africa as far back as 1500 BC. The *Old Testament* also mentions the use of medicinal herb and their cultivation. On the other hand, Ayurveda, the oldest surviving medical system of India about 5000 BC, uses nearly 750 plants like *Aconitum*, *Clitoria*, *Cosinium*, *Shorea*, and many more. Ayurveda is not only the rational use of medicinal plants but the central tenant of Ayur Bijnan (Science of life) is to maintain the harmony of human body and mind with all the elements of the universe, which can help in the management of life-style or microbial diseases including viral fever, meningitis, genital lesions, Amoebiasis, Leishmaniasis, high blood pressure, asthma, diabetes, and even cancer (Jiang et al., 2013). Like Ayurveda and African Traditional Medicine, ancient Chinese Traditional Medicine (CTM) also relies on harmony or balance between the body and soul in addition to uses of herbs. Traditional Chinese healers use herbs as well as other practices like acupuncture, tai-chi, and ai-gong for the treatment and/or prevention of diseases. Acupuncture involves the stimulation of specific nerve points on the body, while tai-chi and ai-gong involve gentle dance-like body movements with mental focus, breathing, and relaxation. Interestingly, Yoga, practiced by Indus-Sarasvati civilization in Northern India over 5000 years ago, mentioned in the Rig Veda, served as mainstream medical practice to maintain health and longevity. Chinese medicine, particularly CTM, also known to treat several dermatological disorders, coronary heart disease, hypertension, stroke, diabetes, atherosclerosis, etc. (Gong et al., 2017). Other schools of traditional medicines includes traditional Korean medicine, Arabic medicine, Haitian folk medicine, Uyghur traditional medicine, Celtic Medicine, Japanese Kampo medicine, and many more.

3.3 ETHNOMEDICINE IN INDIAN CONTEXT

India harbors a rich history of traditional medicine and a great knowledge of medicinal plants. It has been speculated that India's most known traditional medicine "Ayurveda" existed even at the time of Indus valley civilization, and reached its apex during the Vedic period due to its widespread use and significant development. Ayurveda emphasizes to maintain balance between human body, mind, and the nature. With the use of natural resources, mainly plants and minerals, Ayurveda also focuses on exercise, healthy diet, yoga, and meditation to heal any ailments or disease. One of the main pillars of Ayurveda is the medicinal resources used by ethnic communities (Ethnomedicine) and plant-based treatments using different parts of plants including roots, leaves, fruits, bark, or even seeds. The book *Yoga Ratnakara* (1700−1800 CE, unknown author) mentioned the use of opium (*Papaver somniferum*) with minerals as an herbo-mineral formula for diarrhea; while in the book *Bhaisajya Ratnavali* opium and camphor (*Cinnamomum camphora*) are used for acute gastroenteritis, and hemp/marijuana/ganja (*Cannabis indica)* for treating diarrhea. During the British rule the practice of Ayurveda was neglected by the British Indian Government. However, after the independence of India the focus was intensified on Ayurveda and other traditional medical systems. In the past two decades, Ayurveda and other traditional healthcare system such as Yoga, Naturopathy, Unani, Siddha, and Homeopathy (AYUSH) brought under a separate Ministry, and became a part of the Indian National healthcare system with establishment of dedicated colleges and hospitals. Today the AYUSH system of healthcare is running parallel with the mainstream modern medicine in India.

India is inhabited by about 645 indigenous tribes having rich knowledge of wild flora and fauna to manage or cure diseases with other miraculous use. Most of this "traditional wisdom" is undocumented and orally passed over generations. Unfortunately, each time a traditional healer dies without passing their knowledge on to the next generation, the community and the world lose irreplaceable time-tested knowledge about medicinal plants gathered over thousands of years. Indian research policy has turned its attention to validate this untapped knowledge for past few years. However, since 1994 a group of researchers first time initiate a scientific documentation along with validation study to understand why the ancient ethnic communities of Andaman and Nicobar Islands survive, mostly without the help of modern medicine. Five years of rigorous interaction with those communities by establishing personal relationship, the researchers able to unveil some interesting insight of the healthcare culture of Onge, Nicobarese, and Shompen communities through scientific validation of a few useful medicaments which transcribed their traditional "information into modern innovation." Chattopadhyay et al. have identified and validated a number of useful traditional medicaments of Bay Islands used for the management of infection, inflammatory condition, fever, pain, depression, contraception etc., including the antiviral activity of a useful herb *Ophiorrhiza nicobarica* collected from Galathia River basin of Great Nicobar Island, and widely used for skin ailments by the local and tribal communities (Chattopadhyay et al., 2006; Chattopadhyay and Bhattacharya, 2008). Later Bag et al. (2013, 2014) validated its potential and efficacy in cutaneously and vaginally infected animals, which lead to the identification of an alkaloid having potent anti-herpes activity at 1.1−1.5 μg/mL, much lower than acyclovir, by blocking the immediate early transcription of Herpes Simplex Virus types 1 (HSV-1) and 2 (HSV-2). While the extract of *Achyranthes aspera* and its isolated compound Oleanolic acid, used by traditional healers and some tribes of Rajasthan and Gujrat to treat asthma, boils, bronchitis, dysentery, pneumonia, skin diseases, fever, and typhoid (Goyel et al., 2007), was found to inhibit HSV-infection by blocking early stage of virus multiplication (Mukherjee et al., 2013). Similarly the stem bark of *Odina wodier*, a traditional medicine of Jangalmohal, used for curing ulcer, heart diseases and skin infection is found to prevent HSV-1-infected animals by inhibiting the viral multiplication through modulation of host immunity (Ojha et al., 2013), while the isolated chlorogenic acid regulate COX-2-dependent prostaglandin-E2 and TLR 4 signaling pathway (Ojha et al., 2014). Panda et al. (2017) have screened traditional medicinal plants used by tribes of Similipal Biosphere Reserve, Odisha and reported antiviral activity of few plants against Enterovirus-71.

3.4 VIRAL DISEASES—A GLOBAL HEALTH CONCERN

An inefficient virus kills its host, while a clever virus stays with it.

James Lovelock

Viruses are acellular ultra-microscopic metabolically inert obligate intracellular parasites of cellular hosts (Chattopadhyay et al., 2009b), associated with diverse diseases and are threats to human, wildlife, and livestock (Malosh et al., 2017; Marston et al., 2017; Prkno et al., 2017;

Akbari and Elmi, 2017; Vu and Misra, 2018). Many are incurable and have no effective antiviral drugs except a very few (forty) US FDA approved antiviral drugs available for the management of some viral diseases. Complicated life cycle, frequent mutability and unique disease manifestation of each virus are the main reason for the public health concern. Treatment of viral diseases is immensely challenging because of their size, genetic variation, configuration of surface molecules, invasion strategies, mode of transmission, replication and persistent nature with rapid mutability, especially RNA viruses like influenza, severe acute respiratory syndrome (SARS) coronavirus etc., with a very high mutation rate, around 1/genome/replication (Elena and Sanjuan, 2005; Elena et al., 2000). High cost and side effects of present drugs used for treatment are also major contributing factor making antiviral drug development a daunting task. The antiviral drugs Acyclovir against HSV, Zidovudine against human immunodeficiency virus (HIV), Lamivudine against hepatitis B virus (HBV), Ganciclovir against Cytomegalovirus (CMV), Amantadine against influenza, and Pegylated IFN alpha against Hepatitis C virus (HCV) have been proven to be successful at least to some extent. But majority of viruses and viral diseases including Zika, Dengue, Ebola, Marburg, Rabies, Chikungunya, Rotavirus, HIV, etc. do not have any targeted drugs or vaccines till date. The unmet need for successful antivirals can only be fulfilled by thinking *out of the box* and ancient knowledge of ethnomedicine can be a master stroke against these deadly diseases.

3.5 ETHNOMEDICINE AND VIRUS

Ethnomedicines have been found to be effective as antiviral therapy, due to the following perspectives:

- Plants produce thousands of compounds as secondary metabolites than they require for their survival and propagation (primary metabolites). These secondary metabolites, grouped as phenolics (flavonoids, quinones, coumarins, tannins, and anthocyanins), terpenoids (sterols, saponins, essential oils, and cucurbitacins), alkaloids, proteins, peptides, etc., are species specific and widely varied in structure and bioactivity. These compounds are foul-smelling, toxic and are synthesized mainly as weapons of defense against predators and pathogens.
- Plant extracts are being validated for antiviral activity and use in treatments owing to the fact that many viruses are intractable to the typical antivirals. Also the effective life span of most of the antiviral is limited.

- The problems of antiviral resistance, latency, recurrence, and fastidious spread of new or emerging strains, as in the case of HIV and SARS have compelled the virologists to look for better alternatives in nature, especially for people who have very little or no access to the expensive drugs (Chattopadhyay and Naik, 2007).

Plants have the miraculous treasure of numerous compounds with abilities to cure diseases and make our immunity strong. According to WHO fact sheet 2008, about 80% of population in Asia and Africa depend on plant-derived traditional medicine. Beside Asian countries, South America, Australia, and several countries of the European Union have documented ethnomedicine against various diseases including viral diseases.

The screening of antivirals from medicinal plants of traditional use is based on their wisdom, faith, availability and positive results for generations for curing ailments or diseases. Researchers are exploring those traditional plant-based medicaments to identify new source of antivirals. Most of these plants or plant-derived phytocompounds are reported to have antiinflammatory, antioxidant, antipyretic, antihelminthic, antifungal, antibacterial, and antiviral activities (Chattopadhyay et al., 2009a,b). Drug-like activities of plants or plant products, including antiviral activity, have been attributed to the secondary metabolites of plants, mainly alkaloids, flavonoids, saponins, quines, terpenes, lignans, tannins, polysaccharids, steroidal glycoside, thiosulfates, proanthocyanidin, and proteins.

3.5.1 Screening Models for Herbal Antiviral Agents and Their Value in Drug Discovery

- *In vitro primary screening*

  There are different *in vitro* and *in vivo* methods used to study the antiviral activities of plant products, but the most commonly used *in vitro* method for preliminary screening is cytopathic effect reduction (CPE), dye exclusion and plaque reduction assay (PRA) or plaque assay. Another rapid and sensitive *in vitro* procedure of evaluating antiviral agents is based on Spectrophotometric assessment for viability of virus- and mock-infected cells. Detailed methodologies for studying antiviral activities of plant products are available elsewhere (Chattopadhyay et al., 2009a,b, 2015).
- *High throughput screening*

  High-capacity antiviral screening assay in 96-well microtiter plates using tetrazolium dye 3-(4,5-dimethylthiazol-2-yl)-5-(3-carboxymethoxyphenyl)-2-(4-sulfop henyl)-2H-tetrazolium (MTS) or 3-(4,5-dimethylthiazol-2-yl)-5-(3-carboxymethoxyphenyl)-2-(4-sulfop henyl)-2H-tetrazolium (MTT) is convenient as well as cost and time saving. Other methods include TaqMan PCR, and microscopic inspection of the cultures.

3.6 MODE OF ACTION OF PLANT-DERIVED ANTIVIRAL AGENTS

Catechin, present in the globally most popular beverage green tea leaves, when fermented into theaflavins can neutralize bovine rotavirus and coronavir (Lin et al., 1997). Another common herb *Ocimun basilicum* or the sweet Basil of India and China has broad spectrum antiviral activity. The aqueous and ethanolic extract along with purified apigenin, linalool, and ursolic acid showed strong activity against HSV-1 (Bag et al., 2012), adenovirus 8 (ADV-8), Coxsackievirus B type-1 (CVB1) (Chattopadhyay and Naik, 2007). While Isoborneol, a monoterpene of essential oils isolated from Egyptian plant *Melaleuca alternifolia*, exhibited anti-HSV-1 activity by inactivating HSV-1 replication within 30 min of exposure (Armaka et al., 1999). Similarly *Vatica cinerea* from Vietnam is reported to inhibit HIV-1 replication (Zhang et al., 2003). The plant *Cimicifuga racemosa* (black cohosh) having antidepressant activity was also reported to have antiviral activity against human retroviruses by inhibiting HIV-1 reverse transcriptase (Sakurai et al., 2004). It is important to note that the viral diseases caused by Picornavirus and Rhinovirus do not have any drugs till date, while the plant-derived compounds chrysoplenol-C, and its glycoside has virucidal effect against both group of viruses (Wei et al., 2004). Scientists have also explored Himalayan flora used in traditional medicine against viral diseases (Amber et al., 2017). Ethnomedicine from India, Pakistan, China, and Nepal has been explored as source of antivirals against Influenza virus, Rhinovirus, Adenovirus, Coronavirus, and Respiratory Syncytial virus (RSV). Different compounds including monoterpenoids, flavonoids, triterpenoids, iridoid glycosides, sesquiterpenes, benzoic acids, and phenolics have strong antiviral potential. Recent emergence of deadly dengue, which is now a public health concern worldwide, has also been shown to be prevented by plant-derived drug. Lupeol isolated from *Maytenus gonoclada* has shown activity against dengue virus (Silva et al., 2017). Extracts from *Carica papaya* have now been used for the treatment of dengue in hospitals (Ahmad et al., 2011; Dharmarathna et al., 2013) mainly to prevent the reduction on number of platelet due to platelet aggregation, although the anti-dengue or anti-aggregation of platelet activity of papaya extract has not been scientifically proved till date.

In folk medicine, papaya latex is used to cure dyspepsia, external burns and scalds, while its seeds and fruits have excellent antihelminthic and antiamebic activities. However, Chinnappan et al. (2016) have found that the leaf extract of papaya could possess a dengue-specific neutralizing effect on dengue virus-infected plasma that may exert a protective role on platelets. Luteolin, a bioflavonoid isolated from several dietary and medicinal plants, has been shown to have activity against HSV-1 (Ojha et al., 2015), dengue (Peng et al., 2017), Epstein–Barr virus (Wu et al., 2016), Japanese Encephalitis (JE) (Fan et al., 2016), and Chikungunya (Murali et al., 2015). Ginseng, a well-known medicinal herb, has been used in traditional medicine for thousands of years, and was found to be effective for treating influenza (Yoo et al., 2012) and HIV (Park et al., 2014); while in a clinical trial, Ginseng was found to help in curing HBV infection (Choi et al., 2016). Ethnopharmacological use of essential oil extract of three traditional Cretan aromatic plants in Eastern Mediterranean region and Near East claimed to be effective in the prevention and treatment of upper respiratory tract infections of bacterial and viral etiology (Duijker et al., 2015). While an Egyptian plant *Nigella sativa* extract tested against influenza patients showed better activity in people who cannot be treated with interferon-alpha (Barakat et al., 2013). A great deal of scientific research is being conducted to understand the mechanisms by which plant products exert their antiviral effects. Usually, plant-derived compounds exert antiviral effects through diverse mode and mechanism including (1) inhibition by autophagy, (2) generation of reactive oxygen species (ROS), (3) change in viral gene expression, (4) inhibition of viral entry to host cell including attachment and penetration, (5) inhibition of different steps of replication, (6) inhibition of viral release, as well as modulating the host immune parameters, which are briefly presented in Fig. 3.1.

3.7 MECHANISM OF ACTION OF PLANT-DERIVED ANTIVIRAL AGENTS

- *Viral inhibition by autophagy*

  Pentagalloylglucose found in many traditional medicinal herbs exhibits several bioactivities including antiinflammatory, anticancer, antioxidant, and antiviral effects. It has been shown to be effective against HSV-1 by induction of autophagosomes that engulfed HSV-1 virion (Pei et al., 2011). Triterpene glycyrrhizic acid, the major

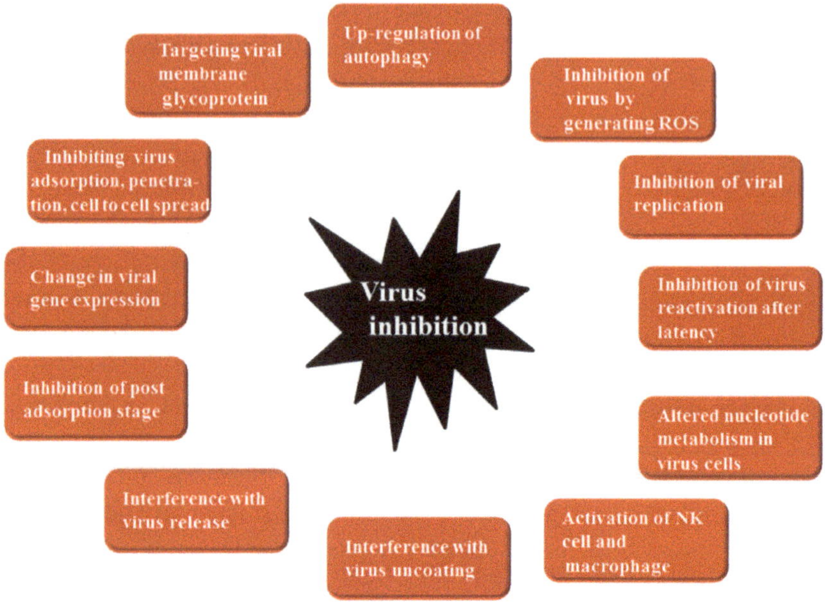

Figure 3.1 Mode of actions of plant-derived antiviral agents.

compound isolated from *Glycyrrhiza glabra*, has also been effective against HSV by promoting autophagy (Laconi et al., 2014).

- *Inhibition of virus replication*

  It has been reported that a cranberry extract known as Oximacro and its purified constituent proanthocyanidins can inhibit HSV-1 and HSV-2 replication *in vitro* and prevent the virus adsorption by targeting the viral envelop glycoprotein gD and gB (Terlizzi et al., 2016). The dichloromethane extract from *Angelica archangelica* L. fruit is a potential antiviral against HSV-1 by inhibiting viral replication (Rajtar et al., 2017). Fig. 3.1 depicts diverse mechanism of action by which the natural compounds are reported to be effective against different viruses. Resveratrol, a natural component of certain foods, such as grapes, has been shown to limit HSV-1 lesion formation by inhibiting viral replication (Docherty et al., 2005). Pentacyclic triterpenes isolated from birch bark (*Betula* species) extract of Eurasian and North America can inhibit HSV-1, especially acyclovir-sensitive and acyclovir-resistant clinical isolates of HSV (Heidary Navid et al., 2014). Sulfated polysaccharide from *Caesalpinia ferrea* can inhibit Poliovirus replication along with virus adsorption, steps after penetration and viral protein synthesis (Lopes et al., 2013). Oxyresveratrol, purified from Thai traditional medicinal plant *Artocarpus lakoocha*, has been shown to possess therapeutic effects in mice infected with HSV-1, by inhibiting early and late phase of viral DNA replication. While the anti-HSV effects of soybean-derived isoflavonoids were reported to be due to the inhibition of viral DNA replication (Argenta et al., 2018).

- *Inhibition of virus by generation of ROS*

  Oligomeric stilbenoids, polyphenolic phytoalexins from some plants including *Cannabis sativa* inhibit viral growth by generating ROS (Chen et al., 2012).

- *Change in virus gene expression*

  The n-Docosanol, a behenyl alcohol (22 carbon saturated fatty alcohol) from various plants, used traditionally as an emollient, emulsifier, and thickener in cosmetics, and nutritional supplement, inhibits herpes labialis and immediate early gene expression of the HSV (Treister and Woo, 2010).

- *Inhibition of viral entry*

  The sulfated galactans from the red seaweeds *Gymnogongrus griffithsiae* and *Cryptonemia crenulata* is reported to inhibit the entry of HSV-1 and HIV-2 (Talarico et al., 2004). *Mentha suaveolens* essential oil and its main component piperitenone oxide exert anti-HSV activity by inhibiting viral adsorption (Civitelli et al., 2014). Moreover, piperitenone oxide from *Mentha longifolia* could interfere with some redox-sensitive cellular pathways of host cell exploited for viral replication and thus inhibits viral growth and proliferation. While the virucidal effect of peppermint oil, the essential oil of *Mentha piperita* against HSV can be attributed to the interaction of peppermint oil with viral adsorption at the host cell surface preventing the virion to enter the host cell for establishing infection (Schuhmacher et al., 2003).

- *Interference with virus release*

  Thai medicinal plant *Dunbariabella prain* interfere with HSV release (Akanitapichat et al., 2006).

3.8 ANTIVIRAL ETHNO-PHARMACOLOGY OF MAJOR CLASSES OF COMPOUNDS

Plants have always been a major source of various bioactive compounds. Many of them have shown anticancer, antibacterial, antiparasitic activities.

Here we will briefly discuss various groups of plant-derived compounds having antiviral activities.

- *Alkaloids*

  Alkaloids are a group of naturally occurring slightly acidic compounds that mostly contain basic nitrogen atoms. More than 10,000 different alkaloids have been discovered in 300 plant families. This group of compounds has been shown to be effective against viruses like hepatitis B (Li et al., 2008), HSV (Ren et al., 2010), dengue (Hishiki et al., 2017), etc.
- *Phenolics*

  Phenols, also called phenolics, are a class of compounds consisting of a hydroxyl group bonded directly to an aromatic hydrocarbon group. Phenolic compounds are classified as simple phenols or polyphenols, based on the number of phenol units in the molecule. This group is an excellent candidate for antimicrobial and antiviral research. These compounds was found to be effective against HSV (Likhitwitayawuid et al., 2005), abies virus (Chavez et al., 2006), Influenzavirus (Ha et al., 2016), and many more.
- *Coumarins*

  Coumarins are organic colorless crystalline benzopyrene, naturally found in many plants. In fact its name comes from a French term for the tonka bean coumarou. Coumarins have been proven to be potent against many viruses like HIV (Lin et al., 2011b), HSV (Ghannadi et al., 2014), Dengue and Chikungunya (Gomez-Calderon et al., 2017).
- *Flavones, flavonoids, and flavonols*

  Flavonoids are a class of secondary metabolites of plants and fungus, while flavones are a class of flavonoids and are synthesized in response to microbial infections. Hence, they are broad spectrum antimicrobial agents. Flavonoids represent an important natural source of antiretroviral agents, especially for AIDS therapy (Pasetto et al., 2014).
- *Terpenoids and essential oils*

  Essential oils are the plant-derived phenolic compounds with a C3 side chain and at a lower level of oxidation without oxygen. The oils that are enriched in isoprene structure are called terpenes; but when they contain additional elements like oxygen they become "terpenoids." These compounds are active against many viruses (Wen et al., 2007; Lu et al., 2015).

- *Quinones*

  Quinones are a class of organic compounds, named after its prototype member 1,4-benzoquinone or cyclohexadienedione, often called as "quinone." It has been shown to be effective against Dengue (Laurent et al., 2005).
- *Tannins*

  Tannins are a group of polymeric plant phenolics. Studies show that tannins are active against many viruses including enveloped viruses like Influenza virus H3N2 and H5N3, HSV, Vesicular Stomatitis virus, Sendai virus, and Newcastle disease virus; as well as against nonenveloped viruses like Poliovirus, Coxsackievirus, Adenovirus, Rotavirus, Feline calicivirus, etc. (Ueda et al., 2013).
- *Lignans*

  Lignans are polyphenols found in plants. These compounds have also shown potent antiviral action (Cui et al., 2014).

Table 3.1 represents a selected list of ethnomedicinal plants and plant-derived compounds that showed promising antiviral activities against some viruses which are of concern in recent time. Dengue, a flavivirus that causes panic in developing countries including India, has sought attention from government and global research communities to look for new treatment. The use of papaya leaf extract in dengue fever, discussed earlier, while other plant-derived compounds including pentacyclic oxindole (Reis et al., 2008), chebulagic acid (Lin et al., 2011a), punicalagin (Lin et al., 2011b), baicalcin (Zandi et al., 2012), and alpha-mangostin (Tarasuk et al., 2017) reported to have anti-dengue activity. These compounds in general can modulate host cell immune system so that the host cell can fight against the dengue virus as the virus attack immune system, invade platelet, and cause postinfection complicacy including cytokine storm, particularly in the presence of preexisting heterologous antibody due to infection by another serotype of the same virus in same individual. Hepatitis caused by HBV, a common concern for liver dysfunction worldwide, has no effective treatment or vaccine. From ethnomedicinal wisdom, compounds such as oxymatrin (Chen et al., 2016) and cinamic acid (Amano et al., 2017) showed inhibitory effect on HBV by inhibiting viral DNA replication. The herpes caused by HSV-1 and HSV-2 in 90% population, via sexual or close contact from infected to uninfected individual are of great concern due to its silent epidemic nature. In spite of available gold-standard antiviral drug acyclovir, the HSV cannot be eradicated or eliminated from the host nor can prevent latency, rather it inhibits virus

Table 3.1 Selected Ethnomedicines and phytocompounds with antiviral activity

Plants	Extract or compound	Virus	Mechanism	Reference
Peucedanum salinum	Methanol extract	Adenovirus (ADV) type 5	Inhibit replication	Gao et al. (2011)
Radix lithospermi	Naphthaquinones Shikonin	Adenovirus (ADV) type 3	Inhibit apoptosis and hexon protein expression of virus	Rajtar et al. (2012)
Uncaria tomentosa	Pentacyclic oxindoleAlkaloid fraction	DENV-2	Immunomodulation	Reis et al. (2008)
Terminalia chebula Retz	Hydrolyzable tanninChebulagic acid and Punicalagin	DENV, HCV, HCMV, HIV, RSV	Inactivate free virus particles and inhibit virus entry	Lin et al. (2011a)
Scutellaria baicalensis root	Flavone Baicalein	DENV-2	By type I interferon induction and virus adsorption	Zandi et al. (2012)
Garcinia mangostana Linn	Xanthones α-mangostin (α-MG)	DENV-2	Immunomodulation	Tarasuk et al. (2017)
Ficus carica Latex extracts	3.4-Dihydroxy benzoic acid, p-OH-phenyl acetic acid, *N*-argenine	ECV-11, HSV-1, and ADV	Replication	Lazreg Aref et al. (2011)
Halenia elliptica Traditional Tibetan medicine (TTM)	2-Methyl chromones 8-methoxy-2-methyl-4H-1-benzopyran-4-one	Hepatitis B virus (HBV)	Inhibiting HBV antigens secretion	Zhang et al. (2015)
Sophora subprostrata	Matrine-type alkaloid Oxymatrine	HBV	Inhibit replication	Yang et al. (2012)

Artemisia scoparia	D-Glucopyran oside Scoparamide A	HBV	Inhibit DNA replication	Geng et al. (2015)
Sophora flavescens root	Matrine-cytisine alkaloid	HBV	Viral growth	Zhang et al. (2016)
Cinnamomum verum bark	Cinnamic acid	HCV	Induction of oxidative stress	Amano et al. (2017)
Ficus benjamina (ethanol extract of leaves)	Quercetin 3-*O*-rutinoside, Kaempferol 3-*O*-rutinoside and Kaempferol 3-*O*-robinobioside	HSV-1	Virus multiplication	Yarmolinsky et al. (2009)
Punica granatum (many plants)	Gallic acid, gallic acid gold nanoparticles	HSV	Attachment and penetration	Halder et al. (2018)
Sophora subprostrata	Propolis	HSV	Pretreated before infection	Nolkemper et al. (2010)
Mallotus peltatus Leaf, bark	Pentacyclic triterpeneUrsolic acid	HSV-1, HSV-2	Inhibiting early stage of multiplication	Bag et al. (2012)
A. aspera	Triterpene acidOleanolic acid	HSV-1 and HSV-2	Modulation of early immunological parameters	Mukherjee et al. (2013)
Ophiorrhiza nicobarica Balkr	Indole alkaloid 7-methoxy-1-methyl-4,9-dihydro-3*H*-pyrido[3,4-*b*] indole	HSV-2	Immediate early transcription by blocking recruitment of LSD-1 by HCF-1	Bag et al. (2013)
Caesalpinia ferrea	Sulfated polysaccharide	HSV, poliovirus	Inhibit adsorption, postpenetration, and synthesis of viral proteins	Lopes et al. (2013)

(*Continued*)

Table 3.1 (Continued)

Plants	Extract or compound	Virus	Mechanism	Reference
Berberis vulgaris, *B. aristata*, *Tinospora cordifolia* *Coptis chinensis*	Isoquinoline alkaloids Berberine from root, stem, bark	HSV	Modulating cellular signaling, including p53 NF-κB, and mitogen-activated protein kinase	Song et al. (2014)
Mentha suaveolens	Essential oilsPiperitenone oxide	HSV	Late phase of replication	Civitelli et al. (2014)
Andrographis paniculata	Diterpenoid lactone 3,19-isopropylidenean drographolide	HSV	Inhibit postentry, early gene expression; suppress ICP8 transcription, DNA replication and gD expression	Kongying yoes et al. (2016)
Curcuma longa rhizome	Curcumin	HCMV	Protein expression of Hsp90α	Lv et al. (2015)
Angelica archangelica	Xanthotoxin, Bergapten, Imperatorin, Phellopterin, Isoimperatorin (isopentenyloxy moiety at C-8 position)	HSV-1 and Coxsackievirus B3	Reduce viral titer	Rajtar et al. (2017)
Rheum root	Anthraquinone derivatives Physcion, Emodin, Rhein	HIV	HIV-1 Reverse Transcriptase (RT)	Esposito et al. (2016)

Rheum officinale baill	Whole extract	HIV-1	RNase H and RT-associated RNA Dependant DNA Polymerase	
Schisandra chinensis	Schisandrin B Deoxyschizandrin	HIV-1	Affect HIV-1 RT	Xu et al. (2015)
Cassia sieberiana *Vitex doniana, Croton megalobotrys*	Methanol extract of root and bark	HIV	Replication	Tietjen et al. (2016)
Cynomorium songaricum Rupr.	Polysaccharide (water soluble)	HIV	Multiplication	Tuvaanjav et al. (2016)
Coffea arabica seed	Polyphenol caffeic acid	Influenza A virus	Inhibits multiplication	Utsunomiya et al. (2014)
Allium sativum Bulb *Plumbago indica*	Allicin and plumbagin	Influenza A (H1N1)	Viral adsorption, nucleoprotein synthesis and polymerase activity	Chavan et al. (2016)
Laggera pterodonta	Pterodontic acid	H1 subtype of Influenza A	Block nuclear export of viral RNP complexes	Guan et al. (2017)
Forsythia suspense (Thunb.) Vahl fruit	Forsythoside A	Influenza A	Reduction of influenza M1 protein	Law et al. (2017)
Dendrobium nobile	Dendrobine	Influenza A	Growth	Li et al. (2017)
Rheum palmatum	Chrysophanol and aloe-emodin	Japanese Encephalitis	IFN-γ triggering host innate immune response	Chang et al. (2014)
CTM hochu-ekkito (HET) or bu-zhong-Žyi-qi-tang (Chinese)	Mixture of 10 plants	MCMV	Induction of IFN-γ and perforin-mediated cytotoxicity by natural killer (NK) cell activation	Hossain et al. (1999)
Coptidis rhizome Rhizome extract	Not detected	RSV	Induction of type I interferon-related signaling	Lee et al. (2017)

HCV, hepatitis C virus; RSV, respiratory syncytial virus; HSV, herpes simplex virus; DENV, dengue virus; HCMV, human cytomegalo virus; MCMV, murine cytomegalo virus; HIV, human immunodeficiency virus; HBV, hepatitis B virus; ADV, adenovirus; JE, Japanese encephalitis virus; ECV, echovirus.

replication by blocking thymidine kinase and/or DNA polymerase of the virus, and thereby promote frequent emergence of drug-resistance strains deficient in thymidine kinase which is also a great concern. HSV is used as a standard model for screening of antivirals and its easy laboratory application made it a choice for researchers to study viruses. A vast number of anti-HSV plant extracts and compounds have been cited. But majority of them are from *in vitro* studies and have been shown to prevent adsorption to cell-to-cell spread. However, latency in HSV is the real twist and we need drugs that can eradicate the virus from brain to prevent its spread. Ursolic acid (Bag et al., 2012), oleanolic acid (Mukherjee et al., 2013), berberine (Song et al., 2014), masilinic acid (Zahmanov et al., 2015), and hermaline (Bag et al., 2013) have been reported to have anti-HSV activity by inhibiting entry, immediate early, early and late replication. Similarly, antiviral plant extracts and compounds have also been reported for Influenzavirus, Poliovirus, JEvirus, and RSV.

3.9 CHALLENGES FOR ETHNOMEDICINES AS ANTIVIRALS

- Many of the viral diseases are incurable and do not have a vaccine yet. Discovery of safe, effective, and inexpensive antivirals from natural source, particularly from ethnomedicinal practices are among the top priorities and challenges in the future.
- Certain natural products show synergistic activities in combination with other natural products or with existing antivirals particularly useful against herpes and retroviral infections. Thus, proper combinatorial techniques to enhance the antiviral activities of these natural products need to be prioritized.
- Recombinant viral vectors mimicking infection and expressing firefly luciferase maker gene are being used in antiviral screening. Therefore, the safety issues involved while screening the antivirals are being met and should be the top most priority in future research as well (Esimone et al., 2005).
- Plants, other than being the sources of natural products, serve as efficient systems for the production of vaccines and pharmaceutical grade peptides/proteins. Several other vaccine antigens have been successfully expressed in plants, after the first subunit vaccine HBsAg was produced in 1992 (Ma Julian et al., 2003; Glenz and Warzecha, 2006).
- Isolation of active ingredient from plants is a challenge and demands continuous up-gradation of technique and laboratory infrastructure

through the years, so that those are available at an affordable cost to the public. In China and India, crude extracts are effectively used in healthcare although it is difficult to get these extracts approved by the FDA. In countries where resources are limited, government sponsored explorations will serve as a gateway for merging modern drug discovery with traditional/conventional medicine.

- Many of these natural products prevent the entry of the virus into the host cell or target specific enzymes of the virus, which may be the molecular basis of viral drug resistance. An alternative mechanism of action can be the potential target to tackle the emerging, reemerging and drug-resistant viral infections. Thus, elucidation of mode as well as mechanism of actions of the potential antivirals is equally important in research.
- Herbal medicines are being used individually or in combination, but their effects on the living systems are not scientifically documented. Results of clinical findings of using combination of plants or plant products like coadministration of kava-kava (*Piper methysticum*) and St. John's wart (*Hypericum perforatum*) lead to hepatotoxicity (Musch et al., 2006). This type of information should be available to the healthcare providers practising traditional medicine.
- It is equally important that the experiments on cytotoxicity of traditionally used plant extracts may be conducted as randomized, double-blind, placebo-controlled multicentric clinical trials so that the incorporation of a particular herbal remedy into the healthcare system is safe, besides being effective.
- Several plant species have become extinct, while many phytocompounds mentioned in the classics may have undergone change with time as well as due to anthropogenic and environmental factors. The nature of the natural compound is complex and requires a flexible standardization process from time to time.
- Screening a potential antiviral from millions of plant species is a tedious and time-taking process. A wide variety of computational techniques will help the chemists to virtually screen a huge library to a manageable size.
- Drug discovery based on ethnomedicine follow "reverse pharmacology," where the already documented traditional plants having healing ability can be further experimented as drug candidates in clinical research. The critical pharmacopoeia tests such as the dissolution time, microbial, pesticide, and heavy metal contamination must be in accordance with the standards globally accepted protocols of antiviral screening (Chattopadhyay et al., 2009b).

3.10 CONCLUSION

Traditional medicines have been increasingly used by diverse communities in many parts of the world, due to its important role in maintaining good health with increasing awareness and research. According to the WHO, the goal of "health for all" cannot be achieved without the incorporation of herbal medicines in primary healthcare system. Traditional medicine has a long history in disease control and public health management. Today, an increasing number of plants used in traditional system of medicine are reported to have diverse activities in infectious diseases, particularly in viral infections, and thus can be new sources of antivirals. The age-old medicinal system is being revisited to tackle the emerging public health issues. Thus, there is a significant need to enhance drug discovery process with respect to natural products, not just for the next 10 or 20 years, but for the next 20 $\pm$ 40 years and beyond, because there are new diseases waiting to discover or invade. Complex structures of phytochemicals, synergistic activity and inadequate validation could not provide the drug which can have the potential to develop as a drug or a drug candidate to inhibit or eradicate the selected viruses alone or in combination with the existing antivirals without much harm to the host. With the advent of new approaches to drug discovery, innovative strategies would be required, that will reveal and contribute the full range of chemical diversity of these valuable natural products to the drug discovery process. It is also important that the development of plants and other natural products for medicinal and other biological purposes should be potentiated on a sustainable and renewable basis in local environment. Thus, pharmaceutical companies should consider developing local resources and decentralizing selected aspects of their research operations globally. Continuous research by using the advanced technology and bowfins can gift us the lead or leads to develop novel antiviral(s) from the treasure of Ethnomedicine.

REFERENCES

Ahmad, N., Fazal, H., Ayaz, M., Abbasi, B.H., Mohammad, I., Fazal, L., 2011. Dengue fever treatment with *Carica papaya* leaves extracts. Asian Pac. J. Trop. Biomed. 1, 330–333.

Akanitapichat, P., Wangmaneerat, A., Wilairat, P., Bastow, K.F., 2006. Anti-herpes virus activity of *Dunbaria bella* Prain. J. Ethnopharmacol. 105, 64–68.

Akbari, M., Elmi, R., 2017. Herpes simplex virus and human papillomavirus co-infections in hyper immunoglobin E syndrome presenting as a conjunctival mass lesion. Case Rep. Med. 2017, 1650841.

Amano, R., Yamashita, A., Kasai, H., Hori, T., Miyasato, S., Saito, S., et al., 2017. Cinnamic acid derivatives inhibit hepatitis C virus replication via the induction of oxidative stress. Antiviral Res. 145, 123–130.

Amber, R., Adnan, M., Tariq, A., Mussarat, S., 2017. A review on antiviral activity of the Himalayan medicinal plants traditionally used to treat bronchitis and related symptoms. J. Pharm. Pharmacol. 69, 109–122.

Argenta, D.F., Bidone, J., Koester, L.S., Bassani, V.L., Simoes, C.M.O., Teixeira, H.F., 2018. Topical delivery of coumestrol from lipid nanoemulsions thickened with hydroxyethylcellulose for antiherpes treatment. AAPS Pharm. Sci. Tech. 19 (1), 192–200.

Armaka, M., Papanikolaou, E., Sivropoulou, A., Arsenakis, M., 1999. Antiviral properties of Isoborneol, a potent inhibitor of herpes simplex virus type 1. Antiviral Res. 43, 79–92.

Bag, P., Chattopadhyay, D., Mukherjee, H., Ojha, D., Mandal, N., Sarkar, M.C., et al., 2012. Anti-herpes virus activities of bioactive fraction and isolated pure constituent of *Mallotus peltatus*: an ethnomedicine from Andaman Islands. Virol. J. 9, 98.

Bag, P., Ojha, D., Mukherjee, H., Halder, U.C., Mondal, S., Chandra, N.S., et al., 2013. An indole alkaloid from a tribal folklore inhibits immediate early event in HSV-2 infected cells with therapeutic efficacy in vaginally infected mice. PLoS One 8, e77937.

Bag, P., Ojha, D., Mukherjee, H., Halder, U.C., Mondal, S., Biswas, A., et al., 2014. A dihydro-pyrido-indole potently inhibits HSV-1 infection by interfering with the viral immediate early transcriptional events. Antiviral Res. 105, 126–134.

Barakat, E.M., Wakeel, L.M.E.L., Hagag, R.S., 2013. Effects of *Nigella sativa* on outcome of hepatitis C in Egypt. World J. Gastroenterol. 19, 2529–2536.

Chang, S.J., Huang, S.H., Lin, Y.J., Tsou, Y.Y., Lin, C.W., 2014. Antiviral activity of *Rheum palmatum* methanol extract and Chrysophanol against Japanese encephalitis virus. Arch. Pharm. Res. 37, 1117–1123.

Chattopadhyay, D., 2009. Ethnomedicinal phytophores in disease management. Internat J. Biomed. Pharmaceut. Sci. 3 (Spl1), 1–125. ISBN: 9784-903313-35-1.

Chattopadhyay, D. (Ed.), 2010. Ethnomedicine: a source of complementary therapeutics a review. Research Signpost Trivandrum, ISBN:978-81-308-0390-6.

Chattopadhyay, D., Bhattacharya, S.K., 2008. Ethnopharmacology: a new search Engine for the development of Antivirals from naturaceuticals. In: Eddouks, M. (Ed.), Handbook of Ethnopharmacology. Research Signpost, India, pp. 129–197. Chapter 5.

Chattopadhyay, D., Naik, T.N., 2007. Antivirals of ethnomedicinal origin: structure–activity relationship and scope. Mini Rev. Med. Chem. 7, 275–301.

Chattopadhyay, D., Arunachalam, G., Mandal, A.B., Bhattacharya, S.K., 2006. Dose dependent therapeutic antiinfectives from Ethnomedicines of Bay Islands. Chemotherapy 52 (3), 151–157.

Chattopadhyay, D., Mukherjee, H., Bag, P., Ghosh, S., Samanta, A., Chakrabarti, S., 2009a. Ethnomedicines in antiviral drug discovery. Int. J. Biomed. Pharmaceut. Sci. 3 (1), 1–25.

Chattopadhyay, D., Sarkar, M.C., Chatterjee, T., Sharma Dey, R., Bag, P., Chakraborti, S., et al., 2009b. Recent advancements for the evaluation of anti-viral activities of natural products. Nat. Biotechnol. 25, 347–368.

Chattopadhyay, D., Ojha, D., Mondal, S., Goswami, D., 2015. Validation of antiviral potential of herbal ethnomedicine. In: Mukherjee, P.K. (Ed.), Evidence-Based Validation of Herbal Medicine. Elsevier Science, USA, pp. 175–200. Chapter 8.

Chavan, R.D., Shinde, P., Girkar, K., Madage, R., Chowdhary, A., 2016. Assessment of anti-influenza activity and hemagglutination inhibition of *Plumbago indica* and *Allium sativum* extracts. Pharmacogn. Res. 8, 105–111.

Chavez, J.H., Leal, P.C., Yunes, R.A., Nunes, R.J., Barardi, C.R., Pinto, A.R., et al., 2006. Evaluation of antiviral activity of phenolic compounds and derivatives against rabies virus. Vet. Microbiol. 116, 53–59.

Chen, X., Qiao, H., Liu, T., Yang, Z., Xu, L., Xu, Y., et al., 2012. Inhibition of herpes simplex virus infection by oligomeric stilbenoids through ROS generation. Antiviral Res. 95, 30–36.

Chen, J.X., Shen, H.H., Niu, M., Guo, Y.M., Liu, X.Q., Han, Y.Z., et al., 2016. Anti-hepatitis B virus effect of matrine-type alkaloid and involvement of p38 mitogen-activated protein kinase and tumor necrosis factor receptor-associated factor 6. Virus Res. 215, 104–113.

Chinnappan, S., Ramachandrappa, V.S., Tamilarasu, K., Krishnan, U.M., Pillai, A.K., Rajendiran, S., 2016. Inhibition of platelet aggregation by the leaf extract of *Carica papaya* during dengue infection: an *in vitro* study. Viral Immunol. 29, 164–168.

Choi, S.H., Yang, K.J., Lee, D.S., 2016. Effects of complementary combination therapy of Korean red ginseng and antiviral agents in chronic hepatitis B. J. Altern. Complement. Med. 22, 964–969.

Civitelli, L., Panella, S., Marcocci, M.E., De Petris, A., Garzoli, S., Pepi, F., et al., 2014. In vitro inhibition of herpes simplex virus type 1 replication by *Mentha suaveolens* essential oil and its main component piperitenone oxide. Phytomedicine 21, 857–865.

Cui, H., Xu, B., Wu, T., Xu, J., Yuan, Y., Gu, Q., 2014. Potential antiviral lignans from the roots of *Saururus chinensis* with activity against Epstein–Barr virus lytic replication. J. Nat. Prod. 77, 100.

Dharmarathna, S.L., Wickramasinghe, S., Waduge, R.N., Rajapakse, R.P., Kularatne, S.A., 2013. Does *Carica papaya* leaf-extract increase the platelet count? An experimental study in a murine model. Asian Pac. J. Trop. Biomed. 3, 720–724.

Docherty, J.J., Fu, M.M., Hah, J.M., Sweet, T.J., Faith, S.A., Booth, T., 2005. Effect of resveratrol on herpes simplex virus vaginal infection in the mouse. Antiviral Res. 67, 155–162.

Duijker, G., Bertsias, A., Symvoulakis, E.K., Moschandreas, J., Malliaraki, N., Derdas, S.P., et al., 2015. Reporting effectiveness of an extract of three traditional Cretan herbs on upper respiratory tract infection: results from a double-blind randomized controlled trial. J. Ethnopharmacol. 163, 157–166.

Elena, S.F., Sanjuan, R., 2005. Adaptive value of high mutation rates of RNA viruses: separating causes from consequences. J. Virol. 79, 11555–11558.

Elena, S.F., Miralles, R., Cuevas, J.M., Turner, P.E., Moya, A., 2000. The two faces of mutation: extinction and adaptation in RNA viruses. IUBMB Life 49, 5–9.

Esimone, C.O., Grunwald, T., Wildner, O., Nchinda, G., Tippler, B., Proksch, P., et al., 2005. *In vitro* pharmacodynamics evaluation of antiviral medicinal plants using a vector-based assay technique. J. Appl. Microbiol. 99, 1346–1355.

Esposito, F., Carli, I., Del vecchio, C., Xu, L., Corona, A., Grandi, N., et al., 2016. Sennoside A, derived from the traditional Chinese medicine plant *Rheum L.*, is a new dual HIV-1 inhibitor effective on HIV-1 replication. Phytomedicine 23, 1383–1391.

Fabricant, D.S., Farnsworth, N.R., 2001. The value of plants used in traditional medicine for drug discovery. Environ. Health Perspect. 109 (Suppl 1), 69–75.

Fan, W., Qian, S., Qian, P., Li, X., 2016. Antiviral activity of luteolin against Japanese Encephalitis virus. Virus Res. 220, 112–116.

Gao, H., Liu, L., Qu, Z.Y., Wei, F.X., Wang, S.Q., Chen, G., et al., 2011. Anti-adenovirus activities of Shikonin, a component of Chinese herbal medicine *in vitro*. Biol. Pharm. Bull. 34, 197–202.

Geng, C.A., Huang, X.Y., Chen, X.L., Ma, Y.B., Rong, G.Q., Zhao, Y., et al., 2015. Three new anti-HBV active constituents from the traditional Chinese herb of Yin-Chen (*Artemisia scoparia*). J. Ethnopharmacol. 176, 109–117.
Ghannadi, A., Fattahian, K., Shokoohinia, Y., Behbahani, M., Shahnoush, A., 2014. Anti-viral evaluation of sesquiterpene coumarins from *Ferula assa-foetida* against HSV-1. Iran. J. Pharm. Res. 13, 523–530.
Glenz, K., Warzecha, H., 2006. New medicinal plants for the production of vaccines. J. für Verbraucherschutz und Lebensmittelsicherheit 1 (Suppl 1), 126–130.
Gomez-Calderon, C., Mesa-Castro, C., Robledo, S., Gomez, S., Bolivar-Avila, S., Diaz-Castillo, F., et al., 2017. Antiviral effect of compounds derived from the seeds of *Mammea americana* and *Tabernaemontana cymosa on* Dengue and Chikungunya virus infections. BMC Complement Altern. Med. 17, 57.
Gong, P., Li, Y., Yao, C., Guo, H., Hwang, H., Liu, X., et al., 2017. Traditional Chinese medicine on the treatment of coronary heart disease in recent 20 years. J. Altern. Complement. Med. 23, 659–666.
Goyel, B.R., Goyal, R.K., Mehta, A.A., 2007. Phytopharmacology of *Achyranthes aspera*: A Review. Pharmacognosy Rev. 1 (1), 143–150.
Guan, W., Li, J., Chen, Q., Jiang, Z., Zhang, R., Wang, X., et al., 2017. Pterodontic acid isolated from *Laggera pterodonta* inhibits viral replication and inflammation induced by influenza A virus. Molecules 22.
Ha, T.K., Dao, T.T., Nguyen, N.H., Kim, J., Kim, E., Cho, T.O., et al., 2016. Antiviral phenolics from the leaves of *Cleistocalyx operculatus*. Fitoterapia 110, 135–141.
Halder, A., Das, S., Ojha, D., Chattopadhyay, D., Mukherjee, A., 2018. Highly monodispersed gold nanoparticles synthesis and inhibition of Herpes Simplex Virus infections. Mater. Sci. Eng. C 89, 413–421.
Heidary Navid, M., Laszczyk-Lauer, M.N., Reichling, J., Schnitzler, P., 2014. Pentacyclic triterpenes in birch bark extract inhibit early step of Herpes Simplex virus type 1 replication. Phytomedicine 21, 1273–1280.
Hishiki, T., Kato, F., Tajima, S., Toume, K., Umezaki, M., Takasaki, T., et al., 2017. Hirsutine, an indole alkaloid of *Uncaria rhynchophylla*, inhibits late step in dengue virus lifecycle. Front. Microbiol. 8, 1674.
Hossain, M.S., Takimoto, H., Hamano, S., Yoshida, H., Ninomiya, T., Minamishima, Y., et al., 1999. Protective effects of hochu-ekki-to, a Chinese traditional herbal medicine against murine cytomegalovirus infection. Immunopharmacology 41, 169–181.
Jiang, G., Xiao, X., Zeng, Y., Nagabhushanam, K., Majeed, M., Xiao, D., 2013. Targeting beta-catenin signaling to induce apoptosis in human breast cancer cells by z-guggulsterone and Gugulipid extract of Ayurvedic medicine plant *Commiphora mukul*. BMC Complement. Altern. Med. 13, 203.
Kongying yoes, B., Priengprom, T., Pientong, C., Aromdee, C., Suebsasana, S., Ekalaksananan, T., 2016. 3,19-Isopropylideneandrographolide suppresses early gene expression of drug-resistant and wild type herpes simplex viruses. Antiviral Res. 132, 281–286.
Krippner, R., Staples, J., 2003. Suspected allergy to artemether-lumefantrine treatment of malaria. J. Travel Med. 10, 303–305.
Laconi, S., Madeddu, M.A., Pompei, R., 2014. Autophagy activation and antiviral activity by a licorice triterpene. Phytother. Res. 28, 1890–1892.
Laurent, D., Baumann, F., Benoit, A.G., Mortelecqe, A., Nitatpattana, N., Desvignes, I., et al., 2005. Structure–activity relationships of dengue antiviral polycyclic quinones. Southeast Asian J. Trop. Med. Public Health 36, 901–905.
Law, A.H., Yang, C.L., Lau, A.S., Chan, G.C., 2017. Antiviral effect of forsythoside A from *Forsythia suspensa* (Thunb.) Vahl fruit against influenza A virus through reduction of viral M1 protein. J. Ethnopharmacol. 209, 236–247.

Lazreg Aref, H., Gaaliche, B., Fekih, A., Mars, M., Aouni, M., Pierre Chaumon, J., et al., 2011. *In vitro* cytotoxic and antiviral activities of *Ficus carica* latex extracts. Nat. Prod. Res. 25, 310–319.

Lee, B.H., Chathuranga, K., Uddin, M.B., Weeratunga, P., Kim, M.S., Cho, W.K., et al., 2017. *Coptidis rhizoma* extract inhibits replication of respiratory syncytial virus *in vitro* and *in vivo* by inducing antiviral state. J. Microbiol. 55, 488–498.

Li, H.L., Han, T., Liu, R.H., Zhang, C., Chen, H.S., Zhang, W.D., 2008. Alkaloids from *Corydalis saxicola* and their anti-hepatitis B virus activity. Chem. Biodivers. 5, 777–783.

Li, R., Liu, T., Liu, M., Chen, F., Liu, S., Yang, J., 2017. Anti-influenza A virus activity of dendrobine and its mechanism of action. J. Agric. Food Chem. 65, 3665–3674.

Likhitwitayawuid, K., Sritularak, B., Benchanak, K., Lipipun, V., Mathew, J., Schinazi, R. F., 2005. Phenolics with antiviral activity from *Millettia erythrocalyx* and *Artocarpus lakoocha*. Nat. Prod. Res. 19, 177–182.

Lin, L.T., Chen, T.Y., Chung, C.Y., Noyce, R.S., Grindley, T.B., Mccormick, C., et al., 2011a. Hydrolyzable tannins (Chebulagic acid and punicalagin) target viral glycoprotein–glycosaminoglycan interactions to inhibit herpes simplex virus 1 entry and cell-to-cell spread. J. Virol. 85, 4386–4398.

Lin, P.H., Ke, Y.Y., Su, C.T., Shiao, H.Y., Hsieh, H.P., Chao, Y.K., et al., 2011b. Inhibition of HIV-1 Tat-mediated transcription by a coumarin derivative, BPRHIV001, through the Akt pathway. J. Virol. 85, 9114–9126.

Lin, Y.M., Anderson, H., Flavin, M.T., Pai, Y.H., Mata-Greenwood, E., Pengsuparp, T., et al., 1997. *In vitro* anti-HIV activity of bioflavonoids isolated from *Rhus succedanea* and *Garcinia multiflora*. J. Nat. Prod. 60, 884–888.

Lopes, N., Faccin-Galhardi, L.C., Espada, S.F., Pacheco, A.C., Ricardo, N.M., Linhares, R.E., et al., 2013. Sulfated polysaccharide of *Caesalpinia ferrea* inhibits Herpes Simplex virus and Poliovirus. Int. J. Biol. Macromol. 60, 93–99.

Lu, H.N., Ma, S.G., Liu, Y.B., Qu, J., Li, Y., Xu, S., et al., 2015. Sesquiterpenes from the roots of *Illicium oligandrum*. J. Asian Nat. Prod. Res. 17, 430–438.

Lv, Y., Gong, L., Wang, Z., Han, F., Liu, H., Lu, X., et al., 2015. Curcumin inhibits human cytomegalovirus by downregulating heat shock protein 90. Mol. Med. Rep. 12, 4789–4793.

Ma Julian, K.C., Drake Pascal, M.W., Christou, P., 2003. The production of recombinant pharmaceutical proteins in plants. Nature Reviews Genetics 4, 794–805.

Malosh, R.E., Martin, E.T., Ortiz, J.R., Monto, A.S., 2017. The risk of lower respiratory tract infection following influenza virus infection: a systematic and narrative review. Vaccine 36 (1), 141–147.

Marston, B.J., Dokubo, E.K., Van Steelandt, A., Martel, L., Williams, D., Hersey, S., et al., 2017. Ebola response impact on public health programs, West Africa, 2014–2017. Emerg. Infect. Dis. 23.

Mukherjee, H., Ojha, D., Bag, P., Chandel, H.S., Bhattacharyya, S., Chatterjee, T.K., et al., 2013. Anti-herpes virus activities of *Achyranthes aspera*: an Indian ethnomedicine, and its triterpene acid. Microbiol. Res. 168 (4), 238–244.

Murali, K.S., Sivasubramanian, S., Vincent, S., Murugan, S.B., Giridaran, B., Dinesh, S., et al., 2015. Anti-chikungunya activity of luteolin and apigenin rich fraction from Cynodon *dactylon*. Asian Pac. J. Trop. Med. 8, 352–358.

Musch, E., Chrissafidou, A., Malek, M., 2006. Acute hepatitis due to kava-kava and St John's Wort: an immune-mediated mechanism? Dtsch. Med. Wochenschr. 131, 1214–1217.

Nolkemper, S., Reichling, J., Sensch, K.H., Schnitzler, P., 2010. Mechanism of herpes simplex virus type 2 suppression by propolis extracts. Phytomedicine 17, 132–138.

Ojha, D., Mukherjee, H., Ghosh, S., Bag, P., Mondal, S., Chandra, N., et al., 2013. Evaluation of anti-infective potential of a Tribal folklore *Odina wodier* Roxb against some selected microbes and Herpes Simplex Virus associated with skin infection. J. Appl. Microbiol. 115, 1317–1328.

Ojha, D., Mukherjee, H., Mondal, S., Jena, A., Dwivedi, V.P., Mondal, K.C., et al., 2014. Anti-inflammatory activity of *Odina wodier* Roxb, an Indian folk remedy, through inhibition of toll-like receptor 4 signaling pathway. PLoS One 9, e104939.

Ojha, D., Das, R., Sobia, P., Dwivedi, V., Ghosh, S., Samanta, A., et al., 2015. *Pedilanthus tithymaloides* inhibits HSV infection by modulating NF-kappaB signaling. PLoS One 10, e0139338.

Panda, S.K., Padhi, L., Leyssen, P., Liu, M., Neyts, J., Luyten, W., 2017. Antimicrobial, antihelminthic, and antiviral activity of plants traditionally used for treating infectious disease in the similipal biosphere reserve, Odisha, India. Front. Pharmacol. 8, 658.

Park, E.H., Yum, J., Ku, K.B., Kim, H.M., Kang, Y.M., Kim, J.C., et al., 2014. Red Ginseng-containing diet helps to protect mice and ferrets from the lethal infection by highly pathogenic H5N1 influenza virus. J. Ginseng Res. 38, 40–46.

Pasetto, S., Pardi, V., Murata, R.M., 2014. Anti-HIV-1 activity of flavonoid Myricetin on HIV-1 infection in a dual-chamber *in vitro* model. PLoS One 9, e115323.

Pei, Y., Chen, Z.P., Ju, H.Q., Komatsu, M., Ji, Y.H., Liu, G., et al., 2011. Autophagy is involved in anti-viral activity of Pentagalloylglucose (PGG) against Herpes simplex virus type 1 infection *in vitro*. Biochem. Biophys. Res. Commun. 405, 186–191.

Peng, M., Watanabe, S., Chan, K.W.K., He, Q., Zhao, Y., Zhang, Z., et al., 2017. Luteolin restricts dengue virus replication through inhibition of the proprotein convertase furin. Antiviral Res. 143, 176–185.

Prkno, A., Hoffmann, D., Goerigk, D., Kaiser, M., Van Maanen, A.C.F., Jeske, K., et al., 2017. Epidemiological investigations of four cowpox virus outbreaks in Alpaca Herds, Germany. Viruses 9.

Rajtar, B., Skalicka-Wozniak, K., Polz-Dacewicz, M., Glowniak, K., 2012. The influence of extracts from *Peucedanum salinum* on the replication of adenovirus type 5. Arch. Med. Sci. 8, 43–46.

Rajtar, B., Skalicka-Wozniak, K., Swiatek, L., Stec, A., Boguszewska, A., Polz-Dacewicz, M., 2017. Antiviral effect of compounds derived from *Angelica archangelica* L. on Herpes simplex virus-1 and Coxsackievirus B3 infections. Food Chem. Toxicol. 109, 1026–1031.

Reis, S.R., Valente, L.M., Sampaio, A.L., Siani, A.C., Gandini, M., Azeredo, E.L., et al., 2008. Immunomodulating and antiviral activities of *Uncaria tomentosa* on human monocytes infected with Dengue Virus-2. Int. Immunopharmacol. 8, 468–476.

Ren, Z., Zhang, C.H., Wang, L.J., Cui, Y.X., Qi, R.B., Yang, C.R., et al., 2010. In vitro anti-viral activity of the total alkaloids from *Tripterygium hypoglaucum* against herpes simplex virus type 1. Virol. Sin. 25, 107–114.

Sakurai, N., Wu, J.H., Sashida, Y., Mimaki, Y., Nikaido, T., Koike, K., et al., 2004. Anti-AIDS agents. Part 57: actein, an anti-HIV principle from the rhizome of *Cimicifuga racemosa* (black cohosh), and the anti-HIV activity of related saponins. Bioorg. Med. Chem. Lett. 14, 1329–1332.

Schuhmacher, A., Reichling, J., Schnitzler, P., 2003. Virucidal effect of peppermint oil on the enveloped viruses Herpes Simplex virus type 1 and type 2 *in vitro*. Phytomedicine 10, 504–510.

Silva, F.C., Rodrigues, V.G., Duarte, L.P., Lula, I.S., Sinisterra, R.D., Vieira-Filho, S.A., et al., 2017. Antidiarrhoeal activity of extracts from *Maytenus gonoclada* and inhibition of Dengue virus by lupeol. An. Acad. Bras. Cienc. 89, 1555–1564.

Song, S., Qiu, M., Chu, Y., Chen, D., Wang, X., Su, A., et al., 2014. Downregulation of cellular c-Jun N-terminal protein kinase and NF-kappaB activation by berberine may

result in inhibition of herpes simplex virus replication. Antimicrob. Agents Chemother. 58, 5068–5078.
Talarico, L.B., Zibetti, R.G., Faria, P.C., Scolaro, L.A., Duarte, M.E., Noseda, M.D., et al., 2004. Anti-herpes simplex virus activity of sulfated galactans from the red seaweeds *Gymnogongrus griffithsiae* and *Cryptonemia crenulata*. Int. J. Biol. Macromol. 34, 63–71.
Tarasuk, M., Songprakhon, P., Chimma, P., Sratongno, P., Na-Bangchang, K., Yenchitsomanus, P.T., 2017. Alpha-mangostin inhibits both dengue virus production and cytokine/chemokine expression. Virus Res. 240, 180–189.
Terlizzi, M.E., Occhipinti, A., Luganini, A., Maffei, M.E., Gribaudo, G., 2016. Inhibition of herpes simplex type 1 and type 2 infections by Oximacro((R)), a cranberry extract with a high content of A-type proanthocyanidins (PACs-A). Antiviral Res. 132, 154–164.
Tietjen, I., Gatonye, T., Ngwenya, B.N., Namushe, A., Simonambanga, S., Muzila, M., et al., 2016. Croton megalobotrys Mull Arg. and Vitex doniana (Sweet): traditional medicinal plants in a three-step treatment regimen that inhibit in vitro replication of HIV-1. J. Ethnopharmacol. 191, 331–340.
Treister, N.S., Woo, S.B., 2010. Topical n-docosanol for management of recurrent herpes labialis. Expert Opin Pharmacother 11 (5), 853–860.
Tuvaanjav, S., Shuqin, H., Komata, M., Ma, C., Kanamoto, T., Nakashima, H., et al., 2016. Isolation and antiviral activity of water-soluble *Cynomorium songaricum* Rupr. Polysaccharides. J. Asian Nat. Prod. Res. 18, 159–171.
Ueda, K., Kawabata, R., Irie, T., Nakai, Y., Tohya, Y., Sakaguchi, T., 2013. Inactivation of pathogenic viruses by plant-derived tannins: strong effects of extracts from persimmon (*Diospyros kaki*) on a broad range of viruses. PLoS One 8, e55343.
Utsunomiya, H., Ichinose, M., Ikeda, K., Uozaki, M., Morishita, J., Kuwahara, T., et al., 2014. Inhibition by caffeic acid of the influenza A virus multiplication in vitro. Int. J. Mol. Med. 34, 1020–1024.
Vu, L., Misra, K., 2018. High burden of HIV, syphilis and HSV-2 and factors associated with HIV infection among female sex workers in Tanzania: implications for early treatment of HIV and pre-exposure prophylaxis (PrEP). AIDS Behav. 22 (4), 1113–1121.
Wei, F., Ma, S.C., Ma, L.Y., But, P.P., Lin, R.C., Khan, I.A., 2004. Antiviral flavonoids from the seeds of *Aesculus chinensis*. J. Nat. Prod. 67, 650–653.
Wen, C.C., Kuo, Y.H., Jan, J.T., Liang, P.H., Wang, S.Y., Liu, H.G., et al., 2007. Specific plant terpenoids and lignoids possess potent antiviral activities against severe acute respiratory syndrome Coronavirus. J. Med. Chem. 50, 4087–4095.
Wu, C.C., Fang, C.Y., Hsu, H.Y., Chen, Y.J., Chou, S.P., Huang, S.Y., et al., 2016. Luteolin inhibits Epstein–Barr virus lytic reactivation by repressing the promoter activities of immediate-early genes. Antiviral Res. 132, 99–110.
Xu, L., Grandi, N., Del Vecchio, C., Mandas, D., Corona, A., Piano, D., et al., 2015. From the traditional Chinese medicine plant *Schisandra chinensis* new scaffolds effective on HIV-1 reverse transcriptase resistant to non-nucleoside inhibitors. J. Microbiol. 53, 288–293.
Yang, J.M., Ip, S.P., Xian, Y., Zhao, M., Lin, Z.X., Yeung, J.H., et al., 2012. Impact of the herbal medicine *Sophora flavescens* on the oral pharmacokinetics of Indinavir in rats: the involvement of CYP3A and P-glycoprotein. PLoS One 7, e31312.
Yarmolinsky, L., Zaccai, M., Ben-Shabat, S., Mills, D., Huleihel, M., 2009. Antiviral activity of ethanol extracts of *Ficus binjamina* and *Lilium candidum in vitro*. Nat. Biotechnol. 26, 307–313.
Yoo, D.-G., Kim, M.-C., Park, M.-K., Park, K.-M., Quan, F.-S., Song, J.-M., et al., 2012. Protective effect of ginseng polysaccharides on influenza viral infection. PLoS

One 7 (3), 1–9. Available from: http://dx.doi.org/10.1371/journal.pone.0033678. e33678.
Zahmanov, G., Alipieva, K., Denev, P., Todorov, D., Hinkov, A., Shishkov, S., et al., 2015. Flavonoid glycosides profiling in dwarf elder fruits (*Sambucus ebulus* L.) and evaluation of their antioxidant and anti-Herpes Simplex activities. Ind. Crop. Prod 63, 58–64.
Zandi, K., Teoh, B.T., Sam, S.S., Wong, P.F., Mustafa, M.R., Abubakar, S., 2012. Novel antiviral activity of baicalein against dengue virus. BMC Complement. Altern. Med. 12, 214.
Zhang, H.J., Tan, G.T., Hoang, V.D., Hung, N.V., Cuong, N.M., Soejarto, D.D., et al., 2003. Natural anti-HIV agents. Part IV. Anti-HIV constituents from *Vatica cinerea*. J. Nat. Prod. 66, 263–268.
Zhang, Y.B., Zhan, L.Q., Li, G.Q., Wang, F., Wang, Y., Li, Y.L., et al., 2016. Dimeric matrine-type alkaloids from the roots of *Sophora flavescens* and their anti-hepatitis B virus activities. J. Org. Chem. 81, 6273–6280.
Zhang, Z., Bian, Q., Luo, P., Sun, W., 2015. Ethnopharmacological, chemical, and pharmacological aspects of *Halenia elliptica*: a comprehensive review. Pharmacogn. Rev. 9, 114–119.

FURTHER READING

Warzecha, H., Mason, H.S., 2003. Benefits and risks of antibody and vaccine production in transgenic plants. J. Plant Physiol. 160, 755–764.
Xiang, Y.F., Qian, C.W., Xing, G.W., Hao, J., Xia, M., Wang, Y.F., 2012. Anti-herpes simplex virus efficacies of 2-aminobenzamide derivatives as novel HSP90 inhibitors. Bioorg. Med. Chem. Lett. 22, 4703–4706.

CHAPTER 4

Plant-Derived Prebiotics and Its Health Benefits

Abdullah Safar Althubiani[1], Saleh Bakheet Al-Ghamdi[2], Samreen[3], Faizan Abul Qais[3], Mohammad Shavez Khan[3], Iqbal Ahmad[3] and Hesham A. Malak[1]

[1]Department of Biology, Faculty of Applied Science, Umm Al-Qura University, Mecca, Kingdom of Saudi Arabia
[2]Biology Department, Faculty of Science, Al Baha University, Al Baha, Kingdom of Saudi Arabia
[3]Department of Agricultural Microbiology, Aligarh Muslim University, Aligarh, Uttar Pradesh, India

4.1 INTRODUCTION

According to an estimate $\sim 10^{14}$ microorganisms that are inhabiting in human body and it is much more similar/closer to that of human cells. Human gastrointestinal (GI) tract is predominantly characterized by these microbes, collectively termed as gut microbiota. Earlier, gut microbiota was known as commensals but now they are considered as symbionts that are beneficial for human beings. Obligate anaerobes like Bacteroidetes and Firmicutes are predominant phyla constituting 90% of total gut microbiome. Other phyla that are found include, Proteobacteria, Actinobacteria, and Tenericutes (Xu et al., 2017). Around 2000 years long ago, a statement "all disease begins in the gut" quoted by Hippocrates and researchers are now beginning to appreciate his statement. Dysbiosis (disruption in microbiota composition) is extensively correlated to not only gut diseases but also related with hypertension, insulin resistance, type 2 diabetes, and obesity. Disequilibrium in gut microbial community affects host's physiological states *via* different mechanisms such as metabolic and immunomodulation. Both host (genetic and gender) and environmental factors (lifestyle, hygiene, diet and antibiotic treatment) shape the intestinal microbiome species diversity (Conlon and Bird, 2014). The functional output of intestinal microbiota plays an important role in homeostasis by controlling metabolic pathways and important modulators of physiological state of GI such as nutrient metabolism and vitamins production. Gut microbiota plays an essential role in human through broad-spectrum metabolites and their diverse

New Look to Phytomedicine
DOI: https://doi.org/10.1016/B978-0-12-814619-4.00004-5

action. Recent studies have revealed that they play significant role in immunomodulation influencing host physiology. Short-chain fatty acids (SCFAs) such as propionate and butyrate induce native T cells differentiation into an inflammatory regulatory. T cells also produce interleukins (IL) (Sridharan et al., 2014).

Human gut microflora makes a complex ecosystem in GI tract and imparts various health benefits to humans. This microecosystem is a consequence of symbiotic association between the host and gut microbiota that is responsible for defense against opportunistic pathogens, metabolize indigestible compounds provide essential nutrients like vitamins, short-chain fatty acid (SCFA) low-molecular-weight molecules. Among the numerous intestinal microbes, those participate to exhibit potential health benefits to their host through modulation of intestinal microbial constitution are regarded as probiotics. Therefore, the live microorganisms, which when administrated in adequate dose exert beneficial effects on the host. Most common species of genera *Lactobacillus* and *Bifidobacterium* have been found beneficial probiotic bacterial strains (Martín et al., 2013). Some of the common inhabitant bacterial strains like *Lactobacillus reuteri*, *Lactobacillus rhamnosus*, *Bifidobacteria*, certain *Enterococcus* such as *Enterococcus faecium* SF68, *Escherichia coli* strain Nissle 1917, some strains of *Bacillus* like *Bacillus licheniformis* exhibit probiotic properties and have pronounced beneficial effect on host physiology.

4.1.1 Definition and Concept of Prebiotics, Probiotics, and Synbiotic

Prebiotics: The pioneering concept of prebiotics was firstly introduced in 1995 by Gibson and Roberfiod as nondigestible components of food that selectively trigger growth or activity in GI microbiota and improve the health status of host. Number of times prebiotic concept was evolved by FAO and UN in 2015. Bindels et al. (2015) gave another concept by requirement of selective fermentation of dietary components, selectively in terms of microbial fermentation is a major concern on this concept. According to them, prebiotics are nondigestible compounds that through metabolization by microorganisms residing in colon modulates gut microbiome composition or activity confer beneficial physiological effects on the host (Gibson et al., 2017).

Among the first prebiotics assessed in human and also used for commercial level are found to stimulate *Lactobacillus* and *Bifidobacterium* especially preventing the growth of pathogens like *E. coli* and members of

Clostridia class. These genera are also used as probiotics, thus there is commodity between probiotics and prebiotics. Prebiotics supplements can be used as an alternative to probiotics or auxiliary support to them. Specific simulation of *Lactobacillus* and *Bifidobacterium* was considered to have diverse functions by stimulating the growth of different endogenous gut inhabitants. Instead of nutraceutical effect of fermentable probiotics, it must have additional values beyond those of daily nutrition. Prebiotics are mostly fermentable fibers that are selectively fermented by the species regarding beneficial to host. Fructooligosaccharide (FOS) and glucooligosaccharide (GOS) are the examples of most common confirmed prebiotics used. Therefore, prebiotics have specified food components that affect the activity of specific bacteria, end product produced after fermentation exert possible beneficial effects on their host. Prebiotics resist digestion or adsorption by their host before fermentation by specific existing strains of gut microbiota (Fanaro et al., 2005).

Probiotics: As prebiotics are nonviable compounds, probiotics are live microorganisms including dominating strains and residing (indigenous) microbiota. The terminology of probiotic is derived from greek language means "for life." The most widely used definition for the development of probiotic concept followed by Fuller is that "the live microbial feed additives that exert beneficial effect on host animal by means of improving intestinal flora balance." Later on FAO and WHO proposed another concept related with probiotics as "live microorganisms which when administered in proper doses exert/impart health benefits to their host." In relation with food the definition is modified by focusing that the beneficial effect is provided by the microorganisms when they are taken as part of food in adequate amount. The most commonly known beneficial effect attributed by probiotics is maintaining intestinal health, stimulation of immune response, decrease in serum cholesterol, and prevention various types of cancer (Kechagia et al., 2013). Probiotics are well known as "health eco-friendly bacteria," they impart lots of benefits some of them are listed above through various mechanisms like decrease in gut pH, production of SCFAs, and competition for nutrients as mucosal barrier function and immunomodulation. In addition, probiotics are also used as therapeutic preparation in dermatological and oral disease and reduction in depression and anxiety through heart brain modulation (Kechagia et al., 2013; Shi et al., 2016). The common reported functions of probiotics includes: (1) reduction in antibiotic or rotavirus-associated diarrhea, (2) prevention in inflammatory disease in GI tract, (3) cancer prevention, and (4) prevention in allergic and atopic reactions in infants (De Vrese and Schrezenmeir, 2008).

In addition to prebiotics, Gibson gave another concept of synbiotic, synbiotic products have both prebiotic and probiotic properties. The term synbiotic originate from the word "synergism," it means in the combinational product, prebiotic components selectivity stimulate growth or activity of probiotic organism residing in colon of GI tract. The combination of prebiotic and probiotics in a single product gave it pronounced activity compared with their activity alone (De Vrese and Schrezenmeir, 2008). The probiotics like *Lactobacilli*, *Bifidobacteria*, and *Bacillus coagulans*, etc. and most commonly prebiotics that comprise FOS, GOS, and XOS, inulin from natural sources like yacon roots and cherry is used in synbiotic formulations.

Synbiotic formulations impart various advantages to the host such as (1) preventing dysbiosis and thus helps in maintaining gut microbiota, (2) reduces incidence of nosocomial infection by preventing translocation of pathogens, (3) modulation of immune responses, and (4) reduced hepatic injuries, etc. (Pandey et al., 2015).

4.1.2 What Are Plant Prebiotics?

Nondigestible oligosaccharides (NDOs) are considered as established prebiotics because these oligosaccharides are unable to digest having anomeric C-atom, in such a configuration, the glycosidic bond resists to mammalian hydrolytic enzymes. NDOs are low caloric value food additives that enhance mineral absorption also potential substrate for fermentation (Boler and Fahey, 2012).

Therefore, prebiotics have been characterized as NDOs and polysaccharides that enhance the growth of beneficial bacteria including *Lactobacilli* and *Bifidobacteria* in the colon of the host gut and also exert antagonistic relationship with *Salmonella* spp., *E. coli*, and *Clostridial* population, thereby limiting their proliferation. The most common prebiotics known to date are inulin, its derivatives like FOSs and GOSs. The consumption of prebiotics improves human immune response, colon metabolism, microbiome balance and eliminate digestion associated complications (Sharma et al., 2102). Apart from that prebiotics also accelerate uptake of various micronutrients like iron, zinc, and calcium and significantly reduces or prevent the chances of colon-associated cancers, cholesterol, and elevated levels of triacylglycerols (Dwivedi et al., 2014).

In this chapter we aimed to provide an overview on the prebiotics from food and other sources, mode of action, chemical nature, and their health effect both on intestinal and extraintestinal.

4.2 SOURCES OF PREBIOTICS

On the basis of fermentability, most of the foods and many ingredients are considered as prebiotics. Various food items contain different types of fiber only few of them are fermentable. Although fermentable fibers exert crucial beneficial effect on their host but prebiotics are not such kind of that. Prebiotics are specific components that are aimed to stimulate specific bacteria in colon, producing broad-spectrum of end products, beneficial for host. Therefore, prebiotics may be dietary fermentable fiber, but all fermentable fibers are not prebiotics (Holscher, 2017). There is significant difference in two terminologies such as dietary fibers are metabolized by majority of residing colon microorganisms but prebiotics are fermented by specifically defined group or strain of microorganisms.

Plant-based sources of prebiotics include FOS, GOS, and inulin. Some of the polysaccharides that are component plant cell wall such as pectins, xylans are available in routine diet (Yoo et al., 2012). In addition, isomaltooligosaccharides (IMOs), chicory root–derived inulin (FOS), xylooligosaccharides (XOSs), soybeanoligosaccharides (SBOSs) lactulose, raffinose and sorbitol are found to have huge applications as prebiotic and exerts lots of health benefits (Patel and Goyal, 2012). Resistant starch of whole grain is also recognized as potential prebiotic carbohydrates since they are unable to be digested by host digestive enzymes and these are not absorbed by the intestine, thus stimulate the growth of beneficial microflora of gut (Fuentes-Zaragoza et al., 2011; Van den Ende, 2013). In addition, fermentability of dietary fibers like fenugreek gum and β-glucans to SCFAs exerts plethora of beneficial health effects (Lin et al., 2011).

4.2.1 Prebiotics From Other Food Sources

Assessment conducted by European Commission or European Food Safety Authority states that any food or food component that has no previous history of safe use are considered as "Novel foods." A number of prebiotics are known to be established as novel foods. In south America Yacon (*Smallanthus sonchifolius*) is used as traditional food. Safety issues of yacon is doubtful among European Union. It has clear history of safe use including prebiotic effects by providing myriads of health benefits. It is free from toxic metabolites and antinutrients (Campos et al., 2012). Prebiotics in the form of carbohydrates are present in number of food crops like vegetables, roots including tuber crops, for example okra, shallot, onion, garlic, artichoke involving gourd family of vegetables like wax ground or bottle guard. Among the fruits like chicory, and yacon are major source of inulin. In addition, dahliya, which is a tuber crop, have prebiotic effects comprised of fructose. Few of the species of mushroom such as *Agaricus bisporus* have been reported in providing potential prebiotic benefits. The fruiting body extract of *Pleurotous* species (pleuran) contain β-glucan have been used in various food supplements imparts immunosuppression and growth of gut probiotic to human health (Synytsya et al., 2009). Beneficial species includes *Pleurotus ostreatus* and *Pleurotus eryngii*. A steroid sapogenin compound called as diosgenin isolated from yam plants stimulate the growth and activity of intestinal lactic acid bacteria (LAB) (Huang et al., 2012). Later on, further studies carried out by this group in murine model to demonstrate the effect of diosgenin has also immunomodulatory effects by intestinal regulatory T cells by means of oral administration (Huang et al., 2017). Grain legumes are also predominant source of dietary fibers like chickpea grain comprises of α-GOSs, which is used in the form of prebiotic in functional foods and lupin kernel derived fibers modulate gut microbial composition specifically enhancing *Bifidobacterium* species and lower the level of *Clostridial* populations like *Clostridial spiroforme*, *Clostridial ramosus*, and prevent colon infection and thereby promote intestinal health benefits in human (Fernando et al., 2010; He et al., 2011).

Spices are well known for their medicinal values and have several therapeutic properties like antimutagenic, antiinflammatory, and antimicrobial activities. Spices of various origin like plants leaves, seeds, bark, and fruits have specific phytoconstituents and also exhibit prebiotic properties and thus influencing the relative abundance of *Bifidobacteria* and *Lactobacilli* within gut microbiota and reduces the number of disease/infection causing pathogens including both Gram positives and Gram negatives (Lu et al., 2017). A bioactive extract of cinnamon, cinnamaldehyde exhibits more potent antibacterial activity against common food borne pathogens having low MIC comparable with their crude extract without any modulation on gut probiotics of mouse model.

Studies carried out by Romo-Vaquero et al. (2014) suggest that Rosemary extract supplementation reduces the progression of *Pediococcus* and *Leuconostoc* population and influence the growth of *Bacteroides* group.

Date seeds/kernel is currently used in the feeding of animals such as cattle, sheep, and camel and in the poultry and fish industries as well. Due to the presence of a large quantity of dietary fiber, they are considered to have potential health benefits for human as prebiotics. In a study conducted by Al-Thubiani and Khan (2017), the effect of two products from date palm (*Phoenix dactylifera* L.), the finely grounded date seed powder and the aqueous extract of date seed powder, was investigated on the growth of the isolated LAB *Lactobacillus paracasei* ssp. *paracasei* as probiotics. *P. dactylifera* L. seed product, i.e. dietary fiber concentrate (DFC) was found as a carbon source for bacterial fermentations. The DFC presented the potential to be applied as novel source of prebiotics, by increasing the population of *L. paracasei* ssp. *paracasei* as probiotics in addition to decreasing the pH values.

Overall the ideal characteristics of prebiotics or selection criteria are (1) they must be nondigested by host metabolic or digestive enzymes and reach to the colon in an unaltered form, (2) they should not be absorbed in small intestine, (3) less or poorly fermented by residing bacteria in the oral cavity, (4) strongly fermented or metabolized by specific group of beneficial bacteria of intestinal microbiome, and (5) not at all or poorly metabolized by pathogenic organisms in gut bowel (Lee and Salminen, 2009; Markowiakand Śliżewska, 2017).

4.3 MODE OF ACTION OF PREBIOTICS

The ultimate focus of prebiotic after administration is to stimulate the growth and activity or an increased count of beneficial bacteria residing in GI tract. In turn, the activated intestinal microbiota imparts health benefits to the host through various mechanisms like antagonism (by producing antimicrobial substances) and competition for nutrients or for the epithelial surface. Some of the beneficial effects of prebiotic-directed functions of probiotics are summarized in the following:

- As a consequence of prebiotic substrate fermentation, *Bifidobacterium* and *Lactobacilli* result in production of inhibitory compounds that act as barrier for GI pathogens as well as reduce the pH of intestine.
- After acidification of gut lumen, *Bifidobacteria* retains high tolerance to the produced SCFAs and also to pH (Grajek et al., 2005).
- Administration of prebiotics also increases the absorption of various nutrients like Mg and Ca.

- *Bifidobacteria* and *Lactobacilli* produce variety of antimutagenic compounds and activation of host immune system for the degradation of carcinogenic enzymes produced by colonic microflora and thereby maintaining the permeability of intestine and microbial balance (Kumar et al., 2010).
- Prebiotics utilized as a source of carbon and energy and influencing the activity and composition of GI microbiota, for example studies carried out with GOS prebiotics using microarray analysis suggest GOS in *Lactobacilli acidophilus* NCFM induces gut lac operon that results in production of galactoside pentose-hexuronide (GPH) permease (Lac S), two cytoplasmic β-galactosidases (LacA and Lac LM) including the enzymes involved in galactose metabolism (Blatchford et al., 2013). These gene clusters also play an important role in acquired combination of gut evolved benefits like tolerance to bile salts and nutrient acquisition.
- *Lactobacilli acidophilus* NCFM also have the potential for utilizing different prebiotics substrate units. Another study of Anderson revealed that in 11 prebiotics consisting of α and β-glucosides result in induced expression of phosphoenol pyruvate-dependent phosphotransferase system (PTS). In addition to Raffinose and stachylose (α-galactosides) induced the expression of ATP-binding transporters (ABC Cassets), while in previous case there is induction GPH permease transporter through GOS (β-galactosides). Cellobiose and gentibiose are regioisomers exhibit differences in their structure only by the type of linkage (β1→4 and β1→6, respectively) were found to be metabolized by two distinct GHI phospho β-glucosides and PTS. Also, the above-mentioned studies suggest that *L. acidophilus* have diverse spectrum of transporters and hydrolases toward distinct prebiotic carbohydrates.

Some of the beneficial effects of prebiotics on modulating the immune system are described in literature as presented in the following:

- Prebiotics have the potential to regulate the expression of hepatic lipogenic enzymes that is influenced by the production of SCFAs, propionic acid, and butyric acid by beneficial colonic flora.
- Enhanced production of SCFAs as a fermentation byproduct result in its histone modulation through acetylation that makes changes in chromatins and permit DNA binding proteins/factors to intact with numerous expressed genes activating transcription and cellulase function.
- Modulate the defense system, a glycoprotein mucin underlying on epithelial cells provide protection from both biological entities (bacterial pathogens/viruses) as well as harmful chemicals.

- It has also been demonstrated that FOS and various other prebiotics result in activation of both acquired and innate immune responses and thus increasing β-lymphocytes/leukocytes in peripheral blood and gut-associated lymphoid tissues (GALTs).
- Induction of phagocytosis and increased secretion of IgA by GALTs in intrainflammatory macrophages activates the host immune responses and production of antimicrobial substances (Schley and Field, 2002).

4.4 CHEMICAL NATURE AND TYPE OF PREBIOTICS

Gut microorganisms have ample opportunity to metabolize and ferment available substances in their environment which they derive either from endogenous sources or through diet. Among various sources, starch is found to be degraded by colonic microorganisms since it is resistant to hydrolytic enzymes and can only be fermented by bacterial enzymes. Other sources that are fermentable include dietary fibers, oligosaccharides, and small fraction of absorbable sugars and sugar alcohol.

The structure of carbohydrate greatly influences their fermentation by intestinal microbiota. Different kind of prebiotics having different structure leads to huge difference in fermentation and their byproducts produced. These substances/prebiotics are degraded by wide range of bacterial enzymes like polysaccharidases, aminopeptidases, proteases, glycosidases, and glycanases (both endo- and exoglycanases) to their monomeric constituents such as sugar or amino acids or oligomers (Blatchford et al., 2013).

Another important factor for fermentation by gut microbiota is the molecular weight of the substrate. For example, the low molecular weight substrates like arabinoxylans and oligodextrans preferably fermented by *Bifidobacteria* compared to higher molecular weight counterparts. Similarly, *L. plantarum* and *L. rhamnosus* have special preference for specific molecular weight substances in a complex mixture of FOSs. These organisms selectively ferment fraction of trisaccharides and tetrasachhrides in a series of oligosaccharides (Rossi et al., 2005).

Besides structure and molecular weight of carbohydrates, branching pattern also influences the fermentation selectivity of the substrates, there is decrease in selectivity of fermentation by *Bifidobacteria* once there is an increase in branching among average molecular weight substances. Since *Bifidobacteria* group produces a series of exo-acting glycosidases, which is not occur or consistent in case of *Bacteroides* of saccharolytic group (Lee and O'Sullivan, 2010).

On the basis of chemical nature, the prebiotics may be classified as (1) disaccharides (Lactulose, lactitol), (2) oligosaccharides such as IMOs, FOSs, XOSs, SBOSs, and (3) polysaccharides (inulin, resistant starches). A brief description of certain prebiotics and their effects are elaborate in the following.

4.4.1 Lectulose

Lectulose of β-galactoside fructose is a synthetic disaccharide that is nonabsorbable and used as purgatives since they are not hydrolyzed by digestive enzymes or absorbed in small intestine. Lactulose has prebiotic potential and it gains attention because of bifidogenic effect. *In vitro* studies of continuous fecal culture system indicated lactulose enhancing *Lactobacilli* and *Bifidobacteria* and significantly causes reduction in *Bacteroides* (Gibson et al., 2010). In case of human trials, lactulose administration as 3 g/day for 14 days to eight volunteers resulted in increased *Bifidobacteria* and significant fall in the member of *Enterobacteria*, *Streptococci*, *Colstridia*, and *Bacteroides*. In addition to this, there is also decrease in detrimental enzymes and metabolites (NH_3, indole, phenol, etc.) supported the beneficial aspects of lactulose (Mao et al., 2014).

In food industry, the prebiotic application of lactulose includes promotion of probiotics of skimmed milk. By using lactulose as ingredient in skim milk, it stimulates the growth of *L. acidophilus*, *L. bulgaris*, and *B. lactis* in coculture with *S. thermophilus*, thereby improving the quality of milk (Oliveira et al., 2011).

According to Rai and colleagues, lactulose increases the life expectancy of patients with cirrhosis who have suffered from minimal level of hepatic encephalopathy (Rai et al., 2015).

4.4.2 Inulin-Type Fructans

These fructans are kind of natural polymers that are most widely used for their prebiotic effect beneficial health properties in functional foods. Inulin type fructans (ITFs) have linear chain of fructose polymer with β2→1 linkages that naturally occur in number of different plant foods like onion, garlic, leek, asparagus, chicory roots, and artichoke are being especially rich (Franco-Robles and López, 2015). Most commonly known ITFs are inulin FOSs and oligofructose. Another candidate fructans include GOSs derived from lactulose and lactose. In fructuns, the chain length of fructosyl residue varies. In case of plants, fructose chain length of degree of polymerization (DP) ranges from 30 to 50 fructosyl residues but sometimes also exceed to 200. In many human trials, fructans (including inulin, FCS, and GCS) are found to be suitable prebiotic candidate as they are nondigestible, minor hydrolysis occur in stomach since they contain β-linkages (Tarini and Wolever, 2010).

These fructans modulate the gut physiology by increasing the count of *Bifidobacteria* and *Lactobacilli* species to provide protection from pathogens and helps in constipation. In addition, acting as prebiotics these fructans improve the level of glucose, modulate lipid metabolism reduce plasma-membrane LPS as well as diacylglycerides. They also result in changes in sugar chain found in glycolipids and glycoproteins adhesins present epithelial cell for the bacterial attachment and thereby exerting antimicrobial effects against pathogens (Ortega-González et al., 2014).

4.4.3 Transgalactooligosaccharides

These oligosaccharides (OSs) are galactose containing found in human and cow's milk. They may be synthetically produced by using bacterial origin β-galactosidases mediated hydrolysis of lactose sirup. These OSs mainly contains galactose but also glucose at reducing terminal align with β1→4 and β1→3, β1-6 linkages. TOS have bifidogenic properties, significantly causes the decrease in breath hydrogen in humans as result of 10 g/day TOS administration where there is increase in human flora associated rats that were fed with 10% (w/w) TOS (Bouhnik et al., 1997).

Another synthetic TOS was developed by using enzyme produced from *Bifidobacterium bifidum* 41171 called Bimuno and its prebiotics effect have been evaluated under *in vitro* conditions (Tzortzis, 2010; Depeint et al., 2008). GOS/TOS also modulates the brain function by increased hippocampal brain-derived neurotrophic factor and *N*-methyl-D-aspartic acid significantly provides a basis for study of the use of prebiotic in mitigating mental health brain-associated disorders (Burokas et al., 2015).

4.4.4 Dietary Components That Are Candidate Prebiotics

Any nonviable food ingredient that specially fermented by residing, beneficial colonic microflora, i.e., *Bifidobacteria* and subsequently results in positive effect, is a candidate prebiotic. Some of the candidate prebiotics are briefly discussed in the following.

4.4.4.1 Polydextrose

Polydextrose is a kind of soluble fiber and has high-molecular weight approximately 80,000 Da. Although it is made from starch but commercialization is expensive because it needs to be passed through refining stages to generate an excellent product. It is a polysaccharide made up of glucose moieties having all types of glycosylic bods (1→6 bonds in majority). Polydextrose also has pronounced prebiotic effect. Substrate fermentation results in increased production of SCFA in large intestine (involving butyrate) followed by the propagation of *Lactobacillus* and *Bifidus* and thus maintaining gut health without any adverse impact (Young and O'Sullivan, 2011).

4.4.4.2 Soybeanoligosaccharides

The oligosaccharides found in soybean include raffinose, sucrose, xylose, as well as various mono and disaccharide linkage components are also present. It is predominantly found in soybean seeds and other legumes and approved as well-established prebiotic or generally recognized as safe ingredient by FDA (Chen et al., 2010).

In the GI tract, SBOS is specially fermented by autochthonous bacteria *Lactobacilli* and *Bifidobacteria*. Once there is incorporation of wheat bran oligosaccharide along SBOS, there is enhanced growth and activity of *Bifidobacteria* and *Clostridium perfringens* compared with that of FOS feeding (Ma et al., 2017). In addition, SBOS also activate immune system by stimulating growth and activity of residing intestinal commensals that are protective in nature and become part of first line of defense in intestinal tract (Boehm et al., 2005). Studies carried out in mice suggest that SBOSs have found to raise SOD levels in blood, increase high level of IgG, modulate the body weight, and immunity agent's functions like proliferation and differentiation splenocytes and increased number of antibody producing cells (Xu et al., 2017). Therefore, SBOS has a positive effect on gut health and modulation of immune system.

4.4.4.3 Isomaltooligosaccharides

IMO comprises of monomeric residues of glucose that are linked together by α1-6 glycosidic linkages. Corn starch hydrolyzed by incubating α-amylase, pullulanase, and α-glucosidases makes a mixture of isomaltotriose, pentose, and isomaltose, which is commercially known as Isomalto-900. The potential of IMCs is prebiotics and has been evaluated in human trials. There was 20 g/day administration of IMC in adult males for 10 days and in elderly men and women for 14 days. The results showed that there was significant increase in beneficial colonic microflora (Gibson et al., 2010)

4.4.4.4 Glucans

Glucans are generally referred to as GOS since they are consist of oligosaccharides and polysaccharides. Enzymatic hydrolysis by glucosyltransferases results in different sized GIOS, the mechanism involved transfer of glucose moiety from sucrose to a mannose residue and leaving fructose residue or sucrose free in mixture. Glucan's used as prebiotic supports the activity of probiotics *Lactobacilli* and significantly reduces the risk of mortality associated with *Aeromonas* (Ngamkala et al., 2010). The studies in calves through feeding glucan along with tylosin have greater influence on specific humeral immunological parameters such as increased γ-immunoglobulins and total protein content (Szymańska-Czerwińska and Bednarek, 2011). Prebiotic potential of glucans has also been seen in human trials. It was concluded by the author that glycans exhibit strong bifidogenic effect, increase the count of *Bifidobacteria* as well as it is well tolerated among older and healthy volunteers in their diet intake (Mitsou et al., 2010).

4.4.4.5 Xylooligosaccharides

XOSs are mixture of oligosaccharides like xylobiose, xylotriose, and xylotetrose containing xylose residues linked by α1−4 bonds, naturally found in various food sources like milk, honey, vegetables, fruits, and bamboos shoots (Lin et al., 2016). XOS has positive effects on consumption, it may lead to nourishment of endogenous *Bifidobacterium* species and production of SCFAs in rat studies (Hsu et al., 2004; Lin et al., 2016). The functional and complete utilization of XOS is based on the activities of verity of enzymes like β-xylosidase, acetyl esterases, α-glucuronides released by different strains of rumen bacteria and producing SCFAs like acetic acid, propionic acid, and butyric acid. XOS administration in rats on high-fat diets causes reduction in lipid level and noticeable drop in low-density lipoprotein (LDL) and total cholesterol and significant increase in high-density lipoprotein (HDL). In addition, stimulation of bifidogenic bacteria and other *Lactobacilli* results in propionate and other acids production, promote insulin synthesis through glucagon-like peptide-1 (GLP-1) for the stimulation and glycogenesis and inhibition of lipolysis (Samanta et al., 2015).

4.5 EXTRAINTESTINAL EFFECTS OF PREBIOTICS

Prebiotics health benefitting agents are not only confined to the gut but also imparts health benefits beyond the intestinal gut. These benefits to extrainessential region can be direct through change in native microbiota, for example in case of mouth or vaginal tract or indirectly influenced by metabolic or compositional changes in the gut microbiota.

Potential prebiotics like GOS and FOS have the capacity to stimulate specific beneficial gut microbial populations that play an important role in treating the incidence of respiratory infections (Arslanoglu et al., 2008), optic dermatitis (Marcobal et al., 2010), and increased absorption of minerals like Mg and Ca (Scholz-Ahrens et al., 2016).

Prebiotics lead to beneficial effects outside the intestinal tract by modulating both innate and adaptive immune mechanisms through involving antiinflammatory metabolites. Like prebiotics, galactose promotes increment in the number of *Bifidobacteria* that ultimately results in production of antiinflammatory cytokines and induce phagocytosis (Vulevic et al., 2008).

It is also observed that the mineral absorption outside the intestinal tract is mediated through fermentation of prebiotics and subsequent enhancement in SCFA production. Individual's gut act as a hub where primary feeders carry out carbohydrate degrading mechanisms, the fermentation byproducts or substrates are utilized by secondary feeders (process called cross feeding). This process is supported to decrease in luminal pH, for example *Bifidobacterium longum* utilizing substrates and produces acetate as a byproduct of feeding and, thus protecting from enteropathogenic infections. In turn, this acetate is consumed by another bacterium *Faecalibacterium prausnitzii* (strong SCFA producer in lumen of gut) (Sokol et al., 2008) that have vital need of this fatty acid under *in vitro* conditions. Therefore, health promoting effects are not only associated with beneficial lactic acid producers such as *Bifidobacterium* and *Lactobacillus* but also another residing SCFA producers.

4.5.1 Extraintestinal Effects of Prebiotic in Adiposity and Its Associated Metabolic Diseases

Studies carried out in humans and in rodents suggest that following the consumption of prebiotics, energy intake is reduced through alterations in satiety hormones like peptide YY (PYY), ghrelin, and GLP-1 (Parnell et al., 2012). One of the mechanisms by which prebiotics reduces adiposity is the decreased expression G-protein—coupled receptor 4-3 in subcutaneous adipose tissue thus in reduced concentration, it is unable to cause lipogenesis and results in lipolysis (Dewulf et al., 2011). In addition, prebiotics also reduce the size of adipocytes because the large-sized adipocytes have increased level of insulin resistance, TNF-α, and high levels of fatty acids.

Obesity may correlate with metabolic diseases like dyslipidaemia, type 2 diabetes, and less known nonalcoholic fatty liver (NAFL) disease. NAFL disease is emerging disease associated with pathogenicity of gut microbiota, it can begin and progress to nonalcoholic steatohepatis and then more severe form cirrhosis and then liver failure. Successful studies have only seen in obese mice that 10% (w/w) prebiotic administration reduces the risk factors coupled with brain-derived neurotrophic factor, there is 40% reduction in liver LPS content in genetically obese rats and 30% reduction in diet-induced obese rats (Payne et al., 2012; Reimer et al., 2012).

4.5.2 Prebiotics as Antiadherence Agent

In addition to activation of probiotics in GU tracts, prebiotics also exert tremendous application in interfering with infection process of bacterial pathogens (Gibson et al., 2005; ShoafSweeney and Hutkkins, 2008; Hotchkiss et al., 2016).

The mechanism of action of prebiotics in infection control is basically based on their structural similarity with receptor site of pathogens located on the surface of host epithelial cells. Most of the enteric pathogens causes infections via attachment through lectin like adhesins to the carbohydrate receptor on the epithelial cells lining of intestinal tract (Kato and Ishiwa, 2015). Commercial prebiotics involving plant-based oligosaccharides and food grade prebiotics have been investigated against variety of enteric pathogens for their antiadherence activity. Like mannooligosaccharides taken from seeds and other plant material or roots that is rich in mannans containing α-linked mannose moieties that are known for their adherence activity against enteric including *E. coli* and *Salmonella* (Sharon, 2006). *In vivo* studies carried out by Ganner and Schatzmayr (2012) with Mnnanoligisaccharide, concluding that there is inhibition of type 1 adhesion of enteric pathogens. Pectin that is compared to galacturonan backbone along with rhamnogalacturonan substituted with GOSs and arabinaooligosaccharides, prebiotic activity is confirmed in oligosaccharides fractions obtained from enzymatic hydrolysis. Ganan et al. (2010) showed that POS inhibited the incidence of *Campylobacter jejuni* invasion on Caco-2 cells suggesting that it could be used as alternative antibiotic for the treatment of *C. jejuni* infections.

Chitin oligosaccharide (CHOS) hydrolysates have DP 4–12 also act as potent antiadherence agent. Fractions of CHOS obtained through enzymatic hydrolysis on the basis of the DP and acetylated residues have profound effect on the inhibition of Enteropathogenic *E. coli* adherence under *in vitro* conditions (Quintero-Villegas et al., 2013).

4.6 SIGNIFICANCE OF PLANT-BASED PREBIOTICS IN DIFFERENT DISEASES/CLINICAL APPLICATIONS IN HUMANS

Prebiotics have been used in various food ingredients providing lots of benefit to human health acting as a substrate for beneficial gut probiotics. Some of the possible functions of prebiotics in mitigating heath complications on various diseases are listed in the following.

4.6.1 Inflammatory Bowel Disease

Inflammatory bowel disease (IBD) is associated with intestinal microbiota pathogenesis. It is characterized as low-grade inflammation, change in gut microflora, abdominal pain, and stool consistency. Although probiotic therapy is most commonly used but administration of prebiotics also has supplementary effect. In case of dextran sulfate sodium model of rats, colitis administration of germinated barley foods that are rich in hemicellulose fibers and glutamine protein has profound effect on reducing the incidence on mucosal injury and associated bloody diarrhea (Kanauchi et al.,1998). Not all the prebiotics established have antiinflammatory properties in curing of IBD but reports suggest that the mixture consisting long chains of inulin and oligosaccharides decreased proinflammatory cytokines IL-1β followed by increase in inflammatory TGF-β, simultaneously increase in *Bifidobacteria* and *Lactobacilli* level (Hoentjen et al., 2005). Fiber intake is found protective in IBD and its associated complications like Crohns's disease, diabetes, and asthma. As the fibers are unable to be digested by human digestive enzymes and can only be fermented by colonic microflora. Thus, resultant propionate, butyrate, and acetate (SCFA) stimulate the antiinflammatory activities of intestinal epithelial cells that raise the IgA level and immunosuppressive cytokines (Trop, 2014).

4.6.2 Role in Cardiovascular Diseases

Epidemiological studies suggest that dietary fibers exert a protective effect against cardiovascular diseases (CVDs) and also coronary heart disease. These fibers have prebiotic potential and alter the biomarkers associated with CVD like C-reactive protein, LDL, and also blood pressure. US FDA had allowed the health claims that oats, psyllium, and barley for reduce bad cholesterol or LDL without affecting the concentration of HDL in blood (Slavin, 2008). A double-blind randomized placebo controlled study in 17 patients who were taking normal diet and did not modify their habits were administrated with 10 g inulin/FOS for 6 months. Inulin/FOS has a significant effect on lipogenesis and plasma triacylglycerol concentration and induces a nonsignificant trend for reduced LDL-C level and plasma total and increased HDL-C concentration. They are involved in decreasing dietary cholesterol, another mechanism includes modulation of lipid metabolism, like propionate (Forcheron and Beylot, 2007).

4.6.3 Appetite Control

Prebiotics dietary fables also play a significant role in satiation and satiety. Increase in time for chewing fiber-rich food is responsible for enhanced satiation. Promoting increased saliva and gastric acid production. Some of viscous fibers get solubilized in water and increase stomach distention as result vagal signals of fulfillment are generated that make satiation during taking meal and satiety in the post meal period. The meal travels from upper to lower GI tract triggering signals that are sent to the brain and releasing various satiety-related hormones like GLP, ghrelin, poly PYY that mediate satiety regulate food and calorie intake and energy balance (Slavin, 2013)

4.6.4 Reduces Risk of Colon Cancer

Prebiotics intake decreases intake of genotoxic enzymes β-glycosidase, β-glucuronidase, and acryl sulfatases. Administration of GOS to the system leads to direct inhibition of *Bacteroides* and *Clostridia* and results in enhancement of *Bifidobacteria* and *Lactobacilli* proportions as they produce low level of such enzymes than former. The prebiotic function of two inulins is from agave and others from chicory have been attributed to *Bifidobacteria* producing butyrate and propionate and these acids also inhibit histone acetylate as well as the growth and proliferation of tumor cells. Through the mechanism involved in the production of multifunction cytokines TNF-α and IL-10. These colonic probiotics are indirectly influences the cell proliferation and differentiation, and cells survival and death (Rivera-Huerta et al., 2017). In case of carcinogen-associated colon cancer, administration of amylase starch has beneficial effect as they may lead to the induction of enzymes for detoxification of compound by butyrate or gut probiotics and thus play an important role in cancer prevention (Liong, 2008).

4.6.5 Antibiotic-Associated Diarrhoea and Traveler's Diarrhoea

Brain-derived neurotrophic factor (AAD) is most comply occurring manifestation of normal microflora disturbance by the use of antibiotics. It is generally if two types: nonspecific AAD or *Clostridium difficalis* AAD. Prebiotic stimulation of gut probiotics mainly *Lactobacilli* play an important role in the prevention of these compilations. A study carried out by Lewis et al. (2005) suggest that daily ingestion of 12 g oligosaccharides prebiotics reduces the incidence of complications of diarrhea including *Clostridium difficile* diarrhea. Another study carried out by working group for probiotics and prebiotics of European Societies with the combination of inulin and FOSs prebiotics. They test patient of 6 months to 14 years taking antibiotic treatment as well as inulin and FOS mix according to their age having maximum dose of 5 g/day ($n = 54$) or a placebo ($n = 54$) for the duration of their antibiotic therapy. It was concluded that prebiotics reduces the overall frequency of diarrhea and its associated complication (Szajewska et al., 2012). Prebiotics also play an important role in prevention and treatment of diarrhea (Floch et al., 2011). It was observed that there is significant reduction in prevalence of diarrhea in prebiotic groups as compared with control counterparts, and less severe complications of diarrhea were also observed. In placebo study of 10 g inulin taking before and after 2 weeks, they were aware to high- or medium-risk destinations for Traveler's diarrhea have reduced evidence of infections caused by Enterooxicgenic *E. coli* (Cummings et al., 2001).

4.6.6 Mineral Absorption and Bone Health

In both rodents and humans, prebiotics boost the mineral metabolism, particularly enhancement of calcium absorption. They may also spur magnesium absorption that plays an important role in architecture of bones, thereby enhancing the bone density. For example, studies carried out in healthy male rats indicate enhanced Ca absorption treating with inulin or GOS as well as in female rats administered with polydextrose for 4 weeks (McCabe et al., 2015). In case of humans, prebiotic plays a potential role in adolescent girls and boys for enhanced Ca absorption, obesity range of different treatment directions starting from 9 days, 3 weeks to 1 year. After 1-year report of adolescent girls with prebiotic treatment also promote the efficiency of prebiotics in enhancing body bones mineral density (Griffin et al., 2002; Abrams et al., 2005). The basic mechanism of mineral absorption is based on decrease in colonic pH through organic avoids produced as a consequence of prebiotic selective fermentation.

4.7 CONCLUSION AND FUTURE DIRECTIONS

Consumption of prebiotics is undoubtedly associated with plethora of health benefits by heir stimulation of growth and activity of gut microflora. In this review, we assessed the generalized idea of prebiotics, prebiotic-directed probiotic activation, and their subsequent beneficial effects, combination of both pre and probiotics, i.e. synbiotic. Plants provide ample source of dietary prebiotics or nondigestible prebiotics. From many past decades, the research on health beneficial effects of prebiotic along with probiotic rocked the sky high. Plant-based healthy dietary components were previously beyond the thinking among researchers but now it is evidenced that they exert potential application to prevent or cure various disorders or diseases like gastroenteritis, obesity, CVD, and fatty liver disease through their prebiotic effect (promoting beneficial colonic microflora).

Prebiotics are no quite identical in terms of structure and function and thus activate different kinds of microorganisms in different individuals that sometimes lead to more worsening situation in disease. Therefore, there is need of better characterization of prebiotic and different fiber as well as human intervention studies to identify their health benefits. The composition of diet has also influence on gut microbial metabolome, modulating microbial composition, thus effecting the disease status of individual. In future, there is of patient's microbiome characterization (composition/function) for disease prevention or treatment. Further studies may explain the mechanism of action of combination of prebiotic and probiotic conferring a beneficial effect on human health.

REFERENCES

Abrams, S.A., Griffin, I.J., Hawthorne, K.M., Liang, L., Gunn, S.K., Darlington, G., et al., 2005. A combination of prebiotic short-and long-chain inulin-type fructans enhances calcium absorption and bone mineralization in young adolescents. Am. J. Clin. Nutr. 82 (2), 471–476.

Al-Thubiani, A.S., Khan, M.S.A., 2017. The prebiotics properties of date palm (*Phoenix dactylifera* L.) seeds in stimulating prebiotic *Lactobacillus*. J. Pure Appl. Microbiol. 11 (4), 1675–1686.

Arslanoglu, S., Moro, G.E., Schmitt, J., Tandoi, L., Rizzardi, S., Boehm, G., 2008. Early dietary intervention with a mixture of prebiotic oligosaccharides reduces the incidence of allergic manifestations and infections during the first two years of life. J. Nutr. 138 (6), 1091–1095.

Bindels, L.B., Delzenne, N.M., Cani, P.D., Walter, J., 2015. Towards a more comprehensive concept for prebiotics. Nat. Rev. Gastroenterol. Hepatol. 12 (5), 303f.

Blatchford, P., Ansell, J., De Godoy, M.R.C., Fahey, G., Garcia-Mazcorro, J.F., Gibson, G.R., et al., 2013. Prebiotic mechanisms, functions and applications—a review. Int. J. Probiot. Prebiot. 8 (4), 109.

Boehm, G., Stahl, B., Jelinek, J., Knol, J., Miniello, V., Moro, G.E., 2005. Prebiotic carbohydrates in human milk and formulas. Acta Paediatr. 94 (s449), 18–21.

Boler, B.M.V., Fahey, G.C., 2012. Prebiotics of plant and microbial origin. Direct-Fed Microbials and Prebiotics for Animals. Springer, New York, NY, pp. 13–26.

Bouhnik, Y., Flourié, B., D'Agay-Abensour, L., Pochart, P., Gramet, G., Durand, M., et al., 1997. Administration of transgalacto-oligosaccharides increases fecalbifidobacteria and modifies colonic fermentation metabolism in healthy humans. J. Nutr. 127 (3), 444–448.

Burokas, A., Moloney, R.D., Dinan, T.G., Cryan, J.F., 2015. Microbiota regulation of the mammalian gut–brain axis, Advances in Applied Microbiology, Vol. 91. Academic Press, Cambridge, MA, pp. 1–62.

Campos, D., Betalleluz-Pallardel, I., Chirinos, R., Aguilar-Galvez, A., Noratto, G., Pedreschi, R., 2012. Prebiotic effects of yacon (*Smallanthussonchifolius* Poepp. & Endl), a source of fructooligosaccharides and phenolic compounds with antioxidant activity. Food Chem. 135 (3), 1592–1599.

Chen, H., Li-jun, L., Jian-jun, Z., Bo, X., Rui, L., 2010. Chemical composition analysis of soybean oligosaccharides and its effect on ATPase activities in hyperlipidemic rats. Int. J. Biol. Macromol. 46 (2), 229–231.

Conlon, M.A., Bird, A.R., 2014. The impact of diet and lifestyle on gut microbiota and human health. Nutrients 7 (1), 17–44.

Cummings, J.H., Christie, S., Cole, T.J., 2001. A study of fructo oligosaccharides in the prevention of travellers' diarrhoea. Aliment. Pharmacol. Ther. 15 (8), 1139–1145.

De Vrese, M., Schrezenmeir, J., 2008. Probiotics, prebiotics, and synbiotics. Food Biotechnology. Springer, Berlin, Heidelberg, pp. 1–66.

Depeint, F., Tzortzis, G., Vulevic, J., I'Anson, K., Gibson, G.R., 2008. Prebiotic evaluation of a novel galactooligosaccharide mixture produced by the enzymatic activity of Bifidobacterium bifidum NCIMB 41171, in healthy humans: a randomized, double-blind, crossover, placebo-controlled intervention study. Am. J. Clin. Nutr. 87 (3), 785–791.

Dewulf, E.M., Cani, P.D., Neyrinck, A.M., Possemiers, S., Van Holle, A., Muccioli, G. G., et al., 2011. Inulin-type fructans with prebiotic properties counteract GPR43 overexpression and PPARγ-related adipogenesis in the white adipose tissue of high-fat diet-fed mice. J. Nutr. Biochem. 22 (8), 712–722.

Dwivedi, S., Sahrawat, K., Puppala, N., Ortiz, R., 2014. Plant prebiotics and human health: biotechnology to breed prebiotic-rich nutritious food crops. Electr. J. Biotechnol. 17 (5), 238–245.

Fanaro, S., Boehm, G., Garssen, J., Knol, J., Mosca, F., Stahl, B., et al., 2005. Galacto-oligosaccharides and long chain fructooligosaccharides as prebiotics in infant formulas: a review. Actapaediatrica 94 (s449), 22–26.

Fernando, W., Hill, J., Zello, G., Tyler, R., Dahl, W., Van Kessel, A., 2010. Diets supplemented with chickpea or its main oligosaccharide component raffinose modify faecal microbial composition in healthy adults. Benef. Microbes 1 (2), 197–207.

Floch, M.H., Walker, W.A., Madsen, K., Sanders, M.E., Macfarlane, G.T., Flint, H.J., et al., 2011. Recommendations for probiotic use—2011 update. J. Clin. Gastroenterol. 45, 168–171.

Forcheron, F., Beylot, M., 2007. Long-term administration of inulin-type fructans has no significant lipid-lowering effect in normolipidemic humans. Metab. Clin. Exp. 56 (8), 1093–1098.

Franco-Robles, E., López, M.G., 2015. Implication of fructans in health: immunomodulatory and antioxidant mechanisms. Sci. World J. 289267.

Fuentes-Zaragoza, E., Sánchez-Zapata, E., Sendra, E., Sayas, E., Navarro, C., Fernández-López, J., et al., 2011. Resistant starch as prebiotic: a review. Starch-Stärke 63 (7), 406–415.

Ganan, M., Collins, M., Rastall, R., Hotchkiss, A.T., Chau, H.K., Carrascosa, A.V., et al., 2010. Inhibition by pectic oligosaccharides of the invasion of undifferentiated and differentiated Caco-2 cells by *Campylobacter jejuni*. Int. J. Food Microbiol. 137 (2–3), 181–185.

Ganner, A., Schatzmayr, G., 2012. Capability of yeast derivatives to adhere enteropathogenic bacteria and to modulate cells of the innate immune system. Appl. Microbiol. Biotechnol. 95 (2), 289–297.

Gibson, G.R., McCartney, A.L., Rastall, R.A., 2005. Prebiotics and resistance to gastrointestinal infections. Br. J. Nutr. 93 (S1), S31–S34.

Gibson, G.R., Scott, K.P., Rastall, R.A., Tuohy, K.M., Hotchkiss, A., Dubert-Ferrandon, A., et al., 2010. Dietary prebiotics: current status and new definition. Food Sci. Technol. Bull. Funct. Foods 7 (1), 1–19.

Gibson, G.R., Hutkins, R., Sanders, M.E., Prescott, S.L., Reimer, R.A., Salminen, S.J., et al., 2017. Expert consensus document: the International Scientific Association for Probiotics and Prebiotics (ISAPP) consensus statement on the definition and scope of prebiotics. Nat. Rev. Gastroenterol. Hepatol. 14 (8), 491.

Grajek, W., Olejnik, A., Sip, A., 2005. Probiotics, prebiotics and antioxidants as functional foods. Acta Biochim. Pol. 52 (3), 665–671.

Griffin, I.J., Davila, P.M., Abrams, S.A., 2002. Non-digestible oligosaccharides and calcium absorption in girls with adequate calcium intakes. Br. J. Nutr. 87 (S2), 187–191.

He, J.Y., Zhang, Y., Li, W., Sun, Y., Zheng, X.X., 2011. Prebiotic function of alphagalactooligosaccharides from chickpea seeds. Food Sci. 32, 94–98.

Hoentjen, F., Welling, G.W., Harmsen, H.J., Zhang, X., Snart, J., Tannock, G.W., et al., 2005. Reduction of colitis by prebiotics in HLA-B27 transgenic rats is associated with microflora changes and immunomodulation. Inflamm. Bowel Dis. 11 (11), 977–985.

Holscher, H.D., 2017. Dietary fiber and prebiotics and the gastrointestinal microbiota. Gut Microbes 8 (2), 172–184.

Hotchkiss, A.T., Nunez, A., Khoo, C., Strahan, G.D., Ocean Spray Cranberries Inc and US Department of Agriculture, 2016. Cranberry xyloglucan oligosaccharide composition. U.S. Patent 9,314,494.

Hsu, C.K., Liao, J.W., Chung, Y.C., Hsieh, C.P., Chan, Y.C., 2004. Xylooligosaccharides and fructooligosaccharides affect the intestinal microbiota and precancerous colonic lesion development in rats. J. Nutr. 134 (6), 1523–1528.

Huang, C.H., Cheng, J.Y., Deng, M.C., Chou, C.H., Jan, T.R., 2012. Prebiotic effect of diosgenin, an immunoactive steroidal sapogenin of the Chinese yam. Food Chem. 132 (1), 428–432.

Huang, C.H., Wang, C.C., Lin, Y.C., Hori, M., Jan, T.R., 2017. Oral administration with diosgenin enhances the induction of intestinal T helper 1-like regulatory T cells in a murine model of food allergy. Int. Immunopharmacol. 42, 59–66.

Kanauchi, O.S.A.M.U., Mitsuyama, K., Saiki, T., Agata, K., Nakamura, T., Iwanaga, T., 1998. Preventive effects of germinated barley foodstuff on methotrexate-induced enteritis in rats. Int. J. Mol. Med. 1 (6), 961–967.

Kato, K., Ishiwa, A., 2015. The role of carbohydrates in infection strategies of enteric pathogens. Trop. Med. Health 43 (1), 41–52.

Kechagia, M., Basoulis, D., Konstantopoulou, S., Dimitriadi, D., Gyftopoulou, K., Skarmoutsou, N., et al., 2013. Health benefits of probiotics: a review. ISRN Nutr. 481651.

Kumar, M., Kumar, A., Nagpal, R., Mohania, D., Behare, P., Verma, V., et al., 2010. Cancer-preventing attributes of probiotics: an update. Int. J. Food Sci. Nutr. 61 (5), 473–496.

Lee, J.H., O'Sullivan, D.J., 2010. Genomic insights into bifidobacteria. Microbiol. Mol. Biol. Rev., 74 (3), 378–416.

Lee, Y.K., Salminen, S., 2009. Handbook of Probiotics and Prebiotics. John Wiley & Sons, Hoboken, NJ978-0-470-13544-0.

Lewis, S., Burmeister, S., Brazier, J., 2005. Effect of the prebiotic oligofructose on relapse of Clostridium difficile-associated diarrhea: a randomized, controlled study. Clin. Gastroenterol. Hepatol. 3 (5), 442–448.

Lin, B., Gong, J., Wang, Q., Cui, S., Yu, H., Huang, B., 2011. In-vitro assessment of the effects of dietary fibers on microbial fermentation and communities from large intestinal digesta of pigs. Food Hydrocol. 25 (2), 180–188.

Lin, S.H., Chou, L.M., Chien, Y.W., Chang, J.S., Lin, C.I., 2016. Prebiotic effects of xylooligosaccharides on the improvement of microbiota balance in human subjects. Gastroenterol. Res. Pract. 2016.

Liong, M.T., 2008. Roles of probiotics and prebiotics in colon cancer prevention: postulated mechanisms and in-vivo evidence. Int. J. Mol. Sci. 9 (5), 854–863.

Lu, Q.Y., Summanen, P.H., Lee, R.P., Huang, J., Henning, S.M., Heber, D., et al., 2017. Prebiotic potential and chemical composition of seven culinary spice extracts. J. Food Sci. 82 (8), 1807–1813.

Ma, Y., Wu, X., Giovanni, V., Meng, X., 2017. Effects of soybean oligosaccharides on intestinal microbial communities and immune modulation in mice. Saudi J. Biol. Sci. 24 (1), 114–121.

Mao, B., Li, D., Zhao, J., Liu, X., Gu, Z., Chen, Y.Q., et al., 2014. In vitro fermentation of lactulose by human gut bacteria. J. Agric. Food Chem. 62 (45), 10970–10977.

Marcobal, A., Barboza, M., Froehlich, J.W., Block, D.E., German, J.B., Lebrilla, C.B., et al., 2010. Consumption of human milk oligosaccharides by gut-related microbes. J. Agric. Food Chem. 58 (9), 5334–5340.

Markowiak, P., Śliżewska, K., 2017. Effects of probiotics, prebiotics, and synbiotics on human health. Nutrients 9 (9), 1021.

Martín, R., Miquel, S., Ulmer, J., Kechaou, N., Langella, P., Bermúdez-Humarán, L.G., 2013. Role of commensal and probiotic bacteria in human health: a focus on inflammatory bowel disease. Microb. Cell Fact. 12 (1), 71.

McCabe, L., Britton, R.A., Parameswaran, N., 2015. Prebiotic and probiotic regulation of bone health: role of the intestine and its microbiome. Curr. Osteopor. Rep. 13 (6), 363–371.
Mitsou, E.K., Panopoulou, N., Turunen, K., Spiliotis, V., Kyriacou, A., 2010. Prebiotic potential of barley derived β-glucan at low intake levels: a randomised, double-blinded, placebo-controlled clinical study. Food Res. Int. 43 (4), 1086–1092.
Ngamkala, S., Futami, K., Endo, M., Maita, M., Katagiri, T., 2010. Immunological effects of glucan and Lactobacillus rhamnosus GG, a probiotic bacterium, on Nile tilapia Oreochromisniloticus intestine with oral Aeromonas challenges. Fish. Sci. 76 (5), 833–840.
Oliveira, R.P.D.S., Florence, A.C.R., Perego, P., De Oliveira, M.N., Converti, A., 2011. Use of lactulose as prebiotic and its influence on the growth, acidification profile and viable counts of different probiotics in fermented skim milk. Int. J. Food Microbiol. 145 (1), 22–27.
Ortega-González, M., de Medina, F.S., Molina-Santiago, C., López-Posadas, R., Pacheco, D., Krell, T., et al., 2014. Fructooligosacharides reduce *Pseudomonas aeruginosa* PAO1 pathogenicity through distinct mechanisms. PLoS One 9 (1), e85772.
Pandey, K.R., Naik, S.R., Vakil, B.V., 2015. Probiotics, prebiotics and synbiotics—a review. J. Food Sci. Technol. 52 (12), 7577–7587.
Parnell, J.A., Raman, M., Rioux, K.P., Reimer, R.A., 2012. The potential role of prebiotic fibre for treatment and management of non-alcoholic fatty liver disease and associated obesity and insulin resistance. Liver Int. 32 (5), 701–711.
Patel, S., Goyal, A., 2012. The current trends and future perspectives of prebiotics research: a review. 3 Biotech 2 (2), 115–125.
Payne, A.N., Chassard, C., Lacroix, C., 2012. Gut microbial adaptation to dietary consumption of fructose, artificial sweeteners and sugar alcohols: implications for host–microbe interactions contributing to obesity. Obes. Rev. 13 (9), 799–809.
Quintero-Villegas, M.I., Aam, B.B., Rupnow, J., Sørlie, M., Eijsink, V.G., Hutkins, R. W., 2013. Adherence inhibition of enteropathogenic Escherichia coli by chitooligosaccharides with specific degrees of acetylation and polymerization. J. Agric. Food Chem. 61 (11), 2748–2754.
Rai, R., Saraswat, V.A., Dhiman, R.K., 2015. Gut microbiota: its role in hepatic encephalopathy. J. Clin. Exp. Hepatol. 5, S29–S36.
Reimer, R.A., Maurer, A.D., Eller, L.K., Hallam, M.C., Shaykhutdinov, R., Vogel, H.J., et al., 2012. Satiety hormone and metabolomic response to an intermittent high energy diet differs in rats consuming long-term diets high in protein or prebiotic fiber. J. Proteome Res. 11 (8), 4065–4074.
Rivera-Huerta, M., Lizárraga-Grimes, V.L., Castro-Torres, I.G., Tinoco-Méndez, M., Macías-Rosales, L., Sánchez-Bartéz, F., et al., 2017. Functional effects of prebiotic fructans in colon cancer and calcium metabolism in animal models. BioMed Res. Int. 9758982.
Romo-Vaquero, M., Selma, M.V., Larrosa, M., Obiol, M., García-Villalba, R., González-Barrio, R., et al., 2014. A rosemary extract rich in carnosic acid selectively modulates caecum microbiota and inhibits β-glucosidase activity, altering fiber and short chain fatty acids fecal excretion in lean and obese female rats. PLoS One 9 (4), e94687.
Rossi, M., Corradini, C., Amaretti, A., Nicolini, M., Pompei, A., Zanoni, S., et al., 2005. Fermentation of fructooligosaccharides and inulin by bifidobacteria: a comparative study of pure and fecal cultures. Appl. Environ. Microbiol. 71 (10), 6150–6158.
Samanta, A.K., Jayapal, N., Jayaram, C., Roy, S., Kolte, A.P., Senani, S., et al., 2015. Xylooligosaccharides as prebiotics from agricultural by-products: Production and applications. Bioact. Carbohydr. Diet. Fibre 5 (1), 62–71.

Schley, P.D., Field, C.J., 2002. The immune-enhancing effects of dietary fibres and prebiotics. Br. J. Nutr. 87 (S2), S221–S230.
Scholz-Ahrens, K.E., Adolphi, B., Rochat, F., Barclay, D.V., de Vrese, M., Açil, Y., et al., 2016. Effects of probiotics, prebiotics, and synbiotics on mineral metabolism in ovariectomized rats—impact of bacterial mass, intestinal absorptive area and reduction of bone turn-over. NFS J. 3, 41–50.
Sharma, S., Agarwal, N., Verma, P., 2012. Miraculous health benefits of prebiotics. Int. J. Pharm. Sci. Res. 3 (6), 1544.
Sharon, N., 2006. Carbohydrates as future anti-adhesion drugs for infectious diseases. Biochim. Biophys. Acta 1760 (4), 527–537.
Shi, L.H., Balakrishnan, K., Thiagarajah, K., Ismail, N.I.M., Yin, O.S., 2016. Beneficial properties of probiotics. Trop. Life Sci. Res. 27 (2), 73.
ShoafSweeney, K.D., Hutkins, R.W., 2008. Adherence, anti-adherence, and oligosaccharides: preventing pathogens from sticking to the host. Adv. Food Nutr. Res. 55, 101–161.
Slavin, J., 2013. Fiber and prebiotics: mechanisms and health benefits. Nutrients 5 (4), 1417–1435.
Slavin, J.L., 2008. Position of the American Dietetic Association: health implications of dietary fiber. J. Am. Diet. Assoc. 108 (10), 1716–1731.
Sokol, H., Pigneur, B., Watterlot, L., Lakhdari, O., Bermúdez-Humarán, L.G., Gratadoux, J.J., et al., 2008. Faecalibacterium prausnitzii is an anti-inflammatory commensal bacterium identified by gut microbiota analysis of Crohn disease patients. Proc. Natl Acad. Sci. 105 (43), 16731–16736.
Sridharan, G.V., Choi, K., Klemashevich, C., Wu, C., Prabakaran, D., Pan, L.B., et al., 2014. Prediction and quantification of bioactive microbiota metabolites in the mouse gut. Nat. Commun. 5, 5492.
Synytsya, A., Míčková, K., Synytsya, A., Jablonský, I., Spěváček, J., Erban, V., et al., 2009. Glucans from fruit bodies of cultivated mushrooms Pleurotus ostreatus and Pleurotus eryngii: Structure and potential prebiotic activity. Carbohydr. Polym. 76 (4), 548–556.
Szajewska, H., Weizman, Z., Abu-Zekry, M., Jaklin Kekez, A., Braegger, C.P., Kolacek, S., et al., 2012. Inulin and fructo-oligosaccharides for the prevention of antibiotic-associated diarrhea in children: report by the ESPGHAN working group on probiotics and prebiotics. J. Pediatr. Gastroenterol. Nutr., 54 (6), 828–829.
Szymańska-Czerwińska, M., Bednarek, D., 2011. Effect of tylosin and prebiotics on the selected humoral immunological parameters in calves. Medycyna Weterynaryjna 67 (4), 275–278.
Tarini, J., Wolever, T.M., 2010. The fermentable fibre inulin increases postprandial serum short-chain fatty acids and reduces free-fatty acids and ghrelin in healthy subjects. Appl. Physiol. Nutr. Metab. 35 (1), 9–16.
Trop, T.K., 2014. Intestinal microbiota, probiotics and prebiotics in inflammatory bowel disease. World J. Gastroenterol. 20 (33), 11505.
Tzortzis, G., 2010. Development and functional properties of Bimuno®: a second-generation prebiotic mixture. Food Sci. Technol. Bull. 6, 81–89.
Van den Ende, W., 2013. Multifunctional fructans and raffinose family oligosaccharides. Front. Plant Sci. 4, 247.
Vulevic, J., Drakoularakou, A., Yaqoob, P., Tzortzis, G., Gibson, G.R., 2008. Modulation of the fecal microflora profile and immune function by a novel trans-galactooligosaccharide mixture (B-GOS) in healthy elderly volunteers. Am. J. Clin. Nutr. 88 (5), 1438–1446.
Xu, X., Jia, X., Mo, L., Liu, C., Zheng, L., Yuan, Q., et al., 2017. Intestinal microbiota: a potential target for the treatment of postmenopausal osteoporosis. Bone Res. 5, 17046.

Yoo, H.D., Kim, D., Paek, S.H., 2012. Plant cell wall polysaccharides as potential resources for the development of novel prebiotics. Biomol. Ther. 20 (4), 371.
Young, N.W.G., O'Sullivan, G.R., 2011. The influence of ingredients on product stability and shelf life. Food and Beverage Stability and Shelf Life. pp. 132–183.

FURTHER READING

Andersen, J.M., Barrangou, R., Hachem, M.A., Lahtinen, S., Goh, Y.J., Svensson, B., et al., 2011. Transcriptional and functional analysis of galactooligosaccharide uptake by lacS in Lactobacillus acidophilus. Proc. Natl Acad. Sci. 108 (43), 17785–17790.
Arslanoglu, S., Moro, G.E., Boehm, G., 2007. Early supplementation of prebiotic oligosaccharides protects formula-fed infants against infections during the first 6 months of life. J. Nutr. 137 (11), 2420–2424.
Azcarate-Peril, M.A., Ritter, A.J., Savaiano, D., Monteagudo-Mera, A., Anderson, C., Magness, S.T., et al., 2017. Impact of short-chain galactooligosaccharides on the gut microbiome of lactose-intolerant individuals. Proc. Natl Acad. Sci. 114 (3), E367–E375.
Bruzzese, E., Volpicelli, M., Salvini, F., Bisceglia, M., Lionetti, P., Cinquetti, M., et al., 2006. Early administration of GOS/FOS prevents intestinal and respiratory infections in infants. J. Pediatr. Gastroenterol. Nutr. 42 (5), E95.
Costalos, C., Kapiki, A., Apostolou, M., Papathoma, E., 2008. The effect of a prebiotic supplemented formula on growth and stool microbiology of term infants. Early Hum. Dev. 84 (1), 45–49.
Cui, X., Ye, L., Li, J., Jin, L., Wang, W., Li, S., et al., 2018. Metagenomic and metabolomic analyses unveil dysbiosis of gut microbiota in chronic heart failure patients. Sci. Rep. 8 (1), 635.
Daubioul, C.A., Horsmans, Y., Lambert, P., Danse, E., Delzenne, N.M., 2005. Effects of oligofructose on glucose and lipid metabolism in patients with nonalcoholicsteatohepatitis: results of a pilot study. Eur. J. Clin. Nutr. 59 (5), 723.
Fadden, K., Owen, R.W., 1992. Faecal steroids and colorectal cancer: the effect of lactulose on faecal bacterial metabolism in a continuous culture model of the large intestine. Eur. J. Cancer Prevent. 1 (2), 113–127.
Garcia, A.L., Otto, B., Reich, S.C., Weickert, M.O., Steiniger, J., Machowetz, A., et al., 2007. Arabinoxylan consumption decreases postprandial serum glucose, serum insulin and plasma total ghrelin response in subjects with impaired glucose tolerance. Eur. J. Clin. Nutr. 61 (3), 334.
Gibson, G.R., Roberfroid, M.B., 1995. Dietary modulation of the human colonic microbiota: introducing the concept of prebiotics. J. Nutr. 125 (6), 1401–1412.
Isolauri, E., Sütas, Y., Kankaanpää, P., Arvilommi, H., Salminen, S., 2001. Probiotics: effects on immunity. Am. J. Clin. Nutr. 73 (2), 444–450.
Kim, Y.S., Tsao, D., Morita, A., Bella, A., 1982. Effect of sodium butyrate and three human colorectal adenocarcinoma cell lines in culture. In: Falk Symposium 31, pp. 317–323.
Kurhekar, J.V., 2013. *Curcuma longa* and *Allium sativum* as prebiotics. Bionano Front. 6 (2), 327–329.
Lindsay, J.O., Whelan, K., Stagg, A.J., Gobin, P., Al-Hassi, H.O., Rayment, N., et al., 2006. Clinical, microbiological, and immunological effects of fructo-oligosaccharide in patients with Crohn's disease. Gut 55 (3), 348–355.
Lu, Z.X., Walker, K.Z., Muir, J.G., O'Dea, K., 2004. Arabinoxylan fibre improves metabolic control in people with Type II diabetes. Eur. J. Clin. Nutr. 58 (4), 621.

Luo, J., Van Yperselle, M., Rizkalla, S.W., Rossi, F., Bornet, F.R., Slama, G., 2000. Chronic consumption of short-chain fructooligosaccharides does not affect basal hepatic glucose production or insulin resistance in type 2 diabetics. J. Nutr. 130 (6), 1572–1577.

Moro, G., Arslanoglu, S., Stahl, B., Jelinek, J., Wahn, U., Boehm, G., 2006. A mixture of prebiotic oligosaccharides reduces the incidence of atopic dermatitis during the first six months of age. Arch. Dis. Childhood 91 (10), 814–819.

Munjal, U., Glei, M., Pool-Zobel, B.L., Scharlau, D., 2009. Fermentation products of inulin-type fructans reduce proliferation and induce apoptosis in human colon tumour cells of different stages of carcinogenesis. Br. J. Nutr. 102 (5), 663–671.

Parnell, J.A., Reimer, R.A., 2009. Weight loss during oligofructose supplementation is associated with decreased ghrelin and increased peptide YY in overweight and obese adults. Am. J. Clin. Nutr. 89 (6), 1751–1759.

Prasad, S., Dhiman, R.K., Duseja, A., Chawla, Y.K., Sharma, A., Agarwal, R., 2007. Lactulose improves cognitive functions and health-related quality of life in patients with cirrhosis who have minimal hepatic encephalopathy. Hepatology 45 (3), 549–559.

Scholtens, P.A., Alliet, P., Raes, M., Alles, M.S., Kroes, H., Boehm, G., et al., 2008. Fecal secretory immunoglobulin A is increased in healthy infants who receive a formula with short-chain galacto-oligosaccharides and long-chain fructo-oligosaccharides. J. Nutr. 138 (6), 1141–1147.

Van Den Elsen, L.W., Poyntz, H.C., Weyrich, L.S., Young, W., Forbes-Blom, E.E., 2017. Embracing the gut microbiota: the new frontier for inflammatory and infectious diseases. Clin. Trans. Immunol. 6 (1).

Welters, C.F., Heineman, E., Thunnissen, F.B., van den Bogaard, A.E., Soeters, P.B., Baeten, C.G., 2002. Effect of dietary inulin supplementation on inflammation of pouch mucosa in patients with an ileal pouch-anal anastomosis. Dis. Colon Rectum 45 (5), 621–627.

Younis, K., Ahmad, S., Jahan, K., 2015. Health benefits and application of prebiotics in foods. J. Food Process. Technol. 6 (4), 1.

SECTION 2A

Biological Activity and Discovery of New Compounds From Herbs, Medicinal Plants, and Herbal Medicine

CHAPTER 5

Moroccan Medicinal Plants as Antiinfective and Antioxidant Agents

Malika Ait-Sidi-Brahim, Mohammed Markouk and Mustapha Larhsini
Laboratory of Biotechnology, Protection and Valorization of Plant Resources; Phytochemistry and Pharmacology of Medicinal Plants Unit, (URAC35 Association Unit) Faculty of Sciences Semlalia, Cadi Ayyad University, Marrakech, Morocco

5.1 INTRODUCTION

Since ancient time, medicinal and aromatic plants have been widely used by populations and constitute the basis for medical treatments worldwide. Medicinal plants as a rich source of phytochemicals have gained a great interest despite the tremendous progress in human medicine because they are easily accessible in the nature and expect low side effects (Al-Adhroey et al., 2010). In addition, medicinal plants are endowed with multiple and various activities like antibacterial, antifungal, antiviral, antioxidant, insecticidal, antitumor, antiinflammatory activities, etc. Natural products derived from aromatic and medicinal plants play a very important therapeutic role in the survival of tribal and ethnic communities. Indeed, enormous variety of natural products isolated from plants are with diverse structures, of which more than 40% constitute a source of alternative drugs used for the treatment of several diseases. Previously, natural products or "secondary metabolites" have been considered waste products because they have no function inside the plant, as opposed to "primary metabolites" which are primarily involved in primary metabolism (plant growth and development). However, with the development of science, it has been shown that these natural products exert important functions in the interaction of the plant with its biotic and abiotic environment (Springob and Kutchan, 2009). These natural products can play several physiological functions at the plant level, for example, they can be involved in defenses against herbivores and pathogens and they can be also flower pigments that attract pollinators, or hormones and signaling

New Look to Phytomedicine
DOI: https://doi.org/10.1016/B978-0-12-814619-4.00005-7

molecules (Springob and Kutchan, 2009). In addition to their plant-level functions, they also have a strong impact on human survival, so they have been used throughout human history as spices, condiments, pigments, and pharmaceuticals. The most important classes of natural products include terpenes, polyphenols, and alkaloids. Many of these compounds have been used as plant extracts or essential oils (EOs) for their medical or nutritional benefits on the human body and playing various pharmacological activities such as antibacterial, antioxidant, antileishmanial, and antitumor.

Infectious diseases caused by bacteria, fungi, viruses, and parasites are a major cause of the death worldwide. With the tremendous progress in human medicine, some of these diseases have been controlled for some decades. However, the indiscriminate and extensive uses of drugs have developed multidrug resistance. These problems have led to the research of new alternative strategies based on natural products derived from medicinal plants to control some disease especially those caused by microorganisms and oxidative stress. In the recent years, natural products from plant origin are also used in combination with some conventional antibiotics to fight infectious disease especially those caused by multiresistant bacteria and *Candida* strains.

Herbal preparations, including plant extracts, can be found in the pharmacopoeias of many countries. Traditional Moroccan pharmacopoeia has been also considered as a rich source of medicinal plants, due to the diversity of the country in climates and biotopes (desert, mountains, coastal areas, etc.). In fact, Morocco possesses a very important floristic system characterized by a great diversity of plants with 4200 species and subspecies (Rankou et al., 2013). Furthermore, Morocco is characterized by a rich and varied aromatic and medicinal flora with high levels of endemic plants belonging to different botanical families (Lamiaceae, Asteraceae, Rosaceae, Chenopodiaceae, Papaveraceae, Caryophyllaceae, Cupressaceae, Rutaceae, Anacardiaceae, Zygophyllaceae, etc.). This endemic flora (22%) contains 879 species and subspecies divided into 55 families and 287 genera (El Midaoui et al., 2011). Such a richness of endemic species places Morocco in an important position among the other Mediterranean countries. Morocco also has a long history of folk medicine and a lot of plants are used by the local population especially in rural areas for the primary health care.

This chapter is dedicated to summarize recent scientific investigations performed to determine the antiinfective and antioxidant activities of plant extracts and EOs from Moroccan medicinal and aromatic plants. Studies performed on the synergistic interaction between Moroccan plant extracts and some antibiotics were also mentioned in this chapter.

5.2 ANTIMICROBIAL ACTIVITY OF MOROCCAN MEDICINAL PLANTS

It has been widely known that Moroccan medicinal plants exhibited important antimicrobial effect. Agar diffusion method (disc or well) and dilution method (agar or liquid broth) were the most common methods used to determine the antibacterial activity of EOs and solvent extracts of Moroccan medicinal plants. Quantitative screening is evaluated using the agar disk and well diffusion methods. The method is easy and not expensive and used in order to determine the inhibition zones (IZs) which allow finding the quantitative inhibition values. However, for plant extracts, this method could not always be a reliable method especially for less polar compounds which diffuse more slowly into the culture medium (Moreno et al., 2006). It is necessary to note that the absence of an IZ did not mean that the compound was inactive. Furthermore, the qualitative inhibition is determined by the dilution method which is used to evaluate minimum inhibitory concentration (MIC) and minimum bactericidal concentration (MBC) values. Comparing to agar method, this method is more precise and reproducible and can increase the sensitivity for small quantities of extract and permit to distinguish between bacteriostatic and bactericidal effects. On the other hand, molecular targets of these natural products are determined using transmission electron microscopy and flow cytometry.

5.2.1 Antimicrobial Activity of Essential Oils

Several studies have revealed the efficiency of EOs as natural antimicrobial agents. Recently, numerous Moroccan aromatic and medicinal plants have been widely investigated for their antimicrobial activities. Table 5.1 summarizes the essential works performed in this context. Thyme and Oregano are the most known Moroccan medicinal plants, frequently used in traditional medicine for the treatment of numerous infections. Several scientific reports evaluated the antimicrobial activity of different species of *Thymus* sp. EOs of endemic Moroccan thyme species (*Thymus maroccanus* and *Thymus broussonetii*) were tested against antibiotic-resistant bacteria involved in nosocomial infections. Both EOs exhibited a high inhibitory

Table 5.1 Antibacterial activity of some Moroccan medicinal plants

Families	Species (families)	Origin	Parts used	Extracts	Major components	Effects	Reference
Apocynaceae	*Periploca laevigata*	Essaouira	Leaves	Essential oil	*n*-Hexadecanoic acid and 4,4,7a-trimethyl-5,6,7,7a-tetrahydro-4H-benzofuran-2-one	ø = 12.50–20 mm, MIC = 0.937–3.75 mg/mL for *Candida* species ø = 10.50–18 mm, MIC = 3.75–7.5 mg/mL for Gram-positive bacteria MIC = 15–30 mg/mL for Gram-negative bacteria	Ait Dra et al. (2017)
Apocynaceae	*Phoenix dactylifera* L.	Southeast Morocco	Fruits	Methanolic rich polyphenol extracts	ND	Bousrdon, Bousthammi, Boufgous, and Jihl varieties showed strong antibacterial activity Majhoul and Bouskri extracts had no activity	Bouhlali et al. (2016)
Asteraceae	*Achillea ageratum*	Demnate	Aerial parts	Essential oil	Artemisyl acetate, yomogi alcohol, and artemisia alcohol	Weak activity; ø = 13.50–20 mm MIC = 4.64 74.32 mg/mL for Gram-positive bacteria MIC = > 108 mg/mL for Gram-negative bacteria	El Abdouni Khayari et al. (2016)
	A. ageratum L. (wild and cultivated)	Wild Demnate Cultivated from Marrakech	Leaves and flowers	Essential oils	Artemisyl acetate, yomogi alcohol, santolina alcohol, and artemisia alcohol	Gram-positive bacteria ø = 24–59.33 mm; MIC = 2.55–7.02 mg/mL. Gram-negative bacteria ø = 8.83–10.67 mm; MIC = 20.40–41.10 mg/mL *Candida* strains: ø = 12.2–34.8 mm; MIC = 5.83–8.42 mg/mL	El Bouzidi et al. (2012)

	Artemisia herba-alba	Tahanaoute	Aerial parts at preflowering stage	Essential oil	Chrysanthenone, camphor, and verbenone	High antibacterial activity MICs obtained for *E. coli* EA27 and *E. aerogenes* AG102, overexpressing efflux system were greater than the MIC obtained for their isogenic strains	Fadli et al. (2016)
	Calendula arvensis	Rabat Khemisset	Flowers	The hexane methanol aqueous extracts	ND	For methanolic and aqueous extracts MIC = 12.5–25 μg/mL for bacteria For hexanolic extracts MICs = 6.25–12.5 μg/mL For methanolic extracts ø = 14 and 20 mm against *Candida tropicalis* and *Candida famata* 1	Abudunia et al. (2017)
	Cotula cinerea	Zagora	Aerial parts	Essential oil	*trans*-Thujone, *cis*-verbenyl acetate, 1,8-cineole and camphor	The oil showed high anticandidal activity against all investigated strains with MIC of 3.2–4.7 mg/mL and CMB of 5.9 mg/mL	El Bouzidi et al. (2011a)
	C. cinerea	Zagora	Aerial parts	Essential oil	*trans*-Thujone, *cis*-verbenyl acetate and santolina triene	Gram-positive bacteria MIC = 3.38–27.00 mg/mL Gram-negative MIC ≥ 108 mg/mL	El Abdouni Khayari et al. (2016)

(*Continued*)

Table 5.1 (Continued)

Families	Species (families)	Origin	Parts used	Extracts	Major components	Effects	Reference
	Pulicaria odora L.	Mrissat (East of Rabat)	Roots	Essential oil and its isolated compounds (2-isopropyl-4-methylphenol and isobutyric acid 2-isopropyl-4-methylphenyleste	2-Isopropyl-4-methylphenol and isobutyric acid 2-isopropyl-4-methylphenylester	The essential oil and the 2-isopropyl-4-methylphenol exhibited a very significant antibacterial and antifungal activity The isobutyric acid 2-isopropyl-4-methylphenylester was inactive for all tested strains	Ezoubeiri et al. (2005)
Cistaceae	*Cistus crispus*	Ouezzane	Leaves	Methanolic, ethanolic, ethyl acetate and *n*-hexanic extracts	ND	Gram-positive bacteria (*L. monocytogenes*, *S. aureus*) are more sensitive than Gram-negative bacteria The highest diameter of inhibition is found with ethanolic extract against *S. aureus* and *L. monocytogenes*	Bouyahya et al. (2017d)
	Cistus villosus	Sous valley (agadir)	Leaves + stem	Hexane; chloroform; ethyl acetate; methanol.	ND	For methanol extract MIC = 0.156 and 1.25 mg/mL and = 2.5 and 5 mg/mL	Talibi et al. (2012)
Cupressaceae	*Juniperus phoenicea L*	Angad (north-eastern part of Morocco)	Aerial parts	EO	Pinene, β-phellandrene and α-terpinyl acetate	The EO showed bacteriostatic and bactericidal activity only against the four Gram-positive strains MIC values ranging from 0.5 to 15 μL/mL and MBC values from 10 to 30 μL/mL	Ait-Ouazzou et al. (2012)

	Cyperus longus L.	Angad (north-eastern part of Morocco)	Aerial parts	EO	β-Himachalene, α-humulene, and γ-himachalene	Bacteriostatic activity only against *S. aureus* (> 0.5 μL/mL) and against *L. monocytogenes* EGD-e and 4b (30 μL/mL)	Ait-Ouazzou et al. (2012)
Ericaceae	*Arbutus unedo* L.	Ouezzane	Leaves	Methanolic, ethanolic, ethyl acetate and *n*-hexanic extracts	ND	For hexane extract ø = 34.42 ± 0.26 mm to 40 ± 0.19 and MICs = 0.255 mg/mL against *S. aureus* and *L. monocytogenes*ø = 19 ± 0.97 mm to 23 ± 0 mm, and MIC = 1–2 mg/mL against *P. aeruginosa* and *E. coli*	Bouyahya et al. (2016b)
Fabaceae	*Ceratonia siliqua*	Sous valley (Agadir)	Leaves	Hexane; chloroform; ethyl acetate; methanol	ND	For methanol extract MIC = 0.156 and 1.25 mg/mL and MFC = 2.5 and 5 mg/mL	Talibi et al. (2012)
Fagaceae	*Quercus suber*	Mâamora	Leaves and bark	Methanolic extracts	ND	The best effect was obtained in the presence of the methanol extract of the bark with MICs = 12.5 and 50 mg/mL for *C. albicans* and *T. rubrum*	Hassikou et al. (2014)
Lamiaceae	*Melissa officinalis*	Hills of the Sefrou City	Fresh leaves	Citronellal, isogeraniol, geraniol acetate, nerol acetate, caryophyllene and b-caryophyllene oxide		Strong antibacterial activity *P. aeruginosa* (16 mm), *K. pneumonia* (13 mm), *S. aureus* (20 mm), and *C. kuseri* (14 mm)	Jalal et al. (2015)

(Continued)

Table 5.1 (Continued)

Families	Species (families)	Origin	Parts used	Extracts	Major components	Effects	Reference
	M. officinalis	Taza (Nord-East of Morocco)	The aerial part	Essential oil	P-mentha-1,2,3-triol, P-menth-3-en-8-ol, pulegone, piperitynone oxide (8.4%) and 2-piperitone oxide	The inhibition rate was 73.2% for *P. expansum* at 0.25 mL/mL and 100% at 1 mL/mL The inhibition rate was 16.27% for *R. stolonifer* at 0.25 mL/mL and 100% at 2 mL/mL The maximum inhibition for B cinerea was 76.81% at 2 mL/mL	El Ouadi et al. (2017)
	Mentha spicate	Settat (Guisser)	Aerial parts	Essential oil	*Carvone*, *Limonène*, *Germacrène-D*	Significant inhibitory activity	Ismaili. et al. (2014)
	M. suaveolens subsp. timija (wild and cultivated)			Essential oils	Menthone, pulegone and isomenthone-Pulegone was more dominant in cultivated mint timija than in wild one, while menthone was more abundant in the wild material	ø = 6.73–30.0 mm and MIC = 0.15–2.33 mg/mL Cultivated essential oil activity > wild	Kasrati et al. (2013)
	Origanum compactum	Taounate	Fresh aerial part (leaves and stems)	Essential oil	Carvacrol, terpinene and thymol	*P. aeruginosa* was the most resistant one, MIC = 4% (v/v)	Ouedrhiri et al. (2016)

	O. compactum Benth	Ouezzane	Leaves	Methanolic, ethanolic, ethyl acetate and *n*-hexanic extracts	ND	All extracts showed strong activity with the largest zone of inhibition of about 34 ± 1.24 mm. The *n*-hexane extract showed the strongest antibacterial activity, while the ethyl acetate extract showed the lowest antibacterial activity	
	officinalis L. *Rosmarinus*	Taounate	Aerial parts	Essential oils	a-Pinene, 1,8-cineole and camphene	Moderate activity with a MIC value of 2% (v/v)	Fadil et al. (2017b)
	Thymus broussonetii L.	Essaouira	Aerial parts at flowering stage	Essential oil	Carvacrol, borneol, bicyclo-germacrene, thymol, and terpinene	Inhibition of efflux pump activity MICs = 0.936–0.468 mL/L for AG102 and EA27 overexpressing the AcrAB pump MICs = 0.234–0.117 mL/L for EAEP294 and AG100A, respectively *P. aeruginosa* PA01 and PA124: MIC = 30 mL/L	Fadli et al. (2011)
	T. broussonetii L.	Essaouira	Aerial parts at flowering stage	Essential oil	Carvacrol, borneol, bicyclo-germacrene; thymol and terpinene	Strong activity CMI = 0.171–0.685 mg/mL Except *P. aeruginosa* CMI = 5.48 mg/mL	Fadli et al. (2012)

(*Continued*)

Table 5.1 (Continued)

Families	Species (families)	Origin	Parts used	Extracts	Major components	Effects	Reference
	T. broussonetii	Essaouirra	Aerial parts	Essential oil	Carvacrol, thymol, γ-terpinene	Strong activity *T. broussonetii* Ø = 49.33–51.17 mm MICs = 0.45 mg/mL	Alaoui Jamali et al. (2012)
	T. ciliatus	Imilchil	Aerial parts	Essential oil	Carvacrol, thymol, p-cymene, γ-terpinene	Great sensitivity Ø = 48–45 mm and MIC = 0.43 mg/mL against *C. albicans* and *C. krusei* and MIC = 0.86 mg/mL against *C. parapsilosis* and *C. glabrata*	Alaoui Jamali et al. (2012)
	T. leptobotrys	Tiznite	Aerial parts	Essential oil	Carvacrol, p-cymene	Strong activity Ø = 0.23–0.46 mm and MICs = 0.23–0.46	Alaoui Jamali et al. (2012)
	T. maroccanus	Ait Ourir	Aerial parts	Essential oil	Carvacrol, γ-terpinene, p-cymene	Strong activity Ø = 50.83–53.33 mm and MICs = 0.46 mg/mL	Alaoui Jamali et al. (2012)
	T. maroccanus L.	Ourika valley	Aerial parts at flowering stage	Essential oil	Carvacrol, bicyclo-germacrene, trans-caryophyllene and o-cymene	Inhibition of efflux pump activity MICs = 0.936–0.468 mL/L for AG102 and EA27 overexpressing the AcrAB pump MICs = 0.234–0.117 mL/L for EAEP294 and AG100A, respectively *P. aeruginosa* PA01 and PA124: MIC = 15 mL/L	Fadli et al. (2011)
	T. maroccanus L.	Ourika valley	Aerial parts at flowering stage	Essential oil	Carvacrol, bicyclo-germacrene, trans-caryophyllene, and o-cymene	Strong activity CMI = 0.086–0.342 mg/mL Except *P.* aeruginosa CMI = 5.52 mg/mL	Fadli et al. (2012)

	T. pallidus	Ait Lkak	Aerial parts	Essential oil	Thymol, γ-terpinene and p-cymene	Moderate anticandidal activity Ø = 37.67–33.67 mm and MICs = 0.90 mg/mL	Alaoui Jamali et al. (2012)
	T. satureioides	Ourika	Aerial parts	Essential oil	Carvacrol, borneol, p-cymene, camphene, and γ-terpinene	Greatest effectiveness ø = 30–11.70 mm Gram-positive bacteria MIC = 2.25–18 mg/mL Gram-negative bacteria MIC = 9– > 18.4	El Abdouni Khayari et al. (2016)
	T. serpyllum	Oukaimeden	Aerial parts	Essential oil	Linalyl acetate, (E)-nerolidol, geranyl acetate	Weak activity Ø = 12–17.33 mm and MICs = 3.52–7.05 mg/mL	Alaoui Jamali et al. (2012)
	T. serpyllum	Taounate	Fresh aerial part (leaves and stems)	Essential oil	p-Cymene, γ-terpinene, and thymol	Strong antibacterial effect *P. aeruginosa* was the most resistant one, the essential oil failed to kill this strain at a concentration of 4% (v/v)	Ouedrhiri et al. (2016)
	T. vulgaris	Tafilelt	Aerial parts	Essential oil	*Thymol, p-Cymène, Gamaterpinène, Bornéol*	Strong activity	Ismaili. et al. (2014)
	T. vulgaris L.	Taounate	Aerial parts	Essential oils	Thymol, p-cymene, c-terpinene, linalool, and carvacrol	Strong antimicrobial activity with MIC value of 0.25% (v/v)	Fadil et al. (2017b)
	T. riatarum	Tazzeka-region	Aerial parts flowering stage	Essential oil	Borneol terpinen-4-ol, and trans-caryophyllene	Strong activity with MICs ranging from 3.57 to 7.5 mL/L *P. aeruginosa* showed the least sensitivity MIC > 7.5 mL/L	Fadli et al. (2016)

(Continued)

Table 5.1 (Continued)

Families	Species (families)	Origin	Parts used	Extracts	Major components	Effects	Reference
	T. riatarum	Tazzeka-region	Aerial parts flowering stage	Essential oil	Borneol terpinen-4-ol, and trans-caryophyllene	Strong inhibition of pump efflux	Fadli et al. (2016)
	T. riatarum L.	Tazzeka	Aerial parts at preflowering stage	Essential oil	Borneol, terpinen-4-ol, and trans-caryophyllene	Noticeable activity MICs = 3.57–7.5 mL/L *P. aeruginosa* least sensitivity MIC > 7.5 mL/L	Fadli et al. (2014)
	Salvia aucheri subsp. *Blancoana*	Ijoukak	Aerial parts	Essential oil	Camphor, camphene, and α-pinene	ø = 18.35–27.50 mm Gram-positive bacteria MIC = 1.71–13.68 mg/mL Gram-negative bacteria MIC ≥ 108 mg/mL	El Abdouni Khayari et al. (2016)
	S. officinalis	Mountain Assoul (Taza)	Aerial part	Essential oil	α-Thujone, 1,8-cineole, trans-pinocarveol, le β-thujone, β-pinene, globulol, α-humulene, trans-caryophyllene	The PF technique demonstrated significant inhibition of the mycelia growth of all strains ($p < 0.05$), with the complete inhibition of *P. expansum* at MIC = 2 μL/mL. The VA assay showed that the essential oil strongly inhibits all three fungi. The complete inhibition of the mycelial growth of *P. expansum* and *R. stolonifer*, was observed respectively at MIQ = 80 and 160 μL/disc	EL Ouadi et al. (2015)
	S. officinalis	Ouezzane	Leaves	Hexane, methanol and ethanol extracts	ND	Good effect observed for the hexane and methanol extracts against *S. aureus*	Et-Touys et al. (2016)

	S. officinalis	Ourika	Aerial parts	Essential oil	*trans*-Thujone, 1,8-cineole, camphor	Weak activity $ø = 10.50-20$ mm Gram-positive bacteria MIC = 3.38–54.00 mg/mL Gram-negative bacteria MIC ≥ 108 mg/mL	El Abdouni Khayari et al. (2016)
Myrtaceae	*Myrtus communis* L.	Ifran	Leaves	Essential oils	α-Pinene, 1,8-cineole myrtenyl acetate, α-terpinol	four chemotypes showed strong antibacterial activity and one showed a moderate activity	Fadil et al. (2017a)
	M. communis L.	Taounate	Aerial parts	Essential oils	Borneol, 1,8-cineole, a-pinene, myrtenyl acetate, trans-pinocarveol, and a-terpineol	Strong activity; MIC = 0.5% (v/v)	Fadil et al. (2017b)
	M. communis (L.)	Ouezzane	Leaves	Methanolic, ethanolic, ethyl acetate, and *n*-hexanic extracts	ND	*n*-Hexane extract showed the strongest activity against *P. aeruginosa* (21 ± 0.77 mm), *L. monocytogenes* (33 ± 0.61 mm), *S. aureus* (29 ± 0.31 mm), and *E. coli* (17 ± 0.12 mm) MIC = 0.25–4 mg/mL and MBC = 0.5–4 mg/mL	Bouyahya et al. (2016c)
Poaceae	*Cymbopogon citrates*	Ourika	Aerial parts at preflowering stage	Essential oil	Geranial, neral, and geraniol	High antibacterial activity MICs obtained for *E. coli* EA27 and *E. aerogenes* AG102, overexpressing efflux system, were greater than the MIC obtained for their isogenic strains	Fadli et al. (2016)

(Continued)

Table 5.1 (Continued)

Families	Species (families)	Origin	Parts used	Extracts	Major components	Effects	Reference
	Pennisetum glaucum			Ethanolic extracts	ND	12.33 ± 1.53–34.00 ± 3.61 mm	Marmouzi et al. (2016)
Rutaceae	*Citrus limonum*	Agadir	Aerial parts	Essential oil	Limonène, Bétapinène, and Gamaterpinène	Significant inhibitory activity	Ismaili. et al. (2014)
Solanaceae	*Withania frutescens* (L.) Pauquy	Hassania (Marrakech)	The roots and leaves	*n*-Hexane and methanol extracts and dichloromethane, ethyl acetate and *n*-butanol fractions	ND	Dichloromethane ø = 14–19 mm, MIC = 0.05 and 0.4 mg/mL *n*-Butanol ø = 10 ± 1.4–17.7 mm, MIC = 2.5–0.4 mg/mL Ethyl acetate fractions ø = 8–13.7 mm, MIC = 0.8–2.5 mg/mL Methanol extract MIC = 0.2–1 mg/mL The dichloromethane fraction exhibited the greatest activity	El Bouzidi et al. (2011b)
Lamiaceae and Apiaceae	*Marrubium vulgare*, *Thymus pallidus*, *Lavandula stoechas* and *Eryngium ilicifolium*	Ourika	Dried Aerial parts	Methanol and aqueous extracts of *Marrubium*	ND	Strong activity has been observed with methanol extracts of the four plants: MIC = 256 μg/mL Aqueous extracts do not show any effect	Warda et al. (2009)

activity against tested bacteria (MIC = 0.086–0.342 and 0.171–0.685 mg/mL, respectively) except *Pseudomonas aeruginosa* (MIC = 5.52 and 5.48 mg/mL, respectively) (Fadli et al., 2012). In addition, *T. maroccanus* L. and *T. broussonetii* L. EOs were tested for their ability to inhibit efflux pump systems of resistant Gram-negative bacteria (*Escherichia coli*, *Enterobacter aerogenes*, *Klebsiella pneumoniae*, *Salmonella enterica* serotype *Typhimurium* and *P. aeruginosa*). The results indicate that these EOs exhibited an interesting activity against tested bacteria by altering efflux pump activity (Fadli et al., 2011). In another investigation, Alaoui Jamali et al. (2012) suggested that biological activity of EOs of the aerial parts of some species of *Thymus* (*T. broussonetii*, *T. ciliatus*, *T. leptobotrys*, *T. maroccanus*, *T. pallidus*, *T. satureioides*, and *T. serpyllum*) collected from different regions in Morocco is in direct relation to the chemical composition. According to the EOs profiles the species are classified into three main groups: a carvacrol group (Group I), including the species *T. maroccanus* and *T. leptobotrys*, a linalyl acetate and (E)-nerolidol group (Group II), represented by *T. serpyllum*, and a thymol and/or carvacrol, γ-terpinene, and p-cymene group (Group III), composed of *T. satureioides*, *T. broussonetii*, *T. ciliatus*, and *T. pallidus*. The EOs were then screened for their anticandidal activity against *Candida albicans* CCMM L and CCMM L5, *Candida krusei* CCMM L10, *Candida glabrata* CCMM L7, and *Candida parapsilosis* CCMM L18, using the agar disc-diffusion and the broth microdilution methods. All tested *Candida* species were inhibited by the Moroccan *Thymus* EOs to a varying degree. With the exception of *T. serpyllum* the EOs of all tested species showed larger IZs (12.00–53.33 mm). The thyme EOs with high amounts of carvacrol (*T. leptobotrys*, *T. broussonetii*, and *T. maroccanus*) proved to be the most active against all tested *Candida*, with MIC values of 0.23, 0.45, and 0.46 mg/mL, respectively. Furthermore, *T. ciliates* EO showed greater sensitivity against *C. albicans* and *C. krusei* (MIC of 0.43 mg/mL) compared with *C. parapsilosis* and *C. glabrata* (MIC of 0.86 mg/mL). The *T. satureioides* and *T. pallidus* EOs exhibited a moderate anticandidal activity (MICs of 0.89–1.78 mg/mL). The weakest activity was obtained for the *T. serpyllum* EO (MICs of 3.52–7.05 mg/mL), which contains only trace amounts of thymol and carvacrol (Alaoui Jamali et al., 2012). In the same way, Kasrati et al. (2014) have noticed that EOs from three wild Moroccan *T. satureioides* populations collected from Northwest–Southeast of Morocco showed a high inhibitory activity against tested bacteria, except *P. aeruginosa*. The most potent activity was observed for the EO obtained from the Er-Rich region (high concentration of carvacrol 45.3%), with MIC and MBC values of 0.14−0.55 mg/mL and 0.28−0.55 mg/mL, respectively. This study confirmed the result obtained above by Alaoui Jamali which shows that the oils rich on carvacrol are the most potent, thus the antimicrobial activity is correlated to the chemical composition. Moreover, the determination of antibacterial activity of another *Thymus* species (*T. riatarum*) EO showed that the oil possessed a noticeable potential of inhibiting the growth of tested strains with MICs ranging from 3.57 to 7.5 mL/L. However, *P. aeruginosa* showed the least sensitivity and was only inhibited with high concentrations (MIC > 7.5 mL/L) (Fadli et al., 2014). In fact, the antimicrobial activity of the thyme EOs is apparently related to their high phenolic content, in particular carvacrol and thymol, which have been tested previously and found to exhibit a significant antimicrobial activity (Manohar et al., 2001; Chami et al., 2005). In the same context, Ouedrhiri et al. (2016) demonstrated that *Origanum compactum*, *Thymus serpyllum*, and *Origanum majorana* (same family; Lamiaceae) EOs exhibited an antibacterial effect against *Staphylococcus aureus* ATCC 29213, *E. coli* ATCC 25922, *Bacillus subtilis* ATCC 3366, and *P. aeruginosa* ATCC 27853 examined by disc-diffusion and microdilution method. In another investigation the EOs from the aerial parts of wild endemic Moroccan *O. compactum* at three phenological stages (vegetative, flowering, and postflowering) were tested against eight bacteria. The results showed that the EOs of vegetative and flowering stage showed remarkable antibacterial activity against *Proteus mirabilis* with an inhibition diameter of 43 and 47 mm, respectively. Moreover, the best bactericidal action was revealed by the EO of flowering stage against *S. aureus* MBLA (MIC = MBC = 0.0312% (v/v)) (Bouyahya et al., 2017a). Comparing the activities of different *Thymus* species EOs, it has been noticed that there is a slightly difference on the antimicrobial activities. This could be explaining by the difference in the chemical composition, the harvest stage, the tested strains, and the experimental methods used.

It has been also reported the remarkable antibacterial and antifungal activities of other species belonging to the same family (Lamaiaceae). For example, *Melissa officinalis* EO was tested against bacteria responsible for nosocomial infections (*P. aeruginosa*, *K. pneumoniae*, *S. aureus*, and *Citrobacter kuseri*) using disc-diffusion method. The studied EO which is rich on citronellal (14.40%), caryophyllene oxide (11.00%), geraniol acetate (10.20%), caryophyllene (8.10%) exhibited a high antibacterial activity against tested strains (Jalal et al., 2015). El Ouadi et al. (2017)

confirmed in another work that *M. officinalis* EO (P-mentha-1,2,3-triol (13.1%), P-menth-3-en-8-ol (8.8%), piperitenone oxide (PEO) (8.4%), and Z-piperitone oxide (PO) (7.3%)) from another region has a significant antifungal activity and inhibited mycelial growth of tested fungi; *Botrytis cinerea*, *Penicillium expansum*, and *Rhizopus stolonifer*. Both *M. officinalis* species showed an antibacterial and antifungal activity while the chemical compositions of the oils were totally different. The two species are collected from two different regions, thus the difference in the chemical compositions could be strongly due to the geographical and environmental factors. Furthermore, an investigation has been conducted on the antibacterial activity of *Mentha pulegium* and *Rosmarinus officinalis* EOs against *E. coli*, *S. aureus*, *Listeria monocytogenes*, *P. mirabilis*, *P. aeruginosa*, and *B. subtilis* using the agar-well diffusion and the broth microdilution assays. The results showed that these two EOs displayed remarkable antibacterial effect. The lowest MIC and MBC values were obtained with *M. pulegium* EO against *S. aureus* MBLA (MIC = MBC = 0.25% (v/v)), while *R. officinalis* EO exhibited a bactericidal effect against *L. monocytogenes* (MIC = MBC = 0.5% (v/v)), *B. subtilis* (MIC = MBC = 1% (v/v)), and *E. coli* (MIC = MBC = 1% (v/v)) (Bouyahya et al., 2017b). The characterization and the antimicrobial activity of three EOs of *Mentha suaveolens* growing in different regions of Morocco revealed that the main aromatic constituents were pulegone, PEO, and PO occurring in different amounts depending on the subspecies. These constituents were tested for their antimicrobial activity against 19 bacteria including Gram-positive and Gram-negative bacteria and three fungi, using solid phase and microtitration assays. Pulegone-rich EO inhibited efficiently all the tested microorganisms with MICs ranging between 0.69 and 2.77 ppm. Additionally, rich PEO EO and rich Pulegone/PEO EO were less effective. These results demonstrated the strong antimicrobial activity of *M. suaveolens* EOs and indicated that the efficacy of EOs depends on their chemical compositions (Oumzil et al., 2002). Similarly, Kasrati et al. (2013) demonstrated the antimicrobial activity of the same species. A comparison of the activity of EOs obtained from the aerial parts of the wild and cultivated *M. suaveolens* subsp. *timija*, an endemic Moroccan medicinal plant showed that the cultivated mint timija EO showed strong antibacterial activity with respect to the wild one. Additionally, the antimicrobial activity of EOs of *M. pulegium* L., *Juniperus phoenicea* L., and *Cyperus longus* L. from Morocco has been tested against seven bacteria of significant importance for food hygiene. The results revealed that EO of

M. pulegium showed the best bacteriostatic and bactericidal effect, followed by *J. phoenicea* and *C. longus* (Ait-Ouazzou et al., 2012). Asdadi et al. (2015) reported that EO extracted from the seeds of *Vitex agnus-castus* L. exhibited a strong antifungal activity against *Candida* strains responsible for nosocomial infection, with IZ diameters and MIC values ranging from 35 to 58 mm and from 0.13 to 2.13 mg/mL, respectively. It is worth quote to note that *Lamiaceae* family is one of the most important families in Morocco and worldwide and the species belonging to this family are known to be one of the most important plants in regards to the production of EOs with antimicrobials (Roldán-Gutiérrez et al., 2010).

Other scientific investigations reported the antimicrobial activities of different Moroccan medicinal plants belonging to different families (Myrtaceae, Asteraceae, Lamiaceae, Lauraceae, etc.). Formerly, Ezoubeiri et al. (2005) evaluated the antibacterial and antifungal activities of *Pulicaria odora* EO and two compounds obtained after fractionation of the EO (2-isopropyl-4-methylphenol (1) and isobutyric acid 2-isopropyl-4-methylphenylester (2). The studied EO and the 2-isopropyl-4-methylphenol exhibited very significant antibacterial and antifungal activities, while the isobutyric acid 2-isopropyl-4-methylphenylester was inactive for all tested strains (Ezoubeiri et al., 2005) (Fig. 5.1).

A scientific investigation determined the antifungal activity of *Thymus vulgaris*, *Mentha spicata*, and *Citrus limonum* EOs against three dermatophyte strains responsible of the superinfection of a dermatitis contact: *Trichophyton mentagrophytes*, *Trichophyton rubrum*, *Epidermophyton floccosum*. The antifungal activity was linked to the inhibition of mycelial growth with particularly high concentrations of EOs and *T. vulgaris* EO showed the greatest activity against dermatophytes compared to the rest of tested EOs (Ismaili et al., 2014). Another plant widely used in Moroccan folk

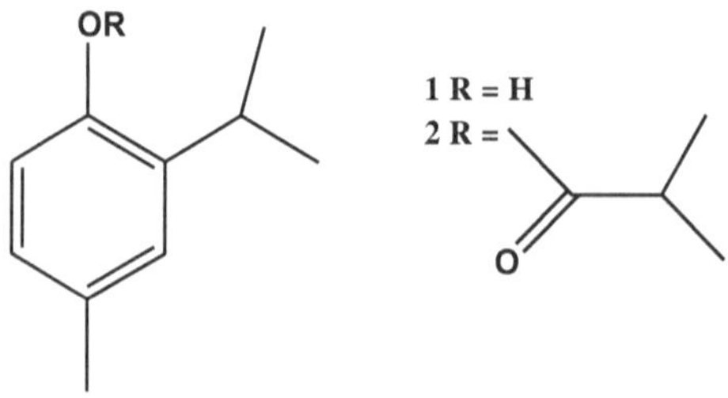

Figure 5.1 Structure of the two isolated compounds from *Pulicaria odora* EO (Ezoubeiri et al., 2005).

medicine for its antiinfective property is *Salvia officinalis*. The in vitro antifungal activity of *S. officinalis* EO using poison food (PF) technique and volatile activity (VA) assay against three phytopathogenic strains causing the deterioration of apple: *Botrytis cinerea*, *Pennicillium expansum*, and *Rhizopus stolonifer* showed complete inhibition of the mycelial growth of *P. expansum* and *R. stolonifer* at MIQ = 80 and 160 μL/disc, respectively (EL Ouadi et al., 2015). In fact, several studies support the antimicrobial effects of *S. officinalis*. *S. officinalis* EO exhibited strong antifungal, bactericidal, and bacteriostatic effects against Gram-positive and Gram-negative bacteria. These effects are due to the presence of terpens and terpenoids compounds (Delamare et al., 2007; Ghorbani and Esmaeilizadeh, 2017). Furthermore, the in vitro evaluation of the antibacterial activity of EOs extracted from the same plant and also from *Mentha viridis*, *Eucalyptus globulus*, and *Myrtus communis* were assessed against eight bacteria strains, using agar well diffusion and microdilution methods. *S. officinalis* EO was the most active on the tested bacteria and exhibited the highest zone of inhibition (23 mm) against *Bacillus subtilis*, while *P. aeruginosa* was the most resistant bacteria. *S. officinalis* and *M. communis* EOs showed an inhibitory effect at MIC = 0.5% (v/v) against *Listeria monocytogenes* and *Proteus mirabilis* (Bouyahya et al., 2017c). On the other hand, *M. communis* L. is also another medicinal plant endemic to Morocco and has been traditionally used for its medicinal purposes since ancient time (Fadil et al., 2017a). Many scientific works reported the antimicrobial activity of Moroccan myrtle. EOs of 20 samples of Moroccan *M. communis* L. collected from the forest of "Ifran" were tested against *B. subtilis* ATCC6633, *Staphylococcus aureus* ATCC 29213, *Salmonella typhimurium* ATCC 14028, and *E. coli* ATCC 2592. The results showed that α-pinene (0.4%–50.3%), 1,8-cineole (8.3%–64.9%), myrtenyl acetate (0%–61.1%), α-terpinolene (0%–20.8%), methyl eugenol (0%–33.6%), and α-terpineol (0.2%–18%) were the main compounds of *M. communis* L. EOs. These oils exhibited an important inhibitor activity against Gram-positive and Gram-negative bacteria except one chemotype which showed a moderate antibacterial activity (Fadil et al., 2017a). The difference in the chemotype activities is due to the difference on the chemical composition. Major compounds are monoterpenes, these compound are known in the literature to possess a potent antimicrobial activity (Yangui et al., 2009). The presence of 1,8-cineole and α-terpineol justify also the antimicrobial activity of *M. communis* EOs. In another study, Fadil et al. (2017b) tested the EOs of *M. communis* L., *T. vulgaris* L., and *R. officinalis*

L. for their antibacterial activity against *S. typhimurium* strain. A strong antibacterial activity was found for *T. vulgaris* L. EO followed by *M. communis* L. EO and *R. officinalis* L. EO.

A study carried out by our team investigated the antimicrobial activity of *Chenopodium ambrosioides* var. *ambrosioides* frequently used in Moroccan folk medicine as vermifuge, analgesic, and antipyretic. *C. ambrosioides* var. *ambrosioides* EO has been tested against a large panel of Gram-positive and Gram-negative bacteria and *Candida* strains; *S. aureus* CCMM B3, *M. luteus* ATCC 10,240, *B. cereus* ATCC 14,579, *E. coli* ATCC 25,922, *P. aeruginosa* ATCC 27,853, and a clinically isolate *K. pneumoniae*, *C. albicans* CCMM L4, *C. glabrata* CCMM L7, *C. krusei* CCMM L10, and *C. parapsilosis* CCMM L18. An inhibitory effect against all tested strains was observed. Gram-positive bacteria were generally found to be more sensitive (ø = 15.33–21.5 mm and MIC = 1.25–5 mg/mL) than Gram-negative ones (ø = 7.17–19.17 mm and MIC = 0.31–20 mg/mL). However, the tested EOs showed a high anticandidal activity, with IZ diameters and MIC values ranging from 14.67 to 20 mm and from 0.075 to 2.5 mg/mL, respectively (Ait Sidi Brahim et al., 2015a). In our laboratory the antimicrobial activities of numerous medicinal plants have been also studied by several researchers. Indeed, El Bouzidi et al. (2011a) tested the EO of Moroccan *Cotula cinerea* aerial parts against a panel of human pathogenic *Candida* strains (*C. albicans* CCMM L4 and CCMM L5, *C. krusei* CCMM L10, *C. glabrata* CCMM L7, and *C. parapsilosis* CCMM L18) using the same methods. The EO showed a high anticandidal activity against all investigated strains with minimal inhibitory concentrations of 3.2–4.7 and 5.9 mg/mL as a minimal candidicidal concentration value (El Bouzidi et al., 2011a). Recently, El Abdouni Khayari et al. (2016) investigated the antibacterial effect of EOs the same plant *C. cinerea* and some other widely used Moroccan aromatic herbs; *T. satureioides*, *Achillea ageratum*, *S. officinalis*, and *Salvia aucheri* subsp. *Blancoana* against a panel of microorganisms. According to the authors the greatest effectiveness was given with *T. satureioides* oil (MIC = 2.25–4.50 mg/mL), while the weakest potency was displayed by the *A. ageratum* EO (MIC = 4.64–74.32 mg/mL) and among the tested EOs, only those obtained from *T. satureioides* and *C. cinerea* were able to control the Gram-negative bacteria except *P. aeroginosa*. This study confirmed the remarkable antimicrobial activities of *T. satureioides* and *C. cinerea* EOs mentioned earlier. El Bouzidi et al. (2012) have investigated the antimicrobial and antifungal activities of the EOs of leaves and flowers of a wild and cultivated rare and threatened Moroccan medicinal species *A. ageratum* L. Both oils obtained from leaves and flowers have interesting action on the growth of tested bacteria and *Candida* strains. The Gram-positive bacteria *M. luteus*, *S. aureus*, *B. subtilis*, and *B. cereus* were more susceptible bacteria than Gram-negative ones with MIC values ranging between 2.55 and 7.02 mg/mL. The MIC values for *Candida* species were found from 5.83 to 8.42 mg/mL.

EOs of some plants belonging to the Asteraceae family were also investigated. Aghraz et al. (2017a, 2017b) studied the EOs of *Cladanthus arabicus* and *Bubonium imbricatum* Cav EOs and demonstrated that *C. arabicus* EO is rich on sabinene (31.1%), β-pinene (16.7%), myrcene (12.3%), and α-pinene (5.3%) and exhibited a strong antibacterial (MIC varied from 0.187 to 3 mg/mL) and antifungal (MIC from 0.187 to 0.75 mg/mL) activities. For *B. imbricatum* Cav, *cis*-chrysanthenyl acetate (31.2%) and thymol isobutyrate (3.4%) were the major components of EO. This EO showed strong antimicrobial activity with MICs values varying from 0.18 to 3 mg/mL and from 0.18 to 0.375 mg/mL, respectively for bacteria and *Candida* strains. These EOs were also tested against Enterobacteriaceae isolates. The results showed that MIC values are varying between 200 and 800 μg/mL for *C. arabicus* and 400–1600 μg/mL for *B. imbricatum*, respectively (Aghraz et al., 2018). In fact the Asteraceae family is known to include numerous species which produce EOs with numerous biological activities especially antimicrobial activities. EOs of this family possesses a broad spectrum of bioactivity, because of the presence of several active compounds which work through various modes of action. *Artemisia herba-alba Asso* is a perennial dwarf shrub belonging to the Asteraceae family and has been extensively used in Moroccan folk medicine as emmenagogue, antidiarrheic or as diuretic agent, antispasmodic. Imelouane et al. (2010) demonstrated that EO of *Artemisia herba-alba* Asso growing in eastern Morocco is rich on camphor (43.07%), camphene (7.2%), 1,8-cineole (7.08%), filifolone (7.04%), borneol (4.88%), and bornyl acetate (3.79%). A strong antimicrobial activity was observed against Gram-positive and Gram-negative bacteria with MIC values ranging from 0.041 to 21.00 mg/mL. *P. aeruginosa* was found to be resistant to this EO. Another investigation reported by Ourid et al. (2016) showed that the EO from harvest samples of *Artemisia herba-alba* is rich on artemisia alcohol, 6-camphenone, cis-hydrate acetate, sabinene, and ß-thujone and possess antibacterial and antifungal activities. The most susceptible microorganisms were *Coniophora puteana* and *Gloeophyllum trabeum*, for wood-destroying fungi, *Aspergillus niger* for molds and *E. coli* for bacterial strains. The evaluation of the antimicrobial activity of another variety (*A. herba-alba hugueii*'s) confirmed that cis-thujone and camphor are the main compound of the ES of this plant. *A. herba-alba hugueii*'s EO exhibited an inhibitory effect against the most tested bacteria (Majdouli et al., 2015). It has been previously reported that the antimicrobial activities of *A. herba-alba* can be attributed to the presence of camphor, 1,8-cineole and thujone which are the major compound of the Moroccan *A. herba-alba*. These three compounds are well documented to possess antibacterial and antifungal effects (Jalsenjak et al., 1987; Sivropoulou et al., 1997). Recently, Fadli et al. (2016) reported also the ability of EOs of *Artemisia herba-alba* and another plant from the Poaceae family *Cymbopogon citratus* EOs to inhibit efflux pumps of some selected antibiotic-resistant Gram-negative bacteria. Antibacterial tests showed that these EOs had high antibacterial activity. For both EOs, the MICs obtained for *E. coli* EA27 and *Enterobacter aerogenes* AG102, overexpressing efflux system were greater than the MIC obtained for their isogenic strains. Therefore these results suggest that the active compounds included in the tested EOs could be substrates of the efflux pumps expressed in these bacteria (Fadli et al., 2016). Additionally the antimicrobial activity of *Periploca laevigata* (Apocynaceae) EO reported by Ait Dra et al. (2017) showed an inhibitory effect against bacteria (MIC = 3.75–30 mg/mL) and *Candida* species (MIC = 0.937–3.75 mg/mL). The MIC values were often equivalent to MBC for bacteria and candida, indicating a bactericidal or fungicidal action of the EO especially on *M. luteus*, *Bacillus cereus*, *B. subtilis*, *Klebsiella pneumonia*, *Pseudomonas aeruginosa*, and *Candida glabrata*.

From all these data, it is worth quote to note that EOs from Moroccan medicinal plants are very effective against a panel of microorganism; bacteria, yeasts, and fungi. In fact, it is well known that EOs composition varied from a species to others and within the same spices. This variability could be attributed to many factors such as the geographical origin, the genetic factors, the plant material, climate, the season at which the plant was collected and the method of preservation and extraction (Bakkali et al., 2008; Sivropoulou et al., 1997). It has been reported that the antimicrobial activity of EOs is not due only to the major component. In fact, EO is mixtures of numerous compounds, thus the activity is due to additive, synergistic and/or antagonistic effects (Ruberto and Baratta, 2000). Many researchers reported that the antimicrobial activity of EOs compounds is ascribed to different mode of action. EOs are characterized by their hydrophobicity that allow them to penetrate through the cell wall and cytoplasmic membrane and then disrupt the cell structure which lead to the eventual death of the bacterial cells (Denyer and Hugo, 1991; Sikkema et al., 1994; Okoh et al., 2010; Fadli et al., 2012). The mechanism of action of EOs also depends on the type of microorganisms. Gram-positive bacteria are more susceptible to EOs than Gram-negative ones, these susceptibilities might be explained by the structure of cell envelope; Gram-negative bacteria possess an extra membrane, named outer membrane, separating the periplasmic space with the cytoplasmic membrane that restricts diffusion of hydrophobic compounds (Tian et al., 2009). The effectiveness of EOs is also influenced by the chemical composition. It was demonstrated that monoterpenes of EOs are able to disrupt and penetrate the lipid structure of the cell wall of bacteria which cause denaturation of protein and destruction of cell membrane leading to cytoplasmic leakage, cell lysis, and eventually cell death (Helander et al., 1998). Moreover, other compounds could act as antimicrobial agents by targeting efflux pump mechanisms, thus restore and modulate drug resistance in strains that overexpress efflux pumps (Fadli et al., 2014). Some also increase the membrane permeability in yeast cells and isolated mitochondria (Andrews et al., 1980; Uribe et al., 1985; Saad et al., 2010).

5.2.2 Antimicrobial Activity of Plant Solvent Extracts

Recently, many studies have been carried out with the aim of highlighting the antimicrobial capacity of many extracts from Moroccan medicinal plants (Table 5.1). Recently, the antibacterial screening of *Cistus. crispus* was studied by Bouyahya et al. (2017d). Data shown that Methanol, ethanol, ethyl acetate, and *n*-hexane extracts had a significant antibacterial activity. Gram-positive bacteria (*L. monocytogenes*, *S. aureus*) are more sensitive than Gram-negative ones. The highest diameter of inhibition was found with ethanolic *C. crispus* extract against *S. aureus* and *L. monocytogenes*. In another work, ethanolic extracts of five medicinal plants, including also two *Cistus* sp. (*Cistus albidus* (L.) and *Cistus monspeliensis*) and *Daphne gnidium* (L.), *Ajuga iva* (L.) *Schreb* and *L. stoechas* (L.), prepared using solid–liquid extraction prevented the growth of the studied bacteria strains by forming significant IZs and *Ajuva iva* ethanolic extracts showed the highest activity (especially against *E. coli* and *S. aureus*) (Bouyahya et a1., 2016a). In an another investigation, methanol and aqueous extracts of *M. vulgare*, *T. pallidus*, *E. ilicifolium*, and *L. stoechas* were tested against *Streptococcus pneumoniae* (responsible for pharyngitis, rhinitis, otitis, and sinusitis infections). A significant activity has been observed with methanol extracts of three plants: *M. vulgare*, *T. pallidus*, and *L. stoechas* with a MIC value of 256 μg/mL (Warda et al., 2009). Talibi et al. (2012) studied the antifungal effect of organic extracts of eight Moroccan plant species against Citrus sour rot agent *Geotrichum candidum*. Authors noticed that among eight plant extracts, only methanol extracts of *Cistus villosus*, *Ceratonia siliqua*, and *Halimiumum bellatum* exhibited strong antifungal and antibacterial effects. Moreover, a strong effect was also obtained with methanolic extract of *Quercus suber* bark with MICs ranging between 12.5 and 50 mg/mL for *C. albicans* and *Trichophyton rubrum*, respectively (Hassikou et al., 2014). Moreover, other studies conducted on the leaves of *Myrtus communis* (L.) and leaves of *Arbutus unedo* L. (Bouyahya et al., 2016b,c) showed that the less polar extracts (hexane) were the most active against *P. aeruginosa* IH, *L. monocytogenes*, *S. aureus* CECT 976 *serovar* 4b CECT 4032 and *E. coli* K12 MBLA with MIC values ranging between 0.25 and 4 mg/mL and 0.25–0.5 mg/mL, respectively. On the other hand, the antibacterial activity of methanol, ethanol, ethyl acetate, and *n*-hexane extracts of *Origanum compactum* Benth was evaluated against the same pathogens using well diffusion method. All samples were proved to be able of inhibiting bacterial strains tested with the largest zone of inhibition of about 34 ± 1.24 mm. The *n*-hexane extract showed the strongest antibacterial activity, while the ethyl acetate extract showed the lowest antibacterial activity. In this study, the Gram-negative bacteria were more resistant than Gram-positive ones (Bouyahya et al., 2017a). In the same context, Et-Touys et al. (2016) evaluated the in vitro antibacterial activity of hexane, methanol, and ethanol extracts of *salvia officinalis* against three reference strains (*S.* aureus, *E. coli*, and *L. monocytogenesserovar*) using the diffusion method and the microdilution assay. The three extracts revealed an inhibitory power against the bacterial strains with a significant difference in the diameters of inhibition. The largest IZ was registered by the hexanic and methanolic extracts against *S. aureus* (22 ± 8 mm), while the weakest one was 11 ± 0.22 mm expressed by the ethanolic extract against *E. coli* and the methanolic extract against *L. monocytogenes*. Furthermore, the lowest inhibitory value was found by methanolic extract against *S. aureus* (MIC = MBC = 2 mg/mL). Abudunia et al. (2017) determined as well the antibacterial and anticandidal activities of the hexanic, methanolic, and aqueous extracts of

Calendula arvensis flowers. All extracts of *C. arvensis* were noticeably effective against *Salmonella aequatoria* and *E. coli* ATCC. Moreover, methanolic extract inhibited only *E. coli* ATCC and *Salmonella braenderup*. However, there were no inhibitory effects of extracts against *Salmonella blockley*, *E. coli* enteropathogenes, and *E. coli MDR*. MIC values of methanolic and aqueous extracts were between 12.5 and 25 μg/mL and were bactericidal for all bacteria. While MICs values of hexanic extracts were between 6.25 and 12.5 μg/mL and were bacteriostatic for all bacteria. However, no effect of extracts was observed on fungi species, except methanolic extracts, which showed moderate activity against two tested strains namely *C. tropicalis* 1 and *C. famata* 1 with IZs of 14 and 20 mm, respectively. In another study, Marmouzi et al. (2016) reported that the ethanolic fraction of *Pennisetum glaucum* exhibited an antibiotic effect on all tested bacteria with IZs diameter varying between 12.33 and 34.00 mm. Furthermore, a study performed by El Bouzidi et al. (2011b) on the antimicrobial activity of *n*-hexane and methanol extracts, dichloromethane, ethyl acetate, and *n*-butanol fractions obtained from the roots and leaves of *Withania frutescens* (L.) Pauquy against multiresistant bacteria involved in nosocomial infections and two *C. albicans* using disc-diffusion and microdilution methods. The dichloromethane fraction exhibited the greatest activity with IZs ranged from 14 to 19 ± 1.4 mm, and MIC values ranging between 50 and 400 μg/mL. However, none of the examined extracts showed anticandidal activity.

In the majority of these studies evaluating the Moroccan plant extracts, no indication of the compounds responsible of the activity was reported, except few studies where the type of the major compounds was indicated. Ait Sidi Brahim et al. (2015b) who worked on the saponins-rich fractions extracted from the aerial parts of *Paronychia argentea* and roots of *Spergularia marginata* were evaluated the antibacterial and anticandidal activities of these saponins-rich fractions using the agar disc-diffusion and broth microdilution methods against Gram-positive and Gram-negative bacteria and *Candida* strains. The result of MIC assay showed that both saponins-rich extracts were found to be active against the majority of *Candida* strains and Gram-positive bacteria with MIC ranging from 8 to 16 mg/mL for both extracts. However, the studied saponin extracts had a weak effect on Gram-negative bacteria. Another investigation by Bouhlali et al. (2016) comparing the effects of methanolic rich polyphenol extracts from the fruit of six Moroccan *Phoenix dactylifera* L. varieties on Gram-positive bacteria and Gram-negative ones showed that the Bousrdon and Jihl varieties were the most active extracts inhibiting tested bacteria with MICs values ranging between 2.5 and 10 mg/mL, whereas Majhoul and Bouskri extracts showed no activity against all tested bacterial strains.

5.3 ANTIMICROBIAL SYNERGETIC INTERACTIONS

The discovery of antifungal drugs had eradicated some infections that ravaged the humankind. But their indiscriminate use has led to the development of multidrug resistant pathogens. One strategy employed to overcome these resistance mechanisms is the use of combination between the EOs of medicinal plants and conventional drugs. Numerous studies have been conducted in order to assess if the combinations between EOs from Moroccan medicinal plants and conventional antibiotics are able to increase the antibiotic susceptibility of multidrug resistant bacteria. An overview of the published studies on interactions between EOs of Moroccan medicinal plants and antibiotics is presented in Table 5.2. In fact, several EOs obtained from plants of *Thymus* sp. have been reported to increase the susceptibility of multiresistant bacteria to conventional antibiotics. Fadli et al. (2012) examined the synergistic interactions between antibiotics (ciprofloxacin, gentamicin, pristinamycin, and cefixime) and EOs of *T. maroccanus* and *T. broussonetii*, and between the two EOs on antibiotic-resistant bacteria responsible for nosocomial infections using the checkerboard test. The combination of *T. maroccanus* with either gentamicin or pristinamycin showed total synergistic effect against all the tested strains, except for *Salmonella* (FICI = 0.751) but the best antibacterial activities were obtained with the combination of *T. maroccanus* and ciprofloxacin with FIC index ranging from 0.09 to 0.37. The combinations between both EOs *T. maroccanus* and *T. broussonetii* provided six total synergism (FICI $\leq$ 0.5) and four partially synergism (FICI 0.56–0.75). The same author investigated the effect of *T. maroccanus* L. and *T. broussonetii* L. EOs in combination with chloramphenicol against *E. aerogenes* (ATCC 13048 and EA27) and *E. coli* (AG100 and AG102) (Fadli et al., 2011). Regarding the strains that overexpress efflux pumps (EA27 and AG102), the MIC decreased by 8–16-fold and 32-fold, respectively. In addition, these EOs greatly increased the chloramphenicol susceptibility of *P. aeruginosa* strain. It has been concluded that the synergy could be attributed to the alteration of efflux pump activity by certain EOs constituents, but also to the ability of carvacrol (76.35% in *T. maroccanus* EO,

Table 5.2 Synergetic interactions between Moroccan medicinal plants EO/extracts and conventional antibiotics

Famille	Species (families)	Origin	Parts used	Extracts	Major components	Antibiotics	Effects	References
Asclepiadaceae	*P. laevigata*	Essaouira	Leaves	Essential oil	*n*-Hexadecanoic acid 6% and 4,4,7a-trimethyl-5,6,7,7a-tetrahydro-4H-benzofuran-2-one	Fluconazole, ciprofloxacin, gentamicin	High synergistic effect. EO/gentamicin against Gram-positive and Gram-negative bacteria: FICi = 0.28–0.50 Essential oil/ciprofloxacin against Gram-positive bacteria: FICi = 0.31–0.38 Essential oil/fluconazole against yeasts except *C. albicans* FICi = 0.25–0.37	Ait Dra et al. (2017)
Asteraceae	*Artemisia herba-alba*	Tahanaoute	Aerial parts at preflowering stage	Essential oil	Chrysanthenone, camphor, and verbenone	Chloramphenicol	Reduction of MIC 2–4-fold, 4-fold, and 8–16-fold, respectively, for EAEP294 (deleted of AcrAB), EA27, and AG102 For *Salmonella* strains, all tested combinations recorded a gain greater than or equal to 64-fold	Fadli et al. (2016)
	Cladanthus arabicus	Ourika	Aerial parts	Essential oil	Sabinene, β-pinene, myrcene, and α-pinene	Amoxicillin and neomycin	Total synergy was obtained with the *B. imbricatum* EO combined with amoxicillin against the majority of tested Enterobacteriaceae isolates (FIC < 0.5) and	Aghraz et al. (2018)

(*Continued*)

Table 5.2 (Continued)

Famille	Species (families)	Origin	Parts used	Extracts	Major components	Antibiotics	Effects	References
							in combination with neomycin against *E. coli* ATCC 25922 and *P. mirabilis*	
	B. imbricatum (Cav.)	Essaouira	Aerial parts	Essential oil	*Cis*-chrysanthenyl acetate and thymol isobutyrate	Amoxicillin and neomycin	Total synergy with amoxicillin against *Salmonella* sp. and *P. mirabilis* A partial synergy against *Enterobacter cloacae*	Aghraz et al. (2018)
Lamiaceae	*O. compactum* + *Thymus serpyllum* + *Origanum majorana*	Taounate	Fresh aerial part (leaves and stems)	Essential oils	Carvacrol, -terpinene, and thymol	None	Strong synergetic effect between the three essential oils *P. aeruginosa* was the most resistant one, the essential oil failed to kill this strain at a concentration of 4% (v/v)	Ouedrhiri et al. (2016)
	T. broussonetii L.	Essaouira	Aerial parts at flowering stage	Essential oil	Carvacrol, borneol, bicyclo-germacrene, thymol, and terpinene	Ciprofloxacin Gentamicin Pristinamycin Cefixime	Ciprofloxacin/EO FICI = 0.14–0.62 total and partial synergism Gentamicin /EO FICI = 0.12–0.62 total and partial synergism Pristinamycin/EO FICI = 0.31–0.75 total and partial synergism Cefixime/EO FICI = 0.5–0.69 total and partial synergism and no effect against *M. luteus*, *K. pneumonia*, and *E. cloacae*	Fadli et al. (2012)

	T. ciliates	Imilchil	Aerial parts	Essential oil	Carvacrol, thymol, p-cymene, γ-terpinene	Cefexime	Total synergistic effect against *Micrococcus luteus, Bacillus cereus* and *P. aeruginosa* (FICIs: 0.370.50), partial synergistic effect against *K. pneumoniae* and *S. aureus* (FICIs: 0.75), but no synergistic effect against *E. coli* (FICI: 1.25)	Alaoui Jamali et al. (2017)
	T. leptobotrys	Tiznite	Aerial parts	Essential oil	Carvacrol, p-cymene	Cefexime	Greatest synergistic effect against all tested bacteria with FICIs between 0.26 and 0.50 and a decrease of the MIC of cefixime by 4–130-fold	Alaoui Jamali et al. (2017)
	T. maroccanus L.	Ourika valley	Aerial parts at flowering stage	Essential oil	Carvacrol, bicyclo-germacrene, trans-caryophyllene and o-cymene	Check board	Ciprofloxacin/EO FICI = 0.09–0.28 total and partial synergism Gentamicin /EO FICI = 0.19–0.75 total and partial synergism Pristinamycin/EO FICI = 0.28–0.75 total and partial synergism Cefixime/EO FICI = 0.18–0.75 total and partial synergism and no effect against *M. luteus, K. pneumonia*, and *E. cloacae*	Fadli et al. (2012)

(*Continued*)

Table 5.2 (Continued)

Famille	Species (families)	Origin	Parts used	Extracts	Major components	Antibiotics	Effects	References
	T. pallidus	Ait Lkak	Aerial parts	Essential oil	Thymol, γ-terpinene, p-cymene	Cefexime	Total synergistic effect against *Micrococcus luteus*, *Bacillus cereus* and *P. aeruginosa* (FICIs: 0.370.50), partial synergistic effect against *K. pneumoniae* and *S. aureus* (FICIs: 0.75), but no synergistic effect against *E. coli* (FICI: 1.25)	Alaoui Jamali et al. (2017)
	T. riatarum L.	Tazzeka	Aerial parts at preflowering stage	EO	Borneol, terpinen-4-ol, and trans-caryophyllene	Chloramphenicol	AG100: fourfold reduction in MIC AG100A: twofold reduction in MIC	Fadli et al. (2014)
	T. saturejoides Coss.	Three regions: Taws, Ourika, and Er-Rich	Blooming stage	Essential oils	Carvacrol, p-cymene, linalool, and borneol in oil from the Er-Rich population (arid site) Carvacrol, borneol, camphene, and -terpinene in oil from Ourika population (medium arid site) Carvacrol, borneol, camphene, and p-cymene in oil from Taws population (less arid site)	Cefexime	67% combination had total synergism, 19% had partial synergistic interaction, and 14% showed no effect Er-Rich EO/cefixime: most potent FIC = 0.29–0.5	Kasrati et al. (2014)

Poaceae	*C. citrates*	Ourika	Aerial parts at preflowering stage	Essential oil	Geranial, neral, and geraniol	Chloramphenicol	Reduction of MIC 2–4-fold, 4-fold, and 8–16-fold, respectively, for EAEP294 (deleted of AcrAB), EA27 and AG102 For *Salmonella* strains, all tested combinations recorded a gain greater than or equal to 64-fold	Fadli et al. (2016)

39.77% in *T. broussonetii* EO) to cause the permeabilization and depolarization of the bacterial cytoplasmic membrane, thus facilitating the passage of the antibiotic to the bacterial cell. Another Moroccan thyme species *T. riatarum* L. EO has been tested in combination with chloramphenicol against *E. coli* AG100 and AG100A (AG100 derivative with AcrAB pump deleted, but resistant to kanamycin). EO used at MIC/4 was able to decrease the MIC values of chloramphenicol against AG100 and AG100A by four- and twofold, respectively. Also, in this case the synergy could be attributed to efflux pump blocking activity which increases the activity of chloramphenicol (Fadli et al., 2014). Synergism against *Candida* strains has been also investigated. The FICI of *T. maroccanus* and *T. broussonetii* EOs combined with amphotericin B and fluconazol, calculated from the checkerboard titer assay, were 0.49, 0.27, 0.37, and 0.3, respectively, indicating a total synergetic effect (Saad et al., 2010). A study investigating the combination between cefexime and EOs (at low concentration MIC/4) of another endemic Moroccan thyme *T. satureioides* collected from three regions indicated that out of 21 combinations tested, 67% showed total synergism, 19% had partial synergistic interaction and 14% showed no effect. Furthermore, the combination between EO from Er-Rich population and cefexime exhibited the highest synergistic effect with FIC index values of 0.29–0.5 (Kasrati et al., 2014). Synergism study of *T. leptobotrys*, *T. pallidus*, or *T. ciliates* EOs and cefixime showed the greatest synergistic effect of *T. leptobotrys* EO against all tested bacteria with FICIs between 0.26 and 0.50 and a decrease of the MIC of cefixime by 4–130-fold. However, the combination of *T. ciliatus* and *T. pallidus* EOs with cefixime demonstrated total synergistic effect against *M. luteus*, *B. cereus*, and *P. aeruginosa* (FICIs: 0.370.50), partial synergistic effect against *K. pneumoniae* and *S. aureus* (FICI: 0.75), but no synergistic effect against *E. coli* (FICI: 1.25) (Alaoui Jamali et al., 2017). The study of in vitro association of EO of *C. ambrosioides* var. *ambrosioides*, a well-known medicinal plant used in Moroccan folk medicine against fever, pain, and gastrointestinal diseases and some commercial antibiotics reported that the total synergistic effect observed for Gram-positive bacteria were obtained with the combination of EO and cefixime with FICi ranging from 0.37 to 0.50, and a decrease of antibiotic MIC of 4–8-fold. However, for Gram-negative bacteria, the best combination was EO-ciprofloxacin with FICi ranging from 0.28 to 0.75, which lead to a remarkable reduction of MIC (2- and 64-fold). Whereas, the association EO-fluconazole showed total synergistic effect for most tested *Candida* and a decrease of the MIC of fluconazole of 8–16-fold (Ait Sidi Brahim et al., 2015a). Recently, a study carried out in our laboratory with leaves of *Periploca laevigata* EO showed a high synergistic effect. Combination with gentamicin exhibited interesting synergistic effect against Gram-positive and Gram-negative bacteria (FICi = 0.28–0.50), followed by EO-ciprofloxacin against Gram-positive bacteria (FICi = 0.31–0.38). Whereas, the association EO-fluconazole showed a total synergistic effect (FICi = 0.25–0.37) against yeasts except *C. albicans* (Ait Dra et al., 2017).

As far as we know, only a few studies have been published on the study of synergetic interactions between plant extracts and antibiotics and interactions between two or three EOs. In this context, we have studied the in vitro association of crude saponins extracted from the aerial parts of *Paronychia argentea* and the roots of *Spergularia marginata* and some commercial antibiotics against a panel of bacteria and *Candida* strains. For bacteria strains, 30 combinations were studied, 17 (56.66%) combinations had total synergism, 7 (23.33%) had partial synergism, 4 (13.33%) had no effect, and 2 (6.66%) had antagonism effect. For *Candida* strains, eight combinations of saponins extracts and fluconazol are tested. All of these combinations (100%) exhibited a total synergism with FICi ranging from 0.31 to 0.50 (Ait Sidi Brahim et al., 2015b).

The combined antibacterial effect of three Moroccan EOs of *O. compactum*, *O. majorana*, and *T. serpyllum* has been investigated by Ouedrhiri et al. (2016). Oregano species beside thyme are commonly used in Moroccan folk medicine to treat infectious disease. The experimental antibacterial activity exhibited by the EOs mixtures depends on the proportion of each EO in the combination and on the target strain. The optimal mixture predicted against *B. subtilis* and *S. aureus* corresponded to 28%, 30%, and 42% of *O. compactum*, *O. majorana*, and *T. serpyllum*, respectively. While the optimal mixture predicted against *E. coli* was composed by *O. compactum* and *O. majorana* EOs at 75% and 25%, respectively. Moreover, to increase the sensibility of *S. typhimurium* strain, a mixture of *T. vulgaris* L., *R. officinalis* L., and *M. communis* L. EOs was used in combined treatment by experimental design methodology (mixture design) (Fadil et al., 2017b). The optimization of mixtures antibacterial activities has highlighted the synergistic effect between *T. vulgaris* L. and *M. communis* L. EOs and a formulation comprising 55% of *T. vulgaris* L. and 45% of *M. communis* L. EOs, respectively, can be considered to increase *S. typhimurium* sensibility. In addition, the association of *T. vulgaris* L. EO with *M. communis* L. EO has more important antibacterial effect with respect to the combination of these two EOs with *R. officinalis* L. EO. This could be due to the fact that the combination of alcoholic monoterpenes (borneol as a major compound in *M. communis* L. EO) with phenolic ones (thymol as a major compound in *T. vulgaris* L. EO) produce a remarkable synergistic effect on several microorganisms. Recently, Aghraz et al. (2018) tested the EOs of *Cladanthus arabicus* and *Bubonium imbricatum* in combination with conventional antibiotics against Enterobacteriaceae isolates. The best synergy was obtained with the *B. imbricatum* EO combined with amoxicillin against the majority of tested Enterobacteriaceae isolates ($FIC < 0.5$) and also in combination with neomycin against *E. coli* ATCC 25922 and *P. mirabilis*. Additionally, *C. arabicus* EO showed a total synergy with amoxicillin against *Salmonella* sp. and *P. mirabilis* and a partial synergy against *Enterobacter cloacae*.

5.4 OTHER ACTIVITIES

Besides antibacterial and antifungal activities, medicinal plants are also an interesting source of other antiinfectious agents interfering on the treatment of leishmanial and viral infections (Table 5.3). It has been reported that *O. compactum* EOs at vegetative, flowering, and postflowering stages showed remarkable cytotoxic activities against *Leishmania major*, *Leishmania tropica*, and *Leishmania infantum* with the high antileishmanial activity observed for the EO of the flowering stage against *L. infantum* with IC_{50} value of 0.02 μg/mL (Bouyahya et al., 2017a). In another study, the same authors demonstrated that *Rosmarinus officinalis* EO exhibited a good antileishmanial effect against *L. tropica*, *L. infantum*, and *L. major*, and the last one was the most sensitive strain ($IC_{50} = 1.2$ mg/mL) (Bouyahya et al., 2017b). *Mentha pulegium* EO was also tested against *L. major*, *L. tropica*, and *L. infantum* using MTT (3-(4.5-dimethylthiazol-2yl)-2.5-diphenyltetrazolium bromide) assay. The obtained results showed that *M. pulegium* EO exhibited a good antileishmanial effect, in particular against *L. major* ($IC_{50} = 1.3 \pm 0.45$ mg/mL) (Bouyahya et al., 2017b). Additionally, the antiparasitic activity of hexanic, methanolic, and ethanolic extracts of *Salvia officinalis* has been tested against *L. major* using MTT assay. The data demonstrated that the studied extracts exhibited a moderate antileishmanial activity with a value of cytotoxicity (IC_{50}) above 1 mg/mL (Et-Touys et al., 2016).

To the best of our knowledge there are a few reports concerning the antiviral activity of Moroccan medicinal plants. One of the work in this regards was conducted by Amzazi et al. (2003) who tested various extracts

Table 5.3 Other antiinfective activities of Moroccan medicinal plant

Families	Species (species)	Origin	Parts used	Extracts	Major components	Effects	References
Lamiaceae	*Mentha pulegium* L.	Ouezzane	Aerial parts	Essential oil	Menthone and pulegone	Remarkable cytotoxic activities The high activity against *L. major* $CI_{50} = 1.3 \pm 0.45$ mg/mL	Bouyahya et al. (2017b)
	Origanum compactum	Ouezzane	Aerial parts of vegetative, flowering, and postflowering	Essential oils	Carvacrol, thymol, p-cymene, and γ-terpinene	Remarkable cytotoxic activities The high antileishmanial activity observed for the essential oil of the flowering stage against *L. infantum* with $IC_{50} = 0.02 \pm 0.004$ μg/mL	Bouyahya et al. (2017a)
	Rosmarinus officinalis	Ouezzane	Aerial parts	Essential oil	α-Pinene, 1,8-cineole and camphor	Strong activity The high antileishmanial activity was observed against *L. infantum* $IC_{50} = 1.2 \pm 0.36$ mg/mL	Bouyahya et al. (2017b)
	Salvia officinalis	Ouezzane	Leaves	Hexane, methanol, and ethanol extracts	ND	The antileishmanial activity was moderate with a value of cytotoxicity (IC_{50}) above 1 mg/mL	Et-Touys et al. (2016)

from *Mentha longifolia* (methanolic, hexanic, ethyl acetate, and butanolic) for their ability to inhibit human immunodeficiency virus type 1 (HIV-1). Nontoxic concentrations (10 μg/mL) of these extracts, especially methanolic and ethyl acetate significantly inhibit HIV-1BaL infection by about 40% and 55%, respectively. However, only ethyl acetate extract shows significant inhibitory activity (50% inhibition) against HIV-1 reverse transcriptase. Furthermore, chemical analysis of these extracts suggests that flavonoids, mainly flavones, may be the major inhibitors of HIV infection. Mouhajir et al. (2001) studied the antiviral activities of methanol extracts of 75 Moroccan endemic plants (64 genera of 35 families), commonly used in the folk medicine to treat infections caused by viruses and microbes against three mammalian viruses: herpes simplex virus, Sindbis virus, and poliovirus, at noncytotoxic concentrations. *Punica granatum* extract, which was the most active, inhibited all tested viruses at a concentration of 1.5 μg/mL. The extracts of *Acacia gummifera*, *Juglans regia*, *Thymus maroccanus*, *Lawsonia inermis*, *Pinus halepensis*, and *Rosa canina* inhibited Sindbis virus at a minimum concentration of 1.5 μg/mL. *Pistacia lentiscus* and *T. maroccanus* were very active against herpes simplex virus. However, the most active extracts against poliovirus were those from *P. halepensis* and *P. granatum* at minimum concentrations of 6.5 μg/mL. These results indicate that these plants are potential potent medicines against viral infections.

5.5 ANTIOXIDANT ACTIVITY

5.5.1 Essential Oils

Many studies support the fact that many redox active/antioxidant phytochemicals are promising compounds to prevent oxidative mechanisms taking place in many pathological diseases. It has been reported that Moroccan medicinal plants contain antioxidant agents, which can act against free radicals a strong antioxidant activity. In this context, several investigations evaluating the antioxidant property of Moroccan medicinal plants have been compiled (summarized in Table 5.4). Antioxidant potential of Thymus species including *T. broussonetii*, *T. ciliatus*, *T. leptobotrys*, *T. maroccanus*, *T. pallidus*, *T. satureioides*, and *T. serpyllum* EOs using 2,2-diphenyl-1-picrylhydrazyl (DPPH) free radical-scavenging ability and the ferric-reducing potential has been evaluated by Alaoui Jamali et al. (2012). It showed that *T. leptobotrys* and *T. maroccanus* EOs (rich in carvacrol) possessed the highest antioxidant activities with IC_{50} values for DPPH assays of 25.37 and 60.72 μg/mL, respectively, while EC_{50} values

Table 5.4 Antioxidant activity of EO/extracts from Moroccan medicinal plants

Families	Species (families)	Origin	Parts used	Extracts	Major components	Effects	References
Apiaceae	*P. verticillata*	Oujda	Aerial parts	Ethyl acetate and diethyl ether extracts	ND	A potent antioxidant activity For ethyl acetate extract $IC_{50} = 0.64$ μg/mL For diethyl ether extract $IC_{50} = 0.92$ μg/mL	El Ouariachi et al. (2011)
	F. vulgare Mill	Gourama, Southeastern Morocco	Seeds	Essential oils obtained from wild and cultivated fennel	ND	Wild plant presented the most important antioxidant power compared to cultivated one	Abdellaoui et al. (2017)
Apocynaceae	*P. laevigata*	Essaouira	Leaves	Essential oil	*n*-Hexadecanoic acid and 4,4,7a-trimethyl-5,6,7,7a-tetrahydro-4H-benzofuran-2-one	A moderate activity with IC_{50} values ranging from 0.69 to 2.86 mg/mL	Ait Dra et al. (2017)
Arecaceae	*P. dactylifera* L.	Southeast Morocco	Fruits	Methanolic rich polyphenol extracts	ND	Strong antioxidant activity for the six varieties Jihl variety was the most active extract with IC_{50} value of 2.05 g/L for DPPH scavenging activity and a ferric-reducing power of 860.89 μmol TE/100 g DW	Bouhlali et al. (2016)
Asteraceae	*C. arabicus*	Ourika	Aerial parts	Essential oil	Sabinene, β-pinene, myrcene, and α-pinene	Strong antioxidant activity $IC_{50} = 55.4 \pm 0.12$ μg/mL and 57.2 ± 0.21 μg/mL, respectively for DPPH and FRAP methods	Aghraz et al. (2017a,b)

(*Continued*)

Table 5.4 (Continued)

Families	Species (families)	Origin	Parts used	Extracts	Major components	Effects	References
Cactaceae	*Opuntia ficus-indica*	Rif region	Flowers	Phenolic rich extracts	ND	Strong antioxidant activity of extracts	
	O. ficus-indica	Rif region	Flowers	Phenolic rich fractions	ND	The ORAC value of ASE methanol/water (V/V) extract was lower (263.72 mol TE/g) than that of maceration methanol extract (326.44 mol TE/g) The 80% acetone/water extract of ASE method was higher (272.84 mol TE/g) than that obtained by maceration method (extract A; 243.68 mol/g)	Benayad et al. (2014)
Caryophyllaceae	*Paronychia argentea*	Oukaimden	Aerial parts	Saponins-rich extract	ND	Using DPPH assay: $IC_{50} = 19.08\ \mu g/mL$ For β-carotene-linoleic acid assay, $IC_{50} = 98.24\ \mu g/mL$ For reducing power assays, the $IC_{50} = 27.22\ \mu g/mL$	Ait Sidi Brahim et al. (2015b)
	Spergularia marginata	Essaouira	Root	Saponins-rich extract	ND	Using DPPH assay, $IC_{50} = 29.65\ \mu g/mL$ For β-carotene-linoleic acid assay, $IC_{50} = 614\ \mu g/mL$ respectively For reducing power assays, $IC_{50} = 61.44\ \mu g/mL$	Ait Sidi Brahim et al. (2015b)
Chenopodiaceae	*C. ambrosioides var. ambrosioides*	55 km from Marrakesh	Aerial parts	Essential oil	γ-Terpinene, ascaridole, p-cymene, neral, geraniol	DPPH assay: moderate antioxidant ($IC_{50} = 4.10^3\ \mu g/mL$) β-Carotene/linoleic acid bleaching: high effect ($IC_{50} = 3.03\ \mu g/mL$) Reducing power assay: high effect ($IC_{50} = 6.02\ \mu g/mL$)	Ait Sidi Brahim et al. (2015a,b)

Cistaceae	*Cistus crispus*	Ouezzane	Leaves	Methanolic, ethanolic, ethyl acetate, and *n*-hexanic extracts	ND	The methanolic extract exhibited the highest antiradical activity ($IC_{50} = 53.95 \pm 7.55$ µg/mL)	Bouyahya et al. (2017d)
	C. monspeliensis L.	Maaziz-Kh'emisset	Aerial parts	Aqueous and hydromethanolic extracts	ND	Potent antioxidant activity using DPPH, ABTS, and FRAP methods	Sayah et al. (2017)
	C. salviifolius L.	Maamoura Forest, Sale	Aerial parts	Aqueous and hydromethanolic extracts	ND	Potent antioxidant activity using DPPH, ABTS, and FRAP methods	Sayah et al. (2017)
Ericaceae	*Arbutus. unedo* L.	Ouezzane	Leaves	Methanolic, ethanolic, ethyl acetate, and *n*-hexanic extracts	ND	All the extracts presented antioxidant capacity, methanol, and hexane extracts were the most active *n*-Hexane $IC_{50} = 73.73$ µg/mL MeOH: $IC_{50} = 95.25$ µg/mL EtOAc $IC_{50} = 276.15$ µg/mL EtOH $IC_{50} = 280.50$ µg/mL	Bouyahya et al. (2016b)
Lamiaceae	*O. compactum* Benth	Ouezzane	Leaves	Methanolic, ethanolic, ethyl acetate, and *n*-hexanic extracts	ND	*n*-Hexane extract is the most active ($IC_{50} = 39.83$ µg/mL)	Bouyahya et al. (2017e)
	O. compactum	Ouezzane	Aerial parts of vegetative, flowering and postflowering	Essential oils	Carvacrol, thymol, p-cymene, and γ-terpinene	Strong antioxidant activity The highest activist is observed for the essential oils from postflowering stage ($IC_{50} = 172.87 \pm 5.38$ µg/mL and $IC_{50} = 53.28 \pm 2.64$ µg/mL for PPH and reducing power assay respectively	Bouyahya et al. (2017a)

(*Continued*)

Table 5.4 (Continued)

Families	Species (families)	Origin	Parts used	Extracts	Major components	Effects	References
	Salvia aucheri subsp. *blancoana* wild and cultivated			Essential oils		Strong activity Cultivated essential oil was more potent than the wild one	El Abdouni Khayari et al. (2014)
	Salvia officinalis	Ouezzane	Flowering tops	Essential oil	ND	Strong activity $IC_{50} = 0.46$ mg/mL	Bouyahya et al. (2017c)
	S. officinalis	Ouezzane	Leaves	Hexane, methanol and ethanol extracts	ND	The methanol extract has significant activity using the radical DPPH test $IC_{50} = 65.655$ μg/mL	Et-Touys et al. (2016)
	Thymus broussonetii	Essaouirra	Aerial parts	Essential oil	Carvacrol, thymol, γ-terpinene	Remarkable activity $IC_{50} = 97.48$ μg/mL for DPPH assays $EC_{50} = 167.86$ μg/mL for ferric-reducing capacity	Alaoui Jamali et al. (2012)
	T. ciliatus	Imilchil	Aerial parts	Essential oil	Carvacrol, thymol p-Cymene, γ-terpinene	Remarkable activity with $IC_{50} = 206.57$ μg/mL for DPPH assays $EC_{50} = 184.05$ μg/mL for ferric-reducing capacity	Alaoui Jamali et al. (2012)
	T. leptobotrys	Tiznite	Aerial parts	Essential oil	Carvacrol, p-cymene	High activity $IC_{50} = 25.37$ μg/mL for DPPH assays $EC_{50} = 19.24$ μg/mL for ferric-reducing capacity	Alaoui Jamali et al. (2012)
	T. maroccanus	Ait Ourir	Aerial parts	Essential oil	Carvacrol, γ-terpinene, p-cymene	High antioxidant activities with IC_{50} values for DPPH 60.721.93 mg/mL EC_{50} values for ferric-reducing capacity were 139.5 μg/mL	Alaoui Jamali et al. (2012)

	T. pallidus	Ait Lkak	Aerial parts	Essential oil	Thymol, γ-terpinene, p-cymene	Remarkable activity with IC_{50} values for DPPH 345.11 μg/mL EC_{50} values for ferric-reducing capacity were 193.69 μg/mL	Alaoui Jamali et al. (2012)
	T. satureioides	Idni	Aerial parts	Essential oil	Carvavrol, Borneol, camphene	Remarkable antioxidant activity IC_{50} = 122.53 μg/mL for DPPH EC_{50} = 177.13 μg/mL for ferric-reducing capacity	Alaoui Jamali et al. (2012)
	Thymus satureioides	Tafilalet Region, Southeast of Morocco	Aerial parts	Aqueous extract, total polyphenols and total flavonoïds extracts	Six phenolic compounds, including caffeic, rosmarinic acids, quercetin, hesperetin, luteolin-7-glycoside and apigenin-7-glycoside	Good antioxidant activity for DPPH radical scavenging activity IC_{50} = 0.418 to 43.891 ± 2.467 mg/mL Form FRAP IC_{50} = 7.03–117.51 mmol trolox/g	Ramchoun et al. (2015)
	Thymus saturejoides Coss.	Three regions: Taws, Ourika, and Er-Rich	Blooming stage	Essential oils	Carvacrol, p-cymene, linalool, and borneol in oil from the Er-Rich population (arid site) Carvacrol, borneol, camphene, and -terpinene in oil from Ourika (medium arid site) Carvacrol, borneol, camphene, and p-cymene in oil from Taws population (less arid site)	High antioxidant activity The highest activity was obtained with essential oil of Er-Rich population IC50 from 19.17 to 44.54	Kasrati et al. (2014)

(*Continued*)

Table 5.4 (Continued)

Families	Species (families)	Origin	Parts used	Extracts	Major components	Effects	References
	T. serpyllum	Oukaimeden	Aerial parts	Essential oil	Linalyl acetate (E)-Nerolidol Geranyl acetate	Not active $CI_{50} > 30{,}000$ μg/mL for bohtesrs	Alaoui Jamali et al. (2012)
	Vitex agnus-castus	southwestern of Morocco	Seeds	Essential oil	1,8-Cineole, sabinene, a-pinene	Low activity	Asdadi et al. (2015)
Myrtaceae	*M. communis*	Ouezzane	Leaves	*Essential oil*	ND	Great efficiency $IC_{50} = 0.24$ mg/mL	Bouyahya et al. (2017c)
Poaceae	*Pennisetum glaucum*		Seeds	Ethanolic extract	ND	Strong antioxidant (208.01 ± 2.54 mg TE/g fDW (DPPH)/ 8.29 ± 0.11 mg TE/g fDW (TEAC)/ 21.20 ± 0.57 mg AAE/g fDW (FRAP)	Marmouzi et al. (2016)
Apiaceae	*Ptychotis verticillata*	Ahfir	Aerial parts	Ethyl acetate and diethyl ether extracts	ND	Potent antioxidant activity	El Ouariachi et al. (2011)
Solanaceae	*Withania frutescens* (L.) Pauquy	Hassania (Marrakech)	Roots and leaves	*n*-Hexane and: methanol extracts and dichloromethane, ethyl acetate and *n*-butanol fractions	ND	The ethyl acetate and *n*-butanol leaf fractions exhibited the highest DPPH radical-scavenging activity with IC_{50} value of 4.53 and 8.49 μg/mL, respectively	El Bouzidi et al. (2011b)

for ferric-reducing capacity were 19.24 and 139.5 μg/mL, respectively. *T. serpyllum* EO was not active. Furthermore, EOs from three wild Moroccan *T. saturejoides* populations collected from the Northwest–Southeast of Morocco showed the highest activity with EO from Er-Rich population (IC_{50} values of 44.54, 22.90, and 19.17 μg/mL, respectively) (Kasrati et al., 2014). A comparison of the aerial parts of wild and cultivated *Salvia aucheri* subsp. *blancoana* EO was evaluated and demonstrated that cultivated Moroccan sage EO showed the highest antioxidant activity comparing to the wild one (El Abdouni Khayari et al., 2014). In another study which also examined the impact of domestication on the antioxidant activities showed that EOs obtained from the wild fennel plant (*Foeniculum vulgare* Mill). presented the most important antioxidant power compared to those extracted from cultivated plant (Abdellaoui et al., 2017). Also, EOs of three phenological stages (vegetative, flowering, and postflowering) of an important wild endemic Moroccan species *Origanum compactum* were extracted by hydrodistillation from the aerial parts and showed a strong antioxidant activity with the highest activity obtained from the EO of the postflowering stage (Bouyahya et al., 2017a). Authors also found that EO of *M. communis* showed great efficiency in reducing the DPPH radical ($IC_{50} = 0.24$ mg/mL), followed by EO of *S. officinalis* ($IC_{50} = 0.46$ mg/mL) (Bouyahya et al., 2017c). The EO obtained from seeds of *Vitex agnus-castus* growing wild in southwestern of Morocco showed a low antioxidant activity ($IC_{50} = 1.072$) comparing with those of BHT and vitamin E (IC50 respectively of 0.5844 mg/mL and 0.4262 mg/mL (Asdadi et al., 2015). In our laboratory, it has been recently shown that the EO of *C. arabicus* aerial parts exhibited a strong antioxidant activity with IC_{50} value of 55.4 and 57.2 μg/mL, respectively for DPPH and FRAP methods (Aghraz et al., 2017a,b). Also, *C. ambrosioides* var. *ambrosioides* EO (Ait Sidi Brahim et al., 2015a,b) and *P. laevigata* EO (Ait Dra et al., 2017) obtained by hydrodistillation exhibited a moderate antioxidant activity as evaluated by DPPH radical method, reducing power and ß-carotene-linoleic acid assays with IC_{50} values ranging from 0.69 to 2.86 mg/mL.

5.5.2 Plant Extracts

Many studies have been also conducted on extracts obtained from Moroccan medicinal plants and showed also an interesting antioxidant activity (Table 5.4). The sequential extracts obtained with ethyl acetate and diethyl ether from the aerial parts of *Ptychotis verticillata* had a potent antioxidant activity using DPPH test with an IC_{50} value of 0.64 and 0.92 μg/mL, respectively (El Ouariachi et al., 2011). El Bouzidi et al. (2011b) determined the antioxidant activity of successive extracts with *n*-hexane and methanol extracts, dichloromethane, ethyl acetate, and *n*-butanol fractions obtained from the roots and leaves of *Withania frutescens* (L.) Pauquy, using the DPPH-free radical scavenging and reducing power methods. The ethyl acetate and *n*-butanol leaf fractions exhibited the highest DPPH radical-scavenging activity with IC_{50} value of 4.53 and 8.49 μg/mL, respectively. It has been demonstrated that the presence of polyphenol and glycowithanolide in the extracts appeared to be responsible for the antioxidant capacity of *W. frutescens* (El Bouzidi et al., 2011b). It is well known in the literature that the antioxidant activity is usually correlated with the presence of phenolic compounds in the extracts. Also, *Opuntia ficus-indica* phenolic rich extracts exhibited a high antioxidant potential (Benayad et al., 2014). Phenolic rich fractions were extracted from the flowers of *O. ficus-indica* using accelerated solvent extraction (ASE) and maceration methods and were then studied for their antioxidant activity using oxygen radical absorbance capacity (ORAC) technique. The extraction methods adopted have great influence on the observed antioxidant activity. Thus the ORAC value of ASE methanol/water (V/V) extract was lower (263.72 mol TE/g) than that of maceration methanol extract (326.44 mol TE/g), and that of the 80% acetone/water extract of ASE method was higher (272.84 mol TE/g) than that obtained by maceration method (extract A; 243.68 mol/g). Aqueous extract, total polyphenols, and total flavonoids extracts of *T. satureioides* showed a good antioxidant activity. IC_{50} values for 1,1-diphenyl-2-picrylhydrazil radical-scavenging activity were 0.480, 0.418, and 43.891 mg/mL for the aqueous extract, total polyphenol, and flavonoïds, respectively. Also the extracts showed ferric-reducing ability and the values were 50.79, 117.51, and 7.03 mmol trolox/g for the aqueous extract, total polyphenol, flavonoids, and trolox, respectively (Ramchoun et al., 2015). Six Moroccan date varieties locally known as Bouskri, Bousrdon, Bousthammi, Boufgous, Jihl, and Majhoul were also screened for their antioxidant activity using DPPH and FRAP tests. Metanolic polyphenol-enriched extracts from these varieties showed strong antioxidant activity and Jihl variety had the highest activity compared to other varieties with an IC_{50} value of 2.05 g/L for DPPH scavenging activity and 860.89 μmol TE/100 g dry weight (DW) for ferric-reducing power assay

(Bouhlali et al., 2016). Furthermore, the hexane, methanol, and aqueous extracts of *Calendula arvensis* were assessed for their antioxidant activity using four different methods (DPPH, FRAP, TEAC, b-carotene bleaching test) and the phenolic and flavonoid contents were also determined. Phytochemical quantification of the methanolic and aqueous extracts revealed that they were rich with flavonoids and phenolics and were found to possess considerable antioxidant activities. However, hexane extract was less potent as antioxidant agent (Abudunia et al., 2017). Additionally, antioxidant evaluation resulted in higher activity for the ethanolic fraction from Moroccan *Pennisetum glaucum* (208.01 mg TE/g fDW (DPPH)/8.29 mg TE/g fDW (TEAC)/21.20 mg AAE/g fDW (FRAP)) using DPPH, ferric-reducing antioxidant power and ABTS assays. Phenolic and flavonoid contents of fractions were varied from 4.19 to 22.78 mg GAE/g fDW, and from 0.75 to 15.60 mg RE/g f DW, respectively (Marmouzi et al., 2016). Bouyahya et al. (2016b) studied the antioxidant activity of methanolic, ethanolic, ethyl acetate, and *n*-hexanic extracts from leaves of *A. unedo* and found that methanol and hexane fractions were the most active with IC_{50} values of 73.73 and 95.25 μg/mL, respectively. This activity has been also correlated with the phenolic content of plant extracts.

Antioxidant potential of methanol, ethanol, ethyl acetate, and *n*-hexane of *Cistus crispus* leave extracts was evaluated using DPPH scavenging method. The methanolic extract exhibited the highest antiradical activity ($IC_{50} = 53.95 \pm 7.55$ μg/mL) comparable to standard antioxidants (ascorbic acid, $IC_{50} = 27.20 \pm 4.3$ μg/mL and Trolox: IC50 = 43.72 ± 7.53 μg/mL) (Bouyahya et al., 2017d). However, according to the same author (Bouyahya et al., 2017a), the quantitative evaluation of the antiradical activity has shown that *n*-hexane extract was the most active for *O. compactum* Benth ($IC_{50} = 39.83$ μg/mL). Indeed, these results confirmed that the presence of effective compounds in the biochemical composition of the plant, particularly polyphenols were the main responsible of the high capacity of extract to reduce DPPH. The antioxidant activity of aqueous and hydromethanolic extracts from the aerial parts of Moroccan *Cistus salviifolius* L. and *C. monspeliensis* were also attributed to their richness of phenolic principles as has been assessed. The extracts of both species exhibited potent antioxidant activity using DPPH, ABTS, and FRAP methods (Sayah et al., 2017).

The antioxidant activity is not only correlated with the presence of phenolic compounds but could be also attributed to other molecules present in plant extracts. Saponins are one of the natural compounds derived from plant extracts which exhibited a strong antioxidant activity. A study undertaken by our team determined the antioxidant activity of crude saponins extracted from two Caryophyllaceae species known for their richness of saponins using the DPPH-free radical, β-carotene-linoleic acid and reducing power assays (Ait Sidi Brahim et al., 2015b). The saponins-rich extract from *P. argentea* aerial parts exhibited a higher antioxidant activity than that from *S. marginata* roots. Using DPPH assay, the IC_{50} values for saponins-rich extracts from *P. argentea* and *S. marginata* were 19.08 and 29.65 μg/mL, respectively. For β-carotene-linoleic acid assay, IC_{50} values were 98.24 and 614 μg/mL respectively for *P. argentea* and *S. marginata*. However, for reducing power assays, the IC_{50} values were respectively 27.22 and 61.44 μg/mL.

5.6 CONCLUSION

Morocco is a rich source of medicinal plants used traditionally to fight against several diseases. Moroccan researchers have exploited scientifically these medicinal plants to extract EOs and plant solvent extracts and determine their antiinfective and antioxidant activities using different experimental methods. This chapter was evidenced to summarize the maximum data presented in this context. From these data, we conclude that EOs and extracts from Moroccan medicinal plants are very effective against Gram-positive and Gram-negative bacteria, *Candida*; fungi; parasites as well as viruses. *Thymus* species are the most studied plant and also the most effective. Some other plants from the Lamiaceae family were found to be also effective. Other species from Asteraceae family for example *Artemisia herba-alba* are very used in Moroccan traditional medicinal and also proved to possess antiinfective activities.

Furthermore, it seems that crude extracts from Moroccan medicinal and aromatic plants are not very well exploited. Many crude extracts studied are effective but we notice that isolation of active principles responsible for the activities is very rare. So, further studies are needed to determine the bioactive molecules especially in the crude solvent extracts as many compounds with interesting antimicrobial and antioxidant properties might be characterized.

Another point that should be mentioned is the diversity of the methods employed for testing at least in the case of antimicrobial activity. Results generated from many of these studies cannot be directly compared due to the lack of a standard method. It is well known that the origin of microorganisms tested, the solvent and the extraction system, the composition of the growth medium greatly influence the activity of the tested extracts or compounds.

REFERENCES

Abdellaoui, M., Bouhlali, E.T., Kasrati, A., El Rhaffari, L., 2017. The effect of domestication on seed yield, essential oil yield and antioxidant activities of fennel seed (*Foeniculum vulgare* Mill) grown in Moroccan oasis. J. Assoc. Arab Univ. Basic Appl. Sci. 24, 107–114.

Abudunia, A.M., Marmouzi, I., Faouzi, M.E.A., Ramli, Y., Taoufik, J., El Madani, N., et al., 2017. Anticandidal, antibacterial, cytotoxic and antioxidant activities of *Calendula arvensis* flowers. J. Mycol. Méd. 27, 90–97.

Aghraz, A., Wanner, J., Schmidt, E., Aitdra, L., Aitsidibrahim, M., Tabanca Ali Abbas, N., et al., 2017a. Chemical composition, *in vitro* antioxidant, antimicrobial and insecticidal activities of essential oil from *Cladanthus arabicus*. J. Essential Oil Bear. Plants 20, 601–609.

Aghraz, A., Wanner, J., Schmidt, E., Aitdra, L., Aitsidibrahim, M., Tabanca, N., et al., 2017b. Chemical composition, antioxidant, antimicrobial and insecticidal activities of essential oil from a Moroccan endemic plant: *Bubonium imbricatum*. J. Nat. Prod. Commun. 11, 1717–1720.

Aghraz, A., Benameur, Q., Gervasi, T., Ait dra, L., Ben-Mahdi, M.H., Larhsini, M., et al., 2018. Antibacterial activity of *Cladanthus arabicus* and *Bubonium imbricatum* essential oils alone and in combination with conventional antibiotics against Enterobacteriaceae isolates. Lett. Appl. Microbiol. Available from: https://doi.org/10.1111/lam.13007.

Ait Sidi Brahim, M., Fadli, M., Hassani, L., Boulay, B., Markouk, M., Bekkouche, K., et al., 2015a. *Chenopodium ambrosioides* var. *ambrosioides* used in Moroccan traditional medicine can enhance the antimicrobial activity of conventional antibiotics. Ind. Crops Prod. 71, 37–43.

Ait Sidi Brahim, M., Fadli, M., Markouk, M., Hassani, L., Larhsini, M., 2015b. Synergistic antimicrobial and antioxidant activity of saponins-rich extracts from *Paronychia argentea* and *Spergularia marginata*. Eur. J. Med.Plants 7, 193–204.

Ait Dra, L., Ait Sidi Brahim, M., Boualy, B., Aghraz, A., Barakate, M., Oubaassine, S., et al., 2017. Chemical composition, antioxidant and evidence antimicrobial synergistic effects of *Periploca laevigata* essential oil with conventional antibiotics. Ind. Crops Prod. 109, 746–752.

Ait-Ouazzou, A., Lorán, S., Arakrak, A., Laglaoui, A., Rota, C., Herrera, A., et al., 2012. Evaluation of the chemical composition and antimicrobial activity of *Mentha pulegium*, *Juniperus phoenicea*, and *Cyperus longus* essential oils from Morocco. Food Res. Int. 45, 313–319.

Al-Adhroey, A.H., Nor, Z.M., Al-Mekhlafi, H.M., Mahmud, R., 2010. Ethnobotanical study on some Malaysian anti-malarial plants A community based survey. Ethnopharmacology 132, 362–364.

Alaoui Jamali, C., El Bouzidi, L., Bekkouche, K., Hassani, L., Markouk, M., Wohlmuth, H., et al., 2012. Chemical composition and antioxidant and anticandidal activities of essential oils from different wild Moroccan thymus species. Chem. Biodivers. 9, 1188–1197.

Alaoui Jamali, C., Kasrati, A., Fadli, M., Hassani, L., Leach, D., Abbad, A., 2017. Synergistic effects of three moroccan thyme essential oils with antibiotic cefixime synergie des huiles essentielles de trois espèces de thyms marocaines avec l'antibiotique Céfixime. Phytothérapie. Available from: https://doi.org/10.1007/s10298-017-1107-2.

Amzazi, S., Ghoulami, S., Bakri, Y., Idrissi, A., Fkih-Tétouani, S., Benjouad, A., 2003. Human immunodeficiency virus type 1 inhibitory activity of *Mentha longifolia* activité anti-VIH des extraits de *Mentha longifolia*. J. Thérapie 58, 531–534.

Andrews, R.E., Parks, L.W., Spence, K.D., 1980. Some effects of Douglas fir terpenes on certain microorganisms. Appl. Environ. Microbiol. 40, 301–304.

Asdadi, A., Hamdouch, A., Oukacha, A., Moutaj, R., Gharby, S., Harhar, H., et al., 2015. Study on chemical analysis, antioxidant and *in vitro* antifungal activities of essential oil from wild *Vitex agnus-castus* L. seeds growing in area of Argan Tree of Morocco against clinical strains of Candida responsible for nosocomial infections. J. Mycol. Méd. 25, e118–e127.

Bakkali, F., Averbeck, S., Averbeck, D., Idaomar, M., 2008. Biological effects of essential oils: a review. Food Chem. Toxicol. 46, 446–475.

Benayad, Z., Martinez-Villaluenga, C., Frias, J., Gomez-Cordoves, C., Es-Safi, N.E., 2014. Phenolic composition, antioxidant and anti-inflammatory activities of extracts from Moroccan *Opuntia ficus-indica* flowers obtained by different extraction methods. Ind. Crops Prod. 62, 412–420.

Bouhlali, E.T., Bammou, M., Sellam, K., Benlyas, M., Alem, C., Filali-Zegzouti, Y., 2016. Evaluation of antioxidant, antihemolytic and antibacterial potential of six Moroccan date fruit (*Phoenix dactylifera* L.) varieties. J. King Saud Univ.-Sci. 28, 136–142.

Bouyahya, A., Abrini, J., El-Baabou, A., Bakri, Y., Dakka, N., 2016a. Determination of phenol content and antibacterial activity of five medicinal plants ethanolic extracts from north-west of Morocco. J. Plant Pathol. Microbiol. 7, 107–111.

Bouyahya, A., El Moussaoui, N., Abrini, J., Bakri, Y., Dakka, N., 2016b. Determination of phenolic contents, antioxidant and antibacterial activities of strawberry tree (*Arbutus unedo* L.) leaf extracts. Br.Biotechnol. J. 14, 1–10.

Bouyahya, A., Abrini, J., El Ouardy, K., Charfi, S., Boujida, N., Harsal, E.L., et al., 2016c. *In vitro* antibacterial activity of organic extracts from north-west Moroccan medicinal plant *Myrtus communis* (L.). Biotechnol. J. Int. 16, 1–9.

Bouyahya, A., Dakka, N., Talbaoui, A., Et-Touys, A., El-Boury, H., Abrini, J., et al., 2017a. Correlation between phenological changes, chemical composition and biological activities of the essential oil from Moroccan endemic Oregano (*Origanum compactum* Benth). Ind. Crops Prod. 108, 729–737.

Bouyahya, A., Et-Touys, A., Bakri, Y., Talbaoui, A., Fellah, H., Abrini, J., et al., 2017b. Chemical composition of *Mentha pulegium* and *Rosmarinus officinalis* essential oils and their antileishmanial, antibacterial and antioxidant activities. Microb. Pathogen. 111, 41–49.

Bouyahya, A., Bakri, Y., Et-Touys, A., Talbaoui, A., Khouchla, A., El Yahyaoui, E.L., et al., 2017c. *In vitro* screening of antibacterial and antioxidant activities of essential oils from four Moroccan medicinal plants. Microbiol. Res. J. Int. 18, 1–10.

Bouyahya, A., Abrini, J., Talbaoui, A., Et-Touys, A., Chatoui, K., Harhar, H., et al., 2017d. Phytochemical screening, antiradical and antibacterial activities of *Cistus crispus* from Morocco. J. Mater. Environ. Sci. 8, 1560–1566.

Bouyahya, A., Guaouguaou, F.E., Dakka, N., Bakri, Y., 2017e. Pharmacological activities and medicinal properties of endemic Moroccan medicinal plant *Origanum compactum* (Benth) and their main compounds. Asian Pacific J. Trop. Dis. 7, 628–640.

Chami, N., Bennis, S., Chami, F., Aboussekhra, A., Remmal, A., 2005. Study of anticandidal activity of carvacrol and eugenol *in vitro* and *in vivo*. Oral Microbiol. Immunol. 20, 106–111.

Delamare, A.P.L., Moschen-Pistorelo, I.V., Artico, L., Atti-Serafini, L., Echeverrigary, S., 2007. Antibacterial activity of the essential oils of *Salvia officinalis* L. and *Salvia triloba* L. cultivated in South Brazil. Food Chem. 100, 603–608.

Denyer, S.P., Hugo, W.B., 1991. Biocide induced damage to the bacterial cytoplasmic membrane. In: Denyer, S.P., Hugo, W.B. (Eds.), Mechanisms of Action of Chemical Biocides, S.A.B. Technical Series, 27th ed Blackwell Scientific Publications, Oxford, London, pp. 171–188.

El Abdouni Khayari, M., Kasrati, A., Alaoui Jamali, C., Zeroual, S., Markouk, M., Bekkouche, K., et al., 2014. Chemical composition, antioxidant and insecticidal properties of essential oils from wild and cultivated *Salvia aucheri* subsp. blancoana (Webb. & Helder), an endemic, threatened medicinal plant in Morocco. Ind. Crops Prod. 57, 106–109.

El Abdouni Khayari, M., Alaoui Jamali, C., Kasrati, A., Hassani, L., Leach, D., Markouk, M., et al., 2016. Antibacterial activity of essential oils of some Moroccan aromatic herbs against selected food-related bacteria. J. Essential Oil Bear. Plants 19, 1075–1085.

El Bouzidi, L., Abbad, A., Fattarsi, K., Hassani, L., Leach, D., Markouk, M., et al., 2011a. Chemical composition and anticandidal properties of the essential oil isolated from aerial parts of *Cotula cinerea*: a rare and threatened medicinal plant in Morocco. Nat. Prod. Commun. 6, 1491–1494.

El Bouzidi, L., Larhsini, M., Markouk, M., Abbad, A., Hassani, L., Bekkouche, K., 2011b. Antioxidant and antimicrobial activities of Withania frutescens 6, 1447–1450.

El Bouzidi, L., Abbad, A., Hassani, L., Fattarsi, K., Leach, D., Markouk, M., et al., 2012. Essential oil composition and antimicrobial activity of wild and cultivated Moroccan *Achillea ageratum* L.: a rare and threatened medicinal species. Chem. Biodivers. 9, 598–605.

El Midaoui, M., Maataoui, A., Benbella, M., Houssa, A.A., Labazi, N., 2011. Ethnobotanical study of some aromatic and medicinal plants in the Middle Atlas mountains of Morocco. Nat. Prod. Commun. 6, 1455–1458.

El Ouadi, Y., Manssouri, M., Bouyanzer, A., Majidi, L., Lahhit, N., Bendaif, H., et al., 2015. Essential oil composition and antifungal activity of *Salvia officinalis* originating from North-East Morocco, against postharvest phytopathogenic fungi in apples. Der Pharm. Chem. 7, 95–102.

El Ouadi, Y., Manssouri, M., Bouyanzer, A., Majidi, L., Bendaif, H., Elmsellem, H., et al., 2017. Essential oil composition and antifungal activity of *Melissa officinalis* originating from north-Est Morocco, against postharvest phytopathogenic fungi in apples. Microb. Pathogen. 107, 321–326.

El Ouariachi, M., Tomi, P., Bouyanzer, A., Hammouti, B., Desjobert, J.M., Costa, J., et al., 2011. Chemical composition and antioxidant activity of essential oils and solvent extracts of *Ptychotis verticillata* from Morocco. Food Chem. Toxicol. 49, 533–536.

Et-Touys, A., Fellah, H., Mniouil, M., Bouyahya, A., Dakka, N., Abdennebi, E.H., et al., 2016. Screening of antioxidant, antibacterial and antileishmanial activities of *Salvia officinalis* L. extracts from Morocco. Br. Microbiol. Res. J. 16, 1–10.

Ezoubeiri, A., Gadhi, C.A., Fdil, N., Benharref, A., Jana, M., Vanhaelen, M., 2005. Isolation and antimicrobial activity of two phenolic compounds from *Pulicaria odora* L. J. Ethnopharmacol. 99, 287–292.

Fadli, M., Chevalier, J., Saad, A., Mezrioui, N.E., Hassani, L., Pages, J.M., 2011. Essential oils from Moroccan plants as potential chemosensitisers restoring antibiotic activity in resistant Gram-negative bacteria. Int. J. Antimicrob. Agents 38, 325–330.

Fadli, M., Saad, A., Sayadi, S., Chevalier, J., Mezrioui, N.E., Pagès, J.M., et al., 2012. Antibacterial activity of *Thymus maroccanus* and *Thymus broussonetii* essential oils against nosocomial infection-bacteria and their synergistic potential with antibiotics. Phytomedicine 19, 464–471.

Fadli, M., Bolla, J.M., Mezrioui, N.E., Pagès, J.M., Hassani, L., 2014. First evidence of antibacterial and synergistic effects of *Thymus riatarum* essential oil with conventional antibiotics. Ind. Crops Prod. 61, 370–376.

Fadli, M., Pagès, J.M., Mezrioui, N.E., Abbad, A., Hassani, L., 2016. *Artemisia herba-alba* Asso and *Cymbopogon citratus* (DC.) Stapf essential oils and their capability to restore antibiotics efficacy. Ind. Crops Prod. 89, 399–404.

Fadil, M., Farah, A., Ihssane, B., Haloui, T., Lebrazi, S., Rachiq, S., 2017a. Intrapopulation variability of *Myrtus communis* L. growing in Morocco: chemometric investigation and antibacterial activity. J. Appl. Res. Med. Aromatic Plants. Available from: https://doi.org/10.1016/j.jarmap.2017.04.006.

Fadil, M., Fikri-Benbrahim, K., Rachid, S., Ihssane, B., Lebrazi, S., Chraibi, M., et al., 2017b. Combined treatment of *Thymus vulgaris* L., *Rosmarinus officinalis* L. and *Myrtus communis* L. essential oils against *Salmonella typhimurium*: optimization of antibacterial activity by mixture design methodology. Eur. J. Pharmaceut. Biopharmaceut. Available from: https://doi.org/10.1016/j.ejpb.2017.06.002.

Ghorbani, A., Esmaeilizadeh, M., 2017. Pharmacological properties of *Salvia officinalis* and its components. J. Trad. Complement. Med. 7, 433–440.

Hassikou, R., Oulladi, H., Arahou, M., 2014. Activité antimycosique des extraits du chêne-liège *Quercus suber* sur *Trichophyton rubrum* et *Candida albicans*. Phytothérapie 12, 206–212.

Helander, I.M., Hanna-Leena, A., Latva-Kala, K., Mattila-Sandholm, T., Pol, I., Smid, E. J., et al., 1998. Characterization of the action of selected essential oil components on Gram-negative bacteria. Agric. Food Chem. 46, 3590–3595.

Imelouane, B., El Bachiri, A., Ankit, M., Khedid, K., Wathelet, J.P., Amhamdi, H., 2010. Essential oil composition and antimicrobial activity of *Artemisiaherba-alba* Asso grown in Morocco. Banat's J. Biotechnol. I, 48–55.

Ismaili, R., Lamiri, A., Moustaid, K., 2014. Etude de l'activité antifongique des huiles essentielles de trois plantes aromatiques marocaines. Int. J. Innov. Sci. Res. 12, 499–505.

Jalal, Z., El Atki, Y., Lyoussi, B., Abdellaoui, A., 2015. Phytochemistry of the essential oil of *Melissa officinalis* L. growing wild in Morocco: preventive approach against nosocomial infections. Asian Pacific J. Trop. Biomed. 5, 458–461.

Jalsenjak, V., Peljnajk, S., Kustrak, D., 1987. Microcapsules of sage oil, essential oils content and antimicrobial activity. Pharmacology 42, 419–420.

Kasrati, A., Alaoui Jamali, C., Bekkouche, K., Lahcen, H., Markouk, M., Wohlmuth, H., et al., 2013. Essential oil composition and antimicrobial activity of wild and cultivated mint timija (*Mentha suaveolens* subsp. *timija* (Briq.) Harley), an endemic and threatened medicinal species in Morocco. Nat. Prod. Res. 27, 1119–1122.

Kasrati, A., Alaoui Jamali, C., Fadli, M., Bekkouche, K., Hassani, L., Wohlmuth, H., et al., 2014. Antioxidative activity and synergistic effect of *Thymus satureoides* Coss. Essential oils with cefixime against selected food-borne bacteria. Ind. Crops Prod. 61, 338–344.

Majdouli, K., Soro N'Dédianhoua, K., Elazzouzi, H., Benmaachou, J., Drioiche, A., Zaïr, T., 2015. Chemical composition and antibacterial activity of *A. herba-alba hugueii's* essential oil from South of Morocco. Int. J. Curr. Res. 7, 14397-1404.

Manohar, V., Ingram, C., Gray, J., Talpur, N.A., Echard, B.W., Bagchi, D., et al., 2001. Antifungal activities of origanum oil against *Candida albicans*. Mol. Cell. Biochem. 228, 111–117.

Marmouzi, I., Kaddafi, A., Harhar, H., Gharby, S., Sayah, K., El Madani, N., 2016. Functional composition, antibacterial and antioxidative properties of oil and phenolics from Moroccan *Pennisetum glaucum* seeds. J. Saudi Soc. Agric. Sci. 17, 229–234.

Moreno, S., Scheyer, T., Romano, C.S., Vojnov, A.A., 2006. Antioxidant and antimicrobial activities of rosemary extracts linked to their polyphenol composition. Free Rad. Res. 40, 223–231.
Mouhajir, F., Hudson, J.B., Rejdali, M., Towers, G.H.N., 2001. Multiple antiviral activities of endemic medicinal plants used by berber peoples of Morocco. Pharmaceut. Biol. 39, 364–374.
Okoh, O.O., Sadimenko, A.P., Afolayan, A.J., 2010. Comparative evaluation of the antibacterial activities of the essential oils of *Rosmarinus officinalis* L. obtained by hydrodistillation and solvent free microwave extraction methods. Food Chem. 120, 308–312.
Ouedrhiri, W., Balouiri, M., Bouhdid, S., Moja, S., Chahdi, F.O., Taleb, M., et al., 2016. Mixture design of *Origanum compactum*, *Origanum majorana* and *Thymus serpyllum* essential oils: optimization of their antibacterial effect. Ind. Crops Prod. 89, 1–9.
Oumzil, H., Ghoulami, S., Rhajaoui, M., Ilidrissi, A., Fkih-Tetouani, S., Faid, M., et al., 2002. Antibacterial and antifungal activity of essential oils of *Mentha suaveolens*. Phytother. Res. 16, 727–731.
Ourid, B., Ghanmi, M., Satrani, B., El Ghadraoui, L., Fatmi, M., Balouiri, M., et al., 2016. Effect of harvest date on yield, chemical composition and antimicrobial activity of *Artemisia herba-alba* essential oil. Int. J. Sci. Eng. Res. 7, 600–606.
Ramchoun, M., Sellam, K., Harnafi, H., Alem, C., Benlyas, M., El Rhaffari, L., et al., 2015. Biological investigations of antioxidant and antimicrobial properties of *Thymus satureioides* collected in tafilalet region, south-east of Morocco. Int. J. Pharm. 5, 339–346.
Rankou, H., Culham, A., Jury, S.L., Christenhusz, M., 2013. The endemic flora of Morocco. Phytotaxa 78, 1–69.
Roldán-Gutiérrez, J.M., Ruiz-Jiménez, J., Luque de Castro, M.D., 2010. Composition and antibacterial activity of essential oils obtained from plants of the Lamiaceae family against pathogenic and benefcial bacteria. Talanta 75, 1369–1375.
Ruberto, G., Baratta, M.T., 2000. Antioxidant activity of selected essential oil components in two lipid model systems. Food Chem. 69, 167–174.
Saad, A., Fadli, M., Bouaziz, M., Benharref, A., Mezrioui, N.E., Hassani, L., 2010. Anticandidal activity of the essential oils of *Thymus maroccanus* and *Thymus broussonetii* and their synergism with amphotericin B and fluconazol. Phytomedicine 17, 1057–1060.
Sayah, K., Marmouzi, I., Naceiri Mrabti, H., Cherrah, Y., Faouzi, M.E., 2017. Antioxidant activity and inhibitory potential of *Cistus salviifolius* (L.) and *Cistus monspeliensis* (L.) aerial parts extracts against key enzymes linked to hyperglycemia. BioMed Res. Int. 2789482. Available from: https://doi.org/10.1155/2017/2789482.
Sikkema, J., De Bont, J.A.M., Poolman, B., 1994. Interactions of cyclic hydrocarbons with biological membranes. Biol. Chem. 269, 8022–8028.
Sivropoulou, A., Nikolaou, C., Papanikolaou, E., Kokkini, S., Lanaras, T., Arsenakis, M., 1997. Antimicrobial, cytotoxic and antiviral activities of *Salviafructicosa* essential oil. Agric. Food Chem. 45, 3197–3201.
Springob, K., Kutchan, T.M., 2009. Introduction to the different classes of natural products. In: Osbourn, A.E., Lanzotti, V. (Eds.), Plant-Derived Natural Products Synthesis, Function, and Application. Springer, New York, pp. 3–50.
Talibi, I., Askarne, L., Boubaker, H., Boudyach, E.H., Msanda, F., Saadi, B., et al., 2012. Antifungal activity of Moroccan medicinal plants against citrus sour rot agent *Geotrichum candidum*. Lett. Appl. Microbiol. 55, 155–161.
Tian, F., Li, B., Ji, B., Yang, J., Zhang, G., Chen, Y., et al., 2009. Antioxidant and antimicrobial activities of consecutive extracts from *Galla chinensis*: the polarity affects the bioactivities. Food Chem. 113, 173–179.
Uribe, S., Ramirez, T., Pena, A., 1985. Effects of β-pinene on yeast membrane functions. Bacteriology 161, 1195–1200.

Warda, K., Markouk, M., Bekkouche, K., Larhsini, M., Abbad, A., Romane, A., et al., 2009. Antibacterial evaluation of selected Moroccan medicinal plants against *Streptococcus pneumoniae*. Afr. J. Pharm. Pharmacol. 3, 101–104.

Yangui, T., Bouaziz, M., Dhouib, A., Sayadi, S., 2009. Potential use of Tunisian Pituranthos chloranthus essential oils as a natural disinfectant. Lett. Appl. Microbiol. 48, 112–117.

CHAPTER 6

Antiinflammatory Properties of Herbs in Oral Infection

Sudhanshu Sharma[1], Vivek Kumar Sharma[1], Sankalp Misra[3], Govind Gupta[2], Deepak Dwivedi[2], Brahma N. Singh[1] and Puneet Singh Chauhan[3]

[1]Pharmacognosy & Ethnopharmacology Division, CSIR-National Botanical Research Institute, Lucknow, Uttar Pradesh, India
[2]Department of Microbiology, Barkatullah University, Bhopal, Madhya Pradesh, India
[3]Division of Plant Microbe Interactions, CSIR-National Botanical Research Institute, Lucknow, Uttar Pradesh, India

6.1 INTRODUCTION

Oral mucosa is innately challenged by highly diverse microorganism load (Paster et al., 2006; Sweeney et al., 2004). Oral cavity has various unique anatomical niches including tongue, teeth, jaws, etc. Oral infections are of significant concern related to overall human health (Gift and Atchison, 1995). Its cases are increasing at an alarming rate and expected to become noteworthy in the upcoming future. Synthetics have been in a common practice over years as pacification agent for dental pathogens (Haffajee et al., 2008). However, the use of such drugs has many clinical and pharmacologic disadvantages along with the emergence of resistance in the pathogens. Therefore search for novel drugs or chemicals to combat oral infection are always warranted. In this context, natural products can play a significant role as a promising alternative that may substitute for synthetic drugs. Plants have presented themselves as rich source of effective and safe medicines since prehistoric times (Hatfield, 2004). As illustrated by all ancient literatures and commonly used by shamans, healers, Vaidya, etc. Plants have a special role in our culture and associated as an integral part of our day to day life. Plants versatility lies in their ability to synthesize range of bioactive chemical constituents as secondary metabolites, namely, alkaloids, flavonoids, tannins, terpenoids, saponins, etc. (Bourgaud et al., 2001; Rates, 2001). All these chemical substances are responsible for mediating special benefits to the host such as role in defense (Bennett and Wallsgrove, 1994). These secondary metabolites are

New Look to Phytomedicine
DOI: https://doi.org/10.1016/B978-0-12-814619-4.00006-9

often perceived as potent drug like identity and biological friendliness than their synthetic counterpart. Medicinal plants are also rich source of precursors and treated as a valued entity in both traditional and modern systems of medicine (Balandrin et al., 1993). Active compounds in the herbs often act in synchronous manner, hence undesirable properties are nullified and desirable properties are augmented (Ernst, 2005).

6.2 INFLAMMATION: OLD FRIEND BUT A DREADFUL FOE

Word "Inflammation" is derived from the Latin "inflammatio, inflammare" meaning "Ignite or burn" (Jain et al., 2015). Its cardinal symptoms were described as early as 1st century BC by Aulus Cornelius Celsus as "calor" (heat), "rubor" (redness), "tumor" (swelling), "dolor" (pain), and "Functio laesa" (loss of function) while last one was added by Galen (Rather, 1971). The seriousness can also be roughly assessed by number of signs it manifests. After initial assault to vascularized tissue, due to physical or chemical means, the process starts automatically. Inflammation is caused by release of mediators like vasoactive amines (mast cells, platelets, and basophils produced) which enhance permeation (e.g., Serotonin, histamine), lipid mediators derived from the phospholipids and polyunsaturated fatty acids [e.g., prostaglandins (PGs), leukotrienes (LTs)], and cytokines (e.g., interleukins, tissue necrosis factor) (Christie and Henderson, 2002; Weissmann et al., 1980). All of them try to protect from obnoxious biological and chemical stimuli by diluting, isolating, or destroying the load (Kjekshus, 2015).

6.3 TYPES OF INFLAMMATION

Inflammation may occur due to aseptic reasons such as chemical irritants, blunt force or septic ones' infections. Broadly speaking, inflammation can be grouped into, i.e., acute and chronic phases. Acute inflammation endures for a short duration from few hours to days, with immediate onset, caused mainly by pathogenic attack and chemical or physical insult (Kumar et al., 2004; Ryan and Majno, 1977). Process is initiated by neutrophils, basophils, and other mononuclear cells. Acute inflammation involves two events, namely, vascular and cellular and results in either complete resolution or pave the way for chronic inflammation. Chronic inflammation persists for a longer period from many months to years with delayed onset caused mainly by autoimmune reaction or persistent bodies, with major involvement of fibroblasts and mononuclear cells (Ingersoll et al., 2011). It is generally trailed by tissue granulation and fibrosis.

6.4 INFLAMMATORY MEDIATORS: KEY PLAYERS IN INFLAMMATION

There are various inflammatory mediators that are released at the time of tissue injury and thereby involved in the whole process, they can be grouped majorly as cell- and plasma-derived mediators. Cell- and plasma-derived mediators work in cumulative manner to stimulate cells by binding to certain receptors, activating cells, engaging cells to sites of injury, and stimulating the release of additional soluble mediators (Mazurek et al., 2003). These mediators have relatively short life span or are generally inhibited by regulated intrinsic mechanisms, effectively switching off their response after the process is resolved. Both mediators differ in their origination, while cell-derived mediators such as histamine, LTs, serotonin, PGs, and platelet-activating factor are liberated at the site of inflammation, whereas plasma-derived mediators are generally originated in the liver and later released in the plasma. While cell derived are preformed plasma derived often require preactivation.

Cell-derived mediators include histamine a well-known cell-derived mediator released during inflammation from cells and initiates vasodilatation and increase vascular permeation on the site of injury. The PGs (arachadonic acid derivatives) are a group of fatty acids type cell-derived mediator produced by cells, and accountable for triggering pain at nerve endings. Whereas mediators derived from plasma includes complement and complement-derived kinins, peptides, and platelets, which are released via the classic or alternative pathways of the complement cascade that aids antibodies. Complement-derived peptides (C3, C5a, and C5b) also augment vascular permeability, leukocytes activation, and stimulate mastocytes degranulation. Bradykinin increases vascular permeability and vasodilatation and, importantly, activates phospholipase A2 to liberate arachidonic acid. Bradykinin is also a major mediator involved in the pain stimulation. Platelets play major contribution in hemostasis and thrombosis during inflammation (Larsen and Henson, 1983; Weissmann, 2013).

6.5 ANTIINFLAMMATORY: A RETALIATION PROCESS

Mainly antiinflammatory sources aim at decreasing PG synthesis, competitively inactivating receptors of histamine (H1 receptors) like Cetirizine (Criado et al., 2010). Antiinflammatory drugs like Aspirin inhibit the cyclo-oxygenase (COX) enzyme both 1 and 2, Cox-1 inhibitors like mofezolac can reduce inflammation, but they may also decrease the natural protective mucus lining of the stomach and intestines, Celecoxib inhibit cox-2 thereby reducing the synthesis of prostanoids (Goto et al., 1998; Hawkey, 2001). However, corticosteroids (Glucocorticoids) work by preventing the formation of both PGs and LTs by initiating the release of lipocortin protein (by binding to receptor and switching on the protein synthesis), which by inhibiting phospholipase A2 subsides arachidonic cascade (Tranchant et al., 1989). But all of these approaches suffer serious side effects in one way or the other (Bjarnason and Hayllar, 1993; Schäcke et al., 2002; Wyngaarden and Seevers, 1951).

6.6 ORAL CAVITY: A DYNAMIC BATTLE GROUND

Our body routinely deals with multifold microorganisms compared with number of cells in our body (Paster et al., 2006). Our oral cavity is a nightmare for our immune system, with dynamic range of antigens from air, water, and food constantly being exposed. Our innate immune system is always on guard. During this complex interaction, it must limit commensal organism as well as sweep off pathogens. And these bouts define the fine line between health and disease (Hollister and Weintraub, 1993). As many systemic inflammatory diseases can be traced back to periodontal diseases or poor oral health, and many oral species can have significant role in infection and inflammation at other sites as emphasized earlier by "theory of focal infection" (Gendron et al., 2000; Gibbons, 1998). Oral bacteria can enter the blood stream through discontinuity of epithelium and cause bacteremia (Poveda-Roda et al., 2008). Vulnerability to oral infection is directly associated with genetics, environmental factors, and personal habits (Tabachnick, 1991). Apart from physical damages from cuts, burns, or allergies, inflammation is caused by infection from pathogens lurking in anaerobic pockets of oral cavity. Generally, oral inflammatory disease involves periodontium tissues (periodontitis) and epithelium (gengivitis) (Page and Kornman, 1997; Ratcliff and Johnson, 1999) (Fig. 6.1). Once colonizing the nooks, they form multilayered biofilm

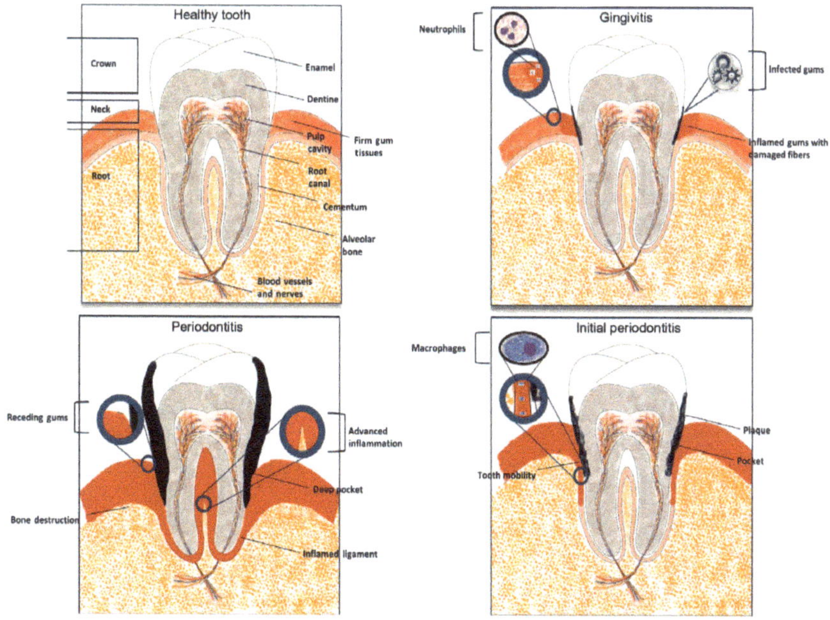

Figure 6.1 Role of inflammation in the development of tooth and gum-related problems such as gingivitis, initial periodontitis, and periodontitis.

and plaque formation (Paster et al., 2001). It gets even worse after neutrophils smite and result in cardinal signs of inflammation (Tonetti et al., 1998). After persistent swelling and redness macrophages joins the strike and launches fierce attack (Hasturk et al., 2012). Inflammatory substances recede gum tissues and can lead to severe damages such as alveolar bone and tooth loose (Cochran, 2008).

6.7 PHYTORESOURCES: NATURAL COMBATANTS

Plants are reported to have gamut of pharmacological active groups and compounds (Balunas and Kinghorn, 2005). Although many of these reports are of academic interest, yet very few will find their entry at any clinical trials (Bent, 2008). Due to multiple active components, herbal formulation's chemical signature is too complex to annotate and many times render them not useful in trials (Bent, 2008). Compilation of such tremendous information resulted from various ancient literatures, traditional medicinal systems, or closely guarded tribal secrets from shamans and healers (Patwardhan et al., 2004; Petrovska, 2012). Table 6.1 lists

Table 6.1 Plant bioresources for the treatment of oral inflammation

Botanical name	Family	Part used	Phytochemical	Mode of action	Model used	References
Azadiracta indica	Meliaceae	Leaves	Azadirachtin, nimbolinin, nimbin, and nimbidin	Suppress the role of macrophage, neutrophils, and inhibit proinflammatory cytokines like NO, PGE_2, and IL-1	Male Wistar rat	Kaur et al. (2004)
Ammi visnaga	Apiaceae	Seed and stem	Khellin and visnagin	Inhibit bacterial strains	*Streptococcus sanguis*, *S. mutans*, and *S. salivarius*	Semyari et al. (2011)
Anacardium occidentale	Anacardiaceae	Bark and leaves	Anacardic acid and cardanol	Inhibit the bacterial strain	*S. mutans*	Akinjogunla et al. (2012)
Anacyclus pyrethrum	Asteraceae	Root	Oxygenated sesquiterpenes	Inhibit bacterial growth	*S. mutans*, *Staphylococcus aureus*, *S. sanguis*, and *Pseudomonas aeruginosa*	Naderi et al. (2012)
Achyranthes aspera	Amaranthaceae	Leaves	Achyranthine and betain	Inhibit growth of oral bacteria	*S. mutans*	Yadav et al. (2016)
Allium cepa	Amaryllidaceae	Whole plant	Thiosulfinates and fructooligosaccharides	Inhibit growth of oral bacteria	*S. mutans*, *S. sorbinus*, and *Porphyromonas gingivalis*	Kim (1997)
Alstonia scholaris	Apocynaceae	Leaves	Picrinine, vallesamine, and scholaricine	Inhibit COX-2 and LOX-5	Mice	Shang et al. (2010)
Barleria prionitis	Acanthaceae	Leaves, bark	Barlerinoside, lupulinoside, and balarenone	Exhibit antibacterial activity	*Lactobacillus rhamnosus*, *S. mutans*, *S. aureus*, *Actinomyces viscoscus*, *S. epidermidis*, *Escherichia coli*, and *Bacillus subtilis*	Diwan and Gadhikar (2012)
Curcuma longa	Zingiberaceae	Rhizome, leaves	Curcumin	Inhibits TNF-α, COX-2, PGE2, NF-κB, and IL-6	Mice	Guimarães et al. (2011)
Camellia sinensis	Theaceae	Leaves	Catechin, EGCG, and CG	Inhibit bacterial strain	*Aggregatibacter actinomycetemcomitans*, *Prevotella intermedia*, and *P. gingivalis*	Araghizadeh et al. (2013)

Centella asiatica and *Punica granatum*	Apiaceae	Leaves	Asiaticosides and brahminoside	Inhibit IL-1β and IL-6	Human	Sastravaha et al. (2005)
Caesalpinia sappan	Leguminoceae	Stem	Xanthone, coumarin, chalcones, and brazilin	Suppress proinflammatory cytokines	Primery human chondrosarcoma cells lines	Wu et al. (2011)
Cinnammomum zeylcanium	Lauraceae	Bark	Cinnamaldehyde and eugenol	Inhibit periodontal bacterial strain	*S. mutans*, *S. mitis*, *S. salivareus*, *P. gingivalis*, and *Actinomycetemcomitans*	Zainal-Abidin et al. (2013)
Calendula officinalis	Asteraceae	Flower	Laurate, myristate, and palmitate	Inhibit TNF-α, IL-1β, IL-6, and COX-2 expression	Mice	Preethi et al. (2009)
Cymbopogon citratus	Poaceae	Essential oil of leaves	Geranniol, myrcene, α-pinene, and linalool	Inhibit the growth of cariogenic bacteria	*S. aureus*, *S. mutans*, *S. epidermidis*, and *C. albicans*	Almeida et al. (2013)
Humulus lupulus	Cannabaceae	Polyphenols	Kaempferol, astragalin, MPPG, and isoquercitrin	Exhibit antimicrobial activity against cariogenic bacterial strain and inhibit PGE-2, IL-1, IL-6, and IL-8 expression	*P. gingivalis* and human gastric epithelial cell line	Inaba et al. (2008)
Krameria lappacea	Krameriaceae	Root	Lignin derivative	Inhibits COX-2, NF-κB, and leukocyte infiltration	Human embryonic kidney cells	Baumgartner et al. (2011)
Matricaria chamomilla	Asteraceae	Flower	Bisabolol, chamazulene, and umbelliferone	Reduce dental plaque and gingivitis	Human volunteers	Pourabbas and Delazar (2010)
Psidium gaujava	Myrtaceae	Leaves	Isoflavonoid, neringenin, gallic acid, rutin, and catechin	Inhibit COX-2, iNOS and NF-kB expression	Mouse RAW264.7 cells	Choi et al. (2008)
Punica granatum	Punicaceae	Fruit	Ascorbic acid, gallic acid, caffeic acid, catechin, rutin, and egallic acid	Inhibit PGE-2, NF-κB, and NO production	Rabbit chondrocytes from articular cartilage	Shukla et al. (2008)

(Continued)

Table 6.1 (Continued)

Botanical name	Family	Part used	Phytochemical	Mode of action	Model used	References
Salvia oficinalis	Lamiaceae	Leaves and sage infusion	1,8-Cineole, borneol, thujone, and rosmarinic acid	Inhibit PMA/I-induced IL-6 and IL-8	Human gingival fibroblast	Ehrnhöfer-Ressler et al. (2013)
Salvadora persica	Salvadoraceae	Stick	Butanediamide, N1,N4-bis (phenylmethyl)-2(S)-hydroxybutanediamide, *N*-benzyl-2-phenylacetamide, *N*-benzylbenzamide and benzylurea	Inhibit bacterial growth	*S. aureus*, *S. mutans*, *S. faecalis*, *S. pyogenis*, *L. acidophilus*, *P. aeruginosa*, and *C. albicans*	Al-Bayati and Sulaiman (2008)
Sanguinaria canadensis	Papaveraceae	Sanguinarine	Sanguinarine	Inhibits the growth of bacterial growth	*Streptococci*, *Actinomyces*, *Enterics*, *Pseudomonads*, *Staphylococci*, and yeast/fungi	Godowski et al. (1995)
Syzygium aromaticum	Myrtaceae	Essential oil of buds	Eugenol and β-caryophyllene	Show antibacterial effect	All strains of *Streptococcus*	Moon et al. (2011)
Vaccinium macrocarpon	Utricaceae	Root, leaves, and stalk	Acetophenone, histamine, chlorogenic acid, and kaempferol	Inhibit cariogenic bacterial strain	*S. mutans* and *S. sobrinus*	Bonifait and Grenier (2010)

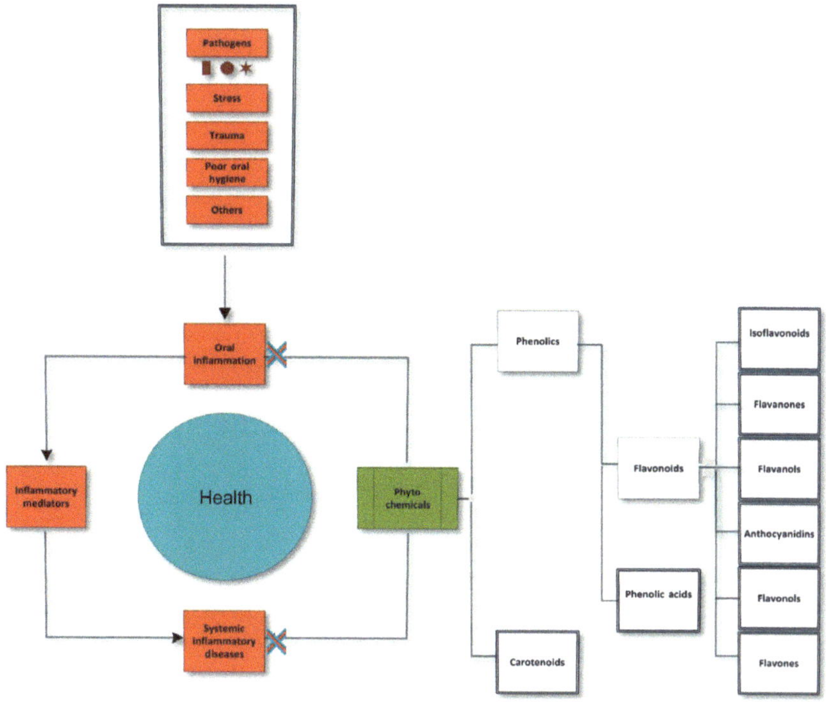

Figure 6.2 Plant-derived therapeutic phytochemicals and their role in the prevention of oral inflammation.

common antiinflammatory plant sources with family and their mode of action. In toxicity studies, the antiinflammatory decoctions from plants show great appeal when compared with modern chemically synthesized similar drugs in terms of lower side effects, higher bioavailability, to name a few (Wachtel-Galor et al., 2011). The discipline of plant-based medicines is significantly much more advance than modern medicine and is generally been practiced from thousands of years with very less changes during the course of time, significance of this art is currently appreciated by whole world (Fig. 6.2).

Garlic (*Allium sativum*) is considered to have its origin in Central Asia and belongs to the Alliacae family. Since ancient times, garlic has been known for its medicinal and therapeutic properties that are evident in Chinese and Indian cultures (Block, 2010). Medicinal use of garlic is now becoming a common practice in both developing and developed countries, 11 inclusively as a remedy for toothache and oral problems (Groppo et al., 2002). *Flos lonicerae*, commonly known as honeysuckle flower belonging to Caprifoliaceae family, is a widely used herb in China for the treatment of infection by exopathogenic wind-heat or epidemic febrile diseases (Ko et al., 2006; Chai et al., 2005). It also has antimicrobial effect against oral pathogens including *Streptococcus mutans*, *Actinomyces viscosus*, and *Bacteroides melaninogenicus* (Wong et al., 2010). Herbal products are being used since long before in the treatment of various oral lesions such as lichen planus, oral submucous fibrosis, leukoplakia, pemphigus vulgaris, aphthous ulcer, candidiasis, and oral infections (Nair and Shiva Prasad, 2017). *Ocimum sanctum* Linn. (Tulsi) is recognized as one of the India's greatest healing herbs. It is also used in preventing tooth decay as well as in reducing various types of oral infections. Carracrol, Tetpene, and Sesquiterpene b-caryophyllene are antibacterial agents present in this plant (Nair and Shiva Prasad, 2017). Addition of these constituents in food additive has also been approved by Food and Drug Administration (FDA). *O. sanctum* leaves also considered as pain reliever due to the presence of eugenol (1-hydroxyl-2-methoxy-4 allyl benzene), which act as COX-2 inhibitor (Nair and Shiva Prasad, 2017). Oil of *O. sanctum* contains eugenol and linalool, which plays an important role in the control of oral candidiasis, caused by *Candida albicans* and *Candida tropicalis*. Tulsi extracts have antimicrobial potential as they are effective against *S. mutans* that causes dental caries. Polyphenol rosmarinic acid present in tulsi can act as a powerful antioxidant, so this property can therapeutically be utilized in treating common oral precancerous lesions and conditions. A number of herbal extracts have been found to show promising antimicrobial effects against oral pathogens both *in vitro* and *in vivo* (Salam et al., 2015). *Salvia officinalis*, commonly known as "Sage," is one of the most commonly used herbs in traditional medicine (Rodrigues et al., 2012). It has been reported that sage exhibited a range of therapeutic activities including antibacterial, antiviral, antifungal, and antioxidant effects (Samuels et al., 2012). Sage extract contains alpha- and beta-thujone, camphor, and cineole along with rosmarinic acid, tannins, and flavonoids owing to which it is effective as antiinflammatory product against gingival inflammation and mouth ulcers.

6.8 CONCLUSION

Oral problems are often neglected, but with increasing evidences we can safely conclude that oral health is a mirror image of your systemic health (Li et al., 2000). Oral symptoms precede and provide the chronic platform for many immunological disasters. Injurious agents are typically destroyed, diluted or walled off during highly regulated process of inflammation. It is one of the most important mechanisms of immune system and it will not be inappropriate to say "organisms' mortality depends on it." Many inflammatory diseases have intermingled roots, which are yet to be fully resolved (Meyer et al., 2008). It is well known that huge chunk of natural products mainly act as starting material for drug discovery and resulted from the diverse structures and intricate carbon skeletons. Plants have colossal potential to meet the needs of future endeavors and recent researches have shown the appreciable potential of natural products in delivering the promised. The major benefits of herbal products are safety, easy availability, increased shelf life, cost effectiveness, and lack of microbial resistance so far. Today is the era of evidence-based medicine—hence any medicine to be used on humans has to undergo extensive research both *in vitro* and *in vivo*. Many researches have enlightened the promising future of herbal products under *in vitro* condition, but preclinical and clinical trials are needed to evaluate the biocompatibility and safety factor before they can conclusively be recommended as natural solutions to various oral problems.

ACKNOWLEDGEMENTS

This chapter is dedicated to our beloved Late Professor Vinod Singh, Head, Department of Microbiology, Barkatullah University, Bhopal. Professor Singh's enthusiasm for teaching and his commitment for development of research in Sciences will always inspire us. As his students and colleagues, we shall carry on not just his love but the spirit with which he worked and lived his life.

REFERENCES

Akinjogunla, O., Adenugba, I., Jumbo, O., 2012. In-vitro antibacterial evaluation of ethanolic stem crude extracts of *Anacardium occidentale* linn.(anacardiaceae) on *Streptococcus mutans* associated with dental caries. Sci. J. Microbiol. 1 (3), 71–81.

Al-Bayati, F.A., Sulaiman, K.D., 2008. In vitro antimicrobial activity of *Salvadora persica* L. extracts against some isolated oral pathogens in Iraq. Turk. J. Biol. 32 (1), 57–62.

Almeida, R., Akisue, G., Cardoso, L., Junqueira, J., Jorge, A., 2013. Antimicrobial activity of the essential oil of *Cymbopogon citratus* (DC) Stapf. on Staphylococcus spp., *Streptococcus mutans* and Candida spp. Rev. Bras. Plantas Med. 15 (4), 474–482.

Araghizadeh, A., Kohanteb, J., Fani, M.M., 2013. Inhibitory activity of green tea (*Camellia sinensis*) extract on some clinically isolated cariogenic and periodontopathic bacteria. Med. Princ. Pract. 22 (4), 368–372.

Balandrin, M.F., Kinghorn, A.D., Farnsworth, N.R., 1993. Plant-Derived Natural Products in Drug Discovery and Development: An Overview. ACS Publications.

Balunas, M.J., Kinghorn, A.D., 2005. Drug discovery from medicinal plants. Life Sci. 78 (5), 431–441.
Baumgartner, L., Sosa, S., Atanasov, A.G., Bodensieck, A., Fakhrudin, N., Bauer, J., et al., 2011. Lignan derivatives from *Krameria lappacea* roots inhibit acute inflammation in vivo and pro-inflammatory mediators in vitro. J. Nat. Prod. 74 (8), 1779–1786.
Bennett, R.N., Wallsgrove, R.M., 1994. Secondary metabolites in plant defence mechanisms. New Phytol. 127 (4), 617–633.
Bent, S., 2008. Herbal medicine in the United States: review of efficacy, safety, and regulation. J. Gen. Intern. Med. 23 (6), 854–859.
Bjarnason, I., Hayllar, J., 1993. Side effects of nonsteroidal anti-inflammatory drugs on the small and large intestine in humans. Gastroenterology 104 (6), 1832–1847.
Block, E., 2010. Garlic and Other Alliums: The Lore and the Science. Royal Society of Chemistry.
Bonifait, L., Grenier, D., 2010. Cranberry polyphenols: potential benefits for dental caries and periodontal disease. J. Can. Dent. Assoc. 76, a130.
Bourgaud, F., Gravot, A., Milesi, S., Gontier, E., 2001. Production of plant secondary metabolites: a historical perspective. Plant Sci. 161 (5), 839–851.
Chai, X.-Y., Li, S.-L., Li, P., 2005. Quality evaluation of *Flos lonicerae* through a simultaneous determination of seven saponins by HPLC with ELSD. J. Chromatogr. A 1070 (1–2), 43–48.
Choi, S.Y., Hwang, J.H., Park, S.Y., Jin, Y.J., Ko, H.C., Moon, S.W., et al., 2008. Fermented guava leaf extract inhibits LPS-induced COX-2 and iNOS expression in Mouse macrophage cells by inhibition of transcription factor NF-κB. Phytother. Res. 22 (8), 1030–1034.
Christie, P.E., Henderson, J.W., 2002. Lipid inflammatory mediators: leukotrienes, prostaglandins, platelet-activating factor. Clin. Allergy Immunol. 16, 233–254.
Cochran, D.L., 2008. Inflammation and bone loss in periodontal disease. J. Periodontol. 79 (8S), 1569–1576.
Criado, P.R., Criado, R.F.J., Maruta, C.W., Machado Filho, C.D.A., 2010. Histamine, histamine receptors and antihistamines: new concepts. Anais Bras. Dermatol. 85 (2), 195–210.
Diwan, P., Gadhikar, Y., 2012. Assessment of phytochemical composition and antibacterial activity of different extracts of *Barleria prionitis* leaves against oral microflora to improve dental hygiene. Asian J. Pharm. Clin. Res. 5 (2), 182–184.
Ehrnhöfer-Ressler, M.M., Fricke, K., Pignitter, M., Walker, J.M., Walker, J., Rychlik, M., et al., 2013. Identification of 1, 8-cineole, borneol, camphor, and thujone as anti-inflammatory compounds in a *Salvia officinalis* L. infusion using human gingival fibroblasts. J. Agric. Food Chem. 61 (14), 3451–3459.
Ernst, E., 2005. The efficacy of herbal medicine – an overview. Fundam. Clin. Pharmacol. 19 (4), 405–409.
Gendron, R., Grenier, D., Maheu-Robert, L.-F., 2000. The oral cavity as a reservoir of bacterial pathogens for focal infections. Microb. Infect. 2 (8), 897–906.
Gibbons, R.V., 1998. Germs, Dr. Billings, and the theory of focal infection. Clin. Infect. Dis. 27 (3), 627–633.
Gift, H.C., Atchison, K.A., 1995. Oral health, health, and health-related quality of life. Med. Care NS57–NS77.
Godowski, K.C., Wolff, E.D., Thompson, D.M., Housley, C.J., Polson, A.M., Dunn, R. L., et al., 1995. Whole mouth microbiota effects following subgingival delivery of sanguinarium. J. Periodontol. 66 (10), 870–877.
Goto, K., Ochi, H., Yasunaga, Y., Matsuyuki, H., Imayoshi, T., Kusuhara, H., et al., 1998. Analgesic effect of mofezolac, a non-steroidal anti-inflammatory drug, against

phenylquinone-induced acute pain in mice. Prostaglandins Other Lipid Mediat. 56 (4), 245–254.

Groppo, F., Ramacciato, J., Simoes, R., Florio, F., Sartoratto, A., 2002. Antimicrobial activity of garlic, tea tree oil, and chlorhexidine against oral microorganisms. Int. Dent. J. 52 (6), 433–437.

Guimarães, M.R., Coimbra, L.S., de Aquino, S.G., Spolidorio, L.C., Kirkwood, K.L., Rossa, C., 2011. Potent anti-inflammatory effects of systemically administered curcumin modulate periodontal disease in vivo. J. Periodont. Res. 46 (2), 269–279.

Haffajee, A., Patel, M., Socransky, S., 2008. Microbiological changes associated with four different periodontal therapies for the treatment of chronic periodontitis. Mol. Oral Microbiol. 23 (2), 148–157.

Hasturk, H., Kantarci, A., Van Dyke, T.E., 2012. Oral inflammatory diseases and systemic inflammation: role of the macrophage. Front. Immunol. 3, 118.

Hatfield, G., 2004. Encyclopedia of Folk Medicine: Old World and New World Traditions. ABC-CLIO.

Hawkey, C., 2001. COX-1 and COX-2 inhibitors. Best Pract. Res. Clin. Gastroenterol. 15 (5), 801–820.

Hollister, M., Weintraub, J., 1993. The association of oral status with systemic health, quality of life, and economic productivity. J. Dent. Educ. 57 (12), 901–912.

Inaba, H., Tagashira, M., Honma, D., Kanda, T., Kou, Y., Ohtake, Y., et al., 2008. Identification of hop polyphenolic components which inhibit prostaglandin E2 production by gingival epithelial cells stimulated with periodontal pathogen. Biol. Pharm. Bull. 31 (3), 527–530.

Ingersoll, M.A., Platt, A.M., Potteaux, S., Randolph, G.J., 2011. Monocyte trafficking in acute and chronic inflammation. Trends Immunol. 32 (10), 470–477.

Jain, P., Pandey, R., Shukla, S.S., 2015. Inflammation: Natural resources and its applications. Springer India, pp. 156.

Kaur, G., Sarwar Alam, M., Athar, M., 2004. Nimbidin suppresses functions of macrophages and neutrophils: relevance to its antiinflammatory mechanisms. Phytother. Res. 18 (5), 419–424.

Kim, J.-H., 1997. Anti-bacterial action of onion (*Allium cepa* L.) extracts against oral pathogenic bacteria. J. Nihon Univ. School Dent. 39 (3), 136–141.

Kjekshus, J., 2015. Inflammation: friend and foe. EBioMedicine 2 (7), 634–635.

Ko, H.-C., Wei, B.-L., Chiou, W.-F., 2006. The effect of medicinal plants used in Chinese folk medicine on RANTES secretion by virus-infected human epithelial cells. J. Ethnopharmacol. 107 (2), 205–210.

Kumar, R., Clermont, G., Vodovotz, Y., Chow, C.C., 2004. The dynamics of acute inflammation. J. Theor. Biol. 230 (2), 145–155.

Larsen, G.L., Henson, P.M., 1983. Mediators of inflammation. Annu. Rev. Immunol. 1 (1), 335–359.

Li, X., Kolltveit, K.M., Tronstad, L., Olsen, I., 2000. Systemic diseases caused by oral infection. Clin. Microbiol. Rev. 13 (4), 547–558.

Mazurek, T., Zhang, L., Zalewski, A., Mannion, J.D., Diehl, J.T., Arafat, H., et al., 2003. Human epicardial adipose tissue is a source of inflammatory mediators. Circulation 108 (20), 2460–2466.

Meyer, M.S., Joshipura, K., Giovannucci, E., Michaud, D.S., 2008. A review of the relationship between tooth loss, periodontal disease, and cancer. Cancer Causes Control 19 (9), 895–907.

Moon, S.-E., Kim, H.-Y., Cha, J.-D., 2011. Synergistic effect between clove oil and its major compounds and antibiotics against oral bacteria. Arch. Oral Biol. 56 (9), 907–916.

Naderi, N.J., Niakan, M., Khodadadi, E., 2012. Determination of antibacterial activity of *Anacyclus pyrethrum* extract against some of the oral bacteria: an in vitro study. J. Dent. Shiraz Univ. Med. Sci. 13 (2), 59–63.
Nair, S.K., Shiva Prasad, B., 2017. Holy herb tulsi as a cure for oral and periodontal disease – a review. EC Dent. Sci. 10, 106–109.
Page, R.C., Kornman, K.S., 1997. The pathogenesis of human periodontitis: an introduction. Periodontology 2000 14 (1), 9–11.
Paster, B.J., Boches, S.K., Galvin, J.L., Ericson, R.E., Lau, C.N., Levanos, V.A., et al., 2001. Bacterial diversity in human subgingival plaque. J. Bacteriol. 183 (12), 3770–3783.
Paster, B.J., Olsen, I., Aas, J.A., Dewhirst, F.E., 2006. The breadth of bacterial diversity in the human periodontal pocket and other oral sites. Periodontology 2000 42 (1), 80–87.
Patwardhan, B., Vaidya, A.D., Chorghade, M., 2004. Ayurveda and natural products drug discovery. Curr. Sci. 789–799.
Petrovska, B.B., 2012. Historical review of medicinal plants' usage. Pharmacog. Rev. 6 (11), 1.
Pourabbas, R., Delazar, A., 2010. The effect of German chamomile mouthwash on dental plaque and gingival inflammation. Iran. J. Pharm. Res. 105–109.
Poveda-Roda, R., Jimenez, Y., Carbonell, E., Gavalda, C., Margaix-Munoz, M., Sarrión-Pérez, G., 2008. Bacteremia originating in the oral cavity. A review. Med. Oral Patol. Oral Cirugia Bucal 13 (6), E355–E362.
Preethi, K.C., Kuttan, G., Kuttan, R., 2009. Anti-inflammatory activity of flower extract of *Calendula officinalis* Linn. and its possible mechanism of action. Indian J. Exp. Biol. 47 (2), 113–120.
Ratcliff, P.A., Johnson, P.W., 1999. The relationship between oral malodor, gingivitis, and periodontitis. A review. J. Periodontol. 70 (5), 485–489.
Rates, S.M.K., 2001. Plants as source of drugs. Toxicon 39 (5), 603–613.
Rather, L., 1971. Disturbance of function (functio laesa): the legendary fifth cardinal sign of inflammation, added by Galen to the four cardinal signs of Celsus. Bull. N Y Acad. Med. 47 (3), 303.
Rodrigues, M.R.A., Kanazawa, L.K.S., das Neves, T.L.M., da Silva, C.F., Horst, H., Pizzolatti, M.G., et al., 2012. Antinociceptive and anti-inflammatory potential of extract and isolated compounds from the leaves of *Salvia officinalis* in mice. J. Ethnopharmacol. 139 (2), 519–526.
Ryan, G.B., Majno, G., 1977. Acute inflammation. A review. Am. J. Pathol. 86 (1), 183.
Salam, R., Sarker, B.K., Haq, M.R., Khokon, J.U., 2015. Antimicrobial activity of medicinal plant for oral health and hygiene. Int. J. Nat. Soc. Sci. 1 (2), 1–12.
Samuels, N., Grbic, J.T., Saffer, A.J., Wexler, I.D., Williams, R., 2012. Effect of an herbal mouth rinse in preventing periodontal inflammation in an experimental gingivitis model: a pilot study. Compend. Contin. Educ. Dent. 33 (3), 204–211.
Sastravaha, G., Gassmann, G., Sangtherapitikul, P., Grimm, W.-D., 2005. Adjunctive periodontal treatment with *Centella asiatica* and *Punica granatum* extracts in supportive periodontal therapy. J. Int. Acad. Periodontol. 7 (3), 70–79.
Schäcke, H., Döcke, W.-D., Asadullah, K., 2002. Mechanisms involved in the side effects of glucocorticoids. Pharmacol. Ther. 96 (1), 23–43.
Semyari, H., Owlia, P., Farhadi, S., Tabrizi, S.M., 2011. Evaluation of antimicrobial effect of *Ammi visnaga* against oral streptococci. J. Microbiol. Antimicrob. 3 (5), 126–129.
Shang, J.-H., Cai, X.-H., Feng, T., Zhao, Y.-L., Wang, J.-K., Zhang, L.-Y., et al., 2010. Pharmacological evaluation of *Alstonia scholaris*: anti-inflammatory and analgesic effects. J. Ethnopharmacol. 129 (2), 174–181.

Shukla, M., Gupta, K., Rasheed, Z., Khan, K.A., Haqqi, T.M., 2008. Bioavailable constituents/metabolites of pomegranate (*Punica granatum* L) preferentially inhibit COX2 activity ex vivo and IL-1beta-induced PGE 2 production in human chondrocytes in vitro. J. Inflammat. 5 (1), 9.

Sweeney, L.C., Dave, J., Chambers, P.A., Heritage, J., 2004. Antibiotic resistance in general dental practice—a cause for concern? J. Antimicrob. Chemother. 53 (4), 567–576.

Tabachnick, W.J., 1991. Genetic control of oral susceptibility to infection of Culicoides variipennis with bluetongue virus. Am. J. Trop. Med. Hyg. 45 (6), 666–671.

Tonetti, M.S., Imboden, M.A., Lang, N.P., 1998. Neutrophil migration into the gingival sulcus is associated with transepithelial gradients of interleukin-8 and ICAM-1. J. Periodontol. 69 (10), 1139–1147.

Tranchant, C., Braun, S., Warter, J., 1989. Mechanism of action of glucocorticoids: role of lipocortins. Rev. Neurol. 145 (12), 813–818.

Wachtel-Galor, S., Benzie, I.F.F., 2011. Herbal medicine: An introduction to its history, usage, regulation, current trends, and research needs. In: Benzie, I.F.F., Wachtel-Galor, S. (Eds.), Herbal Medicine: Biomolecular and Clinical Aspects, 2nd edition CRC Press/Taylor & Francis, Boca Raton (FL), Chapter 1.

Weissmann, G., 2013. Mediators of Inflammation. Springer Science & Business Media.

Weissmann, G., Smolen, J.E., Korchak, H.M., 1980. Release of inflammatory mediators from stimulated neutrophils. N. Engl. J. Med. 303 (1), 27–34.

Wong, R., Hägg, U., Samaranayake, L., Yuen, M., Seneviratne, C., Kao, R., 2010. Antimicrobial activity of Chinese medicine herbs against common bacteria in oral biofilm. A pilot study. Int. J. Oral Maxillofac. Surg. 39 (6), 599–605.

Wu, S.Q., Otero, M., Unger, F.M., Goldring, M.B., Phrutivorapongkul, A., Chiari, C., et al., 2011. Anti-inflammatory activity of an ethanolic *Caesalpinia sappan* extract in human chondrocytes and macrophages. J. Ethnopharmacol. 138 (2), 364–372.

Wyngaarden, J., Seevers, M., 1951. The toxic effects of anti-histaminic drugs. J. Am. Med. Assoc. 145 (5), 277–282.

Yadav, R., Rai, R., Yadav, A., Pahuja, M., Solanki, S., Yadav, H., 2016. Evaluation of antibacterial activity of *Achyranthes aspera* extract against *Streptococcus mutans*: an in vitro study. J. Adv. Pharm. Technol. Res. 7 (4), 149.

Zainal-Abidin, Z., Mohd-Said, S., Adibah, F., Majid, A., Mustapha, W.A.W., Jantan, I., 2013. Anti-bacterial activity of cinnamon oil on oral pathogens. Open Conference Proc. J. 4 (Suppl-2, M4), 12–16.

CHAPTER 7

Bioactive Molecules, Pharmacology and Future Research Trends of *Ganoderma lucidium* as a Cancer Chemotherapeutic Agent

Temitope O. Lawal[1,2], Sheila M. Wicks[3], Angela I. Calderon[4] and Gail B. Mahady[5]

[1]Department of Pharmaceutical Microbiology, University of Ibadan, Ibadan, Nigeria
[2]Schlumberger Faculty for the Future Fellow, College of Pharmacy, WHO Collaborating Centre for Traditional Medicine, Department of Pharmacy Practice, University of Illinois at Chicago, Chicago, IL, United States
[3]Department of Molecular and Cellular Medicine, Rush University, Chicago, IL, United States
[4]Department of Drug Discovery and Development, Harrison School of Pharmacy, Auburn University, Auburn, AL, United States
[5]College of Pharmacy, WHO Collaborating Centre for Traditional Medicine, Department of Pharmacy Practice, University of Illinois at Chicago, Chicago, IL, United States

7.1 INTRODUCTION

Ganoderma mushrooms, commonly known as "Reishi" (Japan) or "Ling-zhi" (Chinese), have been used in traditional Chinese medicine (TCM) for more than 2000 years (Cör et al., 2018; Wachtel-Galor et al., 2011; Wasser et al., 2005; Zeng et al., 2018). Traditional preparations from Reishi were used to enhance stamina, immunity, and to treat inflammatory diseases such as arthritis, asthma, bronchitis, hepatitis, and nephritis, and were also used for the treatment of various cancers (Boh et al., 2007; Cör et al., 2018; Wachtel-Galor et al., 2011). The genus *Ganoderma* is comprised of more than 200 species; however, only *G. lucidum* (W. Curtis Fr.) P. Karst (GL) and *G. sinense* Zhao, Xu et Zhang (GS) are recognized in the Chinese Pharmacopoeia as "Ling-zhi" having similar pharmacological effects (Xie et al., 2012). In TCM, Ling-zhi is used for nourishing and tonifying Qi of Zang organs, as well as nourishing Qi and blood (Meng et al., 2014). Other indications include conditions of a deficiency of Qi and blood, such as stress, weakness, body fatigue, and mental fatigue. *Ganoderma*

New Look to Phytomedicine
DOI: https://doi.org/10.1016/B978-0-12-814619-4.00007-0

preparations are used alone, or combination for the management of amnesia, anxiety, heart palpitations, insomnia, or dream-disturbed sleep (Meng et al., 2014). The pharmacological effects of Ling-zhi include its' use as a tonic to enhance vigor and vital energy, strengthening heart function; improve ageing and cognition (Wachtel-Galor et al., 2011; Zeng et al., 2018).

Reishi mushrooms are unusual as they are primarily used as a medicinal agent rather than a food, and are available as crude drug, powders, dietary supplements, and teas (Wachtel-Galor et al., 2011; Yuen and Gohel, 2005). The pharmaceutical and traditional medicine products are prepared from the mycelia, spores, and fruiting bodies (Yuen and Gohel, 2005). Reishi products are marketed worldwide as traditional medicines, as well as dietary supplements in the United States (Yuen and Gohel, 2005). Reishi products are reported to reduce blood glucose levels and enhance the immune system. They also have antimicrobial activities and have hepatoprotective effects and are also used for the treatment of cancer. The purpose of this work was to review some of the recent studies on GL and its purified bioactive constituents in cancer.

7.2 BIOLOGICALLY ACTIVE POLYSACCHARIDES OF *GANODERMA LUCIDUM*

There are over 400 chemical compounds that have been identified from the mycelia, fruiting bodies, and spores of GL including amino acids, lignans, phenols, polysaccharides, sterols, and terpenes (Cör et al., 2018; Da et al., 2012; Yuen and Gohel, 2005; Zeng et al., 2018). Out of hundreds of compounds identified, to date only the polysaccharides, peptidoglycan, and triterpenes are considered to be the active constituents (see review by Cör et al., 2018; Da et al., 2012; Yuen and Gohel, 2005; Zeng et al., 2018). Many of the chemical constituents isolated from GL have biological activity; therefore it is possible that the bioactive compounds may have additive or synergistic effects that contribute to the overall effects of Reishi.

Interestingly, the two species of *Ganoderma* which are official in the Chinese Pharmacopoeia have different triterpene constituents (Da et al., 2012; Soccol et al., 2016; Xie et al., 2012). A 2012 study investigated 32 batches of commercial Ling-zhi samples from GL and GS to determine the differences in the chemistry of the two species (Da et al., 2012). This study showed that the most striking chemical difference between the two species was a significant lack of many triterpenes in GS. Interestingly, in GL, hundreds of triterpenes have been isolated, and many with anticancer effects (Da et al., 2012). One such group of bioactive triterpenes are the ganoderic acids (GAs) A–K, as well as the ganoderenic acids A–D (Table 7.1). No difference in the polysaccharide constituents was observed, but a difference in the polysaccharide content was seen between the two species (Da et al., 2012). This was confirmed by Xie et al. (2012), who found that the polysaccharides from the two species were similar. Interestingly, polysaccharides are known to stimulate the immune system by activating cellular and humoral components and increasing the function of macrophages, mononuclear cells, and neutrophils (Ren et al., 2012). *Ganoderma* polysaccharides have experimentally shown a wide range of therapeutic effects including antiinflammatory, hypoglycemic, antiulcer, antitumor, and immune-enhancing activities (Bao et al., 2012; Cör et al., 2018; Unlu et al., 2016; Wachtel-Galor et al., 2011; Yuen and Gohel, 2005). Along with immune effects, polysaccharides, such as Ganoderan (MW 20 kDa), which is primarily composed of glucose units, has excellent antitumor activities in tumor-bearing mice, but these polysaccharides may also contain xylose, mannose, galactose, or fucose (Bao et al., 2012; Lee et al., 1995; Ren et al., 2012).

Polysaccharides isolated from GL and GS have been shown to enhance the release of cytokines IL-1α, IL-6, IL-10, as well as TNF-α from RAW 264.7 (abelson murine leukemia virus cells) macrophages (Meng et al., 2014). The results further demonstrated that these polysaccharides increased macrophage function, by enhancing phagocytosis, as well as increasing the release of nitric oxide and the cytokines IL-1α, IL-6, IL-10, and TNF-α (Meng et al., 2014). In addition, the concentration of polysaccharides needed for this effect was 19–300 mg/mL, which corresponds with the commonly used TCM dose of GL (50–300 g/day) in clinical and human studies, further supporting their action (Meng et al., 2014). Currently, a wide range of *Ganoderma* polysaccharides are available as traditional medicines or dietary supplements on the world market for the treatment of cancer, diabetes, and liver disorders.

In addition to the polysaccharide constituents of *Ganoderma*, biologically active peptidoglycans have also been isolated from the extracts, including a proteoglycan with antiviral activity (Cör et al., 2018; Li et al., 2005; Zeng et al., 2018; Zhang et al., 2010). A substance named GLIS (immunomodulating substance of GL) was isolated by Ji et al. (2007); one named PGY, a water-soluble glycopeptide fractionated and purified from

Table 7.1 Structures of selected triterpenes isolated from *G. lucidum* with activity against various cancer cell lines

<table>
<tr><th>Triterpenes isolated from G. lucidum</th><th>Biological activity</th><th>References</th></tr>
<tr><td>
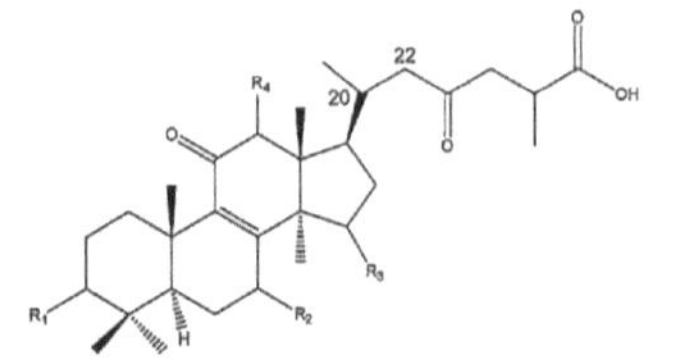

<table>
<tr><th></th><th>R1</th><th>R2</th><th>R3</th><th>R4</th><th>Double Bound</th></tr>
<tr><td>GEC</td><td>β-OH</td><td>β-OH</td><td>α-OH</td><td>H</td><td>$\Delta_{20,22}$</td></tr>
<tr><td>GC2</td><td>β-OH</td><td>β-OH</td><td>α-OH</td><td>H</td><td>-</td></tr>
<tr><td>GG</td><td>β-OH</td><td>β-OH</td><td>=O</td><td>β-OH</td><td>-</td></tr>
<tr><td>GEB</td><td>β-OH</td><td>β-OH</td><td>=O</td><td>H</td><td>$\Delta_{20,22}$</td></tr>
<tr><td>GB</td><td>β-OH</td><td>β-OH</td><td>=O</td><td>H</td><td>-</td></tr>
<tr><td>GEA</td><td>=O</td><td>β-OH</td><td>α-OH</td><td>H</td><td>$\Delta_{20,22}$</td></tr>
<tr><td>GK</td><td>β-OH</td><td>β-OH</td><td>=O</td><td>β-OAc</td><td>-</td></tr>
<tr><td>GA</td><td>=O</td><td>β-OH</td><td>α-OH</td><td>H</td><td>-</td></tr>
<tr><td>GH</td><td>β-OH</td><td>=O</td><td>=O</td><td>β-OAc</td><td>-</td></tr>
<tr><td>GED</td><td>=O</td><td>β-OH</td><td>=O</td><td>H</td><td>$\Delta_{20,22}$</td></tr>
<tr><td>GD</td><td>=O</td><td>β-OH</td><td>=O</td><td>H</td><td>-</td></tr>
<tr><td>GF</td><td>=O</td><td>=O</td><td>=O</td><td>β-OAc</td><td>-</td></tr>
</table>
Twelve triterpenes, namely ganoderic acid C2, ganoderic acid G, ganoderic acid B, ganoderic acid K, ganoderic acid A (GA), ganoderic acid H, ganoderic acid D, ganoderic acid F, and ganoderenic acid C, ganoderenic acid B, ganoderenic acid A, ganoderenic acid D
</td><td>Anticancer and immune enhancing</td><td>Da et al. (2012), Wu et al. (2017)</td></tr>
<tr><td>
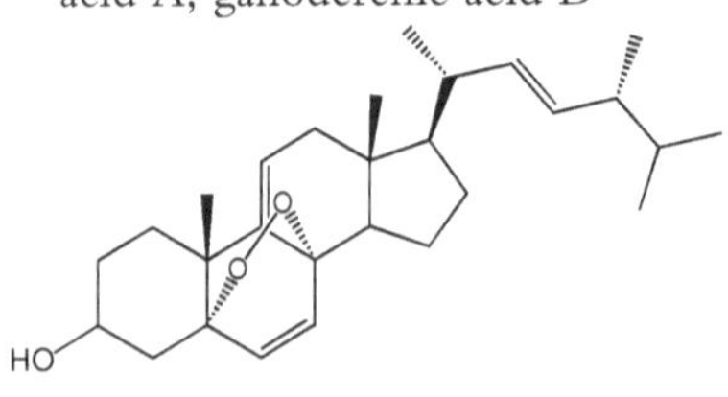

9,11-dehydroergosterol peroxide
</td><td>Inhibits the growth of liver Hep-3B cells</td><td>Chen et al. (2009)</td></tr>
<tr><td>
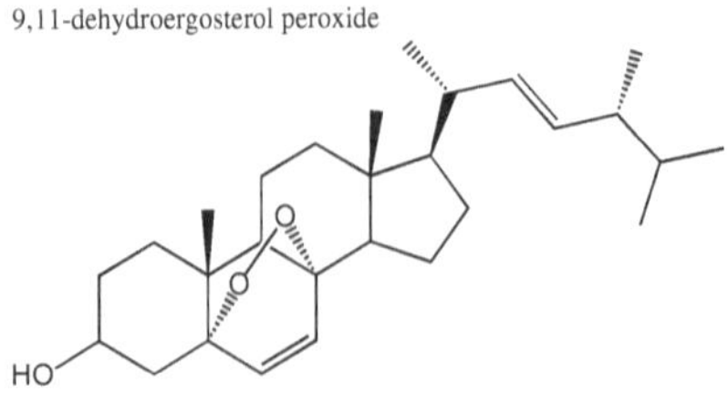

Ergosterol peroxide
</td><td>Inhibits the growth of liver Hep-3B cells</td><td>Chen et al. (2009)</td></tr>
</table>

(*Continued*)

Table 7.1 (Continued)

Triterpenes isolated from *G. lucidum*	Biological activity	References
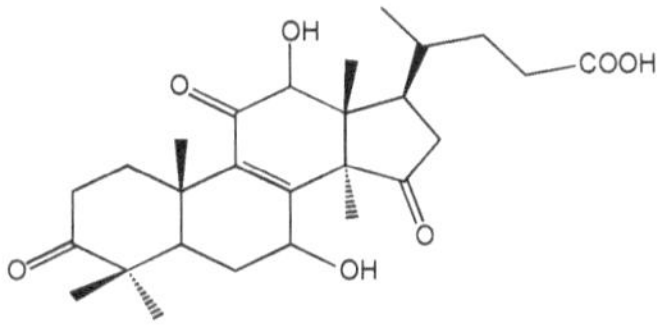 Lucidenic Acid B	Induced apoptosis in HL-60 cells by increasing the release of cytochrome *c*, the activation caspase-9 and caspase-3, and the cleavage of poly(ADP-ribose) polymerase (PARP)	Hsu et al. (2008)
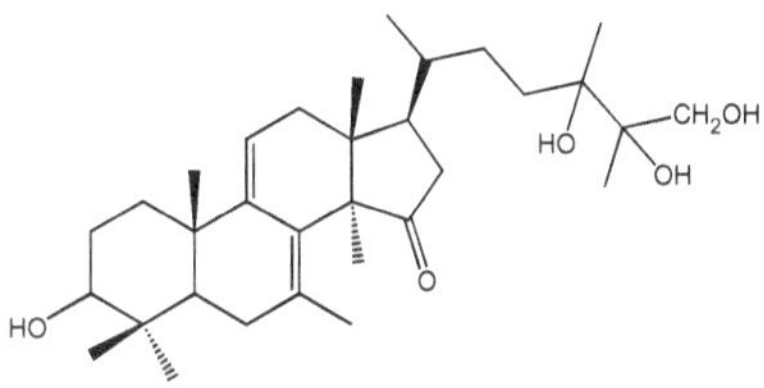 GANODERIOL A	Suppresses migration and adhesion of MDA-MB-231 cells by inhibiting FAK-SRC-paxillin cascade pathway	Wu et al. (2013a, b)
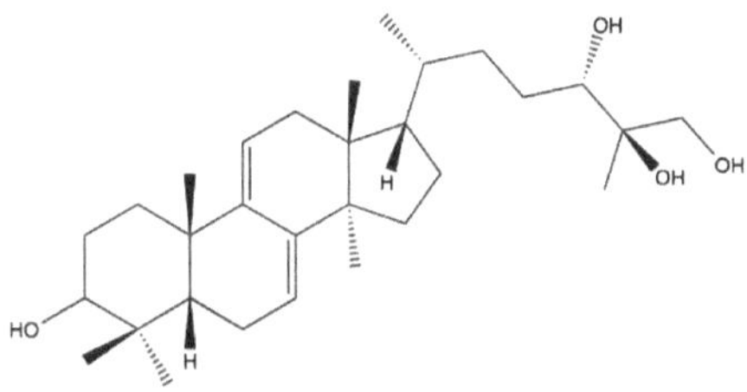 Ganodermanotriol	Cytotoxic in HCT-116 and HT-29 colon cancer cells	Jedinak et al. (2011)
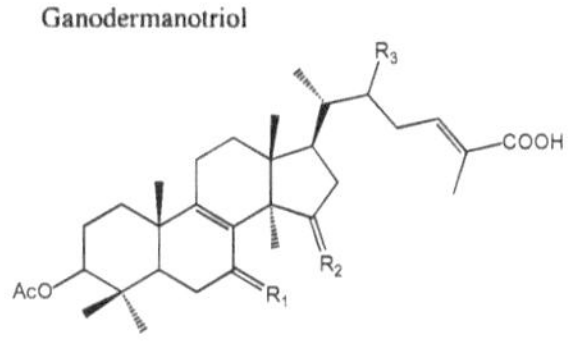 1. R1 = OH, H; R2 = H2; R3 = OAc 3. R1 = OAc, R2 = OH, R3 = OAc 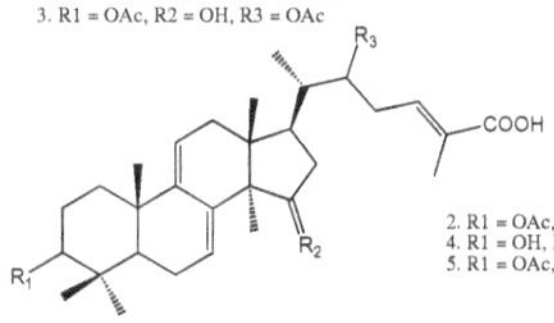 2. R1 = OAc, H; R2 = OH, H; R3 = OAc 4. R1 = OH, H; R2 = H2; R3 = OAc 5. R1 = OAc, H; R2 = OH, H; R3 = H2	Cytotoxic activities against 95D and HeLa tumor cell lines	Li et al. (2013)

(*Continued*)

Table 7.1 (Continued)

Triterpenes isolated from *G. lucidum*	Biological activity	References
3α,22β-Diacetoxy-7α-hydroxyl-5α-lanost-8,24E-dien-26-oic acid (1), together with four known compounds GA-Mk (2), GA-Mc (3), GA-S (4) and GA-Mf (5) 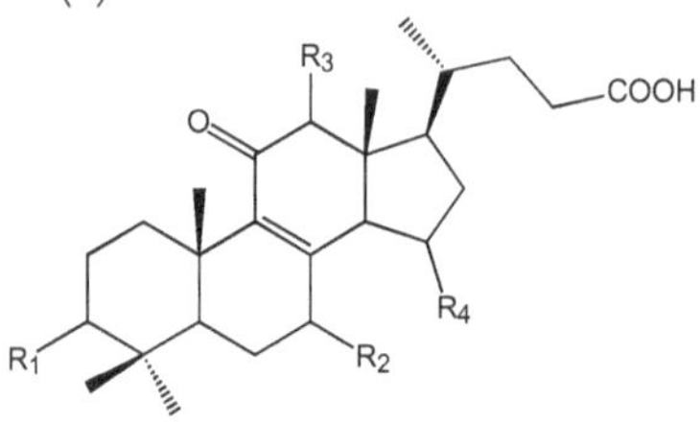 1. R1 = OH; R2 = OH; R3 = H; R4 = OH 2. R1 = OH; R2 = OH; R3 = H; R4 = O 3. R1 = OH; R2 = O; R3 = H; R4 = O 4. R1 = OH; R2 = O; R3 = OAc; R4 = O 5. R1 = H; R2 = OH; R3 = H; R4 = OH (1) 7,15-Trihydroxy-4,4,14-trimethyl-11-oxochol-8-en-24-oic acid; (2) 3,7-dihydroxy-4,4,14-trimethyl-11,15-dioxochol-8-en-24-oic acid; (3) 3-hydroxy-4,4,14-trimethyl-7,11,15-trioxochol-8-en-24-oic acid; (4) 12-acetoxy-3-hydroxy-4,4,14-trimethyl-7,11,15-trioxochol-8-en-24-oic acid (lucidenic acid E); and (5) 7,15-dihydroxy-4,4,14-trimethyl-3,11-dioxochol-8-en-24-oic acid	Synergism with doxorubicin in HeLa cells	Yue et al. (2008)

aqueous extracts of GL fruit bodies (Wu and Wang, 2009); Ganoderma lucidium polysaccharide (GL-PS) peptide (Ho et al., 2007); and F3, a fucose-containing glycoprotein fraction (Chien et al., 2004).

7.3 BIOLOGICALLY ACTIVE TRITERPENES OF *GANODERMA LUCIDUM*

In addition to the polysaccharides, GL contains a group of compounds called triterpenes, a subclass of terpenes with a carbon skeleton of C_{30} and molecular weights of 400–600 kDa (Cör et al., 2018; González et al., 1999; Soccol et al., 2016; Yuen and Gohel, 2005). The triterpenes present in GL are often highly complex and oxidized. Some of the first triterpenes isolated from GL include the GAs A and B (Table 7.1). The GAs, as well as lucidenic acids A and B, are some of the most common triterpenes in GL and are reported to have good antitumor activities (Table 7.1) (Da et al., 2012). Other triterpenes identified include the ganoderals, ganoderiols, and ganodermic acids (Table 7.1; Akihisa et al., 2007; Da et al., 2012; González et al., 1999; Jiang et al., 2008; Ma et al., 2002; Nishitoba et al., 1984; Sato et al., 1986). However, some triterpenes with antitumor effects, the stumbling block has been a lack of biologically active triterpenes being used successfully in clinic as anticancer agents (Da et al., 2012).

7.4 EFFECTS OF *GANODERMA* EXTRACTS AND BIOACTIVE COMPOUNDS ON OVARIAN AND BREAST CANCERS

Ovarian cancer is one of the most serious and difficult to treat gynecological cancers, having a high mortality rate, and most patients will experience cisplatin resistance (Zhao et al., 2011). One study of a GL semipurified polysaccharide extract (GL-P) in IOSE-398, EOC, OV2008, C13, A2780s, and A2780-cp ovarian cancer cell lines demonstrated that this polysaccharide-containing extract inhibited both ovarian cancer cell proliferation and migration. The GL-P extract induced cell apoptosis through the down-regulation of antiapoptotic proteins in the ovarian cancer cells. GL-P increased the sensitivity of the chemoresistant ovarian cancer cells to cisplatin, by up-regulating p53 and down-regulating Akt (Zhao et al., 2011). The effects of a water and ethanol extract of GL on ovarian cancer OVCAR-3 cells showed that both GL extracts reduced OVCAR-3 cell proliferation, but the ethanol extract was more effective with an IC_{50} of 10 μg/mL (Hsieh and Wu, 2011). In addition, the ability of OVCAR-3 cells to form colonies was more significantly inhibited by ethanol extract than the aqueous extracts of GL, with a significant reduction in both the number and the size of colonies (Hsieh and Wu, 2011). The ethanol extract reduced OVCAR-3 cell cycle progression by down-regulating the expression of cyclin D1. Hsieh and Wu (2011) also demonstrated that GL had chemopreventive effects including increasing the activities of superoxide dismutase (SOD), catalase NAD (P) H: quinone oxidoreductase 1 and glutathione S-transferase P1 by mediating the Nrf2 signaling pathway (Hsieh and Wu, 2011). Since the water extract contained primarily polysaccharides, the ethanol extract may have contained both the polysaccharides and some triterpenes, thus presenting the possibility that the combination of the two compounds may have synergistic effects against ovarian cancer.

G. lucidum extracts (GLEs) and purified compounds have also been tested in a variety of breast cancer assays. Thyagarajan et al. (2006) reported that a GL extract reduced the proliferation and invasiveness of metastatic breast cancer cells in vitro. This GL extract contained ~13.5% polysaccharides and 6% triterpenes and appeared to reduce oxidative stress-induced migration of MCF-7 cells (Thyagarajan et al., 2006). A more recent study investigated the effects of a GLE on breast tumor proliferation and breast-to-lung cancer metastasis (Loganathan et al., 2014). Results from this study suggested that treatment of mice implanted with MDA-MB-231 human breast cancer cells with GLE (100 mg/kg/every other day) suppressed the growth of large tumors, but the results were not statistically significant. However, treatment of the mice with GLE did significantly reduce MDA-MB-231 breast-to-lung cancer metastases, and down-regulated the expression of genes associated with invasive behavior (Loganathan et al., 2014). Thus while the extract of GLE did not inhibit the growth of the tumors, it significantly inhibited metastatic disease. A study by Suarez-Arroyo et al. (2013) showed that the same GL extract used in the previous study, inhibited the growth of inflammatory breast cancer (IBC). IBC is an aggressive and very lethal, and current treatment involves the use of nontargeted systemic chemotherapy, surgery, and radiation. In the Suarez-Arroyo study, treatment of the IBC cells with GLE reduced cell growth via the phosphoinositide-3-kinase/AKT/mammalian target of rapamycin (mTOR) signaling pathway in vitro. In an animal study, severe combined immunodeficient mice inoculated with IBC cells treated orally with the GLE at a dose of 28 mg/kg for 13 weeks, showed a 50% reduction in tumor size. In the excised tumors, GLE reduced the expression of E-cadherin, mTOR, eIF4G, p70S6K, and the activity of ERK1/2 (Suarez-Arroyo et al., 2013). The triterpenes appear to be the bioactive constituents of GLE that reduce the proliferation of breast cancer cells. Ganoderiol A (GA; Table 7.1), as well as the dehydrogenated GA and GA isomer, inhibited MDA-MB-231cellular adhesion and migration (Wu et al., 2013a,b). These triterpenes were shown to inhibit focal adhesion kinase (FAK) activity. Treatment of the cells also down-regulated the expression of RhoA, Rac1, and Cdc42 mRNA and reduced the FAK-SRC-paxillin signaling pathway (Cör et al., 2018; Unlu et al., 2016; Wu et al., 2013a; Zeng et al., 2018).

7.5 EFFECTS OF *GANODERMA* POLYSACCHARIDES AND TRITERPENES IN COLORECTAL CANCER

Extracts of GL are reported to be effective for reducing colon cancer risk in a variety of animal models (Cör et al., 2018; Hong et al., 2004; Jedinak et al., 2011; Lu et al., 2003, 2002; Wu et al., 2013b; Zeng et al., 2018). The effects of a polysaccharide containing fraction (GLE-1), and a triterpenoid fraction (without polysaccharides; GLE-2) isolated from a GL extract were investigated in SW 480 human colorectal cancer cells (Xie et al., 2006). Treatment with GLE-1 or GLE-2 significantly inhibited the proliferation of SW 480 cells (Xie et al., 2006). GA T, a triterpene isolated from a GL extract inhibited the growth of HCT-116 cells in vitro (Chen et al., 2008, 2009). In 2011, ganodermanontriol (GNDT; Table 7.1) another triterpene was isolated and identified by Jedinak et al. (2011) was found to inhibit the proliferation of HCT-116 and HT-29 colon cancer cells by reducing the activity of ß-catenin, and cyclin D1 in a concentration-dependent manner (Jedinak et al., 2011). Expression of Cdk-4 and PCNA was also reduced, but expression of Cdk-2, p21, and cyclin E was not. GNDT further induced a dose-dependent increase in protein expression of E-cadherin and ß-catenin in HT-29 cells. GNDT also reduced tumor growth in a mouse xenograft model using the same HT-29 cells (Jedinak et al., 2011). Thus GNDT appears to be active against colon cancer both in vitro and in vivo.

Administration of a triterpene containing extract of GL (GLT) inhibited the proliferation of HT-29 colon cancer cells and reduced the growth of tumors in a rodent xenograph model of colon cancer (Thyagarajan et al., 2010). In vitro treatments inhibited tumor growth, induced cell cycle arrest and stimulated programmed cell death Type II (autophagy). In the cells and tumors, GLT induced the expression of two autophagy proteins, namely Beclin-1 and LC-3, and suppressed the phosphorylation of p38 mitogen-activated protein kinase (MAPK) by ~60%. In the HT-29 xenograph model, GLT also inhibited the growth of colon tumors by the induction of autophagy, indicating a novel mechanism of action (Thyagarajan et al., 2010).

In HCT-116 human colon cancer cells, treatment with a GL polysaccharide (GLP) fraction significantly reduced HCT-116 cell proliferation in a time- and concentration-dependent manner, and also induced cell apoptosis (Liang et al., 2014a). Apoptosis was determined by significant morphological changes, DNA fragmentation, a decrease in the mitochondrial membrane potential, an increase in the S phase, as well as activations of caspases-3 and -9. Inhibition of the c-Jun N-terminal kinase led to a decrease in GLP-induced apoptosis. Western blot analysis further showed that GLP up-regulated Bax/Bcl-2, caspase-3, and poly (ADP-ribose) polymerase (PARP) mRNA expression. The results suggest that GLP-induced apoptosis in human colorectal cancer cells is associated with activation of mitochondrial and MAPK pathways (Liang et al., 2014b). In a follow-up study, these authors further demonstrated that GLP reduced cell migration, and induced cell morphological changes, as well as altered intracellular Ca_2^+ elevation and lactic acid dehydrogenase (LDH) release. Western blot analysis also showed an alteration in Fas and caspase-3 protein expression. The study suggested that GLP inhibited cell proliferation in HCT-116 human colon cancer cell line by increasing intracellular calcium release and death receptor signaling (Liang et al., 2014b). In 2015, the same authors investigated the effects of GLPs on LoVo human colon cancer cells (Liang et al., 2015). GLP-induced cytotoxicity in LoVo cells, as well as inhibited cell migration. GLP also enhanced DNA fragmentation, morphological alterations, and lactate dehydrogenase release. These authors determined that GLP-induced apoptosis in LoVo cells by the activation of caspases-3, -8, and -9. Additionally, treatment with GLPs promoted the expression of Fas and caspase-3 proteins (Liang et al., 2015). These data suggest that GLP induces apoptosis and reduces cell migration in colon cancer cells by the activation of the Fas/caspase-dependent apoptosis pathway is involved in the cytotoxicity of GLPs.

The efficacy of an aqueous GL extract on patients with colorectal cancer was published in 2010 (Oka et al., 2010). The trial was a controlled study in patients with diagnosed colorectal adenomas, determined by colonoscopy. Patients were administered 1.5 g/day of the GLM extract for 12 months, with a follow-up colonoscopy performed after 12 months, in which the size, site, and macroscopic type of all adenomas were recorded. The changes in the number of adenomas up to 12 months increased to 0.66 ± 0.10 (mean $\pm$ SE) in the control group, but significantly decreased in the GLM group to 0.42 ± 0.10 ($P < 0.01$). The total size of adenomas increased to 1.73 ± 0.28 mm in the control group and significantly decreased to 1.40 ± 0.64 mm in the GLM group ($P < 0.01$) (Oka et al., 2010).

7.6 EFFECTS OF *GANODERMA* PREPARATIONS ON ASCITIC AND HEPATOCELLULAR CARCINOMAS

G. lucidum polysaccharide preparations (GLPS) have also shown some therapeutic effects in hepatocellular carcinomas in vivo (Li et al., 2015). In a study by Li et al. (2015), the effects of GLPS administration on the balance of regulatory T cell (Treg) and effector T cell (Teff) were determined in hepatoma-bearing mice. Administration GLPS significantly reduced tumor growth, which was associated with an increase of the ratio of Teffs to Tregs. In addition, treatment of T cells with GLPS also reduced Notch1 and FoxP3 mRNA expression by increasing miR-125b expression (Li et al., 2015). An in vitro study by Chen et al. (2009) investigated the effects of an ethanol extract of *G. lucidium* (GLE) mycelia on human hepatocellular carcinoma cells (Hep-3B). GLE reduced Hep-3B cell proliferation in both a concentration- and time-dependent manner, with an IC_{50} of 156.8, 89.9, and 70.1 μg/mL, after the Hep-3B cells were treated for 24, 48, and 72 h, respectively. Fractionation of the ethanol extract resulted in the isolation and identification of two active compounds, 9,11-dehydroergosterol peroxide [9(11)-DHEP] and ergosterol peroxide (EP) (Table 7.1). The IC_{50} values of 9(11)-DHEP and EP based on the cell viability of Hep-3B were 16.7 and 19.4 μg/mL, respectively (Chen et al., 2009).

Mouse sarcoma S180-induced ascites represents one of the most aggressive transplantable cancers in experimental mouse models. Li et al. (2007) investigated the effects of GLP on the nucleotide content and cell cycle distribution of tumor cells in S180 Ascitic tumor-bearing mice (Li et al., 2007). Mice bearing S180 ascitic tumors were treated with GLP at increasing doses of 100, 200, and 400 mg/kg body weight, orally. Controls included mice treated with normal saline or subcutaneous injection of cyclophosphamide (CTX) at 25 mg/kg. Compared with saline-treated animals, the tumor cells in the 3 GLP groups all showed a significant reduction in RNA and DNA content, at an oral dose of 200 mg/kg. Oral doses of 400, 200, and 100 mg/kg increased the percentage of G2/G2 phase cells and decreased the percentage of G2/M phase tumor cells, while CTX did not ($P = 0.000$). The study results indicated that GLP treatments inhibit DNA and RNA synthesis in the tumor cells and may enhance host immune function thereby interfering with the normal cell cycles (Li et al., 2007). In a 2009 study, sulfated and carboxy-methylated *G. lucidium* polysaccharides (S-GL and CM-GL) were prepared by derivatization of water-insoluble polysaccharides (GL-IV-I) extracted from the fruiting body of GL (Wang et al., 2009). The weight-average molecular mass of GL-IV-I, S-GL, and CM-GL was 13.3×10^4, 10.1×10^4, and 6.3×10^4, respectively. S-GL and CM-GL inhibited the in vitro proliferation of mouse S-180 tumor cells in a dose-dependent manner, with an IC_{50} value of 26 and 38 μg/mL, respectively. These polysaccharides also inhibited the growth of S180 solid tumors implanted in BALB/c mice, with low toxicity to the animals. Flow cytometry analysis also showed that treatment of S180 cells with S-GL and CM-GL induced cell-cycle arrest in the G2/M phase. The expression of Bax mRNA increased, while the expression of Bcl-2 decreased dramatically, suggesting induction of apoptosis (Wang et al., 2009).

GLIS, a proteoglycan isolated and identified from the fruiting bodies of GL, is reported to stimulate the activation of B lymphocytes (Zhang et al., 2010). The immune stimulating effects of GLIS, as well as its antitumor activities were investigated in vitro and in vivo in sarcoma S180 in BALB/c mice (Zhang et al., 2010). Lymphocytes and bone marrow-derived macrophages were isolated from spleen and tibia/femurs, respectively. After treatment with GLIS, spleen-derived B lymphocytes from tumor-bearing mice were activated and produced large amounts of immunoglobulins. Bone marrow-derived macrophages from tumor-bearing mice were also activated after treatment to GLIS, and produced important immunomodulatory substances, such as IL-1β, TNF-α, and nitric oxide. GLIS increased phagocytosis of macrophages, and also increased the macrophage-mediated tumor cytotoxicity. Treatment of mice with GLIS reduced mouse sarcoma S180 tumor growth by 60% in vivo (Zhang et al., 2010).

7.7 LEUKEMIA, FIBROSARCOMA, AND ASTROCYTOMA TUMORS

GLEs have been assessed in melanoma and sarcoma tumors in mice and rats (Unlu et al., 2016; Zeng et al., 2018). Methanol extracts containing total terpenoids (GLT) and a purified methanol extract containing primarily acidic terpenoids (GLAT) were tested. Both GLT and GLAT inhibited tumor growth of B16 mouse melanoma cells inoculated subcutaneously into syngeneic C57BL/6 mice. The extracts also reduced cell proliferation of B16 cells in vitro, but GLT was more effective than GLAT. GLT was also active in two other rodent tumor cell lines,

L929-mouse fibrosarcoma and C6-rat astrocytoma. GLT inhibited cell proliferation and induced caspase-dependent apoptotic cell death mediated via up-regulation of p53, while Bcl-2 expression was inhibited. Furthermore, the antitumor effects of the GLT were associated with the increased production of reactive oxygen species (ROS), whereas the neutralization by the antioxidant, *N*-acetyl cysteine, resulted in partial recovery of cell viability (Cör et al., 2018; Unlu et al., 2016; Zeng et al., 2018).

Triterpenes isolated from an ethyl acetate fraction of the fruiting body of GL, including a new triterpenoid, ethyl 7β-hydroxy-4,4,14α-trimethyl-3,11,15-trioxo-5α-chol-8-en-24-oate (4), named ethyl lucidenates A, along with three known compounds, ganodermanondiol (1), lucidumol B (2), and methyllucidenates A (3) (Table 7.1) were tested using 3-(4,5-Dimethylthiazol-2-yl)-2,5-Diphenyltetrazolium Bromide) (MTT) assay for their cytotoxic activities against two leukemia cell lines K562 and HL-60 (Li et al., 2013). The results showed that compound 4 exhibited good cytotoxicity against HL-60 and CA46 leukemia lines with IC_{50} of 25.98 and 20.42 μg/mL, respectively (Li et al., 2013). The effect of lucidenic acids (A, B, C, and N) isolated from GL was tested in human leukemia cells HL-60 (Hsu et al., 2008). The results showed that the lucidenic acids reduced HL-60 cell growth, and treatment of HL-60 cells with lucidenic acid A, C, and N induced cell cycle arrest in the G1 phase. Lucidenic acid B (LAB; Table 7.1) increased the number of early and late apoptotic HL-60 cells by causing a loss of mitochondria membrane potential. LAB-induced apoptosis in HL-60 cells by increasing the release of cytochrome *c*, the activation caspase-9 and caspase-3, and the cleavage of PARP. Pretreatment of the cells with the caspase-9 inhibitor (Z-LEHD-FMK) and caspase-3 inhibitor (Z-DEVD-FMK) reduced the effects of LAB on HL-60 cells (Hsu et al., 2008).

Ganoderma extracts have been shown to potentiate the effects of chemotherapeutic agents when tested in combination. In a study by Huang et al. (2010), treatment of three urothelial carcinoma cell lines NTUB1, cisplatin-resistant, N/P(14); and arsenic-resistant, N/As(0.5) with a combination of *Ganoderma* polysaccharide fraction 3 (LZP-F3) and cisplatin or arsenic trioxide found that LZP-F3 reversed the chemoresistance of N/P(14) and N/As(0.5) to cisplatin and arsenic, respectively, in a dose-dependent manner. The mechanism(s) involved the activation of p38, as well as down-regulation of Akt and xeroderma pigmentosum, complementation group A (XPA). A dose of 10 μg/mL of LZP-F3 induced significant G1 cell cycle arrest in N/P(14) and N/As(0.5) The mechanisms by which LZP-F3 reduced drug resistance involved significant induction of the death receptor Fas, caspase-3 and -8 activation, Bax and Bad up-regulation, Bcl-2 and Bcl-x(L) down-regulation, and cytochrome *c* release. Thus the combination of LZP-F3 with chemotherapeutic drugs increased cell apoptosis in resistant cell lines (Huang et al., 2010).

In another study, the combination of chemotherapeutic agents and *G. lucidium* triterpenes (GLT) was investigated (Yue et al., 2008). The triterpenes tested included: (1) 3,7,15-trihydroxy-4,4,14-trimethyl-11-oxochol-8-en-24-oic acid; (2) 3,7-dihydroxy-4,4,14-trimethyl-11,15-dioxochol-8-en-24-oic acid (LCN); (3) 3-hydroxy-4,4,14-trimethyl-7,11,15-trioxochol-8-en-24-oic acid; (4) 12-acetoxy-3-hydroxy-4,4,14-trimethyl-7,11,15-trioxochol-8-en-24-oic acid (lucidenic acid E); and (5) 7,15-dihydroxy-4,4,14-trimethyl-3,11-dioxochol-8-en-24-oic acid (Yue et al., 2008; Table 7.1). When used in combination with GLT or LCN, doxorubicin (DOX) had a synergistic interaction in HeLa cells. The molecular mechanism of action of GLT was determined using two-dimensional gel electrophoresis-based comparative proteomics. Proteins with an altered expression after GLT treatment in HeLa cells were identified by matrix-assisted laser desorption/ionization time-of-flight tandem mass spectrometry. GLT treatments enhanced the expression of 14 proteins, which play important roles in cell proliferation, the cell cycle, apoptosis, and oxidative stress. In addition, GLT treatments increased the ROS-producing effect of DOX, which was reversed using an ROS scavenger. The data suggest that the synergism between GTS and DOX may be due to GLT-induced sensitization of cells to chemotherapeutics through enhanced oxidative stress, DNA damage, and apoptosis (Cör et al., 2018; Soccol et al., 2016; Unlu et al., 2016; Yue et al., 2008).

7.8 USE OF CHEMOMETRICS AND BIOCHEMOMETRICS TO IDENTIFY ANTICANCER COMPOUNDS IN *GANODERMA*

One of the primary issues with *Ganoderma* products is the lack of standardization that is necessary for progression to clinical trials. The process of standardization requires knowledge of bioactive and marker compounds present in extracts. The chemistry of *Ganoderma* is very complex and isolation and identification of the biologically active compounds is time consuming, often requiring extensive bioassay-guided fractionation using column chromatography or preparative HPLC. Since bioassay-guided fractionation has many limitations, including (1) overlooking the low abundance of bioactive constituents, (2) the difficulties in the isolation of pure compounds due to a decrease in materials during repetitive separation steps, (3) loss of activity of constituents because of irreversible binding to stationary phases, and (4) degradation during the separation process. As a consequence, newer methodologies are needed to overcome these challenges, including chemometrics and biochemometrics (Britton et al., 2017). These methodologies employ untargeted metabolomics using NMR or MS to detect small molecules in a mixture (the metabolome) without bias, or chemometrics. When metabolite profiles and bioassay data are collected for an extract, and then statically analyzed to create a chemical profile for bioactivity, this process is called biochemometrics (Britton et al., 2017).

Recently, Wu et al. (2017) has employed chemometric analysis to overcome some of the identification and quality control issues for *Ganoderma* extracts. These authors established a fingerprint evaluation system based on similarity analysis, cluster analysis (CA), and principal component analysis (PCA) for the identification of compounds from GL with anticancer activities. GL samples from the Chinese provinces of Hainan, Neimeng, Shangdong, Jilin, Anhui, Henan, Yunnan, Guangxi, and Fujian were analyzed by HPLC-PAD and HPLC-MSn. From these analyses, 47 compounds were detected, of which 42 known triterpenes were identified by comparing their retention times and mass spectrometry data with that of reference compounds and by published literature. GA B (Table 7.1), 3,7,15-trihydroxy-11,23-dioxolanost-8,16-dien-26-oic acid, lucidenic acid A, ganoderic acid G (Table 7.1), and 3,7-oxo-12-acetylganoderic acid DM were used as analytical marker compounds to distinguish the samples with different quality according to both CA and PCA. Chan et al. (2017) also used biochemometrics using liquid chromatography coupled with time-of-flight mass spectrometer (LC-TOF-MS/MS) to identify five new anticancer compounds from *Ganoderma*. Statistical analyses of the datasets generated from LC-TOF-MS/MS were performed to distinguish the chemical difference among different groups, to facilitate the identification of bioactive compounds by comparing the constituents of active and inactive chromatographic fractions from the same plant. In this study, the stripe of GS was fractionated and the anticancer activities were analyzed with an MTT assay using murine breast tumor 4T1 cells. The chemical constituents of five fractions were analyzed by using liquid chromatography/mass spectrometry method with multivariate statistical analysis (Chan et al., 2017). The data were further analyzed using orthogonal partial least-squared discriminant analysis (OPLS-DA) to compare the chemical differences observed in the active and inactive fractions. An S-plot was generated from the OPLS-DA analysis, the ions corresponding to anticancer compounds appeared at the extremity of the S-plot (Chan et al., 2017). Electrospray ionization quadrupole-time-of-flight tandem mass spectrometry provided MS/MS fragmentation pattern along with precursor and product ion masses with low part-per-million accuracy. After comparing the MS/MS fragmentation pattern with the chemical literature and online databases, five potential antitumor compounds were identified (Chan et al., 2017). These new methods for identifying compounds are particularly interesting for use in *Ganoderma* extracts, as there are over 400 known compounds in GL extracts, and using more modern methods rather than strictly bioassay-guided fraction to identify known and unknown compounds with anticancer effects will facilitate the identification of novel compounds for testing. In addition to chemometrics biochemometrics combines the integration of untargeted metabolomics with that of biological activity data to determine bioactive chemical constituents (Britton et al., 2017). Many approaches have been used including PCA, PLS, S-plot and selectivity ratio, to correlate metabolic profiles of extracts with biological data sets (Britton et al., 2017). To date, no examples could be found that integrate both the anticancer biological data with the metabolic profiles of *Ganoderma*. Since there are now good examples of chemometrics for *Ganoderma*, the use of biochemometrics is sure to soon follow. These new methodologies will be very helpful for unraveling the complexities of the chemistry and biology *Ganoderma* extracts and their effects in cancer. They will also be beneficial for overcoming the critical challenge in identifying multiple compounds that may contribute additively, synergistically, or antagonistically to anticancer activities of *Ganoderma* extracts.

7.9 CONCLUSIONS AND FUTURE DIRECTIONS/PROSPECTS

Data from biological and chemical studies suggest that *Ganoderma* species may be of benefit for the treatment and prevention of various cancers. In vitro and in vivo studies suggest that GLEs in particular have an excellent potential to be developed as cancer therapies. However, there is a significant lack of clinical data supporting these preclinical studies. Future clinical trials are urgently needed to assure that the effects observed in animal studies are also valid in humans. However, one of the primary stumbling blocks to effective clinical trials are the significant issues surrounding quality control/assurance of these products. New methodologies including chemometrics and biochemometrics may help to overcome these issues as these methods allow for the correlation of chemistry with biological effects. It is essential that well defined or semipurified extracts be developed for clinical studies to assure reproducible effects. If the safety and efficacy of *Ganoderma* as an adjunct therapy for cancer can be established in clinical trials, then the prospects for *Ganoderma* products worldwide will be enormous.

REFERENCES

Akihisa, T., Nakamura, Y., Tagata, M., Tokuda, H., Yasukawa, K., Uchiyama, E., et al., 2007. Anti-inflammatory and anti-tumor-promoting effects of triterpene acids and sterols from the fungus *Ganoderma lucidum*. Chem. Biodivers. 4, 224–231.

Bao, P.P., Lu, W., Cui, Y., Zheng, Y., Gu, K., 2012. Ginseng and *Ganoderma lucidum* use after breast cancer diagnosis and quality of life: a report from the Shanghai Breast Cancer Survival Study. PLoS One 7 (6), e39343.

Boh, B., Berovic, M., Zhang, J., Zhi-Bin, L., 2007. *Ganoderma lucidum* and its pharmaceutically active compounds. Biotechnol. Ann. Rev. 13, 265–301.

Britton, E., Kellogg, J.J., Kvalheim, O., Cezh, N., 2017. Biochemometrics to identify synergists and additives from botanical medicines: a case study with *Hydrastis canadensis* (goldenseal). J. Nat. Prod. . Available from: https://doi.org/10.1021/acs.jnatprod.7b00654November 1.

Chan, K., Yue, G., Wong, E., 2017. Screening and analysis of potential anti-tumor components from the stipe of *Ganoderma sinense* using high-performance liquid chromatography/time-of-flight mass spectrometry with multivariate statistical tool. J. Chromatogr. A1487, 162–167.

Chen, N.H., Liu, J.W., Zhong, J.J., 2008. Ganoderic acid Me inhibits tumor invasion through down-regulating matrix metalloproteinases 2/9 gene expression. J. Pharmacol. Sci. 108 (2), 212–216.

Chen, Y.K., Kuo, Y.H., Chiang, B.H., Lo, J.M., Sheen, L.Y., 2009. Cytotoxic activities of 9,11-dehydroergosterol peroxide and ergosterol peroxide from the fermentation mycelia of *Ganoderma lucidum* cultivated in the medium containing leguminous plants on Hep 3B cells. J. Agric. Food Chem. 57 (13), 5713–5719.

Chien, C.M., Cheng, J.L., Chang, W.T., Tien, M.H., Tsao, C.M., Chang, Y.H., et al., 2004. Polysaccharides of *Ganoderma lucidum* alter immunophenotypic expression and enhance CD56 + NK-cell cytotoxicity in cord blood. Bioorg. Med. Chem. 12, 5603–5770.

Cör, D., Knez, Ž., Knez Hrnčič, M., 2018. Antitumour, antimicrobial, antioxidant and antiacetylcholinesterase effect of *Ganoderma Lucidum* terpenoids and polysaccharides: a review. Molecules 13, 23–33.

Da, J., Wu, W., Hou, J.J., Long, H., Yao, S., Yang, Z., et al., 2012. Comparison of two official Chinese pharmacopoeia species of *Ganoderma* based on chemical research with multiple technologies and chemometrics analysis. J. Chromatogr. A 1222, 59–70.

González, A.G., Leoén, F., Rivera, A., Padroén, J.I., González-Plata, J., Zuluaga, J.C., et al., 1999. Lanostanoid triterpenes from *Ganoderma lucidum*. J. Nat. Prod. 62, 1700–1701.

Ho, Y.W., Yeung, J.S., Chiu, P.K., Tang, W.M., Lin, Z.B., Man, R.Y., et al., 2007. *Ganoderma lucidum* polysaccharide peptide reduced the production of proinflammatory cytokines in activated rheumatoid synovial fibroblast. Mol. Cell. Biochem. 301, 173–179.

Hong, K.J., Dunn, D.M., Shen, C.L., Pence, B.C., 2004. Effects of *Ganoderma lucidum* on apoptotic and anti-inflammatory function in HT-29 human colonic carcinoma cells. Phytother. Res. 18, 768–770.

Hsieh, T.C., Wu, J.M., 2011. Suppression of proliferation and oxidative stress by extracts of *Ganoderma lucidum* in the ovarian cancer cell line OVCAR-3. Int. J. Mol. Med. 28 (6), 1065–1069.

Hsu, C.L., Yu, Y.S., Yen, G.C., 2008. Lucidenic acid B induces apoptosis in human leukemia cells via a mitochondria-mediated pathway. J. Agric. Food Chem. 56 (11), 3973–3980.

Huang, C.Y., Chen, J.Y., Wu, J.E., Pu, Y.S., Liu, G.Y., Pan, M.H., et al., 2010. Ling-Zhi polysaccharides potentiate cytotoxic effects of anticancer drugs against drug-resistant urothelial carcinoma cells. J. Agric. Food Chem. 58 (15), 8798–8805.

Jedinak, A., Thyagarajan-Sahu, A., Jiang, J., Silva, D., 2011. Ganodermanontriol, a lanostanoid triterpene from *Ganoderma lucidum*, sup- presses growth of colon cancer cells through ß-catenin signaling. Int. J. Oncol. 38 (3), 761–767.

Ji, Z., Tang, Q., Zhang, J., Yang, Y., Jia, W., Pan, Y., 2007. Immuno-modulation of RAW264.7 macrophages by GLIS, a proteo-polysaccharide from *Ganoderma lucidum*. J. Ethnopharmacol. 112 (3), 445–450.

Jiang, J., Grieb, B., Thyagarajan, A., Silva, D., 2008. Ganoderic acids suppress growth and invasive behavior of breast cancer cells by modulating AP-1 and NF-kappaB signaling. Int. J. Mol. Med. 21, 577–584.

Lee, S.S., Wei, Y.H., Chen, C.F., Wang, S.Y., Chen, K.Y., 1995. Antitumor effects of *Ganoderma lucidum*. J. Clin. Med. 6, 1–12.

Li, A., Shuai, X., Jia, Z., Li, H., Liang, X., Su, D., et al., 2015. *Ganoderma lucidum* polysaccharide extract inhibits hepatocellular carcinoma growth by downregulating regulatory T cells accumulation and function by inducing microRNA-125b. J. Transl. Med. 26 (13), 100.

Li, J.J., Lei, L.S., Yu, C.L., Zhu, Z.G., Zhang, Q., Wu, S.G., 2007. Effect of Ganoderma lucidum polysaccharides on tumor cell nucleotide content and cell cycle in S180 ascitic tumor-bearing mice. Nan Fang Yi Ke Da Xue Xue Bao 27 (7), 1003–1005.

Li, Y.B., Liu, R.M., Zhong, J.J., 2013. A new ganoderic acid from *Ganoderma lucidum* mycelia and its stability. Fitoterapia 84, 115–122.

Liang, Z., Guo, Y.T., Yi, Y.J., Wang, R.C., Hu, Q.L., Xiong, X.Y., 2014a. *Ganoderma lucidum* polysaccharides target a fas/caspase dependent pathway to induce apoptosis in human colon cancer cells. Asian Pac. J. Cancer Prev. 15 (9), 3981–3986.

Liang, Z., Yi, Y., Guo, Y., Wang, R., Hu, Q., Xiong, X., 2014b. Chemical characterization and antitumor activities of polysaccharide extracted from *Ganoderma lucidum*. Int. J. Mol. Sci. 15 (5), 9103–9116.

Liang, Z.E., Yi, Y.J., Guo, Y.T., Wang, R.C., Hu, Q.L., Xiong, X.Y., 2015. Inhibition of migration and induction of apoptosis in LoVo human colon cancer cells by polysaccharides from *Ganoderma lucidum*. Mol. Med. Rep. 12 (5), 7629–7636.

Loganathan, J., Jiang, J., Smith, A., Jedinak, A., Thyagarajan-Sahu, A., Sandusky, G.E., et al., 2014. The mushroom *Ganoderma lucidum* suppresses breast-to-lung cancer metastasis through the inhibition of pro-invasive genes. Int. J. Oncol. 44 (6), 2009–2015.

Lu, H., Kyo, E., Uesaka, T., Katoh, O., Watanabe, H., 2002. Prevention of development of *N,N'*-dimethylhydrazine-induced colon tumors by a water-soluble extract from cultured medium of *Ganoderma lucidum* (Reishi) mycelia inmale ICR mice. Int. J. Mol. Med. 9, 113–117.

Lu, H., Kyo, E., Uesaka, T., Katoh, O., Watanabe, H., 2003. A water-soluble extract from cultured medium of *Ganoderma lucidum* (Reishi) mycelia suppresses azoxymethane-induction of colon cancers in male F344 rats. Oncol. Rep. 10, 375–379.

Ma, J.Y., Ye, Q., Hua, Y.J., Zhang, D.C., Cooper, R., Chang, M.N., et al., 2002. New lanostanoids from the mushroom *Ganoderma lucidum*. J. Nat. Prod. 65, 72–75.

Meng, L.Z., Xie, J., Lv, G.P., Hu, D.G., Zhao, J., Duan, J.A., et al., 2014. A comparative study on immunomodulatory activity of polysaccharides from two official species of *Ganoderma* (Lingzhi). Nutr. Cancer . Available from: https://doi.org/10.1080/01635581.2014.948215.

Nishitoba, T., Sato, H., Kasai, T., Kawagishi, H., Sakamura, S., 1984. New bitter C27 and C30 terpenoids from the fungus *Ganodema lucidum* (Reishi). J. Agric. Biol. Chem. 48, 2905–2907.

Oka, S., Tanaka, S., Yoshida, S., Hiyama, T., Ueno, Y., Ito, M., et al., 2010. A water-soluble extract from culture medium of *Ganoderma lucidum* mycelia suppresses the development of colorectal adenomas. Hiroshima J. Med. Sci. 59 (1), 1–6.

Ren, L., Perera, C., Hemar, Y., 2012. Antitumor activity of mushroom polysaccharides: a review. Food Funct. 3, 1118–1130.

Sato, H., Nishitoba, T., Shirasu, S., Oda, K., Sakamura, S., 1986. Gano- deriol-A and ganoderiol-B, new triterpeneoids from the fungus *Ganoderma lucidum* (Reishi). Agric. Biol. Chem. 50 (11), 2887–2890.

Soccol, C.R., Bissoqui, L.Y., Rodrigues, C., Rubel, R., Sella, S.R., Leifa, F., et al., 2016. Pharmacological properties of bio-compounds from Spores of the Lingzhi or Reishi medicinal mushroom *Ganoderma lucidum* (Agaricomycetes): a review. Int. J. Med. Mushrooms 18 (9), 757–767.

Suarez-Arroyo, I.J., Rosario-Acevedo, R., Aguilar-Perez, A., Clemente, P.L., Cubano, L. A., 2013. Anti-tumor effects of *Ganoderma lucidum* (Reishi) in inflammatory breast cancer in vivo and in vitro models. PLoS One 8 (2), e57431.

Thyagarajan, A., Jiang, J., Hopf, A., Adamec, J., Sliva, D., 2006. Inhibition of oxidative stress-induced invasiveness of cancer cells by *Ganoderma lucidum* is mediated through the suppression of interleukin-8 secretion. Int. J. Mol. Med. 18 (4), 657–664.

Thyagarajan, A., Jedinak, A., Nguyen, H., Terry, C., Baldridge, L.A., Jiang, J., et al., 2010. Triterpenes from *Ganoderma lucidum* induce autophagy in colon cancer through the inhibition of p38 mitogen-activated kinase (p38 MAPK). Nutr. Cancer 62 (5), 630–640.

Unlu, A., Nayir, E., Kirca, O., Ozdogan, M., 2016. *Ganoderma Lucidum* (Reishi Mushroom) and cancer. J. Buon. 21 (4), 792–798.

Wachtel-Galor, S., Yuen, J., Buswell, J.A., Benzie, I., 2011. *Ganoderma lucidum* (Lingzhi or Reishi) a medicinal mushroom, Herbal Medicine: Biomolecular and Clinical Aspects, 2nd ed. CRC Press, Boca Raton, FL, p. 101, Chapter 12.

Wang, J., Zhang, L., Yu, Y., Cheung, P.C., 2009. Enhancement of antitumor activities in sulfated and carboxymethylated polysaccharides of *Ganoderma lucidum*. J. Agric. Food Chem. 57 (22), 10565–10572.

Wasser, S.P., Coates, P., Blackman, M., Cragg, G., Levine, M., Moss, J., et al., 2005. Reishi or Ling-zhi (*Ganoderma lucidum*). Encyclopedia of Dietary Supplements. Marcel Dekker, New York, NY, pp. 680–690.

Wu, Y., Wang, D., 2009. A new class of natural glycopeptides with sugar moiety-dependent antioxidant activities derived from *Ganoderma lucidum* fruiting bodies. J. Proteome Res. 8, 436–442.

Wu, G.S., Guo, J.J., Bao, J.L., Li, X.W., Chen, X.P., Lu, J.J., et al., 2013a. Anti-cancer properties of triterpenoids isolated from *Ganoderma lucidum*—a review. Expert Opin. Investig. Drugs 22 (8), 981–992.
Wu, G.S., Song, Y.L., Yin, Z.Q., Guo, J.J., Wang, S.P., Zhao, W.W., et al., 2013b. Ganoderiol A-enriched extract suppresses migration and adhesion of MDA-MB-231 cells by inhibiting FAK-SRC-paxillin cascade pathway. PLoS One 8 (10), e76620.
Wu, L., Liang, W., Chen, W., Li, S., Cui, Y., Qi, Q., et al., 2017. Screening and analysis of the marker components in *Ganoderma lucidum* by HPLC and HPLC-MS with the aid of chemometrics. Molecules 22 (4), E584.
Xie, J., Zhao, X., Hu, D.J., Duan, D.A., Tang, Y., Li, S., 2012. Comparison of polysaccharides from two species of *Ganoderma*. Molecules 17, 740–752.
Xie, J.T., Wang, C.Z., Wicks, S., Yin, J.J., Kong, J., Li, J., et al., 2006. *Ganoderma lucidum* extract inhibits proliferation of SW 480 human colorectal cancer cells. Exp. Oncol. 28 (1), 25–29.
Yue, Q., Xie, F.B., Guan, S.H., Chao, M., Yang, M., Bao-Hong, B.H., et al., 2008. Interaction of *Ganoderma* triterpenes with doxorubicin and proteomic characterization of the possible molecular targets of *Ganoderma* triterpenes. Cancer Sci. 99, 1461–1470.
Yuen, J.W., Gohel, M.D., 2005. Anticancer effects of *Ganoderma lucidum*: a review of scientific evidence. Nutr. Cancer 53 (1), 11–17.
Zhang, J., Tang, Q., Zhou, C., Jia, W., Da Silva, L., Nguyen, L.D., et al., 2010. GLIS, a bioactive proteoglycan fraction from *Ganoderma lucidum*, displays anti-tumour activity by increasing both humoral and cellular immune response. Life Sci. 87 (19–22), 628–637.
Zeng, P., Guo, Z., Zeng, X., Hao, C., Zhang, Y., Zhang, M., et al., 2018. Chemical, biochemical, preclinical and clinical studies of *Ganoderma lucidum* polysaccharide as an approved drug for treating myopathy and other diseases in China. J. Cell Mol. Med. . Available from: https://doi.org/10.1111/jcmm.13613.
Zhao, S., Ye, G., Fu, G., Cheng, J.X., Yang, B.B., Peng, C.Z., 2011. *Ganoderma lucidum* exerts anti-tumor effects on ovarian cancer cells and enhances their sensitivity to cisplatin. Int. J. Oncol. 38 (5), 1319–1327.

CHAPTER 8

Indian Berries and Their Active Compounds: Therapeutic Potential in Cancer Prevention

Mohammad Shavez Khan, Faizan Abul Qais and Iqbal Ahmad
Department of Agricultural Microbiology, Aligarh Muslim University, Aligarh, Uttar Pradesh, India

8.1 INTRODUCTION

Recent reports and projections on cancer epidemiology worldwide show that cancer-linked mortality and morbidity in human population will continue to grow in the coming decades. It was estimated that in 2015, about 8.8 million deaths were caused by cancer globally (WHO, 2017); while in 2012, 14 million new cases of cancer were registered worldwide; and this number is expected to rise by about 70% over next two decades (WHO, 2017). This situation becomes more aggravated in developing countries due to lack of awareness and limited resources for cancer management. The Indian Council of Medical Research (ICMR) has anticipated that over 1.7 million new cases of cancer and 0.8 million deaths related to cancer will occur in India by 2020 (ICMR, 2014). There are several therapeutic approaches for cancer treatment (surgery, chemotherapy, radiotherapy, etc.), however, associated complications in addition to financial constraint are evident. Moreover, strategies to prevent the onset of cancer are still in stage of infancy (Cuzick, 2017). Thus, in view of global scenario of escalation in cancer burden and impediment related to current therapies for cancer treatment the development of novel anticancer drug has become one of the most significant thrust area in modern biological research. Although, synthetic chemistry offers an attractive arena for discovery of lead molecules, trend showed that the success rate of synthetic molecules as potent anticancer agents was too low during last decades (Scannell et al., 2012). Moreover, the transformation period from identification of lead molecule to successful drug development against cancer is very lengthy, cost intensive, and with regulatory hitches (Jonsson and Bergh, 2012). Therefore, researchers focused

New Look to Phytomedicine
DOI: https://doi.org/10.1016/B978-0-12-814619-4.00008-2

their attention on exploration of natural products from medicinal plants/herbal medicine for anticancer compounds/products (Cragg et al., 2009).

In human history, plant products/herbal medicines have been used for treatment of various diseases including cancers and the interest on herbal medicine has increased globally due its recognition for diverse therapeutic efficacy and safety (Graham et al., 2000). In fact, herbal medicines are still considered as primary source of treatment in developing nations (Atanasov et al., 2015). The research on herbal medicine has increased globally and its role in combating various health-related problems is progressively recognized (Ahmad et al., 2006). Concurrently, there is vast scientific interest in identification of drug molecules or effective herbal formulations against cancer as they are considered to be comparatively less-toxic to normal cells with better tolerability. Plant-derived bioactive molecules from root, stem, bark, leaf, flower, and fruits are found to be associated with different pharmacological actions (Singh et al., 2016). In recent years researchers have shown that phytochemicals are able to differentially prevent the progression of cancer at different stages, from initiation to the critical phases, that is, cell proliferation, metastasis, angiogenesis, immortality, immunity, genome stability, inflammation, mutation, and cell metabolism (Lee et al., 2011; González-Vallinas et al., 2013; Maru et al., 2016). Many plants products/molecules are being used as anticancer agents, including vinca alkaloids, taxanes, campothecins, omacetaxine, ingenol mebutate, vincristine, vinblastine, etc. (Amin et al., 2009; Newman and Cragg, 2016).

Numerous studies have shown the relation between functional foods and cancer management (Gonzalez and Riboli, 2010; Wu et al., 2012). Functional foods are food products with additional health benefits, characterized by an array of ingredients or constituents with potential benefits. Globally, there is a growing trend in the use of fruits, vegetables, medicinal herbs, and their extract or specific components as functional food or as a dietary supplements for prevention of various diseases including cancer (Paredes-López et al., 2010; Aghajanpour et al., 2017). Among these, berries are gaining special attention as an important component of healthy diet as they are rich in nutrients and bioactive phytochemicals. Health benefits of berry fruit are due to high levels of polyphenols, antioxidants, vitamins, minerals, and fibers (Jimenez-Garcia et al., 2013). Among those, polyphenols such as flavonoids (anthocyanins, flavanones, isoflavonoids, flavonols, and flavones), tannins, phenolic acid, and stilbenes constitute a major group (Nile and Park, 2014). Berry extracts are widely used in plant-based dietary supplements for potential health benefits. Many in vitro and animal studies have shown that berries have anticancer and antiproliferative properties (Nile and Park, 2014; Baby et al., 2017). Cancer preventive mechanisms of berry's bioactive phytochemicals are complementary and overlapping including the modulation of gene expression, activation of metabolizing enzymes, cell proliferative checkpoints, and apoptosis (Seeram 2008; Gali-Muhtasib et al., 2015). Anticancer mechanisms of different types of berries and their phytocompounds are presented in Table 8.1.

This chapter specifically focuses on Indian berries known for dietary and medicinal uses, highlighting their anticancer potential and recent research on berry's phytochemicals and their in vitro and in vivo studies.

8.2 INDIAN BLACKBERRY

Syzygium cumini L. (synonym: *Syzygium jambolana, Eugenia jambolana, Eugenia cumini*) belong to polyembryonic species of the family Myrtaceae, commonly known as Indian blackberry or Jamun. *S. cumini*, an evergreen tropical tree, is native to Indian subcontinent and naturalized in America, Africa, and Australia (Aqil et al., 2015). The oblong berries having deep purple to violet color with pinkish pulp are widely consumed as fruit. In addition to its nutraceutical value, fruits are used in traditional medicine for treatment of various diseases (Ayyanar and Subash-Babu, 2012). In vitro studies of the fruits extracts showed antiinflammatory (Chaudhuri et al., 1990), antioxidant (Banerjee et al., 2005), antidiabetic (Helmstädter, 2008), and antimutagenic activity (Saxena et al., 2013; Khan et al., 2018), and also act as detoxifier (Abdalla et al., 2011), and protection against radioactivity (Jagetia et al., 2012). Chemical components of the fruit and seed are mainly anthocyanins (in pulp) and other phenolics (Veigas et al., 2007; Benherlal and Arumughan, 2007; Faria et al., 2011; Aqil et al., 2012; Ayyanar and Subash-Babu, 2012).

S. cumini berry's extracts are shown to possess potent anticancer activity in different test systems and against different tumor cell lines (Swami et al., 2012). The selective activity of standardized extracts of Jamun fruit was also evident. It was observed that the extracts were more apoptotic against estrogen-dependent mammary breast cancer cells (MCF-7aro) when compared to estrogen-independent cell lines (MDA-MB-231), while no toxicity was observed when tested against nontumorigenic/normal counter parts MCF-10A (Li et al., 2009). Proapoptotic and antiproliferative

Table 8.1 Anticancer activity of common Indian berries and their phytoconstituents

Name	Family	Anticancer models	Reference	Anticancer active phytocompounds mechanism of action		
				Compounds	Mechanism of action	Reference
Syzygium cumini L.	Myrtaceae	MCF-7, HCT-116, A549 Skin carcinogenesis (Mice), Mammary tumor (Mice),	Li et al., 2009; Parmar et al., 2010; Charepalli et al., 2016; Aqil et al., 2016	Anthocyanins (cyaniding, malvidin, peonidin, petunidin and delphinidin)	1. Oncogenic Notch and WNT pathways 2. Cleavage Bcl-2 and PARP 3. Inhibition of TNFα-induced NF-kappa B activation 4. Modulation of DNA repair genes 5. Affecting transcription factors 6. Modulation of metastatic and angiogenic factors	Stoner et al., 2010; Kausar et al., 2012; Aiyer et al., 2012; Li et al., 2017
				Ellagic acid	1. Induction of apoptosis (elevating p53 and Cip1/p21 and decreasing cyclin D1 and E levels; suppressing Akt, Shh and Notch pathways) 2. DNA protection	Chung et al., 2013; Zhao et al., 2013; Weisburg et al., 2013; Khan et al., 2013; Eskandari et al., 2016; Zahin et al., 2014; Wang et al., 2014; Liu et al., 2017

					3. Inhibition of metastasis (downregulation of MMP2 and MMP9; inhibition of interleukin-8 and signaling pathways) 4. Antiinflammatory responses	
Phyllanthus emblica.	Euphorbiaceae	A549, HepG2, HeLa, MDA-MB-231, SK-OV3, SW620, HCCSCs, Buccal pouch carcinoma Hamster	Ngamkitidechakul et al., 2010; Krishnaveni and Mirunalini, 2012; Zhu et al., 2013; Vadde et al., 2016	Geraniin and isocorilagin	Cytotoxic and Immunomodulatory effects	Liu et al., 2012
				Polyphenolic rich extracts	Induction of apoptosis (Cell cycle arrest, apoptotic proteins viz. Fas, FasI etc are increased)	Zhu et al., 2013
Solanum nigrum	Solanaceae	HepG2, MGC-803; Mice melanoma,	Wang et al., 2010, Yang et al., 2010	Polyphenolic rich extracts	1. Induction of apoptosis (G2/M phase arrest). 2. Suppression of angiogenesis (down expression of CD31 and AKT/mTOR pathway)	Yang et al., 2010; Wang et al., 2011; Yang et al., 2016

(*Continued*)

Table 8.1 (Continued)

Name	Family	Anticancer models	Reference	Anticancer active phytocompounds mechanism of action		
				Compounds	Mechanism of action	Reference
				Solamargine	Induction of apoptosis (Cell cycle arrest at G2/M phase; upregulation of caspase-3 protein)	Ding et al., 2012b
				Steroidal glycoalkaloids	1. Modulation of apoptotic markers (Bax, Bcl-2, mutant p53 and caspase-3) 2. Cell cycle arrest at S phase	Ding et al., 2013
Withania coagulans	Solanaceae	MDA-MB-231, Vero cell	Ahmad et al., 2017	Withacoagulin-H	Inhibition of tumor necrosis factor factor-α (TNF-α) mediated nuclear factor-kappa B (NF-κB) activation	Haq et al., 2013
Morinda citrifolia	Rubiaceae	MCF-7, MDA-MB-231, HEK-293, A549	Sharma et al., 2016; Huang et al., 2016; Günay et al., 2016	Damnacanthal	1. Inhibition of oncogene Ras, p56lck tyrosine kinase, NF-κB pathway 2. Modulation of Tyrosine kinase (Met-C) and secretory system	Abu et al., 2014; García-Vilas et al., 2015; Sukamporn et al., 2016; García-Vilas et al., 2017

					3. Inhibition of Cyclin D1 4. Induction of apoptosis 5. Inhibition of different processes involve in angiogenesis	
Physalis peruviana		HeLa, L929 fibroblast, Caco-2, MCF-7	Mier-Giraldo et al., 2017; Ramadan et al., 2015	4β-Hydroxy-withanolide E	1. Induction of apoptosis 2. Generation of ROS 3. G0/G1 cell cycle arrest 4. Downregulation of Hsp90 (indirect modulation of survival and oncogenic proteins)	Chiu et al., 2013; Park et al., 2016
Zanthoxylum armatum				Flavonoid-rich extract	1. Induction of apoptosis 2. Modulation of apoptotic proteins viz. Bcl-2 and Bax	Alam et al., 2017
				Crude saponins	Induction of apoptosis	Alam et al., 2017

properties of the fruit extracts were evident against human colon cancer cells and colon cancer stem cells. Standardized Jamun pulp extracts suppressed proliferation in HCT-116 cell lines and induced apoptosis in normal cancer cells and cancer stem cells. Moreover, it was evident from colony formation assay that Jamun extract can suppress the colon cancer stem cells (Charepalli et al., 2016).

Protective efficacy of Jamun seed's extracts was evident against DMBA-induced skin carcinogenesis in Swiss albino mice. The number of papilloma and tumor incidence was significantly reduced in animals when their diet was supplemented with Jamun seeds (Parmar et al., 2010). Similarly, when Jamun-supplemented diet was given to female August Copenhagen Irish rats that received 17β-estradiol, a clear attenuation in estrogen-mediated mammary tumor incidence, tumor burden, and tumor multiplicity was observed; continous supplementation resulted into inhibition of overall mammary carcinogenicity-related biomarkers (ER-α, cyclin D1, and candidate miRNAs) when compared to control group (Aqil et al., 2016).

Anthocyanins or anthocyanidins are the major phenolic compounds (flavonoid) present in the fruit pulp of *S. cumini*. These water-soluble pigments are reported to exhibit anticarcinogenic properties such as induction of cell cycle arrest and apoptosis as well as inhibition of tumor formation and growth in animals (Wang and Stoner, 2008; Putta et al., 2017). Anthocyanins-enriched extract of Jamun fruit was found to be active against human lung cancer A549 cells (Aqil et al., 2012). The anticancer effect of mixture of anthocyanidins was found to be significantly higher than individual compounds, suggesting synergistic activity against tumor cell proliferation and metastasis, as well as in modulation of various molecular targets (Kausar et al., 2012). It was presumed that the action of individual anthocyanidins at distinct and overlapping targets, associated with carcinogenesis, was responsible for overall synergistic outcome (Kausar et al., 2012). These active constituents (anthocyanins/anthocyanidins) such as cyaniding, malvidin, peonidin, petunidin, and delphinidin possess multitargeted mechanism against malignant cell survival. Both the anthocyanins and anthocyanidins inhibit cancerous cell survival via modulation of many signaling molecule, including DNA repair genes, transcription factor, and pathways such as Bcl-2, COX-2, Cyclin D1, Notch, Pl3/AKT as well as metastatic and angiogenic mediators VEGF, uPAR, and MMPs (Stoner et al., 2010; Aiyer et al., 2012; Li et al., 2017).

Other important phytochemical constituents having chemopreventive effects are ellagitannins, flavonols, and phenolic acids predominantly found in the seeds of *S. cumini* (Aqil et al., 2015). These polyphenols are known for their protective and antioxidant nature and are extensively investigated for their anticancer activity (Ren et al., 2003; Fresco et al., 2006; León-González et al., 2015), protective ability, and mode of action (Yar Khan et al., 2012; Gali-Muhtasib et al., 2015; Mileo and Miccadei, 2016). However, ellagitannins especially ellagic acid needs special reference as it exerts potent therapeutic effects against different types of cancer including breast, prostate, colon, and skin, as described by Zhang et al. (2014). Ellagic acid acts via different mechanisms including induction of apoptosis pathways in different cell lines (Chung et al., 2013; Zhao et al., 2013; Weisburg et al., 2013; Mishra and Vinayak, 2014), prevention of DNA damage by carcinogens (Zahin et al., 2014), inhibiting metastasis (Pitchakarn et al., 2013; Wang et al., 2014; Liu et al., 2017), and by antiinflammatory actions (Khan et al., 2013; Eskandari et al., 2016). Other indirect mechanisms involved in ellagic acid anticancer action includes radio sensitization, antiviral activity, inhibition of drug resistance protein, and protective effects on heart and liver (Zhang et al., 2014). Fig. 8.1 shows a diagrammatic representation on cancer prevention mechanism of ellagic acid.

8.3 INDIAN GOOSEBERRY

Phyllanthus emblica L. (Synonym: *Emblica officinalis*) is a medium-sized deciduous tree belonging to the family Euphorbiaceae, commonly known as Indian gooseberry, emblic myrobalans, and Amla (in Hindi). The plant species is native to India, also growing in Sri Lanka, Uzbekistan, South East Asia, and China nowadays (Baliga and Dsouza, 2011). This dietary globular fruit, yellowish-green in color with obtusely triangular six-celled nut, is of immense use in various folk and Indian traditional medicinal systems (Mirunalini and Krishnaveni, 2010; Baby et al., 2017). Amla extracts is extensively investigated for different biological activities as reviewed by Variya et al. (2016). Some of the recent in vitro studies revealed antiinflammatory (Golechha et al., 2014), antioxidant (Yamamoto et al., 2016; Packirisamy et al., 2017), cryoprotective (Zhang et al., 2016), antiaging (Pientaweeratch et al., 2016), nephrotoxicity modulation (Malik et al., 2016), antidiabetic (Fatima et al., 2017), and hepatoprotective (Huang et al., 2017) properties.

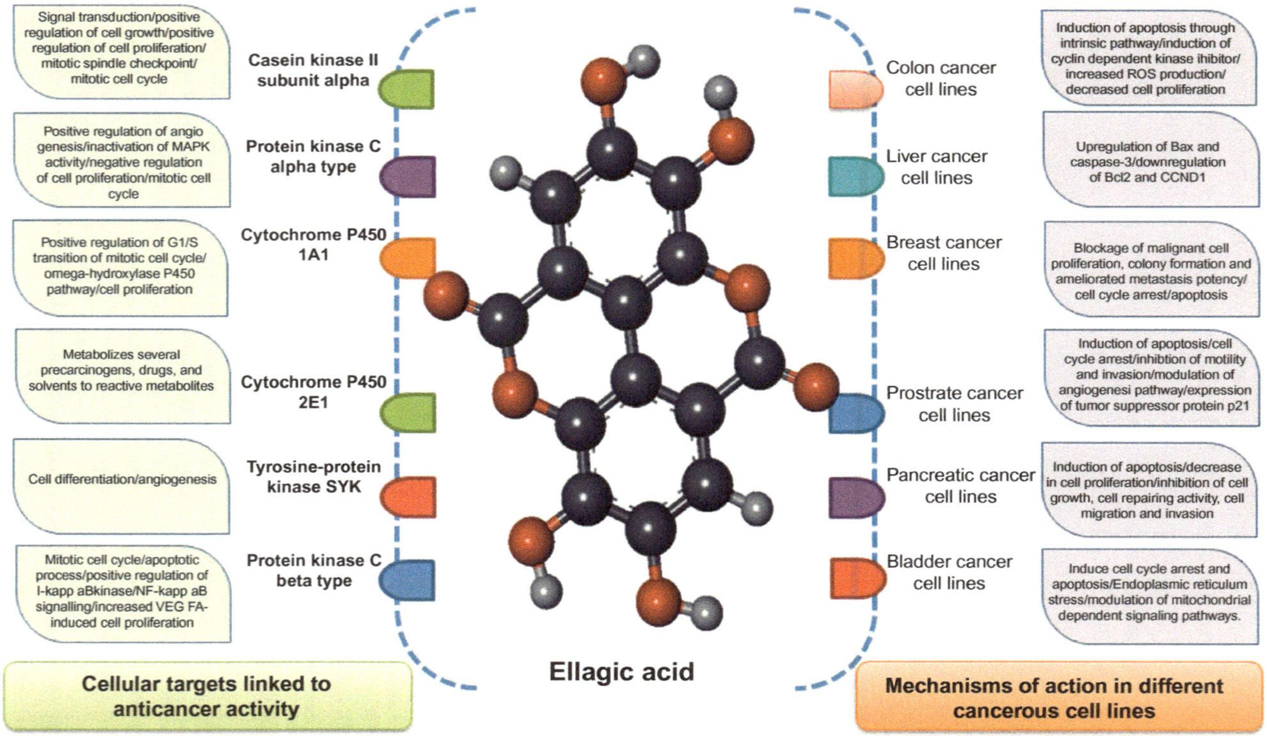

Figure 8.1 Anticancer targets and mechanism of ellagic acid on different malignant cell lines.

The fruit of Amla is rich in vitamin C (ascorbic acid) and contains several bioactive phytochemicals, of which majority are of polyphenols (ellagic acid, chebulinic acid, gallic acid, chebulagic acid, apeigenin, quercetin, corilagin, leutolin, etc.) as described by various workers (Patel et al., 2011; Variya et al., 2016; Yadav et al., 2017). Sugar-substituted phenolics such as flavone glycosides, phenolic glycosides, and flavonol glycosides (Variya et al., 2016), as well as tannins, such as emblicanin A, emblicanin B, phyllaemblicin B, and punigluconoin, are reported in fruit's pulp (Bhattacharya et al., 1999; Yadav et al., 2017).

Amla is largely explored for therapeutic potential against various diseases including cancers (Baliga and Dsouza, 2011; Variya et al., 2016; Baby et al., 2017). Anticancer mechanisms of Amla extracts include free-radical and antioxidant effect, modulation of various enzymes of inflammation and carcinogenesis, modulation of cell cycle proteins, induction of apoptosis in neoplastic cells, and prevention of metastasis as reviewed by Yang and Liu (2014). Reports have also showed considerable degree of antiproliferative activity of extracts of Indian gooseberry (Liu et al., 2012; Zhao et al., 2015). Mahata et al. (2013) studied the effect of Amla extract on transcriptional activity of human papilloma virus (HPV) and on activator protein-1 (AP-1). It was found that Indian gooseberry extract inhibits the DNA binding of activator protein (AP-1) in HPV16- and HPV18-positive cervical cancer cell lines (Mahata et al., 2013). Chemopreventive activity of Amla extract was also evident against buccal pouch carcinoma in hamsters (Krishnaveni and Mirunalini, 2012). It was demonstrated that administration of Amla extract results in improvement of antioxidant biomarkers in pouch and plasma of DMBA-induced tumor in hamsters (Krishnaveni and Mirunalini, 2012).

Preclinical evidence using different cancer cell lines showed that aqueous extract of *P. emblica* berry induced apoptosis at considerably low concentrations (Ngamkitidechakul et al., 2010). Polyphenolic rich extracts of the Amla berry were also shown to induce cell cycle arrest at G2/M phase. Moreover, apoptotic marker protein such as Fas, FasL, and cleaved caspase-8 were shown to increase in HeLa cells treated with Amla extract (Zhu et al., 2013). Cytotoxicity induced by Amla extracts was also reported to be selective for cancerous cell lines while normal cell lines remained least affected (Vadde et al., 2016). De et al. (2013) showed potent anticarcinogenic activity of the extract against ovarian cancer cell line and in vivo model demonstrating that the antiproliferative action of the extract is independent of apoptosis, but mediated through angiogenesis and autophagy.

The antiproliferative activity of Amla extract is thought to be an individual or synergistic effect of its phytoconstituent, especially tannins and phenolics. Geraniin and isocorilagin, among other phenolics reported from Amla berry, were shown to be most potent against MCF-7 and HeLa cell lines (Liu et al., 2012). Anticancer activity and possible mode of action of those phenolic constituents are comprehensively reviewed by León-González et al., 2015. However, hydrolysable tannins are among the major bioactive components of the Amla which are shown to be active against malignant cell lines (Yang and Liu, 2014). These anticancer effects are predominantly due to proliferation inhibition, apoptosis induction, invasion suppression, and angiogenesis inhibition (Fresco et al., 2006; Cai et al., 2017).

8.4 OTHER IMPORTANT INDIAN BERRIES OF MEDICINAL/EDIBLE IMPORTANCE

A brief description on anticancer potential of *Solanum nigrum* (Family: Solanaceae; Common name: Makoi), *Withania coagulans* (Family: Solanaceae; Common name: Paneer phool), *Zanthoxylum armatum* (Family: Rutaceae; Common name: Tejbal), *Morinda citrifolia* (Family: Rubiaceae; Common name: Indian Mulberry, Noni), and *Physalis peruviana* (Family: Solanaceae; Common name: Rasbhari, Cape gooseberry) is included in this section.

S. nigrum (Makoi), a herbaceous plant with small rounded berry fruits used in traditional medicine for treatment of various diseases including cancer (Muthu et al., 2006); different extracts of Makoi are reported for anticancer potential. Overall decrease in tumor incidence and growth following treatment with Makoi extracts has been mechanistically linked with the inhibitory effect of the extract on migrating and invasive abilities of the melanomas (Wang et al., 2010). Similarly, tetradecanoylphobor-13-acetate mediated migration and invasion of HepG2 cells were also found to be attenuated by polyphenol-rich extracts of Makoi berry (Yang et al., 2010). Polyphenolic extracts also induced apoptosis and arrest the cell cycle at G2/M phase in HepG2 carcinoma, providing basis behind complete inhibition of tumor progression in experimental animal (Wang et al., 2011). It was also evident that polyphenolic and aqueous extract of Makoi suppress the angiogenesis, as evident from reduced expression of CD31 (angiogenesis maker) in nude mice bearing tumor xenograft (Yang et al., 2016). Moreover, these extracts were also shown to downregulate signaling pathway (AKT/mTOR pathway), critical for growth and differentiation in human umbilical vascular endothelial cells and HepG2 cells (Yang et al., 2016).

Atanu et al. (2011) described several bioactive phytochemicals including phenolics, alkaloids, saponins, polysaccharide, glycoproteins, steroidal derivatives etc., in Makoi fruit. Polysaccharides, steroidal glycosides, and glycoalkaloids isolated from *S. nigrum* were shown to be highly active against tumors and carcinomas (Ding et al., 2012a; Ding et al., 2013). The antitumor activity of water-soluble polysaccharide could be linked with immunostimulating potential and also with monosaccharide composition (Ding et al., 2012a). Solamargine (steroidal alkaloid glycoside) isolated from *S. nigrum* caused cell cycle arrest at G2/M and upregulate the expression of caspase-3 protein, thus by inducing apoptosis in human hepatoma SMMC-7721 and HepG2 cells (Ding et al., 2012b). Steroidal glycoalkaloids, isolated from Makoi, induced cell cycle arrest at S-phase and modulated apoptotic markers viz., Bax, Bcl-2, mutant p53, and caspase-3 in MGC-803 cells (Ding et al., 2013).

Fruit of *W. coagulans* (Paneer phool), a traditionally used herbal plant is also subjected to anticancer evaluation. Nine bioactive withanoloids were isolated from the fruit of *W. coagulans* via bioactive-guided fractionation for cancer chemoprevention (Haq et al., 2013). All the isolated withanoloids show varying degree of cancer chemopreventive potential, and among isolated compounds the 1-withacoagulin-H (steroidal lactone) was found to be most active in inhibiting tumor necrosis factor-α (TNF-α)-mediated nuclear factor-kappa B (NF-κB) activation. Recently, *W. coagulans* fruit extract was evaluated against human breast cancer (MDA-MB-231) and Vero cell lines by Ahmad et al. (2017). It was observed that the extract was active against human breast cancer while noncancerous Vero cell lines remain unaffected. Further, the DNA fragmentation pattern of the treated cells reveal clear sign of apoptotic-mediated cell toxicity of the berry extract against the human breast carcinoma. Withaferin A, a major constituent of *W. coagulans*, was detected in the bioactive extracts.

M. citrifolia (Noni) fruits are also used for treatment of many health-related problems including cancer, diabetes, hypertension, etc. (Brown, 2012). Noni extract and its phytoconstituents are shown to impart cytotoxic affect in numerous carcinomas (Sharma et al., 2016; Huang et al., 2016). The ethyl acetate extract inhibited the proliferation of MCF-7, MDA-MB-231, and HEK-293 cell lines; increase in apoptosis in MCF-7 and MDA-MB-231 cells; and arrested the cell cycle in the G1/S phase in MCF-7 and G0/G1 phase in MDA-MB-231 cells (Sharma et al., 2016).

Although, 150 bioactive phytochemicals are isolated from Noni berry, majority of them belong to phenolics or alkaloids (Brown, 2012). Damnacanthal, an anthraquinone, initially isolated from Noni roots and fruit is extensively investigated for its antiproliferative activity, which exhibits cell growth arrest and induces caspase activity in colorectal cancer cells (Nualsanit et al., 2012). In vitro studies confirmed that Damnacanthal inhibits oncogene Ras, p56lck tyrosine kinase, Nf-KB pathway, and induces apoptosis (Abu et al., 2014). Cell growth assay, soft agar clonogenic, migration, and invasion assays of HepG2 carcinoma with Damnacanthal revealed the antitumor effects of the anthraquinone. Tyrosine kinase (Met-C) and secretory system in HepG2 were the molecular targets of Damnacanthal, accompanied by apoptosis, antiangiogenesis, and clonogenic properties (García-Vilas et al., 2015). Benzo(a)pyrene-induced interferon levels along with some of the oxidative stress markers were also shown to decrease considerably in the Damnacanthal treated A549 cell line. Moreover, Damnacanthal also induced apoptotic gene expression while downregulating the antiapoptotic gene expression in treated cells (Günay et al., 2016). Cyclin D1, a protein involved in the regulation of cell cycle and overexpressed in carcinomas, was found to be inhibited under the effect of Damnacanthal and its nanoformulation (Sukamporn et al., 2016). Expression of the cyclin protein was not affected, rather the suppression in protein expression was observed at posttranslational levels in the panel of cancerous cells (HCT-116, HT-29, MCF-7, and PC-3) when treated with the compound (Sukamporn et al., 2016). Moreover, in vivo and ex vivo experiments clearly demonstrate the specific effects of Damnacanthal on different steps of angiogenic process including inhibition of tublogenesis, endothelial cell proliferation, survival, migration, and production of extracellular matrix remodeling enzyme (García-Vilas et al., 2017).

P. peruviana (Rasbhary), another edible berry belonging to family Solanacae, is thought to be an important source of bioactive chemicals and is a functional food (Hassanien, 2011). Rasbhary is also used extensively in Peruvian traditional medicine for treatment of various diseases including cancer (Puente et al., 2011). Fruit extracts of Rasbhary were evaluated for cytotoxic activity and immunomodulatory potential. Extracts showed appreciable cytotoxic activity against HeLa cells and L929 fibroblast and also reduced the release of interleukin (IL)-6, IL-8, and MCP-1 (Mier-Giraldo et al., 2017). In another study, polyphenol-rich ethanol extract of berry was evaluated for its anticancer potential against different cell lines. The extract was found to be more active against colon cancer cell line (Caco-2) compared to breast cancer cell line MCF-7 (Ramadan et al., 2015). Rasbhary juice was also shown to exhibit antitumor activity on hepatocellular carcinoma, when coadministered with doxorubicin in experimental mice. Hepatocellular carcinoma-induced rats who received Rasbhary juice showed improvement in all the cancer-linked parameters. The antitumor effect of Rasbhary juice was linked with downregulation of p53 expression and upregulation of Bcl-2 domain (Hassan et al., 2017).

Similar to other berries, *P. peruviana* is also reported to be a rich source of bioactive phytochemicals showing anticancer potential. Withanolides, naturally occurring steroidal lactones, are the most abundantly occurring phytoconstiuent of the genus Physalis and are reported to be associated with anticancer/cytotoxic activity (Lan et al., 2009; Zhang and Tong, 2016). 4β-Hydroxywithanolide E (4HWE), a withanolide isolated from *P. peruviana*, is reported to inhibit proliferation of various human cancer cell lines including breast cancers, pancreatic cancers, oral cancers, liver cancer, and lung cancer (Machin et al., 2010; Zhang et al., 2012; Chiu et al., 2013; You et al., 2014; Gu et al., 2014; Sang-Ngern et al., 2015). Underlying molecular mechanisms for cytotoxic effect of 4HWE include induction of apoptosis, ROS generation, and DNA damage (Chiu et al., 2013). Recently, Park et al. (2016) showed that treatment of HT-29 (human colorectal adenocarcinoma cell lines) with high concentration of 4HWE induce apoptosis, whereas at low concentrations G0/G1 cell cycle arrest was observed. Additionally, compound 4HWE downregulated the Hsp90 client protein expression and epigenetic modulation of growth response genes, thus modulating a relatively complex array of survival and oncogenic proteins.

Z. armatum (Tejbal), another wild berry plant of high medicinal value, is found in Himalayan and sub-Himalayan regions of Indian subcontinent (Singh and Singh, 2011). The flavonoid-rich fruit extract was reported to have apoptotic effect on Ehrlich ascites tumor in Swiss albino mice. The fruit extract was also shown to induce DNA fragmentation and modulation in ratio of apoptotic proteins (Bcl-2 and Bax) (Karmakar et al., 2016). In another study (Alam et al., 2017), crude saponins from fruit of *Z. armatum* showed cytotoxic effect against panel of cancerous cell lines (MDA-MB-468, MCF-7, and Caco-2). It was evident that the extracts induce apoptosis in all the cell lines tested while breast cancer lines showed more sensitivity toward crude saponins.

8.5 CONCLUSION

The role of phytocompounds in prevention and treatment of various types of cancer is subjected to scrutiny since decades. Indian berries as a promising reservoir of different bioactive anticancer phytocompounds has been evident from the present review. Phytocompounds/herbal medicines from berries are emerging as a suitable candidate for anticancer drug development. Some of these identified compounds (anthocyanins, ellagic acid, hydrolysable tannins, solamargine, withanoloids, and damnacanthal) target multiple stages of cancer developmental pathways. However, extensive research needs to be conducted on these phytochemicals/extracts for their efficacy against different types of cancers, both in vitro and *in* vivo to develop multidimensional understanding of their chemotherapeutic utility in cancer preventions.

REFERENCES

Abdalla, F.H., Bellé, L.P., Bitencourt, P.E.R., De Bona, K.S., Zanette, R.A., Boligon, A.A., et al., 2011. Protective effects of *Syzygium cumini* seed extract against methylmercury-induced sistemic toxicity in neonatal rats. Biometals 24 (2), 349–356.

Abu, N., Mohd Ali, N., Yong Ho, W., Keong Yeap, S., Aziz, A., Yusran, M., et al., 2014. Damnacanthal: a promising compound as a medicinal anthraquinone. Anti-Cancer Agents Med. Chem. 14 (5), 750–755.

Aghajanpour, M., Nazer, M.R., Obeidavi, Z., Akbari, M., Ezati, P., Kor, N.M., 2017. Functional foods and their role in cancer prevention and health promotion: a comprehensive review. Am. J. Cancer Res. 7 (4), 740.

Ahmad, I., Aqil, F., Ahmad, F., Owais, M., 2006. Herbal medicines: prospects and constraints. In: Ahmad, I., Aqil, F., Owais, M. (Eds.), Modern Phytomedicine: Turning Medicinal Plants Into Drugs. Wiley-VCH Verlag GmbH & Co. KGaA, Weinheim, Germany, pp. 59–77.

Ahmad, R., Fatima, A., Srivastava, A.N., Khan, M.A., 2017. Evaluation of apoptotic activity of *Withania coagulans* methanolic extract against human breast cancer and Vero cell lines. J. Ayurveda Integr. Med. 8 (3), 177–183.

Aiyer, H.S., Warri, A.M., Woode, D.R., Hilakivi-Clarke, L., Clarke, R., 2012. Influence of berry polyphenols on receptor signaling and cell-death pathways: implications for breast cancer prevention. J. Agric. Food Chem. 60 (23), 5693–5708.

Alam, F., us Saqib, Q.N., Waheed, A., 2017. Cytotoxic activity of extracts and crude saponins from *Zanthoxylum armatum* DC. against human breast (MCF-7, MDA-MB-468) and colorectal (Caco-2) cancer cell lines. BMC Complementary Altern. Med. 17 (1), 368.

Amin, A., Gali-Muhtasib, H., Ocker, M., Schneider-Stock, R., 2009. Overview of major classes of plant-derived anticancer drugs. Int. J. Biomed. Sci. 5 (1), 1.

Aqil, F., Gupta, A., Munagala, R., Jeyabalan, J., Kausar, H., Sharma, R.J., et al., 2012. Antioxidant and anti-proliferative activities of anthocyanins/ellagitannin-enriched extracts from *Syzygium cumini* L. (Jamun, the Indian Blackberry). Nutr. Cancer 64 (3), 428–438.

Aqil, F., Munagala, R., Jeyabalan, J., Joshi, T., Gupta, R.C., Singh, I.P., 2015. The Indian blackberry (Jamun), antioxidant capacity, and cancer protection. In: Preedy, V. (Ed.), Cancer: Oxidative Stress and Dietary Antioxidants. Academic Press, Amsterdam, pp. 101–113.

Aqil, F., Jeyabalan, J., Munagala, R., Singh, I.P., Gupta, R.C., 2016. Prevention of hormonal breast cancer by dietary Jamun. Mol. Nutr. Food Res. 60 (6), 1470–1481.

Atanasov, A.G., Waltenberger, B., Pferschy-Wenzig, E.M., Linder, T., Wawrosch, C., Uhrin, P., et al., 2015. Discovery and resupply of pharmacologically active plant-derived natural products: a review. Biotechnol. Adv. 33 (8), 1582–1614.

Atanu, F.O., Ebiloma, U.G., Ajayi, E.I., 2011. A review of the pharmacological aspects of *Solanum nigrum* Linn. Biotechnol. Mol. Biol. Rev. 6 (1), 1–8.

Ayyanar, M., Subash-Babu, P., 2012. *Syzygium cumini* (L.) Skeels: a review of its phytochemical constituents and traditional uses. Asian Pac. J. Trop. Biomed. 2 (3), 240–246.

Baby, B., Antony, P., Vijayan, R., 2017. Antioxidant and anticancer properties of berries. Crit. Rev. Food Sci. Nutr. Available from: https://doi.org/10.1080/10408398.2017.1329198.

Baliga, M.S., Dsouza, J.J., 2011. Amla (Emblica officinalis Gaertn), a wonder berry in the treatment and prevention of cancer. Eur. J. Cancer Prev. 20 (3), 225–239.

Banerjee, A., Dasgupta, N., De, B., 2005. In vitro study of antioxidant activity of *Syzygium cumini* fruit. Food Chem. 90 (4), 727–733.

Benherlal, P.S., Arumughan, C., 2007. Chemical composition and in vitro antioxidant studies on *Syzygium cumini* fruit. J. Sci. Food Agric. 87 (14), 2560–2569.

Bhattacharya, A., Chatterjee, A., Ghosal, S., Bhattacharya, S.K., 1999. Antioxidant activity of active tannoid principles of *Emblica officinalis* (Amla). Indian J. Exp. Biol. 37 (7), 676–680.

Brown, A.C., 2012. Anticancer activity of *Morinda citrifolia* (Noni) fruit: a review. Phytother. Res. 26 (10), 1427–1440.

Cai, Y., Zhang, J., Chen, N.G., Shi, Z., Qiu, J., He, C., et al., 2017. Recent advances in anticancer activities and drug delivery systems of tannins. Med. Res. Rev. 37 (4), 665–701.

Chaudhuri, A.K., Pal, S., Gomes, A., Bhattacharya, S., 1990. Anti-inflammatory and related actions of *Syzygium cuminii* seed extract. Phytother. Res. 4 (1), 5–10.

Charepalli, V., Reddivari, L., Vadde, R., Walia, S., Radhakrishnan, S., Vanamala, J.K., 2016. *Eugenia jambolana* (Java Plum) fruit extract exhibits anti-cancer activity against early stage human HCT-116 colon cancer cells and colon cancer stem cells. Cancers 8 (3), 29.

Chiu, C.C., Haung, J.W., Chang, F.R., Huang, K.J., Huang, H.M., Huang, H.W., et al., 2013. Golden berry-derived 4β-hydroxywithanolide E for selectively killing oral cancer cells by generating ROS, DNA damage, and apoptotic pathways. PLoS. One. 8 (5), e64739.

Chung, Y.C., Lu, L.C., Tsai, M.H., Chen, Y.J., Chen, Y.Y., Yao, S.P., et al., 2013. The inhibitory effect of ellagic acid on cell growth of ovarian carcinoma cells. Evidence-Based Complementary Altern. Med. 306705.

Cragg, G.M., Grothaus, P.G., Newman, D.J., 2009. Impact of natural products on developing new anti-cancer agents. Chem. Rev. 109 (7), 3012–3043.

Cuzick, J., 2017. Preventive therapy for cancer. Lancet Oncol. 18 (8), 472–482.

De, A., De, A., Papasian, C., Hentges, S., Banerjee, S., Haque, I., Banerjee, S.K., 2013. *Emblica officinalis* extract induces autophagy and inhibits human ovarian cancer cell proliferation, angiogenesis, growth of mouse xenograft tumors. PLoS One 8 (8), e72748.

Ding, X., Zhu, F., Gao, S., 2012a. Purification, antitumour and immunomodulatory activity of water-extractable and alkali-extractable polysaccharides from *Solanum nigrum* L. Food Chem. 131 (2), 677–684.

Ding, X., Zhu, F.S., Li, M., Gao, S.G., 2012b. Induction of apoptosis in human hepatoma SMMC-7721 cells by solamargine from *Solanum nigrum* L. J. Ethnopharmacol. 139 (2), 599–604.

Ding, X., Zhu, F., Yang, Y., Li, M., 2013. Purification, antitumor activity in vitro of steroidal glycoalkaloids from black nightshade (*Solanum nigrum* L.). Food Chem. 141 (2), 1181–1186.

Eskandari, E., Heidarian, E., Amini, S.A., Saffari-Chaleshtori, J., 2016. Evaluating the effects of ellagic acid on pSTAT3, pAKT, and pERK1/2 signaling pathways in prostate cancer PC3 cells. J. Cancer Res. Ther. 12 (4), 1266.

Faria, A.F., Marques, M.C., Mercadante, A.Z., 2011. Identification of bioactive compounds from jambolão (*Syzygium cumini*) and antioxidant capacity evaluation in different pH conditions. Food Chem. 126 (4), 1571–1578.

Fatima, N., Hafizur, R.M., Hameed, A., Ahmed, S., Nisar, M., Kabir, N., 2017. Ellagic acid in *Emblica officinalis* exerts anti-diabetic activity through the action on β-cells of pancreas. Eur. J. Nutr. 56 (2), 591–601.

Fresco, P., Borges, F., Diniz, C., Marques, M.P.M., 2006. New insights on the anticancer properties of dietary polyphenols. Med. Res. Rev. 26 (6), 747–766.

Gali-Muhtasib, H., Hmadi, R., Kareh, M., Tohme, R., Darwiche, N., 2015. Cell death mechanisms of plant-derived anticancer drugs: beyond apoptosis. Apoptosis 20 (12), 1531–1562.

García-Vilas, J.A., Quesada, A.R., Medina, M.A., 2015. Damnacanthal, a noni anthraquinone, inhibits c-Met and is a potent antitumor compound against Hep G2 human hepatocellular carcinoma cells. Sci. Rep. 5.

García-Vilas, J.A., Pino-Ángeles, A., Martínez-Poveda, B., Quesada, A.R., Medina, M.Á., 2017. The noni anthraquinone damnacanthal is a multi-kinase inhibitor with potent anti-angiogenic effects. Cancer Lett. 385, 1–11.

Golechha, M., Sarangal, V., Ojha, S., Bhatia, J., Arya, D.S., 2014. Anti-inflammatory effect of *Emblica officinalis* in rodent models of acute and chronic inflammation: involvement of possible mechanisms. Int. J. Inflammation 178408.

Gonzalez, C.A., Riboli, E., 2010. Diet and cancer prevention: contributions from the European Prospective Investigation into Cancer and Nutrition (EPIC) study. Eur. J. Cancer 46 (14), 2555–2562.

González-Vallinas, M., González-Castejón, M., Rodríguez-Casado, A., Ramírez de Molina, A., 2013. Dietary phytochemicals in cancer prevention and therapy: a complementary approach with promising perspectives. Nutr. Rev. 71 (9), 585–599.

Graham, J.G., Quinn, M.L., Fabricant, D.S., Farnsworth, N.R., 2000. Plants used against cancer—an extension of the work of Jonathan Hartwell. J. Ethnopharmacol. 73 (3), 347–377.

Gu, M., Yu, Y., Gunaherath, G.K.B., Gunatilaka, A.L., Li, D., Sun, D., 2014. Structure-activity relationship (SAR) of withanolides to inhibit Hsp90 for its activity in pancreatic cancer cells. Invest. New Drugs 32 (1), 68–74.

Günay, E., Celik, S., Sarinc-Ulasli, S., Özyürek, A., Hazman, Ö., Günay, S., et al., 2016. Comparison of the anti-inflammatory effects of proanthocyanidin, quercetin, and damnacanthal on benzo (a) pyrene exposed A549 alveolar cell line. Inflammation 39 (2), 744–751.

Hassan, H.A., Serag, H.M., Qadir, M.S., Ramadan, M.F., 2017. Cape gooseberry (*Physalis peruviana*) juice as a modulator agent for hepatocellular carcinoma-linked apoptosis and cell cycle arrest. Biomed. Pharmacother. 94, 1129–1137.

Hassanien, M.F.R., 2011. *Physalis peruviana*: a rich source of bioactive phytochemicals for functional foods and pharmaceuticals. Food Rev. Int. 27 (3), 259–273.

Haq, I.U., Mirza, B., Kondratyuk, T.P., Park, E.J., Burns, B.E., Marler, L.E., Pezzuto, J.M., 2013. Preliminary evaluation for cancer chemopreventive and cytotoxic potential of naturally growing ethnobotanically selected plants of Pakistan. Pharm. Biol. 51 (3), 316–328.

Helmstädter, A., 2008. *Syzygium cumini* (L.) SKEELS (Myrtaceae) against diabetes–125 years of research. Die Pharmazie-An International. J. Pharm. Sci. 63 (2), 91–101.

Huang, C., Wei, Y.X., Shen, M.C., Tu, Y.H., Wang, C.C., Huang, H.C., 2016. Chrysin, abundant in *Morinda citrifolia* fruit water–etoac extracts, combined with apigenin synergistically induced apoptosis and inhibited migration in human breast and liver cancer cells. J. Agric. Food Chem. 64 (21), 4235–4245.

Huang, C.Z., Tung, Y.T., Hsia, S.M., Wu, C.H., Yen, G.C., 2017. The hepatoprotective effect of *Phyllanthus emblica* L. fruit on high fat diet-induced non-alcoholic fatty liver disease (NAFLD) in SD rats. Food Funct. 8 (2), 842–850.

ICMR, 2014. <https://www.icmr.nic.in/Gallery/ncrp/press_release.pdf>.

Jagetia, G.C., Shetty, P.C., Vidyasagar, M.S., 2012. Inhibition of radiation-induced DNA damage by jamun, *Syzygium cumini*, in the cultured splenocytes of mice exposed to different doses of γ-radiation. Integr. Cancer Ther. 11 (2), 141–153.

Jimenez-Garcia, S.N., Guevara-Gonzalez, R.G., Miranda-Lopez, R., Feregrino-Perez, A. A., Torres-Pacheco, I., Vazquez-Cruz, M.A., 2013. Functional properties and quality characteristics of bioactive compounds in berries: biochemistry, biotechnology, and genomics. Food Res. Int. 54 (1), 1195–1207.

Jonsson, B., Bergh, J., 2012. Hurdles in anticancer drug development from a regulatory perspective. Nat. Rev. Clin. Oncol. 9 (4), 236–243.

Karmakar, I., Haldar, S., Chakraborty, M., Chaudhury, K., Dewanjee, S., Haldar, P.K., 2016. Regulation of apoptosis through bcl-2/bax proteins expression and DNA damage by *Zanthoxylum alatum*. Pharm. Biol. 54 (3), 503–508.

Kausar, H., Jeyabalan, J., Aqil, F., Chabba, D., Sidana, J., Singh, I.P., et al., 2012. Berry anthocyanidins synergistically suppress growth and invasive potential of human non-small-cell lung cancer cells. Cancer Lett. 325 (1), 54–62.

Khan, M.K., Ansari, I.A., Khan, M.S., Arif, J.M., 2013. Dietary phytochemicals as potent chemotherapeutic agents against breast cancer: inhibition of NF-κB pathway via molecular interactions in rel homology domain of its precursor protein p105. Pharmacogn. Mag. 9 (33), 51.

Khan, M.S., Qais, F.A., Ahmad, I., Hussain, A., Alajmi, M.F., 2018. Genotoxicity inhibition by *Syzygium cumini* (L.) seed fraction and rutin: understanding the underlying mechanism of DNA protection. Toxicol Res. 7 (2), 156–171.

Krishnaveni, M., Mirunalini, S., 2012. Chemopreventive efficacy of *Phyllanthus emblica* L. (Amla) fruit extract on 7, 12-dimethylbenz (a) anthracene induced oral carcinogenesis–A dose–response study. Environ. Toxicol. Pharmacol. 34 (3), 801–810.

Lan, Y.H., Chang, F.R., Pan, M.J., Wu, C.C., Wu, S.J., Chen, S.L., et al., 2009. New cytotoxic withanolides from *Physalis peruviana*. Food Chem. 116 (2), 462–469.

Lee, K.W., Bode, A.M., Dong, Z., 2011. Molecular targets of phytochemicals for cancer prevention. Nat. Rev. Cancer 11 (3), 211–218.

León-González, A.J., Auger, C., Schini-Kerth, V.B., 2015. Pro-oxidant activity of polyphenols and its implication on cancer chemoprevention and chemotherapy. Biochem. Pharmacol. 98 (3), 371–380.

Li, L., Adams, L.S., Chen, S., Killian, C., Ahmed, A., Seeram, N.P., 2009. *Eugenia jambolana* Lam. berry extract inhibits growth and induces apoptosis of human breast cancer but not non-tumorigenic breast cells. J. Agric. Food Chem. 57 (3), 826–831.
Li, D., Wang, P., Luo, Y., Zhao, M., Chen, F., 2017. Health benefits of anthocyanins and molecular mechanisms: update from recent decade. Crit. Rev. Food Sci. Nutr. 57 (8), 1729–1741.
Liu, X., Zhao, M., Wu, K., Chai, X., Yu, H., Tao, Z., et al., 2012. Immunomodulatory and anticancer activities of phenolics from emblica fruit (*Phyllanthus emblica* L.). Food Chem. 131 (2), 685–690.
Liu, H., Zeng, Z., Wang, S., Li, T., Mastriani, E., Li, Q.H., et al., 2017. Main components of pomegranate, ellagic acid and luteolin, inhibit metastasis of ovarian cancer by down-regulating MMP2 and MMP9. Cancer Biol. Ther. 18 (12), 990–999.
Machin, R.P., Veleiro, A.S., Nicotra, V.E., Oberti, J.C., Padrón, M.J., 2010. Antiproliferative activity of withanolides against human breast cancer cell lines. J. Nat. Prod. 73 (5), 966–968.
Mahata, S., Pandey, A., Shukla, S., Tyagi, A., Husain, S.A., Das, B.C., Bharti, A.C., 2013. Anticancer activity of *Phyllanthus emblica* Linn.(Indian gooseberry): inhibition of transcription factor AP-1 and HPV gene expression in cervical cancer cells. Nutr. Cancer 65 (Suppl. 1), 88–97.
Malik, S., Suchal, K., Bhatia, J., Khan, S.I., Vasisth, S., Tomar, A., et al., 2016. Therapeutic potential and molecular mechanisms of *Emblica officinalis* Gaertn in countering Nephrotoxicity in rats induced by the chemotherapeutic agent Cisplatin. Front. Pharmacol. 7.
Maru, G.B., Hudlikar, R.R., Kumar, G., Gandhi, K., Mahimkar, M.B., 2016. Understanding the molecular mechanisms of cancer prevention by dietary phytochemicals: from experimental models to clinical trials. World J. Biol. Chem. 7 (1), 88.
Mier-Giraldo, H., Díaz-Barrera, L.E., Delgado-Murcia, L.G., Valero-Valdivieso, M.F., Cáez-Ramírez, G., 2017. Cytotoxic and immunomodulatory potential activity of *Physalis peruviana* fruit extracts on cervical cancer (hela) and fibroblast (l929) cells. J. Evidence-Based Complementary Altern. Med. 2156587217718751.
Mileo, A.M., Miccadei, S., 2016. Polyphenols as modulator of oxidative stress in cancer disease: new therapeutic strategies. Oxid. Med. Cell. Longevity. 2016.
Mirunalini, S., Krishnaveni, M., 2010. Therapeutic potential of *Phyllanthus emblica* (Amla): the ayurvedic wonder. J. Basic Clin. Physiol. Pharmacol. 21 (1), 93–105.
Mishra, S., Vinayak, M., 2014. Ellagic acid induces novel and atypical PKC isoforms and promotes caspase-3 dependent apoptosis by blocking energy metabolism. Nutr. Cancer 66 (4), 675–681.
Muthu, C., Ayyanar, M., Raja, N., Ignacimuthu, S., 2006. Medicinal plants used by traditional healers in Kancheepuram district of Tamil Nadu, India. J. Ethnobiol. Ethnomed. 2 (1), 43.
Newman, D.J., Cragg, G.M., 2016. Natural products as sources of new drugs from 1981 to 2014. J. Nat. Prod. 79 (3), 629–661.
Ngamkitidechakul, C., Jaijoy, K., Hansakul, P., Soonthornchareonnon, N., Sireeratawong, S., 2010. Antitumour effects of *Phyllanthus emblica* L.: induction of cancer cell apoptosis and inhibition of in vivo tumour promotion and in vitro invasion of human cancer cells. Phytother. Res. 24 (9), 1405–1413.
Nile, S.H., Park, S.W., 2014. Edible berries: bioactive components and their effect on human health. Nutrition 30 (2), 134–144.
Nualsanit, T., Rojanapanthu, P., Gritsanapan, W., Lee, S.H., Lawson, D., Baek, S.J., 2012. Damnacanthal, a noni component, exhibits antitumorigenic activity in human colorectal cancer cells. J. Nutr. Biochem. 23 (8), 915–923.

Packirisamy, R.M., Bobby, Z., Panneerselvam, S., Koshy, S.M., Jacob, S.E., 2017. Metabolomic analysis and antioxidant effect of amla (*Emblica officinalis*) extract in preventing oxidative stress induced red cell damage and plasma protein alterations: an *in vitro* study. J. Med. Food 21 (1), 81–89.

Paredes-López, O., Cervantes-Ceja, M.L., Vigna-Pérez, M., Hernández-Pérez, T., 2010. Berries: improving human health and healthy aging, and promoting quality life—a review. Plant Foods Hum. Nutr. 65 (3), 299–308.

Park, E.J., Sang-Ngern, M., Chang, L.C., Pezzuto, J.M., 2016. Induction of cell cycle arrest and apoptosis with downregulation of Hsp90 client proteins and histone modification by 4β-hydroxywithanolide E isolated from *Physalis peruviana*. Mol. Nutr. Food Res. 60 (6), 1482–1500.

Parmar, J., Sharma, P., Verma, P., Goyal, P.K., 2010. Chemopreventive action of *Syzygium cumini* on DMBA-induced skin papillomagenesis in mice. Asian Pac. J. Cancer Prev. 11 (1), 261–265.

Patel, J.R., Tripathi, P., Sharma, V., Chauhan, N.S., Dixit, V.K., 2011. *Phyllanthus amarus*: ethnomedicinal uses, phytochemistry and pharmacology: a review. J. Ethnopharmacol. 138 (2), 286–313.

Pientaweeratch, S., Panapisal, V., Tansirikongkol, A., 2016. Antioxidant, anti-collagenase and anti-elastase activities of *Phyllanthus emblica*, *Manilkara zapota* and silymarin: an in vitro comparative study for anti-aging applications. Pharm. Biol. 54 (9), 1865–1872.

Pitchakarn, P., Chewonarin, T., Ogawa, K., Suzuki, S., Asamoto, M., Takahashi, S., et al., 2013. Ellagic acid inhibits migration and invasion by prostate cancer cell lines. Asian Pac. J. Cancer Prev. 14 (5), 2859–2863.

Puente, L.A., Pinto-Muñoz, C.A., Castro, E.S., Cortés, M., 2011. *Physalis peruviana* Linnaeus, the multiple properties of a highly functional fruit: a review. Food Res. Int. 44 (7), 1733–1740.

Putta, S., Yarla, N.S., Peluso, I., Tiwari, D.K., Reddy, G.V., Giri, P.V., et al., 2017. Anthocyanins: possible role as multitarget therapeutic agents for prevention and therapy of chronic diseases. Curr. Pharm. Des. Available from: https://doi.org/10.2174/1381612823666170519151801.

Ramadan, M.M., El-Ghorab, A.H., Ghanem, K.Z., 2015. Volatile compounds, antioxidants, and anticancer activities of Cape gooseberry fruit (*Physalis peruviana* L.): an in-vitro study. J. Arab Soc. Med. Res. 10 (2), 56.

Ren, W., Qiao, Z., Wang, H., Zhu, L., Zhang, L., 2003. Flavonoids: promising anticancer agents. Med. Res. Rev. 23 (4), 519–534.

Sang-Ngern, M., Youn, U.J., Park, E.J., Kondratyuk, T.P., Miklossy, G., Walla, M.M., et al., 2015. Anticancer potential of withanolides and its derivatives from *Physalis peruviana* (POHA). Planta Med. 81 (11), PX81.

Saxena, S., Gautam, S., Sharma, A., 2013. Comparative evaluation of antimutagenicity of commonly consumed fruits and activity-guided identification of bioactive principles from the most potent fruit, Java plum (*Syzygium cumini*). J. Agric. Food Chem. 61 (42), 10033–10042.

Scannell, J.W., Blanckley, A., Boldon, H., Warrington, B., 2012. Diagnosing the decline in pharmaceutical R&D efficiency. Nat. Rev. Drug. Discov. 11 (3), 191–200.

Seeram, N.P., 2008. Berry fruits: compositional elements, biochemical activities, and the impact of their intake on human health, performance, and disease. J. Agric. Food Chem. 56 (3), 627–629.

Sharma, K., Pachauri, S.D., Khandelwal, K., Ahmad, H., Arya, A., Biala, P., et al., 2016. Anticancer effects of extracts from the fruit of *Morinda Citrifolia* (Noni) in breast cancer cell lines. Drug Res. 66 (03), 141–147.

Singh, T.P., Singh, O.M., 2011. Phytochemical and pharmacological profile of *Zanthoxylum armatum* DC.-an overview. Indian J. Nat. Prod. Resour. 2 (3), 275–285.
Singh, S., Sharma, B., Kanwar, S.S., Kumar, A., 2016. Lead phytochemicals for anticancer drug development. Front. Plant Sci. 7, 1667.
Stoner, G.D., Wang, L.S., Sardo, C., Zikri, N., Hecht, S.S., Mallery, S.R., 2010. In: Milner, J.A., Romagnolo, D.F. (Eds.), Bioactive Compounds and Cancer. Nutrition and Health. Humana Press, Totowa, NJ, pp. 703–723.
Sukamporn, P., Rojanapanthu, P., Silva, G., Zhang, X., Gritsanapan, W., Baek, S.J., 2016. Damnacanthal and its nanoformulation exhibit anti-cancer activity via cyclin D1 down-regulation. Life Sci. 152, 60–66.
Swami, S.B., Thakor, N.S.J., Patil, M.M., Haldankar, P.M., 2012. Jamun (*Syzygium cumini* (L.)): a review of its food and medicinal uses. Food Nutr. Sci 3 (8), 1100.
Vadde, R., Radhakrishnan, S., Kurundu, H.E.K., Reddivari, L., Vanamala, J.K., 2016. Indian gooseberry (*Emblica officinalis* Gaertn.) suppresses cell proliferation and induces apoptosis in human colon cancer stem cells independent of p53 status via suppression of c-Myc and cyclin D1. J. Funct. Foods 25, 267–278.
Variya, B.C., Bakrania, A.K., Patel, S.S., 2016. *Emblica officinalis* (Amla): a review for its phytochemistry, ethnomedicinal uses and medicinal potentials with respect to molecular mechanisms. Pharmacol. Res. 111, 180–200.
Veigas, J.M., Narayan, M.S., Laxman, P.M., Neelwarne, B., 2007. Chemical nature, stability and bioefficacies of anthocyanins from fruit peel of *Syzygium cumini* Skeels. Food Chem. 105 (2), 619–627.
Wang, L.S., Stoner, G.D., 2008. Anthocyanins and their role in cancer prevention. Cancer Lett. 269 (2), 281–290.
Wang, H.C., Wu, D.H., Chang, Y.C., Li, Y.J., Wang, C.J., 2010. *Solanum nigrum* Linn. water extract inhibits metastasis in mouse melanoma cells in vitro and in vivo. J. Agric. Food Chem. 58 (22), 11913–11923.
Wang, H.C., Chung, P.J., Wu, C.H., Lan, K.P., Yang, M.Y., Wang, C.J., 2011. *Solanum nigrum* L. polyphenolic extract inhibits hepatocarcinoma cell growth by inducing G2/M phase arrest and apoptosis. J. Sci. Food Agric. 91 (1), 178–185.
Wang, L., Li, W., Lin, M., Garcia, M., Mulholland, D., Lilly, M., et al., 2014. Luteolin, ellagic acid and punicic acid are natural products that inhibit prostate cancer metastasis. Carcinogenesis 35 (10), 2321–2330.
Weisburg, J.H., Schuck, A.G., Reiss, S.E., Wolf, B.J., Fertel, S.R., Zuckerbraun, H.L., et al., 2013. Ellagic acid, a dietary polyphenol, selectively cytotoxic to HSC-2 oral carcinoma cells. Anticancer Res. 33 (5), 1829–1836.
WHO, 2017. <http://www.who.int/mediacentre/factsheets/fs297/en/>.
Wu, Q.J., Yang, Y., Vogtmann, E., Wang, J., Han, L.H., Li, H.L., et al., 2012. Cruciferous vegetables intake and the risk of colorectal cancer: a *meta*-analysis of observational studies. Ann. Oncol. 24 (4), 1079–1087.
Yadav, S.S., Singh, M.K., Singh, P.K., Kumar, V., 2017. Traditional knowledge to clinical trials: a review on therapeutic actions of *Emblica officinalis*. Biomed. Pharmacother. 93, 1292–1302.
Yamamoto, H., Morino, K., Mengistu, L., Ishibashi, T., Kiriyama, K., Ikami, T., et al., 2016. Amla enhances mitochondrial spare respiratory capacity by increasing mitochondrial biogenesis and antioxidant systems in a murine skeletal muscle cell line. Oxid. Med. Cell Longevity 1735841.
Yang, B., Liu, P., 2014. Composition and biological activities of hydrolyzable tannins of fruits of *Phyllanthus emblica*. J. Agric. Food Chem. 62 (3), 529–541.
Yang, M.Y., Hsu, L.S., Peng, C.H., Shi, Y.S., Wu, C.H., Wang, C.J., 2010. Polyphenol-rich extracts from *Solanum nigrum* attenuated PKC α-mediated migration and invasion of hepatocellular carcinoma cells. J. Agric. Food Chem. 58 (9), 5806–5814.

Yang, M.Y., Hung, C.H., Chang, C.H., Tseng, T.H., Wang, C.J., 2016. *Solanum nigrum* suppress angiogenesis-mediated tumor growth through inhibition of the Akt/mTOR pathway. Am. J. Chin. Med. (Gard. City. N. Y). 44 (06), 1273–1288.

Yar Khan, H., Zubair, H., Fahad Ullah, M., Ahmad, A., Mumtaz Hadi, S., 2012. A prooxidant mechanism for the anticancer and chemopreventive properties of plant polyphenols. Curr. Drug Targets 13 (14), 1738–1749.

You, B.J., Wu, Y.C., Lee, C.L., Lee, H.Z., 2014. Non-homologous end joining pathway is the major route of protection against 4β-hydroxywithanolide E-induced DNA damage in MCF-7 cells. Food Chem. Toxicol. 65, 205–212.

Zahin, M., Ahmad, I., Gupta, R.C., Aqil, F., 2014. Punicalagin and ellagic acid demonstrate antimutagenic activity and inhibition of benzo [a] pyrene induced DNA adducts. Biomed. Res. Int. 467465.

Zhang, W.N., Tong, W.Y., 2016. Chemical constituents and biological activities of plants from the genus physalis. Chem. Biodiversity 13 (1), 48–65.

Zhang, H., Samadi, A.K., Cohen, M.S., Timmermann, B.N., 2012. Antiproliferative withanolides from the Solanaceae: a structure–activity study. Pure Appl. Chem. 84 (6), 1353–1367.

Zhang, H.M., Zhao, L., Li, H., Xu, H., Chen, W.W., Tao, L., 2014. Research progress on the anticarcinogenic actions and mechanisms of ellagic acid. Cancer Biol Med. 11 (2), 92.

Zhang, Y., Zhao, L., Guo, X., Li, C., Li, H., Lou, H., et al., 2016. Chemical constituents from *Phyllanthus emblica* and the cytoprotective effects on H2O2-induced PC12 cell injuries. Arch. Pharmacol. Res. 39 (9), 1202–1211.

Zhao, M., Tang, S.N., Marsh, J.L., Shankar, S., Srivastava, R.K., 2013. Ellagic acid inhibits human pancreatic cancer growth in Balb c nude mice. Cancer Lett. 337 (2), 210–217.

Zhao, T., Sun, Q., Marques, M., Witcher, M., 2015. Anticancer properties of *Phyllanthus emblica* (Indian gooseberry). Oxid. Med. Cell Longevity 950890.

Zhu, X., Wang, J., Ou, Y., Han, W., Li, H., 2013. Polyphenol extract of *Phyllanthus emblica* (PEEP) induces inhibition of cell proliferation and triggers apoptosis in cervical cancer cells. Eur. J. Med. Res. 18 (1), 46.

FURTHER READING

Chatterjee, A., Chatterjee, S., Biswas, A., Bhattacharya, S., Chattopadhyay, S., Bandyopadhyay, S.K., 2012. Gallic acid enriched fraction of *Phyllanthus emblica* potentiates indomethacin-induced gastric ulcer healing via e-NOS-dependent pathway. Evidence-Based Complementary Altern. Med. 487380.

D'souza, J.J., D'souza, P.P., Fazal, F., Kumar, A., Bhat, H.P., Baliga, M.S., 2014. Antidiabetic effects of the Indian indigenous fruit *Emblica officinalis* Gaertn: active constituents and modes of action. Food Func. 5 (4), 635–644.

Li, J., Li, Q., Feng, T., Li, K., 2008. Aqueous extract of *Solanum nigrum* inhibit growth of cervical carcinoma (U14) via modulating immune response of tumor bearing mice and inducing apoptosis of tumor cells. Fitoterapia 79 (7), 548–556.

Shivananjappa, M.M., Joshi, M.K., 2012. Influence of *Emblica officinalis* aqueous extract on growth and antioxidant defense system of human hepatoma cell line (HepG2). Pharm. Biol. 50 (4), 497–505.

CHAPTER 9

Prospects of Essential Oils in Controlling Pathogenic Biofilm

Huma Jafri, Firoz Ahmad Ansari and Iqbal Ahmad
Department of Agricultural Microbiology, Aligarh Muslim University, Aligarh, Uttar Pradesh, India

9.1 INTRODUCTION

9.1.1 Biofilm in Disease Development

Biofilm is described as a well organized microbial community enclosed in extra polymeric substances (EPSs) and adhering to biotic and abiotic surfaces (Costerton et al., 1999). Biofilm formation is a multistep phenomenon which includes physical, biological, and chemical processes. These processes are influenced by environmental and hydrodynamic conditions (Bryers, 2008). Biofilm formation comprises a sequence of steps as described by Kostakioti et al. (2013) and Crouzet et al. (2014) as follows (Fig. 9.1).

1. Preconditioning of the solid surface by macromolecules in the medium.
2. Transport of free-living cells from the medium to the surface.
3. Nonspecific adsorption of planktonic cells at the surfaces.
4. Irreversible attachment of bacterial cells at the surface.
5. Communication of cells or production of signaling molecules (quorum sensing).
6. Substrate transport within the biofilms.
7. Metabolism of substrate by the sessile cells and transport of metabolized products leading to cell multiplication, replication, production of EPS.
8. Detachment of biofilm.

Biofilm development constitutes an alternative lifestyle which involves multiple regulatory networks that leads to change in gene expression thereby mediating reorganization of the microbial cells (Kostakioti et al., 2013; Atray and Atray, 2015). These microorganisms may form very dense EPS matrix which constitutes 90% of the biomass (Flemming and

New Look to Phytomedicine
DOI: https://doi.org/10.1016/B978-0-12-814619-4.00009-4

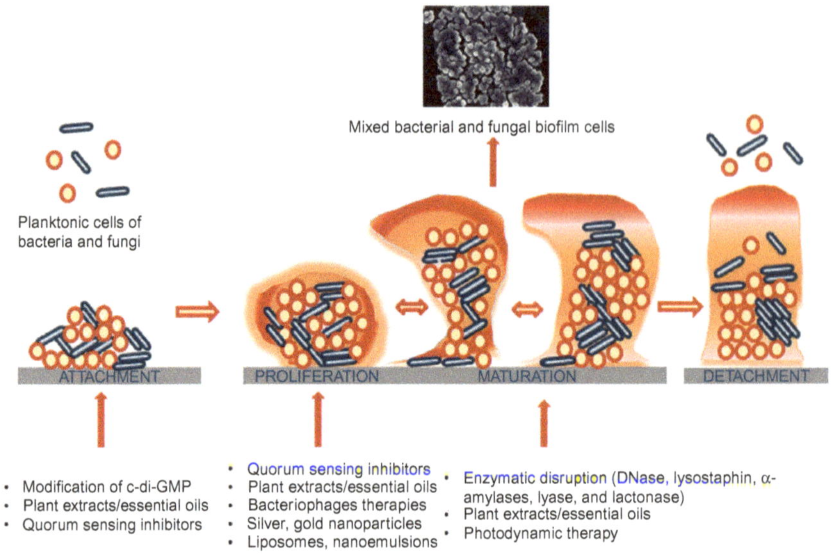

Figure 9.1 Stages of biofilm development and strategies to prevent and eradicate biofilm in vitro.

Wingender, 2010). The chemical composition of EPS matrix constitutes mainly carbohydrate, proteins, extracellular DNA (eDNA), water, and other adhesive fibers (Flemming and Wingender, 2010; Fong and Yildiz, 2015). The biofilm cells have entirely different physiology and physical properties (Barbara et al., 2009). The architecture of biofilms consists of water channel for nutrients transport and the water is efficiently retained within the biofilm with hydrophilic polysaccharides (Flemming and Wingender, 2010). Furthermore, these microorganisms secrete enzymes which modify the composition of EPS in response to nutrient availability thereby making biofilm environment more favorable for microbial community (Flemming, 2016). The structural component of the matrix yielded a highly hydrated structure that help to maintain the microbial cells in close physical proximity, enabling close cell-cell interactions and exchange of genetic material, as well as protecting sessile cells from radiations, desiccation, and other damaging agents (de Carvalho, 2007; Kostakioti et al., 2013). The tough architecture of biofilm is partly due to division of labor within the subpopulation of microbial community which alter the gene expression in response to availability of nutrient and oxygen (Lopez et al., 2009; Hadjifrangiskou et al., 2012). Biofilm gene expression pattern is different than planktonic form. Various workers have reported differential regulation of more than 3000 genes under the planktonic and sessile mode of growth (Harding et al., 2009; Fanning and Mitchell, 2012). Some of these genes are involved in upregulation of secondary metabolic pathways including antimicrobial drug resistance (Blankenship and Mitchell, 2006; Munita and Arias, 2016). It is also reported that metabolically inactive nondividing persister cells within biofilms are also present. These cells are genetically identical to rest of the cells of the biofilm community but showed resistance to many antibiotics (Wood et al., 2013). It is believed that these cells are responsible for reoccurring of biofilms on treatment with antimicrobial drugs in the clinical setting (Fauvart et al., 2011). The matrix also provides protection to biofilm cells from exposure to innate immune defenses (such as opsonization and phagocytosis) and antibiotic treatments within the biofilm niche (Limoli et al., 2015).

9.1.2 Biofilm on Living Tissue (Mucosal Biofilm) and Their Associated Problems

The National Institute of Health (NIH) estimates that biofilm accounts for over 80% of all infections in the body. About 80% of all chronic infections and 65% of all microbial infections are associated with biofilms. Pathogenic biofilms are the 10th leading cause of death in the United States (Harriott and Noverr, 2009). The term "mucosal biofilm" was coined by Garth Ehrlich and denotes biofilms that grow on mucosal surfaces (Costerton et al., 1999). Mucosal biofilm provides protection to microorganisms from innate immune defense which may further contribute to long term survival and invasion into host tissues, resulting in damage of mucosal tissue (Dongari-Bagtzoglou, 2008; Phillips and Schultz, 2012). Majority of the human mucosal surfaces are frequently colonized by opportunistic pathogens which may result in development of infection. The nature of mucosal biofilms may be polymicrobial (Dongari-Bagtzoglou, 2008). Synergistic or antagonistic interactions of bacteria and fungi influence the biofilms. Understanding on the cellular and molecular basis of interactions among pathogens as well as with normal flora will be helpful to explain the pathogenesis of mucosal biofilm infections (Dongari-Bagtzoglou, 2008; Peters et al., 2012).

Various bacteria in oral cavity interact with other bacterial species and host cells associated with dental plaque development which is a well known example of pathogenic biofilm infections (Vasudevan, 2017). *Pseudomonas aeruginosa*, associated pulmonary infection with cystic fibrosis patient, is also a type of a mucosal biofilm. *P. aeruginosa* colonizes the respiratory tract indefinitely through formation of biofilm development (Høiby et al., 2010). Mucosal biofilm is also present in the gut mucosa including inflammatory bowel disease (IBD). The mucosal biofilm diversity is low in relation to high degree of diversity in gut flora (Swidsinski et al., 2005; de Vos, 2015). The mucosal biofilm associated with chronic otitis media is known to involve the mucosal surface of the middle ear (Ehrlich et al., 2002; Akyıldız et al., 2013). Biofilm associated mucosal infections such as tonsillitis, stomatitis, rhinitis, sinusitis, otitis, urethritis, cystitis, vaginitis, and dermatitis are examples of such inflammatory diseases. Thus, variety of pathogenic bacteria are involved in the development of mucosal biofilm. Some of the examples are listed in Table 9.1.

9.1.3 Biofilm on Medical Devices and Their Associated Problems

According to Centre for Disease Control and Prevention Report 2007, nosocomial infections estimated 1.7 million of infections and 99,000 deaths each year (Revelas, 2012). Biofilm associated chronic infections such as upper respiratory tract infections (*P. aeruginosa*), urinary tract infections (UTIs) (*Candida albicans, Candida tropicalis, Streptococcus mutans,* and others), catheter induced and device associated infections (*E. coli, Staphylococcus aureus, Staphylococcus epidermidis, C. albicans,* and others) have been increasingly reported (Desai et al., 2013; Buommino et al., 2014; Falahati et al., 2016; Hogan et al., 2016; Cigana et al., 2017; Sheikh et al., 2017). Nosocomial infections account for 7% in developed and 10% in developing countries (Khan et al., 2017a). Furthermore, 32% of all nosocomial infections are UTIs, 22% are surgical site infections, 15% are lung infections (pneumonia), and 14% are blood stream infections (Bryers, 2008). About 50% of nosocomial infection are associated with implanted medical devices (Paredes et al., 2014). Various medical implants, such as catheters, prosthetic heart valves, orthopedic implants, vascular prosthesis, and contact lenses, etc. can be colonized by pathogenic biofilms (Bryers, 2008; Rodrigues, 2011). The best treatment for foreign body associated biofilm infections is to remove the infected device. However, it is very difficult to remove medical implants. Thus, the increasing uses of invasive medical procedures, infection involving biofilm form an important consideration as a risk factor for complication postoperatively (Eiff et al., 2005; Ribeiro et al., 2012; Tande and Patel, 2014). In some cases (e.g., otitis media, cholesteatoma, tonsillitis), patients

Table 9.1 Infectious diseases caused by biofilm forming microorganisms

Infectious disease	Biofilm forming microorganisms	References
Biliary tract infection	*Escherichia coli*, *Klebsiella pneumoniae*	Newman and Cragg (2007), Gargouri et al. (2015), and Royo-Cebrecos et al. (2017)
Cystic fibrosis	*P. aeruginosa*, *Staphylococcus aureus*, *Haemophilus influenzae*, *Burkholderia cepacia*	Coutinho et al. (2008) and Chmiel et al. (2014)
Dental caries	*Streptococcus mutans*, *Streptococcus oralis*, *Lactobacillus acidophilus*, *Candida albicans*	Freires et al. (2015), Shimazu et al. (2016), and Valdebenito et al. (2017)
Endocarditis	*Streptococcus* species, *S. aureus*, *Enterococcus* species, *Enterococcus faecalis*	Nallapareddy et al. (2006) and Olmos et al. (2017)
Musculoskeletal infection	*S. aureus*, *Streptococcus pyogenes*	Rosenfeld et al. (2017)
Periodontitis	*Treponema denticola*, *Porphyromonas gingivalis*, *Actinobacillus actinomycetemcomitans*, *Eubacterium nodatum*, *Campylobacter rectus*, *Prevotella intermedia*, *Peptostreptococcus micros*	Guthmiller et al. (2002) and Al Yahfoufi (2017)
Prostatitis	*E. coli*, *Klebsiella*, *Proteus*, *Staphylococcus* species, *Chlamydia trachomatis*, *Trichomonas vaginalis*, *Mycoplasma genitalium*, *Neisseria gonorrhoeae*, *Mycobacterium tuberculosis*	Lipsky et al. (2010) and Lee et al. (2016)
Osteomyelitis	*S. aureus*, *P. aeruginosa*, *E. coli*	Howell and Goulston (2011) and Mohiti-Asli et al. (2016)
Otitis media	*S. aureus*, *Pseudomonas* species, *S. epidermidis*, *Klebsiella* species, *E. coli*, *Streptococcus pneumoniae*, *Haemophilus influenzae*, *Moraxella catarrhalis*	Ramakrishnan et al. (2007), Akhtar et al. (2017), and Bergenfelz and Hakansson (2017)

(*Continued*)

Table 9.1 (Continued)

Infectious disease	Biofilm forming microorganisms	References
Wound infections	*Enterococcus faecalis, Streptococcus pyogenes, Candida parapsilosis S. aureus, P. aeruginosa, Proteus mirabilis, E. coli, Corynebacterium* species.	Bessa et al. (2015) and Antonio et al. (2017)

may cured by surgical treatment or by growth related anatomical changes, while many other affected individuals were relegated to intermittent antibiotic therapy for the rest of their lives (Cystic fibrosis, prostatitis) (Costerton et al., 2003). Although antibiotics fail to kill the biofilm but it may reverses the symptoms caused by the detachment of planktonic cells from the biofilm (Aparna and Yadav, 2008; Stoodley et al., 2011). Therefore biofilm infections are recalcitrant, its symptoms may reoccur after cycles of antibiotic therapy until the biofilm cells are surgically removed from the body (Del Pozo et al., 2006; Wu et al., 2015). It is evident that biofilms escape antimicrobial challenges by multiple mechanisms (Costerton et al., 1999; Lebeaux et al., 2014). Occurrence of biofilm by pathogenic bacteria on medical devices has been reported by many workers in recent years. Some of these reports are listed in Table 9.2.

Considering the medical significance of biofilm based infections and problem associated with its treatment a number of strategies have been developed and described (Khan et al., 2014) including the role of plant derived products/essential oils. In this chapter we have briefly listed the various strategies of combating biofilm and special focus on the prospects of essential oils.

9.2 STRATEGIES TO PREVENT/ERADICATE BIOFILM

Increasing scientific research over the last two decades in biofilm enables to understand the possible targets to eradicate or prevent biofilm infections. Various strategies have been introduced which provide potential platform to combat infections in susceptible populations. Antibiofilm strategies that could be employed to combat infection problem have been addressed by many workers as described by Taraszkiewicz et al. (2013), Khan et al. (2014), Sadekuzzaman et al. (2015), and Wu et al. (2015).

Table 9.2 List of type of medical devices involving biofilm infections

Type of medical devices	Biofilm forming microorganisms	References
Arteriovenous shunts	*S. epidermidis, S. aureus*	MacRae et al. (2016)
Biliary stent blockage	*Enterococcus* species, *E. coli, Klebsiella* species	Donelli et al. (2007), Guaglianone et al. (2010), and Kwon and Lehman (2016)
Central venous catheters or dialysis catheters	*S. epidermidis, S. aureus, E. faecalis, K. pneumoniae, P. aeruginosa, C. albicans*	Patil et al. (2011) and Gahlot et al. (2014)
Contact lens	*P. aeruginosa*, Gram-positive bacteria, fungi	Stapleton and Carnt (2012)
Endotracheal tubes	*S. epidermidis, P. aeruginosa*	Hotterbeekx et al. (2016)
Orthopedic devices	*S. epidermidis*	Morgenstern et al. (2016)
Sutures	*S. epidermidis, S. aureus, E. faecalis*	Ming et al. (2008) and Henry-Stanley et al. (2010)
Urinary catheter	*S. aureus, S. epidermidis, E. faecalis, P. aeruginosa, E. coli, K. pneumoniae, Enterobacter* sp., *P. mirabilis, S. agalactiae, C. albicans*	Hola et al. (2010) and Murugan et al. (2016)

A brief description of the common antibiofilm strategies have been given for the convenience of the reader.

1. *Enzymatic disruption of biofilm.* The disruption of biofilm structure could be achieved via the degradation of individual component of matrix by various enzymes (DNase, lysostaphin, α-amylases, lyase, and lactonase) (Mann et al., 2009; Taraszkiewicz et al., 2013).
2. *Quorum sensing inhibitors.* Inhibition of quorum sensing is a promising approach to prevent biofilm development. Biofilm development involves quorum sensing and inhibitors of quorum sensing are considered as potent antibiofilm agent. There are different quorum sensing targets available which include inhibition of signal synthesis, direct degradation of a signal molecule, and/or inhibition of binding of the signal transduction cascade (Sadekuzzaman et al., 2015; Chang et al., 2017).
3. *Nanoparticle as antibiofilm agent.* Nanoparticles, both organic (liposomes, nanoemulsions) and inorganic nanoparticles (silver, gold, selenium

nanoparticle), are reported to have antibacterial and antibiofilm potencies (Ramalingam et al., 2012; Singh et al., 2016). These are also used as surface coating and drug delivery agents and thus offer a very promising alternative to conventional methods of biofilm control. Recently, nanotechnologies have become a promising tool for biofilm prevention and control (Sadekuzzaman et al., 2015).

4. *Bacteriophages therapies.* Phages are currently considered as a potential alternative to antibiotics for biofilm inhibition or disruption. However, this is still in infancy but increased interest by scientific community has reviewed the scope of phages in combating resistance and biofilm problems (Soothill, 2013).
5. *Modification of c-di-GMP.* c-di-GMP was discovered 25 years ago and has been emerged as one of the most common and important bacterial second messengers. Modification of c-di-GMP as target to disperse biofilm infections. c-di-GMP have been shown to play important role in transforming from the planktonic state to sessile state to establish multicellular biofilm communities, and change from the virulent state of acute infections to the less virulent but chronic infections. Therefore, modulating c-di-GMP signaling pathways in bacteria could offer a new way to combat the biofilm formation in medical situations (Hong et al., 2015).
6. *Photodynamic therapy (PDT).* PDT is a medical treatment that utilizes photosensitizer or photosensitizing agent as a drug. This approach has emerged as alternative to antimicrobial agents and mechanical means for eradicating preformed biofilms (Soukos and Goodson, 2011; García-Quintanilla et al., 2013).
7. *Plant derived antimicrobials.* Several natural products such as plant extracts and essential oils have been attributed to antibiofilm properties and these properties have been extensively studied (Sadekuzzaman et al., 2015). Extracts from different parts of the plants have been widely used in modulating drug resistance, biofilm inhibition or eradication and these studies could provide possible directions in the treatment of biofilm-related infections (Yap et al., 2014). Many authors have reported the antibiofilm activity of extracts prepared from different parts of the plants. Ghosh et al. (2017) investigated the antibiofilm activity of *Mussaenda roxburghii* phytoconstituents. Four compounds were characterized as 2α, 3β, 19α, 23-tetra hydroxyurs-12-en-28-oic acid (1), β-sitosterol glucoside (4), lupeol palmitate (5), and myoinositol (6) and tested for antibacterial and antibiofilm activity against

P. aeruginosa. It was found that compound 1 exhibited three times more antibiofilm activity with minimum inhibitory concentration (MIC) at 0.74 mm compared to that of streptomycin. The data also revealed that compound 1 was also noncytotoxic against sheep RBC and murine peritoneal macrophages at sub-MIC doses. Choi et al. (2017) investigated the antibiofilm activity of methanol extracts from 37 Korean medicinal plants against dental pathogens *C. albicans* and *S. mutans*. They also investigated the antimicrobial activity of plant extracts by broth dilution and disk diffusion assay. The antibiofilm and antioxidant activities were evaluated based on the inhibitory effect against glucosyl transferase (GTase) and the DPPH assay, respectively. Among 37 herbs, eight plant extracts showed growth and biofilm inhibitory activity against both organisms. Among all, the methanol extracts (1.0 mg/mL) from *Camellia japonica* and *Thuja orientalis* significantly inhibited the growth of *C. albicans and S. mutans* by over 76% and 83% respectively. MICs of these methanol extracts were found to be 0.5 mg/mL using the disk diffusion assay on solid agar media. Biofilm inhibition was 92.4% and 98% respectively, at same concentration of each extract. Based on above report, *C. japonica* and *T. orientalis* were found potentially useful as antimicrobial and antibiofilm agents in preventing dental diseases. Yan et al. (2017) studied the antibiofilm activity of emodin 1,2,8-trihydroxy-6-methyl anthraquinone, an anthraquinone derivative isolated from *Polygonum cuspidatum* and *Rheum palmatum*, against *S. aureus* CMCC 26003 biofilm cultures in vitro. They also studied the synergistic interaction of emodin and berberine chloride. Emodin significantly reduced *S. aureus* biofilm growth in a dose dependent manner as quantified by crystal violet assay. This was also confirmed by SEM analysis. It was also found that emodin (at sub-MIC concentration) intervened the release of extracellular DNA and inhibited expression of the biofilm-related genes (cidA, icaA, dltB, agrA, sortase A, and sar A) by real time RT-PCR. The studies concluded that emodin can be used to control *S. aureus* biofilm-related infections.

9.3 PLANT ESSENTIAL OILS

Due to increase in antimicrobial resistance among pathogenic bacteria and poor progress in the discovery of new antimicrobial drugs demanded in development of other alternative strategies to combat bacterial infection including herbal products. Plants produced wide variety of secondary metabolites that exhibited antimicrobial activity against variety of pathogens (bacteria, fungi, and viruses) (Hammer et al., 1999; Kavanaugh and Ribbec, 2012; Saviuc et al., 2015). Some plants are known to produce essential oils, especially superior plants, angiosperms and gymnosperms (Saviuc et al., 2015). Essential oils are natural plant derived volatile substances (Sadekuzzaman et al., 2015). It is expected that essential oil effectively kills bacteria without promoting the acquisition of resistance (Kavanaugh and Ribbec, 2012) Essential oils showed little toxicity and degrade quickly making them environment friendly (Jin et al., 2011). The mode of antibacterial/antifungal action of essential oils has been well documented in literature (Lopez-Romero et al., 2015). The common mechanisms include disrupt cell wall and cytoplasmic membrane, leading to lysis and leakage of intracellular compounds (Lopez-Romero et al., 2015). Other biological activity of essential oils reported includes antigenotoxic (Sinha et al., 2011; Thirugnanasampandan et al., 2012). In the recent past role of natural products derived from medicinal plants and other sources for interfering pathogenic biofilms have gained increased attention by the researches. Many excellent articles have been published (Derakhshan et al., 2010; Lang and Buchbauer, 2011; Khan and Ahmad, 2012; de Paula et al., 2014; Mathur et al., 2014; Kim et al., 2015; Almeida et al., 2016). In this chapter we aimed to provide recent scientific reports which have mainly focused on essential oils and its active constituents. Kim et al. (2015) screened the various essential oils for their antibiofilm activity against *P. aeruginosa*. Essential oil from cinnamon bark oil and cinnamaldehyde inhibited growth of *P. aeruginosa* at 0.5% (v/v). Production of pyocyanin and 2-heptyl-3-hydroxy-4 (1H)-quinolone, the swarming motility and the hemolytic activity of *P. aeruginosa* was also inhibited by cinnamon bark oil and eugenol. Interestingly, at 0.01% (v/v) of cinnamon bark oils and active compounds (cinnamaldehyde and eugenol) also inhibited biofilm formation of enterohemorrhagic *E. coli* 0157: H7 (EHEC). Transcriptional analysis also revealed that cinnamon bark oil downregulated the curli genes and Shiga like toxin gene stx2 in EHEC. They concluded that cinnamon bark oil and its constituents, cinnamaldehyde and eugenol decreased the pyocyanin production and pseudomonas quinolone signal (PQS), the swarming motility and the hemolytic activity of *P. aeruginosa* and inhibit EHEC biofilm formation (Table 9.3).

Bogavac et al. (2015) studied the antimicrobial and antibiofilm potential of coriander (*Coriandrum sativum*) and thyme (*Thymus vulgaris*) against

Table 9.3 List of essential oils and their mode of action against biofilm forming microorganisms

Essential oils	Active compounds	Antimicrobial activity against pathogens	Mode of action	References
Allium sativum (Garlic)	Allicin	*C. albicans*, *E. coli*, *S. aureus*, *Listeria monocytogenes*, *Salmonella typhimurium*	Induced leakage from *E. coli* cells	Nejad et al. (2014), Liu et al. (2017), and Marschollek et al. (2017)
Brassica nigra (Mustard oil)	Allyl isothiocyanate	*E. coli*, *Salmonella typhimurium*	Damaged cell shape, loss of intracellular ATP in *E. coli* and *Salmonella typhimurium*	Turgis et al. (2009)
Carum copticum (Ajowan)	Thymol, γ-terpinene, o-cymene, ethylene methacrylate, β-pinene, hexadecane	*S. aureus*, *Streptococcus sanguis*, *Streptococcus salivarius*, *K. pneumonia*	–	Mahboubi and Kazempour (2011) and Talei et al. (2017)
Cinnamomum verum (Cinnamon)	E-cinnamaldehyde, β-caryophyllene, linalool, cinnamyl acetate, eugenol	*Campylobacter jejuni*, *Enterobacter aerogenes*, *E. coli*, *L. monocytogenes*, *P. aeruginosa*, *Salmonella enteritidis*, *S. aureus*	Cell damage, leakage of small electrolytes, proteins and nucleic acid of *E. coli* and *S. aureus* cells	Gupta et al. (2008) and Zhang et al. (2016)
Citrus aurantifolia (Lime)	Limonene, *p*-cymene	*Aspergillus flavus*, *Aspergillus parasiticus*	Destructive alteration of plasma and nucleus membrane, loss of cytoplasm, vacuole fusion and detachment of fibrillar layer in *A. flavus* and *A. parasiticus*	Rammanee and Hongpattarakere (2011)

(*Continued*)

Table 9.3 (Continued)

Essential oils	Active compounds	Antimicrobial activity against pathogens	Mode of action	References
Coriandrum sativum (Coriander)	Linalool, (E)-2-decena l, decanol, (E)-2-decen-1-ol	*Candida* species, *L. monocytogenes*	Increased ionic permeability and caused membrane damage leading to cell death (*Candida* species)	Silva et al. (2011) and Freires et al. (2014)
Cuminum cyminum (Cumin)	Cumin aldehyde, α-terpinen-7-al, γ-terpinene, *p*-cymene	*Bacillus cereus*, *Bacillus subtilis*, *K. pneumoniae*	In *K. pneumoniae*, repression of capsule expression, cell elongation, and inhibition of urease activity	Derakhshan et al. (2010) and Pajohi et al. (2011)
Cymbopogon citratus (Lemon Grass)	Geranial, myrcene, 6-methylhept-5-en-2-one	*Staphylococcus aureus, Bacillus cereus, Bacillus subtilis, Escherichia coli, K. pneumoniae*	–	Naik et al. (2010) and Aiemsaard et al. (2011)
Cymbopogon martini (Palmarosa)	Myrcene, linalool, geraniol, geranyl acetate, dipentene, limonene	*S. aureus, E. coli, Saccharomyces cerevisiae*	Alterations in the *S. cerevisiae* cell membrane composition	Lodhia et al. (2009)
Eucalyptus microtheca (Eucalyptus)	1,8-Cineol, α-pinene, p-cymene, borneol, cryptone, spathulenol, viridiflorol	*Staphylococcus aureus* (MRSA), *Escherichia coli, P. aeruginosa, C. albicans*	Changed in cell wall structure of *S. aureus* (MRSA), *Escherichia coli* and *P. aeruginosa*	Seyyednejad et al. (2014) and Sambyal et al. (2017)
Foeniculum vulgare (Fennel)	*Trans*-anethole, estragole, D-limonene, fenchone	*Streptococcus mutans, E. coli, Shigella dysenteriae, Staphylococcus epidermidis, Staphylococcus saprophyticus, Aspergillus niger, Aspergillus flavus*	Disturbed the permeability of cell membrane of *S. dysenteriae*, leakage of electrolytes, losses of proteins, reducing sugars, etc.	Badgujar et al. (2014); Liu et al. (2017)

Lavandula angustifolia (Lavender)	Camphor, terpinen-4-ol, linalool, linalyl acetate, beta-ocimene, 1,8-cineole	*E. coli, S. aureus*	–	Ali et al. (2015)
Mentha longifolia (Menthol)	Menthol, menthone, pulegone, 1,8-cineole, terpineol-4	*Salmonella typhimurium, E. coli, Micrococcus luteus, S. aureus*	Perturbation of the lipid fraction from plasma membrane, altered membrane permeability, leakage of intracellular material in *S. aureus* and *E. coli*	Trombetta et al. (2005) and Hafedh et al. (2010)
Ocimum basilicum (Basil)	Linalool	*A. niger, Mucor mucedo, Fusarium solani, Botryodiplodia theobromae*	–	Hussain et al. (2008)
Origanum compactum (Oregano)	Carvacrol, thymol, *p*-cymene, γ-terpinene	*P. aeruginosa, S. aureus*	Affected the cell membrane permeability, respiratory activity and failed the membrane potential of *P. aeruginosa*. Significant reduction in membrane potential of *S. aureus*	Bouhdid et al. (2009) and Babacan et al. (2012)
Piper nigrum (Black Pepper)	Piperine	*S. aureus, B. cereus, B. subtilis, E. coli, Salmonella typhi, P. aeruginosa*	–	Liu et al. (2017)
Rosmarinus officinalis (Rosemary)	Carnosic acid, carnosol, rosmadial, genkwanin, rosmarinic acid	*C. albicans, S. cerevisiae*	–	Pérez-Fons et al. (2006) and Tavassoli et al. (2011)

(*Continued*)

Table 9.3 (Continued)

Essential oils	Active compounds	Antimicrobial activity against pathogens	Mode of action	References
Syzygium aromaticum (Clove)	Eugenol, eugenyl acetate, caryophyllene	*C. jejuni*, *E. aerogenes*, *E. coli*, *S. aureus*, *Brochothrix thermosphacta*, *Lactobacillus rhamnosus*, *Pseudomonas fluorescens*, *L. monocytogenes*, *S. enteritidis*, *C. albicans*	Destroyed cell wall and membrane, penetrated the cytoplasmic membrane, and inhibited the normal DNA and protein synthesis in *S. aureus*. Causes disruption to the cell wall of *C. albicans*	Latifah-Munirah et al. (2015), Yadav et al. (2015), Liu et al. (2017), and Sambyal et al. (2017)
Thymus vulgaris (Thyme)	Thymol, *p*-cymene, carvacrol, γ-terpinene	*C. albicans*, *S. cerevisiae*, *B. subtilis*, *C. jejuni*, *Erwinia amylovora*, *E. coli*, *L. innocua*, *L. monocytogenes*, *S. aureus*	Caused changes in outer membrane protein profile of *Erwinia* strains	Rasooli et al. (2006) and Liu et al. (2017)
Zingiber officinale (Ginger)	Pinene, borneol, camphene, linalool	*B. subtilis*, *Bacillus nutto*, *P. aeruginosa*, *Rhodoturola* species, *Salmonella newport*, *S. enteritidis*, *Fusarium* species	—	Liu et al. (2017)

E. coli, Proteus mirabilis, S. aureus, Enterococcus species, *S. aureus* ATCC 25923, *S. aureus* ATCC 6538, and *E. coli* 25922; two standard/wild-type strains of *C. albicans* ATCC 10231 and two clinical strains of *C. albicans* by macro-diffusion (disk, well) and microdilution method in 96-well microtiter plates. MIC of *Coriandrum sativum* was 0.4–45.4 μL/mL against all tested bacteria, except multiple resistant strains of *Enterococcus* species and *Proteus* species. MIC of thyme was found to be 0.11 mg/mL against all *C. albicans* strains. The antibiofilm activity of essential oils was evaluated for two strains of *S. aureus* and *E. coli* as well as *C. albicans* filamentation ability. GC-MS analysis revealed that oxygenated monoterpenes as dominant constituents. The results showed coriander essential oil can be used against *S. aureus, E. coli*, and *C. albicans* vaginal infections in alternative gynecological treatment.

Almeida et al. (2016) demonstrated the anti-*Candida* and antibiofilm efficacy of two essential oils from *Cymbopohon winterianus* (citronella) and *Cinnamon cassia* (cinnamon). MIC was evaluated by broth microdilution assay while antibiofilm activity was determined on acrylic surfaces against matured biofilm. SEM analysis and cell viability assay of biofilm cells were performed at 0 h (immediately after treatment) and 48 h. MICs of cinnamon and citronella oils were 65 and 250 μg/mL respectively. They found that both oils significantly reduced the number of viable microorganisms and accumulation of biofilms at 0 h. However, there was no difference between treated and untreated biofilms at 48 h.

Walmiki and Vittal (2017) determined the antibiofilm activity of three essential oils such as *Syzygium aromaticum, Cuminum cyminum*, and *Piper nigrum* against two Gram negative bacteria *E. coli* MTCC 40 and *Salmonella* spp. MTCC 1163 and one Gram-positive bacteria *S. aureus* MTCC 7443. Antibiofilm activity and inhibition of cell attachment were evaluated by using crystal violet assay. All the three oils were found to be effective in inhibition of cell attachment on the wells of a microtiter plate with the mean value of >50% each. In contrast, inhibition of preformed biofilm was difficult by all the three (*S. aromaticum, C. cyminum*, and *P. nigrum*) oils. Among all the three oils, *S. aromaticum* oil proved to be more significant ($P < 0.05$) on the test bacteria in both the assays and *S. aureus* was proved to be more resistant to oils in both the assays.

Here we list selected antimicrobial essential oils that can eradicate bacteria and fungi within the biofilms with higher efficiency than certain important antimicrobial drugs, making them ideal candidate for the treatment of biofilm-related infections (Fig. 9.2).

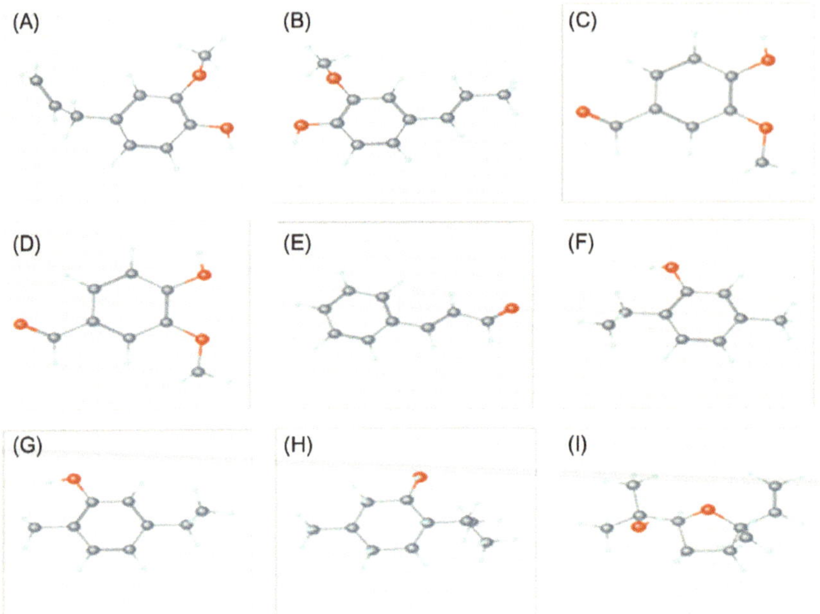

Figure 9.2 Chemical structure of essential oils. The structures are obtained from https://pubchem.ncbi.nlm.nih.gov. (A) Eugenol (CID: 3314), (B) Isoeugenol (CID: 853433), (C) Vanillin (CID: 1183), (D) Safrole (CID: 5144), (E) Cinnamaldehyde (CID: 637511), (F) Thymol (CID: 6989), (G) Carvacrol (CID: 10364), (H) Menthol (CID: 16666), (I) Linalool (CID: 6432254).

9.3.1 Terpenes and Terpenoids

Terpenes are hydrocarbons produced by joining several isoprene units (C_5H_8) (Yadav et al., 2014). They are synthesized by plant cells via the mevalonic acid pathways initiating from acetyl-CoA (Schwab et al., 2008; Singh and Sharma, 2015). Terpenes constitute most diverse class of secondary metabolites. Over 55,000 members are isolated to date. Terpene cyclase enzymes convert the simple, linear hydrocarbon phosphates into an array of chiral, carboxylic structures. (Maimone and Baran, 2007). The major group comprises diterpenes, triterpenes, tetraterpenes as well as hemiterpenes and sesquiterpenes (Pichersky et al., 2006; Nazzaro et al., 2013; Upadhyay et al., 2014; Singh and Sharma, 2015). When the compound contains additional element mostly oxygen they are termed as terpenoids (Upadhyay et al., 2014). Terpenoids are further subdivided into alcohols, esters, aldehydes, ketones, ethers, phenols, and epoxides.

Thymol, carvacrol, linalool, linalyl acetate, citronellal, piperitone, menthol, and geraniol are examples of terpenoids (Nazzaro et al., 2013).

The terpenoids consist of large number of antimicrobial compounds that shows activity against broad spectrum of microorganisms. Thymol and carvacrol identified so far as the most active monoterpenoids. Terpenoids exhibited antimicrobial activity against variety of pathogens which is reported by many workers. Ceylan and Ugur (2015) investigated the antimicrobial and antibiofilm activities and chemical composition of *Thymus sipyleus* BOISS sub-sp. Sipyleus BOISS var. davisinus RONNIGER essential oil. The essential oil was obtained by hydro-distillation and examined by gas chromatography-mass spectrometry. Out of the 14 compounds, thymol (38.31%) and carvacrol (37.95%) were the major components. MICs of oils and components were evaluated by serial dilution method and antibiofilm activity was determined by microtiter plate assay against five Gram-positive (*S. aureus* MU 38, MU 40, MU 46, MU 47, *S. epidermidis* MU 30) and five Gram negative (*P. aeruginosa* MU 187, MU 188, MU 189, *Pseudomonas fluorescence* MU180, MU181) bacteria. MICs of essential oils, thymol, and carvacrol were found to be between 5 and 50 μL/mL, 0.125–0.5 μL/mL, and 0.125–0.5 μL/mL, respectively. The data revealed that the MIC produced greater antibiofilm influence than 0.5, 0.25, and 0.125 MIC. The mean biofilm formation and inhibition value in the presence of essential oils was equal to 67 ± 5.5% and 60% for *P. aeruginosa* MU 188. The results also revealed that carvacrol (MIC) was able to inhibit 72.9 ± 4.1% *S. aureus* (MU 40) biofilm. In case of *P. fluorescence* MU181, there were 68.6 ± 5.3% reduction in biofilm formation. They conclude that of *T. sipyleus* BOISS sub-sp. Sipyleus BOISS var. davisinus RONNIGER essential oil exhibited antimicrobial and antibiofilm activity. Thymol and carvacrol showed exceptional efficiency in the treatment of *Pseudomonas* and *Staphylococcus* biofilms.

9.3.1.1 Antibiofilm Activity of Thymol and Carvacrol

Thymol is the major component of thyme oil which is structurally similar to carvacrol having hydroxyl group at a distinct position on the phenolic ring. It is evident that the interaction of thymol affects membrane permeability which in turn causes loss of membrane potential, leakage of K^+ ions, and ATP and carboxy fluorescein (Xu et al., 2008). In addition to this thymol also interact with membrane bound or periplasmic proteins by hydrophilic and hydrophobic interactions. Di Pasqua et al. (2010) also supported the interaction of thymol with membrane proteins. They exposed the *Salmonella enterica* to sublethal concentration of thymol and observed the accumulation of misfolded outer membrane proteins and upregulation of genes involved in synthesis of outer membrane proteins. Thymol also impaired the citrate metabolic pathway which ultimately affects the synthesis of ATP. The mode of action of thymol also investigated in yeast which showed interaction with cell envelope and intracellular targets. It may disrupt the vesicles and cell membrane and impaired ergosterol biosynthesis in *Candida* strains, which ultimately affected the cell membrane integrity (Cristani et al., 2007; Ahmad et al., 2011).

Carvacrol is a phenolic monoterpenoid and major component of oregano. The antimicrobial activity of carvacrol is similar to that of thymol causing structural and functional damages to the cell membrane. Carvacrol causes disintegration of outer membrane and ultimately release of lipopolysaccharides from Gram negative bacteria (Helander et al., 1998; La Storia et al., 2011). It is also well established that carvacrol enhances the membrane fluidity which in turn increased the permeability of membrane (Di Pasqua et al., 2007). There are also a few reports on the interaction of carvacrol with membrane proteins and periplasmic enzymes. Carvacrol induces the overexpression of outer membrane proteins that may affect the outer membrane protein folding or insertion (Horváth et al., 2009). The antifungal activity of carvacrol showing disruption of Ca^{2+} and H^{+} homeostasis, up and downregulation of transcription, disruption of membrane integrity and impairment of ergosterol biosynthesis in *Candida* strains (Ahmad et al., 2011).

Raei et al. (2017) investigated the effects of carvacrol and thymol on biofilm formation and antimicrobial activity against different carbapenemase-producing Gram negative bacilli. The antimicrobial and antibiofilm effects of carvacrol and thymol were determined against strains harboring different genes related to carbapenemase resistance. Antimicrobial resistance was examined by an agar dilution method and antibiofilm effect was evaluated by microtiter plate assay and staining by crystal violet. They observed that thymol and carvacrol had antibacterial effects ranging from 200 to 1600 and 62 to 250 μg/mL respectively; and antibiofilm effect from 125 to 500 and 400 to 1600 μg/mL respectively. Seoul imipenemase (SIM) producing isolates had the highest sensitivity, and NDM (New Delhi metallo-beta-lactamase) producing isolates had the lowest sensitivity to these components. The finding suggests that carvacrol and thymol could be used in controlling carbapenemase-producing gram negative bacterial infections. These findings helped to develop herbal drugs for replacing antibiotics. Moreover, their antibiofilm effects showed that carvacrol and thymol inhibit biofilm formation of carbapenemase-producing strains. Similar study was conducted by Khan et al. (2017b). They determined the antimicrobial and antibiofilm activities of carvacrol (M-1) and thymol (M-2), purified from *Origanum vulgare* L. along with chlorhexidine digluconate against dental caries causing bacteria *S. mutans*. MIC of carvacrol and thymol was found to be 65 and 54 μg/mL respectively. Live and dead staining and MTT assays showed that 100 μg/mL concentration of these compounds reduced viability and metabolic activity of *S. mutans* by more than 50%. At 100 μg/mL, the biofilm formation in polystyrene plates was significantly reduced by M-1 and M-2 as revealed by SEM and colorimetric analysis. These results were also confirmed by RT-PCR studies. There were 2.2 and 2.4-fold increase in Autolysin gene (AtlE) expression level on exposure to 25 μg/mL of M-1 and M-2 respectively. The superoxide dismutase gene (sod A) activity was also increased by 1.3 and 1.4-fold with the same concentration of M-1 and M-2 respectively. After the treatment with M-1 and M-2, there were increase in the ymcA gene and a decrease in the gtfB gene expression was observed. The findings revealed that carvacrol and thymol isolated from *O. vulgare* L. exhibited strong antibacterial and antibiofilm activity against *S. mutans*. These compounds can be used to control dental caries.

9.3.2 Phenylpropenes

Phenylpropenes are synthesized from amino acid precursor phenylalanine in plants and constitutes various groups of organic compounds called phenyl proponoids. These compounds represent a relatively small portion of essential oils (Tzin and Galili, 2010). The phenylpropenes that are thoroughly studied are eugenol, isoeugenol, vanillin, safrole, and cinnamaldehyde (Chouhan et al., 2017). Several authors have documented the antibiofilm activity of phenylpropenes. Khan and Ahmad (2012) investigated the four phytocompounds (cinnamaldehyde, citral, eugenol, and geraniol) for antibiofilm activity against of *C. albicans* at sub-MICs. The test compounds were analyzed for inhibitory effect on biofilms by XTT reduction assay and microscopic examination. The result showed that *C. albicans* 04 and *C. albicans* SC 5314 formed strong biofilms. *Candida* biofilms exhibited several fold increase in resistance level to antifungal drugs. However, poor resistance increase was noticed against essential oil compounds tested. At 0.5 × MIC, eugenol and cinnamaldehyde showed strong inhibitory activity against biofilm formation. Light and scanning electron microscopy analysis revealed the deformity in biofilm structure at sub-MICs of eugenol and cinnamaldehyde. SEM studies also showed that the cell membrane was one of the target sites of tested compounds in both biofilm and planktonic cell growth.

9.3.2.1 Antibiofilm Activity of Eugenol

Eugenol is present in the essential oil and extracted from many plants specifically the *S. aromaticum* (clove) and its antimicrobial activity is based on its ability to permeabilized the cell membrane and interact with proteins. The nonspecific permeabilization of the cytoplasmic membrane is evidenced with leakage of K^+ and ATP from the cell by eugenol (Hemaiswarya and Doble, 2009). Bacterial cell damage was also reported by eugenol on *E. coli* and *B. thermosphacta* cells (Di Pasqua et al., 2007). The antifungal activity of eugenol is linked with altered cell membrane and cell wall structure resulting in release of cell content (Bennis et al., 2004). de Paula et al. (2014) reported the effect of eugenol on the adherence properties and biofilm formation capacity of *Candida dubiliensis* and *C. tropicalis* isolated from the oral cavity of HIV infected patients. All strains were able to form biofilms on different substrates. Eugenol showed inhibitory activity against planktonic and sessile cells of *Candida* spp. Biofilm cells showed no metabolic activity after 24 h of treatment. SEM studies revealed that eugenol significantly reduced the number of sessile cells on denture material surfaces. There was significant difference in cell surface hydrophobicity after exposure of planktonic cells to eugenol for 1 h. Eugenol also caused a reduction in adhesion of the isolates of *Candida* spp. to HEp-2 cells and to polystyrene. These findings corroborate its potential as an antifungal applied to inhibit the growth of the planktonic and biofilm formation on different surfaces.

Kim et al. (2016) studied the antibiofilm activity of various essential oils and their compound against Enterohemorrhagic *Escherichia coli* 0157: H7 (EHEC). Bay, clove, pimento berry oils, and their major component eugenol at 0.005% (v/v) were found to inhibit EHEC biofilms without affecting planktonic cells growth. The other derivatives of eugenol namely isoeugenol, 2-methoxy-4-propylphenol, and 4-ethyl guaiacol also showed antibiofilm activity. Three essential oils and eugenol did not inhibit biofilm development by three laboratory *E. coli* K-12 strains that reduced curli fimbriae production. The genes involved in biofilm formation, attachment and effacement phenotype [curli genes (csg ABDFG), fimbriae genes (fim CDH), and ler-controlled toxin genes (espD, escJ, escR, and tir)] were downregulated by eugenol as revealed from transcriptional analysis. They concluded that biocompatible poly (lactic-co-glycolic acid) coated with clove oil or eugenol exhibited efficient biofilm inhibition on solid surfaces. In *Caenorhabditis elegans* nematode model, virulence of EHEC was attenuated by clove oil/eugenol.

Zhang et al. (2017) studied the antibacterial and antibiofilm effect and mechanism of eugenol from *S. aromaticum* (L.) MERR. & L.M. Perry (clove) leaf essential oil (CLEO) against oral anaerobe *Porphyromons gingivalis.* Eugenol, with content of 90.84% in CLEO, exhibited antibacterial activity against *P. gingivalis* at a concentration of 31.25 μM. Scanning Electron Microscopy (SEM) analysis showed shrinkage and lysis of cells. The release of macromolecules and uptake of fluorescent dye indicated that the antibacterial activity was due to the ability of eugenol to permeabilize the cell membrane and destroy the integrity of plasmatic membrane irreversibly. Furthermore, eugenol inhibited biofilm formation and reduced preformed biofilm of *P. gingivalis* at different concentrations. The downregulation of virulence factor genes related to biofilm (fimA, hagA, hagB, rgpA, rgpB, kgp) revealed that eugenol suppressed biofilm formation at the initial stage. The study suggested that eugenol and CLEO may be potential additives in food and personal healthcare products as a prophylactic approach to periodontitis.

9.3.2.2 Antibiofilm Activity of Cinnamaldehyde

Cinnamaldehyde is the major active constituents of *Cinnamomum verum.* Cinnamaldehyde has been widely studied for its various biological activities including antimicrobial activities (Zhang et al., 2016). The major mode of action of cinnamaldehyde on fungi is reported to inhibit cell division, cell wall synthesizing enzymes and also act as noncooperative inhibitor of β-(1,3)-glucan synthase and chitin synthase (Bang et al., 2000). It inhibits different enzymes involved in cytokinesis at low concentration. At sublethal concentration, cinamaldehyde, cinnamaldehyde act as an ATPase inhibitor and a lethal concentration, it disturbs the cell membrane (Nazzaro et al., 2013). The antifungal activity of encapsulated preparation of cinnamaldehyde (CNMA) efficacy was determined against *Candida* biofilm by SEM, TEM, and light microscopy and its percent inhibition was determined by XTT and crystal violet assay in multilamellar liposomes (ML) against *C. albicans.* ML-CNMA possessed more fungicidal activity than free CNMA as well as multilamellar liposomal amphotericin B (ML-Aml-B). This was further confirmed by spot test assay and Log-logistic dose response (Khan et al., 2017a,b).

Kot et al. (2015) evaluated the antimicrobials in preventing and inactivating *E. coli* biofilm on urinary catheters. Catheter fragments were inoculated with *E. coli* and treated with trans-cinnamaldehyde, p-coumaric, and ferulic acids (0%, 0.1%, 0.25%, and 0.5%) for 0, 1, 3, and 5 days to test the prevention and inactivation of biofilm formation. They found that all used concentrations of trans-cinnamaldehyde prevented and efficiently inactivated the *E. coli* biofilm formed on urinary catheters fragments. The p-coumaric (0.25% and 0.5%) and ferulic acids (0.5%) showed preventive action on *E. coli* biofilm formation in urinary catheters fragments; and both the agents significantly reduced the number of uropathogenic *E. coli* sessile cells in the lumen of a urinary catheters. In contrast, trans-cinnamaldehyde resulted in complete inactivation of biofilm.

Ramasamy et al. (2017) studied the synthesis of gold nanoparticle complexes using cinnamaldehyde, a potent antibiofilm agent. It contains 0.01% cinnamaldehyde by weight exhibited effective biofilm inhibition of up to >80% against drug resistant Gram-positive and Gram negative bacteria as well as against *C. albicans*. The hyphae formation in *C. albicans* was attenuated by using gold nanoparticle complexes using cinnamaldehyde.

9.4 ANTIBIOFILM ACTIVITY OF OTHER ESSENTIAL OILS

9.4.1 Cumin Oil (*Cuminum cyminum L.*)

Cumin oil has been used as a flavoring agent and in medical formulations. It acts as astringent in digestive system and broncho pulmonary disorders. It has been extensively used in cough treatment and as an analgesic and is known to exhibit antibacterial activity against both Gram-positive and Gram negative bacteria (Lang and Buchbauer, 2011). The mode of action of cumin oil is cell membrane damage and release of intracellular DNA and protein (Dua et al., 2013). Derakhshan et al. (2010) reported the antibiofilm activity of cumin seed essential oil against *Klebsiella pneumoniae* strains. Synergistic interaction of oil with ciprofloxacin was also demonstrated by disk diffusion assay against *K. pneumoniae* isolates. The essential oil reduced the biofilm formation and also increased the activity of the ciprofloxacin disk. The author highlighted the safe exploitation of cumin seed essential oil against *K. pneumoniae* biofilm in vitro.

9.4.2 Tea Tree Oil (*Melaleuca alternifolia* (Maiden & Betche) Cheel)

Tea tree oil is obtained from *Melaleuca alternifolia* and widely explored for its antimicrobial activity (Carson et al., 2006). Biofilm inhibition by tea tree oil against *P. aeruginosa*, *S. aureus*, *C. albicans*, and *Vibrio harveyi* has been demonstrated by Hammer et al. (2008). There were significant amount of reduction in preformed biofilm as well as viability of the microorganisms after treatment with tea tree oil. Crystal violet staining revealed reduction in biomass. The study concluded that tea tree oil have a potential to be used as an antibiofilm agent.

Brady et al. (2010) studied the antimicrobial treatment of cochlear implant biofilm infections. 5% of tea tree oil treatment showed complete eradication of biofilm formed by MSSA isolates.

9.4.3 Eucalyptus Oil (*Eucalyptus globulus* Labill)

The essential oil extracted from *Eucalyptus globulus* is used in the treatment of respiratory diseases, antiinflammatory drug, as an insect repellent, antibacterial agent against *S. aureus* and *E. coli*, and antitumor agent in rats (Djenane et al., 2011).

Mathur et al. (2014) evaluated the antibiofilm effect of the eucalyptus essential oil along with the determination of bioactive components contributing to the biofilm inhibition.

The antibiofilm activity of the oil was determined against *P. mirabilis* ATCC 7002 strain. SEM analysis confirmed the antibiofilm activity of eucalyptus oil. TLC and GC-MS analysis revealed that the terpenes mainly 1,8-cineole (44.22%) and pinene (13.6%) and other components in small fractions were the component of eucalyptus essential oil. SEM studies indicated that the eucalyptus oil had a very strong inhibitory effect on the bacterial biofilm on urinary catheters. The major components of eucalyptus oil were found to be terpenes and related components, which seems to be active agent for biofilm inhibition. The use of eucalyptus oil and their active constituents in biofilm control associated with urinary catheters was highlighted.

9.5 IN VIVO STUDIES

Few in vivo studies have been conducted to demonstrate the effect of essential oils in in vivo conditions (Ohno et al., 2003; Yadav et al., 2015).

Yadav et al. (2015) investigated the antibiofilm activity of eugenol on MRSA and MSSA biofilms in vitro and bacterial colonization in vivo. They studied the colonization of *S. aureus* in the middle ear of rats. The digital images of bulla of rat treated with bacteria and eugenol or DMSO showed clear images as compared to the bulla of rat inoculated with bacteria only. The bulla of rat treated with bacteria only had swollen mucosa, presence of cell debris and matrix. Viability assay and SEM images indicated the reduction in *S. aureus* colonization in rat middle ear. They concluded that in vivo experiments demonstrated the notable activity of eugenol against colonization of *S. aureus.* Further investigation with suitable OM model is needed to study to elucidate the pre-established biofilm eradication activity of eugenol in the middle ear. Similar study was conducted by Manoharan et al. (2017). The antibiofilm activity of cascarilla bark oil, α-longipinene, and linalool against *C. albicans* was evaluated using the nematode *C. elegans* as an in vivo model. The nematode *C. elegans* was infected with *C. albicans*. Fewer than 5% of nematodes were survived in the absence of any treatment after four days, whereas treatment with fluconazole (a commercial antifungal) at 0.01% increased survival to $>90\%$. Interestingly, cascarilla bark oil, α-longipinene, and linalool at 0.01% markedly enhanced survival to the same extent as fluconazole. Even at 0.001%, the survival rate was approximately 60%. SEM images also revealed that most *C. elegans* died in the negative control sample after 4 days, which was possibly due to hyphae piercing the worm's cuticle. However, no hyphal formation was observed in worms treated with cascarilla bark oil, α-longipinene, or linalool at 0.01% (v/v). They confirmed that cascarilla bark oil, α-longipinene, and linalool can prevent hyphal growth of *C. albicans* and effectively diminish its virulence in a nematode model. In addition, the toxicity of α-longipinene and linalool tested against *C. elegans* survival rate was determined after 4 days and it was found that the toxic concentration of these compounds to the nematode would be $>0.5\%$ (v/v), which is 50-fold higher than the concentrations ($<0.01\%$) for antibiofilm activity.

9.6 CONCLUSION

Based on the available information it is concluded that biofilm development by pathogenic microorganisms results in the major problem in the treatment of infections. Among the available antibiofilm agents, chemotherapeutic potential for oral and systemic infections is limited. In vivo studies have demonstrated by various workers on essential oils and its active constituents which have given hope for its potential use in biofilm control by adopting improved strategies. Various essential oils demonstrated their common mode of action/mechanism against biofilm. Role of active compounds such as eugenol, thymol, and carvacrol seems to be promising in biofilm control in oral system. Through solving essential oil solubility, stability, and efficacy by nanotechnology, the future of such compounds are expected to be more promising. However, bioactivity and toxicity studies are limited and further efficacy should be evaluated in suitable animal model to uncover therapeutic value of essential oil against mono and mixed biofilms.

REFERENCES

Ahmad, A., Khan, A., Akhtar, F., Yousuf, S., Xess, I., Khan, L.A., et al., 2011. Fungicidal activity of thymol and carvacrol by disrupting ergosterol biosynthesis and membrane integrity against *Candida*. Eur. J. Clin. Microbiol. Infect. Dis. 30 (1), 41–50.

Aiemsaard, J., Aiumlamai, S., Aromdee, C., Taweechaisupapong, S., 2011. The effect of lemongrass oil and its major components on clinical isolate mastitis pathogens and their mechanisms of action on *Staphylococcus aureus* DMST 4745. Res. Vet. Sci. 91 (3), e31–e37.

Akhtar, N., Datta, P.G., Yasmin, M., Naha, A., Jewel, A.M., Hossen, F., et al., 2017. Microbiology of chronic suppurative otitis media in a tertiary care hospital in Bangladesh. Mymensingh. Med. J. 26 (3), 592–599.

Akyıldız, I., Take, G., Uygur, K., Kızıl, Y., Aydil, U., 2013. Bacterial biofilm formation in the middle-ear mucosa of chronic otitis media patients. Indian J. Otolaryngol. Head Neck Surg. 65 (3), 557–561.

Al Yahfoufi, Z., 2017. Prevalence of periodontal destruction and putative periodontal pathogens in the same Lebanese family. J. Contemp. Dent. Pract. 18 (10), 970–976.

Ali, B., Al-Wabel, N.A., Shams, S., Ahamad, A., Khan, S.A., Anwar, F., 2015. Essential oils used in aromatherapy: a systemic review. Asian Pac. J. Trop. Biomed. 5 (8), 601–611.

Almeida, LdeF., Paula, J.F., Almeida, R.V., Williams, D.W., Hebling, J., Cavalcanti, Y.W., 2016. Efficacy of citronella and cinnamon essential oils on *Candida albicans* biofilms. Acta Odontol. Scand. 74 (5), 393–398.

Antonio, L.F., Lyra, M.R., Saheki, M.N., Schubach, A.O., Miranda, L.F.C., Madeira, M. F., et al., 2017. Effect of secondary infection on epithelialisation and total healing of cutaneous leishmaniasis lesions. Mem. Inst. Oswaldo. Cruz. 112 (9), 640–646.

Aparna, M.S., Yadav, S., 2008. Biofilms: microbes and disease. Braz. J. Infect. Dis. 12 (6), 526–530.

Atray, D., Atray, M., 2015. Correlation between biofilm production and antibiotic resistance pattern in uropathogenic *Escherichia coli* in tertiary care hospital in Southern Rajasthan, India. Int. J. Curr. Microbiol. App. Sci. 4 (7), 640–646.

Babacan, O., Cengiz, S., Akan, M., 2012. Detection of antibacterial effect of oregano plant on various *Salmonella* serotypes. Ank. Univ. Vet. Fak. Derg. 59, 103–106.

Badgujar, S.B., Patel, V.V., Bandivdekar, A.H., 2014. *Foxxeniculum vulgare* mill: a review of its botany, phytochemistry, pharmacology, contemporary application, and toxicology. Biomed. Res. Int. 2014, 842674.
Bang, K.H., Lee, D.W., Park, H.M., Rhee, Y.H., 2000. Inhibition of fungal cell wall synthesizing enzymes by trans-cinnamaldehyde. Biosci. Biotechnol. Biochem. 64 (5), 1061–1063.
Barbara, V.U., Chen, M., Crawford, R.J., Ivanova, E.P., 2009. Bacterial extracellular polysaccharides involved in biofilm formation. Molecules 14 (7), 2535–2554.
Bennis, S., Chami, F., Chami, N., Bouchikhi, T., Remmal, A., 2004. Surface alteration of *Saccharomyces cerevisiae* induced by thymol and eugenol. Lett. Appl. Microbiol. 38 (6), 454–458.
Bergenfelz, C., Hakansson, A.P., 2017. *Streptococcus pneumoniae* otitis media pathogenesis and how it informs our understanding of vaccine strategies. Curr. Otorhinolaryngol. Rep. 5 (2), 115–124.
Bessa, L.J., Fazii, P., Di Giulio, M., Cellini, L., 2015. Bacterial isolates from infected wounds and their antibiotic susceptibility pattern: some remarks about wound infection. Int. Wound. J. 12 (1), 47–52.
Blankenship, J.R., Mitchell, A.P., 2006. How to build a biofilm: a fungal perspective. Curr. Opin. Microbiol. 9, 588–594.
Bogavac, M., Karaman, M., Janjušević, L.J., Sudji, J., Radovanović, B., Novaković, Z., et al., 2015. Alternative treatment of vaginal infections—in vitro antimicrobial and toxic effects of *Coriandrum sativum* L. and *Thymus vulgaris* L. essential oils. J. Appl. Microbiol. 119 (3), 697–710.
Bouhdid, S., Abrinin, J., Zhiri, A., Espuny, M.J., Munresa, A., 2009. Investigation of functional and morphological changes in *Pseudomonas aeruginosa* and *Staphylococcus aureus* cells induced by *Origanum compactum* essential oil. J. Appl. Microbiol. 106, 1558–1568.
Brady, A.J., Farnan, T.B., Toner, J.G., Gilpin, D.F., Tunney, M.M., 2010. Treatment of a cochlear implant biofilm infection: a potential role for alternative antimicrobial agents. J. Laryngol. Otol. 124, 729–738.
Bryers, J.D., 2008. Medical biofilms. Biotechnol. Bioeng. 100 (1), 1–18.
Buommino, E., Scognamiglio, M., Donnarumma, G., Fiorentino, A., D'Abrosca, B., 2014. Recent advances in natural product-based anti-biofilm approaches to control infections. Mini-Rev. Med. Chem. 14 (14), 1169–1182.
Carson, C.F., Hammer, K.A., Riley, T.V., 2006. *Melaleuca alternifolia* (Tea Tree) oil: a review of antimicrobial and other medicinal properties. Clin. Microbiol. Rev. 19 (1), 50–62.
Ceylan, O., Ugur, A., 2015. Chemical composition and anti-biofilm activity of *Thymus sipyleus* BOISS. subsp. sipyleus BOISS. var. davisianus RONNIGER essential oil. Arch. Pharm. Res. 38 (6), 957–965.
Chang, H., Zhou, J., Zhu, X., Yu, S., Chen, L., Jin, H., et al., 2017. Strain identification and quorum sensing inhibition characterization of marine-derived *Rhizobium* sp. NAO1. Royal Soc. Open Sci. 4 (3), 170025.
Chmiel, J.F., Aksamit, T.R., Chotirmall, S.H., Dasenbrook, E.C., Elborn, J.S., LiPuma, J.J., et al., 2014. Antibiotic management of lung infections in cystic fibrosis. I. The microbiome, methicillin-resistant *Staphylococcus aureus*, Gram-negative bacteria, and multiple infections. Ann. Am. Thorac. Soc. 11 (7), 1120–1129.
Choi, H.A., Cheong, D.E., Lim, H.D., Kim, W.H., Ham, M.H., Oh, M.H., Wu, Y., Shin, H.J., Kim, G.J., 2017. Antimicrobial and Anti-Biofilm Activities of the Methanol Extracts of Medicinal Plants against Dental Pathogens *Streptococcus mutans* and *Candida albicans*, J Microbiol Biotechnol. 27 (7), 1242–1248. Available from: https://doi.org/10.4014/jmb.1701.01026.

Chouhan, S., Sharma, K., Guleria, S., 2017. Antimicrobial activity of some essential oils—present status and future perspectives. Medicines (Basel) 4 (3), 58. Available from: https://doi.org/10.3390/medicines4030058.
Cigana, C., Bianconi, I., Baldan, R., De Simone, M., Riva, C., Sipione, B., et al., 2017. *Staphylococcus aureus* impacts *Pseudomonas aeruginosa* chronic respiratory disease in murine models. J. Infect. Dis. . Available from: https://doi.org/10.1093/infdis/jix621.
Costerton, J.W., Stewart, P.S., Greenberg, E.P., 1999. Bacterial biofilms: a common cause of persistent infections. Science 284 (21), 1318–1322.
Costerton, J.W., Veeh, R., Shirtliff, M., Pasmore, M., Post, C., Ehrlich, G., 2003. The application of biofilm science to the study and control of chronic bacterial infections. J. Clin. Invest. 112 (10), 1466–1477.
Coutinho, H.D.M., Falcão-Silva, V.S., Gonçalves, G.F., 2008. Pulmonary bacterial pathogens in cystic fibrosis patients and antibiotic therapy: a tool for the health workers. Int. Arch. Med. 1, 24.
Cristani, M., D'Arrigo, M., Mandalari, G., Castelli, F., Sarpietro, M.G., Micieli, D., et al., 2007. Interaction of four monoterpenes contained in essential oils with model membranes: implications for their antibacterial activity. J. Agric. Food. Chem. 55, 6300–6308.
Crouzet, M., Senechal, C.L., Brözel, V.S., Costaglioli, P., Barthe, C., Bonneu, M., et al., 2014. Exploring early steps in biofilm formation: set-up of an experimental system for molecular studies. BMC Microbiol. 14, 253.
de Carvalho, C.C.C.R., 2007. Biofilms: recent developments on an old battle. Recent Pat. Biotechnol. 1 (1), 49–57.
de Paula, S.B., Bartelli, T.F., Di Raimo, V., Santos, J.P., Morey, A.T., Bosini, M.A., et al., 2014. Effect of eugenol on cell surface hydrophobicity, adhesion, and biofilm of *Candida tropicalis* and *Candida dubliniensis* isolated from oral cavity of HIV-infected patients. Evid. Based. Complement. Alternat. Med. 2014, Article ID 505204.
de Vos, W.M., 2015. Microbial biofilms and the human intestinal microbiome. NPJ Biofilms Microbiomes 15005.
Del Pozo, J.L., Serrera, A., Martinez-Cuesta, A., Leiva, J., Penades, J., Lasa, I., 2006. Biofilm related infections: is there a place for conservative treatment of port-related bloodstream infections? Int. J. Artif. Organs 29 (4), 379.
Derakhshan, S., Sattari, M., Bigdeli, M., 2010. Effect of cumin (*Cuminum cyminum*) seed essential oil on biofilm formation and plasmid Integrity of *Klebsiella pneumoniae*. Pharmacogn. Mag. 6 (21), 57–61.
Desai, J.V., Bruno, V.M., Ganguly, S., Stamper, R.J., Mitchell, K.F., Solis, N., et al., 2013. Regulatory role of glycerol in *Candida albicans* biofilm formation. mBio 4 (2), e00637-12.
Di Pasqua, R., Betts, G., Hoskins, N., Edwards, M., Ercolini, D., Mauriello, G., 2007. Membrane toxicity of antimicrobial compounds from essential oils. J. Agric. Food Chem. 55 (12), 4863–4870.
Di Pasqua, R., Mamone, G., Ferranti, P., Ercolini, D., Mauriello, G., 2010. Changes in the proteome of *Salmonella enterica* serovar Thompson as stress adaptation to sub-lethal concentrations of thymol. Proteomics. 10 (5), 1040–1049.
Djenane, D., Yangela, J., Amrouche, T., Boubrit, S., Boussad, N., Roncals, P., 2011. Chemical composition and antimicrobial effects of essential oils of *Eucalyptus globulus, Myrtus communis* and *Satureja hortensis* against *Escherichia coli* O157:H7 and *Staphylococcus aureus* in minced beef. Food Sci. Technol. Int. 17 (6), 505–515.
Donelli, G., Guaglianone, E., Rosa, R.D., Fiocca, F., Basoli, A., 2007. Plastic biliary stent occlusion: factors involved and possible preventive approaches. Clin. Med. Res. 5 (1), 53–60.

Dongari-Bagtzoglou, A., 2008. Pathogenesis of mucosal biofilm infections: challenges and progress. Expert Rev. Anti-Infect. Ther. 6 (2), 201–208.

Dua, A., Gaurav, G., Balkar, S., Mahajan, R., 2013. Antimicrobial properties of methanolic extract of cumin (cuminum cyminum) seeds. Int. J. Res. Ayurveda Pharm. 4, 1.

Ehrlich, G.D., Veeh, R., Wang, X., Costerton, J.W., Hayes, J.D., Hu, F.Z., et al., 2002. Mucosal biofilm formation on middle-ear mucosa in the chinchilla model of otitis media. J. Am. Med. Assoc. 287 (13), 1710–1715.

Eiff, C.V., Jansen, B., Kohnen, W., Becker, K., 2005. Infections associated with medical devices pathogenesis, management and prophylaxis. Drugs 65 (2), 179–214.

Falahati, M., Farahyar, S., Akhlaghi, L., Mahmoudi, S., Sabzian, K., Yarahmadi, M., et al., 2016. Characterization and identification of candiduria due to *Candida* species in diabetic patients. Curr. Med. Mycol. 2 (3), 10–14.

Fanning, S., Mitchell, A.P., 2012. Fungal biofilms. PLoS Pathog. 8 (4), e1002585.

Fauvart, M., De Groote, V.N., Michiels, J., 2011. Role of persister cells in chronic infections: clinical relevance and perspectives on anti-persister therapies. J. Med. Microbiol. 60 (6), 699–709.

Flemming, H.C., Wingender, J., 2010. The biofilm matrix. Nat. Rev. Microbiol. 8, 623–633.

Flemming, H.C., 2016. EPS—then and now. Microorganisms 4 (4), 41.

Fong, J.N.C., Yildiz, F.H., 2015. Biofilm matrix proteins. Microbiol. Spectr. 3, 2.

Freires, I.A., Murata, R.M., Furletti, V.F., Sartoratto, A., Alencar, S.M.D., Figueira, G.M., et al., 2014. *Coriandrum sativum* L. (Coriander) essential oil: antifungal activity and mode of action on *Candida* spp., and molecular targets affected in human whole-genome expression. PLoS ONE 9 (6), e99086.

Freires, I.A., Denny, C., Benso, B., de Alencar, S.M., Rosalen, P.L., 2015. Antibacterial activity of essential oils and their isolated constituents against cariogenic bacteria: a systematic review. Molecules 20, 7329–7358.

Gahlot, R., Nigam, C., Kumar, V., Yadav, G., Anupurba, S., Gahlot, R., et al., 2014. Catheter-related bloodstream infections. Int. J. Crit. Illn. Inj. Sci. 4 (2), 162–167.

García-Quintanilla, M., Pulido, M.R., López-Rojas, R., Pachón, J., McConnell, M.J., 2013. Emerging therapies for multidrug resistant *Acinetobacter baumannii*. Trends Microbiol. 21, 157–163.

Gargouri, D., Ouakaa-Kchaou, A., Kochlef, A., Elloumi, H., Bibani, N., Trad, D., et al., 2015. Microbiological study and antimicrobial susceptibility of bile in biliarytherapeutic endoscopy. Tunis. Med. 93 (10), 602–605.

Ghosh, R., Das, M.C., Sarkar, A., Das, A., Sandhu, P., Dinda, B., et al., 2017. Exploration of phytoconstituents from *Mussaenda roxburghii* and studies of their anti-biofilm effect. Chem. Biodivers. 14 (10).

Guaglianone, E., Cardines, R., Vuotto, C., Rosa, R.D., Babini, V., Mastrantonio, P., et al., 2010. Microbial biofilms associated with biliary stent clogging. FEMS. Immunol. Med. Microbiol. 59 (3), 410–420.

Gupta, C., Garg, A.P., Uniyal, R.C., Kumari, A., 2008. Comparative analysis of the antimicrobial activity of cinnamon oil and cinnamon extract on some food-borne microbes. Afr. J. Microbiol. Res. 2, 247–251.

Guthmiller, J.M., Karen, F., Novak, K.F., 2002. Periodontal Diseases. ASM Press, Washington, DC (Chapter 8).

Hadjifrangiskou, M., Gu, A.P., Pinkner, J.S., Kostakioti, M., Zhang, E.W., Greene, S.E., et al., 2012. Transposon mutagenesis identifies uropathogenic *Escherichia coli* biofilm factors. J. Bacteriol. 194 (22), 6195–6205.

Hafedh, H., Fethi, B.A., Mejdi, S., Emira, N., Bakhrouf Amin, B., 2010. Effect of *Mentha longifolia* L. ssp. longifolia essential oil on the morphology of four pathogenic bacteria visualized by atomic force microscopy. Afr. Microbiol. J. Res. 4 (11), 1122–1127.

Hammer, K., Carson, C., Riley, T., 1999. Antimicrobial activity of essential oils and other plant extracts. J. Appl. Microbiol. 86, 985–990.
Hammer, K.A., Carson, C.F., Tan, T.J., Riley, T.V., 2008. Effects of tea tree oil on biofilm formation. RIRDC Publication No 08/ 140, Project No PRJ-000451.
Harding, M.W., Marques, L.L., Howard, R.J., Olson, M.E., 2009. Can filamentous fungi form biofilms? Trends Microbiol. 17 (11), 475–480.
Harriott, MM, Noverr, MC, 2009. Candida albicans and Staphylococcus aureus form polymirobial biofilms: effects on antimicrobial resistance. Antimicrob Agents Chemother 53 (9), 3914–3922.
Helander, I.M., Alakomi, H.L., Latva, K., Mattila-Sandholm, T., Pol, I., Smid, E.J., et al., 1998. Characterization of the action of selected essential oil components on Gram-negative bacteria. J. Agric. Food. Chem. 46, 3590–3595.
Hemaiswarya, S., Doble, M., 2009. Synergistic interaction of eugenol with antibiotics against Gram negative bacteria. Phytomedicine 16, 997–1005.
Henry-Stanley, M.J., Hess, D.J., Barnes, A.M.T., Dunny, G.M., Carol, L., Wells, C.L., 2010. Bacterial contamination of surgical suture resembles a biofilm. Surg. Infect. (Larchmt). 11 (5), 433–439.
Hogan, S., Zapotoczna, M., Stevens, N.T., Humphreys, H., O'Gara, J.P., O'Neill, E., 2016. Eradication of *Staphylococcus aureus* catheter-related biofilm infections using ML: 8 and Citrox. Antimicrob. Agents Chemother. 60 (10), 5968–5975.
Høiby, N., Ciofu, O., Bjarnsholt, T., 2010. *Pseudomonas aeruginosa* biofilms in cystic fibrosis. Future Microbiol. 5 (11), 1663–1674.
Hola, V., Ruzicka, F., Horka, M., 2010. Microbial diversity in biofilm infections of the urinary tract with the use of sonication techniques. FEMS. Immunol. Med. Microbiol. 59, 525–528.
Hong, W., Moser, C., Wang, H.Z., Høiby, N., Song, Z.J., 2015. Strategies for combating bacterial biofilm infections. Int. J. Oral Sci. 7, 1–7.
Horváth, G., Kovács, K., Kocsis, B., Kustos, I., 2009. Effect of thyme (*Thymus vulgaris* L.) essential oil and its main constituents on the outer membrane protein composition of *Erwinia* strains studied with microfluid chip technology. Chromatographia. 70, 1645–1650.
Hotterbeekx, A., Xavier, B.B., Bielen, K., Lammens, C., Moons, P., Schepens, T., et al., 2016. The endotracheal tube microbiome associated with *Pseudomonas aeruginosa* or *Staphylococcus epidermidis*. Sci. Rep. 6, 36507.
Howell, W.R., Goulston, C., 2011. Osteomyelitis: an update for hospitalists. Hosp. Pract. (1995) 39 (1), 153–160.
Hussain, A.L., Anwar, F., Hussain Sherazi, S.T., Przybylski, R., 2008. Chemical composition, antioxidant and antimicrobial activities of basil (*Ocimum basilicum*) essential oils depends on seasonal variations. Food Chem. 108 (3), 986–995.
Jin, Y.C., Yuan, K., Zhang, J., 2011. Chemical composition, and antioxidant and antimicrobial activities of essential oil of *Phyllostachys heterocycla* cv. *Pubescens* varieties from China. Molecules 16, 4318–4327.
Kavanaugh, N.L., Ribbec, K., 2012. Selected antimicrobial essential oils eradicate *Pseudomonas* spp. and *Staphylococcus aureus* Biofilms. Appl. Environ. Microbiol. 78 (11), 4057–4061.
Khan, M.S., Ahmad, I., 2012. Antibiofilm activity of certain phytocompounds and their synergy with fluconazole against *Candida albicans* biofilms. J. Antimicrob. Chemother. 67 (3), 618–621.
Khan, M.S.A., Ahmad, I., Sajid, M., Cameotra, S.S., 2014. Current and emergent control strategies for medical biofilms. In: Rumbaugh, K.P., Ahmad, I. (Eds.), Antibiofilm Agents-From Diagnosis to Treatment and Prevention, 8(1). Springer Series on Biofilm, pp. 117–160.

Khan, H.A., Baig, F.K., Mehboob, R., 2017a. Nosocomial infections: epidemiology, prevention, control and surveillance. Asian Pac. J. Trop. Biomed. 7 (5), 478–482.

Khan, S.T., Khan, M., Ahmad, J., Wahab, R., Abd-Elkader, O.H., Musarrat, J., et al., 2017b. Thymol and carvacrol induce autolysis, stress, growth inhibition and reduce the biofilm formation by *Streptococcus mutans*. AMB Express 7 (1), 49.

Kim, Y.G., Lee, J.H., Kim, S.I., Baek, K.H., Lee, J., 2015. Cinnamon bark oil and its components inhibit biofilm formation and toxin production. Int. J. Food Microbiol. 195, 30–39.

Kim, Y.G., Lee, J.H., Gwon, G., Kim, S.I., Park, J.G., Lee, J., 2016. Essential oils and eugenols inhibit biofilm formation and the virulence of *Escherichia coli* O157: H7. Sci. Rep. 6, 36377.

Kostakioti, M., Hadjifrangiskou, M., Hultgren, S.J., 2013. Bacterial biofilms: development, dispersal, and therapeutic strategies in the dawn of the postantibiotic era. Cold Spring Harb. Perspect. Med. 3 (4), a010306.

Kot, B., Wicha, J., Piechota, M., Wolska, K., Gruzewska, A., 2015. Antibiofilm activity of trans-cinnamaldehyde, p-coumaric, and ferulic acids on uropathogenic *Escherichia coli*. Turk. J. Med. Sci. 45 (4), 919–924.

Kwon, C., Lehman, G.A., 2016. Mechanisms of biliary plastic stent occlusion and efforts at prevention. Clin. Endosc. 49 (2), 139–146.

La Storia, A., Ercolini, D., Marinello, F., Di Pasqua, R., Villani, F., Mauriello, G., 2011. Atomic force microscopy analysis shows surface structure changes in carvacrol-treated bacterial cells. Res. Microbiol. 162 (2), 164–172.

Lang, G., Buchbauer, G., 2011. A review on recent research (2008–2010) on essential as antimicrobials and antifungals. A review. Flavour Fragr. J. 27 (1), 13–39.

Latifah-Munirah, B., Himratul-Aznita, W.H., Zain, N.M., 2015. Eugenol, an essential oil of clove, causes disruption to the cell wall of *Candida albicans* (ATCC 14053). Front. Life Sci. 8 (3), 231–240.

Lebeaux, D., Ghigo, J.M., Beloin, C., 2014. Biofilm-related infections: bridging the gap between clinical management and fundamental aspects of recalcitrance toward antibiotics. Microbiol. Mol. Biol. Rev. 78 (3), 510–543.

Lee, D.S., Choe, H.S., Kim, H.Y., Kim, S.W., Bae, S.R., Yoon, B.I., et al., 2016. Acute bacterial prostatitis and abscess formation. BMC Urol. 16 (1), 38.

Limoli, D.H., Jones, C.J., Wozniak, D.J., 2015. Bacterial extracellular polysaccharides in biofilm formation and function. Microbiol. Spectr. 3, 3.

Lipsky, B.A., Byren, I., Hoey, C.T., 2010. Treatment of bacterial prostatitis. Clin. Infect. Dis. 50 (12), 1641–1652.

Liu, Q., Meng, X., Li, Y., Zhao, C.N., Tang, G.Y., Li, H.B., 2017. Antibacterial and antifungal activities of spices. Int. J. Mol. Sci. 18 (6), 1283.

Lodhia, M.H., Bhatt, K.R., Thaker, V.S., 2009. Antibacterial activity of essential oils from palmarosa, evening primrose, lavender and tuberose. Indian J. Pharm. Sci. 71 (2), 134–136.

Lopez, D., Vlamakis, H., Kolter, R., 2009. Generation of multiple cell types in *Bacillus subtilis*. FEMS Microbiol. Rev. 33 (1), 152–163.

Lopez-Romero, J., González-Ríos, H., Borges, A., Simões, M., 2015. Antibacterial effects and mode of action of selected essential oils components against *Escherichia coli* and *Staphylococcus aureus*. Evid. Based Complement. Alternat. Med. 2015, Article ID 795435.

MacRae, J.M., Dipchand, C., Matthew Oliver, M., Moist, L., Yilmaz, S., Lok, C., et al., 2016. Arteriovenous access infection, neuropathy, and other complications. Can. J. Kidney Health Dis. 3, 2054358116669127.

Mahboubi, M., Kazempour, N., 2011. Chemical composition and antimicrobial activity of *Satureja hortensis* and *Trachyspermum copticum* essential oil. Iran. J. Microbiol. 3 (4), 194–200.

Maimone, T.J., Baran, P.S., 2007. Modern synthetic efforts toward biologically active terpenes. Nat. Chem. Biol. 3 (7), 396–407.
Mann, E.E., Rice, K.C., Boles, B.R., Endres, J.L., Ranjit, D., Chandramohan, L., et al., 2009. Modulation of eDNA release and degradation affects *Staphylococcus aureus* biofilm maturation. PLoS ONE 4 (6), e5822.
Manoharan, R.K., Lee, J.H., Kim, Y.G., Kim, S.I., Lee, J., 2017. Inhibitory effects of the essential oils α-longipinene and linalool on biofilm formation and hyphal growth of *Candida albicans*. Biofouling 3 (2), 143–155.
Marschollek, C., Karimzadeh, F., Jafarian, M., Ahmadi, M., Mohajeri, S.M., Rahimi, S., et al., 2017. Effects of garlic extract on spreading depression: in vitro and in vivo investigations. Nutr. Neurosci. 20 (2), 127–134.
Mathur, S., Udgire, M., Khambhapati, A., Paul, D., 2014. Anti-biofilm activity and bioactive component analysis of eucalyptus oil against urinary tract pathogen. Int. J. Curr. Microbiol. Appl. Sci. 3 (5), 912–918.
Ming, X., Rothenburger, S., Nichols, M.M., 2008. In vivo and in vitro antibacterial efficacy of PDS plus (polidioxanone with triclosan) suture. Surg. Infect. (Larchmt). 9 (4), 451–457.
Mohiti-Asli, M., Molina, C., Diteepeng, T., Pourdeyhimi, B., Loboa, E.G., 2016. Evaluation of silver ion-releasing scaffolds in a 3d coculture system of MRSA and human adipose-derived stem cells for their potential use in treatment or prevention of osteomyelitis. Tissue Eng. Part A 22 (21–22), 1258–1263.
Morgenstern, M., Post, V., Erichsen, C., Hungerer, S., Bühren, V., Militz, M., et al., 2016. Biofilm formation increases treatment failure in *Staphylococcus epidermidis* device-related osteomyelitis of the lower extremity in human patients. J. Orthop. Res. 34 (11), 1905–1913.
Munita, J.M., Arias, C.A., 2016. Mechanisms of antibiotic resistance. Microbiol. Spectr. 4 (2).
Murugan, K., Selvanayaki, K., Al-Sohaibani, S., 2016. Urinary catheter indwelling clinical pathogen biofilm formation, exopolysaccharide characterization and their growth influencing parameters. Saudi J. Biol. Sci. 23 (1), 150–159.
Naik, M.I., Fomda, B.A., Jaykumar, E., Bhat, J.A., 2010. Antibacterial activity of lemongrass (*Cymbopogon citratus*) oil against some selected pathogenic bacteria's. Asian Pac. J. Trop. Dis. 3 (7), 535–538.
Nallapareddy, S.R., Singh, K.V., Sillanpää, J., Garsin, D.A., Höök, M., Erlandsen, S.L., et al., 2006. Endocarditis and biofilm-associated pili of *Enterococcus faecalis*. J. Clin. Invest. 116 (10), 2799–2807.
Nazzaro, F., Fratianni, F., De Martino, L., Coppola, R., De Feo, V., 2013. Effect of essential oils on pathogenic bacteria. Pharmaceuticals (Basel) 6 (12), 1451–1474.
Nejad, A., Shabani, S., Bayat, M., Hosseini, S.E., 2014. Antibacterial effect of garlic aqueous extract on *Staphylococcus aureus* in hamburger. Jundishapur J. Microbiol. 7, e13134.
Newman, D.J., Cragg, G.M., 2007. Natural products as sources of new drugs over the last 25 years. J. Nat. Prod. 70 (3), 461–477.
Ohno, T., Kita, M., Yamaoka, Y., Imamura, S., Yamamoto, T., Mitsufuji, S., et al., 2003. Antimicrobial activity of essential oils against Helicobacter pylori. Helicobacter 8, 207–215.
Olmos, C., Vilacosta, I., Fernández-Pérez, C., Bernal, J.L., Ferrera, C., García-Arribas, D., et al., 2017. The evolving nature of infective endocarditis in Spain: a population-based study (2003 to 2014). J. Am. Coll. Cardiol. 70 (22), 2795–2804.
Pajohi, M.R., Tajik, H., Farshid, A.A., Hadian, M., 2011. Synergistic antibacterial activity of the essential oil of *Cuminum cyminum* L. seed and nisin in a food model. J. Appl. Microbiol. 110 (4), 943–951.

Paredes, J., Alonso-Arce, M., Schmidt, C., Valderas, D., Sedano, B., Legarda, J., et al., 2014. Smart central venous port for early detection of bacterial biofilm related infections. Biomed. Microdev. 16 (3), 365–374.

Patil, H.V., Patil, V.C., Ramteerthka, M.N., Kulkarni, R.D., 2011. Central venous catheter-related bloodstream infections in the intensive care unit. Indian J. Crit. Care Med. 15, 213–223.

Pérez-Fons, L., Aranda, F.J., Guillén, J., Villalaín, J., Micol, V., 2006. Rosemary (*Rosmarinus officinalis*) diterpenes affect lipid polymorphism and fluidity in phospholipid membranes. Arch. Biochem. Biophys. 453 (2), 224–236.

Peters, B.M., Jabra-Rizk, M.A., Graeme, A.O., Costerton, J.W., Shirtliff, M.E., 2012. Polymicrobial interactions: impact on pathogenesis and human disease. Clin. Microbiol. Rev. 25 (1), 193–213.

Phillips, P.L., Schultz, G.S., 2012. Molecular mechanisms of biofilm infection: biofilm virulence factors. Adv. Wound Care (New Rochelle) 1 (3), 109–114.

Pichersky, E., Noel, J.P., Dudareva, N., 2006. Biosynthesis of plant volatiles: nature's diversity and ingenuity. Science 311 (5762), 808–811.

Raei, P., Pourlak, T., Memar, M.Y., Alizadeh, N., Aghamali, M., Zeinalzadeh, E., et al., 2017. Thymol and carvacrol strongly inhibit biofilm formation and growth of carbapenemase-producing Gram negative bacilli. Cell Mol. Biol. (Noisy-le-grand) 63 (5), 108–112.

Ramakrishnan, K.K., Sparks, R.A., Berryhill, W.E., 2007. Diagnosis and treatment of otitis media. Am. Fam. Physician. 76 (11), 1650–1658.

Ramalingam, K., Amaechi, B.T., Ralph, R.H., Lee, V.A., 2012. Antimicrobial activity of nanoemulsion on cariogenic planktonic and biofilm organisms. Arch. Oral. Biol. 57 (1), 15–22.

Ramasamy, M., Lee, J.H., Lee, J., 2017. Direct one-pot synthesis of cinnamaldehyde immobilized on gold nanoparticles and their antibiofilm properties. Colloids Surf. B Biointerfaces 160, 639–648.

Rammanee, K., Hongpattarakere, T., 2011. Effects of tropical citrus essential oils on growth, aflatoxin production, and ultrastructure alterations of *Aspergillus flavus* and *Aspergillus parasiticus*. Food Bioprocess Tech. 4 (6), 1050–1059.

Rasooli, I., Rezaei, M.B., Allameli, A., 2006. Ultrastructural studies on antimicrobial efficacy of thyme essential oils on *Listeria monocytogenes*. Int. J. Infect. Dis. 10, 236–241.

Revelas, A., 2012. Healthcare-associated infections: a public health problem. Niger. Med. J. 53 (2), 59–64.

Ribeiro, M., Monteiro, F.J., Ferraz, M.P., 2012. Infection of orthopedic implants with emphasis on bacterial adhesion process and techniques used in studying bacterial-material interactions. Biomatter 2 (4), 176–194.

Rodrigues, L.R., 2011. Novel approaches to avoid microbial adhesion onto biomaterials. J. Biotechnol. Biomat. 1 (4), 1000104e.

Rosenfeld, S.B., Copley, L.A., Mignemi, M., An, T., Benvenuti, M., Schoenecker, J., 2017. Key concepts of musculoskeletal infection. Instr. Course Lect. 66, 569–584.

Royo-Cebrecos, C., Gudiol, C., García, J., Tubau, F., Laporte, J., Ardanuy, C., et al., 2017. Characteristics, aetiology, antimicrobial resistance and outcomes of bacteraemic cholangitis in patients with solid tumours: a prospective cohort study. J. Infect. 74 (2), 172–178.

Sadekuzzaman, M., Yang, S., Mizan, M.F.R., Ha, S.D., 2015. Current and recent advanced strategies for combating biofilms. Compr. Rev. Food Sci. Food Saf. 14, 491–509.

Sambyal, S.S., Sharma, P., Shrivastava, D., 2017. Antibiofilm activity of selected plant essential oils against *Pseudomonas aeruginosa* and *Staphylococcus aureus*. Int. J. Curr. Microbiol. Appl. Sci. 6 (3), 444–450.

Saviuc, C.M., Drumea, V., Olariu, L., Chifiriuc, M.C., Bezirtzoglou, E., Lazar, V., 2015. Essential oils with microbicidal and antibiofilm activity. Curr. Pharm. Biotechnol. 16 (2), 137–151.

Schwab, W., Davidovich-Rikanati, R., Lewinsohn, E., 2008. Biosynthesis of plant-derived flavor compounds. Plant J. 54 (4), 712–732.

Seyyednejad, S.M., Motamedi, H., Najvani, F.D., Hassannejad, Z., 2014. Antibacterial effect of *Eucalyptus microtheca*. Int. J. Enteric Pathog. 2 (2), e16515.

Sheikh, M.A., Shokr, M., Ibrahim, W., Cardozo, S., 2017. Fibrin sheath-associated endovascular infection of the heart: the Trojan horse of indwelling central venous catheters. BMJ Case Rep. 2017.

Shimazu, K., Oguchi, R., Takahashi, Y., Konishi, K., Karibe, H., 2016. Effects of surface reaction-type pre-reacted glass ionomer on oral biofilm formation of *Streptococcus gordonii*. Odontology 104 (3), 310–317.

Silva, F., Ferreira, S., Queiroz, J.O.A., Domingues, F.C., 2011. Coriander (*Coriandrum sativum* L.) essential oil: its antibacterial activity and mode of action evaluated by flow cytometry. J. Med. Microbiol. 60, 1479–1486.

Singh, B., Sharma, R.A., 2015. Plant terpenes: defense responses, phylogenetic analysis, regulation and clinical applications. 3 Biotech. 5 (2), 129–151.

Singh, R., Nadhe, S., Wadhwani, S., Shedbalkar, U., Chopade, B.A., 2016. Nanoparticles for control of biofilms of Acinetobacter species. Materials 9 (5), 383.

Sinha, S., Biswas, D., Mukherjee, A., 2011. Antigenotoxic and antioxidant activities of palmarosa and citronella essential oils. J. Ethnopharmacol. 137 (3), 1521–1527.

Soothill, J., 2013. Use of bacteriophages in the treatment of *Pseudomonas aeruginosa* infections. Expert Rev. Anti-Infect. Ther. 11 (9), 909–915.

Soukos, N.S., Goodson, J.M., 2011. Photodynamic therapy in the control of oral biofilms. Periodontol. 2000 55 (1), 143–166.

Stapleton, F., Carnt, N., 2012. Contact lens-related microbial keratitis: how have epidemiology and genetics helped us with pathogenesis and prophylaxis. Eye (Lond.) 26 (2), 185–193.

Stoodley, P., Ehrlich, G.D., Sedghizadeh, P.P., Hall-Stoodley, L., Baratz, M.E., Altman, D. T., et al., 2011. Orthopaedic biofilm infections. Curr. Orthop. Pract. 22 (6), 558–563.

Swidsinski, A., Weber, J., Loening-Baucke, V., Hale, L.P., Lochs, H., 2005. Spatial organization and composition of the mucosal flora in patients with inflammatory bowel disease. J. Clin. Microbiol. 43 (7), 3380–3389.

Talei, G.R., Mohammadi, M., Bahmani, M., Kopaei, M.R., 2017. Synergistic effect of *Carum copticum* and *Mentha piperita* essential oils with ciprofloxacin, vancomycin, and gentamicin on Gram-negative and Gram-positive bacteria. Int. J. Pharm. Investig. 7 (2), 82–87.

Tande, A.J., Patel, R., 2014. Prosthetic joint infection. Clin. Microbiol. Rev. 27 (2), 302–345.

Taraszkiewicz, A., Fila, G., Grinholc, M., Nakonieczna, J., 2013. Innovative strategies to overcome biofilm resistance. Biomed. Res. Int. 2013, Article ID 150653.

Tavassoli, S.K., Mousavi, S.M., Emam-Djomeh, Z., Razavi, S.H., 2011. Chemical composition and evaluation of antimicrobial properties of *Rosmarinus officinalis* L. essential oil. Afr. J. Biotechnol. 10, 13895–13899.

Thirugnanasampandan, R., Jayakumar, R., Prabhakaran, M., 2012. Analysis of chemical composition and evaluation of antigenotoxic, cytotoxic and antioxidant activities of essential oil of *Toddalia asiatica* (L.) Lam. Asian Pac. J. Trop. Biomed. S1276–S1279.

Trombetta, D., Castelli, F., Sarpietro, M.G., Venuti, V., Cristani, M., Daniele, C., et al., 2005. Mechanisms of antibacterial action of three monoterpenes. Antimicrob. Agents Chemother. 49 (6), 2474–2478.

Turgis, M., Han, J., Caillet, S., Lacroix, M., 2009. Antimicrobial activity of mustard essential oil against *Escherichia coli* O157:H7 and *Salmonella typhi*. Food. Control. 20, 1073–1079.

Tzin, V., Galili, G., 2010. The biosynthetic pathways for shikimate and aromatic amino acids in *Arabidopsis thaliana*. Arabidopsis Book 8, e0132.

Upadhyay, A., Upadhyaya, I., Kollanoor-Johny, A., Venkitanarayanan, K., 2014. Combating pathogenic microorganisms using plant-derived antimicrobials: a mini review of the mechanistic basis. Biomed. Res. Int. 2014, Article ID 761741.

Valdebenito, B., Tullume-Vergara, P.O., González, W., Kreth, J., Giacaman, R.A., 2017. In silico analysis of the competition between *Streptococcus sanguinis* and *Streptococcus mutans* in the dental biofilm. Mol. Oral Microbiol. Available from: https://doi.org/10.1111/omi.12209.

Vasudevan, R., 2017. Dental plaques: microbial community of the oral cavity. J Microbiol. Exp. 4 (1), 00100.

Walmiki, M.R., Vittal, R.R., 2017. Cell attachment inhibition and anti-biofilm activity of *Syzygium aromaticum, Cuminum cyminum* and *Piper nigrum* essential oils against pathogenic bacteria. J. Essent. Oil Bear. Pl. 20 (1), 59–68.

Wood, T.K., Knabel, S.J., Kwan, B.W., 2013. Bacterial persister cell formation and dormancy. Appl. Environ. Microbiol. 79, 237116–237121.

Wu, H., Moser, C., Wang, H.Z., Høiby, N., Song, Z.J., 2015. Strategies for combating bacterial biofilm infections. Int. J. Dent. Oral Sci. 7 (1), 1–7.

Xu, J., Zhou, F., Ji, B.P., Pei, R.S., Xu, N., 2008. The antibacterial mechanism of carvacrol and thymol against *Escherichia coli*. Lett. Appl. Microbiol. 47 (3), 174–179.

Yadav, N., Yadav, R., Goyal, A., 2014. Chemistry of terpenoids. Int. J. Pharm. Sci. Rev. Res. 27 (2), 272–278. Article No. 45.

Yadav, M.K., Chae, S.W., Im, G.J., Chung, J.W., Song, J.J., 2015. Eugenol: a phyto-compound effective against methicillin-resistant and methicillin-sensitive *Staphylococcus aureus* clinical strain biofilms. PLoS ONE 10 (3), e0119564.

Yan, X., Gu, S., Shi, Y., Cui, X., Wen, S., Ge, J., 2017. The effect of emodin on *Staphylococcus aureus* strains in planktonic form and biofilm formation in vitro. Arch. Microbiol. 199 (9), 1267–1275.

Yap, P.S.X., Yiap, B.C., Ping, H.C., Lim, S.H.E., 2014. Essential oils, a new horizon in combating bacterial antibiotic resistance. Open Microbiol. J. 8, 6–14.

Zhang, Y., Liu, X., Wang, Y., Jiang, P., Quek, S.Y., 2016. Antibacterial activity and mechanism of cinnamon essential oil against *Escherichia coli* and *Staphylococcus aureus*. Food. Control. 59, 282–289.

Zhang, Y., Wang, Y., Zhu, X., Cao, P., Wei, S., Lu, Y., 2017. Antibacterial and antibiofilm activities of eugenol from essential oil of *Syzygium aromaticum* (L.) Merr. & LM Perry (clove) leaf against periodontal pathogen *Porphyromonas gingivalis*. Microb. Pathog. 113, 396–402.

FURTHER READING

Ganjewala, D., 2009. Cymbopogon essential oils: chemical compositions and bioactivities. Int. J. Essen. Oil Ther. 3, 56–65.

CHAPTER 10

Anticancer Phytocompounds: Experimental and Clinical Updates

Farrukh Aqil[1,2], Radha Munagala[1,2], Ashish K. Agrawal[2] and Ramesh Gupta[2,3]
[1]Department of Medicine, University of Louisville, Louisville, KY, United States
[2]James Graham Brown Cancer Center, University of Louisville, Louisville, KY, United States
[3]Department of Pharmacology and Toxicology, University of Louisville, Louisville, KY, United States

Cancer is a major public health concern worldwide and is among the leading causes of deaths ranking second after cardiovascular disorders in the developed nations (ACS, 2016). In 2012, there were 14 million new cases and 8.2 million cancer-related deaths worldwide. According to the American Cancer Society, cancer is the second most common cause of death in the United States and accounts for nearly one of every four deaths. More alarming is that number of new cancer cases will raise to 22 million within the next two decades (Global Burden of Disease Cancer et al., 2015). If we examine the geographical statistics worldwide, over 60% of the new cancer cases occur in developing countries in Africa, Asia, and Central and South America. Not only the high cancer incidence but about 70% of the world's cancer deaths also occur in these regions. In the developing countries, one-third of people are at risk of developing cancer during their lifetime (Jemal et al., 2017).

The cancer is a group of diseases characterized by the uncontrolled proliferation and spread of abnormal cell. Cancer in humans is a multistep process, which involves various genetic or epigenetic changes ultimately driving the malignant transformation of the normal cells (Fig. 10.1). In recent years the development of sophisticated proteomic and genomics techniques has provided better understanding of pathways and genetic elements responsible for the development of cancerous phenotypes. The development of early detection techniques has reduced the cancer-related deaths to some extent (ACS, 2016). However, the management for most cancers is still a long way from reality.

New Look to Phytomedicine
DOI: https://doi.org/10.1016/B978-0-12-814619-4.00010-0

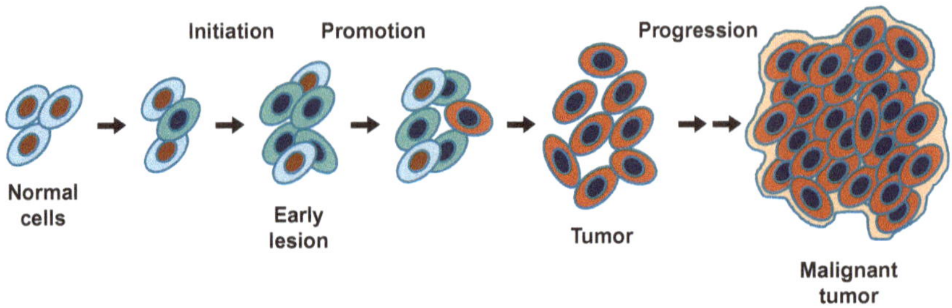

Figure 10.1 Process of cancer cell progression.

10.1 CURRENT PROBLEM IN CANCER THERAPY

Treatment of cancer is often multifaceted involving surgery, radiation therapy, chemotherapy, immunotherapy, hormone, and/or targeted therapy. Surgery is the standard treatment of patients diagnosed with early-stage cancers. However, despite complete resection, 30%–60% patients relapse within 5 years and succumb to metastases (ACS, 2016; Egermann et al., 2002; DeSantis et al., 2014; Siegel et al., 2014).

Systemic therapy employs anticancer drugs given intravenously in general and includes drugs classified as targeted therapy like herceptin (for breast cancer), chemo drug including antimetabolites (e.g., methotrexate), DNA-interactive agents (e.g., cisplatin, doxorubicin), antitubulin agents (taxanes), hormones- and molecular-targeting agents, and cyclophosphamide as adjuvant to surgery, and hormone therapy in women who test positive for hormone receptors breast cancer. Chemo drugs may be used as neo-adjuvant therapy prior to surgery to shrink the tumor or as adjuvant therapy after surgery to kill the tumor cells (Downs-Holmes and Silverman, 2011).

A major reason for failure in cancer treatment is the inherent or acquired drug resistance of the tumor. Emergence of resistance to antitumor chemotherapeutic agents has been associated with either intrinsic or acquired chemo-resistance. The underlying biochemical and genetic reasons of drug resistance are still not completely clear. More than 80% of currently used antitumor agents can be transported by ATP-binding cassette (ABC)-containing drug efflux transporters. Elevated expression of these transporters is frequently found in various cancers and correlations with elevated expression of PgP or MRP1 to chemotherapeutic outcomes have been observed in some cases, suggesting that these transporters may contribute to chemo-resistance (Kuo, 2007). However, attempts to modulate the activities of these transporters using reversal agents have met with limited success. Despite advances in surgery, tailored chemotherapeutic regimens, cytotoxic combinations, developments of molecularly targeted therapies and immune-checkpoint inhibitors and the treatment of cancer remains a formidable clinical challenge due to high rates of relapse, drug resistance, dose-limiting toxicity, and metastasis. An urgent need for alternative strategies to overcome the challenges of chemotherapy has drawn increased attention among researchers for the use of plant bioactives in cancer treatment, due to their potentially wider safety margin and the potential to complement conventional chemotherapeutic drugs.

Natural compounds are good sources for the development of new remedies for different diseases. It has been estimated that two-third of all cancers can be prevented by lifestyle and dietary measures alone. Overall, at least 20% of all cancers can be prevented by consumption of diets rich in vegetables and fruits (Donaldson, 2004; Vyas and Singh, 2014). In this chapter, we highlight the experimental evidence concerning promising plant therapeutics that exhibit protective effect against cancer.

First of all it is important to understand the need of right selection procedures for the chemopreventive agents. Literature is full of several thousand chemical agents and a large number of defined and undefined mixtures, which have epidemiological evidence of preventing cancer. In such a scenario the selection of the chemopreventive agents was largely based on the observational studies built on intake of fruits and vegetables and reduced risk of cancer (Steward and Brown, 2013). Also several randomized clinical trials have been conducted in exposed and control populations even without early phase clinical studies. However, with the time, the approach for selection has completely changed and now it is largely based on preclinical (animal) studies with the doses clinically achievable and evidence of nontoxicity both in vitro and in vivo (Fig. 10.2). Chemopreventive agent development for clinics has exploited a similar model to new drug development in cancer therapy with sequential phase I, II, and III studies (Lippman et al., 1994). Therefore now a days chemoprevention studies are conducted to analyze the appearance of first tumor (tumor incidence), multiplicity and burden for the cancer prevention analyses. Furthermore, studies are also conducted to determine the effect of agents in tumor microenvironment in relevant animal models and in patient-derived tumor xenografts (PDX models) for better clinical relevance. The schematic representation for the selection criteria has been shown in Fig. 10.2, which is adopted from a recent manuscript by Steward and Brown (2013).

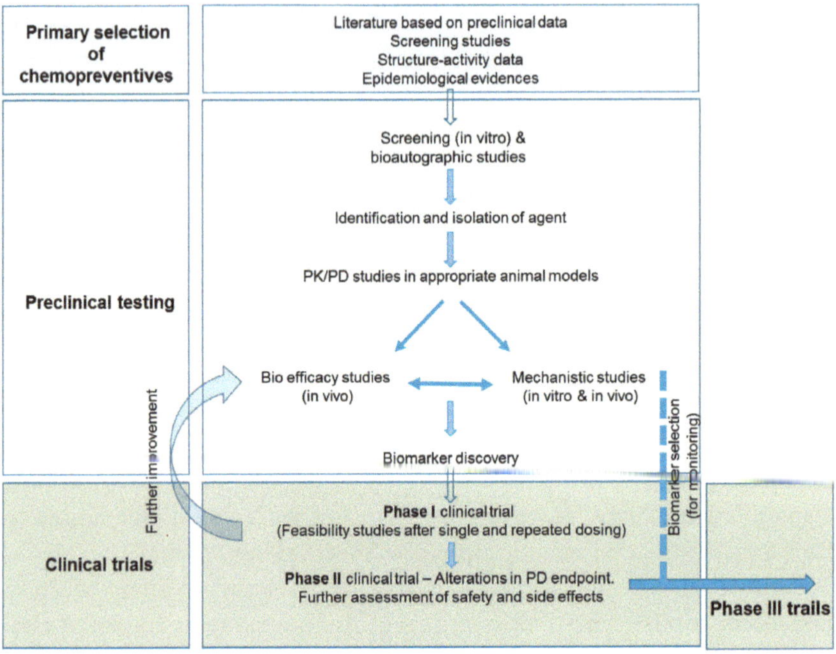

Figure 10.2 Selection criteria for plant bioactives for the development as potential therapeutics.

10.2 PLANT BIOACTIVES IN CANCER PREVENTION/THERAPY

Cancer chemoprevention involves use of a synthetic, natural, or biological agent to lower risk or delay the occurrence of cancer (Sporn, 1976). The concept of cancer chemoprevention appears to have evolved through epidemiological observations. Several studies have reported that cancer incidence is lower in the developing countries specially in the population groups with diets rich in fruits and vegetables (Block et al., 1992). Plant-derived compounds have historically led to some of our most useful cancer drugs such as vinblastine and vincristine, etoposide, paclitaxel (Taxol), docetaxel, topotecan, and irinotecan. These compounds are among the most effective cancer chemo-therapeutics currently available (Cragg and Pezzuto, 2016). In addition to eliciting strong therapeutic activity, dietary phytochemicals have also shown great promise in reducing cancer cell progression to metastasis (Pratheeshkumar et al., 2012). Metastasis involves cancer cell migration, invasion, dissemination through the lymphatics or vasculature, and, ultimately, colonization. Dietary phytochemicals are well documented to block the molecular pathways that lead to metastatic events (Pratheeshkumar et al., 2012).

In the past few decades as chemoprevention has gained much interest, the dietary polyphenolics have become not only important potential chemopreventive but also therapeutic natural agents. Dietary polyphenols gained the interest of the scientific community due to their wide content in a variety of plant-derived foods ("functional foods") and beverages commonly consumed, such as fruits, vegetables, coffee, tea, and cocoa. Importantly, polyphenols have been intensively studied eliciting excellent antioxidant activity, which, in part, contributes to anticancer function of the natural compounds. Natural agents are presumed to offer advantages over synthetic compounds due to their broader range of targets compared to mostly single-target synthetic anticancer compounds. In view of their multitargeting properties along with relatively lower systemic toxicity, plant therapeutics are considered to offer significant therapeutic advantages for prevention and treatment of cancer.

Several polyphenolics have been used extensively in the preclinical studies and are either tested in clinical trials itself or in adjuvant settings. As a proof of concept, tamoxifen and related compounds, such as raloxifene, are well-established agents that are currently in use for the prevention of cancer in high-risk breast cancer patients (Park and Jordan, 2002). Recently, letrozole has also been marketed for the breast cancer prevention (Tremont et al., 2017).

Chemoprevention has led to the isolation and study of many dominant components of dietary materials, such as β-carotene, lycopene, curcumin, catechins, resveratrol, ellagic acid, withaferin A, celastrol, anthocyanins, and anthocyanidins. A couple of clinical trials performed with resounding negative results have disclaimed the entire concept of cancer chemoprevention in the past (Albanes et al., 1996; Potter, 2014). However, with revived interest in the field over 300 clinical trials are currently ongoing (http://www.clinicaltrials.gov), most of which are investigating chemopreventives or their derivatives believed to have some activity. Few compounds, which have been extensively used in chemoprevention studies and their potential, are discussed later.

10.3 CURCUMIN

Curcumin, present in the human environment for centuries, is probably the most studied phytocompounds with thousands of publications demonstrating its potential against several ailments including cancer. Curcumin is a mixture of three curcuminoids (desmethoxy-curcumin,

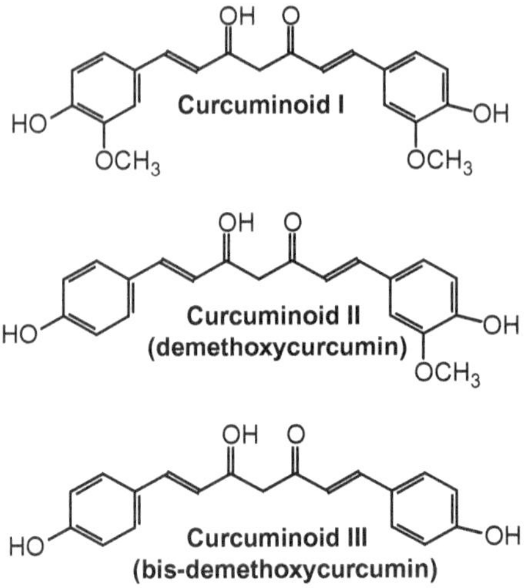

Figure 10.3 Structure of curcumin (curcuminoids I, II, and III).

bis-demethoxycurcumin, and curcumin). It is a major active component of the yellow spice, turmeric (*Curcuma longa*) (Fig. 10.3) representing 3%–5% of turmeric. There is extensive toxicological and preclinical evidence that demonstrate either no or minimal adverse effects of curcumin administration in mice, rats, dogs, and monkeys (Anon., 1996). Phase I and phase II clinical trials have demonstrated the safety of curcumin even at high doses (8–12 g/d) over several months (Gupta et al., 2013b). Numerous studies have demonstrated curcumin to possess antioxidant, antiinflammatory, and anticancerous properties. Several hundred papers have been published on the anticancer properties of curcumin. Most of these studies demonstrated that curcumin inhibits cancer cells by targeting multiple molecular pathways. Numerous studies from the Dr. Aggarwal's laboratory (Gupta et al., 2013a,b, 2012, 2011; Aggarwal et al., 2013) have shown preclinical and clinical data on curcumin.

Over 50 clinical trials of curcumin and its various formulations against various cancers have been conducted so far (Table 10.1). However, due to its known poor solubility and bioavailability, curcumin failed to show its complete potential in the clinical studies even at high doses. The hydrophobic nature of curcumin results in low water solubility, and rapid intestinal/hepatic metabolism limits its oral bioavailability, impeding clinical development of curcumin as a potential therapeutic agent (Anand et al., 2007). Therefore several formulations have been tested to improve its oral bioavailability. Liposomes, polymeric nanoparticles, micelles, and

Table 10.1 Clinical trials of curcumin, curcumin analogs, and its formulations

S. no.	Drug	Cancer	Year	Title	Study center
1.	Curcumin in combination with gemcitabine	Pancreatic	2010	Gemcitabine with curcumin for pancreatic cancer	Rambam Health Care Campus
2.	Curcumin alone and in combination with bioperine	Multiple myeloma	2011	Curcumin (diferuloylmethane derivative) with or without bioperine	U.T.M.D. Anderson Cancer Center Houston, TX, United States
3.	Curcumin C3 complex	Breast	2012	Curcumin for the prevention of radiation-induced dermatitis in breast cancer patients	University of Rochester Medical Center & Wilmot Cancer Center Rochester, NY, United States
4.	Curcumin (dietary supplement)	Colorectal	2012	Curcumin for the prevention of colon cancer	Comprehensive Cancer Center Ann Arbor, MI, United States
5.	Curcumin + bioprine	Chronic obstructive pulmonary disease	2012	Effect of curcumin on lung inflammation	VA Medical Center Houston, TX, United States
6.	Curcumin (dietary supplement)	Prostate	2013	Radio-sensitizing and radio-protective effects of curcumin in prostate cancer	Oncology and radiotherapy department, Besat Hospital Tehran, Iran

(Continued)

Table 10.1 (Continued)

S. no.	Drug	Cancer	Year	Title	Study center
7.	Curcumin C3 tablet	Colorectal	2013	Curcumin biomarkers	UNC Department of Family Medicine, Chapel Hill, NC, United States
8.	Curcumin	Glioblastoma	2013	Curcumin bioavailability in glioblastoma patients	Department of Neurosurgery, Johann Wolfgang Goethe-University, Frankfurt, Hessen, Germany
9.	Curcumin	Uterine cervical dysplasia	2013	Trial on safety and pharmacokinetics of intravaginal curcumin	Emory University Atlanta, GA, United States
10.	Curcumin	Pancreatic neoplasms Adenocarcinoma	2014	Trial of curcumin in advanced pancreatic cancer	MD Anderson Cancer Center Houston, TX, United States
11.	Curcumin/green tea extract/polygonum cuspidatum extract/soybean extract capsule	Healthy, no evidence of disease	2014	A nutritional supplement capsule containing curcumin, green tea extract, *Polygonum cuspidatum* extract, and soybean extract in healthy participants	Barbara Ann Karmanos Cancer Institute Detroit, MI, United States
12.	Surface-controlled water soluble curcumin	Advanced cancers	2015	Phase I: Study of surface-controlled water soluble curcumin (theracurmin CR-011L)	UT MD Anderson Cancer Center Houston, TX, United States

13.	Curcumol	Chemotherapy-induced mucositis	2015	Curcumin for prevention of oral mucositis in children chemotherapy	Hadassah Medical Organization Jerusalem, Israel
14.	Curcumin	Healthy, no evidence of disease	2015	Phase II: A trial of curcumin among patients with prevalent subclinical neoplastic lesions (aberrant crypt foci)	Chao Family Comprehensive Cancer Center Orange, CA, United States
15.	Microgranular curcumin C3 complex	Head and neck	2016	Curcumin biomarker trial in head and neck cancer	LSUHSC-Shreveport and Feist-Weiller Cancer Center, Shreveport, LA, United States
16.	Curcumin as dietary supplement	Endometrial carcinoma	2016	Effect of curcumin addition to standard treatment on tumor-induced inflammation in endometrial carcinoma	University Hospital KU Leuven Campus Gasthuisberg, Leuven, Belgium
17.	Curcumin	Radiation-induced dermatitis	2016	Oral curcumin for radiation dermatitis	University of Rochester, United States
18.	Micellar curcumin	Metabolic syndrome	2016	Micellar curcumin and metabolic syndrome biomarkers	I University of Hohenheim Stuttgart, Baden-Württemberg, Germany

(*Continued*)

Table 10.1 (Continued)

S. no.	Drug	Cancer	Year	Title	Study center
19.	Curcumin (dietary supplement)	Prostate	2017	Comparison of duration of treatment interruption with or without curcumin during the off treatment periods in patients with prostate cancer undergoing intermittent androgen deprivation therapy	Samsung Medical Center
20.	Calcumin (curcumin)	Familial adeno-matous polyposis	2017	Use of curcumin for treatment of intestinal adenomas in familial adeno-matous polyposis (FAP)	University of Puerto Rico Comprehensive Cancer Center San Juan, Puerto Rico
21.	Curcumin	Familial adeno-matous polyposis	2017	Curcumin in treating patients with familial adeno-matous polyposis	Johns Hopkins University/Sidney
22.	Curcumin	–	2017	Curcumin pharmacokinetics	UNC Department of Family Medicine Chapel Hill, NC, United States

Source: http://www.clinicaltrial.gov.

other formulations have provided leads as drug-delivery vehicles; however, they are associated with inherent limitations, such as short circulation time and stability issues when used as unmodified liposomes in vivo (Siddiqui et al., 2009; Sonaje et al., 2007). This subject has recently been reviewed (Bansal et al., 2011a).

To overcome low oral bioavailability issues, our laboratory has demonstrated the use of polymeric implants for slow and sustained systemic delivery of curcumin. We have shown that the curcumin administered via implants resulted in significant reduction in both the tumor multiplicity (60% reduction) and tumor volume (184 vs 280 mm^3). In contrast the dietary administration of curcumin despite higher doses than the implant dose failed to elicit any response. We demonstrated that dietary curcumin increased hepatic CYP1A and CYP1B1 activities without any effect on CYP3A4 activity, whereas curcumin implants increased both CYP1A and CYP3A4 activities but decreased CYP1B1 activity in the presence of 17β-estradiol (Bansal et al., 2014). In another study, we showed that curcumin implants exhibited diffusion-mediated biphasic release pattern with ~twofold higher in vivo release as compared to in vitro. We showed almost eightfold higher curcumin levels in brain compared with diet (Bansal et al., 2012). Sustained release of curcumin and the other curcuminoids has also been reported (Aqil et al., 2012). Although curcumin levels were similar in liver from both the routes, implants were more efficacious in altering hepatic CYP1A1 levels and CYP3A4 activity despite ~28-fold lower doses after 90 days (Bansal et al., 2012). Curcumin delivered via polymeric implants bypasses the oral route and rapid hepatic metabolism of curcumin and achieves higher blood and tissue levels. However, the use of polymeric implants is restricted as it requires surgical grafting (Bansal et al., 2011b).

We have recently demonstrated that the use of exosomes as nanodrug delivery vehicles has tremendous potential for the development of novel therapeutic applications (Munagala et al., 2016). Exosomes, recognized in the past decade, are endogenous nanovesicles, have been suggested as a potential drug carrier (Ren et al., 2016; Batrakova and Kim, 2015; Johnsen et al., 2014), and are an emerging field. Because of their nanosize (30–100 nm), exosomes can readily enter cells and deliver their payloads. We identified bovine milk as a viable biocompatible and scalable source for exosomes that could serve as a potential drug-delivery platform. Exosomal formulation of curcumin (ExoCUR) provides stability to curcumin and achieves significantly higher curcumin tissue levels compared to free curcumin following oral administration. Furthermore, ExoCUR exhibited enhanced biological efficacy measured as antiproliferative, anti-inflammatory, and antitumor activities (Aqil et al., 2017). The exosomal formulation was well tolerated with no adverse effects based on gross toxicity as well as systemic toxicity analyses. Since the bioavailability and stability of curcumin can be enhanced by various means such as implant or via the exosome, such techniques should be used to enhance the utility of curcumin in clinical settings.

10.4 GREEN TEA POLYPHENOLS

Tea is probably the most widely consumed drink, after water. Its consumption can be traced back to 2737 BC as believed by the Chinese (Zink and Traidl-Hoffmann, 2015). It is produced from the *Camellia sinensis* and can generally be divided into several categories based on its processing. Green tea which is unfermented leaves of the plant is generally considered superior to black tea in health benefits. The chemical composition of green tea, with regard to its major components, is similar to that of the fresh leaves of the plant. It contains a unique set of catechins (accounting for about 30% of the dry weight) that demonstrate various activities potentially useful in the prevention and treatment of various diseases, including cancer. The major catechins or polyphenols of green tea include epigallocatechin-3-gallate (EGCG) (Fig. 10.4), epigallocatechin (EGC), epicatechin-3-gallate (ECG), and epicatechin (EC). EGCG accounts for over 50% of the total catechins, and appears to be the most

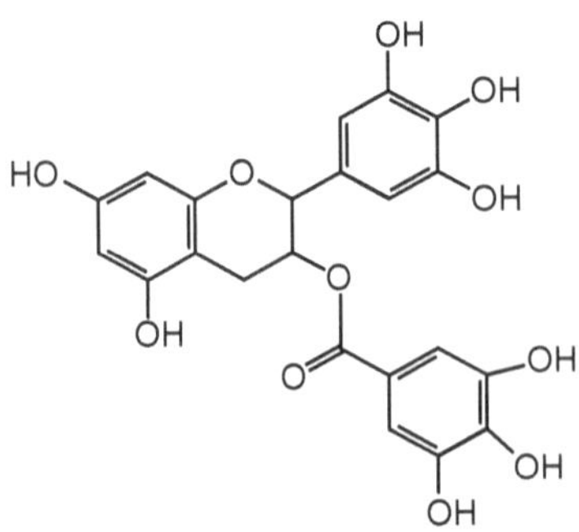

EGCG

Figure 10.4 Structure of epigallocatechin gallate (EGCG), the major constituent of tea.

effective and best-studied constituent of green tea. EGCG has been shown to have anticancer activity against various cancers and holds considerable promise for chemoprevention according to epidemiological, cell culture, and animal and clinical studies (Granja et al., 2016).

Polyphenone E (Poly E), a standardized green tea extract, has been approved by the US FDA for use to treat genital and perianal warts in ointment form (Veregen; MediGene AG, Munich, Germany). Poly E has been investigated in clinical trials for prostate cancer, and a Phase II study has investigated the effects of Poly E in patients with chronic lymphocytic leukemia. Several other clinical studies are conducted worldwide to show its full potential (Table 10.2).

Despite strong antitumor activity of EGCG in various animal models, its anticancer effects in humans and the epidemiological evidences on tea consumption have not yielded conclusive results. Interestingly, continuous administration of green tea polyphenols (GTPs) has demonstrated potential effect to reduce carcinogenesis in premalignant stages of the experiment versus the doses given at the preinitiation stages (Sur and Panda, 2017). Similarly, Chow et al. (2003) showed that the repeated administration of 800 mg of green tea polyphenols once daily for 4 weeks resulted in a 60% increase in the systemic availability of free EGCG, which may be due to the inhibition of presystemic elimination of the catechins. On the other hand, the authors also showed that oral bioavailability of tea polyphenols in humans was low.

In our own studies to administer EGCG continuously, we used polymeric implants to show the anticancer activity of green tea extracts. We showed the continuous released of GTPs from polymeric implants for long durations (Cao et al., 2014). In another study, we demonstrated systemic delivery of GTPs to significantly diminish benzo[*a*]pyrene-induced DNA adducts (50%) analyzed by P^{32}-postlabeling after 1 week of treatment at relatively low doses compared to GTPs given in drinking water (<100-fold; 15.7 vs 1632 mg, respectively). The reduction in the DNA adducts occurred presumably due to known scavenging of carcinogenic antidiolepoxide metabolites of benzo[*a*]pyrene by the Poly E catechins (Cao et al., 2011).

10.5 OTHER CHEMOPREVENTIVE AGENTS

Besides curcumin and green tea polyphenols, there are several other chemopreventive agents that have shown strong cancer chemoprevention

Table 10.2 Clinical trials of green tea and its constituents

S. no.	Drug	Cancer	Year	Title	Study center
1.	Polyphenon E ointment	Condylomata acuminata	2007	Efficacy and safety study of polyphenon E to treat external genital warts	Davis, San Diego, California, United States
2.	Green tea catechin extract	Solid tumor	2010	Green tea extract (polyphenon E) in preventing cancer in healthy participants	Arizona Cancer Center at University of Arizona Health Sciences Center Tucson, Arizona, United States
3.	Polyphenon E (veregen) 15% ointment	Genital warts Perianal warts	2010	Pharmacokinetic study of topically applied veregen 15% compared with oral intake of green tea beverage	Charité Research Organization Berlin, Germany
4.	Polyphenon E (EGCG)	Prostate	2012	Green tea extract and prostate cancer	LSU Health Sciences Center Shreveport, Louisiana, United States
5.	Green tea catechin extract	Lung	2012	Green tea extract in preventing cancer in former and current heavy smokers with abnormal sputum	British Columbia Cancer Agency–Vancouver Cancer Centre Vancouver, British Columbia, Canada
6.	Veregen	Anogenital warts	2012	Systemic exposure of catechins from veregen 15% ointment in patients with external anogenital warts and from oral intake of green tea beverage in healthy volunteers	CardioSec Clinical Research GmbH, Erfurt, Germany

7.	Green tea drink and capsules	Prostate	2013	Lycopene or green tea for men at risk of prostate cancer	Southmead Hospital Bristol, England, United Kingdom
8.	Green tea	Prostate	2013	Diet and psa levels in patients with prostate cancer	Memorial Sloan-Kettering Cancer Center, New York, NY, United States
9.	Green tea and polyphenon E	Lung	2013	Green tea or polyphenon E in preventing lung cancer in former smokers with chronic obstructive pulmonary disease	Virginia G. Piper Cancer Center at Scottsdale Healthcare-Shea Tucson, AZ, United States
10.	Green tea extract	Solid tumor	2013	Green tea extract in treating patients with advanced solid tumors	Memorial Sloan-Kettering Cancer Center New York, NY, United States
11.	Double-brewed green tea	Ovarian	2013	Green tea intake for the maintenance of complete remission in women with advanced ovarian carcinoma	Centre Hospitalier Universitaire de Québec, Quebec, Canada
12.	Polyphenon E	Leukemia	2013	Green tea extract in treating patients with stage 0, stage i, or stage ii chronic lymphocytic leukemia	Mayo Clinic in Arizona Scottsdale, Arizona; Mayo Clinic Rochester, MN, United States

(Continued)

Table 10.2 (Continued)

S. no.	Drug	Cancer	Year	Title	Study center
13.	Green tea catechins extract	Prostate	2014	Defined green tea catechins in treating patients with prostate cancer undergoing surgery to remove the prostate	Arizona Cancer Center-Tucson Tucson, AZ, United States
14.	Green tea extract	Breast	2014	Defined green tea catechins extract in treating women with hormone receptor negative stages i–iii breast cancer	Columbia University Medical Center New York, NY, United States
15.	Polyphenon E	Nonmelanomatous skin cancer	2014	Green tea extract in treating patients with actinic keratosis	Chao Family Comprehensive Cancer Center Orange, CA, United States
16.	Green tea catechins extract	Barrett esophagus	2014	Defined green tea catechin extract in preventing esophageal cancer in patients with Barrett's esophagus	Columbia University Medical Center, New York, NY, United States
17.	Green tea extract	Colorectal	2015	Green tea extracts for the prevention of colorectal adenomas and colorectal cancer	Seoul National University Hospital, South Korea
18.	Polyphenon E and tarceva	Lung (NSCLC)	2015	Study of polyphenon e in addition to erlotinib in advanced nonsmall-cell lung cancer	LSUHSC-Shreveport, Feist-Weiller Cancer Center Shreveport, LA, United States

19.	Polyphenon E and erlotinib	Bladder	2015	Erlotinib and green tea extract (polyphenon E) in preventing cancer recurrence in former smokers who have undergone surgery for bladder cancer	Bladder Cancer Genitourinary Oncology, PC Phoenix, AZ, United States
20.	Green tea catechin extract	Cervical	2015	Green tea extract in preventing cervical cancer in patients with human papilloma virus and low-grade cervical intraepithelial neoplasia	Arizona Cancer Center-Tucson Tucson, AZ, United States
21.	Green tea	Cancer	2015	Green tea anticancer mechanisms in smokers	The Ohio State University Columbus, OH, United States
22.	Green tea extract	Prostate	2016	Green tea extract in treating patients with metastatic prostate cancer that has not responded to hormone therapy	CCOP-Scottsdale Oncology Program Scottsdale, AZ, United States
23.	Green tea extract supplement	Breast	2016	Green tea and reduction of breast cancer risk	Fairview Southdale Breast Center Edina, MN, United States
24.	Green tea	Lung cancer prevention	2016	High tea consumption on smoking-related oxidative stress	University of Arizona, United States
25.	The effect of green tea and vitamin C on skin health	Skin cancer	2016	The effect of green tea and vitamin c on skin health	Salford Royal NHS Foundation Trust Manchester, United Kingdom

(*Continued*)

Table 10.2 (Continued)

S. no.	Drug	Cancer	Year	Title	Study center
26.	Sinecatechins 10%	Carcinoma, basal cell	2016	Topical green tea ointment in treatment of superficial skin cancer	Maastricht University Medical Centre Maastricht, Limburg, The Netherlands
27.	Green tea extract, vitamin E, lycopene, vitamin D3, selenium	Prostate	2017	Study of antioxidants on prostate tumors in men undergoing radical prostatectomy for prostate cancer	University Health Network, Princess Margaret Hospital Toronto, Ontario, Canada
28.	Polyphenon E	Breast	2017	A study of the effect of polyphenon E (green tea extract) on breast cancer progression	LSU Health Sciences Center Shreveport, Louisiana, United States
29.	Epigallocatechin gallate (EGCG)	Colon	2017	Chemopreventive effects of epigallocatechin gallate (EGCG) in colorectal cancer patients	UT Health San Antonio San Antonio, TX, United States
30.	Green tea catechin extract	Bladder	2017	Green tea extract in treating patients with nonmetastatic bladder cancer	University of Alabama at Birmingham Cancer Center Birmingham, AL, United States

Source: http://www.clinicaltrial.gov.

potential in various in vitro and in vivo preclinical and clinical studies. We have summarized a few well-known and some new and emerging agents as follows.

Ellagic acid (Fig. 10.5) has been considered a gold standard in chemoprevention and is a natural antioxidant polyphenol found in numerous fruits and vegetables, in particular pomegranate, persimmon, raspberry, black raspberry (BRB), strawberry, peach, plumes, and nuts (walnuts, almonds). Ellagic acid has been shown to have anticancer activity against bladder (Ceci et al., 2016), colon (Lee et al., 2016), breast (Aqil et al., 2016b), and various other cancers as reviewed in Derosa et al. (2016). We have shown the effect of ellagic acid against breast cancer. Since this compound is notoriously insoluble in aqueous solution, we used punicalagins which are highly water-soluble and bioavailable as a source of ellagic acid. Each molecule of punicalagins contains one molecule of ellagic acid. When punicalagins were delivered in rats by implant route, we could detect ellagic acid and elicited the biological response based on reduction of benzo[*a*]pyrene-induced DNA adducts. This study provides the proof-of-principle for the successful delivery of this potent chemopreventive agent. Ellagic acid has been shown to target various molecular pathways including inducing IGFBP7 in cervical cancer (Guo et al., 2016) and enhance sensitivity of breast cancer cells to γ-radiation (Ahire et al., 2017).

Resveratrol (Fig. 10.5) is a pleiotropic phytochemical belonging to the stilbene family. Resveratrol is present in red grapes. This compound has been tested in various clinical trials specially related to colon cancer as reviewed in Varoni et al. (2016). It has been shown to inhibit multiple molecular pathways that lead to activities such as antioxidant, antiinflammatory, and immunomodulatory, reducing damage induced by oxidative stress (DNA damage, protein oxidation, and lipid peroxidation) and increasing immune on cosurveillance. Resveratrol triggers apoptosis of

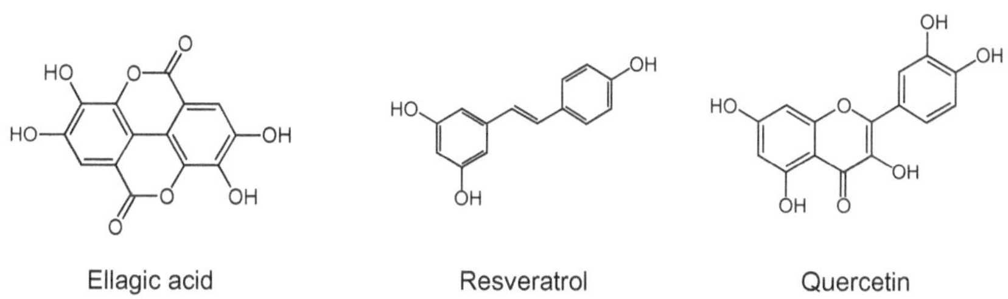

Figure 10.5 Structures of ellagic acid, resveratrol, and quercetin.

activated T cells and suppresses tumor necrosis factor-α, interluekin-17 (IL-17), and other pro-inflammatory molecules, and thus is of benefit in autoimmune diseases (Diaz-Gerevini et al., 2016). Resveratrol alters gut microbiota and influences stem cell proliferation and differentiation. These actions of resveratrol may explain the multitude of its actions and benefits. A major limitation of resveratrol is its poor oral bioavailability.

Quercetin (Fig. 10.5) is a natural flavonoid compound found abundant in the variety of foods including apple skin, berries, brassica vegetables, and grapes, etc. (Kelly, 2011). Epidemiologic studies have demonstrated a negative association between quercetin intake and prostate cancer incidence. The suggested chemopreventive effect of quercetin on prostate cancer has been exhibited in animal experiments (Yang et al., 2015). The major mechanisms of quercetin involve inhibition of the androgen receptor (Xing et al., 2001), phosphatidylinositol 3-kinase (PI3K)/AT signaling pathway (Morgan et al., 2009), inhibition of angiogenesis and insulin-like growth factor (IGF) signaling pathway (Senthilkumar et al., 2010). Quercetin has been shown to inhibit oxidative species-generating enzymes such as xanthine oxidase, LOX, and nicotinamide adenine dinucleotide phosphate oxidase (Day et al., 2000). The clinical and mechanistic details have been previously reviewed (Yang et al., 2015).

Besides the agents summarized in this chapter, there are several other potent plant therapeutics. Describing other well-known chemopreventive and therapeutic agents is beyond the scope of this article. However, we have attempted to review some emerging agents later and their use in clinical studies are shown in Table 10.3.

10.6 EMERGING PHYTOCOMPOUNDS

10.6.1 Withaferin A

Withaferin A (WFA), a triterpenoid (Fig. 10.6), obtained from the herb *Withania somnifera*, popularly known as "Ashwagandha," is a medicinal plant, which has been used for over 3000 years in traditional Ayurvedic and Unani Indian medical systems and has a vast range of therapeutic applications (White et al., 2016). WFA is emerging as a potent therapeutic agent against various cancers, including breast (Stan et al., 2008a,b), uterine cervix (Munagala et al., 2011), prostate (Srinivasan et al., 2007), lung (Aqil et al., 2012), and pancreatic (Yu et al., 2010). Root extract of *W. somnifera* was shown to be effective for prevention of chemically induced cancer in experimental animals (Khazal et al., 2013; Padmavathi

Table 10.3 Clinical trials of other plant compounds and new emerging chemo-therapeutics

S. no.	Drug	Cancer	Year	Title	Study center
1.	Ashwagandha	Autoimmune diseases inflammation Cancer	2010	Ashwagandha: effects on stress, inflammation and immune cell activation	Helfgott Research Institute Portland, OR, United States
2.	Ashwagandha extract and curcumin powder	Osteosarcoma	2011	Pilot study of curcumin formulation and ashwagandha extract in advanced osteosarcoma	Tata Memorial Hospital Mumbai, Maharashtra, India
3.	Blackberries	Healthy volunteers	2011	Blackberry intake and biomarkers of cancer risk	Beltsville Human Nutrition Research Center, Beltsville, MD, United States
4.	Quercetin capsules	Chemotherapy-induced oral mucositis	2012	Effect of quercetin in prevention and treatment of oral mucositis	Oral Medicine Department of Mashhad dental School MAshhad, Khorasan Razavi, Iran
5.	Withania somnifera root extract	Breast cancer	2013	Effect of *Withania somnifera* (ashwagandha) on the development of chemotherapy-induced fatigue and quality of life in breast cancer patients	Department of Nuclear Medicine, Radiotherapy & Oncology, Health Campus, Universiti Sains Malaysia, Malaysia
6.	Berry powder	Lung (NSCLC)	2014	The effect of berries on lung cancer tumors	James Graham Brown Cancer Center, Louisville, KY, United States
7.	Black raspberry confection	Prostate	2016	Absorption and metabolism of lyophilized black raspberry food products in men with prostate cancer undergoing surgery	Ohio State University Medical Center, Columbus, OH, United States

(*Continued*)

Table 10.3 (Continued)

S. no.	Drug	Cancer	Year	Title	Study center
8.	Anthocyanin-rich corn extract	Toxicity due to radiotherapy	2016	Supplementation with dietary anthocyanins and side effects of radiotherapy for breast cancer	Department of Radiotherapy Campobasso, CB, Italy
9.	Quercetin and green tea extract	Prostate	2017	Effect of quercetin on green tea polyphenol uptake in prostate tissue from patients with prostate cancer undergoing surgery	Veterans Administration Los Angeles Healthcare System, Los Angeles, CA, United States
10.	Isoquercetin	Colorectal, pancreatic and nonsmall-cell lung cancer	2017	Cancer associated thrombosis and isoquercetin (CAT IQ)	USC/Norris Comprehensive Cancer Center Los Angeles, CA, United States
11.	Isoquercetin and sunitinib	Renal cell carcinoma Kidney cancer	2017	Isoquercetin as an adjunct therapy in patients with kidney cancer receiving first-line sunitinib: a phase I/II trial	Ospedale Oncologico Regionale-Centro di Riferimento Oncologico di Basilicata U.O. di Oncologia Medica Rionero in vulture, Potenza, Italy
12.	Mirtoselect + meriva	Colorectal adenoma Risk reduction	2017	Anthocyanin extract and phospholipid curcumin in colorectal adenoma	ASL 3, Ospedale Villa Scassi, S. C. Gastroenterologia Genova, Italy
13.	Fast, intermediate and prolong release BRB confection	Healthy volunteers	2017	Black raspberry confection in preventing oral cancer in healthy volunteers	Ohio State University Medical Center Columbus, OH, United States
14.	VitaBlue	Breast	2018	COGNUTRIN in breast cancer survivors	H. Lee Moffitt Cancer Center and Research Institute, Tampa, FL, United States

Source: http://www.clinicaltrial.gov.

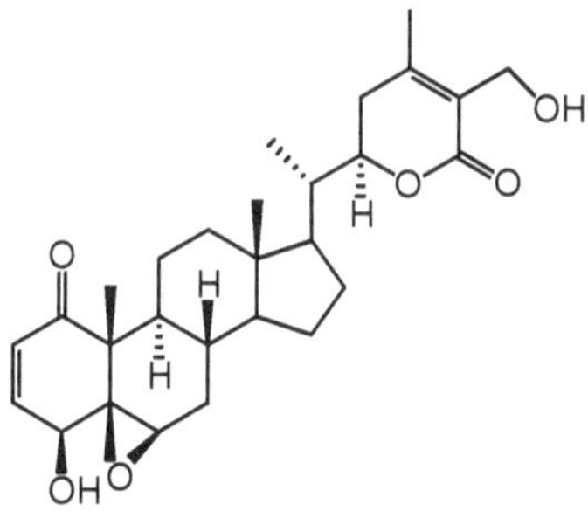

Figure 10.6 Structure of withaferin A (WFA).

et al., 2005). Alleviation of cancer chemotherapy-induced toxicity and fatigue and improvement in quality of life in patients with cancer by administration of Ashwagandha extract (containing 500 mg active ingredients) were also shown (Biswal et al., 2013; Hamza et al., 2008).

The potential mechanisms of action of WFA include: (1) inhibition of NF-κB activation (Maitra et al., 2009); (2) induction of Par-4 (Srinivasan et al., 2007), and inhibition of HSP90 in prostate cancer cells (Yu et al., 2010); and (3) induction of G2/M arrest (Stan et al., 2008b) and FoxO3a and Bim regulation in BC (Stan et al., 2008a). WFA has been reported to induce apoptosis via intrinsic and extrinsic pathways in human prostate, breast (Stan et al., 2008a), leukemia (Mandal et al., 2008), head and neck, and melanoma (Mayola et al., 2011) cancer cells via the reduction of the mitochondrial membrane potential and activation of various caspases and proteases, which trigger degradation of various substrates such as cytoskeletal proteins and poly(ADP-ribose) polymerase cleavage. The first report on the anticancer activity of root extract of *W. somnifera* in animal model was published back in 1967 (Shohat et al., 1967). Since then *Withania* extracts and WFA have been extensively investigated against various cancers, some of the recent studies are described here.

Thaiparambil et al. (2011) demonstrated the antiinvasive and metastasis potential of WFA at subcytotoxic doses in breast cancer. WFA was shown to have weak cytotoxic and apoptotic activity at low doses, but retained potent antiinvasive activity. In a breast cancer metastasis mouse model, WFA showed dose-dependent inhibition of metastatic lung nodules. Limited pharmacokinetics data suggested that WFA reaches peak concentration up to 2 μM in plasma with a half-life of 1.36 h following a single 4 mg/kg dose given intraperitonially.

In another study the antitumor and radio-sensitizing effects of WFA were studied on Ehrlich ascites carcinoma in vivo (Sharada et al., 1996). Swiss Albino mice were inoculated with tumor cells, and WFA was given intraperitonially at different dose fractions (5 or 7.5 mg/kg × 8, 10 mg/kg × 5, 20 or 30 mg/kg × 2) with or without abdominal gamma irradiation (RT, 7.5 Gy) after the first drug dose. WFA inhibited tumor growth and increased survival, which was dependent on the WFA dose per fraction rather than the total dose (Sharada et al., 1996).

Since *W. somnifera* has been shown to possess antibacterial and antiviral properties, we tested WFA against cervical cancer which is caused by human papilloma virus (HPV) expressing E6 and E7 oncoproteins (Munagala et al., 2011). WFA down regulated HPV E6 and E7 oncoproteins and induced tumor suppressor protein p53. WFA intervention resulted in nearly 70% reduction of growth of cervical tumor xenograft. These findings suggest use of WFA as a potent therapeutic agent for the treatment and prevention of cervical cancer without deleterious effects.

In a recent study, WFA shown to provide 100% protection from tumor formation in skin carcinogenesis (Li et al., 2016). WFA suppressed cell proliferation rather than inducing apoptosis during skin carcinogenesis as evident from morphological observations of the skin tissues. In addition to the studies describing modulation of different markers, direct effect on tumor xenografts has been studied. Some of these studies are reviewed in Palliyaguru et al. (2016).

In order to reduce its effective dose, we delivered WFA via subcutaneous polymeric implants. WFA showed inhibition of lung cancer (A549) tumor xenograft when given via polymeric implants while it was ineffective when the same total dose was administered intraperitonially. This was, however, effective when the dose was increased by twofold. WFA implants were also shown to inhibit lung DNA adducts in mice treated with dibenzo[*a*,*l*]pyrene, a highly potent carcinogen, compared with sham treatment; however, the effect was statistically insignificant (Aqil et al., 2012). Recently, we showed that mushroom-shaped polymeric cervical insert could deliver WFA locally in large animals. Cervical inserts of WFA (2% and 4% WFA, by weight) grafted in the cervix of goats, provided sustained dose-dependent release, and WFA was still releasing after 3 months. The drug release was essentially restricted to the target site and adjacent tissues as analyzed by LCMS analysis of tissues collected following euthanasia, providing proof-of-principle for local continuous delivery of therapeutics (Sherwood et al., 2017).

10.6.2 Celastrol

Celastrol is a bioactive compound derived from traditional Chinese medicinal herbs of the Celastraceae family. Celastrol (CEL), a pentacyclic triterpenoid (Fig. 10.7), is isolated from *Tripterygium wilfordii*, and is known to possess antiinflammatory and antioxidant activities and has been shown to have various pharmacological properties. The therapeutic usefulness of CEL has been extensively explored in various diseases such as Alzheimer's disease (Allison et al., 2001; Paris et al., 2010), arthritis rheumatoid (Venkatesha et al., 2011), asthma (Liu et al., 2004), hypertension (Morita, 2010), systemic lupus erythematosus (Li et al., 2005), and various cancer (He et al., 2009; Raja et al., 2011; Abbas et al., 2007; Chang et al., 2003). Besides being a potent inhibitor of NF-κB and proteosomal activation (Yang et al., 2006), CEL is a known inducer of TGFß/Smad signaling, and responsible for its antiinflammatory action (Jin et al., 2002). CEL also induces the activation of heat-shock response via the inhibition of Hsp90 activity and induction of other heat-shock proteins, both of which are known to be essential for attenuation of multiple signaling pathways important for survival and proliferation of cancer cells (Zhang et al., 2009).

In spite of its remarkable pharmacological effects, CEL is restricted from any clinical application due to low water solubility, reduced oral bioavailability, narrow window of dosage, and adverse effects reported in animal models (Song et al., 2011). Oral administration of CEL in vivo (rats) resulted in ineffective absorption into the systemic circulation and showed a bioavailability of only 17% (Zhang et al., 2012). In another report, it was suggested that besides low aqueous solubility in vivo metabolism and/or tissue distribution might also cause this poor bioavailability (Li et al., 2012).

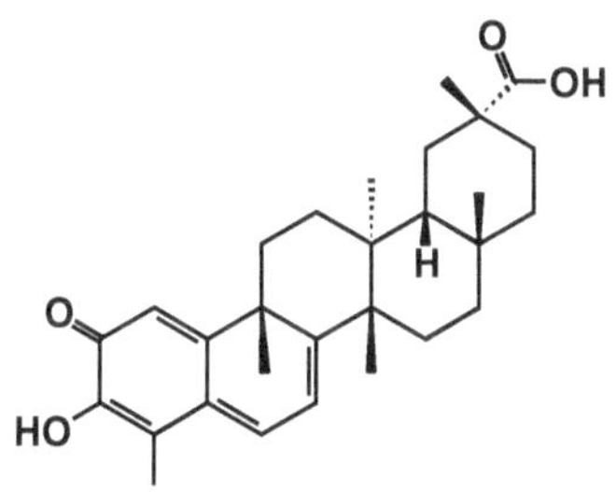

Figure 10.7 Structure of celastrol.

Researchers have used different nanoformulations to overcome these limitations, such as exosomes (Aqil et al., 2016c), lipid nanospheres, nanoencapsulation (Sanna et al., 2015), liposomes (Wolfram et al., 2014), polymeric micelles (Peng et al., 2012), cell-penetrating peptides-coated nanostructured lipid carriers (Yuan et al., 2013), and self-microemulsifying drug-delivery system (Qi et al., 2014). Many of these studies have been recently reviewed by Cascao and colleagues (Cascao et al., 2017) in a recent article, thus not discussed in detail.

We showed that CEL inhibited the proliferation of A549 and H1299 lung cancer cells in a time- and concentration-dependent manner as indexed by MTT assay. Mechanistically, CEL pretreatment of H1299 cells completely abrogated TNFα-induced NF-κB activation and upregulated the expression of ER-stress chaperones Grp 94, Grp78, and pPERK. CEL demonstrated very strong antitumor activity in the nude mice bearing subcutaneous lung cancer xenograft at relatively low doses of CEL in exosomal formulation. ExoCEL achieved nearly 80% tumor inhibition while about 60% inhibition with CEL alone at the same dose suggesting higher bioavailability of CEL in Exo formulation and reduce dose-related toxicity (Aqil et al., 2016c).

10.6.3 Berry Bioactives

Berries are gaining increasing importance lately for their chemopreventive and therapeutic potential against several cancers. Our earlier studies have shown both chemopreventive and therapeutic activities (Aiyer and Gupta, 2010; Jeyabalan et al., 2014; Ravoori et al., 2012; Aiyer et al., 2008) diet supplemented with freeze-dried blueberry (BB) powder against 17β-estradiol (E2)-mediated mammary cancer. There is also large body of data about chemopreventive and therapeutic activity of dietary blackberry and BRB against carcinogen-induced tumors in rat esophagus, colon, and mammary gland and the hamster cheek pouch in preclinical studies (Aiyer et al., 2008; Harris et al., 2001; Casto et al., 2002; Kresty et al., 2001), as well as its significant protective effects against colon and esophageal cancers in pilot clinical studies (reviewed in Weh et al., 2016).

Berries are rich in anthocyanins, which give the characteristic blue and purple colors to the fruits. Work from Dr. Stoner and colleagues has shown anticarcinogenic potential of BRB, blackberry, and strawberry against gastro-intestinal tumors (Stoner et al., 2007; Kresty et al., 2006). In their studies, these berries showed 30%–60% reductions in the tumor indices. Ellagic acid, the key polyphenol present in BRB, is shown to modulate both phase I and phase II enzymes (Aiyer and Gupta, 2010). The anthocyanins present in berries are known to impart anticancer effects by induction of metabolizing enzymes, modulation of gene expression, modulation of cell proliferation, apoptosis, and their downstream signaling pathways and reviewed in several manuscripts (Bishayee et al., 2016; Stoner and Wang, 2013).

Studies from our laboratory with 2.5% BB and BRB diets have shown protection against mammary tumors in rats induced by subcutaneous implantation of E2 (Aiyer et al., 2008). These studies provided insights into the modulation of some of the molecular events involved in E2 metabolism by the berry diets. In another study, we showed dose-dependent effect of BB and BRB against the E2-induced breast cancer. BB and BRB diets delayed the tumor appearance by 24 and 39 days, and tumor burden by 51% and 42%, respectively. BB is effective in downregulating CYP1A1 expression, while BRB down-regulate ERα expression effectively. The distinct anticarcinogenic effects of the two berries correspond to their distinct phytochemical signatures. The BB contains five distinct anthocyanidins while BRB contains almost exclusively cyanidin along with abundance of ellagic acid (Ravoori et al., 2012; Kausar et al., 2012). In another study the Indian blackberry "Jamun" (*Syzygium jambolana*) which contains five distinct anthocyanidins, as in BB and abundance of ellagic acid showed strong inhibition of E2-induced tumourogenesis in August Copenhagen x Irish (ACI) rats. This study also showed the role of berry bioactives in modulation of E2-associated miRNAs (Aqil et al., 2016b).

Since berry bioactives are potent chemopreventive, the therapeutic activity of BB has also been reported (Jeyabalan et al., 2014). For therapeutic activity, Rubel blueberry (5%) which contains almost 10-fold higher anthocyanins was used. Both chemopreventive and therapeutic interventions showed delay of tumor latency by 28 and 37 days, respectively. Tumor volume and multiplicity were also reduced significantly suggesting that berry bioactives are not only chemopreventive but also therapeutic activity (Jeyabalan et al., 2014). BB also showed lung tumor xenograft inhibition and reduced the tumor growth by 45% and 64% at 2.5% and 5% dietary BB, respectively (Aqil et al., 2016a). In another study, we observed that a mixture of BB and BRB provided enhanced growth inhibition compared with individual berries, suggesting potential synergistic activity of distinct phytochemicals in these two berries. The plasma levels of ellagic acid in the rats fed with BRB, and lung tissue levels of mice treated with BB diet have also been reported (Ravoori et al., 2012; Aqil et al., 2016a), suggesting that the berry bioactives can reach outside of the GI tract.

Translatability of dietary berries for therapeutic effects to large population, however, is not practical due to rather large daily doses of berries, besides the costs. Data demonstrate that anthocyanidins—the aglycone forms—are more potent than their glycone counterparts, i.e., anthocyanins (Aqil et al., 2016a) (Fig. 10.8). Berry anthocyanidins (cyanidin, malvidin, peonidin, petunidin, and delphinidin) have increasingly been explored for their anticancer effects; however, their combinatorial effects as a mixture, as present in BB, bilberry and Indian blackberry ("Jamun") remain untested. In vivo studies, both the native mixture of anthocyanidins from bilberry (20 mg/kg) and the most potent anthocyanidins, delphinidin (60 mg/kg) significantly inhibited the growth of H1299 xenografts in nude mice—both by nearly 60%. Since the effective dose of delphinidin, the anthocyanidins mixture was eightfold lower than delphinin alone, further emphasizing synergism (Kausar et al., 2012).

Anthocyanins are highly water-soluble molecules. Glycosylation confers increased stability and water solubility, and acylation of the sugar residues further improves anthocyanin stability. However, the physicochemical and pharmacokinetic limitations of anthocyanidins such as low

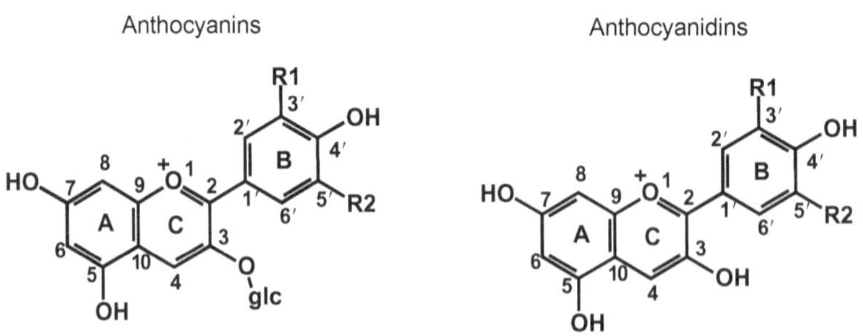

Figure 10.8 Structure of anthocyanins and anthocyanidins. Chemical structure of anthocyanins and anthocyanidins. Depending on the number and position of hydroxyl and methoxyl groups, various anthocyanidins have been described, and of these, six are commonly found in fruits and vegetables. Delphinidin (R1, $-OH$; R2, $-OH$), Cyanidin (R1, $-OH$; R2, $-H$), Peonidin (R1, $-OCH_3$; R2, $-H$), Petunidin (R1, $-OH$; R2, $-OCH_3$), Malvidin (R1, $-OCH_3$; R2, $-OCH_3$), and Pelargonidin (R1, $-H$; R2, $-H$).

permeability and poor oral bioavailability are considered as unfavorable properties for development as drugs. Since we have shown exosomes to enhance the stability and bioavailability, nanoformulation of anthocyanidins was tested against several cancer types. Anthocyanidins showed enhanced antiproliferative and antiinflammatory effects compared with the free anthocyanidins against various cancer cells in vitro and showed enhanced therapeutic response against lung cancer tumor xenograft in nude mice (Munagala et al., 2017). Furthermore, the anthocyanidins showed no signs of gross or systemic toxicity in wild-type mice. Thus exosomal nanoformulation of anthocyanidins can provide an effective alternative that is efficacious, and safe with widely applicable therapeutic agent against several cancers.

10.7 CONCLUSION

The choices of drugs for the cancer treatments are limited and many of them accompany dose-related toxicities. Moreover, in many cases, the etiology of cancer is simply unknown. In such cases the availability of an efficacious chemopreventive agent is imperative and requires continuous efforts from the scientific community. It is clear that many opportunities exist for the development of novel chemopreventive agents and novel strategies which can lead to the identification of plant therapeutics with superior clinical efficacy and decreased toxicity. Chemoprevention studies are generally well received by the nutraceuticals research community, but the long-term safety and tolerability of any chemopreventive compound (natural or synthetic) for human consumption must be considered, since it is likely that the agent will have to be consumed/administered for a long period of time. Historically, only a handful of agents had satisfactory instances of clinical evidence for supporting their anticancer effects. It should be understood that discovery and development of effective plant therapeutics will require close multidisciplinary collaboration, with optimization through the application of combinatorial and medicinal chemistry, their pharmacokinetics aspects and most importantly their delivery to the target site. Therefore multitude of approaches should be used not only for the new compounds but also for the potent known chemopreventive to further develop them as therapeutic agents. These agents can be used as a complementary therapeutics along with current chemotherapy drugs against various types of cancer.

ACKNOWLEDGMENTS

This work was supported from the USPHS grant CA-118114, Kentucky Lung Cancer Research Program, and Agnes Brown Duggan Endowment and Helmsley Funds. Ramesh Gupta holds the Agnes Brown Duggan Chair in Oncological Research.

REFERENCES

Anon., 1996. Clinical development plan: curcumin. J. Cell Biochem. Suppl. 26, 72–85.

Abbas, S., Bhoumik, A., Dahl, R., Vasile, S., Krajewski, S., Cosford, N.D., et al., 2007. Preclinical studies of celastrol and acetyl isogambogic acid in melanoma. Clin. Cancer Res. 13, 6769–6778.

ACS, 2016. Cancer Treatment & Survivorship Facts & Figures 2016–2017. American Cancer Society, Atlanta.

Aggarwal, B.B., Yuan, W., Li, S., Gupta, S.C., 2013. Curcumin-free turmeric exhibits anti-inflammatory and anticancer activities: identification of novel components of turmeric. Mol. Nutr. Food Res. 57, 1529–1542.

Ahire, V., Kumar, A., Mishra, K.P., Kulkarni, G., 2017. Ellagic acid enhances apoptotic sensitivity of breast cancer cells to gamma-radiation. Nutr. Cancer 69, 904–910.

Aiyer, H.S., Gupta, R.C., 2010. Berries and ellagic acid prevent estrogen-induced mammary tumorigenesis by modulating enzymes of estrogen metabolism. Cancer Prev. Res. (Phila.) 3, 727–737.

Aiyer, H.S., Srinivasan, C., Gupta, R.C., 2008. Dietary berries and ellagic acid diminish estrogen-mediated mammary tumorigenesis in ACI rats. Nutr. Cancer 60, 227–234.

Albanes, D., et al., 1996. Alpha-tocopherol and beta-carotene supplements and lung cancer incidence in the alpha-tocopherol, beta-carotene cancer prevention study: effects of base-line characteristics and study compliance. J. Natl. Cancer Inst. 88, 1560–1570.

Allison, A.C., Cacabelos, R., Lombardi, V.R.M., Alvarez, X.A., Vigo, C., 2001. Celastrol, a potent antioxidant and anti-inflammatory drug, as a possible treatment for Alzheimer's disease. Prog. Neuropsychopharmacol. Biol. Psychiatry 25, 1341–1357.

Anand, P., Kunnumakkara, A.B., Newman, R.A., Aggarwal, B.B., 2007. Bioavailability of curcumin: problems and promises. Mol. Pharm. 4, 807–818.

Aqil, F., Jeyabalan, J., Kausar, H., Bansal, S.S., Sharma, R.J., Singh, I.P., et al., 2012. Multi-layer polymeric implants for sustained release of chemopreventives. Cancer Lett. 326, 33–40.

Aqil, F., Jeyabalan, J., Kausar, H., Munagala, R., Singh, I.P., Gupta, R., 2016a. Lung cancer inhibitory activity of dietary berries and berry polyphenolics. J. Berry Res. 6, 105–114.

Aqil, F., Jeyabalan, J., Munagala, R., Singh, I.P., Gupta, R.C., 2016b. Prevention of hormonal breast cancer by dietary jamun. Mol. Nutr. Food Res. 60, 1470–1481.

Aqil, F., Kausar, H., Agrawal, A.K., Jeyabalan, J., Kyakulaga, A.H., Munagala, R., et al., 2016c. Exosomal formulation enhances therapeutic response of celastrol against lung cancer. Exp. Mol. Pathol. 101, 12–21.

Aqil, F., Munagala, R., Jeyabalan, J., Agrawal, A.K., Gupta, R., 2017. Exosomes for the enhanced tissue bioavailability and efficacy of curcumin. AAPS J. 19, 1691–1702.

Bansal, S.S., Goel, M., Aqil, F., Vadhanam, M.V., Gupta, R.C., 2011a. Advanced drug delivery systems of curcumin for cancer chemoprevention. Cancer Prev. Res. (Phila.) 4, 1158–1171.

Bansal, S.S., Kausar, H., Aqil, F., Jeyabalan, J., Vadhanam, M.V., Gupta, R.C., et al., 2011b. Curcumin implants for continuous systemic delivery: safety and biocompatibility. Drug Deliv. Transl. Res. 1, 332–341.

Bansal, S.S., Kausar, H., Vadhanam, M.V., Ravoori, S., Gupta, R.C., 2012. Controlled systemic delivery by polymeric implants enhances tissue and plasma curcumin levels compared with oral administration. Eur. J. Pharm. Biopharm. 80, 571–577.

Bansal, S.S., Kausar, H., Vadhanam, M.V., Ravoori, S., Pan, J., Rai, S.N., et al., 2014. Curcumin implants, not curcumin diet, inhibit estrogen-induced mammary carcinogenesis in ACI rats. Cancer Prev. Res. (Phila.) 7, 456–465.

Batrakova, E.V., Kim, M.S., 2015. Using exosomes, naturally-equipped nanocarriers, for drug delivery. J. Control. Release 219, 396–405.

Bishayee, A., Haskell, Y., Do, C., Siveen, K.S., Mohandas, N., Sethi, G., et al., 2016. Potential benefits of edible berries in the management of aerodigestive and gastrointestinal tract cancers: preclinical and clinical evidence. Crit. Rev. Food Sci. Nutr. 56, 1753–1775.

Biswal, B.M., Sulaiman, S.A., Ismail, H.C., Zakaria, H., Musa, K.I., 2013. Effect of *Withania somnifera* (Ashwagandha) on the development of chemotherapy-induced fatigue and quality of life in breast cancer patients. Integr. Cancer Ther. 12, 312–322.

Block, G., Patterson, B., Subar, A., 1992. Fruit, vegetables, and cancer prevention: a review of the epidemiological evidence. Nutr. Cancer 18, 1–29.

Cao, P., Vadhanam, M.V., Spencer, W.A., Cai, J., Gupta, R.C., 2011. Sustained systemic delivery of green tea polyphenols by polymeric implants significantly diminishes benzo[a]pyrene-induced DNA adducts. Chem. Res. Toxicol. 24, 877–886.

Cao, P., Jeyabalan, J., Aqil, F., Ravoori, S., Gupta, R.C., Vadhanam, M.V., 2014. Polymeric implants for the delivery of green tea polyphenols. J. Pharm. Sci. 103, 945–951.

Cascao, R., Fonseca, J.E., Moita, L.F., 2017. Celastrol: a spectrum of treatment opportunities in chronic diseases. Front. Med. (Lausanne) 4, 69.

Casto, B.C., et al., 2002. Chemoprevention of oral cancer by black raspberries. Anticancer Res. 22, 4005–4015.

Ceci, C., et al., 2016. Ellagic acid inhibits bladder cancer invasiveness and in vivo tumor growth. Nutrients 8.

Chang, F.R., et al., 2003. Antitumor agents. 228. Five new agarofurans, reissantins A–E, and cytotoxic principles from Reissantia buchananii. J. Nat. Prod. 66, 1416–1420.

Chow, H.H., et al., 2003. Pharmacokinetics and safety of green tea polyphenols after multiple-dose administration of epigallocatechin gallate and polyphenon E in healthy individuals. Clin. Cancer Res. 9, 3312–3319.

Cragg, G.M., Pezzuto, J.M., 2016. Natural products as a vital source for the discovery of cancer chemotherapeutic and chemopreventive agents. Med. Princ. Pract. 25 (Suppl. 2), 41–59.

Day, A.J., Bao, Y., Morgan, M.R., Williamson, G., 2000. Conjugation position of quercetin glucuronides and effect on biological activity. Free Radic. Biol. Med. 29, 1234–1243.

Derosa, G., Maffioli, P., Sahebkar, A., 2016. Ellagic acid and its role in chronic diseases. Adv. Exp. Med. Biol. 928, 473–479.

DeSantis, C.E., et al., 2014. Cancer treatment and survivorship statistics, 2014. CA Cancer J. Clin. 64, 252–271.

Diaz-Gerevini, G.T., Repossi, G., Dain, A., Tarres, M.C., Das, U.N., Eynard, A.R., 2016. Beneficial action of resveratrol: how and why? Nutrition 32, 174–178.

Donaldson, M.S., 2004. Nutrition and cancer: a review of the evidence for an anti-cancer diet. Nutr. J. 3, 19.

Downs-Holmes, C., Silverman, P., 2011. Breast cancer: overview & updates. Nurse Pract. 36, 20–26. quiz 7.
Egermann, U., Jaeggi, K., Habicht, J.M., Perruchoud, A.P., Dalquen, P., Soler, M., 2002. Regular follow-up after curative resection of nonsmall cell lung cancer: a real benefit for patients? Eur. Respir. J. 19, 464–468.
Global Burden of Disease Cancer, C., et al., 2015. The global burden of cancer 2013. JAMA Oncol. 1, 505–527.
Granja, A., Pinheiro, M., Reis, S., 2016. Epigallocatechin gallate nanodelivery systems for cancer therapy. Nutrients 8.
Guo, H., Zhang, D., Fu, Q., 2016. Inhibition of cervical cancer by promoting IGFBP7 expression using ellagic acid from pomegranate peel. Med. Sci. Monit. 22, 4881–4886.
Gupta, S.C., Prasad, S., Kim, J.H., Patchva, S., Webb, L.J., Priyadarsini, I.K., et al., 2011. Multitargeting by curcumin as revealed by molecular interaction studies. Nat. Prod. Rep. 28, 1937–1955.
Gupta, S.C., Patchva, S., Koh, W., Aggarwal, B.B., 2012. Discovery of curcumin, a component of golden spice, and its miraculous biological activities. Clin. Exp. Pharmacol. Physiol. 39, 283–299.
Gupta, S.C., Kismali, G., Aggarwal, B.B., 2013a. Curcumin, a component of turmeric: from farm to pharmacy. Biofactors 39, 2–13.
Gupta, S.C., Patchva, S., Aggarwal, B.B., 2013b. Therapeutic roles of curcumin: lessons learned from clinical trials. AAPS J. 15, 195–218.
Hamza, A., Amin, A., Daoud, S., 2008. The protective effect of a purified extract of *Withania somnifera* against doxorubicin-induced cardiac toxicity in rats. Cell Biol. Toxicol. 24, 63–73.
Harris, G.K., et al., 2001. Effects of lyophilized black raspberries on azoxymethane-induced colon cancer and 8-hydroxy-2'-deoxyguanosine levels in the Fischer 344 rat. Nutr. Cancer 40, 125–133.
He, D., et al., 2009. The NF-kappa B inhibitor, celastrol, could enhance the anti-cancer effect of gambogic acid on oral squamous cell carcinoma. BMC Cancer 9.
Jemal, A., et al., 2017. Annual report to the nation on the status of cancer, 1975–2014, featuring survival. J. Natl. Cancer Inst. 109.
Jeyabalan, J., Aqil, F., Munagala, R., Annamalai, L., Vadhanam, M.V., Gupta, R.C., 2014. Chemopreventive and therapeutic activity of dietary blueberry against estrogen-mediated breast cancer. J. Agric. Food Chem. 62, 3963–3971.
Jin, H.Z., Hwang, B.Y., Kim, H.S., Lee, J.H., Kim, Y.H., Lee, J.J., 2002. Antiinflammatory constituents of Celastrus orbiculatus inhibit the NF-kappa B activation and NO production. J. Nat. Prod. 65, 89–91.
Johnsen, K.B., Gudbergsson, J.M., Skov, M.N., Pilgaard, L., Moos, T., Duroux, M., 2014. A comprehensive overview of exosomes as drug delivery vehicles—endogenous nanocarriers for targeted cancer therapy. Biochim. Biophys. Acta Rev. Cancer 1846, 75–87.
Kausar, H., Jeyabalan, J., Aqil, F., Chabba, D., Sidana, J., Singh, I.P., et al., 2012. Berry anthocyanidins synergistically suppress growth and invasive potential of human non-small-cell lung cancer cells. Cancer Lett. 325, 54–62.
Kelly, G.S., 2011. Quercetin. Monograph. Altern. Med. Rev. 16, 172–194.
Khazal, K.F., Samuel, T., Hill, D.L., Grubbs, C.J., 2013. Effect of an extract of *Withania somnifera* root on estrogen receptor-positive mammary carcinomas. Anticancer Res. 33, 1519–1523.
Kresty, L.A., Morse, M.A., Morgan, C., Carlton, P.S., Lu, J., Gupta, A., et al., 2001. Chemoprevention of esophageal tumorigenesis by dietary administration of lyophilized black raspberries. Cancer Res. 61, 6112–6119.

Kresty, L.A., Frankel, W.L., Hammond, C.D., Baird, M.E., Mele, J.M., Stoner, G.D., et al., 2006. Transitioning from preclinical to clinical chemopreventive assessments of lyophilized black raspberries: interim results show berries modulate markers of oxidative stress in Barrett's esophagus patients. Nutr. Cancer 54, 148–156.
Kuo, M.T., 2007. Roles of multidrug resistance genes in breast cancer chemoresistance. Adv. Exp. Med. Biol. 608, 23–30.
Lee, J., Kim, Y.S., Lee, J., Heo, S.C., Lee, K.L., Choi, S.W., et al., 2016. Walnut phenolic extract and its bioactive compounds suppress colon cancer cell growth by regulating colon cancer stemness. Nutrients 8.
Li, H., Zhang, Y.Y., Huang, X.Y., Sun, Y.N., Jia, Y.F., Li, D., 2005. Beneficial effect of tripterine on systemic lupus erythematosus induced by active chromatin in BALB/c mice. Eur. J. Pharmacol. 512, 231–237.
Li, Z., et al., 2012. Celastrol nanoparticles inhibit corneal neovascularization induced by suturing in rats. Int. J. Nanomedicine 7, 1163–1173.
Li, W., et al., 2016. Withaferin A suppresses the up-regulation of acetyl-coA carboxylase 1 and skin tumor formation in a skin carcinogenesis mouse model. Mol. Carcinog. 55, 1739–1746.
Lippman, S.M., Benner, S.E., Hong, W.K., 1994. Cancer chemoprevention. J. Clin. Oncol. 12, 851–873.
Liu, R.L., Liu, Z.L., Li, Q., Qiu, Z.M., Lu, H.J., Yang, Z.M., et al., 2004. The experimental study on the inhibitory effect of tripterine on airway inflammation in asthmatic mice. Zhonghua Jie He He Hu Xi Za Zhi 27, 165–168.
Maitra, R., Porter, M.A., Huang, S., Gilmour, B.P., 2009. Inhibition of NFkappaB by the natural product Withaferin A in cellular models of Cystic Fibrosis inflammation. J. Inflamm. (Lond). 6, 15.
Mandal, C., Dutta, A., Mallick, A., Chandra, S., Misra, L., Sangwan, R.S., et al., 2008. Withaferin A induces apoptosis by activating p38 mitogen-activated protein kinase signaling cascade in leukemic cells of lymphoid and myeloid origin through mitochondrial death cascade. Apoptosis 13, 1450–1464.
Mayola, E., et al., 2011. Withaferin A induces apoptosis in human melanoma cells through generation of reactive oxygen species and down-regulation of Bcl-2. Apoptosis 16, 1014–1027.
Morgan, T.M., Koreckij, T.D., Corey, E., 2009. Targeted therapy for advanced prostate cancer: inhibition of the PI3K/Akt/mTOR pathway. Curr. Cancer Drug Targets 9, 237–249.
Morita, T., 2010. Celastrol: a new therapeutic potential of traditional Chinese medicine. Am. J. Hypertens. 23, 821.
Munagala, R., Kausar, H., Munjal, C., Gupta, R.C., 2011. Withaferin A induces p53-dependent apoptosis by repression of HPV oncogenes and upregulation of tumor suppressor proteins in human cervical cancer cells. Carcinogenesis 32, 1697–1705.
Munagala, R., Aqil, F., Jeyabalan, J., Gupta, R.C., 2016. Bovine milk-derived exosomes for drug delivery. Cancer Lett. 371, 48–61.
Munagala, R., et al., 2017. Exosomal formulation of anthocyanidins against multiple cancer types. Cancer Lett. 393, 94–102.
Padmavathi, B., Rath, P.C., Rao, A.R., Singh, R.P., 2005. Roots of *Withania somnifera* inhibit forestomach and skin carcinogenesis in mice. Evid. Based Complement. Alternat. Med. 2, 99–105.
Palliyaguru, D.L., Singh, S.V., Kensler, T.W., 2016. *Withania somnifera*: from prevention to treatment of cancer. Mol. Nutr. Food Res. 60, 1342–1353.
Paris, D., et al., 2010. Reduction of beta-amyloid pathology by celastrol in a transgenic mouse model of Alzheimer's disease. J. Neuroinflammation. 7, 17.

Park, W.C., Jordan, V.C., 2002. Selective estrogen receptor modulators (SERMS) and their roles in breast cancer prevention. Trends Mol. Med. 8, 82–88.
Peng, X., et al., 2012. Optimized preparation of celastrol-loaded polymeric nanomicelles using rotatable central composite design and response surface methodology. J. Biomed. Nanotechnol. 8, 491–499.
Potter, J.D., 2014. The failure of cancer chemoprevention. Carcinogenesis 35, 974–982.
Pratheeshkumar, P., et al., 2012. Cancer prevention with promising natural products: mechanisms of action and molecular targets. Anticancer Agents Med. Chem. 12, 1159–1184.
Qi, X., Qin, J., Ma, N., Chou, X., Wu, Z., 2014. Solid self-microemulsifying dispersible tablets of celastrol: formulation development, charaterization and bioavailability evaluation. Int. J. Pharm. 472, 40–47.
Raja, S.M., et al., 2011. Anticancer activity of celastrol in combination with ErbB2-targeted therapeutics for treatment of ErbB2-overexpressing breast cancers. Cancer Biol. Ther. 11, 263–276.
Ravoori, S., Vadhanam, M.V., Aqil, F., Gupta, R.C., 2012. Inhibition of estrogen-mediated mammary tumorigenesis by blueberry and black raspberry. J. Agric. Food Chem. 60, 5547–5555.
Ren, J., He, W., Zheng, L., Duan, H., 2016. From structures to functions: insights into exosomes as promising drug delivery vehicles. Biomater. Sci. 4, 910–921.
Sanna, V., Chamcheu, J.C., Pala, N., Mukhtar, H., Sechi, M., Siddiqui, I.A., 2015. Nanoencapsulation of natural triterpenoid celastrol for prostate cancer treatment. Int. J. Nanomedicine 10, 6835–6846.
Senthilkumar, K., Elumalai, P., Arunkumar, R., Banudevi, S., Gunadharini, N.D., Sharmila, G., et al., 2010. Quercetin regulates insulin like growth factor signaling and induces intrinsic and extrinsic pathway mediated apoptosis in androgen independent prostate cancer cells (PC-3). Mol. Cell. Biochem. 344, 173–184.
Sharada, A.C., Solomon, F.E., Devi, P.U., Udupa, N., Srinivasan, K.K., 1996. Antitumor and radiosensitizing effects of withaferin A on mouse Ehrlich ascites carcinoma in vivo. Acta Oncol. 35, 95–100.
Sherwood, L.C., et al., 2017. Development of a goat model for evaluation of withaferin A: cervical implants for the treatment of cervical intraepithelial neoplasia. Exp. Mol. Pathol. 103, 320–329.
Shohat, B., Gitter, S., Abraham, A., Lavie, D., 1967. Antitumor activity of withaferin A (NSC-101088). Cancer Chemother. Rep. 51, 271–276.
Siddiqui, I.A., et al., 2009. Introducing nanochemoprevention as a novel approach for cancer control: proof of principle with green tea polyphenol epigallocatechin-3-gallate. Cancer Res. 69, 1712–1716.
Siegel, R., Ma, J., Zou, Z., Jemal, A., 2014. Cancer statistics, 2014. CA Cancer J. Clin. 64, 9–29.
Sonaje, K., Italia, J.L., Sharma, G., Bhardwaj, V., Tikoo, K., Kumar, M.N., 2007. Development of biodegradable nanoparticles for oral delivery of ellagic acid and evaluation of their antioxidant efficacy against cyclosporine A-induced nephrotoxicity in rats. Pharm. Res. 24, 899–908.
Song, J., Shi, F., Zhang, Z., Zhu, F., Xue, J., Tan, X., et al., 2011. Formulation and evaluation of celastrol-loaded liposomes. Molecules 16, 7880–7892.
Sporn, M.B., 1976. Approaches to prevention of epithelial cancer during the preneoplastic period. Cancer Res. 36, 2699–2702.
Srinivasan, S., Ranga, R.S., Burikhanov, R., Han, S.S., Chendil, D., 2007. Par-4-dependent apoptosis by the dietary compound withaferin A in prostate cancer cells. Cancer Res. 67, 246–253.

Stan, S.D., Hahm, E.R., Warin, R., Singh, S.V., 2008a. Withaferin A causes FOXO3a- and Bim-dependent apoptosis and inhibits growth of human breast cancer cells in vivo. Cancer Res. 68, 7661–7669.
Stan, S.D., Zeng, Y., Singh, S.V., 2008b. Ayurvedic medicine constituent withaferin a causes G2 and M phase cell cycle arrest in human breast cancer cells. Nutr. Cancer 60 (Suppl. 1), 51–60.
Steward, W.P., Brown, K., 2013. Cancer chemoprevention: a rapidly evolving field. Br. J. Cancer 109, 1–7.
Stoner, G.D., Wang, L.S., 2013. Chemoprevention of esophageal squamous cell carcinoma with berries. Top. Curr. Chem. 329, 1–20.
Stoner, G.D., Wang, L.S., Zikri, N., Chen, T., Hecht, S.S., Huang, C., et al., 2007. Cancer prevention with freeze-dried berries and berry components. Semin. Cancer Biol. 17, 403–410.
Sur, S., Panda, C.K., 2017. Molecular aspects of cancer chemopreventive and therapeutic efficacies of tea and tea polyphenols. Nutrition 43–44, 8–15.
Thaiparambil, J.T., et al., 2011. Withaferin A inhibits breast cancer invasion and metastasis at sub-cytotoxic doses by inducing vimentin disassembly and serine 56 phosphorylation. Int. J. Cancer 129, 2744–2755.
Tremont, A., Lu, J., Cole, J.T., 2017. Endocrine therapy for early breast cancer: updated review. Ochsner. J. 17, 405–411.
Varoni, E.M., Lo Faro, A.F., Sharifi-Rad, J., Iriti, M., 2016. Anticancer molecular mechanisms of resveratrol. Front. Nutr. 3, 8.
Venkatesha, S.H., Yu, H., Rajaiah, R., Tong, L., Moudgil, K.D., 2011. Celastrus-derived celastrol suppresses autoimmune arthritis by modulating antigen-induced cellular and humoral effector responses. J. Biol. Chem. 286, 15138–15146.
Vyas, A.R., Singh, S.V., 2014. Molecular targets and mechanisms of cancer prevention and treatment by withaferin a, a naturally occurring steroidal lactone. AAPS J. 16, 1–10.
Weh, K.M., Clarke, J., Kresty, L.A., 2016. Cranberries and cancer: an update of preclinical studies evaluating the cancer inhibitory potential of cranberry and cranberry derived constituents. Antioxidants (Basel) 5.
White, P.T., Subramanian, C., Motiwala, H.F., Cohen, M.S., 2016. Natural withanolides in the treatment of chronic diseases. Adv. Exp. Med. Biol. 928, 329–373.
Wolfram, J., et al., 2014. Evaluation of anticancer activity of celastrol liposomes in prostate cancer cells. J. Microencapsul. 31, 501–507.
Xing, N., Chen, Y., Mitchell, S.H., Young, C.Y., 2001. Quercetin inhibits the expression and function of the androgen receptor in LNCaP prostate cancer cells. Carcinogenesis 22, 409–414.
Yang, H.J., Chen, D., Cui, Q.Z.C., Yuan, X., Dou, Q.P., 2006. Celastrol, a triterpene extracted from the Chinese "Thunder of God Vine," is a potent proteasome inhibitor and suppresses human prostate cancer growth in nude mice. Cancer Res. 66, 4758–4765.
Yang, F., Song, L., Wang, H., Wang, J., Xu, Z., Xing, N., 2015. Quercetin in prostate cancer: chemotherapeutic and chemopreventive effects, mechanisms and clinical application potential (review). Oncol. Rep. 33, 2659–2668.
Yu, Y., et al., 2010. Withaferin A targets heat shock protein 90 in pancreatic cancer cells. Biochem. Pharmacol. 79, 542–551.
Yuan, L., Liu, C., Chen, Y., Zhang, Z., Zhou, L., Qu, D., 2013. Antitumor activity of tripterine via cell-penetrating peptide-coated nanostructured lipid carriers in a prostate cancer model. Int. J. Nanomedicine 8, 4339–4350.
Zhang, T., Li, Y., Yu, Y., Zou, P., Jiang, Y., Sun, D., 2009. Characterization of celastrol to inhibit hsp90 and cdc37 interaction. J. Biol. Chem. 284, 35381–35389.

Zhang, J., Li, C.Y., Xu, M.J., Wu, T., Chu, J.H., Liu, S.J., et al., 2012. Oral bioavailability and gender-related pharmacokinetics of celastrol following administration of pure celastrol and its related tablets in rats. J. Ethnopharmacol. 144, 195–200.
Zink, A., Traidl-Hoffmann, C., 2015. Green tea in dermatology—myths and facts. J. Dtsch. Dermatol. Ges. 13, 768–775.

CHAPTER 11

Plant-Derived Molecules in Managing HIV Infection

Jay Trivedi[1], Anjali Tripathi[1], Debprasad Chattopadhyay[2] and Debashis Mitra[1]
[1]National Centre for Cell Science, Pune University Campus, Pune, Maharashtra, India
[2]ICMR-Virus Unit Kolkata, ID & BG Hospital Campus, Kolkata, West Bengal, India

11.1 DISCOVERY OF HIV

In the late 1970s and the early 1980s in Los Angeles, the United States, several previously healthy young gay men were reported with unusual symptoms including *Pneumocystis carinni* pneumonia or PCP. PCP is a rare opportunistic infection known to be present in the individuals with severely compromised immunity. Soon after this, another group of individuals were reported to have developed similar symptoms in addition to a rare type of skin cancer, Kaposi's sarcoma (KS) (Gottlieb et al., 1981). Many such cases with PCP and KS were reported in subsequent years, and this outbreak was identified as a new disease condition with an unknown cause (Masur et al., 1981). Although the individuals showed variety of opportunistic infections, one striking similarity observed among all the individuals was severely compromised immunity. These reports alarmed the Centre for Disease Control and Prevention (CDC) to find out the cause of these disease conditions. In subsequent months, it was clear that the disease condition affected a set of populations which included homosexual men, intravenous drug users, hemophiliacs, and blood transfusion recipients. A CDC task force was formed in the same year and soon the hunt to find out the causative agent of these disease conditions began. After recognizing the anomalous pattern of symptoms in diseased individuals, the task force named it as acquired immunodeficiency syndrome or AIDS (CDC, 1982).

The pattern of transmission and the population at risk suggested that the disease was transmitted sexually by a novel unidentified pathogen.

New Look to Phytomedicine
DOI: https://doi.org/10.1016/B978-0-12-814619-4.00011-2

The next task for the research community was to identify and characterize the pathogen responsible for AIDS. Several microorganisms were proposed to be responsible for AIDS, but in 1983, three groups individually reported the discovery of the virus responsible for AIDS. In 1983, Montagnier et al. first isolated the causative agent of AIDS from swollen lymph nodes of infected individuals. This was named as lymphadenopathy-associated virus or LAV (Barre-Sinoussi et al., 1983). In the same year, Robert Gallo reported the isolation of the virus from AIDS patients, which was showing striking similarities with human T-cell lymphotropic virus (HTLV) group of viruses. Hence, Gallo et al. (1983) named it as HTLV-III. But the characteristic loss of $CD4^+$ T cells in infected individuals contradicted the hypothesis that the new virus is a member of the HTLV group. Levy in the same year reported the isolation of virus from the circulating blood of AIDS patients and named it as AIDS-related virus or ARV (Levy et al., 1984). Nucleotide sequences of all three reported virus were the same and the International Virus Taxonomy Consortium finally named this virus as human immunodeficiency virus type 1 or HIV-1 (Wain-Hobson et al., 1985). Few years later in 1986, an immunologically distinct variant of HIV-1 was isolated from individuals in the Western region of Africa and was named HIV-2 (Clavel et al., 1986). Out of these two species, HIV-1 is mainly responsible for the AIDS-related complications and deaths, whereas HIV-2 is more confined to Western Africa (Coffin et al., 1986).

11.2 ORIGIN OF HIV

Cross-species transmission of HIV from chimpanzees is hypothesized to be the main cause of introduction of HIV to humans (Sharp and Hahn, 2011). At least three cross-species transmissions may have occurred from monkey species to human which gave rise to three major groups of HIV-1 termed as group M, N, and O (Hayflick, 1992). Multiple mathematical models have predicted that the time of cross-species transmission was the early 1930s but we have failed to trace back the accurate timings of these cross-species transmissions (Gao et al., 1999). Although the oldest HIV-1 evidences dated back to 1959 from preserved tissue samples of individuals from Africa and England independently (Corbitt et al., 1990; Zhu et al., 1998). This gives an idea that HIV-1 might be present in human population even before our predicted timings, but the major outbreak happened not before 1959.

11.3 EPIDEMIOLOGY

According to the UNAIDS report, there are approximately 39 million people living with HIV and the new infections are added to the population at the rate of 2.0–2.5 million infections per year (UNAIDS, 2016). Although the rate of new infections has dropped significantly in recent years, but the number still remains a matter of concern. Due to the extremely fast rate of transcription, lack of proofreading activity, and homologous recombination in host, this virus develops various mutations giving rise to multiple subspecies of virus even in a single host. These subspecies show different replicative behavior as well as different response to the antiviral medications. Initially, subtype-B virus, which was mostly prevalent in the Middle East, was considered to be the main cause for AIDS-related deaths. But recent data suggest that subtype-C virus, which was initially confined to Indian subcontinent and surroundings, has distributed across the globe and it outnumbers subtype-B in terms of AIDS-related deaths (Taylor et al., 2008).

11.4 PATHOGENESIS

HIV-1 primarily infects the vital cells of immune system which includes $CD4^+$ T-cells, monocytes, and macrophages. HIV-1 also infects the progenitor cells and establishes latent reservoirs in these cells which results in the generation of new immune cells with *hidden* provirus (Carter et al., 2010). Infection of these immune cells alters the cellular microenvironment which ultimately results in death of the infected cells by multiple mechanisms over a period of time (Gaiha and Brass, 2014). The loss of immune cells results in compromised immunity which invites opportunistic infections and cancer to thrive. The drop in $CD4^+$ T-cell number is directly correlated with the disease progression and establishment of AIDS. HIV infection and AIDS-related illness is mainly divided into three phases: (1) *acute phase* or the initial infections is the first phase of HIV infection in which the infected individual feels influenza-like symptoms. The normal diagnostic assay may fail to detect the presence of virus in this phase as the viral load is maintained at very low-level due to body's response against the pathogen. (2) The second phase is the *clinical latency* where virus remains in an inactive state for an indefinite period of time. This stage is marked by a gradual loss in $CD4^+$ T-cells, and this period is mainly asymptomatic and varies from 3 years to 20 years. (3) When the

$CD4^+$ T-cell count drops below 200 cells/mm^3 of blood, the stage is referred to as *AIDS*. Due to severely compromised immunity, opportunistic infections are established in infected individuals which ultimately results in death (Stevenson, 2003; Coffin and Swanstrom, 2013; GBD 2013 Mortality and Causes of Death Collaborators, 2015).

11.5 TREATMENT

Since its discovery in 1983 as the causative agent of AIDS, HIV continues to kill humans. Well-studied life cycle and high-resolution structure of various components of virus have enabled us to strengthen the anti-HIV therapeutics over the period of past 35 years. Currently, an arsenal of more than 26 US FDA approved medicines and their analogs are in use to treat HIV- and AIDS-related complications (United States Food and Drug Administration, 2014). With currently practiced highly active antiretroviral therapy (HAART), we have successfully controlled the disease to a great extent and increased the life span of infected individuals but have failed to completely cure AIDS. Many factors limit the efficacy of anti-HIV therapeutics. As mentioned earlier, extremely high replication rate and lack of proofreading activity enables virus to develop the resistant mutants against these drugs in a short span of treatment. Majority of the time, these medications are toxic to the host as they are synthetic molecules and are required at a high concentration to control the HIV replication in body. Inability to detect and destroy the latent viral reservoirs is also a major drawback of these anti-HIV medicines.

Studies are ongoing on many approaches to block HIV replication in host cells. CCR5 knockout hematopoietic stem cell helped us to engineer cells which were resistant to HIV infection to some extent and also tested on HIV-infected individuals (Holt et al., 2010; Tebas et al., 2014). This gene therapy approach gained a lot of attention in scientific community, but the trials failed due to the nonspecific off-target effects (Pandit and de Boer, 2016). Synthesis of novel inhibitors of HIV-1 and their analogs is continuously ongoing to strengthen the current anti-HIV therapeutics, but this approach is both time and resource consuming.

Naturally occurring bioactive products such as plant derivatives and products from marine organisms have always been of great interest because of their various biological activities. Usage of plant product is advantageous over marine organisms because a large number of plant-based bioactive molecules are available as dietary intake and hence are

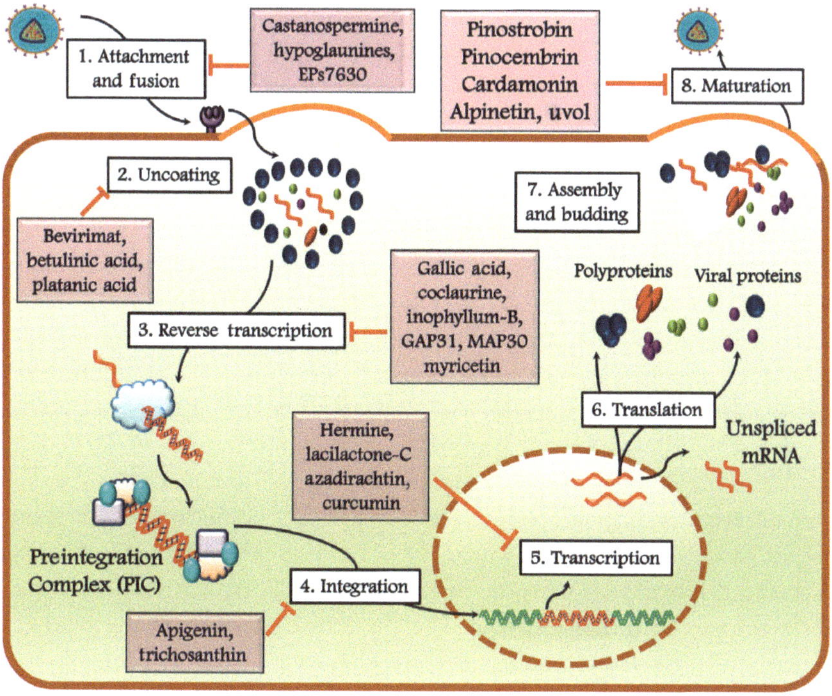

Figure 11.1 Schematic representation of HIV-1 life cycle and stages that are inhibited by plant derivatives.

very less toxic. Apart from that, availability of plants makes it comparatively easy to isolate and study the active molecules and design their analogs to get *better* molecules. Many of such plant derivatives have already been evaluated or are under evaluation for their various biological activities including anti-HIV activity (Asres et al., 2005; Singh et al., 2005; Kurapati et al., 2016). Indian medicinal plants have also been extensively studied for their anti-HIV activity as Indian subcontinent and surroundings are rich in plant diversity and several of them are of great medical importance (Singh and Bodiwala, 2010; Sabde et al., 2011). In the next sections, a detailed discussion is presented on the role of phytocompounds against HIV-1 and their mode of action as schematically highlighted in Fig. 11.1.

11.5.1 Inhibition of Virus Entry

11.5.1.1 Attachment

Life cycle of HIV-1 begins with the attachment of HIV-1 gp120 glycoprotein to CD4 receptor present on $CD4^+$ cells. This attachment leads to several conformational changes in glycoproteins which facilitate virus entry into the host cell. Castanospermine (*Castanospermum australe*) is one of the first plant alkaloids reported to inhibit HIV-1 attachment to the cell. It alters gp120 glycosylation pattern by inhibiting alpha-galactosidase enzyme (Walker et al., 1987). Plant-specific lectins, especially from *Myrianthus holstii*, are also reported to block virus entry at significantly low concentration ($IC_{50} = \sim 150$ nM). Lectins bind to gp120 glycoprotein at the variable region and alter the glycosylation pattern which hampers the interaction of gp120 to CD4 receptor (Akkouh et al., 2015). Root extracts of traditional African medicinal plant *Pelargonium sidoides*, EPs-7630, also blocks virus attachment to the host cells at significantly low concentration and currently in its phase-II trial (Helfer et al., 2014). Several alkaloids including triptonine A/B, wilfordinines A/B/C, and hypoglaunine A/B block HIV-1 attachment to cell by altering glycosylation pattern at very low concentration and with a significantly high therapeutic index (TI $>$ 1000) (Duan et al., 1999). Attachment inhibitors alter the glycosylation pattern on the virus gp120 and hence have very less cytotoxicity to the host cells. But the drawback of these inhibitors is, majority of them fail to block virus at higher viral load (Table 11.1).

11.5.1.2 Fusion

Attachment of gp120 to CD4 receptors leads to the conformational changes which result in unfolding of gp41. This event is referred to as *Fusion*. gp41 interacts with CCR5/CXCR4 chemokine co-receptors which results in 6-helix bundle (6-HB) formation. 6-HB has always been the target of HIV-1 fusion inhibitors as it is the key step for fusion of virus membrane to the cell membrane (Chan and Kim, 1998; Wilen et al., 2012). Derivatives of many plants groups have been reported to block virus entry by blocking the 6-HB bundle formation. Flavonoids such as polyphenol groups of plant metabolites are one of the most diverse chemical groups with various molecules blocking HIV at multiple stages. Michellamine B (*Ancistrocladus korupensis*) blocks HIV-1 fusion by intercalating with the 6-HB (McMahon et al., 1995). In the early 2000s, Baicalin (*Scutellaria baicalensis*) and Taxifolin (*Silybum marianum*) were reported to block HIV-1 entry by blocking conformational changes during the fusion by two independent groups (Li et al., 2000; Min et al., 2002). One of the major advantages of these entry inhibitors is that they block cell-to-cell transmission of the virus as well with equal potential (Table 11.1). But utilization of different chemokine receptors and

Table 11.1 Summary of selected plant derivatives that block HIV-1 entry in the host cells

Plant derivatives	Plant source	Anti-HIV activity	Class	Reference
Castanospermine	*Castanospermum australe*	$IC_{50} = \sim 5\ \mu M$	Alkaloids	Walker et al. (1987)
Michellamine B	*Ancistrocladus Korupensis*	$IC_{50} = 100\ \mu M$	Flavonoids	McMahon et al. (1995)
Triptonine A/B/C	*Tripterygium hypoglaucum*	$IC_{50} = \sim 3\ \mu g/mL$	Alkaloids	Duan et al. (1999)
Wilfordinines A/B/C	*Tripterygium hypoglaucum*	$IC_{50} = < 0.1\ \mu g/mL$	Alkaloids	Duan et al. (1999)
Hypoglaunine A/B	*Tripterygiun hypoglaucum*	$TI = > 1000$	Alkaloids	Duan et al. (1999)
Baicalin	*Scutellaria baicalensis*	$IC_{50} = 4\ \mu M$	Flavonoids	Li et al. (2000)
Taxifolin	*Silybum marianum*	$IC_{50} = 25\ \mu g/mL$	Flavonoids	Min et al. (2002)
EPs-7630	*Pelargonium sidoides*	$TI = > 150$	Flavonoids	Helfer et al. (2014)
M. holstii lectins (MHL)	*Myrianthus holstii*	$IC_{50} = 150\ nM$	Proteins	Akkouh et al. (2015)

different modes of entry in some cases limits the efficiency of these entry blockers (Pollakis and Paxton, 2012; Woodham et al., 2016).

11.5.2 Inhibition of Uncoating

Upon fusion, the cone-shaped capsid of the virus is released into the cytoplasm. The uncoating step removes the capsid and releases the viral genome and other accessory content in the cytoplasm. The pentacyclic triterpenoids, betulinic acid, and platanic acid, found in barks of several plant species including *Betula pubescens*, were reported to have weak anti-HIV activity, but its mechanism of action is not clear (Fujioka et al., 1994). Recently, bevirimat (BVM), a derivative of betulinic acid was proven to be very effective capsid inhibitor. BVM, also referred to as PA-457, DSB, and PMC-4326, very efficiently blocks multiple strains of HIV including several drug-resistant strains at a very low concentration (TI > 2500). BVM is the only capsid inhibitor so far identified and has been tested in humans in phase-II trials (Thenin-Houssier and Valente, 2016) (Table 11.2).

11.5.3 Inhibition of Reverse Transcription

HIV contains two copies of ~10 kb single-stranded RNA genome. Once the capsid content is released in the cytoplasm, with the help of reverse transcriptase (RT) enzyme, this ssRNA genome is converted into cDNA by reverse transcription (Hu and Hughes, 2012). Reverse transcription is one of the most well-characterized steps of HIV life cycle and has always been the prime target for anti-HIV molecules. In fact, first US FDA approved anti-HIV molecule, zidovudine, is a nucleoside analog which blocks HIV-1 reverse transcription (Fischl et al., 1987).

Table 11.2 Summary of selected plant derivatives that block virion capsule uncoating in the host cells

Plant derivatives	Plant source	Anti-HIV activity	Class	Reference
Betulinic acid	*Betula pubescens*	TI = ~100	Terpenes	Fujioka et al. (1994)
Platanic acid	*Betula pubescens*	TI = ~101	Terpenes	Fujioka et al. (1994)
Bevirimat	*Betula pubescens*	TI = > 2500	Terpenes	Thenin-Houssier and Valente (2016)

11.5.3.1 Nonnucleotide Reverse Transcriptase Inhibitors

Several plant derivatives are nonnucleotide RT inhibitors (NNRTIs) that inhibit HIV-1 RT activity. Several *alkaloids* including, michellamine B from *A. korupensis*, triptonin A/B, nitidine from *Toddalia asiatica* (Rashid et al., 1995), benzylisoquinolines including (+)-1(*R*)-coclaurine, (−)-1(*S*)-norcoclaurine, quercetin 3-*O*-β-D-glucuronide isolated from *Nelumbo nucifera* (Kashiwada et al., 2005); several phenanthroindolizidine including (−)-antofine and dehydroantofine of *Cryptocarya chinensis* (Wu et al., 2012) inhibits HIV-1 RT with very high potential. TI of some of these alkaloids is significantly high (>125) and also inhibits not only HIV-1 RT but a range of other viral RT enzymes, as well. Flavonoids, including kaempferol and its analogs isolated from *Securigera securidaca* (Behbahani et al., 2014), quercetin from *Acer Okamotoanum* (Pasetto et al., 2014), myricetin from *Allium cepa* (Pasetto et al., 2014), pinocembrin of *Tunera diffusa* (Rasul et al., 2013), ellagic acid and gallic acid from *Lagerstroemia speciose* (Nutan et al., 2013) bind to HIV-1 RT enzyme and induce conformational changes which suppress enzyme activity. Several subgroup bioflavonoids including amentoflavone, agathisflavone, robustaflavone, hinokiflavone, volkensiflavone, morelloflavone, rhusflavanone, succedaneaflavanone isolated from *Rhus succedanea* have also been reported to significantly inhibit HIV reverse transcription by mechanically binding to the RT enzyme (Lin et al., 1997). Callophyllum is a rich source of various bioactive molecules that belong to coumarin group of plant products. Inophyllum B isolated from *Callophyllum inophyllum* is one of the first reported NNRTI which inhibits HIV-1 RT at nanomolar concentration of IC_{50} = 38 nM (Patil et al., 1993). Protein products isolated from various plants such as GAP31 from *Gelonium multiflorum* (Li et al., 2010), MAP30 and MRK29 from *Momordica charantia* (Lee-Huang et al., 1995) acts as NNRTIs at significantly low concentration.

11.5.3.2 Nucleotide Reverse Transcriptase Inhibitors

Nucleotide analogs calanolide A/B, castatolide, dihydrocostatolide and cordatolide A/B, dipyranocoumarins, and soulattrolide isolated from *Callophyllum lanigerum* efficiently hampers RT activity at the synthesis step (Buckheit et al., 1999; Huerta-Reyes et al., 2004). Swertifrancheside is the first reported flavone−xanthone dimer extracted from the roots of *Swertia franchetiana* that inhibits HIV-1 RT in both cell-based and cell-free assay (Wang et al., 1994).

One of the major advantages of RT inhibitors (Table 11.3) is that they inhibit not only HIV but a range of retroviruses as reverse transcription is

Table 11.3 Summary of selected plant derivatives that block HIV-1 reverse transcription

Plant derivatives	Plant source	Anti-HIV activity	Class	Reference
Inophyllum B	*Callophyllum inophyllum*	$IC_{50} = \sim 38$ nM	Coumarins	Patil et al. (1993)
Swertifrancheside	*Swertia franchetiana*	$IC_{50} = 43$ μM	Xanthone	Wang et al. (1994)
MAP30	*Momordica charantia*	TI = > 10,000	Proteins	Lee-Huang et al. (1995)
Michellamine B	*Ancistrocladus korupensis*	$IC_{50} = 100$ μM	Alkaloids	McMahon et al. (1995)
Nitidine	*Toddalia asiatica*	$IC_{50} = 1$ μM	Alkaloids	Rashid et al. (1995)
Triptonine A/B	*Toddalia asiatica*	$IC_{50} = < 0.1$ μg/mL	Alkaloids	Rashid et al. (1995)
MRK29	*Momordica charantia*	$IC_{50} = 18$ mg/ml	Proteins	Lee-Huang et al. (1995)
Bioflavonoids	*Rhus succedanea*	$IC_{50} = 1$ mM	Flavonoids	Lin et al. (1997)
Amentoflavone				
Agathisflavone				
Robustaflavone				
Hinokiflavone				
Volkensiflavone				
Morelloflavone				
Succedaneaflavanone				
Nucleotide analogs	*Callophyllum lanigerum*	$IC_{50} = 0.2-19$ mM	Coumarins	Buckheit et al. (1999), Huerta-Reyes et al. (2004)
Cordatolide A/B				
Calanolide A/B				
Dihydrocostatolide				
Castatolide				
Dipyranocoumarins				
Soulattrolide				

(+)-1(*R*)-Coclaurine	*Nelumbo nucifera*	TI = > 125	Alkaloids	Kashiwada et al. (2005)
(−)-1(*S*)-Norcoclaurine	*Nelumbo nucifera*	IC_{50} = 1 μM	Alkaloids	Kashiwada et al. (2005)
Quercetin 3-*O*-β-D-glucuronide	*Nelumbo nucifera*	IC_{50} = 1 μM	Alkaloids	Kashiwada et al. (2005)
GAP31	*Gelonium multiflorum*	IC_{50} = 0.2–0.3 nM	Proteins	Li et al. (2010)
(−)-Antofine	*Cryptocarya chinensis*	IC_{50} = 1.88 mg/mL	Alkaloids	Wu et al. (2012)
Dehydroantofine	*Cryptocarya chinensis*	IC_{50} = 1 mg/mL	Alkaloids	Wu et al. (2012)
Pinocembrin	*Turnera diffusa*	IC_{50} = 53.29 mM	Flavonoids	Rasul et al. (2013)
Ellagic acid	*Lagerstroemia speciosa*	TI = 3.4	Flavonoids	Nutan et al. (2013)
Gallic acid	*Lagerstroemia speciosa*	TI = 447	Flavonoids	Nutan et al. (2013)
Kaempferol	*Securigera securidaca*	IC_{50} = 100 mg/mL	Flavonoids	Behbahani et al. (2014)
Quercetin	*Acer okamotoanum*	IC_{50} = 39.26 mM	Flavonoids	Pasetto et al. (2014)
Myricetin	*Allium cepa*	IC50 = 3.23 mM	Flavonoids	Pasetto et al. (2014)

more or less conserved process across the group. But there are few drawbacks of RT inhibitors. NRTI group of inhibitors hampers the normal DNA synthesis at higher concentration and hence they are very toxic to the host, whereas due to extremely fast replication rate HIV develops resistance to NNRTIs faster than any other class of anti-HIV molecules. Hence, more research is needed to get more specific and less toxic molecules to block HIV-1 RT (Cimarelli and Darlix, 2014).

11.5.4 Inhibition of Integration

At the end of reverse transcription, ssRNA genome is converted to dsDNA genome which interacts with several host as well as viral factors and forms a structure referred to as pre-integration complex (PIC). PIC then, with the help of HIV-1 integrase enzyme, gets imported into the nucleus and gets integrated into the host genome which is now referred to as the provirus (Figueiredo and Hope, 2011). Once integrated, it is impossible to identify and destroy the *hidden* provirus and hence blocking integration is of unique importance.

Apigenin isolated from *Chrysanthemum morifolium*, a flavonoid, inhibits interaction of integrase enzyme with that of LEDGF/p75 which is an essential component required for nuclear import of PIC (Lee et al., 2003). Trichosanthin and other plant protein derivatives isolated from *Trichosanthes kirilowii* such as agrostin, gelonin, luffin, momorcharin, and saporin belong to a group of ribosome-inactivating proteins or RIPs (Sha et al., 2013). RIPs are known to hamper 3′-end processing step during strand transfer reaction of integrase enzyme which results in abortive integration. RIPs efficiently block integrase enzyme activity in cell-free assay system, but their efficacy is reduced in cell-based testing (Au et al., 2000).

Targeting integrase enzyme is very challenging as the enzyme is embedded in PIC and is not easily accessible for inhibitor molecules. This limits the therapeutic use of these molecules, but as mentioned above, it is important to block integration as once integrated it is impossible to eradicate the virus from the infected cell (Table 11.4).

11.5.5 Inhibition of Transcription

Once integrated, provirus utilizes cellular transcription machinery to produce new transcripts which get translated and form new functional virus progeny. It is difficult to detect and destroy the provirus at this stage because provirus is now a part of the host genome itself (Karn and Stoltzfus, 2012). HIV transcription has always been of great importance

Table 11.4 Summary of selected plant derivatives that block integration of HIV-1 dsDNA in the host genome

Plant derivatives	Plant source	Anti-HIV activity	Class	Reference
Apigenin	*Chrysanthemum morifolium*	$IC_{50} = 41.86$ mg/mL	Flavonoids	Lee et al. (2003)
Trichosanthin derivatives	*Trichosanthes kirilowii*	TI = >25	Proteins	Au et al. (2000), Sha et al. (2013)
Agrostin				
Gelonin				
Luffin				
Momorcharin				
Saporin				

to eradicate virus in two ways. First is to suppress the virus transcription and second is to reactivate and destroy the latent viral reservoirs.

Almost all major groups of plant derivatives have representative molecules which inhibit the transcription directly or indirectly. Alkaloids including castanospermine from *C. australe*, cepharanthine from *Stephania cepharantha* (Okamoto et al., 1998), and hermine isolated from *Symplocos setchuensis* (Wang et al., 2015) modulate cell signaling that ultimately inhibit the HIV-1 transcription significantly. Andrographolide isolated from *Andrographis paniculata*, a diterpenoid, is reported to inhibit HIV transcription by inhibiting cellular signaling including NF-kB pathway with TI > 51 (Wang et al., 2010). Azadirachtin from *Azadirachta indica*, curcumin from *Curcuma longa* (IC_{90} = 100 nM) inhibits TNF-alpha mediated activation of NF-kB pathway which ultimately blocks the transcription (Atawodi and Atawodi, 2009; Thoh et al., 2010; Prasad and Tyagi, 2015). HIV-1 transcription inhibitors belonging to coumarin group involves imperatorin from *Ferula sumbul* (Sancho et al., 2004) and suksdorfin isolated from *Lomatium suksdorfii* (Lee et al., 1994), which inhibit HIV transcription in an Sp1-dependent manner. Many of the flavonoids are also reported to block HIV-1 transcription at significantly low concentration. Luteolin isolated from *Salvia tomentosa* (Mehla et al., 2011) and triptolide from *Tripterygium wilfordii* (Wan and Chen, 2014) blocks HIV-1 transcription by obstructing the function of trans-activator of transcription (Tat) protein. The protein products such as trichosanthin and TAP29 from *T. kirilowii*, GAP31 from *G. multiflorum* and MAP30 isolated from *M. charantia* and terpene including lasilactone C and its derivatives from *Kadsura lancilimba* block HIV-1 transcription by blocking the formation of transcription complex (Maimone and Baran, 2007). Niruriside isolated from *Phyllanthus niruri*, a carbohydrate derivative, is one of the most interesting plant derivatives as one of the very few reported plant products to block HIV accessory protein Rev. Niruriside blocks HIV-Rev protein's binding to Rev responsive element or RRE (Qian-Cutrone et al., 1996).

However, there are several limitations of transcription inhibitors (Table 11.5). As mentioned earlier, virus utilizes the host machinery for the transcription and hence these inhibitors, at higher concentration, blocks host transcription as well. And also, the inhibitory effect of these molecules is temporary because the virus transcription resumes as soon as the inhibitory force is withdrawn.

11.5.6 Inhibition of Maturation

Upon transcription, HIV-1 LTR produces mRNA transcripts which undergo posttranscriptional modification giving rise to unspliced, single spliced, and multi-spliced forms of mRNAs. After translation, the ssRNA genome, polyproteins, and other machinery assemble to the plasma membrane and bud out of the cell as *immature* virus particle. Proteolytic cleavage mediated by protease enzyme gives rise to *mature* virus particles which are now ready to infect other host cells (Sundquist and Kräusslich, 2012). Protease inhibitors usually bind to the active site of the protease enzyme which blocks the binding of the substrate to the enzyme and, hence, inhibit the maturation step (Eron, 2017; Amin et al., 2018).

Docking studies reveal that several plant alkaloids including jatrorrhizine, magnoflorine, and tinosporide isolated from *Tinospora cordifolia* bind to HIV-1 protease active site with a very high affinity (Van Kiem et al., 2010). Several flavonoids possess very specific inhibitory action on HIV-1 protease. Pinostrobin, pinocembrin, cardamonin, and alpinetin (*Boesenbergia pandurate*) significantly inhibit HIV-1 protease enzyme at nanomolar to picomolar concentration (Tewtrakul and Subhadhirasakul, 2003). Other flavonoid such as gallic acid and its dimer form, ellagic acid isolated from *Lagerstroemia speciousa*, also inhibit HIV-1 protease with TI $>$ 450. Several triterpenes from *Geum japonicum* including ursolic acid, epipomolic acid, maslinic acid, euscaphic acid, and tormentic acid inhibit HIV-1 protease at micromolar concentration in both cell-free as well as cell-based assay system (Xu et al., 1996). Uvol isolated from

Table 11.5 Summary of selected plant derivatives that block HIV-1 provirus transcription and gene expression

Plant derivatives	Plant source	Anti-HIV activity	Class	Reference
Castanospermine	*Castanospermum austral*	$IC_{50} = \sim 5\ \mu M$	Alkaloids	Walker et al. (1987)
Suksdorfin	*Lomatium suksdorfii*	NA	Coumarin	Lee et al. (1994)
MAP30	*Momordica charantia*	$TI = > 10,000$	Proteins	Lee-Huang et al. (1995)
MRK29	*Momordica charantia*	$IC_{50} = 18$ mg/ml	Proteins	Lee-Huang et al. (1995)
Niruriside	*Phyllanthus niruri*	NA	Carbohydrates	Qian-Cutrone et al. (1996)
Cepharanthine	*Stephania cepharantha*	$IC_{50} = \sim 1\ \mu M$	Alkaloids	Okamoto et al. (1998)
Trichosanthin	*Trichosanthes kirilowii*	$TI = > 50$	Proteins	Au et al. (2000), Sha et al. (2013)
Imperatorin	*Ferula sumbul*	NA	Coumarin	Sancho et al., 2004
Lacilactone C	*Kadsura lancilimba*	$TI = 150$	Terpenes	Maimone and Baran (2007)
Azadirachtin	*Azadirachta indica*	NA	Terpenes	Atawodi and Atawodi (2009), Thoh et al. (2010)
Andrographolide	*Andrographis paniculata*	$TI = > 51$	Terpenes	Wang et al. (2010)
GAP31	*Gelonium multiflorum*	$IC_{50} = 0.2–0.3$ nM	Proteins	Li et al. (2010)
Luteolin	*Salvia tomentosa*	NA	Flavonoids	Mehla et al. (2011)
Triptolide	*Tripterygium wilfordii*	$IC_{50} = \sim 3.5$ nM	Flavonoids	Wan and Chen (2014)
Hermine	*Symplocos setchuensis*	$TI = > 100$	Alkaloids	Wang et al. (2015)
Curcumin	*Curcuma longa*	$IC_{90} = 100$ nM	Phenols	Prasad and Tyagi (2015)

Crataegus pinatifida (Byung et al., 1999) Lancilactone C of *Kadusra lancilimba* (Chen et al., 1999) Garciosaterpenes A/B/C from *Garcinia speciose* (Rukachaisirikul et al., 2003) also significantly block HIV-1 protease activity and hence block virus maturation. Crude extracts of *Hoodia gordonii* is also reported to have significant antiviral effect and inhibit HIV-1 protease enzyme activity at low concentration (Kapewangolo et al., 2016). Biological activity of Hoodia derivatives is controversial and hence these products, despite various reported pharmacological properties, have not gained much attention.

One of the major drawbacks of protease/maturation inhibitors (Table 11.6) is that they act at the final stage of HIV life cycle and hence their efficacy is reduced several folds with high viral load (Kurapati et al., 2016). Apart from that, they also fail to block cell-to-cell transmission of HIV-1 which is a major mode of virus transmission. Specificity is also a matter of concern when these inhibitors are used as therapeutics because many of them inhibit cellular proteases as well. Also, they need to be used for long term and hence are very toxic to the host. Hence, maturation inhibitors should be preferred in the case of acute infection over the chronic infection.

11.5.7 Inhibition of Syncytia Formation

The HIV-infected cells fuse with the bystander uninfected cells giving rise to a multinucleated cell called syncytium. Multiple theories have been proposed but still the mechanism of syncytia formation is unclear. Due to excess metabolic load, syncytium, in the end, leads to the cell death. This mode of cell-to-cell transmission of virus is the major mode of HIV spread in the body where more than 80% virus is spread by cell-mediated transmission (Shaw and Hunter, 2012). The fusion of uninfected cell to that of infected cells is more or less similar to fusion event of virus entry. But the involvement of chemokine signaling in cell-to-cell fusion makes it more complex. Several alkaloids which include buchapine derivatives from *Euodia roxburghiana* (Ahmed et al., 2010), castanospermine of *C. australe*, aromoline, and FK-3000 isolated from *S. cepharantha* (Ma et al., 2002) block cell-to-cell transmission and syncytia formation at nanomolar concentration (Table 11.7). Several plant proteins and derivatives such as trichosanthin and TAP29 obtained from *T. kirilowii*, GAP31 from *G. multiflorum*, MAP30 from *M. charantia*, circulins and cyclociolins from *Leonia cymosa* (Hallock et al., 2000), and palicourein of *Palicourea*

Table 11.6 Summary of selected plant derivatives that block HIV-1 maturation

Plant derivatives	Plant source	Anti-HIV activity	Class	Reference
Triterpene derivatives Ursolic acid Epipomolic acid Maslinic acid Euscaphic acid Tormentic acid	*Geum japonicum*	NA	Terpenes	Xu et al. (1996)
Uvol	*Crataegus pinatifida*	$IC_{50} = \sim 250$ nM	Terpenes	Byung et al. (1999)
Lacilactone C	*Kadsura lancilimba*	TI = 150	Terpenes	Chen et al. (1999)
Pinostrobin	*Boesenbergia pandurate*	$IC_{50} = \sim 10$ nM	Flavonoids	Tewtrakul and Subhadhirasakul (2003)
Pinocembrin	*Boesenbergia pandurate*	$IC_{50} = \sim 10$ nM	Flavonoids	Tewtrakul and Subhadhirasakul (2003)
Cardamonin	*Boesenbergia pandurate*	$IC_{50} = \sim 800$ nM	Flavonoids	Tewtrakul and Subhadhirasakul (2003)
Alpinetin	*Boesenbergia pandurate*	$IC_{50} = \sim 800$ nM	Flavonoids	Tewtrakul and Subhadhirasakul (2003)
Garciosaterpenes A/B/C	*Garcinia speciose*	$IC_{50} = \sim 80$ nM	Terpenes	Rukachaisirikul et al. (2003)
Jatrorrhizine derivatives	*Tinospora cordifolia*	NA	Alkaloids	Van Kiem et al. (2010)
Ellagic acid	*Lagerstroemia speciosa*	TI = 3.4	Flavonoids	Nutan et al. (2013)
Gallic acid	*Lagerstroemia speciosa*	TI = 447	Flavonoids	Nutan et al. (2013)
Hoodia gordonii derivatives	*Hoodia gordonii*	$IC_{50} = \sim 70$ μg/mL	NA (crude extracts)	Kapewangolo et al. (2016)

Table 11.7 Summary of selected plant derivatives that block cell-to-cell fusion and syncytia formation

Plant derivatives	Plant source	Anti-HIV activity	Class	Reference
MAP30	*Momordica charantia*	TI = > 10,000	Proteins	Lee-Huang et al. (1995)
TAP29	*Trichosanthes kirilowii*	IC_{50} = 10 nM	Proteins	Lee-Huang et al. (1995)
Cepharanthine	*Stephania cepharantha*	IC_{50} = 5 μM	Alkaloids	Okamoto et al. (1998)
Lacilactone C	*Kadsura lancilimba*	TI = > 150	Terpenes	Chen et al. (1999)
Circulins	*Leonia cymosa*	IC_{50} = 4 μM	Proteins	Hallock et al. (2000)
Cyclociolins	*Leonia cymosa*	IC_{50} = 1 μM	Proteins	Hallock et al. (2000)
Trichosanthin	*Trichosanthes kirilowii*	NA	Proteins	Au et al. (2000), Sha et al. (2013)
Palicourein	*Palicourea condensate*	NA	Proteins	Bokesch et al. (2001)
Aromoline	*Stephania cepharantha*	IC_{50} = 500 nM	Alkaloids	Ma et al. (2002)
FK-3000	*Stephania cepharantha*	IC_{50} = 500 nM	Alkaloids	Ma et al. (2002)
Garciosaterpenes A/B/C	*Garcinia speciose*	TI = > 150	Terpenes	Rukachaisirikul et al. (2003)
Buchapine derivatives	*Eodia roxburghiana*	TI = 4.59–29.47	Alkaloids	Ahmed et al. (2010)
GAP31	*Gelonium multiflorum*	IC_{50} = 0.2–0.3 nM	Proteins	Li et al. (2010)

condensate (Bokesch et al., 2001) are some of the most potent syncytia blockers, some of which act at as low as 0.3 nM concentration and with TI > 10,000. Lancilactone C from *K. lancilimba* and garcisterpene A/B/C of *G. speciose* blocks HIV-1-induced syncytia formation with TI > 150.

11.6 CONCLUDING REMARKS

HIV continues to kill humans from past three and a half decades. Extremely fast replication rate without proofreading activity allows virus to produce diverse strains which are resistant to anti-HIV molecules. Apart from that the toxicity of these drugs also limits their use in patients for long term. With the current HAART approach, we have successfully controlled the diseases and increased the life span of infected individuals significantly but have failed to completely cure the diseases. Hence, an update or modification is required in current therapeutics to fight against HIV and to cure the disease.

Design of transgenic plants rich in nutrients and bioactive medicinal compounds have emerged since the middle of the 19th century. Plants are rich sources of bioactive molecules which are effective against various diseases. An estimate suggests that almost half of the US FDA approved drugs till date are natural products, out of which a quarter is isolated from the plants (Patridge et al., 2016).

Use of plant products as anti-HIV agent is very advantageous. Availability of majority of the medicinal plants makes their use more effective. Many of the plant derivatives are available as daily dietary supplements and hence the toxicity levels of these products are comparatively less. Their availability as dietary supplements makes them orally available which is again more advantageous over other mode of administration of drugs. Extraction of the active ingredient from plants and designing novel analogs has helped us to obtain several molecules with significantly high TI than the parent molecules. In future, design of transgenic plants with an increased source of one or more of these anti-HV molecules might change the face of anti-HIV therapeutics and ultimately antiretroviral medications.

One of the biggest disadvantages of the plant-derived molecules is the lack of study on their mechanism of action. Many of the medicinal plants are used for various purposes for their biological activities since ages and hence lack a proper study on their mechanism of action and also their off-target effects. Although a number of plant-derived molecules have reached the clinical trials, looking at the potential of plant-derived molecules, it seems that researchers in the field need to take their work one step further toward translation by increasing collaboration with virologists. Despite a significant amount of research on HIV and with currently available arsenal of anti-HIV therapeutic regimen, it is a fact that there has been no cure for HIV till date, and we are far from eradicating the virus from human population. So, the quest for identification of novel anti-HIV molecules from plants should be continued as plants provide a huge array of molecular diversity to pursue further toward this goal.

REFERENCES

Ahmed, N., Brahmbhatt, K.G., Sabde, S., Mitra, D., Singh, I.P., Bhutani, K.K., 2010. Synthesis and anti-HIV activity of alkylated quinoline 2,4-diols. Bioorg. Med. Chem. 18, 2872–2879. Available from: https://doi.org/10.1016/j.bmc.2010.03.015.

Akkouh, O., Ng, T.B., Singh, S.S., Yin, C., Dan, X., Chan, Y.S., et al., 2015. Lectins with anti-HIV activity: a review. Molecules 20, 648–668. Available from: https://doi.org/10.3390/molecules20010648 [pii];10.3390/molecules20010648 [doi].

Amin, S.A., Adhikari, N., Bhargava, S., Jha, T., Gayen, S., 2018. Structural exploration of hydroxyethylamines as HIV-1 protease inhibitors: new features identified. SAR QSAR Environ. Res. 29, 385–408. Available from: https://doi.org/10.1080/1062936X.2018.1447511.

Asres, K., Seyoum, A., Veeresham, C., Bucar, F., Gibbons, S., 2005. Naturally derived anti-HIV agents. Phyther. Res 19, 557–581. Available from: https://doi.org/10.1002/ptr.1629.

Atawodi, S.E., Atawodi, J.C., 2009. *Azadirachta indica* (neem): a plant of multiple biological and pharmacological activities. Phytochem. Rev. 601–620. Available from: https://doi.org/10.1007/s11101-009-9144-6.

Au, T.K., Collins, R.A., Lam, T.L., Ng, T.B., Fong, W.P., Wan, D.C.C., 2000. The plant ribosome inactivating proteins luffin and saporin are potent inhibitors of HIV-1 integrase. FEBS Lett. 471, 169–172. Available from: https://doi.org/10.1016/S0014-5793(00)01389-2.

Barre-Sinoussi, F., Chermann, J.C., Rey, F., Nugeyre, M.T., Chamarat, S., Gruest, J., et al., 1983. Isolation of a T-lymphotropic retrovirus from a patient at risk for acquired immune deficiency syndrome (AIDS). Science. (80–) 220 (4599), 868–971. Available from: https://doi.org/10.1126/science.6189183.

Behbahani, M., Sayedipour, S., Pourazar, A., Shanehsazzadeh, M., 2014. In vitro anti-HIV-1 activities of kaempferol and kaempferol-7-*O*-glucoside isolated from *Securigera securidaca*. Res. Pharm. Sci. 9, 463–469.

Bokesch, H.R., Pannell, L.K., Cochran, P.K., Sowder, R.C., McKee, T.C., Boyd, M.R., 2001. A novel anti-HIV macrocyclic peptide from *Palicourea condensata*. J. Nat. Prod. 64, 249–250. Available from: https://doi.org/10.1021/np000372l.

Buckheit, R.W., White, E.L., Fliakas-Boltz, V., Russell, J., Stup, T.L., Kinjerski, T.L., et al., 1999. Unique anti-human immunodeficiency virus activities of the nonnucleoside reverse transcriptase inhibitors calanolide A, costatolide, and dihydrocostatolide. Antimicrob. Agents Chemother. 43, 1827–1834.

Byung, S.M., Hyun, J.J., Jun, S.L., Young, H.K., Song, H.B., Chao, M.M., et al., 1999. Inhibitory effect of triterpenes from *Crataegus pinatifida* on HIV-I protease. Planta Med. 65, 374–375. Available from: https://doi.org/10.1055/s-2006-960792.

Carter, C.C., Onafuwa-Nuga, A., McNamara, L.A., Riddell, J., Bixby, D., Savona, M.R., et al., 2010. HIV-1 infects multipotent progenitor cells causing cell death and establishing latent cellular reservoirs. Nat. Med. 16, 446–451. Available from: https://doi.org/10.1038/nm.2109.

CDC, 1982. Current trends update on acquired immune deficiency syndrome (AIDS)—United States. MMWR 31, 507–508.

Chan, D.C., Kim, P.S., 1998. HIV entry and its inhibition. Cell. 93 (5), 681–684. Available from: https://doi.org/10.1016/S0092-8674(00)81430-0.

Chen, D.F., Zhang, S.X., Wang, H.K., Zhang, S.Y., Sun, Q.Z., Cosentino, L.M., et al., 1999. Novel anti-HIV lancilactone C and related triterpenes from *Kadsura lancilimba*. J. Nat. Prod. 62, 94–97. Available from: https://doi.org/10.1021/np980291d.

Cimarelli, A., Darlix, J.-L., 2014. HIV-1 reverse transcription. Methods Mol. Biol. 1087, 55–70. Available from: https://doi.org/10.1007/978-1-62703-670-2-6.

Clavel, F., Guétard, D., Brun-Vézinet, F., Chamaret, S., Rey, M.A., Santos-Ferreira, M. O., et al., 1986. Isolation of a new human retrovirus from West African patients with AIDS. Science 233, 343–346. Available from: https://doi.org/10.1126/science.2425430.

Coffin, J., Swanstrom, R., 2013. HIV pathogenesis: dynamics and genetics of viral populations and infected cells. Cold Spring Harb. Perspect. Med. 3. Available from: https://doi.org/10.1101/cshperspect.a012526.

Coffin, J., Haase, A., Levy, J.A., Montagnier, L., Oroszlan, S., Teich, N., et al., 1986. Human immunodeficiency viruses. Science 232, 697. Available from: https://doi.org/10.1126/science.3008335.

Corbitt, G., Bailey, A., Williams, G., 1990. HIV infection in Manchester, 1959. Lancet . Available from: https://doi.org/10.1016/0140-6736(90)91561-N.

Duan, H., Takaishi, Y., Bando, M., Kido, M., Imakura, Y., Lee, K.H., 1999. Novel sesquiterpene esters with alkaloid and monoterpene and related compounds from *Tripterygium hypoglaucum*: a new class of potent anti-HIV agents. Tetrahedron Lett. 40, 2969–2972. Available from: https://doi.org/10.1016/S0040-4039(99)00339-1.

Eron, J.J., 2017. HIV-1 protease inhibitors. Clin. Infect. Dis. 30, 160–170. Available from: https://doi.org/10.1086/313853.

Figueiredo, A., Hope, T.J., 2011. KAPs off for HIV-1 integration. Cell. Host. Microbe. 9, 447–448. Available from: https://doi.org/10.1016/j.chom.2011.05.009.

Fischl, M.A., Richman, D.D., Grieco, M.H., Gottlieb, M.S., Volberding, P.A., Laskin, O. L., et al., 1987. The efficacy of azidothymidine (AZT) in the treatment of patients with AIDS and AIDS-related complex. N. Engl. J. Med. 315, 185–191. Available from: https://doi.org/10.1056/NEJM198707233170401.

Fujioka, T., Kashiwada, Y., Kilkuskie, R.E., Cosentino, L.M., Bailas, L.M., Jiang, J.B., et al., 1994. Anti-aids agents, 11. Betulinic acid and platanic acid as anti-HIV principles from Syzigium claviflorum, and the anti-HIV activity of structurally related triterpenoids. J. Nat. Prod. 57, 243–247. Available from: https://doi.org/10.1021/np50104a008.

Gaiha, G.D., Brass, A.L., 2014. The fiery side of HIV-induced T cell death. Science. (80–) 343, 383–384. Available from: https://doi.org/10.1126/science.1250175.

Gallo, R.C., Sarin, P.S., Gelmann, E.P., Robert-Guroff, M., Richardson, E., Kalyanaraman, V.S., et al., 1983. Isolation of human T-cell leukemia virus in acquired immune deficiency syndrome (AIDS). Science 220, 865–867. Available from: https://doi.org/10.1126/science.6601823.

Gao, F., Bailes, E., Robertson, D.L., Chen, Y., Rodenburg, C.M., Michael, S.F., et al., 1999. Origin of HIV-1 in the chimpanzee *Pan troglodytes troglodytes*. Nature 397, 436–441. Available from: https://doi.org/10.1038/17130.

GBD 2013 Mortality and Causes of Death Collaborators, 2013. Global, regional, and national age-sex specific all-cause and cause-specific mortality for 240 causes of death, 1990–2013: a systematic analysis for the Global Burden of Disease Study 2013. Lancet (London, England) . Available from: https://doi.org/10.1016/S0140-6736(14)61682-2.

Gottlieb, M.S., Schroff, R., Schanker, H.M., Weisman, J.D., Fan, P.T., Wolf, R.A., et al., 1981. *Pneumocystis carinii* pneumonia and mucosal candidiasis in previously healthy homosexual men: evidence of a new acquired cellular immunodeficiency. N. Engl. J. Med. 305, 1425–1431. Available from: https://doi.org/10.1056/NEJM198112103052401.

Hallock, Y.F., Sowder, R.C., Pannell, L.K., Hughes, C.B., Johnson, D.G., Gulakowski, R., et al., 2000. Cycloviolins A-D, anti-HIV macrocyclic peptides from *Leonia cymosa*. J. Org. Chem. 65, 124–128. Available from: https://doi.org/10.1021/jo990952r.

Hayflick, L., 1992. Origin of HIV-1. Lancet 340, 484–485. Available from: https://doi.org/10.1016/0140-6736(92)91806-J.

Helfer, M., Koppensteiner, H., Schneider, M., Rebensburg, S., Forcisi, S., Müller, C., et al., 2014. The root extract of the medicinal plant *Pelargonium sidoides* is a potent HIV-1 attachment inhibitor. PLoS One 9. Available from: https://doi.org/10.1371/journal.pone.0087487.

Holt, N., Wang, J., Kim, K., Friedman, G., Wang, X., Taupin, V., et al., 2010. Human hematopoietic stem/progenitor cells modified by zinc-finger nucleases targeted to CCR5 control HIV-1 in vivo. Nat. Biotechnol. 28, 839–847. Available from: https://doi.org/10.1038/nbt.1663.

Hu, W.-S., Hughes, S.H., 2012. HIV-1 reverse transcription. Cold Spring Harb. Perspect. Med. 2, a006882. Available from: https://doi.org/10.1101/cshperspect.a006882.

Huerta-Reyes, M., Basualdo, M.D.C., Abe, F., Jimenez-Estrada, M., Soler, C., Reyes-Chilpa, R., 2004. HIV-1 inhibitory compounds from *Calophyllum brasiliense* leaves. Biol. Pharm. Bull. 27, 1471–1475. Available from: https://doi.org/10.1248/bpb.27.1471.

Kapewangolo, P., Knott, M., Shithigona, R.E.K., Uusiku, S.L., Kandawa-Schulz, M., 2016. In vitro anti-HIV and antioxidant activity of *Hoodia gordonii* (Apocynaceae), a commercial plant product. BMC Complement. Altern. Med. 16, 411–418. Available from: https://doi.org/10.1186/s12906-016-1403-7.

Karn, J., Stoltzfus, C.M., 2012. Transcriptional and posttranscriptional regulation of HIV-1 gene expression. Cold Spring Harb. Perspect. Med. 2. Available from: https://doi.org/10.1101/cshperspect.a006916.

Kashiwada, Y., Aoshima, A., Ikeshiro, Y., Chen, Y.P., Furukawa, H., Itoigawa, M., et al., 2005. Anti-HIV benzylisoquinoline alkaloids and flavonoids from the leaves of *Nelumbo nucifera*, and structure–activity correlations with related alkaloids. Bioorg. Med. Chem. 13, 443–448. Available from: https://doi.org/10.1016/j.bmc.2004.10.020.

Kurapati, K.R. V., Atluri, V.S., Samikkannu, T., Garcia, G., Nair, M.P.N., 2016. Natural products as anti-HIV agents and role in HIV-associated neurocognitive disorders (HAND): a brief overview. Front. Microbiol. Available from: https://doi.org/10.3389/fmicb.2015.01444.

Lee-Huang, S., Huang, P.L., Chen, H.C., Huang, P.L., Bourinbaiar, A., Huang, H.I., et al., 1995. Anti-HIV and anti-tumor activities of recombinant MAP30 from bitter melon. Gene 161, 151–156. Available from: https://doi.org/10.1016/0378-1119(95)00186-A.

Lee, T.T.Y., Kashiwada, Y., Huang, L., Snider, J., Cosentino, M., Lee, K.H., 1994. Suksdorfin: an anti-HIV principle from *Lomatium suksdorfii*, its structure–activity correlation with related coumarins, and synergistic effects with anti-AIDS nucleosides. Bioorg. Med. Chem. 2, 1051–1056. Available from: https://doi.org/10.1016/S0968-0896(00)82054-4.

Lee, J.S., Kim, H.J., Lee, Y.S., 2003. A new anti-HIV flavonoid glucuronide from *Chrysanthemum morifolium*. Planta Med. 69, 859–861. Available from: https://doi.org/10.1055/s-2003-43207.

Levy, J., Hoffman, A., Kramer, S., Landis, J., Shimabukuro, J., Oshiro, L., 1984. Isolation of lymphocytopathic retroviruses from San Francisco patients with AIDS. Science. (80–) 225, 840–842. Available from: https://doi.org/10.1126/science.6206563.

Li, B.Q., Fu, T., Dongyan, Y., Mikovits, J.A., Ruscetti, F.W., Wang, J.M., 2000. Flavonoid baicalin inhibits HIV-1 infection at the level of viral entry. Biochem. Biophys. Res. Commun. 276, 534–538. Available from: https://doi.org/10.1006/bbrc.2000.3485.

Li, H.G., Huang, P.L., Zhang, D., Sun, Y., Chen, H.C., Zhang, J., et al., 2010. A new activity of anti-HIV and anti-tumor protein GAP31: DNA adenosine glycosidase—structural and modeling insight into its functions. Biochem. Biophys. Res. Commun. 391, 340–345. Available from: https://doi.org/10.1016/j.bbrc.2009.11.060.

Lin, Y.M., Anderson, H., Flavin, M.T., Pai, Y.H.S., Mata-Greenwood, E., Pengsuparp, T., et al., 1997. In vitro anti-HIV activity of biflavonoids isolated from *Rhus succedanea* and *Garcinia multiflora*. J. Nat. Prod. 60, 884–888. Available from: https://doi.org/10.1021/np9700275.

Ma, C., Nakamura, N., Miyashiro, H., Hattori, M., Komatsu, K., Kawahata, T., et al., 2002. Screening of Chinese and Mongolian herbal drugs for anti-human immunodeficiency virus type 1 (HIV-1) activity. Phytother. Res. 16, 186–189.

Maimone, T.J., Baran, P.S., 2007. Modern synthetic efforts toward biologically active terpenes. Nat. Chem. Biol. 3, 396–407. Available from: https://doi.org/10.1038/nchembio.2007.1.

Masur, H., Michelis, M.A., Greene, J.B., Onorato, I., Stouwe, R.A., Holzman, R.S., et al., 1981. An outbreak of community-acquired *Pneumocystis carinii* pneumonia: initial manifestation of cellular immune dysfunction. N. Engl. J. Med. 305, 1431–1438. Available from: https://doi.org/10.1056/NEJM198112103052402.

McMahon, J.B., Currens, M.J., Gulakowski, R.J., Buckheit, R.W., Lackman-Smith, C., Hallock, Y.F., et al., 1995. Michellamine B, a novel plant alkaloid, inhibits human immunodeficiency virus-induced cell killing by at least two distinct mechanisms. Antimicrob. Agents Chemother. 39, 484–488. Available from: https://doi.org/10.1128/AAC.39.2.484.Updated.

Mehla, R., Bivalkar-Mehla, S., Chauhan, A., 2011. A flavonoid, luteolin, cripples HIV-1 by abrogation of Tat function. PLoS One 6. Available from: https://doi.org/10.1371/journal.pone.0027915.

Min, B.S., Lee, H.K., Lee, S.M., Kim, Y.H., Bae, K.H., Otake, T., et al., 2002. Anti-human immunodeficiency virus-type 1 activity of constituents from *Juglans mandshurica*. Arch. Pharm. Res. 25, 441–445.

Nutan, Modi, M., Goelb, T., Das, T., Malik, S., Suri, S., et al., 2013. Ellagic acid & gallic acid from *Lagerstroemia speciosa* L. inhibit HIV-1 infection through inhibition of HIV-1 protease & reverse transcriptase activity. Indian J. Med. Res. 137, 540–548.

Okamoto, M., Ono, M., Baba, M., 1998. Potent inhibition of HIV type 1 replication by an antiinflammatory alkaloid, cepharanthine, in chronically infected monocytic cells. AIDS Res. Hum. Retroviruses 14, 1239–1245.

Pandit, A., de Boer, R.J., 2016. HIV-1 CCR5 gene therapy will fail unless it is combined with a suicide gene. Sci. Rep. 5, 180–188. Available from: https://doi.org/10.1038/srep18088.

Pasetto, S., Pardi, V., Murata, R.M., 2014. Anti-HIV-1 activity of flavonoid myricetin on HIV-1 infection in a dual-chamber in vitro model. PLoS One 9. Available from: https://doi.org/10.1371/journal.pone.0115323.

Patil, A.D., Freyer, A.J., Eggleston, D.S., Haltiwanger, R.C., Bean, M.F., Taylor, P.B., et al., 1993. The inophyllums, novel inhibitors of HIV-1 reverse transcriptase isolated from the Malaysian tree, *Calophyllum inophyllum* Linn. J. Med. Chem. 36, 4131–4138. Available from: https://doi.org/10.1021/jm00078a001.

Patridge, E., Gareiss, P., Kinch, M.S., Hoyer, D., 2016. An analysis of FDA-approved drugs: natural products and their derivatives. Drug Discov. Today 21, 204–207. Available from: https://doi.org/10.1016/j.drudis.2015.01.009.

Pollakis, G., Paxton, W.A., 2012. Use of (alternative) coreceptors for HIV entry. Curr. Opin. HIV AIDS 7, 440–449. Available from: https://doi.org/10.1097/COH.0b013e328356e9f3.

Prasad, S., Tyagi, A.K., 2015. Curcumin and its analogues: a potential natural compound against HIV infection and AIDS. Food Funct. 6, 3412–3419. Available from: https://doi.org/10.1039/C5FO00485C.

Qian-Cutrone, J., Huang, S., Trimble, J., Li, H., Lin, P.F., Alam, M., et al., 1996. Niruriside, a new HIV REV/RRE binding inhibitor from *Phyllanthus niruri*. J. Nat. Prod. 59, 196–199. Available from: https://doi.org/10.1021/np9600560.

Rashid, M., Gustafson, K., Kashmani, Y., Cardellina, J., Mcmahon, J., Boyd, M., 1995. Anti-HIV alkaloids from *Toddalia asiatica*. Nat. Prod. Lett. 6, 153–156. Available from: https://doi.org/10.1080/10575639508044104.

Rasul, A., Millimouno, F.M., Ali Eltayb, W., Ali, M., Li, J., Li, X., 2013. Pinocembrin: a novel natural compound with versatile pharmacological and biological activities. Biomed. Res. Int. Available from: https://doi.org/10.1155/2013/379850.

Rukachaisirikul, V., Pailee, P., Hiranrat, A., Tuchinda, P., Yoosook, C., Kasisit, J., et al., 2003. Anti-HIV-1 protostane triterpenes and digeranylbenzophenone from trunk bark and stems of *Garcinia speciosa*. Planta Med. 69, 1141–1146. Available from: https://doi.org/10.1055/s-2003-818006.

Sabde, S., Bodiwala, H.S., Karmase, A., Deshpande, P.J., Kaur, A., Ahmed, N., et al., 2011. Anti-HIV activity of Indian medicinal plants. J. Nat. Med. 65, 662–669. Available from: https://doi.org/10.1007/s11418-011-0513-2.

Sancho, R., Márquez, N., Gómez-Gonzalo, M., Calzado, M.A., Bettoni, G., Coiras, M. T., et al., 2004. Imperatorin inhibits HIV-1 replication through an Sp1-dependent pathway. J. Biol. Chem. 279, 37349–37359. Available from: https://doi.org/10.1074/jbc.M401993200.

Sha, O., Niu, J., Ng, T.B., Cho, E.Y.P., Fu, X., Jiang, W., 2013. Anti-tumor action of trichosanthin, a type 1 ribosome-inactivating protein, employed in traditional Chinese medicine: a mini review. Cancer Chemother. Pharmacol. 71, 1387–1393. Available from: https://doi.org/10.1007/s00280-013-2096-y.

Sharp, P.M., Hahn, B.H., 2011. Origins of HIV and the AIDS pandemic. Cold Spring Harb. Perspect. Med. 1. Available from: https://doi.org/10.1101/cshperspect.a006841.

Shaw, G.M., Hunter, E., 2012. HIV transmission. Cold Spring Harb. Perspect. Med. 2. Available from: https://doi.org/10.1101/cshperspect.a006965.

Singh, I.P., Bodiwala, H.S., 2010. Recent advances in anti-HIV natural products. Nat. Prod. Rep. 27, 1781. Available from: https://doi.org/10.1039/c0np00025f.

Singh, I.P., Bharate, S.B., Bhutani, K.K., 2005. Anti-HIV natural products. Curr. Sci. 89, 269–290.

Stevenson, M., 2003. HIV-1 pathogenesis. Nat. Med. 9, 853–860. Available from: https://doi.org/10.1038/nm0703-853.

Sundquist, W.I., Kräusslich, H.G., 2012. HIV-1 assembly, budding, and maturation. Cold Spring Harb. Perspect. Med. Available from: https://doi.org/10.1101/cshperspect.a006924.

Taylor, B.S., Sobieszczyk, M.E., McCutchan, F.E., Hammer, S.M., 2008. The challenge of HIV-1 subtype diversity. N. Engl. J. Med. 358, 1590–1602. Available from: https://doi.org/10.1056/NEJMra0706737.

Tebas, P., Stein, D., Tang, W.W., Frank, I., Wang, S.Q., Lee, G., et al., 2014. Gene editing of CCR5 in autologous CD4 T cells of persons infected with HIV. N. Engl. J. Med. 370, 901–910. Available from: https://doi.org/10.1056/NEJMoa1300662.

Tewtrakul, S., Subhadhirasakul, S., 2003. HIV-1 protease inhibitory substances from the rhizomes of *Boesenbergia pandurata* Holtt. Songklanakarin J. Sci. Technol 2 (25), 503–508.

Thenin-Houssier, S., Valente, S.T., 2016. HIV-1 capsid inhibitors as antiretroviral agents. Curr. HIV. Res. 14, 270–282.

Thoh, M., Kumar, P., Nagarajaram, H.A., Manna, S.K., 2010. Azadirachtin interacts with the tumor necrosis factor (TNF) binding domain of its receptors and inhibits TNF-induced biological responses. J. Biol. Chem. 285, 5888–5895. Available from: https://doi.org/10.1074/jbc.M109.065847.

UNAIDS, 2016. Global AIDS update 2016. World Heal. Organ. 422, https://doi.org/ISBN 978-92-9253-062-5.

United States Food and Drug Administration, 2014. FDA-Approved HIV Medicines [WWW Document]. aidsinfo.nih.gov. http://aidsinfo.nih.gov/education-materials/fact-sheets/21/58/fda-approved-hiv-medicines (accessed 21.10.17).

Van Kiem, P., Van Minh, C., Dat, N.T., Van Kinh, L., Hang, D.T., Nam, N.H., et al., 2010. Aporphine alkaloids, clerodane diterpenes, and other constituents from *Tinospora cordifolia*. Fitoterapia 81, 485–489. Available from: https://doi.org/10.1016/j.fitote.2010.01.005.

Wain-Hobson, S., Sonigo, P., Danos, O., Cole, S., Alizon, M., 1985. Nucleotide sequence of the AIDS virus, LAV. Cell. 40, 9–17. Available from: https://doi.org/10.1016/0092-8674(85)90303-4.

Walker, B.D., Kowalski, M., Goh, W.C., Kozarsky, K., Krieger, M., Rosen, C., et al., 1987. Inhibition of human immunodeficiency virus syncytium formation and virus replication by castanospermine. Proc. Natl. Acad. Sci. U.S.A 84, 8120–8124. Available from: https://doi.org/10.1073/pnas.84.22.8120.

Wan, Z., Chen, X., 2014. Triptolide inhibits human immunodeficiency virus type 1 replication by promoting proteasomal degradation of Tat protein. Retrovirology. 11, 88. Available from: https://doi.org/10.1186/s12977-014-0088-6.

Wang, J.N., Hou, C.Y., Liu, Y.L., Lin, L.Z., Gil, R.R., Cordell, G.A., 1994. Swertifrancheside, an HIV-reverse transcriptase inhibitor and the first flavone-xanthone dimer, from *Swertia franchetiana*. J. Nat. Prod. 57, 211–217. Available from: https://doi.org/10.1021/np50104a003.

Wang, B., Ge, L., Huang, W., Zhang, H., Qian, H., Li, J., et al., 2010. Synthesis and preliminary anti-HIV activities of andrographolide derivatives. Med. Chem. (Los Angeles) 6, 252–258.

Wang, P., Alvarez-Perez, J.-C., Felsenfeld, D.P., Liu, H., Sivendran, S., Bender, A., et al., 2015. A high-throughput chemical screen reveals that harmine-mediated inhibition of DYRK1A increases human pancreatic beta cell replication. Nat. Med. 21, 383–388. Available from: https://doi.org/10.1038/nm.3820.

Wilen, C.B., Tilton, J.C., Doms, R.W., 2012. HIV: cell binding and entry. Cold Spring Harb. Perspect. Med. 2. Available from: https://doi.org/10.1101/cshperspect.a006866.

Woodham, A.W., Skeate, J.G., Sanna, A.M., Taylor, J.R., Da Silva, D.M., Cannon, P.M., et al., 2016. Human immunodeficiency virus immune cell receptors, coreceptors, and cofactors: implications for prevention and treatment. AIDS Patient Care STDS 30, 291–306. Available from: https://doi.org/10.1089/apc.2016.0100.

Wu, T.-S., Su, C.-R., Lee, K.-H., 2012. Cytotoxic and anti-HIV phenanthroindolizidine alkaloids from *Cryptocarya chinensis*. Nat. Prod. Commun. 7, 725–727. Available from: https://doi.org/10.1016/j.surg.2006.10.010.Use.

Xu, H.X., Zeng, F.Q., Wan, M., Sim, K.Y., 1996. Anti-HIV triterpene acids from *Geum japonicum*. J. Nat. Prod. 59, 643–645. Available from: https://doi.org/10.1021/np960165e.

Zhu, T., Korber, B.T., Nahmias, A.J., Hooper, E., Sharp, P.M., Ho, D.D., 1998. An African HIV-1 sequence from 1959 and implications for the origin of the epidemic. Nature 391, 594–597. Available from: https://doi.org/10.1038/35400.

SECTION 2B

Mechanism of Action Plant Derived Products/Medicine

CHAPTER 12

Current Strategy to Target Bacterial Quorum Sensing and Virulence by Phytocompounds

Fohad Mabood Husain[1], Nasser A. Al-Shabib[1], Saba Noor[2], Rais Ahmad Khan[3], Mohammad Shavez Khan[4], Firoz Ahmad Ansari[4], Mohd Shahnawaz Khan[5], Altaf Khan[6] and Iqbal Ahmad[4]

[1]Department of Food Science and Nutrition, College of Food and Agriculture, King Saud University, Riyadh, Kingdom of Saudi Arabia
[2]Rajiv Gandhi Centre for Diabetes and Endocrinology, Jawaharlal Nehru Medical College and Hospital, Aligarh Muslim University, Aligarh, Uttar Pradesh, India
[3]Department of Chemistry, College of Science, King Saud University, Riyadh, Kingdom of Saudi Arabia
[4]Department of Agricultural Microbiology, Aligarh Muslim University, Aligarh, Uttar Pradesh, India
[5]Department of Biochemistry, College of Science, King Saud University, Riyadh, Kingdom of Saudi Arabia
[6]Central Laboratory Research Center, College of Pharmacy, King Saud University, Riyadh, Kingdom of Saudi Arabia

12.1 INTRODUCTION

12.1.1 Quorum Sensing in Bacteria: An Overview

Bacterial quorum sensing (QS) is density-dependent cell-to-cell communication system that involves production, detection, and response to signaling molecules known as autoinducers (AIs). As the bacterial population density increases, AIs accumulate, bacteria keep track of this information to monitor changes in their cell numbers and alter gene expression collectively (Rutherford and Bassler, 2012). QS controls wide array of genes that direct number of beneficial activities when performed by groups of bacteria acting in synchrony. QS controls important functions like bioluminescence, sporulation, competence, antibiotic production, virulence factor secretion and biofilm formation (Ng and Bassler, 2009; Williams and Camara, 2009). Production of various virulence factors in pathogenic bacteria of different origins is controlled by QS (Table 12.1). Although all known QS systems differ in the regulatory components and molecular mechanisms but are dependent on three basic principles: first, secretion of signaling molecules (AIs); second, detection of AIs by the receptors existing in the cytoplasm or in the membrane;

New Look to Phytomedicine
DOI: https://doi.org/10.1016/B978-0-12-814619-4.00012-4

Table 12.1 Quorum-sensing-regulated virulence factors in bacterial pathogens of humans, animals, and plants

Pathogens	Autoinducers	QS-regulated phenotypes	References
Human pathogens			
Escherichia coli O157:H7	Indole	Biofilm formation, motility	Bansal et al. (2007)
Helicobacter pylori	AI-2	Motility	Rader et al. (2007)
Pseudomonas aeruginosa	OdDHL, BHL, PQS	Elastase, protease, hemolysin, rhamnolipids, virulence	Lee and Zhang (2015)
	Indole	Biofilm formation, motility	Lee et al. (2009)
Vibrio cholera	AI-2, CAI-1	Biofilm formation, protease	Ng and Bassler (2009)
	Indole	Biofilm formation, motility	Mueller et al. (2009)
Animal pathogens			
Aeromonas hydrophila	BHL	Protease, virulence	Natrah et al. (2012)
Photorhabdus luminescens	Dialkylresorcinols	Virulence	Brameyer et al. (2015)
Vibrio harveyi	OH-BHL, AI-2, CAI-1	Protease, type III secretion, siderophore, virulence	Defoirdt et al. (2008)
	Indole	Biofilm formation, motility, virulence	Yang et al. (2017)
Plant pathogens			
Pectobacterium carotovorum	HHL, OHHL, OOHL	Extracellular cell wall–degrading enzymes	Helman and Chernin (2015)
Pseudomonas syringae	OHHL	Extracellular polysaccharides, motility, virulence	Helman and Chernin (2015)
Xanthomonas campestris	DSF	Extracellular polysaccharides, biofilm formation, virulence	Zhou et al. (2017)

AI-2, autoinducer-2; *OdDHL*, *N*-(3-oxododecanoyl)-L-homoserine lactone; *BHL*, *N*-butanoyl-L-homoserine lactone; *PQS*, Pseudomonas quinolone signal; *CAI-1*, cholerae autoinducer-1; *OH-BHL*, *N*-(3-hydroxybutanoyl)-L-homoserine lactone; *HHL*, *N*-hexanoyl-L-homoserine lactone; *OHHL*, *N*-(3-oxohexanoyl)-L-homoserine lactone; *OOHL*, *N*-(3-oxo-octanoyl)-L-homoserine lactone; *DSF*, diffusible signal factor; *QS*, quorum sensing.

and third, activating expression of genes necessary for cooperative behaviors, detection of AIs resulting in AI production (Novick et al., 1995; Seed et al., 1995).

Gram-positive and Gram-negative bacteria use different types of QS systems. Gram-positive bacteria use peptides, called autoinducing peptides (AIPs), as signaling molecules, that are detected by membrane-bound two-component signal transduction system (Havarstein et al., 1995). Once produced in the cells, AIPs are processed and secreted. When the extracellular concentration of AIP reaches a threshold level, it binds to the cognate two-component histidine kinase receptor. This binding usually activates the receptor's kinase activity, it autophosphorylates and passes phosphate to a cognate cytoplasmic response regulator. The phosphorylated response regulator activates transcription of genes in the QS regulon. In some cases, AIPs are transported back to the cell cytoplasm where they interact with transcription factors to modulate the activity of transcription factor and, in turn, modulate gene expression changes (Rutherford and Bassler, 2012). Examples of QS-regulated behaviors in Gram-positive bacteria are competence in *Streptococcus pneumoniae* and *Bacillus subtilis* and sporulation in *B. subtilis* (Kleerebezem et al., 1997). QS also controls virulence factor production in pathogens including *Staphylococcus aureus*, *Listeria monocytogenes*, *Enterococcus faecalis*, and *Clostridium perfringens* (Autret et al., 2003; Podbielski and Kreikemeyer, 2004; Ohtani et al., 2009; Riedel et al., 2009; Thoendel and Horswill, 2010).

Gram-negative bacteria communicate using small molecules as signaling molecules (AIs). These include acyl-homoserine lactones (AHLs), AI-2, quinolones, indole, pyrones, and dialkylresorcinol (LaSarre and Federle, 2013). Gram-negative bacteria employ LuxI/LuxR-type QS systems. In this, LuxI homolog is an AI synthase that catalyzes reaction between *S*-adenosylmethionine (SAM) and an acyl carrier protein (ACP) to produce freely diffusible AHL molecules (Schaefer et al., 1996). At high concentrations, AHL binds to cognate cytoplasmic LuxR-like transcription factor. If the binding does not take place, LuxR-type proteins degrade rapidly, possibly to prevent the bacteria from "short circuiting" their QS systems. AHL binding stabilizes the LuxR-type proteins, allowing them to fold, bind DNA, and activate transcription of the target genes (Zhu and Winans, 1999, 2001). LuxI/LuxR homologs have been identified in most of the Gram-negative bacteria (Case et al., 2008). AHLs produced by different bacteria differ in the side-chain length and side-chain arrangement. Acyl chains ranging from C4 to C18 have identified along with modifications like the presence of carbonyl and hydroxy moieties at C3 position (Fuqua et al., 2001; Ng and Bassler, 2009). This diversity in the AHLs promotes intraspecies-specific cell–cell communication in bacteria. The substrate-binding pockets of various LuxI homologs have correspondingly different shape and sizes, ensuring that only a particular ACP is accommodated for a particular AI (Watson et al., 2002; Gould et al., 2004). Similarly, the AI detecting LuxR homologs possess unique binding sites that accommodate AHL ligands of a particular type (Zhang et al., 2002; Chen et al., 2011).

12.1.2 Quorum-Sensing Interference Strategies

QS being a community-based bacterial behavior and not essentially a survival requirement can be inhibited and interrupted for a desired trait like virulence factor secretion, biofilm formation, and drug resistance. Therefore, development of antipathogenic drug has gained relevance in the possible treatment of bacterial diseases by the attenuation/inhibition of QS in bacteria (Hentzer and Givskov, 2003). There are many ways to inhibit QS in bacteria. QS pathways provide three major points for the QS inhibitor/ligand binding: the signal generator, the signal molecule, and the signal receptor. These three strategies for inhibiting QS are discussed below.

Interference with the signal generation: AHLs are used as signaling molecules in QS pathways by most of the Gram-negative bacteria. The AHL synthase produces the AHL from corresponding acyl chain derived from fatty acid biosynthesis pathway and SAM (Schaefer et al., 1996). Barring the short-chain AHLs which are freely diffusible, the long-chain ones are transported by efflux pumps across the bacterial membrane (Pearson et al., 1999). The QS inhibitor can be designed to block the activity of the efflux pumps, fatty acid biosynthesis, SAM biosynthesis, AHL synthase protein synthesis. Hoang and Schweizer (1999) demonstrated various analogs of SAM, like *S*-adenosylcysteine, *S*-adenosylhomocysteine, and sinefungin, to be potent inhibitors of AHL synthesis catalyzed by the *Pseudomonas aeruginosa* RhlI protein. *S*-adenosylcysteine inhibited activity of the *P. aeruginosa* LuxI homolog RhlI by up to 97% (Parsek et al., 1999).

Degradation of the signal molecules: The QS-based bacterial communication can be inhibited by a reduction in the concentration of active signal molecule in the environment. Complete degradation or inactivation of the AHL signal molecules can be mediated chemically or by enzymatic decay or metabolism. Several enzymes such as AHL-acylase, AHL-lactonase, and paraoxonase enzymes play a role in the degradation of AHL. This kind of enzymatic disruption of QS is typically termed quorum quenching (QQ). Some bacteria for gaining selective advantage block their competitors QS pathways by enzymatic biodegradation, whereas some bacteria such as *P. aeruginosa* PAI-A and *Variovorax paradoxus* can use AHLs as sole source of energy, carbon, and nitrogen (Huang et al., 2003; Bhardwaj et al., 2013).

Interference of the signal reception: QS signal transduction inhibition can be executed by an antagonistic molecule capable of competing or interfering with the AHL signal molecule for binding to the corresponding receptor protein molecule. These can be achieved either by competitive or noncompetitive inhibitors. Several synthetic analogs have been designed by modifications in the acyl side chain or in the lactone ring or in both these moieties of the AHL molecule. The modifications of the natural halogenated furanones are also used and proved to be potential antagonists (Wu et al., 2004).

The AI-2 and AI-3 pathways can also be targeted similarly like inhibition of AI-1. These QS systems provide target for design and development of broad-spectrum inhibitors. Inhibitors targeting QseC, the receptor molecule for AI-3 found in numerous important animal and plant pathogens have potential as broad-spectrum drugs (Rasko et al., 2008; Ni et al., 2009).

12.1.3 Antiquorum Sensing Agents and Their Limitations

Anti-QS agents were first characterized in the red marine alga, *Delisea pulchra*. This alga was investigated for its antifouling properties and was found to contain halogenated furanones, compounds which block AHLs via competitive inhibition and destabilization of LuxR. The structural similarity allows furanones to competitively inhibit the action of AHL signaling molecules (Manefield et al., 2002). A well-studied furanone isolated from *D. pulchra* has been shown to have direct effect on biofilm formation is (5Z)-4-bromo-5-(bromomethylene)-3-butyl-2-(5*H*)-furanone (Fig. 12.1). Initially, Ren et al. (2001) have shown that the furanone inhibited swarming and biofilm formation of *Escherichia coli* XL1-blue on mild steel surfaces with little toxicity to the bacteria. AHL as well as AI-2-based QS systems are inhibited by furanones as these compounds are

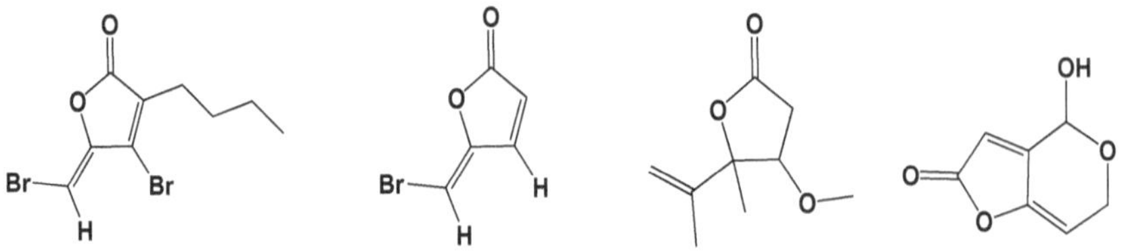

Natural brominated furanone **Furanone 56 (synthetic)** **Penicillic acid** **Patulin**

Figure 12.1 Early known inhibitors of QS from algal and fungal origins. *QS*, quorum sensing.

structural mimics of QS signals (lactones and tetrahydrofuran rings) in both these QS systems (Ren et al., 2001). Though initially it was believed that furanones competitively inhibit the binding of AIs to their receptors, it is now evident that halogenated furanones destabilize and accelerate the turnover of LuxR in *Vibrio fischeri* and *Vibrio harveyi*. This impairs the ability of LuxR to bind DNA and initiate transcription (Lowery et al., 2008).

The improved furanone compound, furanone 4, can increase the susceptibility of *P. aeruginosa* biofilm to tobramycin (Pan and Ren, 2009). Brominated furanones have also been found to be effective against Gram-positive bacteria as well as fungi (Holmstrom and Kjelleberg, 2001; Pan and Ren, 2009). While natural furanones have shown considerable Quorum sensing inhibitory (QSI) activity in some species, a number of synthetic analogs have been shown to inhibit biofilm formation. One such compound called furanone 56 (Fig. 12.1) was first demonstrated by Hentzer et al. (2002) to inhibit QS in *P. aeruginosa*. Furanone 56, unlike many other active furanones, lacks both a side chain and bromine on the furanone ring. It was shown to have little effect on bacterial growth, protein synthesis, and early biofilm formation. However, it was demonstrated that it could penetrate the biofilm matrix and interfere with biofilm maturation, presumable by disrupting QS gene expression. The effectiveness of furanone 56, as well as that of a related furanone, for the treatment of lung infections in mice models has also been demonstrated (Wu et al., 2004). Both compounds have been shown to disrupt AHL QS, ultimately resulting in accelerated lung clearance and prolonged survival time of the mice. Since most of these furanones contain halogens, therefore, are unsuitable for human use. The furanones investigated are too reactive and may be too toxic for treatment of bacterial infections in human (Hentzer and Givskov, 2003; Kociolek, 2009).

Fungi are known to produce secondary metabolites like antibiotics. Around 33 *Penicillium* spp. have been found to produce QS inhibitory compounds—patulin and penicillic acid (Fig. 12.1) (Rasmussen et al., 2005). In mouse pulmonary infection model, the use of patulin could significantly reduce the infections caused by *P. aeruginosa* although not suitable for human use due to toxicity issues.

12.2 DISCOVERY AND EXPLORATION OF QUORUM-SENSING INHIBITORS FROM MEDICINAL PLANTS

Medicinal plants and their secondary metabolites are known to be effective against infection causing pathogens even at low concentrations (Ahmad and Beg, 2001). Mechanism of action is well understood, and these metabolites act on various target sites of pathogen including the cell membrane, cell wall, and respiratory pathways (Bouhdid et al., 2009). The emergence of drug-resistant pathogens has prompted the research community to assess the action of plants and their products against QS in order to limit this resistance. Several findings have reported that targeting QS is an effective strategy for combating infections caused by biofilm formation (Musk and Hergenrother, 2006). Researchers are increasingly focusing on the medicinal plants and their phytoconstituents to search for novel antipathogenic drugs that could be exploited as nontoxic QS inhibitors and, thus, control persistent infections (Hentzer and Givskov, 2003).

Extracts of various plant parts have been found to inhibit QS because of the similarity in their chemical structure to those of AI molecules (AHL) and also because of their ability to degrade signal receptors (LuxR/LasR) (Vattem et al., 2007; Teplitski et al., 2011). Substances with anti-QS properties are either present in the whole plant or in one of the plant parts like leaves, root, seeds, bark, and flowers. Some of the medicinals possessing anti-QS property are enlisted in Table 12.2.

Total of 24 Indian medicinal plants were screened for their anti-QS property, out of which extracts of *Hemidesmus indicus* (root), *Holarrhena antidysenterica* (bark), *Mangifera indica* (leaves), *Punica granatum* (Pericarp), and *Psoralea corylifolia* (seed) demonstrated varying levels of violacein inhibition in the reporter strains (*Chromobacterium violaceum* 12472 and *C. violaceum* CVO26). Significant reduction in the swarming motility of *P. aeruginosa* PAO1 treated with these extracts was also recorded (Zahin et al., 2010). QS inhibition has also been reported in various species of

Table 12.2 Anti-quorum activity of some medicinal plants

Plant species (family)	Products	Major compounds	Strains tested	Effects with target	References
Glycyrrhiza glabra (Fabaceae)	Methanol extract	Flavonoids (licoricone, glycyrin, gylzyrin)	*Acinetobacter baumannii*	Diminution of biofilms formation and motility	Bhargava et al. (2015)
Terminalia chebula (Combretaceae)	Fruit extract	Ellagic acid (benzoic acid)	*Burkholderia cepacia*	Reduction of biofilm	Huber et al. (2003)
Commiphora leptophloeos (Burseraceae)	Extract (stem bark)	ND	*Staphylococcus epidermidis*	Inhibition of virulence and biofilm	da Silva Trentin et al. (2011)
Pityrocarpa moniliformis (Leguminosae)	Extract (leaves)	ND	*Staphylococcus epidermidis*	Inhibition of virulence and biofilm	da Silva Trentin et al. (2011)
Bauhinia acuruana (Leguminosae)	Extract (branches, fruits)	ND	*Staphylococcus epidermidis*	Inhibition of virulence and biofilm.	da Silva Trentin et al. (2011)
Terminalia catappa (Combretaceae)	Methanolic extract (leaf)	ND	*Chromobacterium violaceum and Pseudomonas aeruginosa*	Inhibition of QS-controlled violacein production	Taganna et al. (2011)
Rubus rosaefolius (Rosaceae)	Phenolic extracts	ND	*Chromobacterium violaceum, Aeromonas hydrophila, and Serratia marcescens*	Inhibition of violacein production, swarming motility, and biofilms formation	Oliviera et al. (2016)
Centella asiatica (Apiaceae)	Flavonoid-rich fraction	ND	*Pseudomonas aeruginosa* PAO1 and *Chromobacterium violaceum* ATCC12472	Inhibition of violacein, pyocyanin production, elastolytic and proteolytic activities, swarming motility, and biofilms formation	Vasavi et al. (2016)

Sclerocarya birrea (Anacardiaceae)	Methanolic extract (stem bark)	ND	*Pseudomonas aeruginosa*	Reduction of swarming, motility, and virulence factors production	Sarkar et al. (2014)
Ocimum sanctum (Lamiaceae) *Ananas comosus* (Bromeliaceae) *Musa paradisiaca* (Musaceae) *Manilkara zapota* (Sapotaceae)	Aqueous extracts	ND	*Chromobacterium violaceum and Pseudomonas aeruginosa*	Inhibition violacein, pyocyanin pigment, protease, elastase production, and biofilm formation	Musthafa et al. (2010)
Panax notoginseng (Araliaceae)	Extract (flower and root)	ND	*Chromobacterium violaceum and Pseudomonas aeruginosa*	Interference with violacein production and swarming motility, suppression of LasA and LasB production downregulation of AHLs molecules production	Nyila et al. (2012)
Amomum tsaoko (Zingiberaceae)	Ethanol extract	ND	*Salmonella typhimurium, S. aureus, and P. aeruginosa*	Reduction of biofilm formation	Rahman et al. (2017)
Vernonia blumeoides (Asteraceae)	Hexane, dichloromethane, ethyl acetate, and methanol extracts	-2-(octadeca-9Z, 12Z-dienyloxy); -bufa-20,22-dienolide, catechol; -3,5-stigmastadien-7-one	*Chromobacterium violaceum and Agrobacterium tumefaciens*	Inhibition of violacein production and signal synthesis	Aliyu et al. (2016)

(*Continued*)

Table 12.2 (Continued)

Plant species (family)	Products	Major compounds	Strains tested	Effects with target	References
Nymphaea tetragona (Nymphaeaceae)	Aqueous extract	ND	*Chromobacterium violaceum and Pseudomonas aeruginosa*	Inhibition of violacein production, swarming motility, reduction in pyocyanin production and LasA protease	Hossain et al. (2015)
Cecropia pachystachya (Cecropiaceae)	Aqueous extract	*C*-Glycosyl flavonoids	*Chromobacterium violaceum and Escherichia coli*	Inhibition of QS	Brango-Vanegas et al. (2014)
Mangifera indica	Methanol extract	ND	*Chromobacterium violaceum, P. aeruginosa, and A. hydrophila*	Violacein, elastase, protease, chitinase, EPS, swarming motility, and biofilm inhibition	Husain et al. (2017)
Allium cepa	Ethyl acetate fraction (peel)	Quercetin 4′-*O*-β-D glucopyranoside	*Chromobacterium violaceum, P. aeruginosa*, and *A. hydrophila*	Violacein, elastase, protease, chitinase, EPS, swarming motility, and biofilm inhibition	Al-Yousef et al. (2017)
Schinus terebinthifolia	Flavone-rich fruit extract	ND	*S. aureus*	*agr*-Quenching, inhibition of toxin production	Muhs et al. (2017)
Senegalia nigrescens	Extract	Flavonoids and terpenoids	*C. violaceum*	Violacein inhibition	Bodede et al. (2018)

QS, Quorum sensing; *AHL*, acyl-homoserine lactone; *EPS*, exopolysaccharide; *ND*, Not Detected.

plants including pea seedling (Fatima et al., 2010) and roots of *Panax ginseng, Areca catechu*, and *Panax notoginseng* (Song et al., 2010; Koh and Tham, 2011). Aqueous extracts of edible plants and fruits such as *Ananas comosus*, *Musa paradisiaca*, *Manilkara zapota*, and *Ocimum sanctum* have proved to be effective QSIs by inhibiting violacein production in *C. violaceum* and pyocyanin pigment, staphylolytic protease, elastase production, and biofilm formation in *P. aeruginosa* PAO1 (Musthafa et al., 2010). Hexane extract of clove bud (*Syzygium aromaticum*) demonstrated significant anti-QS activity against two strains of *P. aeruginosa* (Krishnan et al., 2012). Fruit extract of *Lagerstroemia speciosa* is reported to attenuate QS-related genes (*las* and *rhl*) and inhibit virulence factors: LasA protease, LasB elastase, and pyoverdin production in PAO1 (Singh et al., 2012). Methanolic extract of *Rhizophora annamalayana* (bark) exhibited reduction in QS-dependent production of violacein and bioluminescence in *C. violaceum* and aquatic pathogen *V. harveyi*, respectively (Musthafa et al., 2013). Ethanol fractions of medicinal plants like *Adhatoda vasica, Bauhinia purpurea, Lantana camara, Myoporum laetum, Piper longum*, and *Taraxacum officinale* showed significant reduction in pigment produced by *C. violaceum* strain (Zaki et al., 2013). In another study, methanolic extract of *Terminalia chebula* fruits showed QS inhibitory activity using *Agrobacterium tumefaciens* bioreporter strain, and the extract downregulated the expression of LasI/R and rhlI/R genes (Sarabhai et al., 2013).

Extracts of three east European medicinal plants, namely, *Quercus robur* (oak) cortex, *Betula verrucosa* (birch) buds, and *Eucalyptus viminalis* (Manna Gum) leaves showed QS inhibition in *C. violaceum* 31532 and its derived bioreporter strain *C. violaceum* NCTC 13274 (Tomlacheva et al., 2014). Standardized methanol extract of *Sclerocarya birrea* (stem) significantly disrupted the QS-mediated production of biofilm and reduced motility of the PAO1 cells. The extract displayed a regulatory role on the secretion of protease and pyoverdin, two QS-dependent pathogenic factors found in *P. aeruginosa* (Sarkar et al., 2014). *Trigonella foenum-graceum* (fenugreek) is reported to inhibit AHL-regulated production of virulence factors and biofilm in *P. aeruginosa* PAO1 and *Aeromonas hydrophila*. Application of the extract resulted in reduced AHL levels and subsequent downregulation of *lasb* gene. In vivo studies on *Caenorhabditis elegans* nematode model showed enhanced survival after treatment with 1 mg/mL concentration of *T. foenum-graceum* extract (Husain et al., 2015a,b). QS modulatory activity of *Nymphaea tetragona* (methanol extract) was demonstrated against *C. violaceum* and *P. aeruginosa*. The extract at sub-MICs reduced violacein production significantly, while in more than 70% reduction in swarming motility of *P. aeruginosa* was also recorded. Biofilm formation in PAO1 was reduced considerably and the extract was found to be nontoxic using in vitro and in vivo cytotoxicity assays (Hossain et al., 2015). In another study, 12 tannin-rich crude extracts of Indian medicinal plants were assessed for QS inhibition activity. Extracts of *Phyllanthus emblica*, *Terminalia bellirica*, *T. chebula*, *P. granatum*, *Syzygium cumini*, and *M. indica* (flower) were found to interfere with the QS system of Gram-negative (*C. violaceum*) and Gram-positive (*S. aureus*) bacteria effectively over a wide range of subinhibitory concentrations (Shukla and Bhathena, 2016). Bacha et al. (2016) demonstrated that methanol extract of root of *Albizia schimperiana* and petroleum ether extract of *Justicia schimperiana* (seed) quenched QS in *E. coli* reporter strain AI1-QQ.1. Five fractions of *M. indica* (leaves) were screened for their anti-QS property against CV12472 bioreporter strain. Methanol fraction was found to be most active against CV12472 and findings were confirmed using mutant strain CVO26. *M. indica* (leaves) extract demonstrated concentration-dependent reduction in virulence functions of *P. aeruginosa* and *A. hydrophila*. QS-regulated biofilm formation in both the bacterial pathogens was inhibited significantly. In vivo antipathogenic potential of the methanol fraction was also determined and considerably reduced mortality of nematode infected with PAO1 was observed (Husain et al., 2017). Ethyl acetate fraction of onion peel extract (ONPE) inhibited the production of QS-mediated virulence factors such as violacein in *C. violaceum* and elastase, pyocyanin in *P. aeruginosa*. ONPE also reduced QS-controlled biofilm formation, exopolysaccharide (EPS) production and swarming motility in *P. aeruginosa* and *A. hydrophila* (Al-Yousef et al., 2017). In a study conducted on flavone-rich extract of *Schinus terebinthifolia* (Brazilian Peppertree), it reported that the extract inhibited all *S. aureus* accessory gene regulator (*agr*) alleles without affecting the growth of the pathogen. IC_{50} values ranging from 2 to 32 μg/mL were observed for *agr*-quenching activity in transcriptional reporters (Muhs et al., 2017).

12.3 PHYTOCOMPOUNDS IDENTIFIED AS QUORUM-SENSING INHIBITORS

Ideally, QS inhibitors should be molecules with low molecular mass causing significantly reduced expression of QS-regulated genes and should be specific for specific QS regulator and at the same time should not be toxic

(Rasmussen and Givskov, 2006). In addition, these inhibitors should not be degraded by the host metabolic system (Kalia, 2013). Considering the anti-QS properties of array of plants, it is expected that these plants contain compounds with the abovementioned properties. Therefore, QS inhibitors from phytoconstituents can be classified into following classes.

12.3.1 Alkaloids and Derivatives

Using the *C. violaceum* test system for screening anti-QS activity, Norizan et al. (2013) observed that caffeine inhibited violacein production. Further, they found that caffeine was also active in interfering with QS in *P. aeruginosa* PAO1 by suppressing the production of AHL molecules. Husain et al. (2015a,b) also reported similar findings with caffeine against QS-regulated virulence functions of *C. violaceum*, *P. aeruginosa*, and *A. hydrophila*. In another study, Monte et al. (2014) found that indole alkaloid indole-2-carbinol reduced pigment production in the biosensor strain *C. violaceum*.

Tomatidine, a steroidal alkaloid, was assessed for its QS inhibition activity against *S. aureus*. This alkaloid was found to block the expression of various genes that are regulated by QS accessory gene regulator (*agr*) system, leading to altered production of virulence factors like hemolysis production (Mitchell et al., 2011). Another alkaloid capsaicin significantly reduced the production of α-toxin by methicillin-resistant *S. aureus (MRSA)* (Qiu et al., 2012). Treatment of MRSA cells with capsaicin resulted in decreased expression of RNAIII. This observation suggested that the reduced production of α-toxin might be due to the inhibition of *agr* system.

12.3.2 Organosulfur Compounds and Derivatives

Violacein inhibition in a specific *C. violaceum* strain was observed with allyl isothiocyanate, benzyl isothiocyanate, and 2-phenyl isothiocyanate. Although these compounds reduced AHL production, these were cytotoxic against mouse lung fibroblast (Borges et al., 2014).

Leng et al. (2011) demonstrated that allicin reduced the production of α-hemolysin by *S. aureus*. The compound reduces α-hemolysin production by interfering with *agr* QS system of the pathogen.

Iberin, at subinhibitory concentrations, inhibited the expression of *lasB-gfp* fusion in *P. aeruginosa* and showed competence with AHL molecules of regulator molecules. Iberin treatment resulted in significant downregulation of 49 genes controlled by QS (Jakobsen et al., 2012a). Another study conducted by the same group showed that QS disruption in *P. aeruginosa* by garlic extract was due to ajoene (an organosulfur compound). DNA microarray analysis revealed that QS-regulated virulence factors like LasA protease, chitinase, and rhamnolipids were attenuated in a concentration-dependent manner (Jakobsen et al., 2012b).

Ganin et al. (2013) demonstrated that sulforaphane inhibited LasR activation plausibly by binding itself with LasR. Erucin also attenuated QS but the effects of this compound were less pronounced.

12.3.3 Other Aliphatic and Cyclic Compounds

Three cinnamaldehyde analogs, namely, *trans*-2-nonenal, *trans*-3-decen-2-one, and *trans*-3-nonen-2-one were reported to interfere with AI-2 QS; in *Vibrio* spp., *trans*-2-nonenal and *trans*-3-decen-2-one inhibited the AI-2-based QS system by reducing the DNA-binding ability of LuxR, leading to reduced production of QS-regulated virulence functions such as biofilm formation, matrix production, and protease production (Brackman et al., 2011). Ahmad et al. (2015) reported the violacein inhibitory activity of *cis*-3-nonen-1-ol in *C. violaceum*. They also assayed the anti-QS property of several compounds found in essential oil using *C. violaceum* and *P. aeruginosa* bioreporter strains. Two AHL mimics, estragole and *p*-anisaldehyde, were found to inhibit violacein production.

12.3.4 Phenolics

12.3.4.1 Coumarins

Anti-QS activity of coumarin was analyzed using three model bioreporter strains, viz., *Serratia marcescens* for short-chain AHLs, *C. violaceum* for medium-length AHLs and *A. tumefaciens* for long-chain AHLs. Coumarin demonstrated broad-spectrum AHL inhibition and reduced the expression of *pqsA* and *rhlI* genes. Reduction in *Aliivibrio fischeri* bioluminescence, a function regulated by AI-2 QS system, was also observed after treatment with coumarin (Gutiérrez-Barranquero et al., 2015).

Drop in bioluminescence of two *V. harveyi* strains was recorded after treatment with furocoumarins, dihydroxybergamottin, and bergamottin (Girennavar et al., 2008). Derivatives of coumarin like dimethylesculetin and 7-hydroxycoumarin have also been reported as inhibitor of QS-regulated violacein production in *C. violaceum* (Truchado et al., 2012; Monte et al., 2014).

12.3.4.2 Flavonoids

Vandeputte et al. (2011) screened flavonoids for QS inhibition and found that chalcone *trans*-benzylideneacetophenone reduced expression of all QS genes except *lasI*. Three other compounds of the same class, 2,3,5-trihydroxy-4,6,3-trimethoxychalcone, 2,3-dihydroxy-4,6-dimethoxychalchone, and 2,4,4-trihydroxy-3,6-dimethoxychalchone, demonstrated significant interference with QS signaling of *V. harveyi* at low concentrations.

Investigations on the leaves and bark extract of *Combretum albiflorum* revealed that a flavanol catechin inhibited violacein in *C. violaceum*. This compound also reduced the expression of QS-controlled genes such as *lasB*, *rhlA*, *lasI*, *lasR*, *rhlI*, and *rhlR*. The author suggested that catechin might inhibit QS by affecting the detection of AHL signal by RhlT (Vandeputte et al., 2010). In other studies, related compounds such as epicatechin, epigallocatechin, and epigallocatechin gallate demonstrated violacein inhibition, reduced AI-2 signaling, and AHL inhibition, respectively in different QS test system (Huber et al., 2003; Borges et al., 2014; Castillo et al., 2015).

Citrus flavonoids naringenin, naringin, neohesperedin, neoeriocitrin, apigenin, sinensetin, quercetin, kaempferol, and rutin were found to inhibit either HAI-1 (AHL interspecies signal) or AI-2-regulated bioluminescence in *V. harveyi* biosensor strain. Naringin, neohesperedin demonstrated strong inhibition of HAI-1 signaling, while sinensetin reduced bioluminescence effectively (Vikram et al., 2010).

Licochalcone A and E reduced toxin secretion in *S. aureus* by inhibiting *agrA* in a concentration-dependent manner (Qiu et al., 2010a; Zhou et al., 2012). Similarly, luteolin and farrerol caused sixfold reduction in the expression of *agr* locus suggesting that α-toxin production in *S. aureus* is affected by interference with *agr* regulatory system (Qiu et al., 2011a, 2011b).

Truchado et al. (2012) described the QS inhibition by isoflavone daidzein in *C. violaceum*. Rutin, a flavanol reduced the AHL levels in *Yersinia enterocolitica* and *Erwinia carotovora* by inhibiting their synthesis. Brango-Vanegas et al. (2014) isolated *C*-glycosyl flavones isovexitin, vixetin, isoorientin, orientin, and rutin from *Cecropia pachystachya*. Rutin was found to be active in both *E. coli* and *C. violaceum* in QS test system, while vixetin and orientin were most active against *C. violaceum* and *E. coli*, respectively. Flavonoids licoricone, glycyrin, and glyzarin were identified for reduction in QS-mediated virulence of *Acinetobacter baumannii*, and they acted by downregulating the expression of AI synthase, *abaI* which reduced the levels of 3-OH-C12-HSL (Bhargava et al., 2015).

The flavonoids quercetin, quercetin-3-*O*-arabinoside and quercetin 4′-*O*-β-D glucopyranoside, have been reported to inhibit QS-regulated virulence functions in *C. violaceum* and *P. aeruginosa* (Vasavi et al., 2014; Gopu et al., 2015a; Al-Yousef et al., 2017).

Molecular docking analysis revealed malvidin as the phytoconstituent responsible for the QS inhibitory activity of *S. aromaticum*. Docking studies showed that interaction between LasR receptor protein and malvidin were hydrophobic. These findings of the in silico study were confirmed in vitro using *C. violaceum* test system (Gopu et al., 2015b).

Maisuria et al. (2016) demonstrated anti-QS activity of cranberry extract rich in proanthocyanidins against *P. aeruginosa* in *Drosophila melanogaster* model. It decreased the production of AHL-regulated virulence and protected *D. melanogaster* from infection. Proanthocyanidin-rich extract showed reduced production of AHLs, and AHL synthases LasI/RhlI and QS transcriptional regulators LasR/RhlR genes were inhibited and antagonized, respectively.

12.3.4.3 Phenolic Acids

QS inhibition by phenolic acids such as gallic acid and vanillin has been described by Truchado et al. (2012). In another study, gallic acid reduced pigment production but these concentrations were found toxic to mouse lung fibroblast (Borges et al., 2014). In another study, Ponnusamy et al. (2009) examined anti-QS activity of vanillin in *C. violaceum* and *A. tumefaciens*. They observed that vanillin inhibited both short-chain and long-chain AHLs. Another phenolic acid, salicylic acid, has also been reported for QS inhibition in *C. violaceum*, *E. coli*, and *P. aeruginosa* bioreporter strains (Bandara et al., 2006; Chang et al., 2014).

Chong et al. (2011) isolated a resorcinol known as malabaricone C from the *Myristica cinnamomea* bark extract. Malabaricone C demonstrated anti-QS activity by reducing the production of violacein pigment in *C. violaceum*.

Methyl gallate suppressed both the synthesis and activity of AHL in *C. violaceum*. It reduced the biofilm formation and other QS-associated virulence factor of *P. aeruginosa*. Concentration-dependent decrease in the expression of *lasI/R*, *rhlI/R*, and *pqsA* of *P. aeruginosa* was recorded. In vitro studies showed it to be nontoxic (Hossain et al., 2017).

12.3.4.4 Phenylethanoids and Phenylpropanoids

Borges et al. (2014) evaluated the effects of oleoropein glucoside, caffeic acid, and ferulic acid on violacein production in C. *violaceum* bioreporter strain. The compounds inhibited the pigment production, but oleoropein glucoside and caffeic acid were found toxic during in vivo studies conducted on mouse fibroblast cell lines. Chlorogenic acid is also reported to inhibit QS using *C. violaceum* strain (Truchado et al., 2012; Brango-Vanegas et al., 2014). Truchado et al. (2012) observed that chlorogenic acid inhibited violacein production at the highest tested concentration, while Brango-Vanegas et al. (2014) witnessed slight inhibition of violacein and no inhibition of bioluminescence in *E. coli*. It is very vital to discuss that the amount of compound used by Brango-Vanegas et al. was lower than that used in earlier studies.

Cinnamaldehyde was evaluated to assess its potential as a QSI using *C. violaceum*, *Y. enterocolitica*, and *E. carotovora*. Cinnamaldehyde inhibited violacein production and reduced AHL concentrations (Truchado et al., 2012). Another study demonstrated that cinnamaldehyde specifically targets short-chain AHL synthase (RhlI) and inhibits AHL production by RhlI (Chang et al., 2014). Cinnamaldehyde and its analogs 4-methoxycinnamaldehyde, 2-methoxicinnamaldehyde, and 4-dimethylaminocinnamaldehyde inhibited AI-2-based QS system of *V. harveyi* in a dose-dependent manner (Brackman et al., 2008). In another study, conducted by the same group, it was reported that cinnamaldehyde, cinnamic acid, 4-methoxicinnamaldehyde, 4-dimethylaminocinnamaldehyde, and 4-phenyl-2-butanone were effective against AI-2-regulated QS of *Vibrio* spp. (Brackman et al., 2011).

Eugenol, an active constituent found in essential oils, has been reported to inhibit QS in pathogenic bacteria. Zhou et al. (2013) demonstrated reduction in violacein production in *C. violaceum* treated with eugenol. They also observed modulation of lasB and pqsA in *E. coli*, suggesting that eugenol inhibits Las and pseudomonas quinolone signal (PQS)−controlled transcription. Recently, it was reported that eugenol demonstrated significant anti-QS activity in mutant strain *C. violaceum* CVO26 and also reduced the QS-regulated production of elastase, protease, chitinase, pyocyanin, and EPS in *P. aeruginosa* PAO1 at subinhibitory concentrations considerably (Al-Shabib et al., 2017). Methyl eugenol, an eugenol derivative, was observed as the chief phytoconstituent responsible for the anti-QS activity of *Cuminum cyminum* extract using molecular docking studies (Packiavathy et al., 2012). In Gram-positive pathogen *S. aureus*, eugenol demonstrated antivirulence property by reducing exotoxin production. This decreased production of exotoxin is induced by the repression of *agrA* transcription and reduced transcription of QS-controlled exoproteins (Qiu et al., 2010b).

Phenolics such as 6-gingerol, zingerone, and 6-shogaol exhibited QSI in *C. violaceum* bioassay (Kumar et al., 2014). In another study, these authors demonstrated that zingerone decreased *N*-(3-oxododecanoyl)-L-homoserine lactone molecules and PQS production. Docking analysis revealed that the compound may interact with active site of QS receptors, blocking the downstream signaling pathway. Therefore, blockage of receptor–ligand binding seems to be the mechanism by which zingerone reduces the expression of QS-regulated virulence genes (Kumar et al., 2015). QS inhibition by 6-gingerol was confirmed with two biosensor strains, *C. violaceum* (reduction in violacein production) and *A. tumefaciens* (fading cyan color) (Kim et al., 2015).

Phenolic component of *Curcuma longa* root/rhizome, curcumin, reduced QS signal molecule leading to decreased production of virulence factors (Rudrappa and Bais, 2008). Curcumin interfered with the expression QS-regulated genes involved in transcriptional regulation and in type III secretion factors. Several other studies have been conducted to explore the QS inhibition by curcumin, Packiavathy et al. (2013, 2014) observed reduced violacein production in *C. violaceum* and inhibition of bioluminescence in *V. harveyi* after treatment with sub-MICs of curcumin.

12.3.4.5 Quinones

Molecular docking analysis was conducted to screen putative QSIs of *A. tumefaciens* from a database of known compounds used in Chinese traditional medicine. Computational studies identified six potential QSIs; however, only anthraquinone emodin induced proteolysis of the TraR QS signal receptor (Ding et al., 2011). 10′(Z),13′(E)-Heptadecadienylhydroquinone was observed to inhibit QS-regulated swarming motility in *Proteus mirabilis* wild type and *rppA* mutant strains. This effect was not witnessed in *rcsB* mutant indicating that this compound inhibits motility through the RcsB-dependent pathway (Liu et al., 2012).

12.3.4.6 Stilbenoids

Resveratrol was reported to decrease QS-regulated violacein production in *C. violaceum*. In a separate study, it was observed to reduce the AHL concentrations in *Y. enterocolitica* and *E. carotovora* in addition to violacein inhibition (Alvarez et al., 2012; Truchado et al., 2012). Resveratrol is also reported to interfere with virulence factor production in *P. mirabilis* through RsbA (Wang et al., 2006).

12.3.4.7 Tannins

Well-explored tannin ellagic acid interfered with QS of *C. violaceum* and reduced synthesis of AHL synthesis in *Y. enterocolitica* and *E. carotovora* without exerting any bactericidal effect (Truchado et al., 2012). Ellagic acid derivatives from *T. chebula* (fruit) extract demonstrated inhibition of QS-regulated production of extracellular virulence in *P. aeruginosa*. These showed downregulation of AI synthase (*lasI* and *rhlI*) and their cognate receptor (*lasR* and *rhlR*) (Sarabhai et al., 2013). 3-O-Methyl ellagic acid was observed to decrease violacein production significantly and reduced QS-regulated biofilm formation, prodigiosin production, and protease production in *S. marcescens* (Salini and Pandian, 2015).

Kiran et al. (2008) identified hamamelitanin as an inhibitor of RNAIII. In all *S. aureus* and *Staphylococcus epidermidis* tested, hamamelitanin reduced virulence factor production in vitro by downregulating production of RNAIII, a component of *agr* QS system.

Tannic acid, a hydrolysable tannin, is shown to antagonize QS system by inhibiting violacein production. It was also demonstrated to reduce biofilm in a concentration-dependent manner in *P. mirabilis* (Jones et al., 2009). Punicalagin, another well-known tannin, showed reduced production of violacein at sub-MICs in *C. violaceum*. Moreover, punicalagin reduced the production of two QS-related genes (*sidA* and *srgE*) (Li et al., 2014).

12.3.4.8 Terpenoids and Derivatives

Monoterpene, carvacrol, was shown to inhibit *cviI* gene expression in *C. violaceum*; this gene encodes for AHL synthase. Carvacrol reduced the production of violacein and chitinase in *C. violaceum* (Burt et al., 2014). In another report, Kerekes et al. (2013) reported significant drop in violacein production after treatment with terpinen-4-ol, linalool, and α-pinene.

Thymol and menthol also demonstrated decreased production of α-hemolysin, enterotoxin A and B, and toxic shock syndrome in *S. aureus*. At the highest concentration tested, thymol and menthol reduced the transcription levels of *agrA* (Qiu et al., 2010c, 2011c). Menthol was effective in reducing QS-regulated virulence and biofilm formation in *P. aeruginosa* and *A. hydrophila*. Methanol directly inhibited Las and PQS-controlled transcription in *E. coli* (Husain et al., 2015a,b).

Ahmad et al. (2015) reported varying level of QS inhibition by monoterpenes α-phellandrene, *p*-cimene, thymol, carvacol, geraniol, menthone, linalool, camphene, camphor and the sesquiterpenes farnesol, nerolidol, and nerol using *C. violaceum* and *P. aeruginosa* strains.

Cugini et al. (2007) demonstrated that sesquiterpenes farnesol and related compounds geranyllinalool and farnesyl acetate inhibited PQS and pyocyanin production without affecting growth of the bacteria. Decrease in *pqsA* transcript in *P. aeruginosa*, that interferes with the PqsR-mediated transcriptional activation, facilitates the reduction in PQS production.

α-Toxin production in *S. aureus* is encoded by *hla* gene under the control of *agr*. Sesquiterpene lactone isoalantolactone, α-cyperone, and triperpene glycyrrhetinic acid inhibited α-toxin production in *S. aureus* as described by various workers (Qiu et al., 2011d; Luo et al., 2012; Li et al, 2012).

12.4 CONCLUSION

The data presented in this document clearly shows that plants and their metabolites could be exploited in the discovery of novel drugs in order to combat drug-resistant bacterial infections. The different classes of phytocompounds having diverse antivirulence properties are an attractive alternative to bacteria resistant to conventional antibiotics. Numerous reports have suggested that phytocompounds demonstrate bacteriostatic or bactericidal activity against wide array of pathogenic bacteria using different mechanisms. In addition to the antibacterial activity, they cause site-specific modulation of bacterial virulence targets, as shown by their interference in enzymes, toxins, and signal receptors.

QS has been a leading anti-infective drug target, although complexity of the system allows distinct cascades to bypass the inhibited pathway. QS system coordinates many functions in the cells, therefore, any interference with the QS of the pathogen is bound to alter bacterial fitness. Current review brings to the fore the QS inhibitory potential of several compounds of plant origin in different QS systems. These QSIs are safe, stable, and nontoxic and include low risk of resistance development. Although these compounds offer great hope in the fight against drug resistance, but still there is a long way to go. Further, in depth, analysis is required on bioavailability and pharmacodynamics, and studies demonstrating the clinical efficacy of promising compounds should be conducted. It is very important to understand the synergism between conventional antibiotics and QSIs to make progress in this field. Thus, in conclusion, it can be said that the phytocompounds are the promising candidates in the development of antipathogenic drugs, although applicability of this approach remains to be investigated and validated.

ACKNOWLEDGMENTS

The authors wish to thank the Pharmacy Research Centre, College of Pharmacy, and the Deanship of Scientific Research, King Saud University, Riyadh, KSA for funding this research.

REFERENCES

Ahmad, I., Beg, A.Z., 2001. Antimicrobial and phytochemical studies on 45 Indian medicinal plants against multidrug resistant human pathogens. J. Ethnopharmacol. 74, 113–123.

Ahmad, A., Viljoen, A.M., Chenia, H.Y., 2015. The impact of plant volatiles on bacterial quorum sensing. Lett. Appl. Microbiol. 60 (1), 8–19.

Aliyu, A.B., Koorbanally, N.A., Moodley, B., Singh, P., Chenia, H.Y., 2016. Quorum sensing inhibitory potential and molecular docking studies of sesquiterpene lactones from *Vernonia blumeoides*. Phytochemistry 126, 23–33.

Al-Shabib, N.A., Husain, F.M., Ahmad, I., Baig, M.H., 2017. Eugenol inhibits quorum sensing and biofilm of toxigenic MRSA strains isolated from food handlers employed in Saudi Arabia. Biotechnol. Biotechnol. Equip. 31 (2), 387–396.

Alvarez, M.V., Moreira, M.R., Ponce, A., 2012. Antiquorum sensing and antimicrobial activity of natural agents with potential use in food. J. Food Saf. 32 (3), 379–387.

Al-Yousef, H.M., Ahmed, A.F., Al-Shabib, N.A., Laeeq, S., Khan, R.A., Rehman, M.T., et al., 2017. Onion peel ethylacetate fraction and its derived constituent quercetin 4′-O-β-D glucopyranoside attenuates quorum sensing regulated virulence and biofilm formation. Front. Microbiol. 8, 1675.

Autret, N., Raynaud, C., Dubail, I., Berche, P., Charbit, A., 2003. Identification of the agr locus of *Listeria monocytogenes*: role in bacterial virulence. Infect. Immun. 71, 4463–4471.

Bacha, K., Tariku, Y., Gebreyesus, F., Zerihun, S., Mohammed, A., Weiland-Bräuer, N., et al., 2016. Antimicrobial and anti-Quorum Sensing activities of selected medicinal plants of Ethiopia: implication for development of potent antimicrobial agents. BMC. Microbiol. 16 (1), 139.

Bandara, M.B., Zhu, H., Sankaridurg, P.R., Willcox, M.D., 2006. Salicylic acid reduces the production of several potential virulence factors of *Pseudomonas aeruginosa* associated with microbial keratitis. Invest. Ophthalmol. Visual Sci. 47 (10), 4453–4460.

Bansal, T., Englert, D., Lee, J., Hegde, M., Wood, T.K., Jayaraman, A., 2007. Differential effects of epinephrine, norepinephrine, and indole on *Escherichia coli* O157: H7 chemotaxis, colonization, and gene expression. Infect. Immun. 75 (9), 4597–4607.

Bhardwaj, A.K., Vinothkumar, K., Rajpara, N., 2013. Bacterial quorum sensing inhibitors: attractive alternatives for control of infectious pathogens showing multiple drug resistance. Recent Pat Antiinfect. Drug Discovery 8, 68–83.

Bhargava, N., Singh, S.P., Sharma, A., Sharma, P., Capalash, N., 2015. Attenuation of quorum sensing-mediated virulence of *Acinetobacter baumannii* by *Glycyrrhiza glabra* flavonoids. Future Microbiol. 10 (12), 1953–1968.

Bodede, O., Shaik, S., Chenia, H., Singh, P., Moodley, R., 2018. Quorum sensing inhibitory potential and in silico molecular docking of flavonoids and novel terpenoids from *Senegalia nigrescens*. J. Ethnopharmacol. 214, 134–146.

Borges, A., Serra, S., Cristina Abreu, A., Saavedra, M.J., Salgado, A., Simões, M., 2014. Evaluation of the effects of selected phytochemicals on quorum sensing inhibition and in vitro cytotoxicity. Biofouling 30 (2), 183–195.

Bouhdid, S., Abrini, J., Zhiri, A., Espuny, M.J., Manresa, A., 2009. Investigation of functional and morphological changes in *Pseudomonas aeruginosa* and *Staphylococcus aureus* cells induced by Origanum compactum essential oil. J. Appl. Microbiol. 106, 1558–1568.
Brackman, G., Defoirdt, T., Miyamoto, C., Bossier, P., Van Calenbergh, S., Nelis, H., et al., 2008. Cinnamaldehyde and cinnamaldehyde derivatives reduce virulence in *Vibrio* spp. by decreasing the DNA-binding activity of the quorum sensing response regulator LuxR. BMC Microbiol. 8 (1), 149.
Brackman, G., Celen, S., Hillaert, U., Van Calenbergh, S., Cos, P., Maes, L., et al., 2011. Structure-activity relationship of cinnamaldehyde analogs as inhibitors of AI-2 based quorum sensing and their effect on virulence of *Vibrio* spp. PLoS One 6 (1), e16084.
Brameyer, S., Kresovic, D., Bode, H.B., Heermann, R., 2015. Dialkylresorcinols as bacterial signaling molecules. Proc. Natl Acad. Sci. U.S.A. 112 (2), 572–577.
Brango-Vanegas, J., Costa, G.M., Ortmann, C.F., Schenkel, E.P., Reginatto, F.H., Ramos, F.A., et al., 2014. Glycosylflavonoids from *Cecropia pachystachya* Trécul are quorum sensing inhibitors. Phytomedicine 21 (5), 670–675.
Burt, S.A., Ojo-Fakunle, V.T., Woertman, J., Veldhuizen, E.J., 2014. The natural antimicrobial carvacrol inhibits quorum sensing in *Chromobacterium violaceum* and reduces bacterial biofilm formation at sub-lethal concentrations. PLoS One 9 (4), e93414.
Case, R.J., Labbate, M., Kjelleberg, S., 2008. AHL-driven quorum-sensing circuits: their frequency and function among the Proteobacteria. ISME J. 2, 345–349.
Castillo, S., Heredia, N., García, S., 2015. 2 (5*H*)-Furanone, epigallocatechin gallate, and a citric-based disinfectant disturb quorum-sensing activity and reduce motility and biofilm formation of *Campylobacter jejuni*. Folia Microbiol. (Praha) 60 (1), 89–95.
Chen, G., Swem, L.R., Swem, D.L., Stauff, D.L., O'Loughlin, C.T., Jeffrey, P.D., et al., 2011. A strategy for antagonizing quorum sensing. Mol Cell 42, 199–209.
Chong, Y.M., Yin, W.F., Ho, C.Y., Mustafa, M.R., Hadi, A.H.A., Awang, K., et al., 2011. Malabaricone C from *Myristica cinnamomea* exhibits anti-quorum sensing activity. J. Nat. Prod. 74 (10), 2261–2264.
Chang, C.Y., Krishnan, T., Wang, H., Chen, Y., Yin, W.F., Chong, Y.M., et al., 2014. Non-antibiotic quorum sensing inhibitors acting against *N*-acyl homoserine lactone synthase as druggable target. Sci. Rep. 4, 7245.
Cugini, C., Calfee, M., Farrow, J.M., Morales, D.K., Pesci, E.C., Hogan, D.A., 2007. Farnesol, a common sesquiterpene, inhibits PQS production in *Pseudomonas aeruginosa*. Mol. Microbiol. 65 (4), 896–906.
da Silva Trentin, D., Giordani, R.B., Zimmer, K.R., Da Silva, A.G., Da Silva, M.V., dos Santos Correia, M.T., et al., 2011. Potential of medicinal plants from the Brazilian semi-arid region (Caatinga) against *Staphylococcus epidermidis* planktonic and biofilm lifestyles. J. Ethnopharmacol. 137 (1), 327–335.
Defoirdt, T., Boon, N., Sorgeloos, P., Verstraete, W., Bossier, P., 2008. Quorum sensing and quorum quenching in *Vibrio harveyi*: lessons learned from in vivo work. ISME J. 2 (1), 19.
Ding, X., Yin, B., Qian, L., Zeng, Z., Yang, Z., Li, H., et al., 2011. Screening for novel quorum-sensing inhibitors to interfere with the formation of *Pseudomonas aeruginosa* biofilm. J. Med. Microbiol. 60 (12), 1827–1834.
Fatima, Q., Zahin, M., Khan, M.S.A., Ahmad, I., 2010. Modulation of quorum sensing controlled behavior of bacteria by growing seedling, seed and seedling extracts of leguminous plants. Indian J. Microbiol. 50, 238–242.
Fuqua, C., Parsek, M.R., Greenberg, E.P., 2001. Regulation of gene expression by cell-to-cell communication: acyl-homoserine lactone quorum sensing. Annu. Rev. Genet. 35, 439–468.

Ganin, H., Rayo, J., Amara, N., Levy, N., Krief, P., Meijler, M.M., 2013. Sulforaphane and erucin, natural isothiocyanates from broccoli, inhibit bacterial quorum sensing. MedChemComm 4 (1), 175–179.

Girennavar, B., Cepeda, M.L., Soni, K.A., Vikram, A., Jesudhasan, P., Jayaprakasha, G.K., et al., 2008. Grapefruit juice and its furocoumarins inhibits autoinducer signaling and biofilm formation in bacteria. Int. J. Food Microbiol. 125 (2), 204–208.

Gopu, V., Meena, C.K., Shetty, P.H., 2015a. Quercetin influences quorum sensing in food borne bacteria: in-vitro and in-silico evidence. PLoS One 10 (8), e0134684.

Gopu, V., Kothandapani, S., Shetty, P.H., 2015b. Quorum quenching activity of *Syzygium cumini* (L.) skeels and its anthocyanin malvidin against *Klebsiella pneumoniae*. Microb. Pathog. 79, 61–69.

Gould, T.A., Schweizer, H.P., Churchill, M.E., 2004. Structure of the *Pseudomonas aeruginosa* acyl-homoserine lactone synthase LasI. Mol. Microbiol. 53, 1135–1146.

Gutiérrez-Barranquero, J.A., Reen, F.J., McCarthy, R.R., O'Gara, F., 2015. Deciphering the role of coumarin as a novel quorum sensing inhibitor suppressing virulence phenotypes in bacterial pathogens. Appl. Microbiol. Biotechnol. 99 (7), 3303–3316.

Havarstein, L.S., Coomaraswamy, G., Morrison, D.A., 1995. An unmodified heptadecapeptide pheromone induces competence for genetic transformation in *Streptococcus pneumoniae*. Proc. Natl Acad. Sci. U.S.A. 92, 11140–11144.

Helman, Y., Chernin, L., 2015. Silencing the mob: disrupting quorum sensing as a means to fight plant disease. Mol. Plant Pathol. 16 (3), 316–329.

Hentzer, M., Givskov, M., 2003. Pharmacological inhibition of quorum sensing for the treatment of chronic bacterial infections. J. Clin. Invest. 112, 1300–1307.

Hentzer, M., Riedel, K., Rasmussen, T.B., Heydorn, A., Andersen, J.B., Parsek, M.R., et al., 2002. Inhibition of quorum sensing in *Pseudomonas aeruginosa* biofilm bacteria by a halogenated furanone compound. Microbiology 148 (1), 87–102.

Hoang, T.T., Schweizer, H.P., 1999. Characterization of *Pseudomonas aeruginosa* enoyl-acyl carrier protein reductase (FabI): a target for the antimicrobial triclosan and its role in acylated homoserine lactone synthesis. J. Bacteriol. 181, 5489–5497.

Holmstrom, G.P.C., Kjelleberg, S., 2001. Inhibition of fungi. WO2001068091A1.

Hossain, M.A., Lee, S.J., Park, J.Y., Reza, M.A., Kim, T.H., Lee, K.J., et al., 2015. Modulation of quorum sensing-controlled virulence factors by *Nymphaea tetragona* (water lily) extract. J. Ethnopharmacol. 174, 482–491.

Hossain, M.A., Lee, S.J., Park, N.H., Mechesso, A.F., Birhanu, B.T., Kang, J., et al., 2017. Impact of phenolic compounds in the acyl homoserine lactone-mediated quorum sensing regulatory pathways. Sci. Rep. 7 (1), 10618.

Huang, J.J., Han, J.I., Zhang, L.H., Leadbetter, J.R., 2003. Utilization of acylhomoserine lactone quorum signals for growth by a soil pseudomonad and *Pseudomonas aeruginosa* PAO1. Appl. Environ. Microbiol. 69, 5941–5949.

Huber, B., Eberl, L., Feucht, W., Polster, J., 2003. Influence of polyphenols on bacterial biofilm formation and quorum-sensing. Z. Naturforsch. [C] 58 (11–12), 879–884.

Husain, F.M., Ahmad, I., Khan, M.S., Ahmad, E., Tahseen, Q., Khan, M.S., et al., 2015a. Sub-MICs of *Mentha piperita* essential oil and menthol inhibits AHL mediated quorum sensing and biofilm of Gram-negative bacteria. Front. Microbiol. 6, 420.

Husain, F.M., Ahmad, I., Khan, M.S., Al-Shabib, N.A., 2015b. *Trigonella foenum-graceum* (Seed) extract interferes with quorum sensing regulated traits and biofilm formation in the strains of *Pseudomonas aeruginosa* and *Aeromonas hydrophila*. Evid. Based Complement. Alternat. Med. 2015, 1–10.

Husain, F.M., Ahmad, I., Al-thubiani, A.S., Abulreesh, H.H., AlHazza, I.M., Aqil, F., 2017. Leaf extracts of *Mangifera indica* L. inhibit quorum sensing–regulated production of virulence factors and biofilm in test bacteria. Front. Microbiol. 8, 727.

Jakobsen, T.H., Bragason, S.K., Phipps, R.K., Christensen, L.D., van Gennip, M., Alhede, M., et al., 2012a. Food as a source for QS inhibitors: iberin from horseradish revealed as a quorum sensing inhibitor of *Pseudomonas aeruginosa*. Appl. Environ. Microbiol. AEM-05992.

Jakobsen, T.H., van Gennip, M., Phipps, R.K., Shanmugham, M.S., Christensen, L.D., Alhede, M., et al., 2012b. Ajoene, a sulfur-rich molecule from garlic, inhibits genes controlled by quorum sensing. Antimicrob. Agents Chemother. 56 (5), 2314–2325.

Jones, S.M., Dang, T.T., Martinuzzi, R., 2009. Use of quorum sensing antagonists to deter the formation of crystalline *Proteus mirabilis* biofilms. Int. J. Antimicrob. Agents 34 (4), 360–364.

Kalia, V.C., 2013. Quorum sensing inhibitors: an overview. Biotechnol. Adv. 31 (2), 224–245.

Kerekes, E.B., Deák, É., Takó, M., Tserennadmid, R., Petkovits, T., Vágvölgyi, C., et al., 2013. Anti-biofilm forming and anti-quorum sensing activity of selected essential oils and their main components on food-related micro-organisms. J. Appl. Microbiol. 115 (4), 933–942.

Kim, H.S., Lee, S.H., Byun, Y., Park, H.D., 2015. 6-Gingerol reduces *Pseudomonas aeruginosa* biofilm formation and virulence via quorum sensing inhibition. Sci. Rep. 5, 8656.

Kiran, M.D., Adikesavan, N.V., Cirioni, O., Giacometti, A., Silvestri, C., Scalise, G., et al., 2008. Discovery of a quorum-sensing inhibitor of drug-resistant staphylococcal infections by structure-based virtual screening. Mol. Pharmacol. 73, 1578–1586.

Kleerebezem, M., Quadri, L.E., Kuipers, O.P., de Vos, W.M., 1997. Quorum sensing by peptide pheromones and two-component signal-transduction systems in Gram-positive bacteria. Mol. Microbiol. 24, 895–904.

Kociolek, M.G., 2009. Quorum-sensing inhibitors and biofilms. Antiinfect. Agents J. Med. Chem. 8, 315–326.

Koh, K.H., Tham, F.Y., 2011. Screening of traditional Chinese medicinal plants for quorum sensing inhibitors activity. J. Microbiol. Immunol. Infect. 44, 144–148.

Krishnan, T., Yin, W.F., Chan, K.G., 2012. Inhibition of quorum sensing-controlled virulence factor production in *Pseudomonas aeruginosa* PAO1 by ayurveda spice clove *(Syzygium aromaticum)* bud extract. Sensors 12, 4016–4030.

Kumar, N.V., Murthy, P.S., Manjunatha, J.R., Bettadaiah, B.K., 2014. Synthesis and quorum sensing inhibitory activity of key phenolic compounds of ginger and their derivatives. Food Chem. 159, 451–457.

Kumar, L., Chhibber, S., Kumar, R., Kumar, M., Harjai, K., 2015. Zingerone silences quorum sensing and attenuates virulence of *Pseudomonas aeruginosa*. Fitoterapia 102, 84–95.

LaSarre, B., Federle, M.J., 2013. Exploiting quorum sensing to confuse bacterial pathogens. Microbiol. Mol. Biol. Rev. 77, 73–111.

Lee, J., Zhang, L., 2015. The hierarchy quorum-sensing network in *Pseudomonas aeruginosa*. Protein Cell 6 (1), 26–41.

Lee, J., Attila, C., Cirillo, S.L., Cirillo, J.D., Wood, T.K., 2009. Indole and 7-hydroxyindole diminish *Pseudomonas aeruginosa* virulence. Microb. Biotechnol. 2 (1), 75–90.

Leng, B.F., Qiu, J.Z., Dai, X.H., Dong, J., Wang, J.F., Luo, M.J., et al., 2011. Allicin reduces the production of α-toxin by *Staphylococcus aureus*. Molecules 16 (9), 7958–7968.

Li, H.E., Qiu, J.Z., Yang, Z.Q., Dong, J., Wang, J.F., Luo, M.J., et al., 2012. Glycyrrhetinic acid protects mice from *Staphylococcus aureus* pneumonia. Fitoterapia 83 (1), 241–248.

Li, G., Yan, C., Xu, Y., Feng, Y., Wu, Q., Lv, X., et al., 2014. Punicalagin inhibits Salmonella virulence factors and has anti-quorum-sensing potential. Appl. Environ. Microbiol. 80 (19), 6204–6211.

Liu, M.C., Lin, S.B., Chien, H.F., Wang, W.B., Yuan, Y.H., Hsueh, P.R., et al., 2012. 10′(Z), 13′(E)-heptadecadienylhydroquinone inhibits swarming and virulence factors and increases polymyxin B susceptibility in *Proteus mirabilis*. PLoS One 7 (9), e45563.
Lowery, C.A., Dickerson, T.J., Janda, K.D., 2008. Interspecies and interkingdom communication mediated by bacterial quorum sensing. Chem. Soc. Rev. 37, 1337–1346.
Luo, M., Qiu, J., Zhang, Y., Wang, J., Dong, J., Li, H., et al., 2012. α-Cyperone alleviates lung cell injury caused by *Staphylococcus aureus* via attenuation of α-hemolysin expression. J. Microbiol. Biotechnol. 22, 1170–1176.
Maisuria, V.B., Lopez-de Los Santos, Y., Tufenkji, N., Déziel, E., 2016. Cranberry-derived proanthocyanidins impair virulence and inhibit quorum sensing of *Pseudomonas aeruginosa*. Sci. Rep. 6, 30169.
Manefield, M., Rasmussen, T.B., Henzter, M., 2002. Halogenated furanones inhibit quorum sensing through accelerated LuxR turnover. Microbiology 148 (4), 1119–1127.
Mitchell, G., Lafrance, M., Boulanger, S., Séguin, D.L., Guay, I., Gattuso, M., et al., 2011. Tomatidine acts in synergy with aminoglycoside antibiotics against multiresistant *Staphylococcus aureus* and prevents virulence gene expression. J. Antimicrob. Chemother. 67 (3), 559–568.
Monte, J., Abreu, A.C., Borges, A., Simões, L.C., Simões, M., 2014. Antimicrobial activity of selected phytochemicals against *Escherichia coli* and *Staphylococcus aureus* and their biofilms. Pathogens 3 (2), 473–498.
Mueller, R.S., Beyhan, S., Saini, S.G., Yildiz, F.H., Bartlett, D.H., 2009. Indole acts as an extracellular cue regulating gene expression in *Vibrio cholerae*. J. Bacteriol. 191 (11), 3504–3516.
Muhs, A., Lyles, J.T., Parlet, C.P., Nelson, K., Kavanaugh, J.S., Horswill, A.R., et al., 2017. Virulence inhibitors from Brazilian peppertree block quorum sensing and abate dermonecrosis in skin infection models. Sci. Rep. 7, 42275.
Musk, D.J., Hergenrother, P.J., 2006. Chemical counter measures for the control of bacterial biofilms: effective compounds and promising targets. Curr. Med. Chem. 13, 2163–2177.
Musthafa, K.S., Ravi, A.V., Annapoorani, A., Packiavathy, I.S.V., Pandian, S.K., 2010. Evaluation of anti-quorum-sensing activity of edible plants and fruits through inhibition of the *N*-acyl-homoserine lactone system in *Chromobacterium violaceum* and *Pseudomonas aeruginosa*. Chemotherapy 56, 333–339.
Musthafa, K.S., Sahu, S.K., Ravi, A.V., Kathiresan, K., 2013. Anti-quorum sensing potential of the mangrove *Rhizophora annamalayana*. World J. Microbiol. Biotechnol. 10, 1851–1858.
Natrah, F.M.I., Alam, M.I., Pawar, S., Harzevili, A.S., Nevejan, N., Boon, N., et al., 2012. The impact of quorum sensing on the virulence of *Aeromonas hydrophila* and *Aeromonas salmonicida* towards burbot (*Lota lota* L.) larvae. Vet. Microbiol. 159 (1–2), 77–82.
Ng, W.L., Bassler, B.L., 2009. Bacterial quorum-sensing network architectures. Annu. Rev. Genet. 43, 197–222.
Ni, N., Li, M., Wang, J., Wang, B., 2009. Inhibitors and antagonists of bacterial quorum sensing. Med. Res. Rev. 29, 65–124.
Norizan, S.N.M., Yin, W.F., Chan, K.G., 2013. Caffeine as a potential quorum sensing inhibitor. Sensors 13 (4), 5117–5129.
Novick, R.P., Projan, S.J., Kornblum, J., Ross, H.F., Ji, G., Kreiswirth, B., et al., 1995. The agr P2 operon: an autocatalytic sensory transduction system in *Staphylococcus aureus*. Mol. Gen. Genet. 248, 446–458.
Nyila, M.A., Leonard, C.M., Hussein, A.A., Lall, N., 2012. Activity of South African medicinal plants against *Listeria monocytogenes* biofilms, and isolation of active compounds from Acacia karroo. S. Afr. J. Bot. 78, 220–227.

Ohtani, K., Yuan, Y., Hassan, S., Wang, R., Wang, Y., Shimizu, T., 2009. Virulence gene regulation by the agr system in *Clostridium perfringens*. J. Bacteriol. 191, 3919–3927.
Oliveira, B.D.Á., Rodrigues, A.C., Cardoso, B.M.I., Ramos, A.L.C.C., Bertoldi, M.C., Taylor, J.G., et al., 2016. Antioxidant, antimicrobial and anti-quorum sensing activities of *Rubus rosaefolius* phenolic extract. Ind. Crops Prod. 84, 59–66.
Packiavathy, I.A.S.V., Agilandeswari, P., Musthafa, K.S., Pandian, S.K., Ravi, A.V., 2012. Antibiofilm and quorum sensing inhibitory potential of *Cuminum cyminum* and its secondary metabolite methyl eugenol against Gram negative bacterial pathogens. Food Res. Int. 45 (1), 85–92.
Packiavathy, I.A.S.V., Sasikumar, P., Pandian, S.K., Ravi, A.V., 2013. Prevention of quorum-sensing-mediated biofilm development and virulence factors production in *Vibrio* spp. by curcumin. Appl. Microbiol. Biotechnol. 97 (23), 10177–10187.
Packiavathy, I.A.S.V., Priya, S., Pandian, S.K., Ravi, A.V., 2014. Inhibition of biofilm development of uropathogens by curcumin – an anti-quorum sensing agent from *Curcuma longa*. Food Chem. 148, 453–460.
Pan, J., Ren, D., 2009. Quorum sensing inhibitors: a patent overview. Expert Opin. Ther. Pat 19, 1581–1601.
Parsek, M.R., Val, D.L., Hanzelka, B.L., Cronan, J.E., Greenberg Jr, E.P., 1999. Acyl homoserine-lactone quorum-sensing signal generation. Proc. Natl Acad. Sci. U.S.A. 96, 4360–4365.
Pearson, J.P., Van Delden, C., Iglewski, B.H., 1999. Active efflux and diffusion are involved in transport of *Pseudomonas aeruginosa* cell-to-cell signals. J. Bacteriol. 181, 1203–1210.
Podbielski, A., Kreikemeyer, B., 2004. Cell density-dependent regulation: basic principles and effects on the virulence of Gram-positive cocci. Int. J. Infect. Dis. 8, 81–95.
Ponnusamy, K., Paul, D., Kweon, J.H., 2009. Inhibition of quorum sensing mechanism and *Aeromonas hydrophila* biofilm formation by vanillin. Environ. Eng. Sci. 26 (8), 1359–1363.
Qiu, J., Jiang, Y., Xia, L., Xiang, H., Feng, H., Pu, S., et al., 2010a. Sub-inhibitory concentrations of licochalcone A decrease alpha-toxin production in both methicillin-sensitive and methicillin-resistant *Staphylococcus aureus* isolates. Lett. Appl. Microbiol. 50 (2), 223–229.
Qiu, J., Feng, H., Lu, J., Xiang, H., Wang, D., Dong, J., et al., 2010b. Eugenol reduces the expression of virulence-related exoproteins in *Staphylococcus aureus*. Appl. Environ. Microbiol. 76 (17), 5846–5851.
Qiu, J., Wang, D., Xiang, H., Feng, H., Jiang, Y., Xia, L., et al., 2010c. Subinhibitory concentrations of thymol reduce enterotoxins A and B and α-hemolysin production in *Staphylococcus aureus* isolates. PLoS One 5 (3), e9736.
Qiu, J., Li, H., Meng, H., Hu, C., Li, J., Luo, M., et al., 2011a. Impact of luteolin on the production of alpha-toxin by *Staphylococcus aureus*. Lett. Appl. Microbiol. 53 (2), 238–243.
Qiu, J., Xiang, H., Hu, C., Wang, Q., Dong, J., Li, H., et al., 2011b. Sub-inhibitory concentrations of farrerol reduce α-toxin expression in *Staphylococcus aureus*. FEMS Microbiol. Lett. 315 (2), 129–133.
Qiu, J., Luo, M., Dong, J., Wang, J., Li, H., Wang, X., et al., 2011c. Menthol diminishes *Staphylococcus aureus* virulence-associated extracellular proteins expression. Appl. Microbiol. Biotechnol. 90 (2), 705–712.
Qiu, J., Luo, M., Wang, J., Dong, J., Li, H., Leng, B., et al., 2011d. Isoalantolactone protects against *Staphylococcus aureus* pneumonia. FEMS Microbiol. Lett. 324 (2), 147–155.
Qiu, J., Niu, X., Wang, J., Xing, Y., Leng, B., Dong, J., et al., 2012. Capsaicin protects mice from community-associated methicillin-resistant *Staphylococcus aureus* pneumonia. PLoS ONE 7 (3), e33032.

Rader, B.A., Campagna, S.R., Semmelhack, M.F., Bassler, B.L., Guillemin, K., 2007. The quorum-sensing molecule autoinducer 2 regulates motility and flagellar morphogenesis in *Helicobacter pylori*. J. Bacteriol. 189 (17), 6109–6117.

Rahman, M.R.T., Lou, Z., Yu, F., Wang, P., Wang, H., 2017. Anti-quorum sensing and anti-biofilm activity of *Amomum tsaoko* (Amommum tsao-ko Crevost et Lemarie) on foodborne pathogens. Saudi J. Biol. Sci. 24 (2), 324–330.

Rasko, D.A., Moreira, C.G., Li, R., Reading, N.C., Ritchie, J.M., Waldor, M.K., et al., 2008. Targeting QseC signaling and virulence for antibiotic development. Science 321, 1078–1080.

Rasmussen, T.B., Skindersoe, M.E., Bjarnsholt, T., Phipps, R.K., Christensen, K.B., Jensen, P.O., et al., 2005. Identity and effects of quorum-sensing inhibitors produced by Penicillium species. Microbiology 151, 1325–1340.

Rasmussen, T.B., Givskov, M., 2006. Quorum sensing inhibitors: a bargain of effects. Microbiology 152 (4), 895–904.

Ren, D., Sims, J.J., Wood, T.K., 2001. Inhibition of biofilm formation and swarming of *Escherichia coli* by (5Z)-4-bromo-5-(bromomethylene)-3-butyl-2(5*H*)-furanone. Environ. Microbiol. 3 (11), 731–736.

Riedel, C.U., Monk, I.R., Casey, P.G., Waidmann, M.S., Gahan, C.G., Hill, C., 2009. AgrD-dependent quorum sensing affects biofilm formation, invasion, virulence and global gene expression profiles in *Listeria monocytogenes*. Mol. Microbiol. 71, 1177–1189.

Rudrappa, T., Bais, H.P., 2008. Curcumin, a known phenolic from *Curcuma longa*, attenuates the virulence of *Pseudomonas aeruginosa* PAO1 in whole plant and animal pathogenicity models. J. Agric. Food. Chem. 56 (6), 1955–1962.

Rutherford, S.T., Bassler, B., 2012. Bacterial quorum sensing: its role in virulence and possibilities for its control. Cold Spring Harb. Perspect. Med. 2, a012427.

Salini, R., Pandian, S.K., 2015. Interference of quorum sensing in urinary pathogen *Serratia marcescens* by *Anethum graveolens*. Pathog. Dis. 73 (6), ftv038.

Sarabhai, S., Sharma, P., Capalash, N., 2013. Ellagic acid derivatives from *Terminalia chebula Retz*. downregulate the expression of quorum sensing genes to attenuate *Pseudomonas aeruginosa* PAO1 virulence. PLoS One 8, e53441.

Sarkar, R., Chaudhary, S., Sharma, A., Yadav, K., Nema, N., Sekhoachad, M., et al., 2014. Anti-biofilm activity of Marula—a study with the standardized bark extract. J. Ethnopharmacol. 154, 170–175.

Schaefer, A.L., Val, D.L., Hanzelka, B.L., Cronan, J.E., Greenberg, E.P., 1996. Generation of cell-to-cell signals in quorum sensing: acyl homoserine lactone synthase activity of a purified *Vibrio fischeri* LuxI protein. Proc. Natl Acad. Sci. U.S.A. 93, 9505–9509.

Seed, P.C., Passador, L., Iglewski, B.H., 1995. Activation of the *Pseudomonas aeruginosa* lasI by LasR and the *Pseudomonas* autoinducer PAI: an autoinduction regulatory gene hierarchy. J. Bacteriol. 177, 654.

Shukla, V., Bhathena, Z., 2016. Broad spectrum anti-quorum sensing activity of tannin-rich crude extracts of Indian medicinal plants. Scientifica 2016, 5823013.

Singh, B.N., Singh, H.B., Singh, A., Singh, B.R., Mishra, A., Nautiyal, C.S., 2012. *Lagerstroemia speciosa* fruit extract modulates quorum sensing-controlled virulence factor production and biofilm formation in *Pseudomonas aeruginosa*. Microbiology 158, 529–538.

Song, Z., Kong, K.F., Wu, H., 2010. *Panax ginseng* has anti-infective activity against opportunistic pathogen *Pseudomonas aeruginosa*by inhibiting quorum sensing, a bacterial communication process critical for establishing infection. Phytomedicine 17, 1040–1046.

Taganna, J.C., Quanico, J.P., Perono, R.M.G., Amor, E.C., Rivera, W.L., 2011. Tannin-rich fraction from *Terminalia catappa* inhibits quorum sensing (QS) in *Chromobacterium violaceum* and the QS-controlled biofilm maturation and LasA staphylolytic activity in *Pseudomonas aeruginosa*. J. Ethnopharmacol. 134 (3), 865–871.

Teplitski, M., Mathesius, U., Rambaugh, K.P., 2011. Perception and degradation of *N*-acyl homoserine lactone quorum sensing signals by mammalian and plant cells. Chem. Rev. 111, 100–116.

Thoendel, M., Horswill, A.R., 2010. Biosynthesis of peptide signals in gram-positive bacteria. Adv. Appl. Microbiol. 71, 91–112.

Tomlacheva, A.A., Rogozhin, E.A., Deryabin, D.G., 2014. Antibacterial and quorum sensing regulatory activities of some traditional Eastern-European medicinal plants. Acta Pharm. 64, 173–186.

Truchado, P., Tomás-Barberán, F.A., Larrosa, M., Allende, A., 2012. Food phytochemicals act as quorum sensing inhibitors reducing production and/or degrading autoinducers of *Yersinia enterocolitica* and *Erwinia carotovora*. Food Control. 24 (1–2), 78–85.

Vandeputte, O.M., Kiendrebeogo, M., Rajaonson, S., Diallo, B., Mol, A., El Jaziri, M., et al., 2010. Identification of catechin as one of the flavonoids from *Combretum albiflorum* bark extract that reduces the production of quorum-sensing-controlled virulence factors in *Pseudomonas aeruginosa* PAO1. Appl. Environ. Microbiol. 76 (1), 243–253.

Vandeputte, O.M., Kiendrebeogo, M., Rasamiravaka, T., Stevigny, C., Duez, P., Rajaonson, S., et al., 2011. The flavanone naringenin reduces the production of quorum sensing-controlled virulence factors in *Pseudomonas aeruginosa* PAO1. Microbiology 157 (7), 2120–2132.

Vasavi, H.S., Arun, A.B., Rekha, P.D., 2014. Anti-quorum sensing activity of *Psidium guajava* L. flavonoids against *Chromobacterium violaceum* and *Pseudomonas aeruginosa* PAO1. Microbiol. Immunol. 58 (5), 286–293.

Vasavi, H.S., Arun, A.B., Rekha, P.D., 2016. Anti-quorum sensing activity of flavonoid-rich fraction from *Centella asiatica* L. against *Pseudomonas aeruginosa* PAO1. J. Microbiol. Immunol. Infect. 49 (1), 8–15.

Vattem, D.A., Mihali, K.K., Crixell, S.H., Mclean, R.J.C., 2007. Dietary phytochemicals as quorum sensing inhibitors. Fitoterapia 78, 302–310.

Vikram, A., Jayaprakasha, G.K., Jesudhasan, P.R., Pillai, S.D., Patil, B.S., 2010. Suppression of bacterial cell–cell signalling, biofilm formation and type III secretion system by citrus flavonoids. J. Appl. Microbiol. 109 (2), 515–527.

Wang, W.B., Lai, H.C., Hsueh, P.R., Chiou, R.Y.Y., Lin, S.B., Liaw, S.J., 2006. Inhibition of swarming and virulence factor expression in *Proteus mirabilis* by resveratrol. J. Med. Microbiol. 55 (10), 1313–1321.

Watson, W.T., Minogue, T.D., Val, D.L., von Bodman, S.B., Churchill, M.E., 2002. Structural basis and specificity of acyl-homoserine lactone signal production in bacterial quorum sensing. Mol. Cell 9, 685–694.

Williams, P., Camara, M., 2009. Quorum sensing and environmental adaptation in *Pseudomonas aeruginosa*: a tale of regulatory networks and multifunctional signal molecules. Curr. Opin. Microbiol. 12, 182–191.

Wu, H., Song, Z., Hentzer, M., Andersen, J.B., Molin, S., Givskov, M., et al., 2004. Synthetic furanones inhibit quorum-sensing and enhance bacterial clearance in *Pseudomonas aeruginosa* lung infection in mice. J. Antimicrob. Chemother. 53, 1054–1061.

Yang, Q., Pande, G.S.J., Wang, Z., Lin, B., Rubin, R.A., Vora, G.J., et al., 2017. Indole signalling and (micro) algal auxins decrease the virulence of *Vibrio campbellii*, a major pathogen of aquatic organisms. Environ. Microbiol. 19 (5), 1987–2004.

Zahin, M., Hasan, S., Aqil, F., Khan, M.S.A., Husain, F.M., Ahmad, I., 2010. Screening of Indian medicinal plants for their anti-quorum sensing activity. Indian J. Exp. Biol. 48, 1219–1224.

Zaki, A.A., Shaaban, M.I., Hashish, N.E., Amer, M.A., Lahloub, M.F., 2013. Assessment of anti-quorum sensing activity for some ornamental and medicinal plants native to Egypt. Sci. Pharm. 81, 251–258.

Zhang, R.G., Pappas, K.M., Brace, J.L., Miller, P.C., Oulmassov, T., Molyneaux, J.M., et al., 2002. Structure of a bacterial quorum sensing transcription factor complexed with pheromone and DNA. Nature 417, 971–974.

Zhou, T., Deng, X., Qiu, J., 2012. Antimicrobial activity of licochalcone E against *Staphylococcus aureus* and its impact on the production of staphylococcal alpha-toxin. J. Microbiol. Biotechnol. 22 (6), 800–805.

Zhou, L., Zheng, H., Tang, Y., Yu, W., Gong, Q., 2013. Eugenol inhibits quorum sensing at sub-inhibitory concentrations. Biotechnol. Lett. 35 (4), 631–637.

Zhou, L., Zhang, L.H., Cámara, M., He, Y.W., 2017. The DSF family of quorum sensing signals: diversity, biosynthesis, and turnover. Trends Microbiol. 25 (4), 293–303.

Zhu, J., Winans, S.C., 1999. Autoinducer binding by the quorum-sensing regulator TraR increases affinity for target promoters in vitro and decreases TraR turnover rates in whole cells. Proc. Natl Acad. Sci. U.S.A. 96, 4832–4837.

Zhu, J., Winans, S.C., 2001. The quorum-sensing transcriptional regulator TraR requires its cognate signaling ligand for protein folding, protease resistance, and dimerization. Proc. Natl Acad. Sci. U.S.A. 98, 1507–1512.

CHAPTER 13

Understanding Biochemical and Molecular Mechanism of Complications of Glycation and Its Management by Herbal Medicine

Faizan Abul Qais[1], Mohammad Shavez Khan[1], Abdullah Safar Althubiani[2], Saleh Bakheet Al-Ghamdi[3] and Iqbal Ahmad[1]
[1]Department of Agricultural Microbiology, Aligarh Muslim University, Aligarh, Uttar Pradesh, India
[2]Department of Biology, Faculty of Applied Science, Umm Al-Qura University, Mecca, Kingdom of Saudi Arabia
[3]Biology Department, Faculty of Science, Al Baha University, Al Baha, Kingdom of Saudi Arabia

13.1 INTRODUCTION

Diabetes and its related complications is a global health issue which is going to become world's largest contributor of mortality and morbidity (Jang et al., 2011; Singh et al., 2014). In the next two decades, diabetes is expected to increase two to fivefold in many countries including India, China, and the United States (Alberti and Zimmet, 2013). In adults, diabetes is one of the leading causes of renal failure, heart disease, blindness, and limb amputations (Ahola et al., 2010; Antonetti et al., 2012; Martin et al., 2006). The underlying reason for adverse effects and complications in diabetes is persistent elevation in plasma glucose concentrations. The elevated expression of glucose transporter 1 (GLUT 1) makes it difficult to regulate intracellular glucose levels and hence become more prone to hyperglycemia-induced damages (Singh et al., 2014). Overexpression of GLUT 1 in renal mesangial cells leads to diabetic phenotypes such as increased extracellular matrix (ECM) synthesis and activation of the polyol pathway (Heilig et al., 1995). The renal complications might be major drivers of increased cardiovascular mortality and morbidity (Afkarian et al., 2013). The exact series of proceedings and its mechanism leading to cellular malfunction because of hyperglycemia is not fully understood. One of such events is reaction of glucose with proteins to

New Look to Phytomedicine
DOI: https://doi.org/10.1016/B978-0-12-814619-4.00013-6

form advanced glycation end products (AGEs) (Fraser and Hansen, 2005). The process is mainly nonenzymatic and called glycation. Elevated levels of glucose induce the glycation of numerous structural and functional proteins such as collagen and plasma proteins (Negre-Salvayre et al., 2009). The products of nonenzymatic reaction exert several deleterious effects including platelet activation, alteration in drug binding in the plasma, generation of oxygen-free radicals, impairment in immune system regulation, and impaired fibrinolysis (Helou et al., 2014; Negre-Salvayre et al., 2009). Furthermore, alteration in osteoblast differentiation is caused by the structural modification in collagen which leads to skeletal fragility and bone remodeling (Alikhani et al., 2007; Saito et al., 2006). In diabetic patients, autoantibodies produced against serum AGEs are capable of forming AGE—immune complexes that may play a key role in atherogenesis (Turk et al., 2001). The free radicals produced by glycation are capable of oxidizing nucleic acids and lipids and causing protein fragmentation (Baynes, 1991). Amino groups present in adenine and guanine bases of DNA are also vulnerable to glycation and AGEs formation (Baynes, 2002; Bohlooli et al., 2016).

13.2 DIABETES: A GLOBAL HEALTH PROBLEM

Diabetes has become a leading cause of death globally and warrants effective management of this disease (Zimmet et al., 2014). It is estimated by International Diabetes Federation that there are more than 380 million people affected by this epidemic and expected to reach 600 million people by 2035 as presented in Fig. 13.1. Majority of affected patients belong to low- and middle-income countries, out of which 60% live in Asia and approximately one-third in China (Guariguata et al., 2014). There has been significant increase in both type 1 and type 2 diabetes in last few decades, but type 2 is much more prevalent worldwide (Nanditha et al., 2016). The most accelerating rate of disease outbreak is in developing regions, mainly South Asia and Western Pacific region (Chan et al., 2009; Ramachandran et al., 2014, 2010). The accelerating incidence of diabetes is ascribed to certain interrelated factors including urbanization, industrialization as well as the changes in recent lifestyle (Chan et al., 2009; Ramachandran et al., 2010). The epigenetic and environmental changes also contributed to the increased risk of incidence of type 2 diabetes (Zimmet et al., 2014). The occurrence of type 2 diabetes is increasingly even in adolescents and children. The gestational diabetes

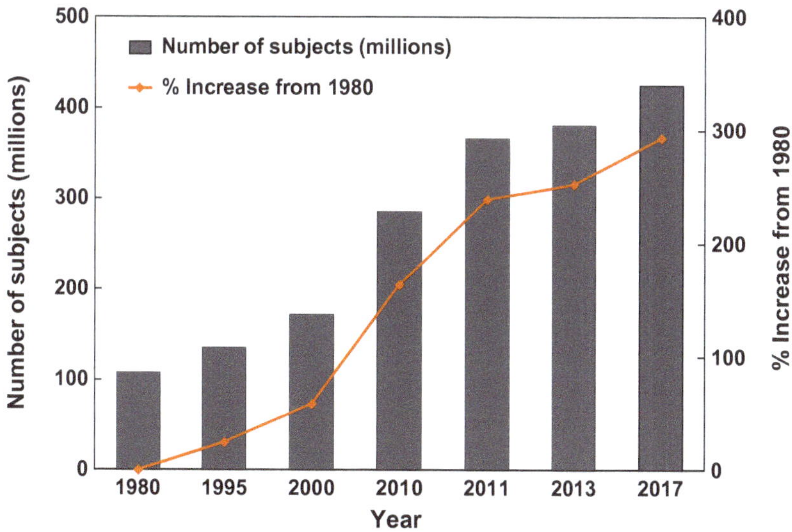

Figure 13.1 Prevalence of diabetes worldwide.

mellitus imparts a higher risk of diabetes in women along with long-term negative consequences on the health of offsprings (Chan et al., 2014; Tutino et al., 2014). The prevalence of prediabetes is higher than diabetes in certain Western Pacific region countries (Nanditha et al., 2016).

With the increased incidence of diabetes, the changing epidemiology of risk factors involved in type 2 diabetes is quite clear. Traditional risk factors include adiposity, age, sedentary lifestyle, and adverse dietary patterns. The increase in these traditional risk factors are driven by urbanization in developing countries; doing less physical activity for daily living, food is easier to acquire and is more likelihood to survive to older age (Herman and Zimmet, 2012). The epidemiologic transition of people from traditional health risks to modern lifestyle health risks is also an important factor to consider (WHO, 2009). Some of the common causes involved in risk factors of higher incidence of diabetes are briefly described below.

13.2.1 Age

The occurrence of diabetes increases with age, and it is expected that there will be higher burden of type 2 diabetes. A recent increase in number of diabetic cases observed in adolescents and children is also a new concern. Still there is lack of more precise data, but the elevated glucose and adiposity will only add to the prevalence of diabetes as the population ages in near future (Shay et al., 2013). Based on the data, in the United States only, there will be 2.3% increase per year in diabetic cases in persons younger than 20 years (Imperatore et al., 2012).

13.2.2 Adiposity

Since 1980, adiposity has been regarded as strong risk factor in incidence of diabetes as obesity and overweight in adults has increased substantially (Finucane et al., 2011). The prevalence of obesity increased from 6.4% to 12% between 1980 and 2008, and prevalence of overweight people has increased to 34.4% from 24.6% during the same time. The countries where there is a largest increment in the number of overweight people are Brazil, China, Mexico, and the United States (Stevens et al., 2012). The increase in number of overweight cases in children and infants have been observed, which again increase the risk of diabetes (de Onis et al., 2010; The et al., 2013). Therefore, the increase in obesity will accumulate more metabolically harmful adipose and contribute to the increased global burden of diabetes.

13.2.3 Physical Activity and Diet

The adoption of the modern Western lifestyle shifted to a lower quality diet and increased access to food and a relatively sedentary lifestyle that resulted in excess adiposity (Danaei et al., 2013). In this shift of lifestyle, carbohydrates taken as whole grains are replaced with lower quality carbohydrates such as sugar-sweetened beverages, and the consumption of fruits and vegetables has decreased (Mattei et al., 2015). Consumption of processed red meat has also appeared to raise the risk of diabetes as found in many observational studies in Japan, the United States, France, and Rotterdam (Kurotani et al., 2013; Lajous et al., 2012; Pan et al., 2013; van Woudenbergh et al., 2012).

13.2.4 Environmental Exposures

Environmental exposures include change in climate, contamination of water, and food supply (by chemicals used in agriculture, plasticizers used in the packaging of foods, and drugs in food animals) as well as AGEs in prepared food (Maruthur, 2013). According to insulin hypersecretion hypothesis, food additives have the strongest impact on increased insulin secretion (Corkey, 2012). A study indicated the possibility that air pollution (as estimated by residential NO_2 levels) may increase diabetes risk (Andersen et al., 2012). Other factors contributing to incidence of diabetes include smoking, socioeconomic factors such as urbanizing, genetic risk, and certain medications in chronic diseases.

13.3 AN OVERVIEW OF COMPLICATIONS ASSOCIATED WITH DIABETES AND ADVANCED GLYCATION END PRODUCTS

The elevated sugar levels, or hyperglycemia, are the major driving force for the emergence of numerous complications such as atherosclerosis, cardiac disease, cataract, neuropathy, retinopathy, and many other complications (Brownlee and Cerami, 1981; Halliwell and Gutteridge, 2006; Valko et al., 2007). Therefore, there is an established link between diabetic complications and chronic hyperglycemia in which oxidative stress is the major reinforcement (Brownlee, 2005; Bandeira et al., 2013). Oxidative stress triggers the several diabetes-linked complications (Kiritoshi et al., 2003). For instance, elevated activity of polyol pathway accelerates oxidative stress susceptibility by excessive consumption of NADPH, an essential cofactor of this pathway, and has a vital role in regeneration of reduced glutathione (GSH) (Ho et al., 2006). This pathway is the major entry route of reactive oxygen species (ROS) caused by increased sugar levels in retina (Lee and Chung, 1999). Higher concentrations of intracellular glucose are reduced to sorbitol and further to fructose by aldose reductase, an enzyme that converts toxic aldehydes into inactive alcohols; this enzyme has been found to be directly proportional to cataract associated with diabetes (Lee and Chung, 1999). There is evidence of involvement of polyol pathway in diabetic neuropathy also, leading to abnormalities in the peripheral nervous tissues (Ho et al., 2006). Production of AGEs and activation of receptor for advanced glycation end products (RAGE) is one of the most prominent mechanisms of diabetic complications (Brownlee, 2005).

13.4 BIOCHEMICAL MECHANISM OF GLYCATION

The process of glycation is a series of complex reactions between free amino group of a protein and the carbonyl group of a reducing sugars. The biochemical reaction involved is a free-radical chain reaction that propagates in several stages (Wu et al., 2011). The nonenzymatic nature of this reaction makes it a slow process that usually takes several weeks to complete. Apart from reducing sugars (glucose, ribose, fructose, and galactose), certain metabolites of polyol pathway (fructose or fructose-3-phosphate) and intermediates of glucose metabolism (glucose-6-phosphate, ribose-5-phosphate, fructose-6-phosphate, desoxyribose-5-phosphate, and glyceraldehyde) also react with free amino groups of biological macromolecules to form AGEs (Dunn et al., 1989). Of note, on comparing to other reducing monosaccharides, glucose is less efficient to undergo glycation (Aragno and Mastrocola, 2017).

Glycation is a concentration-dependent and spontaneous reaction that starts with the nucleophilic addition of a carbonyl group from a reducing sugar to free amino group (of proteins or other molecules), resulting in formation of a reversible Schiff base. The Schiff base is highly susceptible to oxidation and free-radical generation to form reactive carbonyl intermediates (Bonnefont-Rousselot, 2002). Amadori products are the rearranged product of the Schiff base. In the presence of transition-metal ions, glycoxidation of Amadori adducts might get fragmented, resulting in formation of short-chain reactive compounds including methylglyoxal and glyoxal (Negre-Salvayre et al., 2009). Glyoxal and methylglyoxal are also produced during lipid peroxidation and accessory glycolytic pathways. Auto-oxidation of glucose itself causes the formation of keto aldehydes and hydrogen peroxide in presence of transition-metal ions which subsequently fasten the formation of AGEs (Wolff and Dean, 1987). These reactive intermediates or compounds further react with amino groups of biological molecules (mainly proteins), consequently generating a wide range of adducts and cross-links called AGEs (Ahmed and Thornalley, 2007; Peyroux and Sternberg, 2006; Thorpe and Baynes, 2003). Earlier it was thought that glycation only targets proteins, but later it was found that aminophospholipids and DNA also undergo glycation process (Baynes, 2000; Pamplona et al., 2000). Glycation proceeds by the reactions of Amadori products with the amino acids to form carboxymethyl lysine (CML) while glycosyl group gets dehydrated to form deoxyglucosone (Wu et al., 2011). Finally, the highly reactive deoxyglucosone reacts with lysine residues of the protein to form pyrraline. Similarly, pentose reacts with arginine and lysine residues of protein to produce pentosidine and other end products. As a result, intra- or intermolecular heterocyclic cross-linking and fragmentation occur in a protein molecule, causing irreversible damage including protein denaturation. Moreover,

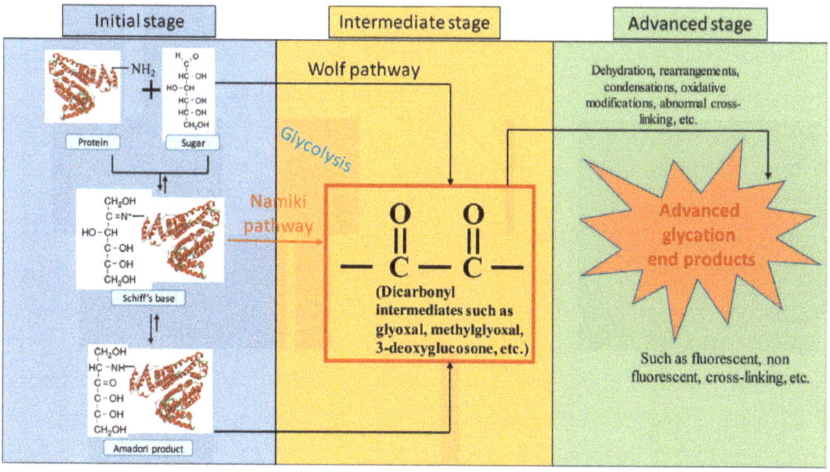

Figure 13.2 Different stages in formation of advanced glycation end products.

other irreversible products of glycation are carboxyethyl lysine, glyoxal-lysine dimer (GOLD), 2-(2-furoyl)-4(5)-(2-furanyl)-1*H*-imidazole, methylglyoxal-lysine dimer, and many other compounds still to be identified. The complete event of glycation has been grouped in three stages (Fig. 13.2).

13.4.1 Initial Phase: Formation of Amadori Products

The process of glycation is initiated by covalent condensation of carbonyl group of a reducing sugar with nitrogen atom of an ε-amino group or terminal α-amino group of lysine and/or arginine residue of protein, nucleic acids, or phospholipids by nucleophilic attack. The step is concentration dependent which takes place at an enhanced rate in patients with hyperglycemia or diabetes (Singh et al., 2001a). The product, an unstable aldimine structure, of this reaction is called the Schiff base. Due to the unstable nature of aldimine functional group, the formation of Schiff base is a reversible reaction. The rearrangement of Schiff base intermediate results in formation of a relatively stable ketosamine structure, called as Amadori products or keto-amines; for example, HbA1c (Stirban et al., 2014). The rate of formation of Amadori products from the Schiff base is comparatively faster than the reverse reaction. This process is believed to be facilitated by acid–base catalysis if there is a lysine or histidine side chain within 5 Å from the target amino residue (Acosta et al., 2000).

13.4.2 Intermediate Phase: Formation of Advanced Glycation End Product Precursors

Formation and reorganization of Amadori products leads to accumulation of reactive intermediates mainly dicarbonyl compounds also called as oxoaldehydes. The major products of these carbonyl intermediates are glyoxal, 3-deoxyglucosone, and methylglyoxal (Baynes, 1991; Suzuki et al., 1999). Under in vivo conditions, these AGE precursors are formed via two major pathways, that is, the metabolic degradation of glucose and fragmentation of the Schiff base and Amadori products. Although the physiologically prevalent form of glucose (i.e., cyclic confirmation), which is approximately 90% of blood glucose, is one of the least reactive forms of reducing sugars to undergo protein glycation and hence are least susceptible to nucleophilic attack by primary amino group (Bunn et al., 1977). However, other sugars and the dicarbonyl compounds, such as glyoxal, methylglyoxal, and 3-deoxyglucosone, undergo glycation at a relatively much faster rate (Li et al., 2008, 2007). Formation of glyoxal occurs through different routes: metal ions catalyzed autoxidation of glucose (Wolff pathway) or lipid peroxidation via the acetol pathway and oxidative degradation of the Schiff base via the Namiki pathway (Dyer et al., 1993; Glomb and Monnier, 1995; Wolff et al., 1991; Wolff and Dean, 1987). 3-Deoxyglucosone is formed by the nonoxidative hydrolysis and fragmentation of Amadori products and also from fructose-3-phosphate which is an intermediate of polyol pathway where aldose reductase reduces glucose to form sorbitol. The reaction proceeds and 3-deoxyglucosone is degraded to form glyceraldehyde and methylglyoxal through retro-aldol condensation (Martins et al., 2003; Thornalley et al., 1999). Despite being derived through nonoxidative mechanisms, they exert intracellular oxidative stress and cause cellular apoptosis (Okado et al., 1996). Among these glycation intermediates, methylglyoxal is one of the most reactive glycating agents, which have affinity to react with several amino acids including arginine and lysine to produce dicarbonyl-derived AGEs, altering the secondary conformation of a protein, browning, denaturation, and finally loss of its function (Meade et al., 2003; Okado et al., 1996; Shinohara et al., 1998). The oxidative and carbonyl stress is the key factor in many pathological complications such as uremia and diabetes, further accelerating the vascular damage (Raj et al., 2000).

13.4.3 Advanced or Last Phase: Formation of Advanced Glycation End Products

AGEs are formed both in vivo and in vitro through a number of pathways such as reactions between AGE precursors and Amadori products, modification by dicarbonyl compounds and direct degradation of the Schiff base or Amadori products. In the last stage of glycation, reactive dicarbonyl compounds combine with lysine and arginine residues of different proteins resulting in formation of a diverse class of irreversible adducts, regarded as AGEs (Jahan and Choudhary, 2015). Consequently, proteins after being altered by active dicarbonyl glycating agents form protein cross-links and induce fragmentation, a situation which is prevalent in several diabetic complications. The AGEs formed are group of heterogeneous compounds such as nonfluorescent cross-linking products including GOLD, fluorescent cross-linking structures such as pentosidine and methylglyoxal lysine, and noncross-linking nonfluorescent compounds such as CML (Ahmed et al., 1986; Wu et al., 2011). There are three major types of AGEs:

1. Cross-linking fluorescent AGEs.
2. Nonfluorescent cross-linking AGE.
3. Nonfluorescent noncross-linking AGEs.

13.5 MECHANISMS OF COMPLICATIONS INDUCED BY GLYCATION

Progression and development of many diabetic complications such as retinopathy, nephropathy, and neuropathy are mainly as result of formation of AGEs (Singh et al., 2014). In smokers, levels of AGEs in tissues and serum are more along with elevated inflammatory markers (Nicholl et al., 1998). Animal studies have found that exogenous AGEs also contribute significantly to development of vascular and renal complications (Zheng et al., 2002). Intracellular accumulation of AGEs results in the formation of glucose-derived dicarbonyl precursors (Brownlee, 2001; Giardino et al., 1994). Such intracellular AGEs activate the intracellular signaling pathways and modify the function of intracellular proteins (Brownlee, 1995). Major sites of accumulation of AGEs are retina, kidney, and atherosclerotic plaques (Hammes et al., 1999). The normal functioning of proteins is hampered as a result of glycation that interferes with receptor recognition, alters enzymatic activity, reduces degradation capacity, and disrupts molecular conformation (Hsieh et al., 2007). The biochemical mechanism by which glycation alters the cell functions includes organopathy as a result of accumulation of AGEs in tissue, denaturation and functional deterioration of the target proteins, generation of both carbonyl stress and oxidative stress, and activation of receptor-dependent signal pathway in cells (Yonekura et al., 2005). AGEs cause the intermolecular cross-linking of collagen which leads to increased diastolic dysfunction, diminished myocardial and arterial compliance, and increased vascular stiffness and systolic hypertension (Cooper et al., 2001). Autoantibodies are formed against serum AGEs accomplishes AGE–immune complex formation that also has a role to play in atherogenesis (Turk et al., 2001). In DNA, amino groups of guanine and adenine bases are vulnerable glycation and AGEs formation (Baynes, 2002). Moreover, AGEs play an important role in pathogenesis of diabetes and its associated complications including cataract, retinopathy, neuropathy, cardiomyopathy, and nephropathy (Abdullah et al., 2017; Singh et al., 2014).

13.6 ACCUMULATION OF ADVANCED GLYCATION END PRODUCTS IN DIABETES AND ITS ASSOCIATED COMPLICATIONS

AGEs are primarily generated via nonenzymatic glycation reaction or Millard reaction which is three-stage process giving rise to so-called endogenous AGEs (Chilelli et al., 2013). A similar category of molecules are advanced lipoxidation end products (ALEs) which are produced from oxidation of fatty acids along with the formation of lipid hydroperoxides. Lipid peroxidation products of ALEs react with proteins to form stable cross-links in which some are fluorescent in nature (Pugliese, 2008). Many other molecules such as methylglyoxal are produced via hexose-monophosphate or glycolysis by enzymatically catalyzed reactions (Kalapos, 2013).

Another way of accumulation of AGEs in diabetic subjects is due to the defects in real excretion of AGEs, e.g. in the case of diabetic nephropathy, as the renal clearance is inversely linked to the levels of AGEs in plasma (Gugliucci and Bendayan, 1996; Tessitore et al., 2004). It further increases the circulating pool of plasma AGEs, which alters the structure or shape of certain proteins at glomerulus level and also interacts with specific receptors. Such interaction of AGEs impairs clearance of glycoxidation product and further worsens chronic kidney disease (Lapolla et al., 2006; Thomas et al., 2005).

Exogenous AGEs, harmful products of the Maillard reaction present in various foods, also accumulate in vivo. Collectively exogenous and endogenous AGEs contribute to the majority of glycation-free adducts which account for a large amount of circulating AGEs in both the diabetic and nondiabetic subjects (Kaňková, 2008). Many modern food-processing techniques, sterilizing, heating, microwaves, etc., also generate exogenous AGEs which increase the nonenzymatic reaction of free NH_2 groups of proteins and lipids to nonreducing sugars (Chilelli et al., 2013). Exogenous AGEs elevate the plasma levels of AGEs in diabetic nephropathic and healthy individuals, inducing oxidative and inflammatory processes which is the pathogenesis of many chronic diseases (Uribarri et al., 2011). It has been found that restriction of exogenous AGEs stops the progression of renal complications, atherosclerosis injury and also improves insulin resistance (Yamagishi and Matsui, 2010).

Progression and pathogenesis of microvascular complications are the most prominent complications caused by AGE pathway in diabetes. AGEs have both intracellular and extracellular targets which are linked together. In the case of increased glycemia, there is glycation of respiratory chain proteins of mitochondria owing to more production of ROS (Rosca et al., 2005). The role of biochemical pathways in the microvascular pathobiology of diabetes confirms the role of AGEs in the progression of diabetic retinopathy, neuropathy, and nephropathy (Chilelli et al., 2013).

There is increasing evidence that formation of AGEs under hyperglycemia plays a key role in the development and progression of diabetes-related complications (Sun et al., 2013; Wan et al., 2015). AGEs have harmful effect that promotes vascular stiffness by causing the cross-linking of long-lived proteins and alters vascular functions. Moreover, the studies have shown the presence of approximately double amount of AGEs in terminal-stage renal diabetic subjects (Galler et al., 2003). Interaction of AGEs with RAGE encourages the intracellular signaling pathway, causing the enhanced oxidative stress, and in turn convolutes prosclerotic and proinflammatory cytokines (Sharma et al., 2012). Apart from accelerated rate of AGEs formation, reduced clearance rate is also responsible for accumulation of AGEs in diabetic patients especially with diabetic nephropathy (Genuth et al., 2005;

Monnier et al., 1999). The AGE–RAGE axis seems to be responsible for continued inflammation, retinal microvascular dysfunction, and neurodegeneration in hyperglycemic conditions (Zong et al., 2011). Accumulation of AGEs on ECM proteins alters cell–matrix as well as matrix–matrix interactions, impedes their degradation by matrix metalloproteinases, and ultimately results in mesangial expansion and membrane thickening (Brownlee and Lilly, 1993; Mott et al., 1997; Sun et al., 2013). These are the hallmarks of diabetic nephropathy (Wan et al., 2015). Also, accumulation of AGEs in subendothelium triggers migration of monocyte across endothelium cell monolayer, leading to activation of nuclear factor-kB (NF-kB) (Morigi et al., 1998). It has been surmised that activation of NADPH oxidase is a primary mechanism of induction of AGEs–RAGE-mediated oxidative stress, resulting in cytokine formation and activation of NF-kB and proinflammatory pathways. Prolonged AGEs–RAGE interaction in pericytes, endothelial cells, and retinal pigment epithelial cells commences multiple cellular events, which eventually leads to retinal damage in diabetic retinopathy (Yamagishi et al., 2012).

13.6.1 Contribution of Advanced Glycation End Products in Diabetic Nephropathy

Nephropathy means damage to kidneys, and in severe cases it might lead to kidney failure. The loss of kidney functions has been found to be correlated with the elevated circulating levels of AGEs in diabetic patients (Genuth et al., 2005). In diabetic nephropathy, there is not only augmented formation of serum AGEs but also their limited clearance (Yamagishi and Matsui, 2010). The disruption of renal architecture is due to cross-linking of matrix proteins by AGEs besides the activation of downstream signaling and ultimately impairs the kidney function (Brownlee and Lilly, 1993). Glycation of laminin and type IV collagen alters their function, leading to increased vascular permeability of albumin. A histological feature of diabetic nephropathy is thickening of basement membrane, caused by the structural changes in other ECM proteins by AGEs, impairing its degradation by metalloproteinases (Thomas et al., 2005). Another important consequence of glycation is mesangium in which increased vascular endothelial growth factor expression and pericyte apoptosis leads to hyperfiltration which is an early kidney dysfunction in diabetes (Wendt et al., 2003).

13.6.2 Contribution of Advanced Glycation End Products in Diabetic Retinopathy

Retinopathy is retinal vascular disease or damage to the retina that may cause vision impairment. It involves structural and functional changes in the retinal vasculature along with development of abnormal vessel causing hemorrhage, infarction, and ischemia which are mainly contributed by AGEs. Apart from production of ROS, accumulation of AGEs leads to vessel thickening, endothelial dysfunction, hypertension, and loss of pericytes (Chilelli et al., 2013). AGEs reduce the survival of platelet and increase its aggregation leading to ischemia and triggers growth factors with angiogenesis and neovascularization (Antonetti et al., 2006). Neuronal damage may also contribute to progression and pathogenesis of retinopathy. There are evidences from animal studies that diabetes severely affects neurosensory retina, increasing neuronal apoptosis besides altering neuro-retinal supporting cell metabolism (Méndez et al., 2010). In hyperglycemia, due to accumulation of AGEs, mullermacroglia play a crucial role in retinal physiology and are severely dysfunctional, increasing glial fibrillary acidic protein expression and ultimately contributing in retinal neuron excitotoxicity (Genuth et al., 2005).

13.6.3 Contribution of Advanced Glycation End Products in Diabetic Neuropathy

Neuropathy is several kinds of abnormalities caused by damage to one or several peripheral nerves. Both peripheral and autonomic nerves are affected in diabetic neuropathy; peripheral nerve damage is associated with atherosclerosis which is a major determinant of lower limb morbidity while autonomic damage may lead to sudden death (Chilelli et al., 2013). The commencement of these complications is governed by nonenzymatic glycation and formation of AGEs. There are evidences of accumulation of methylglyoxal-derived hydroimidazolone, carboxymethyl lysine, frucosyl-lysine, and 3-deoxyglucosone in the peripheral nerves of streptozocin-induced diabetic rats and diabetic patients (Karachalias et al., 2003; Sugimoto et al., 1997). In human subjects, CML was detected in pericytes, vascular endothelial cells, axons, basement membrane, and Schwann cells. The structural proteins found to be modified are neurofilament, tubulin, axonal Na/K ATPase, and myelin proteins. Also, ECM proteins were found modified by AGEs that might be severely affecting cellular function. Glycation of ECM proteins mainly modifies arginine residues, resulting in structural distortions by loss of charge, decreasing integrin-binding affinity and thickening of basement membrane (Duran-Jimenez et al., 2009). Such AGEs-related protein modifications lead to structural changes and impairing peripheral nerve function. Moreover, incubation of Schwann cells and neuronal cells with AGEs induced apoptosis and in vitro (Sekido et al., 2004). Modifications in tubulin and neurofilament intersect axonal transport, contributing to the development of nerve fiber degeneration and atrophy (Wada and Yagihashi, 2005). Nerve fiber demyelination might also be caused by modifications in myelin protein of Schwann cells (Sekido et al., 2004). It has been demonstrated in streptozocin-induced diabetic murine model that glycation of fibronectin and laminin (ECM proteins) causes preconditioned neurite outgrowth and reduces neurotrophin-stimulated neurite outgrowth, a possible mechanism of failure of axonal regeneration (Duran-Jimenez et al., 2009). In rats and mice with diabetic neuropathy, extra-pancreatic proinsulin-producing bone-marrow-derived cells were found in sciatic nerve and neurons of dorsal root ganglion (Chan et al., 2011).

13.6.4 Contribution of Advanced Glycation End Products in Diabetic Cardiomyopathy

In cardiomyopathy, heart muscle becomes rigid or enlarged, causing the worsening of heart by pumping lesser amount of blood through the body. Diabetic cardiomyopathy leads to diastolic dysfunction and is characterized by myocardial fibrosis and myocellular hypertrophy which is associated with a higher rate of heart failure in diabetic subjects (Singh et al., 2014). Diastolic dysfunction is prevalent up to 50%–60% of type 2 diabetic patients. It likely that accumulation of AGEs in myocardium leads to diastolic dysfunction which is correlated to levels of HbA1c (Bell, 2003). Studies in experimental animal models have revealed that aminoguanidine was found to be effective in preventing arterial stiffening and ultimately cardiac hypertrophy, highlighting the pathogenicity of AGEs in this complication (Montagnani, 2009). Methylglyoxal-derived AGEs have been found to upregulate cardiac RAGE mRNA, a triggering factor of cardiomyocyte contractile dysfunction. In nutshell, there are enough evidences that both RAGE and AGE play an important role in the development cardiac dysfunction in diabetes (Petrova et al., 2002).

13.7 CONTROL OF DIABETES BY HERBAL MEDICINE OR PLANTS-BASED MEDICINES

Diabetes is a well-known disease from ancient times, and different civilizations developed their own management practices as per that time. Most of the therapies available earlier were directly herbs or derived from herbs. With the advancement in science, researchers revealed the underlying causes and pathophysiological complications associated with diabetes. This led to the identification of major phytocompounds in herbs responsible for managing diabetes as well as synthesis of novel moieties.

A summary conducted by review literature on antidiabetic activity of various medicinal plants is listed in Table 13.1. The literature indicated that these plants belong to more than 20 families, and many plants are traditionally used in herbal medicine. The major mechanisms of antidiabetic activity documented include reduction in fasting and serum glucose level, increase in glucose metabolism, increase in plasma insulin level, increase in insulin binding to its receptor β-cell rejuvenation, regeneration and stimulation, free-radical scavenging, etc.

In literature, a variety of medicinal and food plants have been documented to exhibit anti-glycation and antidiabetic activities. Excellent review articles are available addressing this issue (Balogun et al., 2016; Bnouham et al., 2006; Mamun-or-Rashid et al., 2014; Mishra et al., 2010; Oghogho Rosalie and Ekpe, 2016; Patel et al., 2012). However, certain plants are common in various herbal medicines developed for diabetic control in traditional and complementary medicine. The details of some such plants are briefly described below.

13.7.1 Gymnema sylvestre

This plant is also known as *gurmar* (sugar destroyer) in Hindi whose leaves are used traditionally to treat diabetes, cholesterol, and obesity in Ayurvedic medicine (Bone, 2002). Leaf extract of this plant produced antihyperglycemic effects and stimulated β-cell regeneration in alloxan-induced diabetic rats (Ahmed et al., 2010; Kang et al., 2012). Gymnemic acid along with some saponins is believed to be the active phytocompound, but the exact mechanism of action is still unclear (Porchezhian and Dobriyal, 2003). Extract of this plant resulted in significant improvements in HbA1c and fasting blood glucose (FBG) levels in patients with type 1 and type 2 diabetes (Baskaran et al., 1990). These promising results in murine model and clinical subjects warrant further study for its judicious use and to understand its exact mode of action.

Table 13.1 List of plants/extracts with reported antidiabetic property (Mishra et al., 2010)

Family	Botanical name	Parts used	Mechanism of action	Chemical constituents	Reference
Acanthaceae	*Andrographis paniculata*	Entire plant	Increase glucose metabolism	Diterpenoid lactone and rographoloid	Yu et al. (2003) and Zhang and Tan (2000)
Anacardiaceae	*Anacardium occidentale*	Entire plant		Flavonols, terpenoid, caumarin	Kamtchouing et al. (1998)
	Mangifera indica	Leaves	Reduction of intestinal absorption of glucose	Mangiferin	Aderibigbe et al. (1999)
Annonaceae	*Annona squamosa*	Leaves	Antihyperglycemic effect, increased plasma insulin level	Acetogenins, squamosin B, squamosamid	Shirwaikar et al. (2004)
Apiaceae	*Carum carvi*	Fruits		V.oil, resin, carvone, fixed oil	Singh et al. (2001b)
Apocynaceae	*Catharanthus roseus*	Leaves, twig, and flower	Increase metabolization of glucose	Indole alkaloid, vincristine, vinblastine	Eddouks et al. (2004)
	Vinca rosea	Leaves	Beta-cell rejuvenation, regeneration, and stimulation	Vincristine, vinblastine	Ghosh and Suryawanshi (2001)
Araliaceae	*Panax ginseng*	Root and entire plant	Lowering blood sugar level	Glycans, panaxans I, J, K, and L	Oshima et al. (1985)

Asclepiadaceae	*Gymnema sylvestre*	Leaves	Lowers plasma glucose level	Gymnemic acid, quercital	Ghalap and Kar (2005, 2003), Shanmugasundaram et al. (1983), and Sugihara et al. (2000)
Asclepidaceae	*Cryptolepis sanguinolenta*	Entire plant	Increase glucose uptake by 3T3-L1 cells	Cryptolepine	Luo et al. (1998)
Asteraceae	*Artemisia pallens*	Aerial parts	Hypoglycemic, increases peripheral glucose utilization	Essential oil, davanone	Subramoniam et al. (1996)
	Bidens pilosa	Aerial parts		Polyacetylenic glucoside	Ubillas et al. (2000)
	Eclipta alba	Leaves	Decrease activity of glucose-6-phosphatase and fructose-1−6, bisphasphatase	Ecliptin alkaloid	Ananthi et al. (2003)
Bixaceae	*Bixa orellana*	Entire plant	Increase plasma insulin conc. and increase insulin binding on insulin receptor	Oleoresin	Russell et al. (2005)
Brassicaceae	*Brassica juncea*	Leaves and seed	Food adjuvants for diabetic patients	Isothiocyanate glycoside singrin, protein, fixed oil	Grover et al. (2003)
Capparidaceae	*Capparis deciduas*	Powder	Hypoglycemic, antioxidant, hypolipidemic		Yadav et al. (1997)
Chenopodiaceae	*Beta vulgaris*	Leaves	Reduce blood glucose level by regeneration of β cells		Bolkent et al. (2000)

(*Continued*)

Table 13.1 (Continued)

Family	Botanical name	Parts used	Mechanism of action	Chemical constituents	Reference
Convolvulaceae	*Ipomoea aquatica*	Leaves	Reduce fasting blood sugar level and serum glucose level	Carotene	Malalavidhane et al. (2003)
Cucurbitaceae	*Luffa aegyptiaca*	Seed	Galactigogue activity	Fatty oil	El-Fiky et al. (1996)
	Momordica charantia	Fruit	Reduce blood glucose level	Momordicine alkaloid, ascorbic acid	Chaturvedi et al. (2004)
Euphorbiaceae	*Embellica officinalis*	Fruits	Reduce 5-hydroxymethylfurfural, creatinine albumin level	Vit. C, tannin	Rao et al. (2005)
	Phyllanthus amarus	Entire plant	Decrease blood glucose level	Alkaloids	Srividya and Periwal (1995)
Fabaceae	*Lupinus albus*	Seed	Lower serum glucose level	Alkaloid, fatty oil, asparagines	Tsiodras et al. (1999)
	Acacia arabica	Seed	Initiate release of insulin	Arabin	Wadood et al. (1989)
	Glycerrhiza glabra	Root	Lowers plasma glucose level	Triterpenoid, saponin, glycerrhizin	Swanston-Flatt et al. (1990)
	Trigonella foenum-graceum	Seed	Decrease blood glucose concentration	Protein, fat, V. oil, fixed oil, carbohydrate	Ajabnoor and Tilmisany (1988)

Gentianaceae	*Enicostemma littorale*	Entire plant	Decrease glycosylated Hb and glucose-6-phosphatase	Swertiamarin glycoside	Maroo et al. (2003), Upadhyay and Goyal (2004), Vijayvargia et al. (2000)
	Swertia chirayata	Entire plant	Stimulates insulin release from islets	Xanthone mangiferin, gentianine, swerchirin	Saxena et al. (1996)
Lamiaceae	*Ocimum sanctum*	Leaves	Lowering blood sugar level	Alkaloid, tannin, ascorbic acid	Chatopadhyay (1993)
Lauraceae	*Cinnamomum zeylanicum*	Bark	Elevation in plasma insulin	Tannin, mannitol, oxalate	Satheesh and Pari (2004)
Liliaceae	*Allium sativum*	Roots	Antihyperglycemic and antinociceptive effect	Allin, allicin	Kumar and Reddy (1999)
	Allium cepa	Bulb	Stimulating effects on glucose utilization and antioxidant enzyme	Vit. A, B, C, allyl propyl disulphide	Kumari and Augusti (2002)
	Aloe vera	Entire plant		Aloin glycoside	Okyar et al. (2001) and Rajasekaran et al. (2004)
	Aloe barbadensis	Leaves	Stimulating synthesis and/or release of insulin	Barbaloin, isobarbaloin, resin	Ajabnoor (1990)
Malvaceae	*Hibiscus rosa-sinensis*	Entire plant	Stimulate insulin secretion from beta cells	Vit. B, C, fat	Sachdewa and Khemani (1999)

(*Continued*)

Table 13.1 (Continued)

Family	Botanical name	Parts used	Mechanism of action	Chemical constituents	Reference
Meliaceae	*Azadirachta indica*	Leaves	Glycogenolytic effect due to epinephrine action was blocked	Nimbidin, Nimbin, Nimbidol, Nimbosterol	Chatopadhyay (1996)
Menispermaceae	*Coscinium fenestratum*	Stem	Increase enzymatic antioxidants	Berberine, glycoside, saponin	Shirwaikar et al. (2005)
	Tinospora cardifolia	Root	Decrease blood glucose and brain lipid	Berberine, starch	Stanely et al. (2000)
Moraceae	*Ficus religiosa*	Entire plant	Initiating release of insulin	Tannin	Wadood et al. (2003)
Myrtaceae	*Eucalyptus globulus*	Leaves	Increase insulin secretion from clonal pancreatic beta line	Essential oil, cineol	Gray and Flatt (1998)
	Myrtus communis	Leaves	Lower blood glucose level	V.oil mirtii oleum	Sepici et al. (2004)
Nyctaginaceae	*Boerhaavia diffusa*	Leaves and entire plant	Increase activity of glucose metabolic enzymes, increase plasma insulin level	Alkaloid punarnavine, punarnavoside	Satheesh and Pari (2004)
Nymphaeaceae	*Nelumbo nucifera*	Rhizome	Reduce blood sugar level	Nuciferin, nornuciferin	Khan et al. (1995)
Oleaceae	*Olea europia*	Leaves	Induced insulin released and increase peripheral uptake of glucose	Oleuropeoside	Gonzalez et al. (1992)

Pandanaceae	*Pandanus odorus*	Root	Decrease plasma glucose level	Essential oil	Peungvicha et al. (1996)
Pinaceae	*Abies pindrow*	Entire plant	Insulin secretagogue activity	Volatile oil	Hussain et al. (2004)
Punicaceae	*Punica granatum*	Seed	Reduce blood sugar level	Vit. C, protein, tannin, gallic acid, pelletierine	Das et al. (2001)
Rhamnaceae	*Zizyphus sativa*	Leaves	Dose-dependent reduction in blood glucose level	Tannin	Anand et al. (1989)
Scrophulariaceae	*Picrorrhiza kurroa*	Entire plant	Decrease serum glucose	Picrorrhizin, kutkin	Joy and Kuttan (1999)
Solanaceae	*Capsicum frutescens*	Entire plant	Increase insulin secretion	Capsaicin, pritein	Talan et al. (2001)
	Withania somnifera	Root	Decrease blood sugar level	Withanine, withaferine, withanolide	Andallu and Radhika (2000)
Sterculiaceae	*Abroma augusta*	Roots and leaves	Lowering blood sugar	Fixed oil, alkaloid	Halim (2003)
	Helicteres isora	Root	Decrease plasma triglyceride level and insulin sensitizing activity	Saponin, tannin, lignin	Chakrabarti et al. (2002)

(*Continued*)

Table 13.1 (Continued)

Family	Botanical name	Parts used	Mechanism of action	Chemical constituents	Reference
Theaceae	*Camellia sinensis*	Leaves	Increase insulin secretion	Polyphenolic constituents (EGCG)	Koyama et al. (2004)
Urticaceae	*Urtica dioica*	Leaves	Increase insulin secretion	Fatty	Farzami et al. (2003)
Zingiberaceae	*Zingiber officinale*	Rhizome	Increase insulin level and decrease fasting glucose level	Sesquiterpene	Akhani et al. (2004)
Zygophyllaceae	*Tribulus terrestris*	Saponin	Decrease serum glucose	Harmine	Li et al. (2002)

13.7.2 Cinnamon

Cinnamomum verum is a small tree whose bark is commonly used as spice. *Cinnamon* is also traditionally used for many centuries for the treatment of diabetes and other conditions. The aqueous extract of this plant is known to enhance the glycogen synthase activity and activate the insulin receptor by diverse mechanisms (Cao et al., 2007; Imparl-Radosevich et al., 1998; Jarvill-Taylor et al., 2001; Qin et al., 2003). Many clinical trials had shown that Cinnamon has a substantial effect on FBG and HbA1c levels in diabetic subjects (Baker et al., 2008; Khan et al., 2003; Mang et al., 2006). Cinnamic acid, a known constituent of *Cinnamomum* plant, has been found to inhibit glycation (Qais et al., 2016). Till date, no major adverse effect of this plant has been reported in trials (Nahas and Moher, 2009).

13.7.3 Camellia sinensis

The plant is commonly known as green tea and has numerous health benefits which are mainly due to polyphenol catechins, predominantly epigallocatechin gallate (Higdon and Frei, 2003; Nagle et al., 2006). These phytocompounds are reported to reduce β-cell damage and improve insulin sensitivity in animal and in vitro studies (Anderson and Polansky, 2002; Wu et al., 2004). Initial administration of caffeine impairs glucose metabolism, but long-term dosage increases basal energy expenditure, stimulates lipolysis, and mobilizes muscle glycogen (Higdon and Frei, 2003; Robinson et al., 2004). Trial studies have found that partly administered oxidized tea to patients for 4 weeks resulted in 30% decrease in FBG levels (Chuang et al., 1992). There is little evidence in support of use of green tea to control hyperglycemia. Therefore, further research is warranted on this plant to be used in modern therapeutic system as antidiabetic drug.

13.7.4 Trigonella foenum-graecum

Trigonella foenum-graecum, commonly known as fenugreek, is cultivated for its medicinal properties for many centuries Mediterranean and Asian and cultures. The seed and leaf of this plant is used for the treatment of diabetes in many traditional systems including Ayurvedic medicine (Nahas and Moher, 2009). The most studied phytocompound is 4-hydroxyisoleucine, which is known to increase pancreatic insulin secretion and inhibit α-amylase activity (Ajabnoor and Tilmisany, 1988; Sauvaire et al., 1998).

A clinical trial has found that seeds of this plant improved FBG levels without any noticeable adverse effects (Sharma and Raghuram, 1990).

13.7.5 Momordica charantia

Momordica charantia, also known as bitter melon, is tropical plant whose fruit is used in Ayurvedic medicine. Active phytoconstituents of this plant are charantin, vicine, and polypeptide (Basch et al., 2003). These ingredients are thought to alter hepatic glucose metabolism and stimulate insulin secretion (Welihinda et al., 1986; Yeh et al., 2003). Although many clinical studies did not find any significant effect on HbA1c or FBG levels, it has a widespread use in traditional system of medicine (Dans et al., 2007; John et al., 2003).

13.8 CONCLUSION

Due to prolonged hyperglycemia, there is an accelerated rate of nonenzymatic glycation. Formation and accumulation of AGEs are drivers of diabetes-related complications. Numerous compounds are generated at elevated sugar levels that activate intracellular signaling pathways, further enhancing diabetic complications. It is well known that RAGE play an important role in pathogenesis of diabetic complications; therefore, the underlying molecular mechanism is needed to be investigated in much detail. The possible strategies for reducing the rate of glycation and accumulation of AGEs is either by lowering the plasma sugar level or blocking AGE binding to RAGE. Many other alternatives are also possible, but no remarkable success has been achieved till date. Therefore, natural products such medicinal plants/herbs should be tested at biochemical and molecular level for their possible therapeutic potential.

ACKNOWLEDGMENT

FAQ is thankful to UGC for providing fellowship.

REFERENCES

Abdullah, K.M., Qais, F.A., Ahmad, I., Naseem, I., 2017. Inhibitory effect of vitamin B_3 against glycation and reactive oxygen species production in HSA: an in vitro approach. Arch. Biochem. Biophys. 627, 21–29.

Acosta, J., Hettinga, J., Flückiger, R., Krumrei, N., Goldfine, A., Angarita, L., et al., 2000. Molecular basis for a link between complement and the vascular complications of diabetes. Proc. Natl Acad. Sci. U.S.A. 97, 5450–5455.

Aderibigbe, A.O., Emudianughe, T.S., Lawal, B.A.S., 1999. Antihyperglycaemic effect of *Mangifera indica* in rat. Phytother. Res. 13, 504–507.

Afkarian, M., Sachs, M.C., Kestenbaum, B., Hirsch, I.B., Tuttle, K.R., Himmelfarb, J., et al., 2013. Kidney disease and increased mortality risk in type 2 diabetes. J. Am. Soc. Nephrol. 24, 302–308.

Ahmed, N., Thornalley, P.J., 2007. Advanced glycation endproducts: what is their relevance to diabetic complications? Diabetes Obes. Metab. 9, 233–245.

Ahmed, M.U., Thorpe, S.R., Baynes, J.W., 1986. Identification of *N* epsilon-carboxymethyllysine as a degradation product of fructoselysine in glycated protein. J. Biol. Chem. 261, 4889–4894.

Ahmed, A.B.A., Rao, A.S., Rao, M.V., 2010. In vitro callus and in vivo leaf extract of *Gymnema sylvestre* stimulate β-cells regeneration and anti-diabetic activity in Wistar rats. Phytomedicine 17, 1033–1039.

Ahola, A.J., Saraheimo, M., Forsblom, C., Hietala, K., Sintonen, H., Groop, P.H., 2010. Health-related quality of life in patients with type 1 diabetes—association with diabetic complications (the FinnDiane Study). Nephrol. Dial. Transplant. 25, 1903–1908.

Ajabnoor, M.A., 1990. Effect of aloes on blood glucose levels in normal and alloxan diabetic mice. J. Ethnopharmacol. 28, 215–220.

Ajabnoor, M.A., Tilmisany, A.K., 1988. Effect of *Trigonella foenum graceum* on blood glucose levels in normal and alloxan-diabetic mice. J. Ethnopharmacol. 22, 45–49.

Akhani, S.P., Vishwakarma, S.L., Goyal, R.K., 2004. Anti-diabetic activity of *Zingiber officinale* in streptozotocin-induced type I diabetic rats. J. Pharm. Pharmacol. 56, 101–105.

Alberti, K.G.M.M., Zimmet, P., 2013. Global burden of disease—where does diabetes mellitus fit in? Nat. Rev. Endocrinol. 9, 258–260.

Alikhani, M., Alikhani, Z., Boyd, C., MacLellan, C.M., Raptis, M., Liu, R., et al., 2007. Advanced glycation end products stimulate osteoblast apoptosis via the MAP kinase and cytosolic apoptotic pathways. Bone 40, 345–353.

Anand, K.K., Singh, B., Grand, D., Chandan, B.K., Gupta, V.N., 1989. Effect of *Zizyphus sativa* leaves on blood glucose levels in normal and alloxan-diabetic rats. J. Ethnopharmacol. 27, 121–127.

Ananthi, J., Prakasam, A., Pugalendi, K.V., 2003. Antihyperglycemic activity of *Eclipta alba* leaf on alloxan-induced diabetic rats. Yale J. Biol. Med. 76, 97–102.

Andallu, B., Radhika, B., 2000. Hypoglycemic, diuretic and hypocholesterolemic effect of winter cherry (*Withania somnifera* Dunal) root. Indian J. Exp. Biol. 38, 607–609.

Andersen, Z.J., Raaschou-Nielsen, O., Ketzel, M., Jensen, S.S., Hvidberg, M., Loft, S., et al., 2012. Diabetes incidence and long-term exposure to air pollution: a cohort study. Diabetes Care 35, 92–98.

Anderson, R.A.R., Polansky, M.M., 2002. Tea enhances insulin activity. J. Agric. Food Chem. 50, 7182–7186.

Antonetti, D.A., Barber, A.J., Bronson, S.K., Freeman, W.M., Gardner, T.W., Jefferson, L.S., et al., 2006. Diabetic retinopathy: seeing beyond glucose-induced microvascular disease. Diabetes 55, 2401–2411.

Antonetti, D.A., Klein, R., Gardner, T.W., 2012. Diabetic retinopathy. N. Engl. J. Med. 366, 1227–1239.

Aragno, M., Mastrocola, R., 2017. Dietary sugars and endogenous formation of advanced glycation endproducts: emerging mechanisms of disease. Nutrients 9, 385.

Baker, W.L., Gutierrez-Williams, G., White, C.M., Kluger, J., Coleman, C.I., 2008. Effect of cinnamon on glucose control and lipid parameters. Diabetes Care 31, 41–43.

Balogun, F.O., Tshabalala, N.T., Ashafa, A.O.T., 2016. Antidiabetic medicinal plants used by the Basotho tribe of Eastern Free State: a review. J. Diabetes Res. 2016, 1–13.

Bandeira, de M., da Fonseca, L.S., Guedes, da S.G., Rabelo, L., Goulart, M., Vasconcelos, S., 2013. Oxidative stress as an underlying contributor in the development of chronic complications in diabetes mellitus. Int. J. Mol. Sci. 14, 3265–3284.

Basch, E., Gabardi, S., Ulbricht, C., 2003. Bitter melon (*Momordica charantia*): a review of efficacy and safety. Am. J. Heal. Pharm. 60, 356–359.

Baskaran, K., Ahamath, B.K., Shanmugasundaram, K.R., Shanmugasundaram, E.R.B., 1990. Antidiabetic effect of a leaf extract from *Gymnema sylvestre* in non-insulin-dependent diabetes mellitus patients. J. Ethnopharmacol. 30, 295–305.

Baynes, J.W., 1991. Role of oxidative stress in development of complications in diabetes. Diabetes 40, 405–412.

Baynes, J.W., 2000. From life to death—the struggle between chemistry and biology during aging: the Maillard reaction as an amplifier of genomic damage. Biogerontology 1, 235–246.

Baynes, J.W., 2002. The Maillard hypothesis on aging: time to focus on DNA. Ann. N.Y. Acad. Sci. 959, 360–367.

Bell, D.S.H., 2003. Diabetic cardiomyopathy. Diabetes Care 26, 2949–2951.

Bnouham, M., Ziyyat, A., Mekhfi, H., Tahri, A., Legssyer, A., 2006. Medicinal plants with potential antidiabetic activity—a review of ten years of herbal medicine research (1990–2000). Int. J. Diabetes Metab. 14, 1–25.

Bohlooli, M., Miri, M., Khajeh, M., Ghamari, F., Khatibi, A., Ghaffari-Moghaddam, M., et al., 2016. The inhibitory influence of 3-β-hydroxybutyrate on calf thymus DNA glycation by glucose. RSC Adv. 6, 83880–83884.

Bolkent, Ş., Yanardağ, R., Tabakoğlu-Oğuz, A., Özsoy-Saçan, Ö., 2000. Effects of chard (*Beta vulgaris* L. var. cicla) extract on pancreatic B cells in streptozotocin-diabetic rats: a morphological and biochemical study. J. Ethnopharmacol. 73, 251–259.

Bone, K., 2002. Gymnema: a key herb in the management of diabetes. (Phytotherapy Review & Commentary.). Townsend Lett. Dr. Patients. 233, 28–31.

Bonnefont-Rousselot, D., 2002. Glucose and reactive oxygen species. Curr. Opin. Clin. Nutr. Metab. Care 5, 561–568.

Brownlee, M.D.M., 1995. Advanced protein glycosylation in diabetes and aging. Annu. Rev. Med. 46, 223–234.

Brownlee, M., 2001. Biochemistry and molecular cell biology of diabetic complications. Nature 414, 813–820.

Brownlee, M., 2005. The pathobiology of diabetic complications: a unifying mechanism. Diabetes 54, 1615–1625.

Brownlee, M., Cerami, A., 1981. The biochemistry of the complications of diabetes mellitus. Annu. Rev. Biochem. 50, 385–432.

Brownlee, M., Lilly, L., 1993. Glycation and diabetic complications. Diabetes 43, 836–841.

Bunn, H., Gabby, K., Gallop, P., 1977. Posttranslational covalent modification of proteins. Science 200, 21–27.

Cao, H., Polansky, M.M., Anderson, R.A., 2007. Cinnamon extract and polyphenols affect the expression of tristetraprolin, insulin receptor, and glucose transporter 4 in mouse 3T3-L1 adipocytes. Arch. Biochem. Biophys. 459, 214–222.

Chakrabarti, R., Vikramadithyan, R.K., Mullangi, R., Sharma, V., Jagadheshan, H., Rao, Y., et al., 2002. Antidiabetic and hypolipidemic activity of *Helicteres isora* in animal models. J. Ethnopharmacol. 81, 343–349.

Chan, J., Malik, V., Jia, W., Kadowaki, T., Yajnik, C.C.S.C., Yoon, K.K.-H.K., et al., 2009. Diabetes in Asia: epidemiology, risk factors, and pathophysiology. JAMA 301, 2129–2140.

Chan, L., Terashima, T., Urabe, H., Lin, F., Kojima, H., 2011. Pathogenesis of diabetic neuropathy: bad to the bone. Ann. N.Y. Acad. Sci. 1240, 70–76.

Chan, J., Yeung, R.R., Luk, A., 2014. The Asian diabetes phenotypes: challenges and opportunities. Diabetes Res. Clin. Pract. 105, 135–139.

Chatopadhyay, R., 1993. Hypoglycemic effect of *Ocimum sanctum* leaf extract in normal & streptozotocin diabetic rats. Indian J. Exp. Biol. 31, 891–893.

Chatopadhyay, R.R., 1996. Possible mechanisms of antihyperglycemic effect of *Azadirachta indica* leaf extract part. Gen. Pharmacol. 27, 431–434.

Chaturvedi, P., George, S., Milinganyo, M., Tripathi, Y.B., 2004. Effect of *Momordica charantia* on lipid profile and oral glucose tolerance in diabetic rats. Phyther. Res. 18, 954–956.

Chilelli, N.C., Burlina, S., Lapolla, A., 2013. AGEs, rather than hyperglycemia, are responsible for microvascular complications in diabetes: a "glycoxidation-centric" point of view. Nutr. Metab. Cardiovasc. Dis. 23, 913–919.

Chuang, L.M., Jou, T.S., Yang, W.S., Wu, H.P., Huang, S.H., Tai, T.Y., et al., 1992. Therapeutic effect of guar gum in patients with non-insulin-dependent diabetes mellitus. J. Formos. Med. Assoc. 91, 15–19.

Cooper, M.E., Bonnet, F., Oldfield, M., Jandeleit-Dahm, K., 2001. Mechanisms of diabetic vasculopathy: an overview. Am. J. Hypertens. 14, 475–486.

Corkey, B.E., 2012. Diabetes: have we got it all wrong?: insulin hypersecretion and food additives: cause of obesity and diabetes? Diabetes Care 35, 2432–2437.

Danaei, G., Singh, G.M., Paciorek, C.J., Lin, J.K., Cowan, M.J., Finucane, M.M., et al., 2013. The global cardiovascular risk transition: associations of four metabolic risk factors with national income, urbanization, and western diet in 1980 and 2008. Circulation 127, 1493–1502.

Dans, A.M.L., Villarruz, M.V.C., Jimeno, C.A., Javelosa, M.A.U., Chua, J., Bautista, R., et al., 2007. The effect of *Momordica charantia* capsule preparation on glycemic control in Type 2 Diabetes Mellitus needs further studies. J. Clin. Epidemiol. 60, 554–559.

Das, A.K., Mandal, S.C., Banerjee, S.K., Sinha, S., Saha, B.P., Pal, M., 2001. Studies on the hypoglycaemic activity of *Punica granatum* seed in streptozotocin induced diabetic rats. Phyther. Res. 15, 628–629.

de Onis, M., Blossner, M., Borghi, E., 2010. Global prevalence and trends of overweight and obesity among preschool children. Am. J. Clin. Nutr. 92, 1257–1264.

Dunn, J.A., Patrick, J.S., Thorpe, S.R., Baynes, J.W., 1989. Oxidation of glycated proteins: age-dependent accumulation of *N* epsilon-(carboxymethyl)lysine in lens proteins. Biochemistry 28, 9464–9468.

Duran-Jimenez, B., Dobler, D., Moffatt, S., Rabbani, N., Streuli, C.H., Thornalley, P.J., et al., 2009. Advanced glycation end products in extracellular matrix proteins contribute to the failure of sensory nerve regeneration in diabetes. Diabetes 58, 2893–2903.

Dyer, D.G., Dunn, J.A., Thorpe, S.R., Bailie, K.E., Lyons, T.J., McCance, D.R., et al., 1993. Accumulation of Maillard reaction products in skin collagen in diabetes and aging. J. Clin. Invest. 91, 2463–2469.

Eddouks, M., Lemhadri, A., Michel, J.-B., 2004. Caraway and caper: potential anti-hyperglycaemic plants in diabetic rats. J. Ethnopharmacol. 94, 143–148.

El-Fiky, F.K., Abou-Karam, M.A., Afify, E.A., 1996. Effect of *Luffa aegyptiaca* (seeds) and *Carissa edulis* (leaves) extracts on blood glucose level of normal and streptozotocin diabetic rats. J. Ethnopharmacol. 50, 43–47.

Farzami, B., Ahmadvand, D., Vardasbi, S., Majin, F., Khaghani, S., 2003. Induction of insulin secretion by a component of *Urtica dioica* leave extract in perifused islets of Langerhans and its in vivo effects in normal and streptozotocin diabetic rats. J. Ethnopharmacol. 89, 47–53.

Finucane, M.M., Stevens, G.A., Cowan, M.J., Danaei, G., Lin, J.K., Paciorek, C.J., et al., 2011. National, regional, and global trends in body-mass index since 1980: systematic analysis of health examination surveys and epidemiological studies with 960 country-years and 9·1 million participants. Lancet 377, 557–567.

Fraser, D.A., Hansen, K.F., 2005. Making sense of advanced glycation end products and their relevance to diabetic complications. Inter Diabetes Monit. 17, 1–7.

Galler, A., Müller, G., Schinzel, R., Kratzsch, J., Kiess, W., Münch, G., 2003. Impact of metabolic control and serum lipids on the concentration of advanced glycation end products in the serum of children and adolescents with type 1 diabetes, as determined by fluorescence spectroscopy and nepsilon-(carboxymethyl)lysine ELISA. Diabetes Care 26, 2609–2615.

Genuth, S., Sun, W., Cleary, P., Sell, D.R., Dahms, W., Malone, J., et al., 2005. Glycation and carboxymethyllysine levels in skin collagen predict the risk of future 10-year progression of diabetic retinopathy and nephropathy in the diabetes control and complications trial and epidemiology of diabetes interventions and complications participants with type 1 diabetes. Diabetes 54, 3103–3111.

Ghalap, S., Kar, A., 2003. Effect of *Inula racemosa* root and *Gymnema sylvestre* leaf extract in the regulation of corticosteroid induced diabetes mellitus: involvement of thyroid hormone. Pharmazie 58, 413–415.

Ghalap, S., Kar, A., 2005. Gymnemic acid from *Gymnema sylvestre* potentially regulates dexamethasone induced hyperglycemia in mice. Pharm. Biol. 43, 192–195.

Ghosh, S., Suryawanshi, S., 2001. Effect of *Vinca rosea* extract in treatment of alloxan diabetes in male albino rats. Indian J. Exp. Biol. 39, 748–759.

Giardino, I., Edelstein, D., Brownlee, M., 1994. Nonenzymatic glycosylation in vitro and in bovine endothelial cells alters basic fibroblast growth factor activity. A model for intracellular glycosylation in diabetes. J. Clin. Invest. 94, 110–117.

Glomb, M.A., Monnier, V.M., 1995. Mechanism of protein modification by glyoxal and glycolaldehyde, reactive intermediates of the Maillard reaction. J. Biol. Chem. 270, 10017–10026.

Gonzalez, M., Zarzuelo, A., Gamez, M., Utrilla, M., Jimenez, J., Osuna, I., 1992. Hypoglycemic activity of olive leaf. Planta Med. 58, 513–515.

Gray, A.M., Flatt, P.R., 1998. Antihyperglycemic actions of *Eucalyptus globulus* (*Eucalyptus*) are associated with pancreatic and extra-pancreatic effects in mice. J. Nutr. 128, 2319–2323.

Grover, J., Yadav, S., Vats, V., 2003. Effect of feeding *Murraya koenigii* and *Brassica juncea* diet kidney functions and glucose levels in streptozotocin diabetic mice. J. Ethnopharmacol. 85, 1–5.

Guariguata, L., Whiting, D.R.R., Hambleton, I., Beagley, J., Linnenkamp, U., Shaw, J.E. E., 2014. Global estimates of diabetes prevalence for 2013 and projections for 2035. Diabetes Res. Clin. Pract. 103, 137–149.

Gugliucci, A., Bendayan, M., 1996. Renal fate of circulating advanced glycated end products (AGE): evidence for reabsorption and catabolism of AGE-peptides by renal proximal tubular cells. Diabetologia 39, 149–160.

Halim, E., 2003. Lowering of blood sugar by water extract of *Azadirachta indica* and *Abroma augusta* in diabetes rats. Indian J. Exp. Biol. 41, 636–640.

Halliwell, B., Gutteridge, J.M., 2006. Free radicals in biology and medicine. 4th ed. Clarendon Press, Oxford.

Hammes, H.-P., Alt, A., Niwa, T., Clausen, J.T., Bretzel, R.G., Brownlee, M., et al., 1999. Differential accumulation of advanced glycation end products in the course of diabetic retinopathy. Diabetologia 42, 728–736.

Heilig, C.W., Concepcion, L.A., Riser, B.L., Freytag, S.O., Zhu, M., Cortes, P., 1995. Overexpression of glucose transporters in rat mesangial cells cultured in a normal glucose milieu mimics the diabetic phenotype. J. Clin. Invest. 96, 1802–1814.

Helou, C., Marier, D., Jacolot, P., Abdennebi-Najar, L., Niquet-Léridon, C., Tessier, F.J., et al., 2014. Microorganisms and Maillard reaction products: a review of the literature and recent findings. Amino Acids 46, 267–277.
Herman, W.H., Zimmet, P., 2012. Type 2 diabetes: an epidemic requiring global attention and urgent action. Diabetes Care 35, 943–944.
Higdon, J.V., Frei, B., 2003. Tea catechins and polyphenols: health effects, metabolism, and antioxidant functions. Crit. Rev. Food Sci. Nutr. 43, 89–143.
Ho, E.C.M., Lam, K.S.L., Chen, Y.S., Yip, J.C.W., Arvindakshan, M., Yamagishi, S.-I., et al., 2006. Aldose reductase-deficient mice are protected from delayed motor nerve conduction velocity, increased c-Jun NH_2-terminal kinase activation, depletion of reduced glutathione, increased superoxide accumulation, and DNA damage. Diabetes 55, 1946–1953.
Hsieh, C.-L., Yang, M.-H., Chyau, C.-C., Chiu, C.-H., Wang, H.-E., Lin, Y.-C., et al., 2007. Kinetic analysis on the sensitivity of glucose- or glyoxal-induced LDL glycation to the inhibitory effect of *Psidium guajava* extract in a physiomimic system. Biosystems 88, 92–100.
Hussain, Z., Waheed, A., Qureshi, R.A., Burdi, D.K., Verspohl, E.J., Khan, N., et al., 2004. The effect of medicinal plants of Islamabad and Murree region of Pakistan on insulin secretion from INS-1 cells. Phyther. Res. 18, 73–77.
Imparl-Radosevich, J., Deas, S., Polansky, M.M., Baedke, D.A., Ingebritsen, T.S., Anderson, R.A., et al., 1998. Regulation of PTP-1 and insulin receptor kinase by fractions from cinnamon: implications for cinnamon regulation of insulin signalling. Horm. Res. Paediatr. 50, 177–182.
Imperatore, G., Boyle, J.P., Thompson, T.J., Case, D., Dabelea, D., Hamman, R.F., et al., 2012. Projections of type 1 and type 2 diabetes burden in the United States population aged <20 years through 2050: dynamic modeling of incidence, mortality, and population growth. Diabetes Care 35, 2515–2520.
Jahan, H., Choudhary, M.I., 2015. Glycation, carbonyl stress and AGEs inhibitors: a patent review. Expert Opin. Ther. Pat. 25, 1267–1284.
Jang, C., Lim, J.H., Park, C.W., Cho, Y.-J., 2011. Regulator of calcineurin 1 isoform 4 (RCAN1.4) is overexpressed in the glomeruli of diabetic mice. Korean J. Physiol. Pharmacol. 15, 299–305.
Jarvill-Taylor, K.J., Anderson, R.A., Graves, D.J., 2001. A hydroxychalcone derived from cinnamon functions as a mimetic for insulin in 3T3-L1 adipocytes. J. Am. Coll. Nutr. 20, 327–336.
John, A.J.J., Cherian, R., Subhash, H.S.S., Cherian, A.M.M., 2003. Evaluation of the efficacy of bitter gourd (*Momordica charantia*) as an oral hypoglycemic agent—a randomized controlled clinical trial. Indian J. Physiol. Pharmacol. 47, 363–365.
Joy, K.L., Kuttan, R., 1999. Anti-diabetic activity of *Picrorrhiza kurroa* extract. J. Ethnopharmacol. 67, 143–148.
Kalapos, M.P., 2013. Where does plasma methylglyoxal originate from? Diabetes Res. Clin. Pract. 99, 260–271.
Kamtchouing, P., Sokeng, S.D., Moundipa, P.F., Watcho, P., Jatsa, H.B., Lontsi, D., 1998. Protective role of *Anacardium occidentale* extract against streptozotocin-induced diabetes in rats. J. Ethnopharmacol. 62, 95–99.
Kang, M.-H., Lee, M.S., Choi, M.-K., Min, K.-S., Shibamoto, T., 2012. Hypoglycemic activity of *Gymnema sylvestre* extracts on oxidative stress and antioxidant status in diabetic rats. J. Agric. Food Chem. 60, 2517–2524.
Kaňková, K., 2008. Diabetic threesome (hyperglycaemia, renal function and nutrition) and advanced glycation end products: evidence for the multiple-hit agent? Proc. Nutr. Soc. 67, 60–74.

Karachalias, N., Babaei-Jadidi, R., Ahmed, N., Thornalley, P.J., 2003. Accumulation of fructosyl-lysine and advanced glycation end products in the kidney, retina and peripheral nerve of streptozotocin-induced diabetic rats. Biochem. Soc. Trans. 31, 1423–1425.
Khan, B.A., Abraham, A., Leelamma, S., 1995. Hypoglycemic action of *Murraya koenigii* (curry leaf) and Brassica juncea (mustard): mechanism of action. Indian J. Biochem. Biophys. 32, 106–108.
Khan, A., Safdar, M., Ali Khan, M.M., Khattak, K.N., Anderson, R.A., 2003. Cinnamon improves glucose and lipids of people with type 2 diabetes. Diabetes Care 26, 3215–3218.
Kiritoshi, S., Nishikawa, T., Sonoda, K., Kukidome, D., Senokuchi, T., Matsuo, T., et al., 2003. Reactive oxygen species from mitochondria induce cyclooxygenase-2 gene expression in human mesangial cells: potential role in diabetic nephropathy. Diabetes 52, 2570–2577.
Koyama, Y., Abe, K., Sano, Y., Shizaki, L., 2004. Effect of green tea on gene expression of hepatic gluconeogenic enzyme in vivo. Planta Med. 70, 1100–1102.
Kumar, G.R., Reddy, K.P., 1999. Reduced nociceptive responses in mice with alloxan induced hyperglycemia after garlic (*Allium sativum* Linn.) treatment. Indian J. Exp. Biol. 37, 662–666.
Kumari, K., Augusti, K., 2002. Antidiabetic and antioxidant effect of *S*-methyl cysteine sulfoxide isolated from onion (*Allium cepa* Linn) as compared to standard drug in alloxan diabetic rats. Indian J. Exp. Biol. 40, 1005–1009.
Kurotani, K., Nanri, A., Goto, A., Mizoue, T., Noda, M., Oba, S., et al., 2013. Red meat consumption is associated with the risk of type 2 diabetes in men but not in women: a Japan Public Health Center-based Prospective Study. Br. J. Nutr. 110, 1910–1918.
Lajous, M., Tondeur, L., Fagherazzi, G., de Lauzon-Guillain, B., Boutron-Ruaualt, M.-C., Clavel-Chapelon, F., 2012. Processed and unprocessed red meat consumption and incident type 2 diabetes among French women. Diabetes Care 35, 128–130.
Lapolla, A., Fedele, D., Seraglia, R., Traldi, P., 2006. The role of mass spectrometry in the study of non-enzymatic protein glycation in diabetes: an update. Mass Spectrom. Rev. 25, 775–797.
Lee, A.Y., Chung, S.S., 1999. Contributions of polyol pathway to oxidative stress in diabetic cataract. FASEB J. 13, 23–30.
Li, M., Qu, W., Wang, Y., Wan, H., Tian, C., 2002. Hypoglycemic effect of saponin from *Tribulus terrestris*. J. Chinese Med. Mater. 25, 420–422.
Li, Y., Dutta, U., Cohenford, M.A., Dain, J.A., 2007. Nonenzymatic glycation of guanosine 5′-triphosphate by glyceraldehyde: an in vitro study of AGE formation. Bioorg. Chem. 35, 417–429.
Li, Y., Cohenford, M.A., Dutta, U., Dain, J.A., 2008. The structural modification of DNA nucleosides by nonenzymatic glycation: an in vitro study based on the reactions of glyoxal and methylglyoxal with 2′-deoxyguanosine. Anal. Bioanal. Chem. 390, 679–688.
Luo, J., Fort, D.M., Carlson, T.J., Noamesi, B.K., nii-Amon-Kotei, D., King, S.R., et al., 1998. *Cryptolepis sanguinolenta*: an ethnobotanical approach to drug discovery and the isolation of a potentially useful new antihyperglycaemic agent. Diabet. Med. 15, 367–374.
Malalavidhane, T.S., Wickramasinghe, S.M.D.N., Perera, M.S.A., Jansz, E.R., 2003. Oral hypoglycaemic activity of *Ipomoea aquatica* in streptozotocin-induced, diabetic Wistar rats and type II diabetics. Phyther. Res. 17, 1098–1100.
Mamun-or-Rashid, A., Shamim Hossain, M., Hassan, N., Kumar Dash, B., Ashrafuzzaman Sapon, M., Kumer Sen, M., 2014. A review on medicinal plants with antidiabetic activity. J. Pharmacogn. Phytochem. 3, 149–159.

Mang, B., Wolters, M., Schmitt, B., Kelb, K., Lichtinghagen, R., Stichtenoth, D.O., et al., 2006. Effects of a cinnamon extract on plasma glucose, HbA1c, and serum lipids in diabetes mellitus type 2. Eur. J. Clin. Invest. 36, 340–344.
Maroo, J., Vasu, V.T., Gupta, S., 2003. Dose dependent hypoglycemic effect of aqueous extract of *Enicostemma littorale* Blume in alloxan induced diabetic rats. Phytomedicine 10, 196–199.
Martin, C.L., Albers, J., Herman, W.H., Cleary, P., Waberski, B., Greene, D.A., et al., 2006. Neuropathy among the diabetes control and complications trial cohort 8 years after trial completion. Diabetes Care 29, 340–344.
Martins, S.I.F., Marcelis, A.T., van Boekel, M.A.J., 2003. Kinetic modelling of Amadori *N*-(1-deoxy-D-fructos-1-yl)-glycine degradation pathways. Part I—Reaction mechanism. Carbohydr. Res. 338, 1651–1663.
Maruthur, N.M., 2013. The growing prevalence of type 2 diabetes: increased incidence or improved survival? Curr. Diab. Rep. 13, 786–794.
Mattei, J., Malik, V., Wedick, N.M., Hu, F.B., Spiegelman, D., Willett, W.C., et al., 2015. Reducing the global burden of type 2 diabetes by improving the quality of staple foods: the Global Nutrition and Epidemiologic Transition Initiative. Global. Health 11, 23.
Meade, S.J., Miller, A.G., Gerrard, J.A., 2003. The role of dicarbonyl compounds in non-enzymatic crosslinking: a structure–activity study. Bioorg. Med. Chem. 11, 853–862.
Méndez, J.D., Xie, J., Aguilar-Hernández, M., Méndez-Valenzuela, V., 2010. Molecular susceptibility to glycation and its implication in diabetes mellitus and related diseases. Mol. Cell. Biochem. 344, 185–193.
Mishra, S., Rao, C., Ojha, S., Vijayakumar, M., Verma, A., 2010. An analytical review of plants for anti diabetic activity with their phytoconstituent and mechanism of action. Int. J. Pharm. Sci. Res. 1, 1647–1652.
Monnier, V.M., Bautista, O., Kenny, D., Sell, D.R., Fogarty, J., Dahms, W., et al., 1999. Skin collagen glycation, glycoxidation, and crosslinking are lower in subjects with long-term intensive vs conventional therapy of type 1 diabetes: relevance of glycated collagen products vs HbA1c as markers of diabetic complications. DCCT Skin Collagen Ancillary Study Group. Diabetes Control and Complications Trial. Diabetes 48, 870–880.
Montagnani, M., 2009. Diabetic cardiomyopathy: how much does it depend on AGE? Br. J. Pharmacol. 154, 725–726.
Morigi, M., Angioletti, S., Imberti, B., Donadelli, R., Micheletti, G., Figliuzzi, M., et al., 1998. Leukocyte-endothelial interaction is augmented by high glucose concentrations and hyperglycemia in a NF-kB-dependent fashion. J. Clin. Invest. 101, 1905–1915.
Mott, J.D., Khalifah, R.G., Nagase, H., Shield, C.F., Hudson, J.K., Hudson, B.G., 1997. Nonenzymatic glycation of type IV collagen and matrix metalloproteinase susceptibility. Kidney Int. 52, 1302–1312.
Nagle, D.G., Ferreira, D., Zhou, Y.-D., 2006. Epigallocatechin-3-gallate (EGCG): chemical and biomedical perspectives. Phytochemistry 67, 1849–1855.
Nahas, R., Moher, M., 2009. Complementary and alternative medicine for the treatment of type 2 diabetes. Can. Fam. Physician 55, 591–596.
Nanditha, A., Ma, R.C.W.W., Ramachandran, A., Snehalatha, C., Chan, J.C.N.N., Chia, K.S., et al., 2016. Diabetes in Asia and the Pacific: implications for the Global Epidemic. Diabetes Care 39, 472–485.
Negre-Salvayre, A., Salvayre, R., Augé, N., Pamplona, R., Portero-Otín, M., 2009. Hyperglycemia and glycation in diabetic complications. Antioxid. Redox Signal. 11, 3071–3109.

Nicholl, I.D., Stitt, A.W., Moore, J.E., Ritchie, A.J., Archer, D.B., Bucala, R., 1998. Increased levels of advanced glycation endproducts in the lenses and blood vessels of cigarette smokers. Mol. Med. 4, 594–601.

Oghogho Rosalie, I., Ekpe, E., 2016. Antidiabetic potentials of common herbal plants and plant products: a glance. Int. J. Herb. Med. IJHM 4, 90–97.

Okado, A., Kawasaki, Y., Hasuike, Y., Takahashi, M., Teshima, T., Fujii, J., et al., 1996. Induction of apoptotic cell death by methylglyoxal and 3-deoxyglucosone in macrophage-derived cell lines. Biochem. Biophys. Res. Commun. 225, 219–224.

Okyar, A., Can, A., Akev, N., Baktir, G., Sütlüpinar, N., 2001. Effect of *Aloe vera* leaves on blood glucose level in type I and type II diabetic rat models. Phyther. Res. 15, 157–161.

Oshima, Y., Konno, C., Hikino, H., 1985. Isolation and hypoglycemic activity of panaxans I, J, K and L, glycans of *Panax ginseng* roots. J. Ethnopharmacol. 14, 255–259.

Pamplona, R., Portero-Otín, M., Ruiz, C., Gredilla, R., Herrero, A., Barja, G., 2000. Double bond content of phospholipids and lipid peroxidation negatively correlate with maximum longevity in the heart of mammals. Mech. Ageing Dev. 112, 169–183.

Pan, A., Sun, Q., Bernstein, A.M., Manson, J.E., Willett, W.C., Hu, F.B., 2013. Changes in red meat consumption and subsequent risk of type 2 diabetes mellitus. JAMA Intern. Med. 173, 1328.

Patel, D., Prasad, S., Kumar, R., Hemalatha, S., 2012. An overview on antidiabetic medicinal plants having insulin mimetic property. Asian Pac. J. Trop. Biomed. 2, 320–330.

Petrova, R., Yamamoto, Y., Muraki, K., Yonekura, H., Sakurai, S., Watanabe, T., et al., 2002. Advanced glycation endproduct-induced calcium handling impairment in mouse cardiac myocytes. J. Mol. Cell. Cardiol. 34, 1425–1431.

Peungvicha, P., Thirawarapan, S.S., Watanabe, H., 1996. Hypoglycemic effect of water extract of the root of *Pandanus odorus* RIDL. Biol. Pharm. Bull. 19, 364–366.

Peyroux, J., Sternberg, M., 2006. Advanced glycation endproducts (AGEs): pharmacological inhibition in diabetes. Pathol. Biol. 54, 405–419.

Porchezhian, E., Dobriyal, R.M., 2003. An overview on the advances of *Gymnema sylvestre*: chemistry, pharmacology and patents. Int. J. Pharm. Sci. 58, 5–12.

Pugliese, G., 2008. Do advanced glycation end products contribute to the development of long-term diabetic complications? Nutr. Metab. Cardiovasc. Dis. 18, 457–460.

Qais, F.A.F.A., Alam, M.M.M., Naseem, I., Ahmad, I., 2016. Understanding the mechanism of non-enzymatic glycation inhibition by cinnamic acid: an in vitro interaction and molecular modelling study. RSC Adv. 6, 65322–65337.

Qin, B., Nagasaki, M., Ren, M., Bajotto, G., Oshida, Y., Sato, Y., 2003. Cinnamon extract (traditional herb) potentiates in vivo insulin-regulated glucose utilization via enhancing insulin signaling in rats. Diabetes Res. Clin. Pract. 62, 139–148.

Raj, D.S.C., Choudhury, D., Welbourne, T.C., Levi, M., 2000. Advanced glycation end products: a nephrologist's perspective. Am. J. Kidney Dis. 35, 365–380.

Rajasekaran, S., Sivagnanam, K., Ravi, K., Subramanian, S., 2004. Hypoglycemic effect of *Aloe vera* Gel on streptozotocin-induced diabetes in experimental rats. J. Med. Food 7, 61–66.

Ramachandran, A., Wan, Ma, R.C., Snehalatha, C., 2010. Diabetes in Asia. Lancet 375, 408–418.

Ramachandran, A., Snehalatha, C., Ma, R.C.W., 2014. Diabetes in South-East Asia: an update. Diabetes Res. Clin. Pract. 103, 231–237.

Rao, T.P., Sakaguchi, N., Juneja, L.R., Wada, E., Yokozawa, T., 2005. Amla (*Emblica officinalis* Gaertn.) extracts reduce oxidative stress in streptozotocin-induced diabetic rats. J. Med. Food 8, 362–368.

Robinson, L.E., Savani, S., Battram, D.S., McLaren, D.H., Sathasivam, P., Graham, T.E., 2004. Caffeine ingestion before an oral glucose tolerance test impairs blood glucose management in men with type 2 diabetes. J. Nutr. 134, 2528–2533.

Rosca, M.G., Mustata, T.G., Kinter, M.T., Ozdemir, A.M., Kern, T.S., Szweda, L.I., et al., 2005. Glycation of mitochondrial proteins from diabetic rat kidney is associated with excess superoxide formation. Am. J. Physiol. Physiol. 289, F420–F430.

Russell, K.R.M., Morrison, E.Y.S.A., Ragoobirsingh, D., 2005. The effect of annatto on insulin binding properties in the dog. Phyther. Res. 19, 433–436.

Sachdewa, A., Khemani, L.D., 1999. A preliminary investigation of the possible hypoglycemic activity of *Hibiscus rosa-sinensis*. Biomed. Environ. Sci. 12, 222–226.

Saito, M., Fujii, K., Mori, Y., Marumo, K., 2006. Role of collagen enzymatic and glycation induced cross-links as a determinant of bone quality in spontaneously diabetic WBN/Kob rats. Osteoporos. Int. 17, 1514–1523.

Satheesh, M., Pari, L., 2004. Antioxidant effect of *Boerhaavia diffusa* L. in tissue of alloxan induced diabetic rat. Indian J. Exp. Biol. 42, 989–992.

Sauvaire, Y., Petit, P., Broca, C., Manteghetti, M., Baissac, Y., Fernandez-Alvarez, J., et al., 1998. 4-Hydroxyisoleucine: a novel amino acid potentiator of insulin secretion. Diabetes 47, 206–210.

Saxena, A.M., Murthy, P.S., Mukherjee, S.K., 1996. Mode of action of three structurally different hypoglycemic agents: a comparative study. Indian J. Exp. Biol. 34, 351–355.

Sekido, H., Suzuki, T., Jomori, T., Takeuchi, M., Yabe-Nishimura, C., Yagihashi, S., 2004. Reduced cell replication and induction of apoptosis by advanced glycation end products in rat Schwann cells. Biochem. Biophys. Res. Commun. 320, 241–248.

Sepici, A., Gürbüz, I., Çevik, C., Yesilada, E., 2004. Hypoglycaemic effects of myrtle oil in normal and alloxan-diabetic rabbits. J. Ethnopharmacol. 93, 311–318.

Shanmugasundaram, K.R., Panneerselvam, C., Samudram, P., Shanmugasundaram, E.R. B., 1983. Enzyme changes and glucose utilisation in diabetic rabbits: the effect of *Gymnema sylvestre*, R.Br. J. Ethnopharmacol. 7, 205–234.

Sharma, R.D., Raghuram, T.C., 1990. Hypoglycaemic effect of fenugreek seeds in non-insulin dependent diabetic subjects. Nutr. Res. 10, 731–739.

Sharma, Y., Saxena, S., Mishra, A., Saxena, A., Natu, S.M., 2012. Advanced glycation end products and diabetic retinopathy. J. Ocul. Biol. Dis. Infor. 5, 63–69.

Shay, C.M., Ning, H., Daniels, S.R., Rooks, C.R., Gidding, S.S., Lloyd-Jones, D.M., 2013. Status of cardiovascular health in United States Adolescents: prevalence estimates from the National Health and Nutrition Examination Surveys (NHANES) 2005–2010. Circulation 127, 1369–1376.

Shinohara, M., Thornalley, P.J., Giardino, I., Beisswenger, P., Thorpe, S.R., Onorato, J., et al., 1998. Overexpression of glyoxalase-I in bovine endothelial cells inhibits intracellular advanced glycation endproduct formation and prevents hyperglycemia-induced increases in macromolecular endocytosis. J. Clin. Invest. 101, 1142–1147.

Shirwaikar, A., Rajendran, K., Kumar, C., 2004. Oral antidiabetic activity of *Annona squamosa* leaf alcohol extract in NIDDM rats. Pharm. Biol. 42, 30–35.

Shirwaikar, A., Rajendran, K., Punitha, I.S.R., 2005. Antidiabetic activity of alcoholic stem extract of *Coscinium fenestratum* in streptozotocin–nicotinamide induced type 2 diabetic rats. J. Ethnopharmacol. 97, 369–374.

Singh, R., Barden, A., Mori, T., Beilin, L., 2001a. Advanced glycation end-products: a review. Diabetologia 44, 129–146.

Singh, S.N., Vats, P., Suri, S., Shyam, R., Kumria, M.M., Ranganathan, S., et al., 2001b. Effect of an antidiabetic extract of *Catharanthus roseus* on enzymic activities in streptozotocin induced diabetic rats. J. Ethnopharmacol. 76, 269–277.

Singh, V.P., Bali, A., Singh, N., Jaggi, A.S., 2014. Advanced glycation end products and diabetic complications. Korean J. Physiol. Pharmacol. 18, 1–14.

Srividya, N., Periwal, S., 1995. Diuretic, hypotensive and hypoglycaemic effect of *Phyllanthus amarus*. Indian J. Exp. Biol. 33, 861–864.
Stanely, P., Prince, M., Menon, V.P., 2000. Hypoglycaemic and other related actions of *Tinospora cordifolia* roots in alloxan-induced diabetic rats. J. Ethnopharmacol. 70, 9–15.
Stevens, G.A., Singh, G.M., Lu, Y., Danaei, G., Lin, J.K., Finucane, M.M., et al., 2012. National, regional, and global trends in adult overweight and obesity prevalences. Popul. Health Metr. 10, 22.
Stirban, A., Gawlowski, T., Roden, M., 2014. Vascular effects of advanced glycation end-products: clinical effects and molecular mechanisms. Mol. Metab. 3, 94–108.
Subramoniam, A., Pushpangadan, P., Rajasekharan, S., Evans, D.A., Latha, P.G., Valsaraj, R., 1996. Effects of *Artemisia pallens* Wall. on blood glucose levels in normal and alloxan-induced diabetic rats. J. Ethnopharmacol. 50, 13–17.
Sugihara, Y., Nojima, H., Matsuda, H., Murakami, T., Yoshikawa, M., Kimura, I., 2000. Antihyperglycemic effects of gymnemic acid IV, a compound derived from *Gymnema sylvestre* leaves in streptozotocin-diabetic mice. J. Asian Nat. Prod. Res. 2, 321–327.
Sugimoto, K., Nishizawa, Y., Horiuchi, S., Yagihashi, S., 1997. Localization in human diabetic peripheral nerve of *N* ε-carboxymethyllysine-protein adducts, an advanced glycation endproduct. Diabetologia 40, 1380–1387.
Sun, Y.-M., Su, Y., Li, J., Wang, L.-F., 2013. Recent advances in understanding the biochemical and molecular mechanism of diabetic nephropathy. Biochem. Biophys. Res. Commun. 433, 359–361.
Suzuki, D., Miyata, T., Saotome, N., Horie, K., Inagi, R., Yasuda, Y., et al., 1999. Immunohistochemical evidence for an increased oxidative stress and carbonyl modification of proteins in diabetic glomerular lesions. J. Am. Soc. Nephrol. 10, 822–832.
Swanston-Flatt, S.K., Day, C., Bailey, C.J., Flatt, P.R., 1990. Traditional plant treatments for diabetes. Studies in normal and streptozotocin diabetic mice. Diabetologia 33, 462–464.
Talan, I., Ragoobirsingh, D., Morrison, E., 2001. The effect of capsaicin on blood glucose plasma insulin level and insulin binding in dog models. Phyther. Res. 15, 391–394.
Tessitore, N., Lapolla, A., Arico, N.C., Poli, A., Gammaro, L., Bassi, A., et al., 2004. Effect of protein leaking BK-F PMMA-based hemodialysis on plasma pentosidine levels. J. Nephrol. 17, 707–714.
The, N.S., Richardson, A.S., Gordon-Larsen, P., 2013. Timing and duration of obesity in relation to diabetes: findings from an ethnically diverse, nationally representative sample. Diabetes Care 36, 865–872.
Thomas, M.C., Forbes, J.M., Cooper, M.E., 2005. Advanced glycation end products and diabetic nephropathy. Am. J. Ther. 12, 562–572.
Thornalley, P.J., Langborg, A., Minhas, H.S., 1999. Formation of glyoxal, methylglyoxal and 3-deoxyglucosone in the glycation of proteins by glucose. Biochem. J. 344 (Pt 1), 109–116.
Thorpe, S.R., Baynes, J.W., 2003. Maillard reaction products in tissue proteins: new products and new perspectives. Amino Acids 25, 275–281.
Tsiodras, S., Shin, R.K., Christian, M., Shaw, L.M., Sass, D.A., 1999. Anticholinergic toxicity associated with lupine seeds as a home remedy for diabetes mellitus. Ann. Emerg. Med. 33, 715–717.
Turk, Z., Ljubic, S., Turk, N., Benko, B., 2001. Detection of autoantibodies against advanced glycation endproducts and AGE-immune complexes in serum of patients with diabetes mellitus. Clin. Chim. Acta 303, 105–115.
Tutino, G.E., Tam, W.H., Yang, X., Chan, J.C.N., Lao, T.T.H., Ma, R.C.W., 2014. Diabetes and pregnancy: perspectives from Asia. Diabet. Med. 31, 302–318.

Ubillas, R.P., Mendez, C.D., Jolad, S.D., Luo, J., King, S.R., Carlson, T.J., et al., 2000. Antihyperglycemic acetylenic glucosides from *Bidens pilosa*. Planta Med. 66, 82–83.
Upadhyay, U.M., Goyal, R.K., 2004. Efficacy of *Enicostemma littorale* in type 2 diabetic patients. Phyther. Res. 18, 233–235.
Uribarri, J., Cai, W., Ramdas, M., Goodman, S., Pyzik, R., Chen, X., et al., 2011. Restriction of advanced glycation end products improves insulin resistance in human type 2 diabetes: potential role of AGER1 and SIRT1. Diabetes Care 34, 1610–1616.
Valko, M., Leibfritz, D., Moncol, J., Cronin, M.T.D., Mazur, M., Telser, J., 2007. Free radicals and antioxidants in normal physiological functions and human disease. Int. J. Biochem. Cell Biol. 39, 44–84.
van Woudenbergh, G.J., Kuijsten, A., Tigcheler, B., Sijbrands, E.J.G., van Rooij, F.J.A., Hofman, A., et al., 2012. Meat consumption and its association with C-reactive protein and incident type 2 diabetes: the Rotterdam study. Diabetes Care 35, 1499–1505.
Vijayvargia, R., Kumar, M., Gupta, S., 2000. Hypoglycemic effect of aqueous extract of *Enicostemma littorale* Blume (chhota chirayata) on alloxan induced diabetes mellitus in rats. Indian J. Exp. Biol. 38, 781–784.
Wada, R., Yagihashi, S., 2005. Role of advanced glycation end products and their receptors in development of diabetic neuropathy. Ann. N.Y. Acad. Sci. 1043, 598–604.
Wadood, A., Wadood, N., Shah, S.A., 1989. Effects of *Acacia arabica* and *Caralluma edulis* on blood glucose levels of normal and alloxan diabetic rabbits. J. Pak. Med. Assoc. 39, 208–212.
Wadood, N., Wadood, A., Nisar, M., 2003. Effect of *Ficus relegiosa* on blood glucose and total lipid levels of normal and alloxan diabetic rabbits. J. Ayub. Med. Coll. Abbottabad 15, 40–42.
Wan, T.-T., Li, X.-F., Sun, Y.-M., Li, Y.-B., Su, Y., 2015. Recent advances in understanding the biochemical and molecular mechanism of diabetic retinopathy. Biomed. Pharmacother. 74, 145–147.
Welihinda, J., Karunanayake, E.H., Sheriff, M.H.H., Jayasinghe, K.S.A., 1986. Effect of *Momordica charantia* on the glucose tolerance in maturity onset diabetes. J. Ethnopharmacol. 17, 277–282.
Wendt, T.M., Tanji, N., Guo, J., Kislinger, T.R., Qu, W., Lu, Y., et al., 2003. RAGE drives the development of glomerulosclerosis and implicates podocyte activation in the pathogenesis of diabetic nephropathy. Am. J. Pathol. 162, 1123–1137.
WHO, 2009. Department of Health Statistics and Informatics in the Information Evidence and Research Cluster of the World Health Organization: Global Health Risks: Mortality and Burden of Disease Attributable to Selected Major Risks. World Health Organization Press, Geneva.
Wolff, S.P., Dean, R.T., 1987. Glucose autoxidation and protein modification. The potential role of "autoxidative glycosylation" in diabetes. Biochem. J 245, 243–250.
Wolff, S.P., Jiang, Z.Y., Hunt, J.V., 1991. Protein glycation and oxidative stress in diabetes mellitus and ageing. Free Radic. Biol. Med. 10, 339–352.
Wu, L.-Y., Juan, C.-C., Hwang, L.S., Hsu, Y.-P., Ho, P.-H., Ho, L.-T., 2004. Green tea supplementation ameliorates insulin resistance and increases glucose transporter IV content in a fructose-fed rat model. Eur. J. Nutr. 43, 116–124.
Wu, C.-H., Huang, S.-M., Lin, J.-A., Yen, G.-C., 2011. Inhibition of advanced glycation endproduct formation by foodstuffs. Food Funct. 2, 224–234.
Yadav, P., Sarar, S., Bhatnagar, D., 1997. Action of *Capparis decidua* against alloxan-induced oxidative stress and diabetes in rat tissues. Pharmacol. Res. 36, 221–228.
Yamagishi, S.I., Matsui, T., 2010. Advanced glycation end products, oxidative stress and diabetic nephropathy. Oxid. Med. Cell. Longev. 3, 101–108.

Yamagishi, S., Maeda, S., Matsui, T., Ueda, S., Fukami, K., Okuda, S., 2012. Role of advanced glycation end products (AGEs) and oxidative stress in vascular complications in diabetes. Biochim. Biophys. Acta - Gen. Subj. 1820, 663–671.

Yeh, G.Y.G.Y., Eisenberg, D.M.D.M., Kaptchuk, T.J.T.J., Phillips, R.S., 2003. Systematic review of herbs and dietary supplements for glycemic control in diabetes. Diabetes Care 26, 1277–1294.

Yonekura, H., Yamamoto, Y., Sakurai, S., Watanabe, T., Yamamoto, H., 2005. Roles of the receptor for advanced glycation endproducts in diabetes-induced vascular injury. J. Pharmacol. Sci. 97, 305–311.

Yu, B.C., Chen, W.C., Cheng, J.T., 2003. Antihyperglycemic effect of andrographolide in streptozotocin-induced diabetic rats. Planta Med. 69, 1075–1079.

Zhang, X.F., Tan, B.K., 2000. Anti-diabetic property of ethanolic extract of *Andrographis paniculata* in streptozotocin-diabetic rats. Acta Pharmacol. Sin. 21, 1157–1164.

Zheng, F., He, C., Cai, W., Hattori, M., Steffes, M., Vlassara, H., 2002. Prevention of diabetic nephropathy in mice by a diet low in glycoxidation products. Diabetes. Metab. Res. Rev. 18, 224–237.

Zimmet, P.Z., Magliano, D.J., Herman, W.H., Shaw, J.E., 2014. Diabetes: a 21st century challenge. Lancet Diabetes Endocrinol. 2, 56–64.

Zong, H., Ward, M., Stitt, A.W., 2011. AGEs, RAGE, and diabetic retinopathy. Curr. Diab. Rep. 11, 244–252.

CHAPTER 14

Insights of Phyto-Compounds as Antipathogenic Agents: Controlling Strategies for Inhibiting Biofilms and Quorum Sensing in *Candida albicans*

Mohd Sajjad Ahmad Khan[1], Mohd Musheer Altaf[2] and Mohammad Sajid[3]

[1]Department of Basic Sciences, Biology Unit, Health Track, Imam Abdulrahman Bin Faisal University, Dammam, Kingdom of Saudi Arabia

[2]Department of Agricultural Microbiology, Aligarh Muslim University, Aligarh, Uttar Pradesh, India

[3]Cell Biology and Immunology Laboratory, Institute of Microbial Technology, Chandigarh, Punjab, India

14.1 INTRODUCTION

Candida albicans resides in the gastrointestinal, vaginal, and oral tracts as a part of normal flora of healthy individuals. However, under certain unfavorable conditions, *Candida* is able to raid the mucosal surface and propagate into the bloodstream causing systemic diseases (Nucci et al., 2010; Barriuso, 2015). *C. albicans* is a dimorphic fungi being able to switch from yeast-to-hyphal growth under certain conditions, which is critical to its pathogenicity. This shifting in morphology is directed by many environmental signals acting through numerous well-established signaling cascades (Lo et al., 1997), controlled by a cell density–dependent behavior termed quorum sensing (QS), which regulate many processes that encompass cell–cell interactions, viz. group motility, secretion of virulence factors, and biofilms establishment (Daniels et al., 2004; Parsek and Greenberg, 2005). QS regulation assists host-associated microbes to delay its detection until an effective population (quorum) has achieved in the appropriate niche within the host. The small diffusible signaling molecules termed as quorum-sensing molecules (QSMs) are contributing to a better adaptation of the metabolic activity of microbial cells, by synchronizing their different genes expression in a particular cell density, to their tolerance to antimicrobials and synthesis of virulence

New Look to Phytomedicine
DOI: https://doi.org/10.1016/B978-0-12-814619-4.00014-8

factors including biofilm. Microbial biofilms are surface-associated three-dimensional organized structures embedded within an exo-polymeric matrix. Infections associated with biofilms formation are difficult to eradicate because these structures are particularly resistant to antimicrobial agents and host immune factors (Deveau and Hogan, 2011). *C. albicans* not only establishes biofilms on host tissue but also forms biofilms on medical devices either pre- or postimplantation leading to recurring infections and even death in some cases (Chandra et al., 2012).

The fungi, like bacteria, also use quorum regulation to affect biofilm formation and pathogenesis in their population and it appears to be prevalent in diverse fungal species (Hogan, 2006). The regulation of the synchronization of the expression of virulence genes as a function of the population density is a crucial step in its pathogenesis and adaptation to host tissues (San-Blas et al., 2000; Miller and Bassler, 2001). This coordinated expression of virulence factors during infection of a host probably constitutes a significant survival advantage by enhancing the chances of establishing infection and escaping the immune response (Bassler and Losick, 2006). In addition, some QSMs can be considered virulence factors by themselves since they are toxic to host cells and also can modulate host immunity (Winzer and Williams, 2001).

This chapter aims to highlight the progress in the research on fungal QS with emphasis on *C. albicans* presenting well-known QSMs in fungi and their physiological effects relating to the ecology of QS. Moreover, we provide a discussion of QS networks responsible for biofilms, the potential targets of phyto-compounds toward discovering newer antifungal therapies.

14.2 BASIC ELEMENTS OF QUORUM-SENSING REGULATION IN *CANDIDA ALBICANS*

In both bacteria and fungi, QS is facilitated by QSMs that accumulate in the extracellular environment. These signal molecules are generally species or strain specific with a high degree of structural diversity among the signaling molecules produced by different microorganisms (Visick and Fuqua, 2005). The mechanism by which a signal accumulates in the medium depends on the system, passive diffusion across the membrane, the action of efflux pumps, and specific transporters (Hui and Morrison, 1991; Pearson et al., 1999). When a signal accumulates to a sufficiently high concentration, the cognate response regulator is activated within the local population of cells, leading to coordinated gene expression. The morphological transitions in *C. albicans* are mediated by two signal molecules with opposite effects, viz farnesol and tyrosol.

14.2.1 Physiological Effects of Quorum-Sensing Molecules in *Candida albicans*

14.2.1.1 Farnesol

QS in eukaryotic organisms was unknown until the discovery of farnesol as a QSM in the pathogenic fungus *C. albicans* (Hornby et al., 2001). *E*, *E*-Farnesol is a sesquiterpine alcohol, which is an alcohol made up of three isoprene units. It is an intermediate in sterol biosynthesis, but the subsequent pathways to secrete the molecule are unknown. Farnesol has been shown to play various roles in *C. albicans* physiology as a signaling molecule and induces damaging effects on host cells and other microbes. Another analogue compound farnesoic acid has been isolated only in a strain of *C. albicans* that does not produce farnesol and its effects are less strong (Nickerson et al., 2006).

14.2.1.1.1 Effects on Filamentation

The first described effects of farnesol as a QSM were on the regulation of *C. albicans* filamentation. Hornby et al. (2001) showed that commercial and conditioned media purified farnesol inhibited the yeast-to-mycelium conversion. However, farnesol showed no effects on cells already committed to mycelial development (Mosel et al., 2005).

14.2.1.1.2 Biofilm Formation

Ramage et al. (2002) evaluated the effects of farnesol on biofilm development and observed that in addition to its role in regulating *C. albicans* morphology, farnesol also inhibited biofilm formation. Microarray analysis of biofilms treated with farnesol exhibited that genes related to drug resistance, cell wall maintenance, cell surface hydrophilicity, iron transport, and heat shock proteins are influenced in addition to the genes associated with hyphaeformation (Cao et al., 2005).

14.2.1.1.3 Oxidative Stress

Farnesol has been identified to deliver protection against oxidative stress induced by hydrogen peroxide and by the superoxide anion-generating agents menadione and plumbagin. In a gene expression experiment conducted by Westwater et al. (2005) showed no induction of anoxidative stress response in response to commercial farnesol (35 μM) in *C. albicans*, suggesting a role for farnesol as an antioxidant in addition to a signaling molecule. Shirtliff et al. (2009) similarly observed that farnesol concentrations of 40 or 100 μM were able to induce up-regulation of *C. albicans* proteins involved in protection against oxidative stress.

14.2.1.1.4 Modulation of Drug Efflux

In a study conducted by Sharma and Prasad (2011), it was shown that farnesol can specifically modulate *C. albicans* drug efflux mediated by ABC multidrug transporters without affecting the multidrug extrusion pump protein CaMdr1p from the major facilitator super family. In their study, they also noticed the potentiation of azoles and polyenes by farnesol through increasing reactive oxygen species (ROS) levels in *C. albicans*.

14.2.1.1.5 Effects on Other Microbes

Farnesol has been reported to deliver detrimental effects on many microbes including bacteria and fungi, such as *Staphylococcus aureus*, *Saccharomyces cerevisiae*, *Aspergillus* sp., *Paracoccidioides brasiliensis*, *Mycobacterium smegmatis*, and *Pseudomonas aeruginosa* (Machida et al., 1999; Jabra-Rizk et al., 2006; Semighini et al., 2006; Derengowski et al., 2009; Jin et al., 2010). *C. albicans* and *P. aeruginosa* are commonly found in mixed opportunistic infections, and both microbes involve in complex interactions involving their QS systems (Cugini et al., 2007). Hogan et al. (2004) found that a homoseryl lactone produced by *P. aeruginosa* inhibited filamentation in *C. albicans* without compromising fungal growth. Additionally, *P. aeruginosa* was able to form biofilms on hyphae of *C. albicans* and subsequently killing it, but not on *C. albicans* yeast cells. This suggests that the morphological switch occurs in *C. albicans* by sensing the presence of *P. aeruginosa*. But upon addition of farnesol to *P. aeruginosa* cultures, it was found that biosynthesis of *Pseudomonas* quinolone signal (PQS) was inhibited (Cugini et al., 2007). However, they further observed that in mixed biofilms, where the concentrations of both PQS and farnesol were high, the overall effect was an increased production of phenazines. The group suggested that farnesol effects are due to the induction of ROS since the effects are decreased in the presence of antioxidants and are similar to the *P. aeruginosa* response to hydrogen peroxide addition (Cugini et al., 2010).

14.2.1.2 Tyrosol

The other known fungal QSMs are all alcohols derived from aromatic amino acids tyrosine (tyrosol), phenylalanine (phenylethanol), and tryptophan (tryptophol). Tyrosol being related to tyrosine requires the biosynthetic pathways of the aromatic amino acid. Tyrosol in *C. albicans* decreases the length of the lag phase of growth and stimulates the filamentation and biofilm formation. These effects are suppressed in the presence of farnesol, suggesting a fine QS-mediated control (Chen et al., 2004).

14.2.1.2.1 Growth Effects

Chen et al. (2004) studied the lag phase time for diluted cultures of *C. albicans* to resume exponential growth. They observed that the addition of conditioned medium reduced this lag phase, and tyrosol is found to be the active molecule. Tyrosol also stimulates filamentation in opposition to effect of farnesol.

14.2.1.2.2 Biofilm Formation

It has been reported in a study by Alem et al. (2006) that the addition of tyrosol at early stages of biofilm formation stimulated hyphal growth in *C. albicans*. They also noticed that tyrosol concentration correlates with the increase of biomass of both planktonic and sessile cells. When biofilms were treated with 50 μM of farnesol, the tyrosol (0.1−1 mM) eliminated the filamentation-inhibiting effects of the farnesol. In biofilms, treated with higher concentrations of farnesol, the addition of tyrosol resulted in biofilms containing mostly yeast cells. This observation suggested that tyrosol could not counteract the effects of higher concentrations of farnesol. They also noticed that tyrosol effects exceed those of farnesol after 14 h of biofilm formation and suggested that tyrosol acts primarily in the early and intermediate stages of biofilm development. However, in mature biofilms farnesol activity and concentration surpass tyrosol and possibly have a critical role on the release of yeast cells for biofilm dispersal (Ramage et al., 2002; Nickerson et al., 2006).

14.2.1.2.3 Oxidative Stress Response

Cremer et al. (1999) highlighted that tyrosol from *C. albicans*-conditioned medium inhibited the respiratory burst of human neutrophils, while Westwater et al. (2005) failed to get any protective effect of exogenous tyrosol in *C. albicans* cells treated with hydrogen peroxide.

14.2.2 Role of Quorum-Sensing Molecules in Virulence

Polymorphism between yeast, hyphae, and pseudohyphae forms in *C. albicans* is critical for its virulence and corresponds to an adaptive response to environmental changes (Saville et al., 2003). This morphological switch is reported to be triggered by temperature, pH, CO_2 concentration, and the presence of mammalian serum (Sudbery et al., 2004). A study in a mouse model of disseminated candidiasis, Park et al. (2005) demonstrated that locking of *C. albicans* into either the yeast or hyphal morphology resulted in severely attenuated virulence. Farnesol prevents the differentiation from yeast-to-hyphal growth and therefore the synthesis and release of this molecule could play a role in pathogenesis (Hornby et al., 2001; Chen et al., 2004). Hornby and Nickerson (2004) demonstrated that *Candida* treated with sublethal concentrations of azoles result in cells which are predominantly in the yeast form and have increased levels of farnesol production suggesting a role of azole action on modulation of morphogenesis. Interestingly, when *C. albicans* cells pretreated with azoles allowed to challenge a mouse in an infection model, the strains exhibited strong increase in their virulence (Navarathna et al., 2005). Furthermore, they also examined the role of farnesol and its effect on cytokine expression to address farnesol's role in enhancing *C. albicans* survival in the host (Navarathna et al., 2007). Using a mouse model, challenged with farnesol alone, *Candida*, or *C. albicans* and exogenous farnesol, they did not observe any change in IL-2 or IL-4 levels in mice challenged with *Candida* or farnesol alone. In mice challenged with *Candida* alone, TNF-α levels were uninfluenced but a high increase is observed after 48 h in mice challenged with farnesol. On the other hand, tyrosol excite germ-tube formation and hyphal expansion in the early and intermediate stages of *C. albicans* growth (Chen et al., 2004). Biofilms formation in *Candida* species is promoted by external addition of farnesol augmenting their susceptibility to antifungals (Cordeiro et al., 2015).

14.2.3 Molecular Mechanism of Quorum-Sensing-Associated Virulence and Biofilm in *Candida albicans*

A number of transcriptional regulators playing important role in regulating yeast-to-hyphal shift have been characterized, but several of these target genes have not yet been identified. The morphological changes in *C. albicans* are regulated at different levels by signal dependent transduction pathways. The signaling cascades controlling the expression of genes under QS regulation in *C. albicans* are still poorly understood. Langford et al. (2009) reviewed signaling pathways in farnesol-mediated QS in *C. albicans*. Hyphal development can be regulated through the modulation of the Cyclic adenosine monophosphate (cAMP)/protein kinase A (PKA) pathway. It is suggested that farnesol can affect filamentation by inhibiting the cAMP pathway by inhibiting the activity of the *Candida* adenylyl cyclase, Cyr1p (Hall et al., 2011). Activation of the cAMP/PKA pathway down-regulates the expression of NRG1, which is the major repressor of hyphal development. Farnesol is blocks the protein degradation of Nrg1, repressing the expression of Sok1, the kinase required for Nrg1 degradation (Lu et al., 2014). The transcription repressor Cup9 is also responsible for the regulation of SOK1 expression in response to farnesol inhibition. The *CEK1* mitogen-activated protein kinase pathway is also involved in response to farnesol (Klengel et al., 2005). Farnesol also modulates the transcription levels of *tup1*, which is a major repressor of the morphological transition in *C. albicans* (Kebaara et al., 2008; Hall et al., 2011; Langford et al., 2013). Kruppa et al. (2004) found that farnesol was not able to inhibit filamentation or biofilm formation in *C. albicans* Chk1p histidine kinase mutants. Their study has suggested that this two-component signaling protein has involvement in QS. These studies in altogether highlighted that farnesol can act through different regulatory pathways in *C. albicans* which are involved in several other physiological processes (Hall et al., 2009).

In a study conducted by Cao et al. (2005), a total of 274 responsive genes were affected in farnesol-treated *C. albicans*. Out of which 104 were up-regulated and 170 were down-regulated. Important changes were detected in the hyphal formation-associated genes (*tup1*, *crk1*, and *pde2*), and a number of other genes with roles related to drug resistance, cell wall maintenance, iron transport, and genes encoding heat shock proteins.

14.3 WHY IS NEED OF STRATEGIES ALTERNATIVE TO EXISTING CHEMOTHERAPEUTIC AGENTS?

Microbial cells in biofilm can produce persister cells which do not grow or die in the presence of antibiotics and are able to avoid killing by drugs. Therefore once biofilm is established, *C. albicans* presents a significant clinical problem, because currently available chemotherapeutic agents are severely limited by the intrinsic tolerance of such fungal biofilms

(Desai et al., 2014). Although, combination therapies involving antibacterial and antifungal drugs have shown some success. The emergence of drug resistance has been continuously reported by investigators, and biofilms are being considered as prominent cause. It has led to decreased efficiency of currently available antifungal agents to combat rapid increase in multidrug-resistant microorganisms (Cooper and Shlaes, 2011). Therefore there is an urgent need to envisage new antiinfective strategies that do not target essential processes in the pathogen rather virulence. Since the inhibition of biofilms in such organisms is a major clinical challenge, the increased understanding of microbial signaling networks to control virulence and biofilm could be potential approach to combat such fungal infections (Bink et al., 2011).

14.4 HOW TO COMBAT BIOFILM AND VIRULENCE IN *CANDIDA ALBICANS*?

Any microorganism's pathogenicity is dependent on its ability to sense the environment including all signals produced by itself and from different cells of the host organism (Davies et al., 1998). Microorganisms receive signals and accordingly react to activate their virulence genes in order to get escaped from host defense mechanisms. They also utilize host factors for their own benefit. Here is the role of QSMs that after being accumulated according to the cellular density of a microbial community or biofilm and subsequently provide their tolerance to antimicrobials and synthesis of virulence factors. It is now evident from literature that resistance of biofilm to antimicrobial agents is acquired through the process of QS (Miller and Bassler, 2001; Alina and Veronica, 2011). A schematic presentation of strategies being used to target QS in *C. albicans* by plant compounds is given in Fig. 14.1.

14.4.1 Blocking Quorum Sensing

Interfering with the mechanism of cell–cell communication can provide a novel approach to prevent biofilm formation and hereby lowering the pathogenicity. The microbial cell produces and secrets out low molecular weight compounds termed as autoinducers. The acyl-homoserine lactones (AHLs) are the most studied autoinducers reported (Wang and Ma, 2014). In addition, others are like autoinducer-2, a boron-bearing compound (Pereira et al., 2013), bradyoxetin (Loh et al., 2002), several diketopiperazines, e.g., farnesol (Albuquerque and Casadevall, 2012),

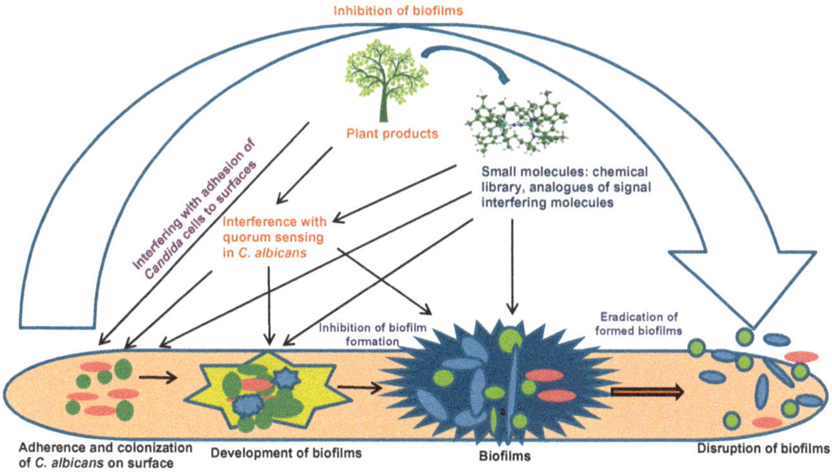

Figure 14.1 Schematic presentation of approaches used to inhibit biofilm formation in *C. albicans* targeting quorum sensing.

cis-2-alkenoic acids (Deng et al., 2011), and a variety of peptides (Monnet et al., 2016), many of which are cyclic compounds (Novick and Geisinger, 2008). AHLs mediate QS within same microbial species, but autoinducer-2, in addition, also mediates QS between phylogenetically distantly related species. Therefore interference in this communication could provide control over mixed biofilm communities (Thompson et al., 2015).

The autoinducers not only initiate biofilm formation but also accountable for maintenance of preestablished biofilms. Several studies have indicated that inhibition of the production of these autoinducers or blocking their receptor results in biofilms as less organized and more susceptible to the host immune system. Therefore blocking QS by targeting these autoinducers has become an vital approach for the search of biofilm modulating compounds for the control of biofilm-associated infections. In this regards, several autoinducers being produced in varying amounts have been identified for regulation of the cellular responses (Scutera et al., 2014). This is a high standards for drugs to block QS effectively and reliably. However, the majority of fungal autoinducers, as well as their molecular mechanisms of action has remained unknown. Understanding how QS systems affect fungal physiology can lead to a better understanding of pathology, thus providing knowledge to develop new systems for control (De Sordi and Muhlschlegel, 2009). The use of reporter strains for the detection of QS inhibition allows high-throughput screening of large compound libraries. QS could be addressed and interfered to prevent biofilm formation or the dispersal of already established biofilms. Since the patients have already developed biofilms before showing any clinical symptoms, the biofilm dissolving approach would be a high priority strategy to combat biofilm infections (Abraham, 2016).

14.5 QUORUM-SENSING INHIBITORS

The application of QS inhibitors (QSIs) is one of the promising techniques for disrupting QS and biofilm formation. Bromated furanones obtained from the red alga *Delisea pulchra* is considered as the first such QSIs. It is reported to inhibit bacterial colonization on the algae. Afterward a series of similar halogenated-furanone compounds have been synthesized. Among them, furanones C-30 and C-56 have been extensively studied as they can inhibit QS-controlled behaviors, in particular the production of extracellular virulence factors and the development of biofilms in microorganisms (Lonn-Stensrud et al., 2009). A large number of compounds, antagonistic to AHLs, have been reported by researchers (Jiang and Li, 2013). Also, many compounds were identified using large chemical libraries and subsequent optimization of the hits obtained from their high-throughput screening (Estrela et al., 2009). Interestingly, a number of well-known phyto-compounds like eugenol, curcumin, cinnamic aldehyde, and ajoene have also been reported to block receptors of AHLs (Rudrappa and Bais, 2008; Jakobsen et al., 2012; Zhou et al., 2013). Many chemical analogs to such natural compounds have been investigated and are found to be inhibiting the autoinducer-2 signaling pathway such as butyl- and isobutyl-4,5-dihydroxy-2,3-pentanedione (Roy et al., 2010; Brackman et al., 2011; Abraham, 2016).

All living organisms including commensal bacteria, probiotics, algae, lichens, plants, and animals synthesize one or more types of QSIs as their defense mechanism against infectious pathogens (Ditu et al., 2011; Chifiriuc et al., 2009). Now-a-days, screening for QSI compounds can be done in high-throughput screenings because several assays targeting the formation of autoinducers or their receptors are available. For the application of an antibiofilm compound, it is important to consider that it should be active against the pathogen and nondamaging. Furthermore, the evaluation of a novel drug for its cytotoxicity is mandatory. Although the natural compounds especially from plants are being considered as safe and displaying variety of targets, the screening for these bioactive compounds is more laborious and slower. However, investigations have highlighted many newer compounds with new targets and new mechanisms, viz. *cis*-2-decenoic acid, ursolic acid (Rudrappa and Bais, 2008; Jakobsen et al., 2012; Zhou et al., 2013; Roy et al., 2010; Brackman et al., 2011; Abraham, 2016). Some of the compounds have several targets allowing the control of a broader spectrum of pathogens. Combining these biofilm controlling compounds with established or novel antibiotics will add another dimension to the treatment of biofilm infections. Here, we have summarized an overview on natural compounds from plant origin modulating biofilms beyond a mere QS interference.

14.6 PHYTO-COMPOUNDS: POTENTIAL INHIBITORS OF *CANDIDA ALBICANS* BIOFILMS

Plants produce two types of metabolites, viz primary and secondary. The primary metabolites such as carbohydrates, amino acids, peptides, proteins, enzymes, lipids, purine, and pyrimidine derivatives are mainly involved in physiological functions and less diverse in chemical structure, whereas the secondary metabolites are produced as a result of plant interactions with environment to enhance its survival and defense against pathogen attack. These compounds do not play a direct role in plant physiology but are very diverse in structure as being synthesized from few basic building blocks such as acetate, isoprenoid, and phenylpropanoids (Gurib-Fakim, 2006). These classes of secondary metabolites are categorized as terpenes/terpenoids, phenylpropanoids, polyphenols, flavonoids, alkaloids, quinones, and phenolics. Many of these pharmacologically active phytomolecules exhibit interesting anti-*Candida* activities and are attractive candidates for drug discovery against biofilms targeting pathogenicity or QS.

14.6.1 Terpenes

Xanthorrhizol, a sesquiterpenoid isolated from the rhizome of *Curcuma xanthorrhiza* Roxb, has been known to exert significant anti-*Candida* activity (Rukayadi et al., 2011). Bakuchiol, a meroterpene, found to be an active component of *Psoralea glandulosa* leaves, and *Psoralea corylifolia* seeds have shown bioactivity against various *Candida* species associated with oral biofilm (Nordin et al., 2015). This compound is nontoxic to human cells (HepG2 cells) up to 5000 μg/mL concentration (Rukayadi et al., 2011). Therefore it could be a better choice over currently available antifungal drugs which have substantial side effects at high concentrations.

Bakuchiol act by altering cell surface properties of *Candida* resulting in decreased adhesion and subsequently reduced biofilm formation and virulence. Gong et al. (2011) isolated Pseudolaric acid B, a diterpene acid, from the bark of *Pseudolarix kaempferi* and reported its biofilm inhibitory activity against *C. albicans*. Casbane, a diterpene, isolated from the ethanolic extract of *Croton nepetaefolius* stalk has been shown to exhibit biofilm preventive effects on several pathogenic bacteria and yeasts mainly *C. albicans*, *Candida glabrata*, and *Candida tropicalis* (Carneiro et al., 2010; Vasconcelos et al., 2014). The mechanism of action is expected to be due to interaction of this diterpene with lipid bilayer and destabilization of membrane phospholipids. Xie et al. (2015) investigated ethyl acetate soluble fraction of the methanol extract of the plant *Carpesium macrocephalum* against *Candida* biofilms.

In a study conducted by Nakamura et al. (2010) for *C. albicans* adhesion and colonization on oral epithelial cells, it was prevented by 0.25 mM of Hinokitiol (α-thujaplicin), isolated from plants in Cupressaceae family. In a comprehensive study, Raut et al. (2013a) have evaluated efficacy of 28 major terpenoids against biofilms of *C. albicans*. Citral, citronellal, and citronellol were found to suppress biofilm formation. Geraniol, a major component of geranium and lemongrass oil (*Cymbopogon citratus*), Thymol, a major constituent of *Origanum vulgarum*, and carvacrol from *Origanum virens* and *Satureja hortensis* and, linalool, a terpenoid alcohol found in *Croton cajucara*, *Coriandrum sativum*, and *Lavandula angustifolia*, were reported to possess antibiofilm activity. These two molecules exerted their anti-*Candida* activity through cell cycle arrest (Dalleau et al., 2008; Zore et al., 2011; Raut et al., 2013a). Menthol, camphene, and myrcene are active principles in many plant extracts and essential oils and are strong antibiofilm compound against *C. albicans*. Their mechanism of action is appeared to be the disruption of cell membrane. Also, nerol, carvacrol, eugenol, and α-thujone were found as efficient as farnesol which is a biofilm preventive QSM in *C. albicans* (Pauli, 2006).

Terpenoids are expected to prevent yeast to hyphae morphogenetic switching and also being antiadhesion, and are considered as strong inhibitor of *Candida* biofilms. However, they are also known to damage the cell membrane and therefore responsible for extenuation of mature biofilms (Silva et al., 2008). Geraniol, being structurally similar to farnesol, results in activation of transcription suppressors (Tup1, Nrg1, and Rfg1), hence inhibits morphogenesis (Kebaara et al., 2008). The CSH of

Candida is responsible for adherence and subsequent biofilm formation (Borghi et al., 2011), and *CSH1* gene encodes a CSH-associated protein which is supposed to contribute to adhesion of *C. albicans* (Singleton et al., 2001). Farnesol is found to be affecting CSH through downregulation of *CSH1* gene, leading to suppression of biofilm growth (Cao et al., 2005). A study from Marcos-Arias et al. (2011) exposed that terpenoids mediated changes in permeability and fluidity of cell membrane results in degradation of cell wall. This leads to altered adherence and subsequently preventing biofilm development in *C. albicans*.

14.6.2 Phenylpropanoids

Raut et al. (2014) evaluated the antifungal properties of several phenylpropanoids against *C. albicans* biofilms. In their study, anisyl alcohol, anisaldehyde, eugenol, salicylaldehyde, and cinnamaldehyde resulted in significant inhibition of biofilm formation. Our group has also shown that highly resistant biofilm cells are susceptible to 0.1 mg/mL concentrations of cinnamaldehyde (Khan and Ahmad, 2012). Eugenol, a major component of clove (*Eugenia caryophyllis*) and *O. vulgarum*, exerts morphogenesis inhibitory and fungicidal activities against *C. albicans*. Disruption of cell membrane is primary mechanism of action of phenylpropanoids. These molecules interact with proteins in lipoprotein bilayer and inactivate them to cause membrane dysfunction. Inhibition of morphogenesis by phenylpropanoids is also one of the reason to prevent biofilm growth (Chami et al., 2005; Sung and Lee, 2010; Raut et al., 2014). Eugenol and cinnamaldehyde aim the cell membrane of sessile cells to disrupt biofilm network (Khan and Ahmad, 2012). Other studies have revealed that mechanism of action of phenylpropanoids is by disrupting the Ca^{++} and H^{+} ion activities, interference in cell cycle progress, and perturbation of amino acid permeases in the cytoplasmic membrane (Rao et al., 2010; Darvishi et al., 2013).

14.6.3 Polyphenols

Polyphenols constitute a diverse group of aromatic alcohols consisting of flavonoids, quinones, tannins, and coumarins, and show remarkable antibiofilm activity against *C. albicans*. Allicin, an important constituent of garlic, was found to be one of the most potent inhibitors of *Candida* biofilms. Allicin is reported to act through inhibition of *HWP*1 gene, which is important for biofilm formation (Khodavandi et al., 2011). Therefore allicin could be appraised as a lead molecule for antibiofilm drug discovery. A study by Ali et al. (2010) exhibited that hydroxy-chavicol also act as a potential inhibitor of oral *Candida* biofilm through membrane destabilizing activity. Moreover, curcumin, pyrogallol, and pyrocatechol have also been shown to hold antibiofilm activity against *Candida* spp. (Shahzad et al., 2014). Zosteric acid, a phenolic compound from seagrass *Zostera marina*, also reported to possess antibiofilm activity in *Candida* (Villa et al., 2011). Epigallocatechin-3-gallate a major component of polyphenols extracted from green tea is reported to inhibit biofilm formation by *C. albicans*. In vivo, the effect is mediated through impairment of the yeast proteasomal activity, which results in reduced cellular metabolism and structural disruptions of sessile cells (Evensen and Braun, 2009). β-Asarone and two derivatives of asaronaldehyde, obtained from *Acorus* sp., are also able to inhibit *Candida* biofilms through targeting ergosterol biosynthesis (Rajput and Karuppayil, 2013).

14.6.4 Flavonoids

Baicalein, from the roots of a Chinese herb, *Scutellaria baicalensis* is a potential inhibitor of preformed/developing biofilms in *C. albicans*. The mechanism of action is expected to be lowering of CSH by down-regulating the CSH1 gene expression in *C. albicans*. Moreover, flavonoids primarily act through interaction with membrane proteins leading to changes in cell permeability and cell surface properties (Cao et al., 2008). A prenylated flavanoid compound isolated from the roots of *Dalea elegans* plant (Fabaceae), named 2′,4′-dihydroxy-5- (1″,1″-dimethylallyl)-8-prenylpinocembrin, had shown potential antibiofilm activity against drug resistant strains of *Candida* (Peralta et al., 2015). Authors suggested that accumulation of ROS and reactive nitrogen intermediates in sessile cells after treatment of this compound leading to increase in oxidative stress is the principal mechanism of action of inhibition.

14.6.5 Alkaloids

In a study from Zhao et al. (2013), an alkaloid tetrandrine isolated from *Stephania tetrandra* was found to exert strong antibiofilm activity in *Candida* by suppressing the biofilm formation through interference in Ras/cAMP pathway. They found significant lowering in the expression of genes involved in adhesion and filamentation. Caffeine inhibits biofilm formation in *C. albicans* by modulating target of rapamycin pathway

(Raut et al., 2013b). Berberine, an alkaloid present in several medicinal plants, including *Hydrastis canadensis* (goldenseal), *Coptis chinensis* (coptis or golden thread), *Berberis aquifolium* (Oregon grape), and other *Berberies* sp., has been identified to reduce preformed biofilms in *C. albicans* (Wei et al., 2011).

14.6.6 Anthraquinone, Tannin, and Phytoalexin

Quinones exhibit strong antibiofilm activity by their interaction and complex formation with nucleophilic amino acids leading to protein inactivation and loss of cell function. In a study conducted by Tsang et al. (2012), purpurin, a natural red pigment found in madder root (*Rubia tinctorum* L.), inhibited developing and preformed *C. albicans*. Mechanism of action of this compound is the downregulation of filamentation-associated genes *ALS3*, *ECE1*, *HWP1*, and *HYR1* and hyphal regulator protein *RAS1*. Tannins exercise antibiofilm effects through inactivation of microbial adhesins and transport proteins. Bakkiyaraj et al. (2013) investigated a tannin compound, ellagic acid, isolated from pomegranate fruit peel and found it to be strong antibiofilm candidate against *C. albicans*. Tannins alter the cell surface properties of *Candida* and thereby interfering with adherence leading to disruption of QS and, therefore, prevention of biofilm growth. In a study carried out by Upadhyay et al. (2014) showed that a hydroxylated derivatives of coumarins, i.e., phytoalexins, exhibited potential antibiofilm effects in *C. albicans*. Pterostilbene isolated from plants including *Pterocarpus marsupium* (the Indian kino tree), *Pterocarpus santalinus* (red sandalwood), *Vitis vinifera* (common grape vine), and *Vaccinium ashei* (rabbiteye blueberry) inhibited biofilm formation and extinguished the mature biofilms at concentrations $<64\,\mu g/mL$ in *C. albicans* (Li et al., 2014). Pterostilbene altered the expression of genes involved in morphological transition, ergosterol biosynthesis, oxidoreductase activity, and cell surface and heat shock proteins. Genes related to filamentation which are regulated by the Ras/cyclic AMP (cAMP) pathway were down-regulated.

14.6.7 Peptides and Lectins

Some primary metabolites such as peptides and lectins also exhibit antimicrobial properties. Their mechanism of action is believed to be through the formation of ion channels in cell membrane. Lectins such as ConA from *Canavalia ensiformis* and ConBr isolated from *Canavalia brasiliensis* have been shown to retard the biofilm development in *C. tropicalis* (Coelho, 2011). Mandal et al. (2011) have evaluated that plant defensins, cysteine-rich peptides, showed antibiofilm activity against *Candida*. They further isolate a novel antifungal peptide Tn-AFP1 (MW1230 Da) from plant *Trapa natans* that exhibited strong biofilm inhibitory activity in *C. tropicalis*. Similarly, datucin, a peptide from pericarp of *Datura stramonium*, has also exhibited fungicidal effects by weakening of mature biofilm of *C. albicans* (Mandal et al., 2012).

14.7 CONCLUSION

The ability of microorganisms to communicate within the microbial population imparts higher fitness in the environment. The polymorphic fungus *C. albicans* is an opportunistic human pathogen able to regulate virulence traits through this behavior by the production of at least two QS signal molecules: farnesol and tyrosol. The ability to undergo morphological switching and biofilms formation are the most important pathogenic characteristics of *C. albicans* regulated by QS and are of clinical relevance. One of the most important consequences of biofilm formation is that the biofilm cells are more resistant to both host defenses and conventional doses of antibiotics. In fact, traditional antimicrobial approaches are often unsuccessful toward these features and also have led to development of resistance. However, bio-compatibility is an issue for drugs applied for the prevention of biofilms on implants. Local drug release may be an attractive option especially for biofilm-preventing compounds on implants. Although fungal QS research is still in its infancy, its discovery has changed our views about the fungal kingdom and could eventually lead to the development of new antifungal therapeutics. But, there is a concern that disruption of QS may results in dispersion of biofilm leading to dissemination of disease. However, more studies of biofilms would be appreciated of targeting this QS pathway for *Candida* biofilm treatment. Further investigations on bioactive phyto-compounds are needed to concentrate on the molecular mechanisms and the genetic regulation of these phenomena in order to recognize alleged novel therapeutic opportunities.

ACKNOWLEDGMENT

We acknowledge scientific research support received from Imam Abdulrahman Bin Faisal University, Dammam for completing this work.

REFERENCES

Abraham, W.R., 2016. Going beyond the control of quorum-sensing to combat biofilm infections. Antibiotics (Basel) 5 (1), 3.

Albuquerque, P., Casadevall, A., 2012. Quorum sensing in fungi—a review. Med. Mycol. 50 (4), 337–345.

Alem, M.A.S., Oteef, M.D.Y., Flowers, T.H., Douglas, L.J., 2006. Production of tyrosol by *Candida albicans* biofilms and its role in quorum sensing and biofilm development. Eukaryot. Cell 5 (10), 1770–1779.

Ali, I., Khan, F.G., Suri, K.A., Gupta, B.D., Satti, N.K., Dutt, P., et al., 2010. In vitro antifungal activity of hydroxychavicol isolated from *Piper betle* L. Ann. Clin. Microbiol. Antimicrob. 9, 7.

Alina, H., Veronica, L., 2011. Inter-kingdom cross-talk: the example of prokaryotes–eukaryotes communication. Biointerface Res. Appl. Chem. 1 (3), 95–110.

Bakkiyaraj, D., Nandhini, J.R., Malathy, B., Pandian, S.K., 2013. The antibiofilm potential of pomegranate (*Punica granatum* L.) extract against human bacterial and fungal pathogens. Biofouling 29 (8), 929–937.

Barriuso, J., 2015. Quorum sensing mechanisms in fungi. AIMS Microbiol. 1 (1), 37–47.

Bassler, B.L., Losick, R., 2006. Bacterially speaking. Cell 125 (2), 237–246.

Bink, A., Pellens, K., Cammue, B.P.A., Thevissen, K., 2011. Anti-biofilm strategies: how to eradicate Candida biofilms. Open Mycol. J. 5, 29–38.

Borghi, E., Sciota, R., Biassoni, C., Cirasola, D., Cappelletti, L., Vizzini, L., et al., 2011. Cell surface hydrophobicity: a predictor of biofilm production in *Candida* isolates? J. Med. Microbiol. 60 (5), 689–690.

Brackman, G., Celen, S., Hillaert, U., van Calenbergh, S., Cos, P., Maes, L., et al., 2011. Structure–activity relationship of cinnamaldehyde analogs as inhibitors of AI-2 based quorum sensing and their effect on virulence of *Vibrio* spp. PLoS One 6, 3.

Cao, Y.Y., Cao, Y.B., Xu, Z., Ying, K., Li, Y., Xie, Y., et al., 2005. cDNA microarray analysis of differential gene expression in *Candida albicans* biofilm exposed to farnesol. Antimicrob. Agents Chemother. 49 (2), 584–589.

Cao, Y., Dai, B., Wang, Y., Huang, S., Xu, Y., Cao, Y., et al., 2008. In vitro activity of baicalein against *Candida albicans* biofilms. Int. J. Antimicrob. Agents 32 (1), 73–77.

Carneiro, V.A., Santos, H.S., Arruda, F.V., Bandeira, P.N., Albuquerque, M.R., Pereira, M.O., et al., 2010. Casbane diterpene as a promising natural antimicrobial agent against biofilm-associated infections. Molecules 16 (1), 190–201.

Chami, N., Bennis, S., Chami, F., Aboussekhra, A., Remmal, A., 2005. Study of anticandidal activity of carvacrol and eugenol in vitro and in vivo. Oral Microbiol. Immunol. 20 (2), 106–111.

Chandra, J., Mukherjee, P.K., Ghannoum, M.A., 2012. *Candida* biofilms associated with CVC and medical devices. Mycoses 55 (s1), 46–57.

Chen, H., Fujita, M., Feng, Q., Clardy, J., Fink, G.R., 2004. Tyrosol is a quorum sensing molecule in *Candida albicans*. Proc. Natl. Acad. Sci. U.S.A. 101 (14), 5048–5052.

Chifiriuc, M.C., Dițu, L.M., Oprea, E., Lițescu, S., Bucur, M., Măruțescu, L., Enache, G., Saviuc, C., Burlibașa, M., Trăistaru, T., Tănăse, G., Lazăr, V., 2009. In vitro study of the inhibitory activity of usnic acid on dental plaque biofilm. Roum Arch Microbiol Immunol. 68 (4), 215–222.

Coelho, F.A., 2011. Effect of Diocleinae Lectins on Bacteria and Fungi Planktonic and Sessile Cells. M.Sc. Dissertation Thesis. Universidade do Minho, (http://hdl.handle.net/1822/20301).

Cooper, M.A., Shlaes, D., 2011. Fix the antibiotics pipeline. Nature 472 (7341), 32.

Cordeiro, A., Teixeira, C.E., Brilhante, R.S., Castelo-Branco, D.S., Alencar, L.P., de Oliveira, J.S., et al., 2015. Exogenous tyrosol inhibits planktonic cells and biofilms of

Candida species and enhances their susceptibility to antifungals. FEMS Yeast Res. 15 (4), fov012.
Cremer, J., Vatou, V., Braveny, I., 1999. 2,4-[Hydroxyphenyl]-ethanol, an antioxidative agent produced by *Candida* spp., impairs neutrophilic yeast killing in vitro. FEMS Microbiol. Lett. 170 (2), 319–325.
Cugini, C., Calfee, M.W., Farrow III., J.M., Morales, D.K., Pesci, E.C., Hogan, D.A., 2007. Farnesol, a common sesquiterpene, inhibits PQS production in *Pseudomonas aeruginosa*. Mol. Microbiol. 65 (4), 896–906.
Cugini, C., Morales, D.K., Hogan, D.A., 2010. *Candida albicans*-produced farnesol stimulates *Pseudomonas* quinolone signal production in LasR defective *Pseudomonas aeruginosa* strains. Microbiology 156 (Pt10), 3096–3107.
Dalleau, S., Cateau, E., Berges, T., Berjeaud, J.M., Imbert, C., 2008. In vitro activity of terpenes against *Candida* biofilms. Int. J. Antimicrob. Agents 31 (6), 572–576.
Daniels, R.J., Vanderleyden, J., Michiels, J., 2004. Quorum sensing and swarming migration in bacteria. FEMS Microbiol. Rev. 28 (3), 261–289.
Darvishi, E., Omidi, M., Bushehri, A.A., Golshani, A., Smith, M.L., 2013. The antifungal eugenol perturbs dual aromatic and branched-chain amino acid permeases in the cytoplasmic membrane of yeast. PLoS One 8 (10), e76028.
Davies, D.G., Parsek, M.R., Pearson, J.P., Iglewski, B.H., Costerton, J.W., Greenberg, E. P., 1998. The involvement of cell-to-cell signals in the development of a bacterial biofilm. Science 280 (5361), 295–298.
De Sordi, L., Muhlschlegel, F.A., 2009. Quorum sensing and fungal-bacterial interactions in *Candida albicans*: a communicative network regulating microbial coexistence and virulence. FEMS Yeast Res. 9 (7), 990–999.
Deng, Y., Wu, J., Tao, F., Zhang, L.H., 2011. Listening to a new language: DSF-based quorum sensing in Gram-negative bacteria. Chem. Rev. 111 (1), 160–173.
Derengowski, L.S., De-Souza-Silva, C., Braz, S.V., Mello-De-Sousa, T.M., Bao, S.N., Kyaw, C.M., et al., 2009. Antimicrobial effect of farnesol, a *Candida albicans* quorum sensing molecule, on *Paracoccidioides brasiliensis* growth and morphogenesis. Ann. Clin. Microbiol. Antimicrob. 8, 13.
Desai, J.V., Mitchell, A.P., Andes, D.R., 2014. Fungal biofilms, drug resistance, and recurrent infection. Cold Spring Harb. Perspect. Med. 4, a019729.
Deveau, A., Hogan, D.A., 2011. Linking quorum sensing regulation and biofilm formation by *Candida albicans*. Methods Mol. Biol. 692, 219–233.
Ditu, L.M., Chifiriuc, CM, Bezirtzoglou, E., Voltsi, C., Bleotu, C., Pelinescu, D., Mihaescu, G., Lazar, V., 2011. Modulation of virulence and antibiotic susceptibility of EPEC strains by Enterococcus faecium probiotic strain culture fractions. Anaerobe. 17, 448–451.
Estrela, A.B., Heck, M.G., Abraham, W.R., 2009. Novel approaches to control biofilm infections. Curr. Med. Chem. 16 (12), 1512–1530.
Evensen, N.A., Braun, P.C., 2009. The effects of tea polyphenols on *Candida albicans*: inhibition of biofilm formation and proteasome inactivation. Can. J. Microbiol. 55 (9), 1033–1039.
Gong, Y.J., Liu, H., Feng, S.Y., Zhou, Y.H., Sun, M.J., 2011. Effect of ten traditional chinese medicine on planktonic and biofilm *Candida albicansin vitro*. Chin. J. Exp. Trad. Med. Form 23, 039.
Gurib-Fakim, A., 2006. Medicinal plants: traditions of yesterday and drugs of tomorrow. Mol. Asp. Med. 27 (1), 1–93.
Hall, R.A., Cottier, F., Muhlschlegel, F.A., 2009. Molecular networks in the fungal pathogen *Candida albicans*. Adv. Appl. Microbiol. 67, 191–212.
Hall, R.A., Turner, K.J., Chaloupka, J., Cottier, F., De Sordi, L., Sanglard, D., et al., 2011. The quorum sensing molecules farnesol/homoserine lactone and dodecanol

operate via distinct modes of action in *Candida albicans*. Eukaryot. Cell 10 (8), 1034–1042.

Hogan, D.A., 2006. Talking to themselves: autoregulation and quorum sensing in fungi. Eukaryot. Cell 5 (4), 613–619.

Hogan, D.A., Vik, A., Kolter, R.A., 2004. *Pseudomonas aeruginosa* quorum sensing molecule influences *Candida albicans* morphology. Mol. Microbiol. 54 (5), 1212–1223.

Hornby, J.M., Nickerson, K.W., 2004. Enhanced production of farnesol by *Candida albicans* treated with four azoles. Antimicrob. Agents Chemother. 48 (6), 2305–2307.

Hornby, J.M., Jensen, E.C., Lisec, A.D., Tasto, J.J., Jahnke, B., Shoemaker, R., et al., 2001. Quorum sensing in the dimorphic fungus *Candida albicans* is mediated by farnesol. Appl. Environ. Microbiol. 67 (7), 2982–2992.

Hui, F.M., Morrison, D.A., 1991. Genetic transformation in *Streptococcus pneumoniae*: nucleotide sequence analysis shows comA, a gene required for competence induction, to be a member of the bacterial ATP dependent transport protein family. J. Bacteriol. 173 (1), 372–381.

Jabra-Rizk, M.A., Meiller, T.F., James, C.E., Shirtliff, M.E., 2006. Effect of farnesol on *Staphylococcus aureus* biofilm formation and antimicrobial susceptibility. Antimicrob. Agents Chemother. 50 (4), 1463–1469.

Jakobsen, T.H., van Gennip, M., Phipps, R.K., Shanmugham, M.S., Christensen, L.D., Alhede, M., et al., 2012. Ajoene, a sulfur-rich molecule from garlic, inhibits genes controlled by quorum sensing. Antimicrob. Agents Chemother. 56 (5), 2314–2325.

Jiang, T., Li, M., 2013. Quorum sensing inhibitors: a patent review. Expert Opin. Ther. Pat. 23 (7), 867–894.

Jin, J., Zhang, J.Y., Guo, N., Sheng, H., Li, L., Liang, J.C., et al., 2010. Farnesol, a potential efflux pump inhibitor in *Mycobacterium smegmatis*. Molecules 15 (11), 7750–7762.

Kebaara, B.W., Langford, M.L., Navarathna, D.H., Dumitru, R., Nickerson, K.W., Atkin, A.L., 2008. *Candida albicans* Tup1 is involved in farnesol-mediated inhibition of filamentous-growth induction. Eukaryot. Cell 7 (6), 980–987.

Khan, M.S., Ahmad, I., 2012. Antibiofilm activity of certain phytocompounds and their synergy with fluconazole against *Candida albicans* biofilms. J. Antimicrob. Chemother. 67 (3), 618–621.

Khodavandi, A., Harmal, N.S., Alizadeh, F., Scully, O.J., Sidik, S.M., Othman, F., et al., 2011. Comparison between allicin and fluconazole in *Candida albicans* biofilm inhibition and in suppression of HWP1 gene expression. Phytomedicine 19 (1), 56–63.

Klengel, T., Liang, W.J., Chaloupka, J., Ruoff, C., Schroppel, K., Naglik, J.R., et al., 2005. Fungal adenylyl cyclase integrates CO_2 sensing with cAMP signaling and virulence. Curr. Biol. 15 (22), 2021–2026.

Kruppa, M., Krom, B.P., Chauhan, N., Bambach, A.V., Cihlar, R.L., Calderone, R.A., 2004. The two-component signal transduction protein Chk1p regulates quorum sensing in *Candida albicans*. Eukaryot. Cell 3 (4), 1062–1065.

Langford, M.L., Atkin, A.L., Nickerson, K.W., 2009. Cellular interactions of farnesol, a quorum-sensing molecule produced by *Candida albicans*. Future Microbiol. 4 (10), 1353–1362.

Langford, M.L., Hargarten, J.C., Patefield, K.D., Marta, E., Blankenship, J.R., Fanning, S., et al., 2013. *Candida albicans* Czf1 and Efg1 coordinate the response to farnesol during quorum sensing, white-opaque thermal dimorphism, and cell death. Eukaryot. Cell 12 (9), 1281–1292.

Li, D.D., Zhao, L.X., Mylonakis, E., Hu, G.H., Zou, Y., Huang, T.K., et al., 2014. *In vitro* and *in vivo* activities of pterostilbene against *Candida albicans* biofilms. Antimicrob. Agents Chemother. 58 (4), 2344–2355.

Lo, H.J., Köhler, J.R., DiDomenico, B., Loebenberg, D., Cacciapuoti, A., Fink, G.R., 1997. Nonfilamentous *C. albicans* mutants are avirulent. Cell 90 (5), 939–949.

Loh, J., Carlson, R.W., York, W.S., Stacey, G., 2002. Bradyoxetin, a unique chemical signal involved in symbiotic gene regulation. Proc. Natl. Acad. Sci. U.S.A. 99 (22), 14446–14451.
Lonn-Stensrud, J., Landin, M.A., Petersen, F.C., Benneche, T., Scheie, A.A., 2009. Furanones, potential agents for preventing *Staphylococcus epidermidis* biofilm infections? J. Antimicrob. Agents Chemother. 63 (2jiang), 309–316.
Lu, Y., Su, C., Unoje, O., Liu, H., 2014. Quorum sensing controls hyphal initiation in *Candida albicans* through Ubr1-mediated protein degradation. Proc. Natl. Acad. Sci. U.S.A. 111 (5), 1975–1980.
Machida, K., Tanaka, T., Yano, Y., Otani, S., Taniguchi, M., 1999. Farnesol-induced growth inhibition in *Saccharomyces cerevisiae* by a cell cycle mechanism. Microbiology 145 (Pt2), 293–299.
Mandal, S.M., Migliolo, L., Franco, O.L., Ghosh, A.K., 2011. Identification of an antifungal peptide from *Trapa natans* fruits with inhibitory effects on *Candida tropicalis* biofilm formation. Peptides 32 (8), 1741–1747.
Mandal, S.M., Migliolo, L., Das, S., Mandal, M., Franco, O.L., Hazra, T.K., 2012. Identification and characterization of a bactericidal and proapoptotic peptide from *Cycas revoluta* seeds with DNA binding properties. J. Cell Biochem. 113 (1), 184–193.
Marcos-Arias, C., Eraso, E., Madariaga, L., Quindos, G., 2011. *In vitro* activities of natural products against oral *Candida* isolates from denture wearers. BMC Compl. Alternat. Med. 11, 119.
Miller, M.B., Bassler, B.L., 2001. Quorum sensing in bacteria. Annu. Rev. Microbiol. 55, 165–199.
Monnet, V., Juillard, V., Gardan, R., 2016. Peptide conversations in Gram-positive bacteria. Crit. Rev. Microbiol. 42 (3), 339–351.
Mosel, D.D., Dumitru, R., Hornby, J.M., Atkin, A.L., Nickerson, K.W., 2005. Farnesol concentrations required to block germ tube formation in *Candida albicans* in the presence and absence of serum. Appl. Environ. Microbiol. 71 (8), 4938–4940.
Nakamura, M., Fujibayashi, T., Tominaga, A., Satoh, N., Kawarai, T., Shinozuka, O., et al., 2010. Hinokitiol inhibits *Candida albicans* adherence to oral epithelial cells. J. Oral Biosci. 52 (1), 42–50.
Navarathna, D.H.M.L.P., Hornby, J.M., Hoerrmann, N., Parkhurst, A.M., Duhamel, G. E., Nickerson, K.W., 2005. Enhanced pathogenicity of *Candida albicans* pre-treated with subinhibitory concentrations of fluconazole in a mouse model of disseminated candidiasis. J. Antimicrob. Chemother. 56, 1156–1159.
Navarathna, D.H.M.L.P., Nickerson, K.W., Duhamel, G.E., Jerrels, T.R., Petro, T.M., 2007. Exogenous farnesol interferes with the normal progression of cytokine expression during candidiasis in a mouse model. Infect. Immun. 75 (8), 4006–4011.
Nickerson, K.W., Atkin, A.L., Hornby, J.M., 2006. Quorum sensing in dimorphic fungi: farnesol and beyond. Appl. Environ. Microbiol. 72 (6), 3805–3813.
Nordin, M.A., Abdul Razak, F., Himratul-Aznita, W.H., 2015. Assessment of antifungal activity of bakuchiol on oral-associated *Candida* spp. Evid. Based Compl. Alternat. Med. 2015. Available from: https://doi.org/10.1155/2015/918624.
Novick, R.P., Geisinger, E., 2008. Quorum sensing in staphylococci. Annu. Rev. Genet. 42, 541–564.
Nucci, M., Queiroz-Telles, F., Tobon, A.M., Restrepo, A., Colombo, A.L., 2010. Epidemiology of opportunistic fungal infections in Latin America. Clin. Infect. Dis. 51 (5), 561–570.
Park, H., Myers, C.L., Sheppard, D.C., Sanchez, A.A., Edwards Jr., J.E., Filler, S.G., 2005. Role of the fungal Ras-protein kinase A pathway in governing epithelial cell interactions during oropharyngeal candidiasis. Cell Microbiol. 7 (4), 499–510.

Parsek, M.R., Greenberg, E.P., 2005. Sociomicrobiology: the connections between quorum sensing and biofilms. Trends Microbiol. 13 (1), 27–33.

Pauli, A., 2006. Anticandidal low molecular compounds from higher plants with special reference to compounds from essential oils. Med. Res. Rev. 26 (2), 223–268.

Pearson, J.P., Van Delden, C., Iglewski, B.H., 1999. Active efflux and diffusion are involved in transport of *Pseudomonas aeruginosa* cell-to-cell signals. J. Bacteriol. 181 (4), 1203–1210.

Peralta, M.A., da Silva, M.A., Ortega, M.G., Cabrera, J.L., Paraje, M.G., 2015. Antifungal activity of a prenylated flavonoid from *Dalea elegans* against *Candida albicans* biofilms. Phytomedicine 22 (11), 975–980.

Pereira, C.S., Thompson, J.A., Xavier, K.B., 2013. AI-2-mediated signalling in bacteria. FEMS Microbiol. Rev. 37 (2), 156–181.

Rajput, S.B., Karuppayil, S.M., 2013. β-Asarone, an active principle of Acorus calamus rhizome, inhibits morphogenesis, biofilm formation and ergosterol biosynthesis in *Candida albicans*. Phytomedicine 20 (2), 139–142.

Ramage, G., Saville, S.P., Wickes, B.L., Lopez-Ribot, J.L., 2002. Inhibition of *Candida albicans* biofilm formation by farnesol, a quorum-sensing molecule. Appl. Environ. Microbiol. 68 (11), 5459–5463.

Rao, A., Zhang, Y., Muend, S., Rao, R., 2010. Mechanism of antifungal activity of terpenoid phenols resembles calcium stress and inhibition of the TOR pathway. Antimicrob. Agents Chemother. 54 (12), 5062–5069.

Raut, J.S., Shinde, R.B., Chauhan, N.M., Karuppayil, S.M., 2013a. Terpenoids of plant origin inhibit morphogenesis, adhesion, and biofilm formation by *Candida albicans*. Biofouling 29 (1), 87–96.

Raut, J.S., Chauhan, N.M., Shinde, R.B., Karuppayil, S.M., 2013b. Inhibition of planktonic and biofilm growth of *Candida albicans* reveals novel antifungal activity of caffeine. J. Med. Plant Res. 7, 777–782.

Raut, J.S., Shinde, R.B., Chauhan, N.M., Karuppayil, S.M., 2014. Phenylpropanoids of plant origin as inhibitors of biofilm formation by *Candida albicans*. J. Microbiol. Biotechnol. 24 (9), 1216–1225.

Roy, V., Smith, J.A.I., Wang, J., Stewart, J.E., Bentley, W.E., Sintim, H.O., 2010. Synthetic analogs tailor native AI-2 signaling across bacterial species. J. Am. Chem. Soc. 132 (32), 11141–11150.

Rudrappa, T., Bais, H.P., 2008. Curcumin, a known phenolic from *Curcuma longa*, attenuates the virulence of *Pseudomonas aeruginosa* PAO1 in whole plant and animal pathogenicity models. J. Agric. Food Chem. 56 (6), 1955–1962.

Rukayadi, Y., Han, S., Yong, D., Hwang, J.K., 2011. *In vitro* activity of xanthorrhizol against *Candida glabrata*, *C. guilliermondii*, and *C. parapsilosis* biofilms. Med. Mycol. 49 (1), 1–9.

San-Blas, G., Travassos, L.R., Fries, B.C., Goldman, D.L., Casadevall, A., Carmona, A. K., et al., 2000. Fungal morphogenesis and virulence. Med. Mycol. 38 (Suppl. 1), 79–86.

Saville, S.P., Lazzell, A.L., Monteagudo, C., Lopez-Ribot, J.L., 2003. Engineered control of cell morphology *in vivo* reveals distinct roles for yeast and filamentous forms of *Candida albicans* during infection. Eukaryot. Cell 2 (5), 1053–1060.

Scutera, S., Zucca, M., Savoia, D., 2014. Novel approaches for the design and discovery of quorum-sensing inhibitors. Expert Opin. Drug Discov. 9 (4), 353–366.

Semighini, C.P., Hornby, J.M., Dumitru, R., Nickerson, K.W., Harris, S.D., 2006. Farnesol-induced apoptosis in *Aspergillus nidulans* reveals a possible mechanism for antagonistic interactions between fungi. Mol. Microbiol. 59 (3), 753–764.

Shahzad, M., Sherry, L., Rajendran, R., Edwards, C.A., Combet, E., Ramage, G., 2014. Utilising polyphenols for the clinical management of *Candida albicans* biofilms. Int. J. Antimicrob. Agents 44 (3), 269–273.

Sharma, M., Prasad, R., 2011. The quorum-sensing molecule farnesol is a modulator of drug efflux mediated by ABC multidrug transporters and synergizes with drugs in *Candida albicans*. Antimicrob. Agents Chemother. 55 (10), 4834–4843.

Shirtliff, M.E., Krom, B.P., Meijering, R.A., Peters, B.M., Zhu, J., Scheper, M.A., et al., 2009. Farnesol-induced apoptosis in *Candida albicans*. Antimicrob. Agents Chemother. 53 (6), 2392–2401.

Silva, C.D., Guterres, S.S., Weisheimer, V., Schapoval, E.E., 2008. Antifungal activity of the lemongrass oil and citral against *Candida* spp. Braz. J. Inf. Dis. 12 (1), 63–66.

Singleton, D.R., Masuoka, J., Hazen, K.C., 2001. Cloning and analysis of a *Candida albicans* gene that affects cell surface hydrophobicity. J. Bacteriol. 183 (12), 3582–3588.

Sudbery, P., Gow, N., Berman, J., 2004. The distinct morphogenic states of *Candida albicans*. Trends Microbiol. 12 (7), 317–324.

Sung, W.S., Lee, D.G., 2010. Antifungal action of chlorogenic acid against pathogenic fungi, mediated by membrane disruption. Pure Appl. Chem. 82 (1), 219–226.

Thompson, J.A., Oliveira, R.A., Djukovic, A., Ubeda, C., Xavier, K.B., 2015. Manipulation of the quorum sensing signal AI-2 affects the antibiotic-treated gut microbiota. Cell Rep. 10 (11), 1861–1871.

Tsang, P.W., Bandara, H.M., Fong, W.P., 2012. Purpurin suppresses *Candida albicans* biofilm formation and hyphal development. PLoS One 7 (11), e50866.

Upadhyay, A., Upadhyaya, I., Kollanoor-Johny, A., Venkitanarayanan, K., 2014. Combating pathogenic microorganisms using plant-derived antimicrobials: a minireview of the mechanistic basis. BioMed. Res. Int. 2014. Available from: https://doi.org/10.1155/2014/761741.

Vasconcelos, M.A., Arruda, F.V., Santos, H.S., Rodrigues, A.S., Bandeira, P.N., Albuquerque, M.R.J.R., et al., 2014. Effect of a casbane diterpene isolated from *Croton nepetaefolius* on the prevention and control of biofilms formed by bacteria and *Candida* species. Ind. Crops Prod. 61, 499–509.

Villa, F., Pitts, B., Stewart, P.S., Giussani, B., Roncoroni, S., Albanese, D., et al., 2011. Efficacy of zosteric acid sodium salt on the yeast biofilm model *Candida albicans*. Microb. Ecol. 62 (3), 584–598.

Visick, K.L., Fuqua, C., 2005. Decoding microbial chatter: cell–cell communication in bacteria. J. Bacteriol. 187 (16), 5507–5519.

Wang, Y., Ma, S., 2014. Small molecules modulating AHL-based quorum sensing to attenuate bacteria virulence and biofilms as promising antimicrobial drugs. Curr. Med. Chem. 21 (3), 296–311.

Wei, G.X., Xu, X., Wu, C.D., 2011. *In vitro* synergism between berberine and miconazole against planktonic and biofilm *Candida* cultures. Arch. Oral Biol. 56 (6), 565–572.

Westwater, C., Balish, E., Schofield, D.A., 2005. *Candida albicans*-conditioned medium protects yeast cells from oxidative stress: a possible link between quorum sensing and oxidative stress resistance. Eukaryot. Cell 4 (10), 1654–1661.

Winzer, K., Williams, P., 2001. Quorum sensing and the regulation of virulence gene expression in pathogenic bacteria. Int. J. Med. Microbiol. 291 (2), 131–143.

Xie, C., Sun, L., Meng, L., Wang, M., Xu, J., Bartlam, M., et al., 2015. Sesquiterpenes from *Carpesium macrocephalum* inhibit *Candida albicans* biofilm formation and dimorphism. Bioorg. Med. Chem. Lett. 25 (22), 5409–5411.

Zhao, L.X., Li, D.D., Hu, D.D., Hu, G.H., Yan, L., Wang, Y., et al., 2013. Effect of tetrandrine against *Candida albicans* biofilms. PLoS One 8 (11), e79671.

Zhou, L., Zheng, H., Tang, Y., Yu, W., Gong, Q., 2013. Eugenol inhibits quorum sensing at sub-inhibitory concentrations. Biotechnol. Lett. 35 (4), 631–637.

Zore, G.B., Thakre, A.D., Rathod, V., Karuppayil, S.M., 2011. Evaluation of anti-*Candida* potential of geranium oil constituents against clinical isolates of *Candida albicans* differentially sensitive to fluconazole: inhibition of growth, dimorphism and sensitization. Mycoses 54 (4), e99–e109.

CHAPTER 15

Neem Leaf Glycoprotein in Cancer Immunomodulation and Immunotherapy

Anamika Bose and Rathindranath Baral
Department of Immunoregulation and Immunodiagnostics, Chittaranjan National Cancer Institute (CNCI), Kolkata, West Bengal, India

15.1 INTRODUCTION

The dreaded disease cancer is attempted to control by primary treatment, i.e., surgery that generally follows chemotherapy and/or radiotherapy. The subject in these three branches of cancer treatment is advanced extensively worldwide; however, its impact on overall 5 years survival is debatable. These treatment modalities could not offer the expected benefits that may be due to following reasons: (1) inability of correction of dysregulated self-defense (immunity) of human body, with imposing impairment on the immune system; (2) unable to prevent recurrence as memory response is not generated; (3) metastasis, a unique character of cancer, cannot be prevented; (4) minimum effect on cancer stem cells. In this scenario, neem leaf glycoprotein (NLGP) holds some promise as it has shown positive impact on all three above-mentioned issues, while fourth issue is yet to be tested. All of these studies are performed in our laboratory and no other group in the world studied on this topic yet. Thus discussions made in this review are solely based on the published work of our group.

15.2 THE STORY BEHIND NEEM LEAF GLYCOPROTEIN RESEARCH

The world has been learned on natural management of several diseases from Indian Ayurveda (Diasio and LoBuglio, 1996; Atal et al., 2006). Among several natural resources in India, neem (*Azadirachta indica*) is known by the name of "*Sarboroganibarani*," i.e., component that can cure all forms of diseases (Biswas et al., 2002). Different components of the neem tree like leaves, bark, flowers, twigs, gum, seeds, pulp, and oil were

New Look to Phytomedicine
DOI: https://doi.org/10.1016/B978-0-12-814619-4.00016-1

found to be beneficiary for the mankind. It possesses antifungal, antihelminthic, antibacterial, antiviral, antidiabetic, and anticancer effects (Paul et al., 2011; Veitch et al., 2008; Puri, 1999; Bodduluru et al., 2014). Myriad of herbal products obtained from the neem tree contributed significantly to both prophylactic and therapeutic treatment regimens (Paul et al., 2011). Most of these functions of neem plants are proved in daily life, not in laboratories.

The neem tree has attracted considerable research attention as a rich source of limonoids that have potent antioxidants and anticancer properties. Researches evaluate chemo-preventive potential of neem limonoids azadirachtin and limbolide on in vitro antioxidant assays and in vivo inhibitory effects on 7,12-dimethylbenz[a]anthracene (DMBA)-induced hamster buccal pouch carcinogenesis. Both azardirachtin and nimbolide exhibited antiradical scavenging activity and parallel administration of both, inhibited development of DMBA-induced carcinomas by several mechanisms, like prevention of pro-carcinogen activation and oxidative DNA damage, upregulation of antioxidant, carcinogen detoxification enzymes, and tumor invasion, angiogenesis. Nimbolide was more potent antioxidant and chemopreventive agent, thus a promising candidate for multitargeted prevention and treatment of cancer (Priyadarsini et al., 2009; Arumugam et al., 2014). In a recent study, antiangiogenic effect of ethanol extract of neem leaves on Human Umbilical vein Endothelial Cells (HUVECs) exhibited inhibition of vascular endothelial growth factor (VEGF)-induced angiogenic response in vitro and in vivo (Mahapatra et al., 2011). Nimbolide retards tumor cell migration, invasion, and angiogenesis by downregulating MMP-2/9 expression in colon cancer cells (Babykutty et al., 2012). Neem components known to have immune-modulatory and immune-stimulatory activities as examined in vivo models (Paul et al., 2011); however, limited information in this area opens the enormous scope of research on neem and immune system. Based on the available literature on neem plants, in 1992, US National Academy of Science designated this tree as "A tree solving global problem" (1992).

Prophylactic use of an aqueous preparation of neem leaf was reported to prevent the growth of murine carcinoma and melanoma (Baral and Chattopadhyay, 2004). Since then a series of evidences have been published favoring immune-editing functions of such neem leaf preparation that effectively restrict the murine tumor growth (Haque and Baral, 2006; Mandal-Ghosh et al., 2007; Sarkar et al., 2008; Bose et al., 2009a,b;

Goswami et al., 2010). Therapeutic efficacy of this preparation in murine tumor models has been tested after identification of the active principal as a glycoprotein, thus named it as NLGP. Although complete chemical characterization of NLGP is under process, it has appeared single peak in HPLC (Goswami et al., 2010). NLGP has carbohydrate moiety consisting of arabinose, galactose, and glucose. Exposure of NLGP to the adverse temperature (100°C), pH (5.7), and enzymes (papain, neuraminidase) resulted complete disappearance of the tumor growth restricting property (Baral et al., 2010), prove that the bioactivity of NLGP depends on both glycol- and protein-part of the molecule.

15.3 NEEM LEAF GLYCOPROTEIN IS NONTOXIC FOR HUMAN USE

The toxicity profile of NLGP on different physiological systems of Swiss mice and Sprague–Dawley rats was studied (Mallick et al., 2013a,b). NLGP injection (25 μg/mice/injection), even in higher doses (200 μg/mice/injection) than effective concentration (25 μg/mice/injection) caused neither behavioral changes in animals nor death. NLGP increased the body weight of mice slightly without any change in organ weights; and showed no adverse effect on the hematological system except little hemato-stimulation. Histological assessment of different organs like brain, liver, kidney, spleen, and lymph nodes revealed no alterations in the organ microstructure in mice and rats after the NLGP treatment. Histological normalcy of liver and kidney was further confirmed by the assessment of liver enzymes alkaline phosphatase, Serum Glutamic Oxaloacetic Transaminase (SGOT), Serum Glutamic Pyruvic Transaminase (SGPT), and nephrological products urea, and creatinine. However, NLGP has no apoptotic effect on immune cells including T cells but induces proliferation of mononuclear cells collected from mice and rats. Accumulated evidences strongly suggest the nontoxic nature of NLGP (Fig. 15.1), thus may safe for human use in anticancer therapy (Mallick et al., 2013a,b).

15.4 IMMUNO-EDITING BY NEEM LEAF GLYCOPROTEIN TARGETS IMMUNE EVASION STRATEGIES OF TUMOR

15.4.1 Neem Leaf Glycoprotein as a Therapeutic Vaccine

After obtaining sufficient data on NLGP as a prophylactic vaccine (Baral and Chattopadhyay, 2004; Haque and Baral, 2006), the role of NLGP as a

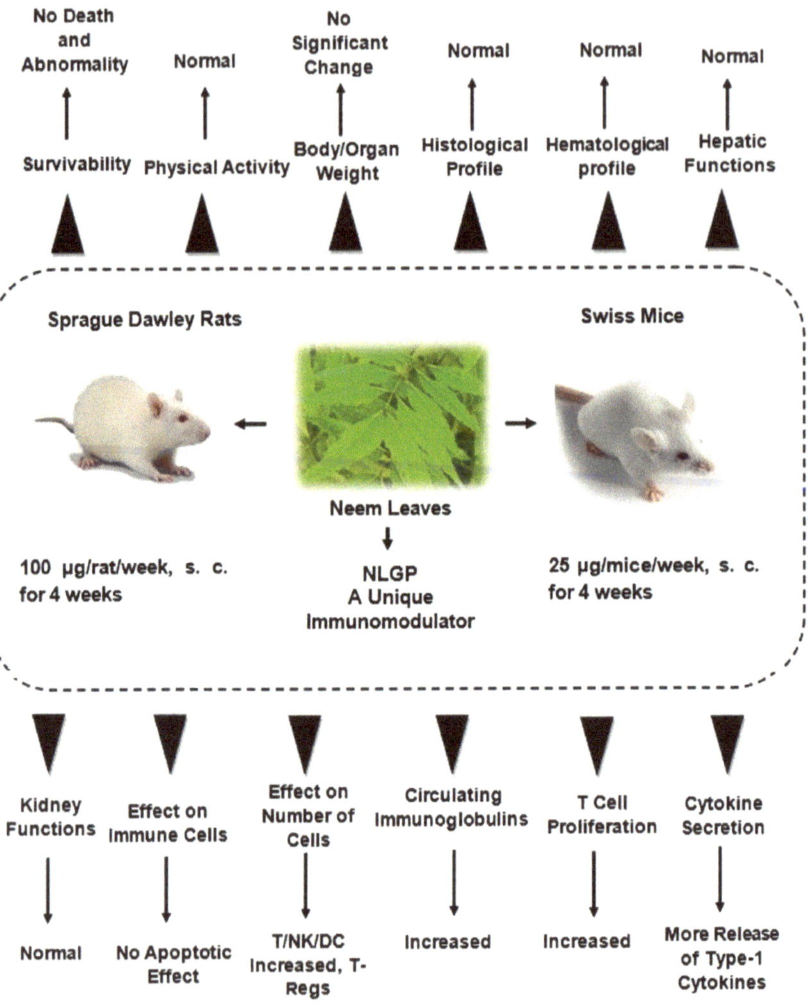

Figure 15.1 Toxicity testing of NLGP in Swiss mice and Sprague–Dawley rats. Physical, histological, biochemical, and immunological assessment revealed NLGP is completely nontoxic.

therapeutic vaccine was investigated. Treatment of mice bearing established sarcoma with NLGP (25 µg/mice/week subcutaneously for 4 weeks) resulted in tumor regression or dormancy (tumor free/regressor, 13/24 (NLGP), 4/24 (PBS)) (Mallick et al., 2013a,b). In addition to sarcoma, therapy with NLGP inhibits murine B16-melanoma in vivo and improves survivability (Barik et al., 2015). NLGP is also effective to reduce solid carcinoma growth in Swiss mice (Bhuniya et al., 2016) (Fig. 15.2A and B).

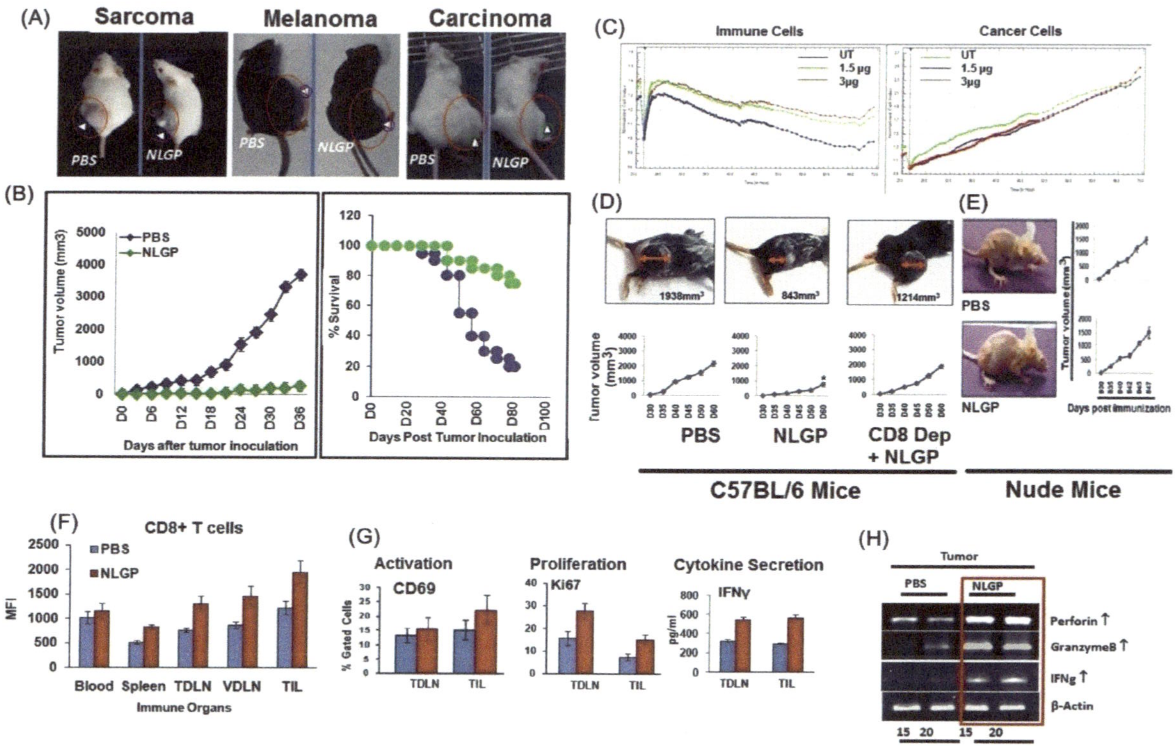

Figure 15.2 Therapeutic property of NLGP in different mouse tumors. (A) NLGP restricts mouse carcinoma, sarcoma, and melanoma. (B) Tumor growth and survival curve in NLGP-treated sarcoma bearing mice showing therapeutic power of NLGP. (C) NLGP has no cytotoxic effect on either immune cells or tumor cells. (D) NLGP restricts solid melanoma growth, but fails when $CD8^+$ T cells are depleted. (E) NLGP has no tumor restricting effect in immune-compromised athymic nude mice. (F) $CD8^+$ T cells are upregulated in different immune compartments from NLGP-treated mice, as determined by flow cytometry. (G) NLGP-stimulated $CD8^+$ T cells are active, proliferative, and IFNγ secreting. (H) RT-PCR analysis for cytotoxic molecules from mRNA extracted from NLGP-stimulated $CD8^+$ T cells.

15.4.2 Therapeutic Potential of Neem Leaf Glycoprotein is Immune Cell Dependent

In order to find out the mechanism of tumor growth restriction by NLGP, as observed in murine melanoma, sarcoma, and carcinoma models in therapeutic settings, a series of investigation was performed. Previous study confirmed that NLGP has no toxic effect on either cancer or normal cells (Fig. 15.2C) (Bose et al., 2007; Chakraborty et al., 2011a,b), thus this molecule might use some different mode of action for tumor restriction. With established knowledge on immunomodulatory power of NLGP, the participation of immune system on observed tumor inhibition was examined by using drug-induced immunosuppressive mice and immunocompromised athymic nude mice models. In first model, mice were immunosuppressed with an immunosuppressive drug, cyclosporine before NLGP treatment. In cyclosporine group, NLGP failed to restrict tumor growth as immune system was not optimally functional. Corroborate, failure to restrict tumor growth by NLGP in cyclosporine treated mice, was partially recovered by adoptive transfer of splenic T cells from NLGP-immunized mice (Banerjee et al., 2014). These findings might indicate that NLGP-mediated tumor restriction is immune dependent. To further validate the influence of NLGP-conditioned immune system to restrain tumor growth, tumor bearing immune-compromised athymic nude mice was treated with NLGP, where no tumor growth inhibition was detected (Fig. 15.2E).

15.4.3 $CD8^+$ T Cell Dependence of Antitumor Action of Neem Leaf Glycoprotein

After confirming the immune dependence of NLGP's antitumor action, involvement of various immune competent cells, like $CD4^+$, $CD8^+$ T cells, NK cells, B cells, etc., was studied. Results revealed a significant increment in $CD8^+$ T cells in blood, spleen, tumor draining lymph node (TDLN), vaccine draining lymph node (VDLN), and tumor in NLGP-treated mice. Elevated expression of CD69, CD44, and Ki67 on $CD8^+$ T cells indicated their state of activation and proliferation induced by NLGP (Fig. 15.2F and G). Depletion of $CD8^+$ T cells in NLGP-treated mice resulted disappearance of NLGP-mediated tumor regression (Fig. 15.2D). An expansion of $CXCR3^+$ and $CCR5^+$ T cells was observed in the TDLN and tumor, along with their corresponding ligands. Enhanced secretion/expression of IFNγ was noted after NLGP therapy. In vitro culture of $CD8^+$ T cells with IL-2 and sarcoma antigen resulted in significant enhancement in cytotoxic efficacy that was further supported by consistent higher expression of CD107a on $CD8^+$ T cells from tumors (Mallick et al., 2013a,b).

This upregulation of $CD8^+$ T cell activity in NLGP-treated mice was also studied in relation to tumor-microenvironment (TME) in B16 melanoma model. Within TME from NLGP-treated mice $CD8^+$ T cell activity was enhanced with dominance of type 1 cytokine/chemokine network with downregulation of suppressive cellular functions. Exposure of $CD8^+$ T cells to NLGP-TME in vitro, but not to PBS-microenvironment, resulted higher perforin and granzyme B (killer molecules) expression with greater in vitro cytotoxicity against B16 melanoma. These $CD8^+$ T cells showed proportionally lower FasR expression, which prevents them from activation-induced cell death. Collectively, these findings support a paradigm in which NLGP dynamically orchestrates the activation, expansion, and recruitment of $CD8^+$ T cells into established tumors to operate significant tumor cell lysis. Therefore NLGP normalizes TME to allow $CD8^+$ T cells to perform optimally to inhibit tumor growth (Barik et al., 2015, 2013).

15.4.4 Neem Leaf Glycoprotein Influences Antigen-Presenting Cells to Optimize $CD8^+$ T Cell Functions

NLGP-mediated murine tumor growth restriction depends on $CD8^+$ T cells. For optimum T cell functions, matured type 1 dendritic cells (DC1) are required. Unfortunately, in cancer, DCs remain in an immature state, thus cannot impart proper assistance to T cells regarding antigen presentation. In a study, $CD14^+$ human monocytes were differentiated to immature DCs with Granulocyte Macrophage Colony Stimulating Factor (GM-CSF)/IL-4 and treated with NLGP either in normal and tumor conditions for maturation. NLGP-matured DCs (NLGP-DCs) showed upregulated expression of CD83, CD80, CD86, CD40, and MHCs, in a comparable extent to the control, i.e., Lipopolysaccharide (LPS). NLGP-DCs secrete high amount of IL-12p70 with low IL-10. NLGP upregulates the expression of crucial transcription factor, ikaros, indicating maturation toward DC1 phenotype. Increased expression of CD28 and CD40L on T cells following coculture with NLGP-DCs was noticed to promote DC-T cell interactions. As a result, low IL-4 and high amounts of IFNγ were released in culture to produce antitumor type 1 immune microenvironment (Goswami et al., 2010).

In vivo efficacy of NLGP-matured DCs (NLGP-DCs) for antigen presentation to $CD8^+$ T cells, thereby enhancement of colon cancer cell killing, was examined. These colon cancer cells express carcinoembryonic antigen (CEA). Vaccination with NLGP-matured CEA pulsed DCs enhances antigen-specific humoral and cellular immunity against tumor antigen, CEA, and restricts the growth of CEA^+ murine colon tumors. NLGP helps in better CEA uptake, and processing and presentation to T and B cells. This vaccination (NLGP-matured DC pulsed with CEA-DCNLGPCEA) elicits mitogen-induced and CEA-specific T cell proliferation, IFNγ secretion and induces specific cytotoxic reactions to CEA^+ colon tumor cells. In addition to T cell response, Dendritic Cell-Neem Leaf Glycoprotein-Carcinoembryonic Antigen (DCNLGPCEA) vaccine generates anti-CEA antibody (principally IgG2a) response, which participates in cytotoxicity of CEA^+ cells in antibody-dependent manner. This strong anti-CEA cellular and humoral immunity protects mice from tumor development and these mice remained tumor free following second tumor inoculation, indicating generation of effector memory response. Evaluation of underlying mechanism suggests vaccination and generates strong CEA-specific cytotoxic T lymphocyte (CTL) and antibody response that can completely prevent the tumor growth following adoptive transfer (Sarkar et al., 2010; Das et al., 2014).

The observation on the efficacy of NLGP-matured murine DCs to prevent CEA^+ tumors was extended with DCs generated with human monocytes obtained from stage IIIB cervical cancer (CaCx IIIB) patients. These DCs showed dysfunctional maturation with dysregulated antitumor T cell functions. With an objective to optimize these dysregulated immune functions, NLGP was used to induce optimum maturation of immature DCs (iDCs) obtained from CaCx IIIB patients. Results showed an upregulated expression of various cell surface markers (CD40, CD83, CD80, CD86, and HLA-ABC), which indicates DC maturation. Consequently, NLGP-matured DCs displayed balanced cytokine secretions, with type 1 bias and noteworthy functional properties. These DCs displayed substantial T cell allo-stimulatory capacity and promoted the generation of CTLs, effective to kill human cervical cancer cells in vitro (Roy et al., 2011).

15.4.5 Neem Leaf Glycoprotein Polarizes Type 1 Immune Microenvironment

To explain the immunological basis of NLGP-mediated tumor growth restriction, NLGP-driven immune activation and NLGP-associated immune polarization were studied, where upregulation of early activation marker CD69 on lymphocytes, monocytes, and DCs was observed. Activation is also denoted by CD45RO enhancement, with a decrease in CD45RA phenotype and CD62L (L-selectin). NLGP-activated T cells secrete greater amount of signature T-helper type 1 cytokine IFNγ and a lower amount of the type 2 cytokine IL-4. Similar type 1 polarization is also observed in antigen-presenting monocytes and DCs by upregulation of IL-12, TNFα, and downregulation of IL-10. Creation of the type 1 microenvironment is also assisted by NLGP-induced downregulation of $FoxP3^+$ regulatory T (Treg) cells. A type 1-specific transcription factor, T-bet, is upregulated in circulating immune cells after their stimulation with NLGP. In the creation of type 1 immune network, increased phosphorylation of STAT1 and STAT4 with decreased phosphorylation of STAT3 might have significance. NLGP may be effective in maintaining normal immune homeostasis by upregulating type 1 response in immunosuppressed hosts, which may have significant role in the induction of host protective antitumor functions (Bose et al., 2009a,b).

15.4.6 Neem Leaf Glycoprotein Reduces Frequency and Suppressive Properties of Regulatory-Immune Cells

In addition to the presence of effector cells, some suppressor cells, like Tregs, tumor-associated macrophages (TAMs), myeloid-derived suppressor cells (MDSCs), are present in a physiological system that suppresses the over-activation of immune cells to prevent autoimmunity. In cancer, suppressive functions of these cells are amplified to paralyze the effector immune functions. Presence of Tregs in tumors is associated with compromised tumor-specific immune responses and has a clear negative impact on survival of cancer patients. Thus downregulation of Tregs is considered as a promising cancer immunotherapeutic approach. NLGP downregulates $CD4^+CD25^+Foxp3^+$ Tregs within tumors (Fig. 15.3A). Treg migration was also inhibited by NLGP in association with the downregulation of CCR4 along with its ligand CCL22. NLGP is not apoptotic to Tregs but significantly downregulates the expression of Treg markers, like Foxp3, CTLA4, and GITR. It also reverses the Treg-induced functional impairment of T-effector cells, in terms of IFNγ secretion, cellular proliferation, and tumor cell cytotoxicity. Interaction between Foxp3, p-NFATc3, and p-Smad2/3, needed for successful Treg function, is also inhibited by NLGP (Chakraborty et al., 2011a,b).

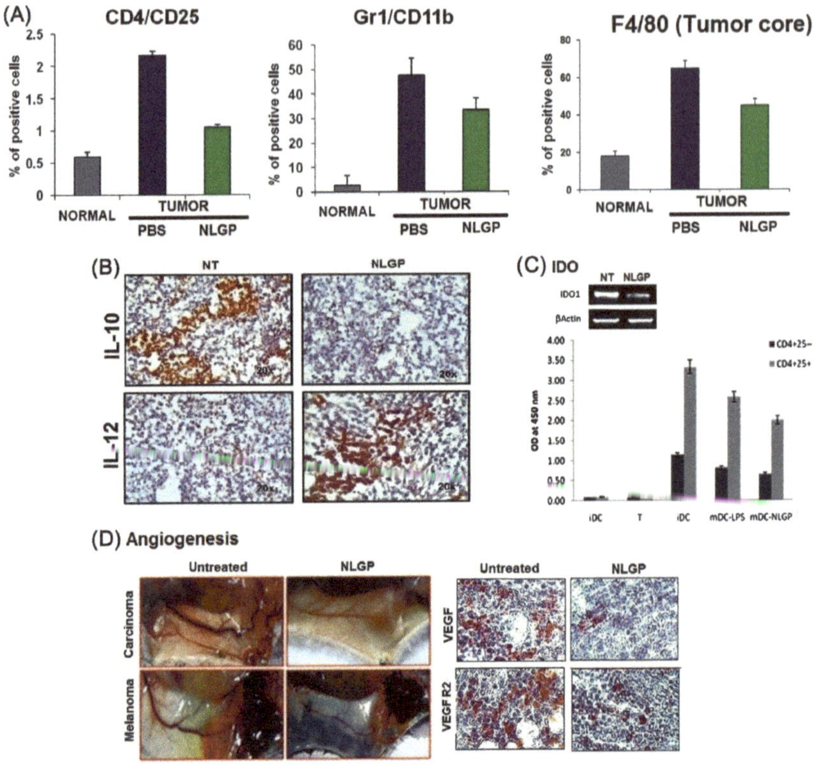

Figure 15.3 Status of immune-suppressive cells and protumor molecules in NLGP-treated mice. (A) Regulatory T cells ($CD4^+CD25^+$), myeloid-derived suppressor cells ($Gr1^+CD11b^+$), and tumor-associated macrophages ($F4/80^+$) are downregulated in NLGP-treated mice. (B) Expression of IL-10, but not IL-12, is downregulated, in tumor core from NLGP-treated mice. (C) Tryptophan degrading enzyme, IDO, is downregulated in NLGP-matured dendritic cells. (D) NLGP normalizes aberrant angiogenesis in carcinoma and melanoma bearing mice. Angiogenic molecule VEGF and its receptor are also downregulated within tumor from NLGP-treated mice (immunohistochemical analysis).

Tregs suppress the tumoricidal functions of $CD14^+CD68^+$ monocytes/macrophages (MO/Mφ) from human peripheral blood and such suppression is reversed by NLGP. Cytotoxic efficacy of MO/Mφ toward macrophage sensitive cells, U937, is decreased in the presence of Tregs (induced); however, it was increased further by NLGP supplementation in culture. Associated Treg-mediated inhibition of nitric oxide release from MO/Mφ was normalized by NLGP. Altered status of signature cytokines, like IL-12, IL-10, IL-6, and TNFα from MO/Mφ under the influence of Tregs, is also corrected by NLGP. In addition to tumoricidal functions, antigen-presenting ability of MO/Mφ is hampered by Treg-induced downregulation of CD80, CD86, and HLA-ABC. NLGP upregulates these molecules in MO/Mφ. Treg-mediated contact dependent inhibition of MO/Mφ chemotaxis was also normalized partially by NLGP, where the participation of CCR5 was documented. Overall results suggest that Treg-influenced pro-tumor MO/Mφ functions are rectified in a significant extent by NLGP to create an antitumor immune environment (Chakraborty et al., 2011a,b).

Macrophages residing in TME or TAMs, preferentially M2 pro-umor macrophages, skewed to promote tumor growth, angiogenesis, invasion, and/or metastasis, thus reversal to M1 type (antitumor) macrophages are desired. In vitro immunomodulatory potential of NLGP in reprogramming of Stage III supraglottic laryngeal tumor cell lysate (SLTCL)-induced M2 TAMs to their classical antitumor shape (M1) was studied. Data generated from this study support that NLGP is effective in preventing the SLTCL-induced generation ($CD68^+CD206^+IL\text{-}10^{high}$ to $CD68^+CD206^-IL\text{-}10^{low}$ TAMs) and functions (NO^{low} to NO^{high}, $MHC\text{-}I^{low}$ to $MHC\text{-}I^{high}$, $CD80^{low}$ to $CD80^{high}$) of protumorous M2 macrophages, which in turn is associated with sustained antitumor effector functions by promoting cytotoxic T cell activities and suppressing Tregs. Furthermore, NLGP prevents M2 skewness of TAMs by downregulating phosphorylation of targeted STAT3 (Goswami et al., 2014).

Another important class of suppressor cells is MDSCs. This group of cells suppresses immunity in cancer by promoting conversion of T cells to regulatory T cells and M1 to M2 macrophages. Interestingly, NLGP downregulates $CD11b^+Gr1^+$ MDSCs in periphery of tumor host and also in tumors (Fig. 15.3A). NLGP helps to withdraw the suppressive effects of MDSC toward effector T cells ($CD8^+$ T cells) by increasing their proliferation, activation, and inhibiting T cell anergy. NLGP can partially rectify the conversion of $CD4^+$ T cells to Tregs by modulating the regulatory suppressive cytokines, TGFβ, and IL-10. In surgically tumor (sarcoma)-removed mice, NLGP prevents recurrence of tumor by activating host immune system by $CD8^+$ T cell-mediated killing of MDSCs using Fas-FasL pathway and by minimizing the suppressive effect of MDSCs. NLGP in TME triggers the apoptosis of MDSCs which downregulates mainly granulocytic MDSCs in TME, along with downregulation of suppressive molecules present on MDSCs that ultimately withdraws the MDSCs' suppression (Sarkar et al., 2017).

15.4.7 Neem Leaf Glycoprotein Protects $CD8^+$ T Cells From Anergy

In tumor growth restriction, prevention of T cell anergy by NLGP plays a key role by maximizing the efficacy of T cells within TME. $CD8^+$ T cells can work properly in TME from NLGP-treated mice (NLGP-TME). To elucidate the role of anergic process in T cell activation and its modulation by NLGP, the status of anergy-related genes by RT-PCR during exposure of normal $CD8^+$ T cells to NLGP-TME and its control counterpart was studied. Results obtained from RT-PCR analysis demonstrated that purified $CD8^+$ T cells, exposed to NLGP-TME, have shown negligible expression of cbl-b, egr2, egr3, itch, GRAIL, and DGK1α compared to T cells exposed to control TME. Further, $CD8^+$ T cells were purified using magnetic sorting technology (MACS) from day 20 tumors from NLGP and PBS-treated mice to obtain mRNA profile. Similar trend of results was obtained as seen in case of TME-exposed T cells. Western blotting data showed the presence of greater amount of pNFAT in NLGP-TME exposed $CD8^+$ T cells, in comparison to PBS-TME and ionomycin-exposed control cells. These data clearly provide evidence on NLGP protection within TME by preventing T cells from tumor-induced anergy (Barik et al., 2015, 2013).

15.4.8 Neem Leaf Glycoprotein Maintains Optimum Tryptophan Supply to T Cells by Inhibiting Indoleamine 2,3-Dioxygenase Secretion From Tolerogenic Dendritic Cells

Another mechanism of tumor immune evasion is deprivation of T cells from tryptophan by abundant production of indoleamine 2,3-dioxygenase (IDO) from tolerogenic DCs. IDO may be cooperatively induced in DCs by Tregs and various DC maturation agents to hamper antitumor immunity. Consequently, a major focus of current immunotherapeutic strategies in cancer is to minimize IDO, which is possible by reducing Tregs and using various IDO inhibitors. As Tregs are inducer of IDO in DCs and hyperactive Treg is a hallmark of cancer, NLGP might abrogate IDO induction in DCs by inhibiting Tregs (Fig. 15.3C). The evidence was presented using immune cells from CaCx-IIIB patients, which shows inhibition of IDO induction in DCs by NLGP with curtailing the over expression of CTLA4 on Tregs and concomitant induction of optimal DC maturation. This finding suggests the reduction of tolerogenecity of DCs in CaCx-IIIB patients by reducing the IDO pool using NLGP (Roy et al., 2013).

15.4.9 Neem Leaf Glycoprotein Promotes Central and Effector Memory Cells in Prophylactic and Therapeutic Settings

During NLGP-mediated $CD8^+$ T cell dependent therapeutic tumor growth restriction, tumor free status was maintained in majority of mice after reinoculation of sarcoma cells in NLGP-treated tumor regressed mice. This is correlated with the increment of $CD44^{hi}CD62L^{hi}$ central memory T cells. Role of the NLGP in generation, functions, and persistence of memory $CD8^+$ T cells was also examined. Under prophylactic settings, sarcoma antigen immunization resulted in better $CD8^+$ memory T cell generation against sarcoma tumor challenge in the presence of NLGP than antigen alone. Dual generation of central memory cells in lymph nodes and effector memory cells in spleen protected immunized mice from future tumor inoculation (Ghosh et al., 2016). In therapeutic settings, NLGP treatment–generated memory $CD8^+$ T cells were superior to those generated after metronomic cyclophosphamide treatment. No tumor recurrence in mice after surgical removal of primary tumor was observed in NLGP-treated mice but tumor recurred in cyclophosphamide-treated mice (Ghosh et al., 2017).

15.5 HYPOXIA-REGULATING AND ANTIANGIOGENIC PROPERTIES OF NEEM LEAF GLYCOPROTEIN WITHIN TUMOR

In continuation with the fact that prophylactic as well as therapeutic administration of NLGP induces significant restriction of solid tumor growth in mice, the effect of such treatment (25 μg/mice; weekly four times) in regulation of tumor angiogenesis, an obligate factor for tumor progression, was investigated. The NLGP pretreatment results in vascular normalization in melanoma and carcinoma bearing mice along with downregulation of CD31, VEGF, and VEGFR2 and facilitates profound infiltration of $CD8^+$ T cells within hypoxic tumor parenchyma, which subsequently regulates VEGF-VEGFR2 signaling in $CD31^+$ vascular endothelial cells and tumor cells to prevent aberrant neo-vascularization (Fig. 15.3D). Pericyte stabilization, inhibition of VEGF-dependent VEC proliferation, and subsequent vascular normalization are also experienced. Studies in immune compromised mice confirmed that these vascular and intratumoral changes in angiogenic profile are dependent upon active adoptive immunity particularly those mediated by $CD8^+$ T cells (Banerjee et al., 2014).

More importantly, tumor tissue obtained from NLGP-treated mice showed significantly reduced expression of HIF1α, a signature molecule of hypoxia. In vitro study suggested that NLGP can modulate HIF1α downstream signaling cascade in immune cells as well as in tumor cells which subsequently reduce VEGF.

15.6 ANTIMETASTATIC PROPERTIES OF NEEM LEAF GLYCOPROTEIN

Metastasis attributes 90% mortality in cancer patients; hence, new therapeutic strategies to prevent metastasis is still in search. NLGP therapeutically restricts primary tumor growth but its role in relation to metastasis was unknown. To ascertain the antimetastatic role of NLGP, spontaneous and experimental metastasis models were established with carcinoma (LLC) and melanoma (B16F10) cells in C57BL/6J mice (Bhuniya et al., 2016), where significant attenuation of metastasis by NLGP was observed along with prolonged mice survival. Metastasis inhibition is due to the reduction of migration or alteration of tumor cells, but metastasis cascade requires colonization which needs neo-angiogenesis and immune-evasion (Fig. 15.4). Normalized angiogenesis along with reduced angiogenic

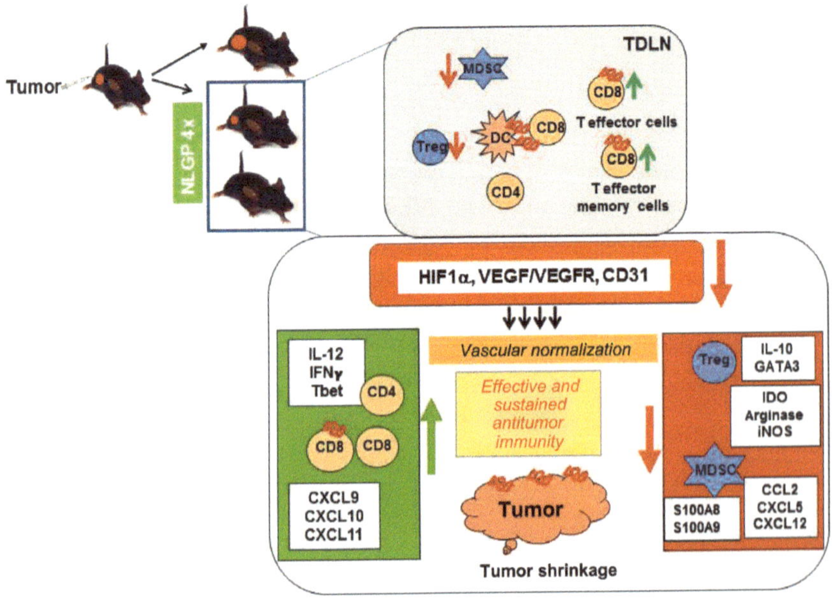

Figure 15.4 Schematic representation on mechanism of action of NLGP to restrict murine tumor growth by immunomodulation.

factors (VEGF, TGFβ) in lungs of NLGP-treated mice was noticed with a high infiltration of activated $CD8^+$ T cells with maximized IFNγ secretion that inhibits colonization. Antimetastatic potential of NLGP might be helpful to propose NLGP as an arsenal against cancer in clinical settings (Bhuniya et al., 2016). The overall mechanism of action of NLGP in tumor shrinkage is depicted in Fig. 15.4.

15.7 CONCLUSION

Anticancer therapy mainly focuses on direct killing of tumor cells. However, history of anticancer therapy confirms such approach can offer temporary benefits without complete eradication of the disease. Other than conventional therapies, immunotherapy proposes other way of thought by either stimulating suppressed immune system or inhibiting suppressor cells or modulating TME. Immunotherapy with NLGP fulfils most of the hallmark criteria of such therapy including elimination of tumor cells in antigen-specific manner, creation of type 1 immune environment with suppression of the suppressor system, generation of memory response, normalization of aberrant angiogenesis, and prevention of metastasis. Fulfilment of one or two among many criterions would be rarely successful to reach the ultimate goal of cancer eradication because other paths are enough to promote cancer. Exceptional antitumor immunomodulatory functions of NLGP warrant clinical translation of this unique natural molecule.

ACKNOWLEDGMENTS

Within the span of 2002–16, the following investigators contributed in this NLGP work and their valuable contributions are acknowledged: Dr. Ishita Mandal-Ghosh, Dr. Enamul Haque, Dr. Koustav Sarkar, Dr. Krishnendu Chakraborty, Dr. Shyamal Goswami, Dr. Tathagata Chakraborty, Dr. Soumyabrata Roy, Dr. Subhasis Barik, Dr. Atanu Mallick, Dr. Saptak Banerjee, Dr. Kuntal Kanti Goswami, Dr. Bipasa Mandal, Dr. Arnab Das, Dr. Sarbari Ghosh, Ms. Tithi Ghosh, Ms. Ipsita Guha, Mr. Avishek Bhuniya, Ms. Madhurima Sarkar, Ms. Shayani Dasgupta, Ms. Akata Saha, Mr. Partha Nandi, Ms. Juhina Das, Mr. Neelanjan Ganguly.

We would like to thank the Director, CNCI, for providing necessary facilities. We also acknowledge our collaborators Prof. Subrata Laskar, Dr. Smarajit Pal, Prof. Walter J. Storkus, Dr. Utpala Chattopadhyay, Dr. Mahesh Kulkarni, Dr. Bhaskar Saha, Prof. Pinak Chakraborti, and others for their generous contribution to NLGP work time to time. Financial assistance from ICMR, CSIR, UGC, DST is acknowledged.

REFERENCES

Atal, C.K., Sharma, M.L., Kaul, A., Khajuria, A., 2006. Immunomodulating agents of plant origin I, preliminary screening. J. Ethnopharmacol. 18, 133–141.

Arumugam, A., Agullo, P., Boopalan, T., 2014. Neem leaf extract inhibits mammary carcinogenesis by altering cell proliferation, apoptosis, and angiogenesis. Cancer Biol. Ther. 15, 26–34.

Babykutty, S., Priya, P.S., Nandini, R.J., 2012. Nimbolide retards tumor cell migration, invasion, and angiogenesis by downregulating MMP-209 expression via inhibiting ERK1/2 and reducing DNA-binding activity of NF-κB in colon cancer cells. Mol. Carcinog. 51, 475–490.

Banerjee, S., Ghosh, T., Barik, S., Das, A., Ghosh, S., Bhuniya, A., et al., 2014. Neem leaf glycoprotein prophylaxis transduces immune dependent stop signal for tumor angiogenic switch within tumor microenvironment. PLoS One 9, e110040.

Baral, R., Chattopadhyay, U., 2004. Neem *(Azadirachta indica)* leaf mediated immune activation causes prophylactic growth inhibition of murine Ehrlich carcinoma and B16 melanoma. Int. Immunopharmacol. 4, 355–366.

Baral, R.N., Sarkar, K., Mandal Ghosh, T., Bose, A., 2010. Neem leaf glycoprotein as a new vaccine adjuvant for cancer immunotherapy. In: Gupta, VK (Ed.), Comprehensive Bioactive Natural Products; Immune-modulation & Vaccine Adjuvants, Vol. 5. Studium Press LLC, India, pp. 21–45, Chapter 2.

Barik, S., Banerjee, S., Mallick, A., Goswami, K.K., Roy, S., Bose, A., et al., 2013. Normalization of tumor microenvironment by neem leaf glycoprotein potentiates effector T cell functions and therapeutically intervenes in the growth of mouse sarcoma. PLoS One 8, e66501.

Barik, S., Banerjee, S., Sarkar, M., Bhuniya, A., Roy, S., Bose, A., et al., 2015. Neem leaf glycoprotein optimizes effector and regulatory functions within tumor microenvironment to intervene therapeutically the growth of B16 melanoma in C57BL/6 mice. Trials Vaccinol. 4, e80–e87.

Bhuniya, A., Guha, I., Ghosh, T., Ghosh, S., Banerjee, S., Bose, A., et al., 2016. Neem leaf glycoprotein attenuated carcinoma and melanoma metastasis by editing DC-CD8 + T cell interaction and angiogenesis. J. Carcinog. 15, S283.

Biswas, K., Chattopadhyay, I., Banerjee, R.K., Bandyopadhyay, U., 2002. Biological activities and medicinal properties of neem (*Azadirachta indica*). Curr. Sci. 82, 1336–1345.

Bose, A., Haque, E., Baral, R., 2007. Neem leaf preparation induces apoptosis of tumor cells by releasing cytotoxic cytokines from human peripheral blood mononuclear cells. Phytother. Res. 21, 914–920.

Bodduluru, L.N., Kasala, E.R., Thota, N., Barua, C.C., Sistla, R., 2014. Chemopreventive and therapeutic effects of nimbolide in cancer: the underlying mechanisms. Toxicol In Vitro 28, 1026–1035.

Bose, A., Chakraborty, K., Sarkar, K., Chakraborty, T., Goswami, S., Pal, S., et al., 2009a. Neem leaf glycoprotein induces perforin mediated tumor cell killing by T and NK cells through differential regulation of IFNγ. J. Immunother. 32, 42–53.

Bose, A., Chakraborty, K., Sarkar, K., Goswami, S., Haque, E., Chakraborty, T., et al., 2009b. Neem leaf glycoprotein directs T-bet associated type 1 immune commitment. Hum. Immunol. 70, 6–15.

Chakraborty, T., Bose, A., Barik, S., Goswami, K., Banerjee, S., Goswami, S., et al., 2011a. Neem leaf glycoprotein inhibits $CD4^{+}CD25^{+}Foxp3^{+}$ regulatory T cells to restrict murine tumor growth. Immunotherapy 3, 749–769.

Chakraborty, T., Bose, A., Goswami, K.K., Goswami, S., Chakraborty, K., Baral, R., 2011b. Neem leaf glycoprotein suppresses regulatory T cell mediated suppression of monocyte/macrophage functions. Int. Immunopharmacol. 12, 326–333.

Das, A., Barik, S., Banerjee, S., Bose, A., Sarkar, K., Biswas, J., et al., 2014. A monoclonal antibody against neem leaf glycoprotein recognizes carcinoembryonic antigen (CEA) and restricts CEA expressing tumor growth. J. Immunother. 37, 394–406.

Diasio, R.B., LoBuglio, A.F., 1996. Immunomodulators: immunosuppressive agents and immunostimulants. The Pharmacological Basis of Therapeutics. Goodman and Gilman's. McGraw-Hill, New York.

Ghosh, S., Sarkar, M., Ghosh, T., Guha, I., Bhuniya, A., Saha, A., et al., 2016. Neem leaf glycoprotein promotes dual generation of central and effector memory $CD8^+$ T cells against sarcoma antigen vaccine to induce protective anti-tumor immunity. Mol. Immunol. 71, 42–53.

Ghosh, S., Sarkar, M., Ghosh, T., Guha, I., Bhuniya, A., Saha, A., et al., 2017. Neem leaf glycoprotein generates superior tumor specific central memory CD8 + T cells than cyclophosphamide that averts post-surgery solid sarcoma recurrence. Vaccine 35, 4421–4429.

Goswami, S., Bose, A., Sarkar, K., Roy, S., Chakraborty, T., Baral, R., 2010. Neem leaf glycoprotein matures myeloid derived dendritic cells and optimizes anti-tumor T cell functions. Vaccine 28, 1241–1252.

Goswami, K.K., Barik, S., Sarkar, M., Bhowmick, A., Biswas, J., Bose, A., et al., 2014. Targeting STAT3 phosphorylation by neem leaf glycoprotein prevents immune evasion exerted by supra-glottic laryngeal tumor induced M2 macrophages. Mol. Immunol. 59, 119–127.

Haque, E., Baral, R., 2006. Neem *(Azadirachta indica)* leaf preparation induces prophylactic growth inhibition of murine Ehrlich carcinoma in Swiss and C57BL/6 by activation of NK cells and NK-T cells. Immunobiology 211, 721–731.

Mahapatra, S., Young, C.Y.F., Kohli, M., 2011. Antiangiogenic effects and therapeutic targets of *Azadirachta indica* leaf extract in endothelial cells. Evid. Based Complement. Alternat. Med. 2012, 303019.

Mallick, A., Barik, S., Goswami, K.K., Banerjee, S., Ghosh, S., Sarkar, K., et al., 2013a. Neem leaf glycoprotein activates $CD8^+$ T cells to promote therapeutic anti-tumor immunity inhibiting the growth of mouse sarcoma. PLoS One 8, e47434.

Mallick, A., Ghosh, S., Banerjee, S., Majumder, S., Das, A., Mondal, B., et al., 2013b. Neem leaf glycoprotein is nontoxic to physiological functions of Swiss mice and Sprague Dawley rats: histological, biochemical and immunological perspectives. Int. Immunopharmacol. 15, 73–83.

Mandal-Ghosh, I., Chattopadhyay, U., Baral, R., 2007. Neem leaf preparation enhances Th1 type immune response and anti-tumor immunity against breast tumor associated antigen. Cancer Immun. 7, 8–17.

Neem: A Tree for Solving Global Problems. National Research Council, Washington, DC: National Academic Press, 1992.

Paul, R., Prasad, M., Sah, N.K., 2011. Anticancer biology of *Azadirachta indica* L. (neem): a minireview. Free Radic. Res. 12, 467–476.

Priyadarsini, R.V., Manikandan, P., Kumar, G.H., 2009. The neem limnoids azadirachtin and nimbolide inhibit hamster cheek pouch carcinogenesis by modulating xenobiotic-metabolizing enzymes, DNA damage, antioxidants, invasion and angiogenesis. Free Radic. Res. 43, 492–504.

Puri, H.S., 1999. Neem: The Divine Tree *Azadirachta indica*, 1st ed. CRC Press, Boca Raton, FL.

Roy, S., Goswami, S., Bose, A., Chakraborty, T., Chakraborty, K., Pal, S., et al., 2011. Neem leaf glycoprotein partially rectifies suppressed dendritic cell functions and associated T cell efficacy in stage IIIB cervical cancer patients. Clin. Vac. Immunol. 18, 571–579.

Roy, S., Barik, S., Banerjee, S., Pal, S., Basu, P.S., Biswas, J., et al., 2013. Neem leaf glycoprotein overcomes indoleamine 2, 3 dioxygenase mediated tolerance in dendritic cells by attenuating hyperactive regulatory T cells in cervical cancer stage IIIB patients. Hum. Immunol. 74, 1015–1023.

Sarkar, K., Bose, A., Chakraborty, K., Haque, E., Ghosh, D., Goswami, S., et al., 2008. Neem leaf glycoprotein helps to generate carcino-embryonic antigen specific antitumor immune responses utilizing macrophage mediated antigen presentation. Vaccine 26, 4352–4362.

Sarkar, K., Goswami, S., Roy, S., Mallick, A., Chakraborty, K., Bose, A., et al., 2010. Neem leaf glycoprotein enhances carcinoembryonic antigen presentation of dendritic cells to T and B cells for induction of antitumor immunity by allowing generation of immune effector/memory response. Int. Immunopharmacol. 10, 865–887.

Sarkar, M., Ghosh, S., Bhuniya, A., Ghosh, T., Guha, I., Barik, S., et al., 2017. Neem leaf glycoprotein prevents post-surgical sarcoma recurrence in Swiss mice by differentially regulating cytotoxic T and myeloid-derived suppressor cells. PLoS One 12, e0175540.

Voitch, G.E., Boyer, A., Ley, S.V., 2008. The Azadirachtin story. Angew Chem. Int. Ed. Engl. 47, 9402–9429.

CHAPTER 16

Role of Phytomedicine in Diabetes and Cardiovascular Diseases

Parul Tripathi[1], Govind Gupta[2] and Puneet Singh Chauhan[3]
[1]Department of Microbiology, Government Medical College, Ambedkar Nagar, Uttar Pradesh, India
[2]Department of Microbiology, Barkatullah University, Bhopal, Madhya Pradesh, India
[3]Division of Plant Microbe Interactions CSIR-National Botanical Research Institute, Lucknow, Uttar Pradesh, India

16.1 INTRODUCTION

Plants have always been important sources of medicines since days of yore. The fact that plant-derived drugs came into use in modern medicine must have stemmed from natural reasoning that spontaneously became integral to domestic cure in traditional systems of medicine. Phytomedicine is a plant-based traditional medicinal practice that uses various plant materials in preventive and therapeutic processes (Wallis, 2005; Ali, 2008). Plants are now being globally considered as biosynthetic laboratory of naturally occurring bioactive compounds with possible biological response modifier abilities. These bioactive molecules or phytochemicals therefore have gained much importance as prophylactic and therapeutic agents in curing several human ailments (Akerele, 1993; Kamboj, 2000).

Phytochemicals are found in all plants and are classified according to their chemical structures and functional properties (Tiwari et al., 2012). There are a number of epidemiological studies and experimental evidences to establish the relationship between the phytochemicals and immune-modulation in both humans and animals. An ideal desirable phytochemical candidate would be one that corrects or regulates the specific part of the immune response that causes the disease. Therefore, if stimulation of the immune response is desired in immune-compromised individuals, suppression of the same is required for others such as transplant recipients or patient with autoimmune or inflammatory diseases. It is apparent that desired overall balance in immune response is thought to be

New Look to Phytomedicine
DOI: https://doi.org/10.1016/B978-0-12-814619-4.00017-3

regulated and manifested at the relevant genetic level. And the latter is modulated to a large extent by the therein physiological milieu like the cytokines, chemokines, etc. This unique potential of phytochemicals as a source of immune-modifier bioactive molecules (Aridoğan, 2006) that can help to maintain the homeostasis of disturbed immune function is hence of great advantage in managing clinical conditions as diabetes and associated cardiovascular diseases (CVDs). Moreover, from an immunological point of view, the biological responses in these diseases involve a rapid upsurge of inflammation, oxidative stress, resulting in progressive disabling symptoms (Martín-Timón et al., 2014).

Though past few decades have observed a significant progress in managing diabetes and associated CVDs. However, results of treatment in patients have been far from perfect and new challenges are being constantly faced. Further, several disadvantages including drug resistance, side effects, and even toxicity have set new hurdles in the path of disease management. Management of CVD in diabetics is of great challenge for the physicians and cardiologists. Since persistent low-grade chronic inflammation has been shown to be central in mediating the underlying pathophysiology of both diabetes and CVDs (Stern, 1995; Dokken, 2008). It is high time that plants with potent immune-modulatory potential should go for well-planned clinical trials and then ultimately to pharmacy without any further delay. Simple and cost-effective interventions implemented timely can effectively prevent these diseases. This will not only help in reducing the socioeconomic burden in the developing countries owing to their minimal side effects, low cost, and abundance. They can also be used as a powerful tool, as part of a customized treatment plan, particularly in the developed countries that are facing the double disease burden due to globalization and changing lifestyles.

This chapter is an attempt to consolidate and highlight the unique immune-modulating ability of plant-based drugs in diabetes and related CVDs. It is clear that from a clinical perspective, there is an urgent need to incorporate these bioactive components into integrated antidiabetic and cardioprotective therapies. Keeping this in mind, we have consciously discussed herein that how it is imperative that out of the several plant-based treatments blindly used for regulating metabolic dysfunctions, selections should be strictly made based upon their ability to effectively ameliorate the underlying diseases-specific inflammatory balances. Finally, the need of the hour is to evolve a systematic holistic treatment approach and to develop well-designed methodologies to bring these novel biomolecules to bedsides of the masses.

16.2 DIABETES AND CVDS: CAUGHT IN A VISCOUS LOOP

16.2.1 Global Burden and Threat

Diabetes is a metabolic disease of pandemic presence. Its ever-increasing prevalence is threatening to cause significant damage both to individuals and society as a whole (Herman, 2017). The precise cause is not always nailed down; however, an underlying genetic predisposition cannot be ruled out. Other factors like age, diet, physical activity, ethnic background, environmental, psychological factors, and hormonal disorders also have their respective and cumulative roles. Diabetes is a persistent systemic disease that is consistently linked with the development of CVDs (Fig. 16.1) (Laakso, 2010). The incidence of diabetes and related CVDs has dramatically increased over the past years with an alarming increase in mortality due to CVDs than diabetes (Amos et al., 1997; Frankel et al., 2008).

Although significant progress has been made, the precise reasons of underlying pathophysiology still remains clouded. We have witnessed an

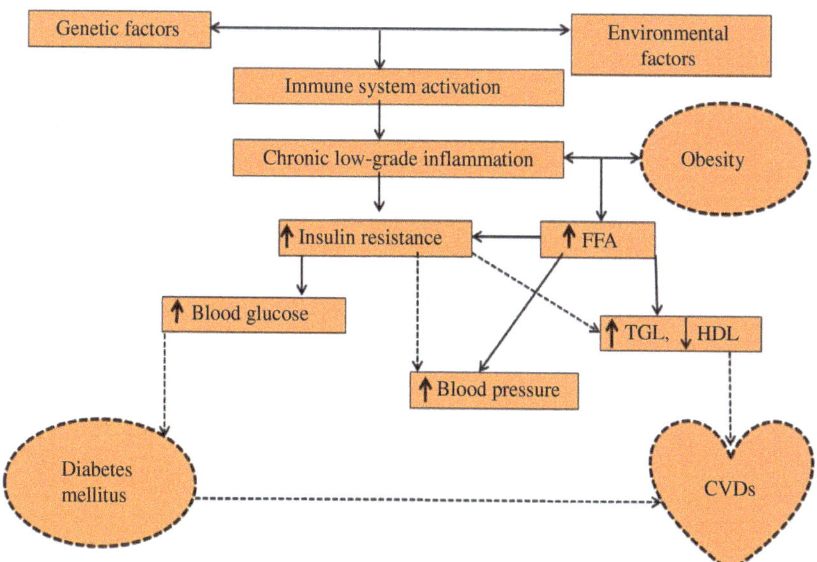

FFA: free fatty acid; TGL: triglycerides; HDL: high-density lipoprotein

Figure 16.1 Cardio-metabolic interrelationship.

explosion of research focused on the role and use of natural plant/food-based products for managing several metabolic diseases with limited long-term effective success (Newman and Cragg, 2007; Simaan 2009; World Health Organization, 2008). Diabetes and related clinical conditions have been at the epicenter of this clinical and scientific volcanic burst (IDF, 2015). It is precisely for this reason that in this chapter, we have consciously restricted the content to current thinking with regards to the role of plant-based components in regulating the underlying disease mechanisms. We do understand that how on one hand, it is helpful to understand, absorb, and analyze the observations accumulated from an array of studies based on these natural components. It is all the more important to question and discuss whether the accumulated data have actually helped to elucidate the complex interplay in diabetes and CVDs. In any case, it is urgent to find new hits and lead compounds.

Ideally, the first step supporting to reduce this double disease burden (Boutayeb, 2006; Mazzone et al., 2008) begins with patient education, awareness, and maintaining a healthy lifestyle (Pradeepa et al., 2012). As mentioned earlier, diabetic milieu supports an increase and persistence in low-grade systemic inflammation and when pushed to extremes can produce several dysfunctions like CVDs (Stern, 1995; Dokken, 2008; Frankel et al., 2008). So a more clinically relevant issue would be to target and identify factors regulating this overlap. While ultimately all forms of diabetes type I, II, and gestational (occurring during pregnancy) are due to the beta cells of the pancreas being unable to produce sufficient insulin to prevent hyperglycemia, the causes are different (Expert Committee Report on Diabetes, 2003; ADA, 2016). Out of the types I and II, the former is caused by autoimmune destruction of β-cells, secondary to environmental triggers like toxins and viruses. Therefore, treatment of type I diabetes depends on exogenous insulin and accounts for nearly 10% of all diabetic patients. Type II diabetes, on the other hand, is a more common form of diabetes constituting 90% of the diabetic population. Patients suffering from type II diabetes are unable to respond to insulin and can be managed with dietary changes, exercise, and medication (Holleman, 2014; Dean and McEntyre, 2004). In addition, as discussed earlier, uncontrolled diabetes leads to several chronic CVD complications (Mazzone et al., 2008). Therefore, considering the nature of the disease, an integrative approach involving plant-based components should be followed.

16.2.2 Disease Obstacles

Despite recent advances in genome-based association studies, the available experimental data aiming to bridge diabetes and CVDs to therapeutics has failed to add important pieces to our understanding of underlying mechanisms (Qazi and Malik, 2013), though it is known that diabetes involves a complex interplay between genetic and environmental factors. Wherein the genetic factors that underlie individual susceptibility are amplified in the presence of certain environmental triggers (Brownlee, 2005). This understanding has not been used to maximum advantage in devising strategies to combat the increasing burden of diabetes and CVDs. Huge body of available literature accumulated from different trials and studies are unanimously conclusive of the poor reality and limited efficacy observed with both conventional and non-pharmacological interventions (Beckman et al., 2013; Paneni et al., 2013). The drugs used for management of diabetes and CVDs are potent but frequently associated with serious, even life-threatening, side effects and are awaiting precise optimization (Bae, 2016; Moller, 2001). At present no drug is able to arrest the progressive loss of pancreatic cells which occurs in type II diabetes mellitus (Goldfine, 2008; Bailey, 2013; Bloomgarden, 2007; Alberti et al., 2009). Even in well-managed patients, daily insulin administration cannot parallel the natural timing and dosing of insulin secretion from the pancreas in response to hyperglycemia, resulting in severe complications (Cersosimo et al., 2014).

Further, although animal models and human studies have elucidated several pathologic features of diabetes and CVDs (Sasase et al., 2013, 2015), the resultant observations have not been studied rigorously with regards to prevention or treatment of diabetic CVDs. As a consequence, no specific recommendations can be made for the use of these models and conclusions with respect to diabetes and associated CVDs. The heterogeneity of types of human diabetes and the lack of exact replicas among non-primate animals often require efficacy studies in more than one model, especially to investigate the mode of action. Currently, a standard model of experimental diabetes to study effects of drugs, which could help in preventing the progressive loss of pancreatic islet function, remains to be established (King, 2012).

Most of the available animal models apparently share similar characteristic features of type II diabetes and have allowed experimentation that would be impossible in humans. Although none of the known single species is exactly equivalent to human diabetes, still each model acts as an essential tool for investigating genetic, endocrine, metabolic, morphologic changes and underlying etiopathogenic mechanisms that could also operate during the evolution of type II diabetes in humans (Cefalu, 2006; Srinivasan and Ramarao, 2007). However, care must be taken in interpretation and extrapolation of the results obtained from these animal models to humans. In addition, in the screening program of antidiabetic compounds, it is particularly important to note that some animal models are better suited to screen particular class of antidiabetic compounds. The use of smaller animal models such as mice will also reduce the expense of producing test materials, while some advanced efficacy studies or toxicological examinations which require invasive procedures and large blood and tissue samples may be facilitated by using animals with large body size such as rat or other nonrodents. Further, selection of particular animal model also depends on the investigator's choice (whether to use inbred or out-bred) and other factors like availability of particular strain, aim of scientific strategy, type of drug being sought, infrastructure and logistics of the concerned institute, etc. contribute to diabetes and associated CVDs research and pharmaceutical drug discovery and development program.

16.3 TARGETING THE INFLAMMATORY NETWORK: TIME TO LET THE CAT OUT OF THE BAG!

Recent studies and systematic reviews point toward the role of inflammation in the pathogenesis of diabetes and associated complications like CVDs (Ziegler, 2005; Williams and Nadler, 2007; Festa and Steven, 2005; Gleissner et al., 2007; Garcia et al., 2010). Inflammation is a short-term adaptive response of the body to harmful stimuli such as allergens, pathogens, any injury to tissues. Inflammation, if not regulated, may disrupt homeostasis (Hotamisligi, 2006). Over the past couple of decades, studies have established persistence of chronic low-grade systemic inflammation in diabetes leading to systemic downstream repercussions (Duncan et al., 2003; de Rooij et al., 2009). However, the debate is still ongoing as to the cause of underlying persistent inflammatory chaos. One school of thought advocates that obesity (that has been clearly shown to be associated to type II diabetes) is the driving perpetuator behind existent chronic inflammation (Mokdad et al., 2001; Van Gaal et al., 2006). As a result, the adipocytes that enlarge beyond healthy limits often die, triggering an immune response. The adipose tissues have been shown to undergo molecular and cellular alterations that subsequently affect systemic metabolism (Haslam, 2010). There is an over expression of proinflammatory cytokines (Antuna-Puente et al., 2008; Ohashi et al., 2014), resulting in localized adipose tissue inflammation that propagates an overall systemic inflammation associated with the development of obesity-related comorbidities (Calle and Fernandez, 2012; Hotamisligi, 2006). It is evident that such an environment activates cytokines, which in turn enhance insulin resistance in adipose and other tissues, thereby increasing the risk for diabetes. Therefore, the role of activated inflammatory markers cannot be ignored, and the risk factors further catalyze the inflammatory odyssey (Fig. 16.2). In fact, patients with diabetes display typical features of an inflammatory process characterized by the presence of cytokines, immune cell infiltration, impaired function, and tissue destruction. In addition, signs of inflammation are also seen in individuals with CVDs, another ailment linked to obesity (Wang and Nakayama, 2010; Sowers, 2003). Other studies have shown that low-grade chronic inflammation may actually drive insulin resistance by stimulating proteins that block

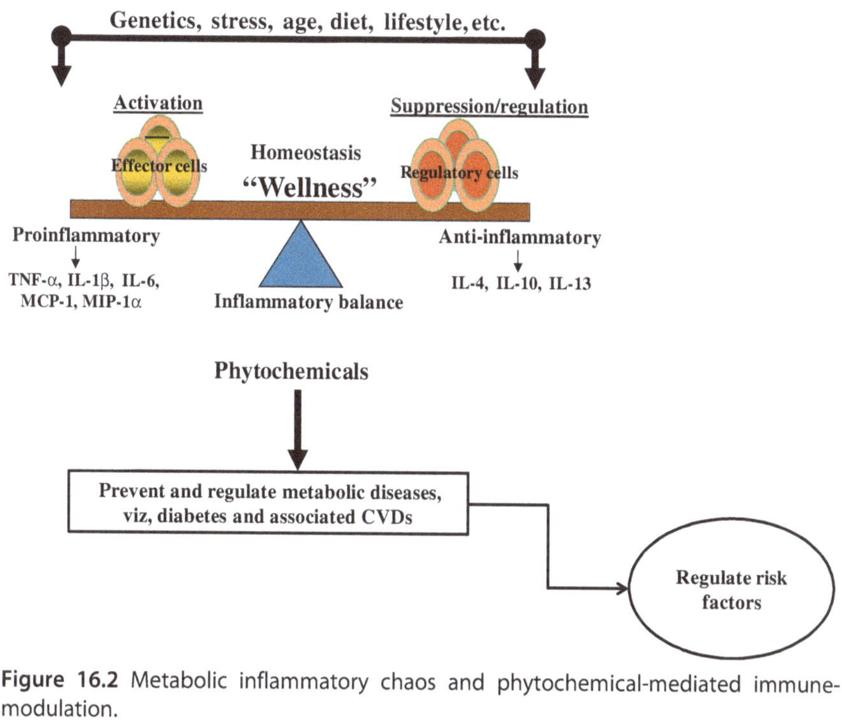

Figure 16.2 Metabolic inflammatory chaos and phytochemical-mediated immune-modulation.

insulin action. The fat and muscle cells require insulin as a signal to absorb glucose from the blood, while any insulin resistance causes cells to ignore insulin leading to type II diabetes (Hotamisligi, 2006; Haffner, 2003; Schmidt et al., 1999).

Growing clinical and experimental evidences suggest that targeting immune system dysfunction and inflammatory signaling overlap may play instrumental roles in designing treatment strategies for both diabetes and CVDs (Donath and Shoelson, 2011; DeClercq et al., 2008; Rajala and Scherer, 2003). Under conditions of metabolic challenge as in diabetes, this integration of metabolism and immunity becomes deleterious and the resultant chronic inflammation poses life-threatening conditions (Hotamisligi, 2006; Donath and Shoelson, 2011). However, as mentioned earlier, the question of how the inflammatory response begins cannot be reliably answered at the present time. An array of known triggers like stress, environmental chemicals, infections, allergies, unhealthy food, poor physical activity, conditions like excess abdominal fat and high blood pressure, etc. all seem to contribute (Dokken, 2008; Herman, 2017). It is still not known which comes first, the inflammation or the fat? This query is obvious as it is being increasingly recognized and accepted that inflammation is central to cardio-metabolic pathophysiology. It is well known that obesity is one of the risk factors for both diabetes and CVDs (Fig. 16.1), as the fat cells can produce chemicals that lead to inflammation. However, the possibility of pancreatic cells being the trigger point is most likely as they are the first cells affected by the development of diabetes (Amos et al., 1997).

Further, a second important mechanism may involve oxidative stress (Giacco and Brownlee, 2010; Csányi and Miller, 2014; Cervantes et al., 2017). Thus, increased production of reactive oxygen species (ROS) in mitochondria can also activate the inflammatory signaling pathways. Endothelial injury in the local tissues might attract inflammatory cells such as macrophages to this site and further exacerbate the local inflammation. The ability to store excess energy, which has been preserved through the course of evolution, is the crux of the interdependency of the metabolic and immune pathways (Hotamisligi, 2006; Kayama et al., 2015). Therefore, on the one hand, regulating inflammation together with capturing the contribution of genetic factors and their interaction with environmental covariates could be one strong strategy for disease prevention/management. In addition, using mediators of inflammation as lead markers could help to provide a phenotypic profile which could be used as sensitive and specific biomarkers independently from established risk factors. However, before applying these biomarkers, protocols for their assessment should be standardized, and laboratory reference intervals need to be used in decision-making processes.

16.4 PHYTOMEDICINE RENAISSANCE IN DIABETES AND CVDS

16.4.1 Promising Implications and Need for Standardization

Phytomedicine is increasingly becoming one of the most important aspects of speedily flourishing global commercial health enterprise. A lot has been written, discussed, investigated regarding the role of these wellness-promoting ingredients (Snyderman and Weil, 2002; Pan et al., 2014; Sharma and Shrivastava, 2016; Tripathi 2010). Further, there are convincing evidences to suggest that consuming foods rich in phytochemicals may progressively reduce the risk of diabetes and CVDs by gradually modulating immune-inflammatory markers (Suroowan and Mahomoodally, 2015; Corson and Crews, 2007; Barnes et al., 2009). In addition, the presence of strong chemical defense system in these plants constantly retains and rekindles the interest of enthusiastic researchers keen to identify novel therapeutics (Sofowora et al., 2013).

World Health Organization has strongly recommended detailed evaluation and adoption of traditional antidiabetic and cardioprotective plant-based treatments (World Health Organization, 2008, 2011). A scientific validation of several plant species has repeatedly proved the efficiency of botanicals as potent immune-modulating agents (Birdee et al., 2016; Suroowan and Mahomoodally, 2015; Jia et al., 2003; Grover et al., 2002; Ceylan-Isik et al., 2008; Mukesh and Namita, 2013; Chauhan et al., 2010). In addition, owing to cognizance and experiences made from reports on potential effectiveness against diabetes and CVDs, an increasing number of health experts recommend and support the addition of secondary metabolites of plants into conventional drugs (Baldwin, 2002). However, it is also relatively important to understand and adapt that no supplementation can be absolutely beneficial until combined with regular exercise and a healthy balanced diet. At the same time, it is also important to consider that risks and complications attributed to plant-based treatments should not be underestimated by both healthcare professionals and consumers.

With this background, it is evident that phytomedicine has clearly shifted and revolutionized the focus on the overall well-being of the individual, rather than on a particular ailment or disease. Accumulated knowledge and experiences from available and ongoing studies is being increasingly applied in a holistic approach. Therefore, it appears that days are not far when the use of naturally occurring novel bioactive molecules shall become central part of all systems of medicine.

16.4.2 Reaping the Benefits of Phytomedicine Prominence

It is evident that currently, there is no known permanent cure for diabetes (World Health Organization, 2011). At the same time, it is also well understood that considering the nature of the disease, treatment for type II diabetes necessitates a multifaceted approach integrating diet, nutrition, exercise, stress management, and required medications. Ironically, plant-derived drugs have continued to be incorporated into the antidiabetic armamentarium for centuries. Thousands of plants have been used in antidiabetic and cardioprotective plant-based formulations, and amongst them, more than hundred have been scientifically validated (Makheswari et al., 2011; Kazi, 2014; Bathaie et al., 2012; Dey et al., 2002). However, largely due to lack of proper standardization methodologies, till date no single approved plant-based drug is available for mass usage (Gupta and De, 2012; Mirhoseini et al., 2013). Plants contain mixture of complex bioactive compounds like flavonoids, terpenoids, phenolic acids, etc., and many of these have been reported to have several promising biological activities including anti-inflammatory, anti-hepatotoxic, antidiabetic, anti-microbial, and immune-modulatory effects (Baldwin, 2002; Filho et al., 2006; Wink and Zhou, 2015; Bahmani et al., 2014). Following a comprehensive literature review of plant-based antidiabetic and cardioprotective treatment modalities, in majority, the lead bioactive molecules have not been followed up with clinical studies, and long-term effects have not been consistent (Sofowora et al., 2013). Despite the hiccups, we can no longer ignore the worldwide exponential surge in public enthusiasm for disease management using these bioactive phytochemicals.

The list of principal antidiabetic and cardioprotective plants is endless. Time has come to retain, utilize, and bring together the important secondary metabolites that clearly are the functional strength of phytomedicines. The added benefit of the latter is bringing about an increase in their overall ability to survive and overcome local challenges. This regulatory effect is then mirrored in the form of an improvement of overall immune response. For instance, alkaloids such as aconitine, anisodamine, charantine, leurosine have been shown to exhibit antidiabetic activity (Li et al., 2004). Berberine has been shown to lower elevated blood glucose as effectively as conventional drugs (Yin et al., 2008). Terpenes are being used both for the regulation of type II diabetes and associated risks (Wink and Zhou, 2015). Gallic acid, a type of phenolic acid, is being effectively used to treat diabetes and albuminuria (Bahmani et al., 2014). Recognizing and overcoming the loopholes, focusing on the novelties, would definitely pave a way for a successful antidiabetic and cardioprotective treatment strategy.

Once the active component in a potential bioactive molecule in a plant is indicated and defined, the next fundamental step should be to extrapolate the mechanism by which they act or lead to mediating observed disease regulating effects. Since numerous scientific reviews and experimental studies have very clearly and specifically described a relationship between inflammation and diabetes (Gothai et al., 2016; Calle and Fernandez, 2012; Donath and Shoelson, 2011). Therefore, this positive correlation between underlying inflammatory chaos and the ability of phytochemicals to rescue from therein biological dysfunction is definitely a light at the end the tunnel. What is required is a corroborative therapeutic intervention through well-designed, randomized, double blind, placebo-controlled clinical trials involving a significant number of subjects. Therefore, all plants and there active components that bring about prevention/regulation/correction of metabolic inflammatory imbalance should be consolidated and further evaluated.

There are a huge number of active medicinal plants and their bioactive molecules that have already been tested in different populations for potential antidiabetic and cardioprotective roles with fairly promising results (Gautami et al., 2015; Makheswari et al., 2011; Suroowan and Mahomoodally, 2015; Birdee and Yeh, 2010). Our aim is not to list plant-based treatments for diabetes and CVDs. Therefore, in Tables 16.1 and 16.2, we have concentrated/restricted on those plants and herbs for which there is some evidence for supporting their antidiabetic and cardioprotective value through immune-modulation. It is needless to mention that the market for phytomedicines is undoubtedly booming with several plant components, showing protection against the pathology of diabetes mellitus through the attenuation of inflammatory mediators (Fürst and Zündorf, 2014; Bohn et al., 2015; Yatoo et al., 2017). In addition, several

Table 16.1 Medicinal plants used for the management of diabetes mellitus (provide reference and mechanism of action with target compounds if available)

Medicinal plant (common name)	Family	Part of plant	Active ingredient	Mode of action
Aloe vera (aloe) Arumugam et al. (2013), Eidi et al. (2009)	Asphodelaceae	Leaf, pulp, and gel	Phytosterols, polysaccharides mannans, lophenol, lectins	Hypoglycemic effect, stimulate synthesis, and/or release of insulin
Aegle marmelos (holy fruit tree) Maity et al. (2009)	Rutaceae	Fruit extract, aqueous seed, and leaf extract	Alkaloids, tannins, cardiac glycosides, terpenoids	Hypoglycemic potential, normalizes glycosylated hemoglobin, blood glucose, and insulin levels, elevates both serum insulin and liver glycogen
Allium cepa (onion) Roman-Ramos et al. (1995), Kumari et al. (1995), Mathew and Augusti (1975)	Amaryllidaceae	Whole onion bulbs, ether fractions, oils	Quercetin	Potent antidiabetic activity (significant hypoglycemic and hypolipidemic effects)
Allium sativum (garlic) Modak et al. (2007), Augusti and Shella (1996)	Amaryllidaceae	Garlic gloves, ethanol extract, oils	Allicin	Potent antidiabetic activity, normalizes fasting blood glucose, serum triglycerides, total cholesterol, urea, creatinine, AST, and ALT levels
Aralia elata (Japanese angelica tree) Singab et al. (2014)	Araliaceae	Root cortex	Elatosides E together with oleanolic acid and its derivatives	Antidiabetic activity (normalizes serum glucose levels)
Azadirachta indica (neem) Sunarwidhi et al. (2014), Satyanarayana et al. (2015)	Meliaceae	Leaves, stem bark, and seeds	Quercetin, rutin, nimbidin, azadirachtin	Regulates hypoglycemia, hyperinsulinemia, and undesirable weight gain, delays onset of diabetes

Brassica nigra (Indian black mustard) Anand et al. (2007, 2009)	Brassicaceae	Aqueous seed extract, seed powder	Isothiocyanate glycosidesingrin	Stimulates insulin release from pancreas and normalizes the effects of glucose metabolizing enzyme, directs and regulates glycogenolysis and gluconeogenesis
Cinnamomum zeylanicum (cinnamon) Goel et al. (2012), Shen et al. (2010)	Lauraceae	Bark, aqueous extract	Cinnamaldehyde	Enhance insulin action, increase glucose uptake and glycogen synthesis
Curcuma longa (turmeric) Olatunde et al. (2014)	Zingiberaceae	Root powder	Curcumin, demethoxycurcumin, bisdemethoxycurcumin, and ar-turmerone	Potent anti-inflammatory activity
Eugenia jambolana (jamun/black plum) Patel et al. (2012)	Myrtaceae	Fruit pulp extract, seed kernels, dried alcohol, and flavonoid-rich extracts	Jamboline	Potent antidiabetic activity
Fiscus bengalensis (sacred fig)	Moraceae	Leaves and bark	Glycoside	Increase insulin secretion
Grewia asiatica (phalsa) Parveen et al. (2012), Goyal (2012)	Tiliaceae	Ethanol fruit extract, stem bark, and leaves	Grewinol, quercetin	Antihyperglycemic effect
Lawsonia inermis (henna) Pritesh et al. (2012), Chaudhary et al. (2010)	Lythraceae	Leaves, whole plant ethanolic and methanolic extracts	Tannins, terpenoids, alkaloids, quinones, xanthones, coumarins	Potent hypoglycemic and hypolipidemic activities
Morus alba (white mulberry) Singab et al. (2005)	Moraceae	Root bark alcohol extract	Morusin, cyclomorusin, neocyclomorusin, and kuwanon E	Antidiabetic activity

(*Continued*)

Table 16.1 (Continued)

Medicinal plant (common name)	Family	Part of plant	Active ingredient	Mode of action
Murraya koeingii (curry leaf tree)	Rutaceae	Whole leaves, leaf aqueous extract	Mahanimbine, carbazole	Antidiabetic potential
Kesari et al. (2005), Jain et al. (2012)				
Ocimum sanctum (holy basil/tulsi)	Labiatae	Leaves, leaf powder	Pectins, eugenol	Antidiabetic agent, stimulate insulin secretion
Prakash and Gupta (2005)				
Panax ginseng (ginseng)	Araliaceae	Roots and leaves	Ginsenosides	Antidiabetic potential
Painter (2009), Attele et al. (2002)				
Psidium guajava (guava)	Myrtaceae	Leaf aqueous and methanolic extract, stem bark ethanol extract	Flavonoid glycosides exemplified by pedunculagin, isostrictinin, and strictinin	Antidiabetic potential
Oh et al. (2005), Mukhtar et al. (2006)				
Trigonella foenumgraecum (fenugreek seeds)	Fabaceae	Defatted seed, seed fiber	Trigonelline	Increase glucose-induced insulin secretion, enhancement of peripheral insulin action
Basch et al. (2003), Basch et al. (2003)				
Zingiber officinale (ginger)	Zingiberaceae	Rhizome juice, extract	Gingerol	Antidiabetic potential
Sekiya et al. (2004)				

Table 16.2 List of cardioprotective compounds

Medicinal plant (common name)	Family	Part of the plant	Active components	Mode of action
Allium sativum (garlic) Augusti (1996), Neha Nausheen et al. (2014)	Amaryllidaceae	Leaves, flowers and cloves, extracts	Allicin, polysulfides, *S*-allylcysteine, AMS, and DAS	Controls and regulates risk factors like hypertension and total cholesterol, strong antioxidant properties
Asparagus racemosus (satavari) Choudhary and Sharma (2014)	Asparagaceae	Roots, flowers and leaves	Saponins, shatavarins	Hypocholesterolemic effect, decreases lipid peroxidation, regulates total lipids, total cholesterol, and triglycerides
Anacardium occidentale (cashew) Dreher et al. (1996), Kris-Etherton et al. (1999)	Anacardiaceae	Stem bark extract	Flavonoids, carotenoids	Regulates cholesterol, improves endothelium-dependent vasodilation in hypercholesterolemic subjects
Antiaris toxicaria (bark cloth tree) Shi Li-Shian et al. (2010)	Moraceae	Ethanolic extract of trunk bark	Cardiac glycosides (cardenolides), e.g., α, β, and γ-antiarin	Circulatory and cardiac stimulant
Camellia sinensis (tea) Moghaddam et al., 2013	Theaceae	Leaves	Catechins	Antioxidant abilities, reverse endothelial dysfunction, reduce superoxide radical and alkyl peroxy radicals, increase endothelium-dependent dilatation
Cynara cardunculus var. *scolymus* (globe artichoke) Llorach et al. (2002), Wang et al. (2003)	Asteraceae	Dried leaves, lower part of the flower, aqueous, and ethanolic extracts	Sesquiterpenes, cynaropicrin, aguerine B, and grosheimin	Hypolipidemic activity, prevents hypercholesterolemia
Commiphora mukul (guggul) Ojha et al. (2011)	Burseraceae	Treated resin	Sterols, and E and Z guggulsterone	Potent antioxidant and antiperoxidative activity

(Continued)

Table 16.2 (Continued)

Medicinal plant (common name)	Family	Part of the plant	Active components	Mode of action
Crataegus species (hawthorn) Wang et al. (2013)	Rosacea	Decoction of leaves and unripe fruits	Biflavonoids	Antioxidant and anti-inflammatory ability, anticardiac remodeling effect
Cactus grandiflorus (queen of the night) Khan (2015), Haque et al. (2015)	Cactaceae	Young succulent stems and flowers	Betacyanins and flavonol glycosides	Antihypertensive and antiarrhythmic activity, exerts inotropic effect in the myocardium
Cinnamomum tamala (Malabar leaf) Nagaraju et al. (2016)	Lauraceae	Bark, leaves, flowers, fruits, and roots	Cinnamaldehyde	Free radical scavenging and antioxidant activity
Digitalis purpurea (lady's glove/fox glove) Ojha et al. (2016), Liu et al. (2009)	Scrophulariaceae	Flowers and leaves	Cardiac glycosides, digitoxin, and digitonin	Positive inotrope, antiarrhythmic agent
Tinospora cordifolia (guduchi) Sharma et al. (2011)	Menispermaceae	Alcoholic extract	Alkaloids, diterpenoid lactones, glycosides, steroids, sesquiterpenoid, phenolics, aliphatic compounds, and polysaccharides	Free radical scavenging activity, enhances endogenous antioxidant levels
Vitis vinifera (grapevine) Berti et al. (2003)	Vitaceae	Fruits	Procyanidin	Free radical scavenging capability, antioxidant impacts

AMS, Allyl methyl sulfide; *DAS*, Diallyl sulfide.

phytochemicals have been shown to exhibit combinational effect of indicated bioactivities, for example, chili peppers, bitter melon, ginseng, turmeric, and tea extracts. Therefore, the combinational intake of these foods or synergistic efficacy of these phytochemicals should be seriously considered as future research area in diabetic and CVDs.

It is apparent that there has been a widespread increase in the use of phytomedicine; however, a major caveat has been a lack of up-to-date documentation. We have therefore aimed to emphasize the front foot requirements of phytochemicals that can contribute to a wellness-based approach in combating the metabolic dysfunctions. However, for an effective antidiabetic and cardioprotective plant-based approach, it is important to include bioactive components that only moderately affect/attenuate the immune-inflammatory balance. This would be crucial to reap long-term health benefits with minimum side effects. Therefore, it would be worthwhile to explore the possibility of a customized successful treatment strategy to suit the need of an individual with inflammation as an important corrective parameter. Targeting phytochemicals that sync with existing disease immunology, i.e., reciprocal interaction between the immune system, inflammation and the magnitude of the disease would help to provide further clues to the future of antidiabetic and cardioprotective drug discovery targets. However, several factors including genetics, environment, severity and nature of the disease, etc. have to be kept in mind.

16.5 CONCLUSIONS

Majority of population in the developing countries still continues to use traditional medicine for their primary medical problems. There are several challenges to our current understanding of the role and contribution of phytomedicine in combating diseases like diabetes and CVDs. Whether phytochemicals absolutely regulate/alter/correct the disturbed underlying mechanistic mechanisms still needs to be fully elucidated. How, when, and why one unique immune-modulatory role becomes central to regulating the effectiveness needs explanation. The debate is still ongoing in support of secondary metabolites present in plants. It is speculated with mixed evidences that these bioactive molecules might exert an inhibitory control on some cell populations like T cells, macrophages, etc. probably through the stimulation of other cell types. It seems reasonable to suggest that the final outcome of the different inhibitory/stimulatory inputs exerted by these components is that of a control aimed to prevent overshooting and excessive activation of the immune system.

Further, it is apparent that not every method, diagnostic technique, or intervention has solid clinical evidences. Therefore, whether medicinal plants can completely combat CVDs and diabetes still needs to be aggressively and elaborately explored! What also needs to be regulated is the ever-increasing perception that these medications are completely safe and do not have adverse effects. The latter is not only incorrect but also misleading as several plant-derived therapies have been shown to elicit adverse and life-threatening consequences. Therefore, it would be challenging yet interesting to find a balance between traditional practice and modern treatments resulting in a state of homeostasis or holistic well-being, and why there is an urgent need for strong interdisciplinary collaboration to bridge the gap between traditional and modern biochemical medicine to be used as an effective tool for the management of diabetes and CVDs.

In view of the above, an investigation of the effect of significant phytochemicals on the balance between the molecular signatures of the immune and metabolic states should also be included as a frontline research area. This should provide long-term benefits, while having little impact on the immune status. It is high time to translate clinical and epidemiologic findings into practice that could help to curb the burden of diabetes worldwide. Amidst this inflammatory chaos, phytomedicine comes across as a powerful tool in managing and correcting underlying metabolic imbalance only if approached in the right direction.

ACKNOWLEDGMENTS

The authors extend a deep sense of gratitude to Late Professor Vinod Singh, Head, Department of Microbiology, Barkatullah University, Bhopal. Prof. Singh succumbed to cancer few months back, and this is a huge loss to the whole scientific fraternity. He has truly been an inspirational pathfinder "The Guru" and a pivotal backbone in molding our respective professional achievements. Though he is no more amongst us, his teachings and philosophies shall always remain imbibed within us.

REFERENCES

Akerele, O., 1993. Nature's medicinal bounty: don't throw it away. World Health Forum 14, 390–395.

Alberti, K.G., Eckel, R.H., Grundy, S.M., Zimmet, P.Z., Cleeman, J.I., Donato, K.A., et al., 2009. Harmonizing the metabolic syndrome: a joint interim statement of the

International Diabetes Federation Task Force on Epidemiology and Prevention; National Heart, Lung, and Blood Institute; American Heart Association; World Heart Federation; International Atherosclerosis Society; and International Association for the Study of Obesity. Circulation 120, 1640–1645.

Ali, M., 2008, Textbook of Indian Book Publishing House. Indian Book Publishing House, New Delhi, India.

American Diabetes Association, 2016. Classification and diagnosis of diabetes. Diabetes. Care 39, S13–S22.

Amos, A.F., McCarty, D.J., Zimmet, P., 1997. The rising global burden of diabetes and its complications: estimates and projections to the year 2010. Diabetic Med. S1–85.

Anand, P., Murali, K.Y., Tandon, V., Chandra, R., Murthy, P.S., 2007. Preliminary studies on antihyperglycemic effect of aqueous extract of *Brassica nigra* (L.) Koch in streptozotocin induced diabetic rats. Indian J. Exp. Biol. 45, 696–701.

Anand, P., Murali, Y.K., Tandon, V., Murthy, P.S., Chandra, R., 2009. Insulinotropic effect of aqueous extract of *Brassica nigra* improves glucose homeostasis in streptozotocin induced diabetic rats. Exp. Clin. Endocrinol. Diabetes 117, 251–256.

Antuna-Puente, B., Feve, B., Fellahi, S., Bastard, J.P., 2008. Adipokines: the missing link between insulin resistance and obesity. Diabetes Metab. 34 (1), 2–11.

Aridoğan, B.C., 2006. Immunomodulatory effects of phytocompounds', in modern phytomedicine: turning medicinal plants into drugs. In: Ahmad, I., Aqil, F., Owais, M. (Eds.), Wiley-VCH Verlag GmbH& Co. KGaA, Weinheim, Germany.

Arumugam, G., Manjula, P., Paari, N., 2013. A review: anti diabetic medicinal plants used for diabetes mellitus. J. Acute Dis. 2, 196–200.

Attele, A.S., Zhou, Y.P., Xie, J.T., Wu, J.A., Zhang, L., et al., 2002. Antidiabetic effects of *Panax ginseng* berry extract and the identification of an effective component. Diabetes 51, 1851–1858.

Augusti, K.T., 1996. Therapeutic values of onion (*Allium cepa* L.) and garlic (*Allium sativum* L.). Indian J. Exp. Biol. 34 (7), 634–640.

Augusti, K.T., Shella, C.G., 1996. Antiperoxide effect of *S*-allyl cysteine sulfoxide, an insulin secretagogue in diabetic rats. Experientia 52, 115–120.

Bae, J.C., 2016. Diabetes drugs and cardiovascular safety. Endocrinol. Metab. 31 (2), 239–244.

Bahmani, M., Golshahi, H., Saki, K., Rafieian-Kopaei, M., Delfan, B., Mohammadi, T., 2014. Medicinal plants and secondary metabolites for diabetes mellitus control. Asian Pac. J. Trop. Dis. 4 (2), S687–S692.

Bailey, C.J., 2013. Interpreting adverse signals in diabetes drug development programs. Diabetes Care 36, 2098–2106.

Baldwin, I.T., 2002. Unraveling the function of secondary metabolites, In: L. Taiz, E.Z. (Eds.), Plant Physiology (3, pp. Essay 13.1). Sunderland, Ma: Sinauer.

Barnes, P.M., Bloom, B., Nahin, R.L., 2009. Complementary and alternative medicine use among adults and children: United States, 2007. Nat. Health Stat. Rep. 12, 1–23.

Basch, E., Ulbricht, C., Kuo, G., Szapary, P., Smith, M., 2003. Therapeutic applications of fenugreek. Altern. Med. Rev. 8, 20–27.

Bathaie, S., Mokarizade, N., Shirali, S., 2012. An overview of the mechanisms of plant ingredients in the treatment of diabetes mellitus. J. Med. Plants 4 (44), 1–24.

Beckman, J.A., Paneni, F., Cosentino, F., Creager, M.A., 2013. Diabetes and vascular disease: pathophysiology, clinical consequences, and medical therapy: part II. Eur. Heart J. 34, 2444–2452.

Berti, F., Manfredi, B., Mantegazza, P., Rossoni, G., 2003. Procyanidins from *Vitis vinifera* seeds display cardioprotection in an experimental model of ischemia-reperfusion damage. Drugs. Exp. Clin. Res. 207–216.

Birdee, G.S., Yeh, G., 2010. Complementary and alternative medicine therapies for diabetes: a clinical review. Clin. Diabetes 28 (4), 147–155.

Bloomgarden, Z.T., 2007. The Avandia debate. Diabetes. Care 30, 2401–2408.

Bohn, T., McDougall, G.J., Alegría, A., Alminger, M., Arrigoni, E., Aura, A.M., et al., 2015. Mind the gap—deficits in our knowledge of aspects impacting the bioavailability of phytochemicals and their metabolites—a position paper focusing on carotenoids and polyphenols. Mol. Nutr. Food. Res. 59, 1307–1323.

Boutayeb, A., 2006. The double burden of communicable and non-communicable diseases in developing countries. Trans. R. Soc. Trop. Med. Hyg. 100 (3), 191–199.

Brownlee, M., 2005. The pathobiology of diabetes complications: a unifying mechanism. Diabetes 54, 1615–1625.

Calle, M.C., Fernandez, M.L., 2012. Inflammation and type 2 diabetes. Diabetes Metab. 38 (3), 183–191.

Cefalu, W.T., 2006. Animal models of type 2 diabetes: clinical presentation and pathophysiological relevance to the human condition. ILAR J. 47 (3), 186–198.

Cersosimo, E., Solis-Herrera, C., Trautmann, M.E., Malloy, J., Triplitt, C.L., 2014. Assessment of pancreatic β-cell function: review of methods and clinical applications'. Curr. Diabetes Rev. 10 (1), 2–42.

Cervantes, Gracia, K., Llanas-Cornejo, D., Husi, H., 2017. CVD and oxidative stress. J. Clin. Med. 6(2), 22.

Ceylan-Isik, A.F., Fliethman, R.M., Wold, L.E., Ren, J., 2008. Herbal and traditional Chinese medicine for the treatment of cardiovascular complications in diabetes mellitus. Curr. Diabetes. Rev. 4 (4), 320–328.

Chaudhary, G., Goyal, S., Poonia, P., 2010. *Lawsonia inermis* Linnaeus: a phytopharmacological review. Int. J. Pharm. Sci. Drug Res. 2, 91–98.

Chauhan, A., Sharma, P.K., Srivastava, P., Kumar, N., Dudhe, R., 2010. Plants having potential anti-diabetic activity: a review. Der Pharmacia Lettre 2 (3), 369–387.

Choudhary, D., Sharma, D.A., 2014. Phytopharmacological review on *Asparagus racemosus*. Int. J. Sci. Res. 3, 742–746.

Corson, T.W., Crews, C.M., 2007. Molecular understanding and modern application of traditional medicines: triumphs and trials. Cell 130 (5), 769–774.

Csányi, G., Miller, F.J., 2014. Oxidative stress in cardiovascular disease. Int. J. Mol. Sci. 15 (4), 6002–6008.

de Rooij, S.R., Nijpels, G., Nilsson, P.M., Nolan, J.J., Gabriel, R., Bobbioni-Harsch, E., et al., 2009. Low-grade chronic inflammation in the relationship between insulin sensitivity and cardiovascular disease (RISC) population: associations with insulin resistance and cardiometabolic risk profile. Diabetes Care 32 (7), 1295–1301.

Dean, L., McEntyre, J., 2004. The Genetic Landscape of Diabetes. National Center for Biotechnology Information (US).

DeClercq, V., Taylor, C., Zahradka, P., 2008. Adipose tissue: the link between obesity and cardiovascular disease. Cardiovasc. Hematol. Disord. Drug. Targets. 8 (3), 228–237.

Dey, L., Attele, A.S., Yuan, C.S., 2002. Alternative therapies for type 2 diabetes. Altern. Med. Rev. 7, 45–58.

Dokken, B.B., 2008. The pathophysiology of cardiovascular disease and diabetes: beyond blood pressure and lipids. Diabetes Spectr. 21 (3), 160–165.

Donath, M.Y., Shoelson, S.E., 2011. Type 2 diabetes as an inflammatory disease. Nat. Rev. Immunol. 11 (2), 98–107.

Dreher, M.L., Maher, C.V., Kearney, P., 1996. The traditional and emerging role of nuts in healthful diets. Nutr. Rev. 54, 241–245.

Duncan, B.B., Schmidt, M.I., Pankow, J.S., Ballantyne, C.M., Couper, D., Vigo, A., et al., 2003. Low-grade systemic inflammation and the development of type 2 diabetes. Diabetes 52 (7), 1799–1805.

Eidi, A., Eidi, M., Darzi, R., 2009. Antidiabetic effect of *Olea europaea* L. in normal and diabetic rats. Phytother. Res. 23, 347–350.

Expert Committee on the Diagnosis and Classification of Diabetes Mellitus, 2003. Report of the expert committee on the diagnosis and classification of diabetes mellitus. Diabetes. Care 26, S5–S20.

Festa, A., Steven, M.H., 2005. Inflammation and cardiovascular disease in patients with diabetes. Circulation 111, 2414–2415.

Filho, J.M.B., Piuvezam, M.R., Moura, M.D., Silva, M.S., Lima, K.V.B., da-Cunha, E.V. L., et al., 2006. Anti-inflammatory activity of alkaloids: a twenty century review. Rev. Bras. Farmacogn. 16 (1).

Frankel, D.S., Meigs, J.B., Massaro, J.M., Wilson, P.W., O'Donnell, C.J., D'Agostino, R. B., et al., 2008. Von Willebrand factor, type 2 diabetes mellitus, and risk of cardiovascular disease: the Framingham Offspring Study. Circulation 118, 2533–2539.

Fürst, R., Zündorf, I., 2014. Plant-derived anti-inflammatory compounds: hopes and disappointments regarding the translation of preclinical knowledge into clinical progress. Mediators Inflamm.

Garcia, C., Feve, B., Ferré, P., Halimi, S., Baizri, H., Bordier, L., et al., 2010. Diabetes and inflammation: fundamental aspects and clinical implications. Diabetes Metab. 36 (5), 327–338.

Gautami, J., Sandhya, M., Revathi, B., 2015. Anti diabetic medicinal plants. J. Pharmacogn. Phytochem. 3 (1).

Giacco, F., Brownlee, M., 2010. Oxidative stress and diabetic complications. Circ. Res. 107 (9), 1058–1070.

Gleissner, C.A., Galkina, E., Nadler, J.L., Ley, K., 2007. Mechanisms by which diabetes increases cardiovascular disease. Drug. Discov. Today Dis. Mech. 4 (3), 131–140.

Goel, R., Bhatia, D., Gilani, S.J., Katiyar, D., 2012. Medicinal plants as antidiabetics: a review. Int. Bull. Drug Res. 1, 100–107.

Goldfine, A.B., 2008. Assessing the cardiovascular safety of diabetes therapies. N. Engl. J. Med. 359, 1092–1095.

Gothai, S., Ganesan, P., Park, S.Y., Fakurazi, S., Choi, D.K., Arulselvan, P., 2016. Natural phyto-bioactive compounds for the treatment of type 2 diabetes: inflammation as a target. Nutrients 8, 461.

Goyal, P.K., 2012. Phytochemical and pharmacological properties in the genus Grewia: a review. Int. J. Pharm. Pharm. Sci. 4 (4), 72–78.

Grover, J.K., Yadav, S., Vats, V., 2002. Medicinal plants of India with antidiabetic potential. J. Ethnopharmacol. 81, 81–100.

Gupta, P.D., De, A., 2012. Diabetes mellitus and its herbal treatment. Int. J. Res. Pharm. Biomed. Sci. 3 (2), 706–721.

Haffner, S.M., 2003. Pre-diabetes, insulin resistance, inflammation and CVD risk. Diabetes Res. Clin. Pract. 61, S9–S18.

Haque, S.E., Verma, R.K., Khan, V., Sharma, S., 2015. *Cactus grandiflorus*: a homeopathic remedy for cardiac ailment. Indian J. Pharm. Pharmacol. 2 (1), 74–80. January–March.

Haslam, D., 2010. Obesity and diabetes: the links and common approaches. Primary Care Diabetes 4 (2), 105–112.

Herman, W.H., 2017. The global burden of diabetes: an overview. In: Dagogo-Jack, S. (Ed.), Diabetes Mellitus in Developing Countries and Underserved Communities. Springer, Cham.

Holleman, F., Introduction to Diabetes Mellitus. Diapedia 1104085113 rev. no. 55. From https://doi.org/10.14496/dia.1104085113.55 (viewed August 2014).
Hotamisligi, G.S., 2006. Inflammation and metabolic disorders. Nature 444 (7121), 860–867.
International Diabetes Federation, IDF Diabetes Atlas, seventh ed. From http://www.diabetesatlas.org (viewed January 2015). ISBN: 978-2-930229-81-2.
Jain, V., Momin, M., Laddha, K., 2012. *Murraya koenigii*: an updated review. Int. J. Ayur. Herb. Med. 2, 607–627.
Jia, W., Gaoz, W., Tang, L., 2003. Antidiabetic herbal drugs officially approved in China. Phytother. Res. 17 (10), 1127–1134.
Kamboj, V.P., 2000. Herbal medicine. Curr. Sci. 78, 35–39.
Kayama, Y., Raaz, U., Jagger, A., Adam, M., 2015. Diabetic cardiovascular disease induced by oxidative stress. Int. J. Mol. Sci. 16, 25234–25263.
Kazi, S., 2014. Use of traditional plants in diabetes mellitus. Int. J. Pharm. 4 (4), 283–289.
Kesari, A.N., Gupta, R.K., Watal, G., 2005. Hypoglycemic effects of *Murraya koenigii* on normal and alloxan-diabetic rabbits. J. Ethnopharmacol. 97, 247–251.
Khan, V., 2015. *Cactus grandiflorus*: A homeopathic remedy for cardiac ailment. Indian J. Pharm. Pharmacol. 2, 74–80.
King, A.J., 2012. The use of animal models in diabetes research. Br. J. Pharmacol. 166 (3), 877–894.
Kris-Etherton, P.M., Yu-Poth, S., Sabate, J., Ratcliffe, H.E., et al., 1999. Nuts and their bioactive constituents: effects on serum lipids and other factors that affect disease risk. Am. J. Clin. Nutr. 70, 504S–511S.
Kumari, K., Mathew, B.C., Augusti, K.T., 1995. Antidiabetic and hypolipidaemic effects of *S*-methyl cysteine sulfoxide, isolated from *Allium cepa* Linn. Ind. J. Biochem. Biophys. 32, 49–54.
Laakso, M., 2010. Cardiovascular disease in type 2 diabetes from population to man to mechanisms. Diabetes Care 33 (2), 442–449.
Li, W.L., Zheng, H.C., Bukura, J., Kimpe, N.D., 2004. Natural medicines used in traditional Chinese medicines for diabetes mellitus. J. Ethnopharmacol. 92, 1–21.
Liu, Z., Song, X.-D., Xin, Y., et al., 2009. Protective effect of chrysoeriol against doxorubicin-induced cardiotoxicity in vitro. Chin. Med. J. 122 (21), 2652–2656.
Llorach, R., Espin, J.C., Tomas-Barberan, F.A., Ferreres, F., 2002. Artichoke (*Cynara scolymus* L.) byproducts as a potential source of health-promoting antioxidant phenolics. J. Agric. Food. Chem. 50, 3458–3464.
Maity, P., Hansda, D., Bandyopadhyay, U., Mishra, D.K., 2009. Biological activities of crude extracts and chemical constituents of Bael, *Aegle marmelos* (L.) Corr. Indian J. Exp. Biol. 47, 849–861.
Makheswari, M., Uma, Sudarsanam, D., 2011. Phytomedicine for diabetes mellitus: an overview. Res. Pharm. 1 (4), 28–37.
Martín-Timón, I., Cristina, S.C., Amparo, S.G., Francisco, J.C.G., 2014. Type 2 diabetes and cardiovascular disease: have all risk factors the same strength? World J. Diabetes 5 (4), 444–470.
Mathew, P.T., Augusti, K.T., 1975. Hypoglycemic effects of onion, *Allium cepa* Linn. on diabetes mellitus—a preliminary report. Ind. J. Physiol. Pharmacol. 19, 213–217.
Mazzone, T., Chait, A., Plutzky, J., 2008. Cardiovascular disease risk in type 2 diabetes mellitus: insights from mechanistic studies. Lancet 371, 1800–1809.
Mirhoseini, M., Baradaran, A., Rafieian-Kopaei, M., 2013. Medicinal plants, diabetes mellitus and urgent needs. J. Herbmed Pharmacol. 2 (2), 53–54.
Modak, M., et al., 2007. Indian herbs and herbal drugs used for the treatment of diabetes. J. Clin. Biochem. Nutr. 40 (3), 163–173. *PMC*. Web. 12 Mar. 2018.

Moghaddam, et al., 2013. Evaluation of cardioprotective effects of *Camellia sinensis* on isoproterenol induced myocardial infarction. IJPSR 1056–1063.

Mokdad, A.H., Ford, E.S., Bowman, B.A., Dietz, W.H., Vinicor, F., Bales, V.S., et al., 2001. Prevalence of obesity, diabetes, and obesity-related health risk factors. JAMA 289 (1), 76–79.

Moller, D.E., 2001. New drug targets for type 2 diabetes and the metabolic syndrome. Nature 414 (6865), 821–827.

Mukesh, R., Namita, P., 2013. Medicinal plants with antidiabetic potential—a review. Am-Eurasian J. Agric. Environ. Sci. 13 (1), 81–94.

Mukhtar, H.M., Ansari, S.H., Bhat, Z.A., Naved, T., Singh, P., 2006. Antidiabetic activity of an ethanol extract obtained from the stem bark of *Psidium guajava* (Myrtaceae). Die Pharmazie—Int. J. Pharm. Sci. 61, 725–727.

Nagaraju, B., Vidhyadhara, S., Kumar, A., Vikas, S., Suryanarayana, D., 2016. Evaluation of cardioprotective activity of ethanolic extract of dried leaves of *Cinnamomum tamala* in rats. Int. J. Biomed. Adv. Res. 7 (4), 181–186.

Neha Nausheen, Q., Ayaz Ali, S., Subur, K., 2014. Cardioprotective and antioxidant activity of onion (*Allium cepa*) leaves extract in doxorubicin induced cardiotoxicity in rats. Ann. Exp. Biol. 2 (2), 37–42.

Newman, D.J., Cragg, G.M., 2007. Natural products as sources of new drugs over the last 25 years. J. Nat. Prod. 70, 461–477.

Oh, W.K., Lee, C.H., Lee, M.S., Bae, E.Y., Sohn, C.B., et al., 2005. Antidiabetic effects of extracts from *Psidium guajava*. J. Ethnopharmacol. 96, 411–415.

Ohashi, K., Shibata, R., Murohara, T., Ouchi, N., 2014. Role of anti-inflammatory adipokines in obesity-related diseases. Trends Endocrinol. Metab. 25 (7), 348–355.

Ojha, S., Bhatia, J., Arora, S., Golechha, M., Kumari, S., Arya, D.S., 2011. Cardioprotective effects of *Commiphora mukul* against isoprenaline-induced cardiotoxicity: a biochemical and histopathological evaluation. J. Environ. Biol. 32 (6), 731–738.

Ojha, S., Al Taee, H., Goyal, S., et al., 2016. Cardioprotective potentials of plant-derived small molecules against doxorubicin associated cardiotoxicity. Oxid. Med. Cell Longev. .

Olatunde, A., Luka, C.D., Tijjani, H., Obidola, S.M., Joel, E., 2014. Anti-diabetic activity of aqueous extract of *Curcuma longa* (Linn) rhizome in normal and alloxan-induced diabetic rats. Researcher. 6, 58–65.

Painter, F., 2009. *Panax ginseng*. Monograph. Altern. Med. Rev. 14, 172–176.

Pan, S.Y., Litscher, G., Gao, S.H., Zhou, S.F., Yu, Z.L., Chen, H.Q., et al., 2014. Historical perspective of traditional indigenous medical practices: the current renaissance and conservation of herbal resources. Evid. Based Complement. Alternat. Med. 2014, 525340.

Paneni, F., Beckman, J.A., Creager, M.A., Cosentino, F., 2013. Diabetes and vascular disease: pathophysiology, clinical consequences, and medical therapy: part I. Eur. Heart J. 34, 2436–2443.

Parveen, A., Irfan, M., Mohammad, F., 2012. Antihyperglycemic activity in *Grewia asiatica*, a comparative investigation. Int. J. Pharm. Pharm. Sci. 4, 210–213.

Patel, D., Prasad, S., Kumar, R., Hemalatha, S., 2012. An overview on antidiabetic medicinal plants having insulin mimetic property. Asian Pac. J. Trop. Biomed. 2 (4), 320–330.

Pradeepa, R., Prabhakaran, D., Mohan, V., 2012. Emerging economies and diabetes and cardiovascular disease. Diabetes Technol. Ther. 14 (S1), S59–S67.

Prakash, P., Gupta, N., 2005. Therapeutic uses of *Ocimum sanctum* Linn (Tulsi) with a note on eugenol and its pharmacological actions: a short review. Indian J. Phys. Pharmacol. 49, 125.

Pritesh, P., Pinal, H., Jagath, P., Nilesh, D., Bhagirath, P., 2012. Antidiabetic herbal drugs a review. Pharmacophore 3, 18–29.

Qazi, M.U., Malik, S., 2013. Diabetes and cardiovascular disease: original insights from the Framingham Heart Study. Global heart 8 (1), 43–48.

Rajala, M.W., Scherer, P.E., 2003. Minireview: the adipocyte—at the crossroads of energy homeostasis, inflammation, and atherosclerosis. Endocrinology 144 (9), 3765–3773.

Roman-Ramos, R., Flores-Saenz, J.L., Alaricon-Aguilar, F.J., 1995. Antihyperglycemic effect of some edible plants. J. Ethnopharmacol. 48, 25–32.

Sasase, T., Pezzolesi, M.G., Yokoi, N., Yamada, T., Matsumoto, K., 2013. Animal models of diabetes and metabolic disease. J. Diabetes Res. .

Sasase, T., Yokoi, N., Pezzolesi, M.G., Shinohara, M., 2015. Animal models of diabetes and metabolic disease. J. Diabetes Res. .

Satyanarayana, K., Sravanthi, K., Shaker, I.A., Ponnulakshmi, R., 2015. Molecular approach to identify antidiabetic potential of *Azadirachta indica*. J. Ayurveda Integr. Med. 6, 165.

Schmidt, M.I., Duncan, B.B., Sharrett, A.R., Lindberg, G., Savage, P.J., Offenbacher, S., et al., 1999. Markers of inflammation and prediction of diabetes in adults (atherosclerosis risk in communities study): a cohort study. Lancet 353, 1649–1652.

Sekiya, K., Ohtani, A., Kusano, S., 2004. Enhancement of insulin sensitivity in adipocytes by ginger. Biofactors 22, 153–156.

Sharma, S., Shrivastava, N., 2016. Renaissance in phytomedicines: promising implications of NGS technologies. Planta 244 (1), 19–38.

Sharma, A.K., Kishore, K., Sharma, D., et al., 2011. Cardioprotective activity of alcoholic extract of *Tinospora cordifolia (Willd.) Miers* in calcium chloride-induced cardiac arrhythmia in rats. J. Biomed. Res. 280–286.

Shen, Y., Fukushims, M., Ito, Y., Muraki, E., Hosono, T., et al., 2010. Verification of the antidiabetic effects of cinnamon (*Cinnamomum zeylanicum*) using insulin-uncontrolled type 1 diabetic rats and cultured adipocytes. Biosci. Biotechnol. Biochem. 74, 2418–2425.

Shi Li-Shian, et al., 2010. Cardiac Glycosides from Antiaris *Toxicaria* with potent cardiotonic activity. J. Nat. Prod. 73 (7), 1214–1222.

Simaan, J.A., 2009. Herbal medicine, what physicians need to know. Lebanese Med. J. 57, 215–217.

Singab, A.N.B., El-Beshbishy, H.A., Yonekawa, M., Nomura, T., Fukai, T., 2005. Hypoglycemic effect of Egyptian *Morus alba* root bark extract: effect on diabetes and lipid peroxidation of streptozotocin-induced diabetic rats. J. Ethnopharmacol. 100, 333–338.

Singab, A.N., Youssef, F.S., Ashour, M.L., 2014. Medicinal plants with potential antidiabetic activity and their assessment. Med. Aromat. Plants 3, 151.

Snyderman, R., Weil, A.T., 2002. Integrative medicine: bringing medicine back to its roots. Arch. Intern. Med. 162, 395–397.

Sofowora, A., Ogunbodede, E., Onayade, A., 2013. The role and place of medicinal plants in the strategies for disease prevention. Afr. J. Tradit. Complement. Altern. Med. 10 (5), 210–229.

Sowers, J.R., 2003. Obesity as a cardiovascular risk factor. Am. J. Med. 115 (8), 37–41.

Srinivasan, K., Ramarao, P., 2007. Animal models in type 2 diabetes research: an overview. Indian J. Med. Res. 125 (3), 451–472.

Stern, M.P., 1995. Diabetes and cardiovascular disease. The "common soil" hypothesis. Diabetes 44, 369–374.

Sunarwidhi, L., Anggit, L., Sudarsono, S., Agung, E.N., 2014. Hypoglycemic effect of combination of *Azadirachta indica* A. Juss. and *Gynura procumbens* (Lour.) Merr.

ethanolic extracts standardized by rutin and quercetin in alloxan-induced hyperglycemic rats. Adv. Pharm. Bull. 2, 613.
Suroowan, S., Mahomoodally, F., 2015. Common phyto-remedies used against cardiovascular diseases and their potential to induce adverse events in cardiovascular patients. Clin. Phytosci. 1, 1.
Tiwari, B.K., Brunton, N., Brennan, C.S., 2012. Handbook of Plant Food Phytochemicals: Sources, Stability and Extraction. Wiley, Hoboken, N.J.
Tripathi, Y., 2010. Natural product research: where have we been—where are we now. ENVIS For. Bull. 10, 34–49.
Van Gaal, L.F., Mertens, I.L., De Block, C.E., 2006. Mechanisms linking obesity with cardiovascular disease. Nature 444, 875–880.
Wallis, T.E., 2005. Textbook of Pharmacognosy. CBS Publishers and Distributors, New Delhi, Bangalore, India.
Wang, Z., Nakayama, T., 2010. Inflammation, a link between obesity and cardiovascular disease. Mediators Inflamm. 2010, 535918.
Wang, M., Simon, J.E., Aviles, I.F., He, K., Zheng, Q.Y., Tadmor, Y., 2003. Analysis of antioxidative phenolic compounds in artichoke (*Cynara scolymus* L.). J. Agric. Food. Chem. 51, 601–608.
Wang, J., Xiong, X., Feng, B., 2013. Effect of *Crataegus* usage in cardiovascular disease prevention: an evidence-based approach. Evid. Based Complement. Alternat. Med.: eCAM. 2013.
Williams, M.D., Nadler, J.L., 2007. Inflammatory mechanisms of diabetic complications. Curr. Diab. Rep. 7 (3), 242–248.
Wink, M., Zhou, S., 2015. Modes of action of herbal medicines and plant secondary metabolites. Medicines (Basel) 2 (3), 251–286.
World Health Organization, 2008. Fact Sheet: Traditional Medicine. Geneva.
World Health Organization, 2011. Diabetes: Key Facts. Geneva, Switzerland.
Yatoo, M.I., Saxena, A., Gopalakrishnan, A., Alagawany, M., Dhama, K., 2017. Promising antidiabetic drugs, medicinal plants and herbs: an update. Int. J. Pharmacol. 13 (7), 732–745.
Yin, J., Xing, H., Ye, J., 2008. Efficacy of berberine in patients with type 2 diabetic mellitus. Metabolism 57 (5), 712–717.
Ziegler, D., 2005. Type 2 diabetes as an inflammatory cardiovascular disorder. Curr. Mol. Med. 5 (3), 309–322.

FURTHER READING

Ali, L., Azed Khan, A., Hassan, Z., Mosihuzzaman, M., Nahar, N., et al., 1995. Characterization of the hypoglycemic effects of *Trigonella foenumgraecum* seed. Planta Med. 61, 358–360.
Esser, N., Legrand-Poels, S., Piette, J., Scheen, A.J., Paquot, N., 2014. Inflammation as a link between obesity, metabolic syndrome and type 2 diabetes. Diabetes Res. Clin. Pract. 105 (2), 141–150.
Fakim, A.G., 2006. Medicinal plants: traditions of yesterday and drugs of tomorrow. Mol. Aspects Med. 27 (1), 1–93.
Zeng, Y., Li, Y., Yang, J., et al., 2017. Therapeutic role of functional components in alliums for preventive chronic disease in human being. Evid. Based Complement. Alternat. Med. Volume 2017.

CHAPTER 17

Plant-Derived Immunomodulators

Arathi Nair[1], Debprasad Chattopadhyay[2,3] and Bhaskar Saha[1,2]
[1]National Centre for Cell Science, Pune, Maharashtra, India
[2]ICMR-National Institute of Traditional Medicine, Belagavi, Karnataka, India
[3]ICMR-Virus Unit Kolkatta, ID & BG Hospital Campus, Kolkata, West Bengal, India

17.1 INTRODUCTION

The concept of "immunomodulation" was conceived with Edward Jenner's discovery of the small pox vaccine in 1796 that certain attenuated infectious agents can boost human immune system to fight against subsequent infections with the same or closely related infectious agents. Our body's multitier immune system tends to maintain a steady state of homeostasis, but it is constantly exposed to invading pathogens such as viruses, bacteria, protozoan parasites, allergens, etc. In addition, chronic stress, illness, environmental and lifestyle changes adversely affect the immune system and impair the host's immune homeostasis. With the discovery of antibiotics and conventional chemotherapy, the homeostasis can be restored, but the use of chemical agents as drugs have further detrimental effects on the immune system. Considering an increased awareness about the adverse effects of chemotherapy, the use of phytomedicines has increased considerably in last few decades. Prophylactic and treatment modalities by "Phytomedicines" and natural immunomodulators offer a safer alternative (Tzianabos, 2000).

17.1.1 Immunomodulators

Certain natural and synthetic compounds that can modulate immune responses in a positive or negative manner are known as "Immunomodulators." Based on their effect on immune system, these agents are categorized as a "suppressor," or a "stimulant," or an "adjuvant." In autoimmune disorders, a hyperactive immune system fails to recognize self from non-self and leads to the destruction of self-entities. Here, immunosuppressors play a pivotal role in suppressing the immune

New Look to Phytomedicine
DOI: https://doi.org/10.1016/B978-0-12-814619-4.00018-5

system to restore normalcy. Immunostimulators are used to replenish the deficiency in the immune system as observed in the treatment of diseases like AIDS (Fig. 17.1).

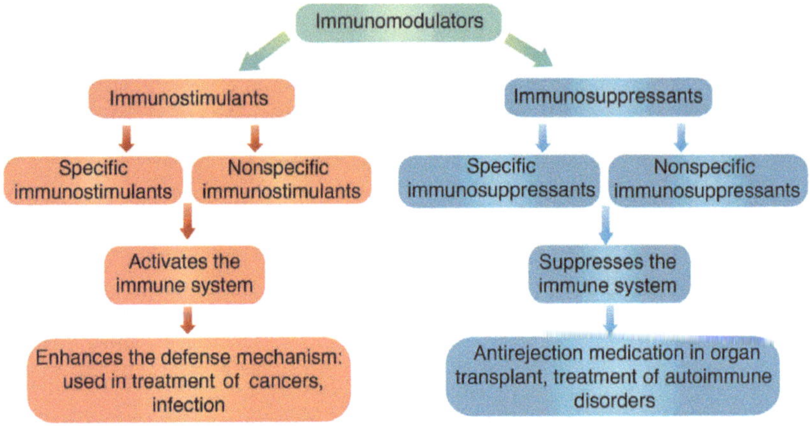

Figure 17.1 An overview of the action of immunomodulators.

Immunoadjuvants on the other hand can enhance the efficacy of vaccines, for example, Freud's adjuvant. Immunomodulators can modulate various cellular events such as apoptosis, protein synthesis, antigen presentation, etc. and target various transcription factors and immune mediators (Fig. 17.2).

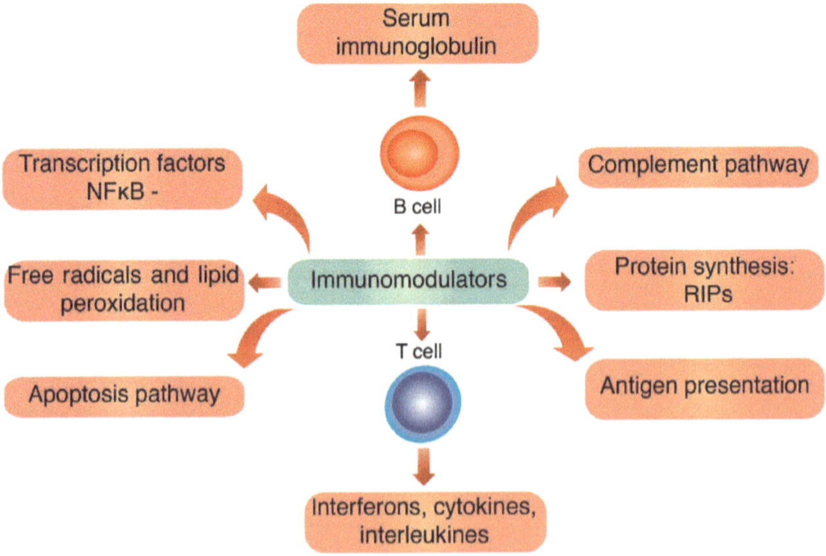

Figure 17.2 An overview of various targets and basic principle of functioning of immunomodulators.

Attempts are being made to generate adjuvants that would enable eliciting selective immune responses such as cellular or humoral, as well as IgE or IgG response (Song et al., 2016).

17.1.2 Plant-Derived Immunomodulators or Phytoimmunomodulators

Since ages, plants have been used for the prevention and cure of various ailments including microbial and lifestyle diseases (Kalia, 2011). According to the World Health Organization (WHO), approximately three quarter of the world's population relies on herbal medicine (Kumar et al., 2012). Plant extracts and phytocompounds are found to fortify the host's immune system, and numerous plants have been listed in this category (Thatte and Dahanukar, 1986). Phytoimmunomodulatory agents can increase the body's immune-responsiveness against pathogens by activating the immune system in a specific or a non-specific manner that includes both the innate and adaptive immune systems (Agarwal and Singh, 1999). Immunomodulatory plants play a pivotal role in the treatment of infection (Jandú et al., 2017), inflammation (Schulze-Koops et al., 1999), and in immunodeficiencies (Ziauddin et al., 1996) by their effect on various cell types via cytokines and interleukines. The mode of action could be as immunostimulators, immunosuppressors, or immunoadjuvants (Billiau and Matthys, 2001) to boost antigen-specific immune response. The liquid endosperm of green coconuts, used as beverage since antiquity for hydrating, health promoting, and disease-preventing abilities, along with its active component shikimic acid protects against oxidative stress through suppression of NF-kB and activation of Nrf2 pathway (Manna et al., 2014). Interestingly, *Odina wodier*, a folk medicine of Jangalmahal, West Bengal showed in vivo and in vitro anti-inflammatory activity through the inhibition of toll-like receptor 4 (TLR4) signaling pathway (Ojha et al., 2014), while *Pedilanthus tithymaloides* inhibits herpes simplex virus (HSV) infection via the modulation of NF-kB signaling pathway (Ojha et al., 2015).

Plant products are widely considered as immunopotentiators and collectively known as biological response modifiers (BRMs). Many dietary phytomolecules, vitamins, and minerals have a protective role in cellular nutrition and in the management of diabetic complications (Chattopadhyay and Eddouks, 2012). The immunomodulatory function attributed to BRMs is possibly through modulation of the immune cells of the body like the macrophages, the lymphocytes (B-cells and T-cells), dendritic cells (DCs), etc. For example, Concanavalin A lectin, a carbohydrate-binding protein from *Canavalia ensiformis* can cross-link glycoproteins like TCR/CD3 and thus activate T-lymphocytes (Quintans et al., 1989). Similarly, *Shorea robusta*, a well-known traditional medicine of India can modulate nitric oxide, prostaglandins E2, TNF-α, iNOS expression with anti-inflammatory (Chattopadhyay et al., 2012) and wound-healing activity (Mukherjee et al., 2013), while one of its major phytochemical Bergenin enhance T helper-1 responses affording antimycobacterial immunity by activating the MAP kinase pathway in macrophages (Dwivedi et al., 2017).

17.1.2.1 Ayurveda: The Ancient Indian Medicinal System and "Rasayanas"

The earliest mention of use of plants as medicine is found in ancient scriptures like Charak Samhita, Atharvaveda, and Sushrut Samhita (Chulet and Pradhan, 2009). The Indian traditional plant-based system of medicine or "Ayurveda" describes the use of several medicinal plants to strengthen the body's defense and contribute to general physical and mental well-being. The use of plants for immunomodulation can be traced back to ancient Ayurveda of 6000 BC (Mukherjee et al., 2012). This system describes medicinal plants as "rasayanas" having rejuvenating properties in terms of fortifying the immune system against various diseases (Dahanukar and Thatte, 1997). Currently, 34 plants have been identified as immunomodulators in rasayana (Table 17.1).

Table 17.1 List of Rasayana herbs (Doshi et al., 2013; Balasubramani et al., 2011)

S. no.	Name of the plant	S. no.	Name of the plant
1.	*Acorus calamus*	14.	*Convolvulus pluricaulis*
2.	*Allium sativum*	15.	*Cynodon dactylon*
3.	*Aloe vera*	16.	*Curculigo orchioides*
4.	*Anacyclus pyrethrum*	17.	*Curcuma longa*
5.	*Asparagus racemosus*	18.	*Glycyrrhiza glabra*
6.	*Azadirachta indica*	19.	*Hemidesmus indicus*
7.	*Bacopa monnieri* Linn	20.	*Phyllanthus emblica*
8.	*Boerhavia diffusa*	21.	*Piper longum*
9.	*Butea monosperma*	22.	*Sida spinosa*
10.	*Centella asiatica* Linn	23.	*Terminalia bellirica*
11.	*Chlorophytum borivilianum*	24.	*Tinospora cordifolia*
12.	*Clitoria ternatea*	25.	*Withania somnifera*
13.	*Commiphora mukul*		

In addition to rasayanas, plants categorized as "non-rasayanas" also exhibit immunostimulatory, antipathogenic, and anti-inflammatory activities. Certain plants that exhibit anti-allergic properties like those that stimulate the T-suppressor cells are also categorized as immunomodulators (Mahajani and Kulkarni 1977; Dorsch et al., 1991). Many herbs with immunomodulatory activities belong to the category of non-rasayanas and are presented in Table 17.2.

Table 17.2 List of non-rasayana herbs (Doshi et al., 2013)

S. no.	Name of the plant	S. no.	Name of the plant
1.	*Andrographis paniculata*	14.	*Mangifera indica*
2.	*Aristolochia indica*	15.	*Melia azadirachta*
3.	*Bauhinia variegata*	16.	*Mentha spicata*
4.	*Bombax malabarium*	17.	*Nardostachys jatamansi*
5.	*Butea monosperma*	18.	*Ocimum canum*
6.	*Butea superba*	19.	*Ocimum sanctum*
7.	*Calotropis procera*	20.	*Picrorhiza kurroa*
8.	*Catharanthus roseus*	21.	*Piper betle*
9.	*Gymnema sylvestre*	22.	*Saraca indica*
10.	*Hibiscus esculentus*	23.	*Vitex negundo*
11.	*Jasminum sambac*	24.	*Viscum album*
12.	*Lawsonia inermis*	25.	*Zingiber officinale*
13.	*Luffa cylindrica*		

17.1.2.2 Properties of Immunomodulators

Immunomodulators are not true antigens but are mitogens or antigenomimetics (mimics the action of real antigens). These can have specific as well as non-specific modes of action. Antigenomimetics require booster doses for efficient functioning. Like drug metabolism in the body, the efficacy of immunomodulators is dependent on various factors like the dosage, time, mode and duration of administration, as well on the immune-responsiveness of system.

17.1.3 High Throughput Screening for Plants and Bioactive Compound

A number of medicinal plants are currently under high throughput screening for a quick assessment of pharmacologically important hit molecules that could be utilized as a lead molecule in drug development. Some of the medicinally important plants currently under research for their immunomodulatory utility are given in Table 17.3.

Table 17.3 List of plants with immunomodulatory potential (Wagner, 1999)

S. no.	Name of the plant	Parts used	Compounds isolated	Reference(s)
1.	*Acanthopanax senticosus*	Root, berry, leaf	Saponins, lignans, coumarins, flavones, syringin, eleutheroside syringaresinol, phenylpropane, beta-sitosterol, daucosterol	Huang et al. (2011)
2.	*Achyranthes bidentata*	Root, leaf, seed	Triterpenoid, saponins, ketosteroids, sterols, alkaloids, flavonoids, anthraquinones, organic acids	He et al. (2017)
3.	*Achyrocline satureioides*	Flower	Luteolin, flavonoids quercetin, 3-*O*-methylquercetin	De Souza et al. (2007)
4.	*Aconitum carmichaelii*	Root	Unesterified-diterpene alkaloids, diester-diterpene alkaloids, monoester-diterpene alkaloids	Wu et al. (2018a,b)
5.	*Actinidia arguta*	Berry	Phenolic compounds, alkaloids	Latocha (2017)
6.	*Aeginetia indica*	Whole plant	Glycoproteins, alkylamides, cichoric acid, polysaccharides, alkaloid, terpenoids, macrocyclic lactones, quinones	Auttachoat et al. (2004)
7.	*Albizia julibrissin*	Bark, flower	Aromatic ethyl esters, monocarboxylic, and dicarboxylic esters	Lv et al. (2011)
8.	*Aloe vera*	Leaf	Glucomannans, amino acids, anthraquinones, glycosides, lipids, sterols, vitamins	Sharrif Moghaddasi (2011)
9.	*Alsophila spinulosa*	Stem, leaf	Diploptene, beta-sitosterol, glucopyranose, caffeic acid, astragalin	Kao et al. (1994), Chiang et al. (1994)
10.	*Angelica acutiloba*	Root, aerial part	Ferulic acid, tokiaerialide, Z-ligustilide, falcarindiol, bergaptol	Shinjyo et al. (2018)
11.	*Angelica sinensis*	Root	Polysaccharides, Z-ligustilide, ferulic acid	Wu and Hsieh (2011)
12.	*Aralia mandshurica*	Root	Aralosides (A, B, C), terpenoid acids, flavonoids, polysaccharides, terpenoid, saponins, sterols	Kochetkov et al. (1962), Shikov et al. (2016)
13.	*Arnica montana*	Flower	Coumarins, phenolic acids, lignans, alkaloids, oligosaccharides, sesquiterpene lactones, flavonoids, carotenoids, essential oils, diterpenes, arnidiol, pyrrolizidine	Kriplani et al. (2017)
14.	*Artemisia capillaris*	Leaf	Capillin, chlorogenic acid, esculetin, isochlorogenic acid, scoparone, scopoletin, artepillin, hyperin, isorhamnetin-1,6-diglucoside	Kim et al. (2018)

15.	*Artemisia iwayomogi*	Aerial parts	Scopoletin, esculetin 6-methylether, scopolin, β-sitosterol, chlorogenic acid, quebrachitol, essential oils, fatty acids, sesquiterpene lactones, flavonoids	Lee et al. (2017a,b), Yan et al. (2014)
16.	*Asarum europaeum*	Root, leaf	Trans-isoasarone, trans-isoeugenol, methyl ether	Gracza (1983)
17.	*Astragalus membranaceus*	Whole plant	Formononetin, Astragaloside IV	Zhang et al. (2018a,b), Qin et al. (2017)
18.	*Atractylodes lancea*	Root	Phenolic acids, sesquiterpenoids, monoterpenes, polyacetylenes, steroids, atractylodin, β-eudesmol, hinesol, atractyloside III	Xu et al. (2016), Na-Bangchang et al. (2017)
19.	*Azadirachta indica*	Whole plant	Azadirachtin, nimbin, quercetin, ß-sitosterol, nimbidin, nimbolide, limonoids	Saleem et al. (2018), Alzohairy (2016)
20.	*Benincasa cerifera*	Fruit	Isovitexin, lupeol, β-sitosterol, cucurbitacin B	Arora and Kaushik (2016)
21.	*Bryonia dioica*	Root	Carbohydrates, polyphenols, triterpenes, alkaloids, *c*-heterosides, saponins, sterols	Benarba et al. (2012)
22.	*Bupleurum chinense*	Root	Polysaccharides	Wang et al. (2018a,b,c)
23.	*Caesalpinia sappan*	Heartwood, seed	Brazilin, secang, brazilein, diterpene caesalsappanin	Hwang and Shim (2018), Handayani et al. (2017), Zhu et al. (2017)
24.	*Calendula officinalis*	Flower	Quercetin (1), isorhamnetin (2)	Lima et al. (2017), Olennikov et al. (2017)
25.	*Camellia sinensis*	Leaf	Theophylline, catechins, flavonoids, saponins, epigallocatechin-3-*O*-gallate	Tanwar et al. (2017), Chi et al. (2017), Kim et al. (2016a,b)
26.	*Carthamus tinctorius*	Flower, seed	Safflower polysaccharide, flavonoids, alkaloids, quinochalcones, steroids, phenolic compounds, moschamine	Wang and Yang (2007), Jo et al. (2017)

(*Continued*)

Table 17.3 (Continued)

S. no.	Name of the plant	Parts used	Compounds isolated	Reference(s)
27.	*Caulophyllum thalictroides*	Root, rhizome	Triterepenes saponins, alkaloids	Lee et al. (2012)
28.	*Chelidonium majus*	Whole plant	Isoquinoline alkaloids, sparteine, coptisine, stylopine, chelidonine, chelerythrine, sanguinarine, bebeerine	Dobrucka et al. (2018), Colombo and Bosisio (1996)
29.	*Choerospondias axillaris*	Folium, fruit peel	Flavones, proanthocyanidins	Qiu et al. (2016), Li et al. (2016)
30.	*Centaurea macrocephala*	Flower	Anthocyanins, cyanidin, cyanidin glycoside, flavonoids, patuletin, patuletin apigenin, kaempferol glycoside, quercetagetin	Mishio et al. (2015)
31.	*Cimicifuga simplex*	Root	Triterpenoid glycosides, phenolic compounds	Su et al. (2017), Lun et al. (2015)
32.	*Cinnamomum cassia*	Twig	Coumarin, cinnamate, cinnamaldehyde, cinnamic acid, cinnamyl alcohol	Park et al. (2018a,b), Rao and Gan (2014), Wu et al. (2018a,b)
33.	*Cistanche salsa*	Whole plant	Iridoid glycosides, phenylethanoid glycosides, phenylpropanoid-substituted diglycosides (Cistansalside A)	Kartbaeva et al. (2015), Ahn et al. (2017)
34.	*Cnidium officinale*	Root, rhizome	Falcarindiol	Hong et al. (2017)
35.	*Coffea arabica*	Berry, leaf	Caffeine, 5-caffeoylquinic acid, ferulic acid, protocatechuic acids, caffeic acid, p-coumaric, vanillic, hydroxycinnamic acid, terpenoids, carotenoids	Patay et al. (2016)
36.	*Combretum micranthum*	Leaf	Kinkéloids A, B, C, and D, inositol, sorbitol, vitexin, isovitexin, orientin, homoorientin, myricetin, stachydrine choline	D'Agostino et al. (1990), Welch et al. (2017), Olajide et al. (2003)

37.	*Cordyceps sinensis*	Whole plant	Polysaccharides, nucleosides, sterols, proteins, amino acids, polypeptides	Li and Ren (2017), Xiao et al. (2018)
38.	*Curcuma longa*	Root	Curcuminoids	Kadri et al. (2018), Wu and Lin (2018)
39.	*Daucus carota*	Root, seed	Anthocyanin, Cyaniding 3-xylosyl galactoside, β-cyclodextrin	Soares et al. (2018), Karkute et al. (2018)
40.	*Echinacea purpurea*	Whole plant	Alkamides, caffeic acid derivatives, polysaccharides	Park et al. (2018a,b), Clifford et al. (2002), Manayi et al. (2015)
41.	*Echinosophora koreensis*	Root	Prenylated flavonoids, alkaloids	Choi et al. (2009), Sohn et al. (2004)
42.	*Epimedium alpinum*	Root, rhizome	Glycosides, flavonoid, alkaloids, lignans, terpenoids, polysaccharides	Čolić et al. (2007), Liang et al. (1997), Kovačević et al. (2006)
43.	*Eupatorium cannabinum*	Aerial parts	Flavonoids, terpenoids, pyrrolizidine, alkaloids, sesquiterpene lactones, benzofurans, dihydrobenzofurans, polysaccharides	Herz (2001), Judzentiene et al. (2016)
44.	*Fagopyrum cymosum*	Rhizome	Flavonoids, phenolic compounds, luteolin, β-sitosterol, hecogenin, *p*-coumaric acid, ferulic acid, protocatechuic acid, dimeric procyanidin, glutinone, epicatechin, shakuchirin	Liu et al. (1983), Ge et al. (2017), Shen et al. (2013)
45.	*Forsythia koreana*	Fruit, leaf	Lignans (arctigenin)	Kim et al. (2016a,b), Lee and Kim (2010)
46.	*Geranium macrorrhizum*	Leaf	Gallic acid, quercetin, ellagic acid	Radulović et al. (2012), Jurkstiene et al. (2007)
47.	*Glycyrrhiza glabra*	Root	Glycyrrhetinic acid, glycyrrhizin, glabridin, flavonoids, beta-chalcones, isoflavones, triterpenoid saponins	Ahn et al. (2013), Yavuz Kocaman and Güzelkokar (2018)

(*Continued*)

Table 17.3 (Continued)

S. no.	Name of the plant	Parts used	Compounds isolated	Reference(s)
48.	*Gymnema sylvestre*	Leaf	Triterpene, flavonoids, saponins, tannins, coumarins, phytosterol	Vaghela et al. (2018), Arora and Sood (2017)
49.	*Gynostemma pentaphyllum*	Whole plant	Saponins, polysaccharides, flavones	Hong et al. (2018)
50.	*Hedysarum polybotrys*	Radix	Heteropolysaccharide, proanthocyanidin	Huang et al. (2013)
51.	*Herpestis monniera*	Whole plant	Glycosides (bacosides A and B)	Sairam et al. (2001)
52.	*Morus alba*	Stem, berry, leaf	Flavones, oxyresveratrol, Anthocyanin, flavonoids, lignans, pyrrole alkaloids, polyphenols, fatty acids	Wang et al. (2018a,b,c), Yiemwattana et al. (2018), Chen et al. (2018a,b)
53.	*Nyctanthes arbor-tristis*	Flower, leaf	Ursolic acid, phenol, flavonoids	Saini et al. (2014), Khanapur et al. (2014)
54.	*Ocimum sanctum*	Whole plant	Carvacrol, ursolic acid, eugenol, eugenic acid, polysaccharides, anthocyans, sitosterol, linalool, limatrol, estragol	Jayanti et al. (2018), Pattanayak et al. (2010)
55.	*Ophiopogon japonicus*	Root	Galactan, steroidal saponins, flavonoids	Gu et al. (2018)
56.	*Paeonia albiflora*	Root	Monoterpenes, glucosides	Duan et al. (2009)
57.	*Panax ginseng*	Whole plant	Ginsenosides, saponins, polyacetylenes	Shin et al. (2018), Yu et al. (2018)
58.	*Petiveria alliacea*	Leaf	*S*-Propyl propanethiosulfinate, *S*-benzyl phenylmethanethiosulfinate	Gutierrez and Hoyo-Vadillo (2017), Santander et al. (2012)
59.	*Phellodendron amurense*	Bark, root	Alkaloids, phenolic compounds, limonoids	Akihisa et al. (2017), Jung et al. (2017)

60.	*Picrorhiza kurroa*	Rhizome	Iridoid glycoside (picroside I, picroside II)	Sharma et al. (2017)
61.	*Pinellia ternata*	Tuber	Ferulic acid, coniferin, lectins, *p*-coumaryl alcohol, dihydroxy-cinnamyl alcohol, lariciresinol, erythro guaiacylglycerol-beta-*O*-4′-sinapyl ether, dehydrodiconiferyl alcohol, isolariciresinol, sachaliside 1	Kim et al. (2017), Xu et al. (2018), Han et al. (2007)
62.	*Pinus strobus*	Whole plant	Dihydrobenzofurans, xanthenes	Richardson et al. (2015)
63.	*Polygala tenuifolia*	Roots	Polysaccharides	Yao et al. (2017), Wang et al. (2017)
64.	*Potentilla tormentilla*	Rhizome, root	Procyanidin, polyphenols, monogalloylquinic acids	Bos et al. (1996), Subbotina et al. (2003), Fecka et al. (2015)
65.	*Pseudostellaria heterophylla*	Whole plant	Pectic polysaccharide, saponins, peptides, amino acids	Chen et al. (2018a,b), Choi et al. (2017)
66.	*Quillaja saponaria*	Bark	Triterpenoid saponins	Tafaghodi et al. (2016), Ahmed Abdel-Reheim et al. (2017)
67.	*Rehmannia glutinosa*	Whole plant	Polysaccharides, acetoside	Wang et al. (2018a,b,c), Gan et al. (2018)
68.	*Sapium sebiferum*	Leaf	Gallic acid, methyl gallate, kaempferol, quercetin, β-sitosterol glycoside, astragalin, 6-*O*-galloyl-*d*-glucose, methyl-3,4,5-trihydroxybenzoate	Liu et al. (1988), Fu et al. (2013, 2015a,b)
69.	*Schisandra chinensis*	Fruit	Schisandrin A, B, C, lignans	Chang et al. (2013), Ran et al. (2018), Park et al. (2013), Zhang et al. (2018a,b)
70.	*Serenoa repens*	Whole plant	Linoleic acid, lauric acid, oleic acid, myristic acid	Saidi et al. (2017), Chua et al. (2014)
71.	*Solenostemma argel*	Leaf	Solenoside A, kaempferol-3-*O*-glucoside, kaempferol-3-*O*-rutinoside, Stemmosides E–K	Innocenti et al. (2005), Plaza et al. (2005)

(*Continued*)

Table 17.3 (Continued)

S. no.	Name of the plant	Parts used	Compounds isolated	Reference(s)
72.	*Taraxacum platycarpum*	Aerial parts	Sesquiterpene glycoside, triterpenes, desacetylmatricarin, polysaccharides	Warashina et al. (2012), Ho et al. (1998), Jeong et al. (1991)
73.	*Tinospora cordifolia*	Stem, root, leaf	Polysaccharides, peptides, alkaloids, steroids, glycosides	Philip et al. (2018), Singh et al. (2003), Alsuhaibani and Khan (2017)
74.	*Trichosanthes kirilowii*	Fruit peel	Polysaccharide, cucurbitacin B	Zhang et al. (2017), Ding et al. (2017), Lee et al. (2017a,b)
75.	*Tripterygium wilfordii*	Root	Glucosides, nerolidol-type sesquiterpene, tripterygosidols A-D, triptolide, tripdiolide, triptonide, celastrol	Ni et al. (2018), Han et al. (2012), Zhou et al. (2018), Allison et al. (2001)
76.	*Viscum album*	Whole plant	Viscotoxins, lectins, polysaccharides, phenolic acids, phenylpropanoids, phytosterols, flavonoids, triterpene	Estko et al. (2015), de Oliveira Melo et al. (2018), Steinborn et al. (2017)
77.	*Uncaria tomentosa*	Bark, leaf	Procyanidin, propelargonidin dimers, hydroxybenzoic acid, hydroxycinnamic acids, flavanols, flavalignans, cinchonains, alkaloids	Navarro-Hoyos et al. (2018), Keplinger et al. (1998), Azevedo et al. (2018)
78.	*Zea mays*	Seed	Diterpenoids, phenolic compounds, flavonoids	Qi et al. (2018)
79.	*Zingiber officinale*	Rhizome	6-Gingerol	Kun-Hua et al. (2011)
80.	*Ziziphus jujuba*	Aerial parts	Phenolics, polysaccharides, vitamin C, flavonoids, triterpenic acids	Molina-Romo et al. (2018)

17.1.4 Screening Protocols for Immunomodulatory Constituents From Plants

In order to isolate the active phytoconstituents, the plant extracts are subjected to chromatographic separation using thin-layer chromatography (TLC). Separation is followed by various purification processes like sequential fractionation using column chromatography, and each isolated fraction is further validated for its immunomodulatory action, a process known as bioactivity-guided fractionation by various in vitro and in vivo assays followed by pharmacological testing on animal models to evaluate its efficacy and toxicity. Several in vitro methods are devised for the pharmacological screening of medicinal plants that involves high throughput screening of a large number of putative medicinal plants, as it gives some insight into the possible mode of action. Validation of the in vitro results should be performed in animal (in vivo) models. Multiple in vitro assays are required to validate various parameters of efficacy of a compound to be used as an immunomodulator. The commonly used animal models for infection and immunosuppression, and the ability of active phytoconstituents to reverse or antagonize the mild, moderate, or severe infection or to enable recovery is assessed in an immunosuppressed system (Fig. 17.3). Some results obtained

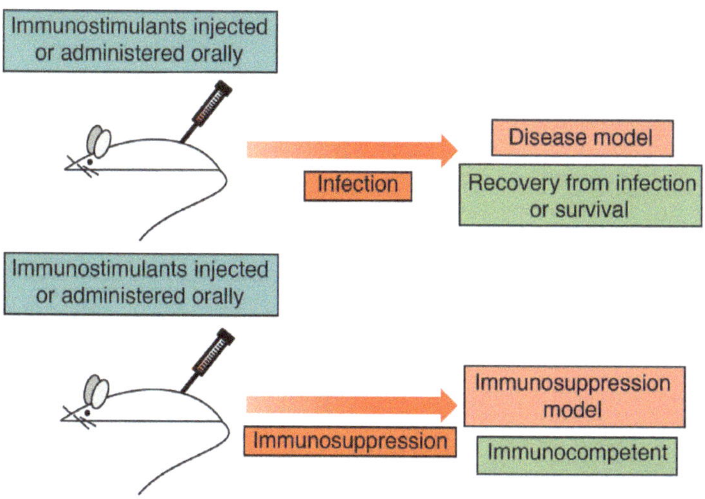

Figure 17.3 Animal models for screening immunostimulants include the infection model and the immunosuppression model. The efficacy of the immunostimulant is based on its ability to induce recovery from infection or to reverse immunosuppression.

in vivo are likely not to be obtained in vitro owing to the absence of a combinatorial effect of various cell populations and mediators.

An overview of in vivo and in vitro methods used for assessment of immunomodulatory potential is summarized in Table 17.4. The most commonly used in vitro assays determine the effect of the compound on

Table 17.4 In vitro and in vivo methods of assessment of immunomodulatory activity (Nagarathna et al., 2013)

Methods for assessment of immunomodulatory activity	
In vitro methods	**In vivo methods**
Inhibition of histamine release from mast cells	Spontaneous autoimmune diseases in animals
Mitogen-induced lymphocyte proliferation	Acute systemic anaphylaxis in rats
Inhibition of T-cell proliferation	Antianaphylactic activity (Schultz–Dale reaction)
Chemiluminescence in macrophages	Passive cutaneous anaphylaxis
PFC (plaque-forming colony) test	Arthus type immediate hypersensitivity
Inhibition of dihydro-orotate dehydrogenase	Delayed-type hypersensitivity
MTT assay	Reversed passive arthus reaction
Neutrophil adhesion test	Adjuvant arthritis in rats
Hemagglutination antibody (HA) titer	Collagen type II induced arthritis in rats
Inducible nitric oxide synthase (iNOS) activity	Proteoglycans-induced progressive polyarthritis in mice with experimental autoimmune thyroiditis
	Coxsackievirus B3-induced myocarditis
	Porcine cardiac myosin-induced autoimmune myocarditis in rats
	Experimental allergic encephalomyelitis
	Acute graft versus host disease (GVHD) in rats
	Influence on SLE-like disorder in MRL/lpr mice
	Prevention of experimentally induced myasthenia gravis in rats
	Glomerulonephritis induced by antibasement membrane antibody in rats
	Autoimmune uveitis in rats
	Inhibition of allogeneic transplant rejection

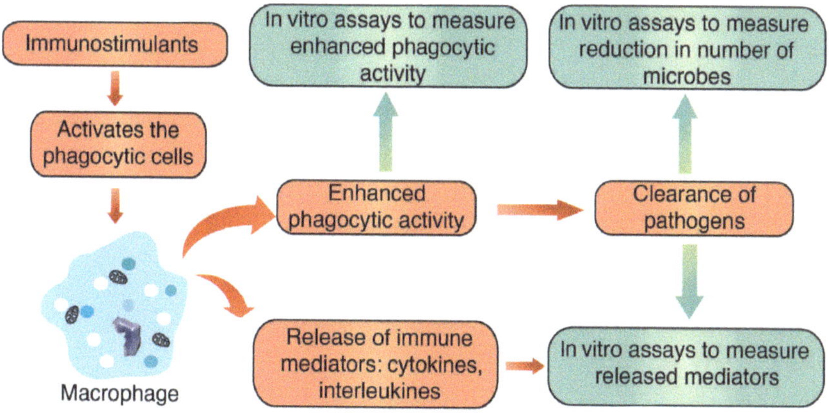

Figure 17.4 In vitro assessment of the effect of phytocompounds on phagocytes using various readouts such as cytokines, interleukines, and chemokines; phagocytic potential of the phagocytic cells; and clearance of microbes.

the mononuclear phagocytic system, complements, and on the proliferation of T-lymphocytes. Multitudes of phagocytic cells including the polymorphonuclear granulocytes, mononuclear cells, peritoneal macrophages, bone marrow–derived macrophages, and of skin (Langerhans cells) and liver (Kupffer cells) phagocytose to eliminate the pathogens. Polymorphonuclear leukocytes eliminate the pathogens by induction of oxidative stress via reactive oxygen species that can be determined in vitro using chemiluminescence assay. Alternatively, Smear test (Pappenheim test) is used to determine the phagocytic activity of granulocytes, macrophages, and monocytes in presence of the test compound by incubating granulocytes fraction with baker's yeast, test compound, and serum for opsonization. The phagocytosis index of the number of viable cells is determined by trypan blue staining (Fig. 17.4).

For in vivo studies, the plant decoction/extract (aqueous or ethanolic), and the chemically defined fractions (isolated secondary metabolites) are administered intragastrically or orally (decoction and crude powdered form) and subcutaneously or intraperitoneally (for isolates) in animal models. Mostly, the treatment with the test compound resulted in an increase in the number of PMNs, macrophages, lymphocytes, natural killer (NK) cells, and leukocytes with a concomitant increase in the cytokines and antibody production. In many animal models, the phytoconstituents were found to protect the animals against infection like those with *Listeria*, *Candida*, *Mycobacteria*, and *Herpesvirus* (Swamy et al., 2016; Dwivedi et al., 2017).

In vitro and in vivo studies are followed by spectroscopic techniques like mass spectrometry (MS), nuclear magnetic resonance (NMR), etc., to determine the structure of the active compound. Following structural analysis, total or partial synthesis of the original compound or its derivatives is done wherein the biological activity can be modulated to suit the needs. This is followed by the large-scale isolation of the compound and its preclinical and clinical trials (Hostettmann et al., 1991).

17.2 CLASSIFICATION OF IMMUNOMODULATORS BASED ON MOLECULAR WEIGHT

A number of factors play a deciding role in determining efficacy of an immunomodulatory compound such as its structural conformation, molecular weight, and solubility. Based on the molecular weight, immunomodulators are classified into low molecular weight and high molecular weight compounds (Table 17.5), both exhibiting immunostimulatory properties.

17.2.1 Low Molecular Weight Bioactive Compounds

A number of low molecular weight compounds have immunomodulatory properties. Alkaloids, heterocyclic compounds containing nitrogen, like Aristolochic acid obtained from the plant *Aristolochia clematitis* was the first alkaloid found to exhibit immunostimulatory property (Kumar et al., 2012). Aristolochic acid enhanced the phagocytic activity of peritoneal macrophages and leukocytes, but due to its carcinogenic property, its use as an immunostimulant is limited (Hoang et al., 2016). Other bioactive immunostimulatory alkaloids were isolated from *Actinidia macrosperma*,

Table 17.5 Low and high molecular weight compounds with immunomodulatory properties (Wagner, 1999)

Low molecular weight compounds	High molecular weight compounds
Alkylamides	Proteins
Phenolic compounds	Peptides
Alkaloids	Polysaccharides
Quinones	Glycolipids
Saponins	Glycoproteins
Sesquiterpenes, di- and triterpenoids	
Tryptamine	
Phytoestrogens	

Cissampelos pareira, *Achillea millefolium*, and *Murraya koenigii*. Cepharanthine isolated from *Stephania cepharantha* and Vincristine from *Catharanthus roseus* is found to stimulate the production of antibodies (Yamanoi et al., 2017). Cepharanthine also counteracts the effect of cytostatic agents on hematopoiesis (Seubwai et al., 2010). Small molecular weight compounds such as Vincristine (Jantan et al., 2015) and Staurosporine exhibit immunomodulation in a dose-dependent manner, lower doses acting as an immunostimulant while higher doses as an immunosuppressor. Alkaloids isolated from *Uncaria tomentosa* increased the number of granulocytes (Farias et al., 2011). A beta-carboline indole alkaloid harmine, isolated from *Ophiorrhiza nicobarica*, inhibit lysine-specific demethylase-1 during immediate early transcription of herpes simplex virus 1 and 2 in vitro and in infected animal. The interference on early transcription, a decisive factor for HSV lytic cycle or latency, reveals an epigenetic target that may help to develop a nonnucleotide antiherpesvirus drug (Bag et al., 2013, 2014). Thiosulfinate obtained from *Allium hirtifolium* are potent adaptogens and immunomodulators (Jafarian et al., 2010). Naphthoquinones also showed similar effects on lymphocytes and granulocytes (Wagner et al., 1988). Plumbagin, a quinoid compound isolated from the roots of *Plumbago zeylanica*, was found to inhibit the growth of hormone refractory prostate cancers (Hafeez et al., 2015) and limits the growth of *Staphylococcus aureus* (Nair et al., 2016). Terpenes and its oxygenated derivatives terpenoids, for example, Amyrine from *Bauhinia variegata* (Nadkarni and Nadkarni, 1954), Eugenol from *Ocimum sanctum* (Vaghasiya et al., 2010), diterpene from *Andrographis paniculata, Achillea millefolium, Alternanthera tenella* and pentacyclic triterpene from *Cecropia telenitida* act as immunomodulators (Peláez et al., 2013). The sapogenins triterpenoids and diterpenoids from *Gymnema sylvestre* have diverse immunomodulatory potentials. Immunostimulatory phorbol esters (derivatives of tetracyclic diterpenoids phorbol) are anticancer drugs at lower doses (Goel et al., 2007). Cichoric acid isolated from *Echinacea purpurea* activated phagocytic cells both in vitro and in vivo (Manayi et al., 2015).

Plant-derived glycosidase are organic sugar ethers that yield sugar on acid or enzymatic hydrolysis. These include iridoid glycosides of *Picrorhiza scrophulariiflora* and anthraquinone glycosides of *Andrographis paniculata*. The Chinese medicinal plant *Dendrobium nobile* yields sesquiterpene glycosides Dendroside A, Dendronobilosides A and B that exerts proliferative effects on lymphocytes. Flavonoids are another class of phytocompounds that exerts immunomodulatory effects. Lasure et al. (1994) reported the effect of flavonoids on the activation of complements using hemolytic assay and found that quercetin, quercitrin, rutin, myricetin, taxifolin, pelargonidin chloride, and cyanidin chloride inhibited the classical pathway, while hyperoside, myricetin, baicalein, and pelargonidin chloride inhibited the alternative pathway in a dose-dependent manner. Other flavonoids like apigenin, anthocyanidins (Bhattacharya and Muruganandam, 2003), flavones, isoflavonoids, and oligomeric proanthocyanidins (Davis and Kuttan, 2000) found in plants like *Terminalia arjuna* show immunomodulatory potential. In addition to this phenolic compounds isolated from *Euphorbiaceae* family also inhibit the classical pathway of complement activation. Coumarin glycosides isolated from *Achillea millefolium*, *Citrus natsudaidai*, and *Heracleum persicum* revealed immunomodulatory properties. Hydroxycoumarins like Esculin, Esculetin, and Scopoletin enhance complement-mediated hemolysis. Esculentin a class of 6,7-dihydroxycoumarin isolated from *Euphorbia lathyris*, *Citrus limonia*, and *Artemisia capillaris* have been attributed with a wide range of immunomodulatory properties (Leung et al., 2005) like free radical scavenging, protecting DNA against oxidative damage. It also shows cancer chemopreventive, antitumor, lipoxygenase-inhibitory activity, and tyrosinase-inhibitory activity. Certain plants and plant-derived compounds like polysaccharides and lipopolysaccharides can activate the alternative pathway of complement activation and hence play a key role in the regulation of inflammation. Moreover, plants like *Rosmarinus officinalis* L. containing the bioactive component Rosmarinic acid can inhibit the complement system via blocking both the classical and alternative pathways of complement activation (Sahu et al., 1999) (Fig. 17.5).

17.2.2 High Molecular Weight Compounds

High molecular weight compounds like polysaccharides exert immunomodulatory effects mainly on the innate arm of the immune system and specifically by enhancing the phagocytosis of granulocytes and affecting the macrophages functions. Polysaccharide derived from the plant *Cistanche deserticola* enhanced the proliferation of murine thymus lymphocytes. High molecular weight compounds isolated from *Salicornia herbacea* show anticancer properties via activating monocytic cells and inducing

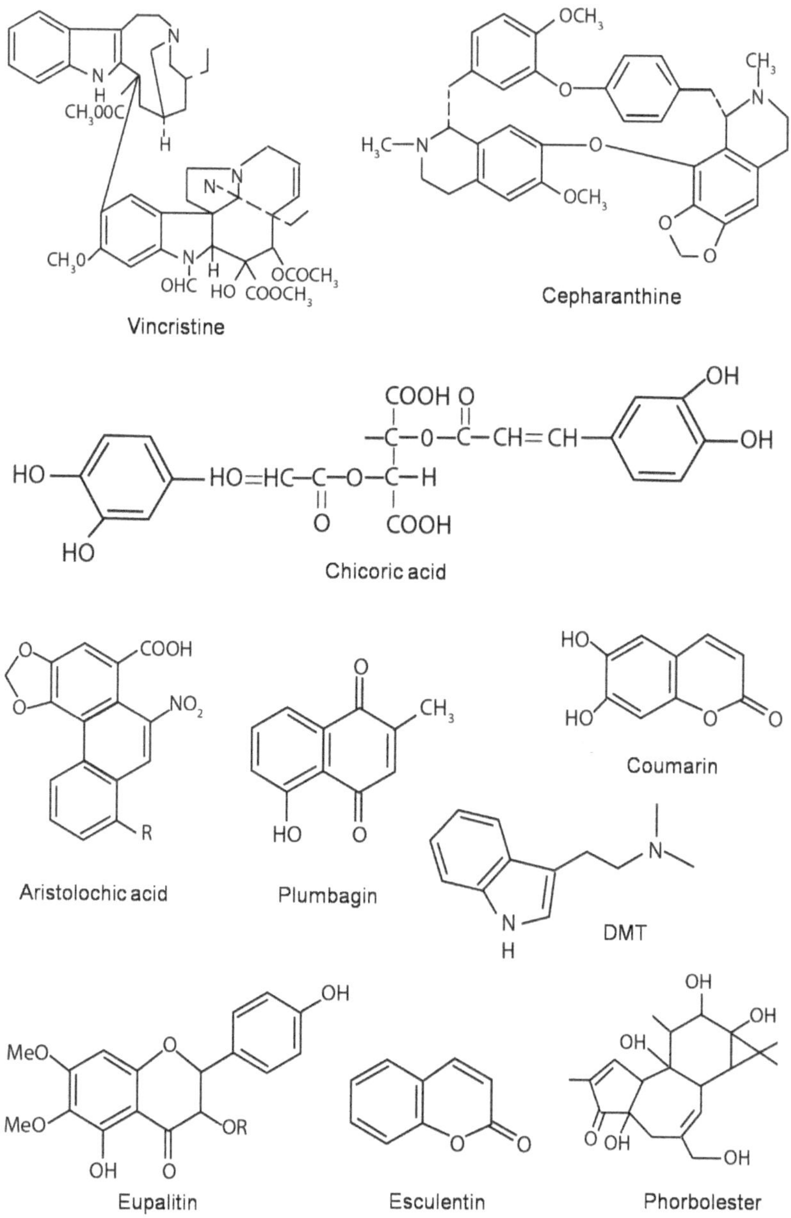

Figure 17.5 Structure of common low molecular weight immunomodulatory compounds.

differentiation of monocytes into macrophages (Alamgir and Uddin, 2010; Ríos, 2010). Axillary mode of action includes induced interferons, interleukines, and complement activation. The binding sites for various polysaccharides to immune components might differ. Xyloglucans (do Rosário et al., 2011), glucuronic acid containing arabinogalactan and methyl glucuronoxylans, play a pivotal role in activating the phagocytes (Luettig et al., 1989). The complement-activating polysaccharides comprise acidic polygalacturonans isolated from *E. purpurea.* Polygalacturonans are also effective in activating macrophages to counteract tumor cells and to destroy intracellular pathogens like *Candida albicans, Leishmania enrietti*, and *Listeria cytogenes.* Glycanogalacturonans from *Achyrocline satureioides* enhance the phagocytic potential of macrophages and granulocytes and show strong anti-complement and anti-inflammatory activities. Polysaccharides are also known to enhance chemotaxis, for instance, the rhizome of *Urtica dioica* yields α-glucan that can stimulate leukocyte migration. Lectins isolated from *Lens culinaris, Canavalia ensiformis, Ricinus communis, Viscum album, Phaseolus vulgaris*, and *Phytolacca americana* induced mitosis (Satoshi et al., 1982), inhibited protein synthesis (Stirpe et al., 1980), bind to the lymphocytes, and agglutinate malignant cells. Certain lectins having interferon-inducing properties were isolated from *Viscum album* and *Urtica dioica* (Peumans et al., 1984).

Not only lectins but also proteins capable of inducing interferons were isolated and analyzed from *Artemisia princeps. N*-Glycosidase that recognizes a conserved stem-loop structure in 23S/25S/28S rRNA and irreversibly block protein translation are known as Ribosome-Inactivating Proteins (RIPs). RIPs have been reported from over 50 plant species including Cucurbitaceae (Zhang and Halaweish, 2007), Euphorbiaceae (Wu et al., 2015), and Poaceae (Loss-Morais et al., 2013). RIPs extracted from *Trichosanthes kirilowii* (Trichosanthin) was found to decrease the retroviral protein and RNA levels in acutely infected T lymphoblast and monocyte/macrophage (Shaw et al., 2005). Other known RIPs are isolated from plants such as *Cucurbita pepo, Ecballium elaterium, Bryonia cretica*, etc. Saponins derived from the plant *Quillaja saponaria* of the family Rosaceae are potent immunoadjuvants. Quil A, a purified form of Quillaja extract, is used in many veterinary vaccines (e.g., rabies, and foot and mouth disease), but its adverse reactions makes it unfit for use in humans (Rajput et al., 2007) (Fig. 17.6).

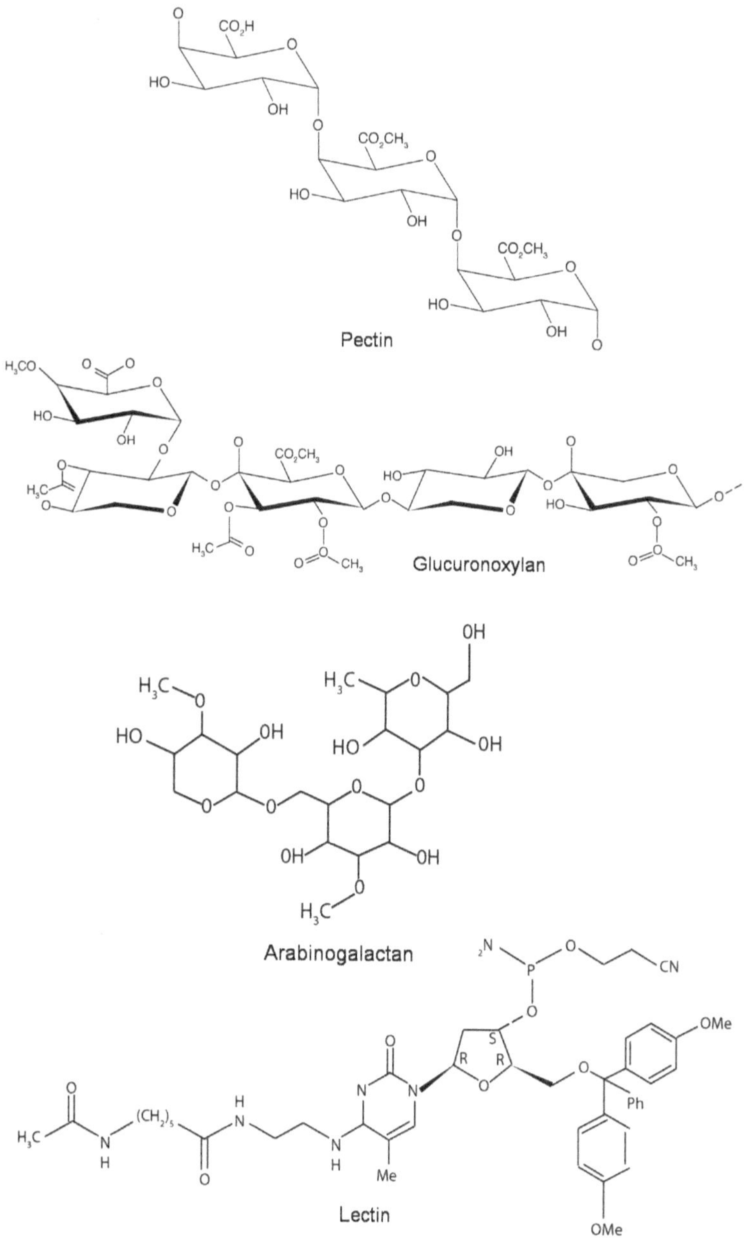

Figure 17.6 Structure of common high molecular weight immunomodulatory compounds.

17.2.3 Formulation Policy

Another deciding factor in the efficacy of an immunomodulatory compound is its formulation. Bioactive compounds present in the plants have certain natural stabilizers and enhancers associated with it, which are crucial for their potency and proper functioning. In herbal formulations, all these naturally evolved components are together and exert its effect in a combinatorial manner. On purification, the natural stabilizer or the activity enhancers are lost. Hence, in a purified fraction, a given compound may or may not exhibit the same level of immunomodulatory potential. Therefore, not only the active compounds but also the stabilizers and the enhancers are required to be identified and mixed in different proportions to make chemically defined natural phytomedicine.

Thus, the mode of immunomodulation by a bioactive compound can be diverse. Immunomodulatory compounds can affect both the innate and the adaptive immune system. Some immunomodulatory compounds scavenge the free radical produced during oxidative stress and hence protect the cells against detrimental effects such as lipid peroxidation and apoptosis. It can also target various components of the apoptosis pathway. Immunomodulatory compounds can affect B-cells as well as T-cells and hence alter antibody production, antigen presentation, and production of immune meditators. Moreover, the immunomodulation can also be achieved via inhibiting the protein synthesis and targeting the components of complement pathway.

17.3 COMMON IMMUNOMODULATORY PLANTS

Numerous studies have been conducted to elucidate bioactive compounds from medicinal plants and to study their effects on the immune system. These include rasayana as well as non-rasayana plants. Some important medicinal plants, their derived phytoimmunomodulators, and its mode of action are as follows:

17.3.1 *Acacia catechu/Senegalia catechu* (Distribution: Asia; Family: Fabaceae; Common Names: Cutch Tree, Black Catechu, Cachou)

The heartwood and bark are used in traditional medicine to treat sore throat and diarrhea. The earliest clinical studies involve the use of *Acacia catechu* to treat lepromatous leprosy (Ojha et al., 1969). A number of bioactive compounds are isolated from *A. catechu* like free radical scavenging catechin (polyhydroxylated benzoic acid), rutin, isorhamnetin (Li et al., 2011), and other compounds like epicatechin, epicatechin-3-*O*-gallate, epigallocatechin-3-*O*-gallate (Stohs and Bagchi, 2015), 4-hydroxybenzoic acid, ophioglonin, quercetin, afzelechin, kaempferol, 3,4′,7-trihydroxyl-3′,5-dimethoxy-flavone, epiafzelechin, mesquitol, aromadendrin, and phenol (Li et al., 2010). Naik et al. (2003) demonstrated that the aqueous extracts of *A. catechu*, in rat liver microsomal preparation, could inhibit radiation-induced lipid peroxidation. Flavonoids isolated from *A. catechu* reduce the production of proinflammatory eicosanoids (Burnett et al., 2007). Of note, 70% methanolic extracts of *A. catechu* is found to have DNA protective properties. The acetone, ethyl acetate, and methanolic extracts of the heartwood, leaves, and bark of *A. catechu* not only scavenges free radicals but also protects DNA against strand breaks (Guleria et al., 2011) and has antimicrobial properties (Negi and Dave, 2010).

A. catechu shows its immunomodulatory effect on both cell-mediated and humoral immunity. In a study, the aqueous extract of *A. catechu* was orally administered (5 and 50 mg/kg), and it was observed that the treated mice showed an increase in the neutrophil adhesion to the nylon fibers, produced a significant increase in the phagocytic index, and a significant protection against cyclophosphamide induced neutropenia indicating its effect on cell-mediated immunity. *A. catechu* extract produced a significant increase in the serum immunoglobulin levels, increase in the hemagglutination titer values, and decrease in the mortality ratio in mice, suggesting enhanced humoral immunity (Ismail and Asad, 2009).

A. catechu bark (methanolic and hexane) extracts have antiproliferative, cytotoxic, and anticancer properties against various cancer cell lines but does not show any effect on human peripheral lymphocytes; thus, this property can be used in designing safe anticancer drug (Nadumane and Nair, 2011). Methanolic extract of *A. catechu* heartwood showed 50% cytotoxic activity in breast adenocarcinoma cell line MCF-7, which is due to the enhancement in Bax/Bcl2 ratio leading to the activation of caspases and subsequent cleavage of polyadeno ribose polymerase (Ghate et al., 2014). The aqueous extracts of heartwood also show antidiabetic and antinociceptive action in a dose-dependent manner (Rahmatullah et al., 2013). Various extract of *A. catechu* have chemo-protective role in chemical (carbon tetrachloride, *t*-butyl hydrogen peroxide,

7,12-dimethylbenz[*a*]anthracene, DMBA) induced hepatocytic damage, breast and squamous cells cancers (Monga et al., 2011). Moreover, various parts of *A. catechu* extract show antimicrobial properties against a number of pathogens including *Salmonella typhi* (Rani and Khullar, 2004), *Pseudomonas aeruginosa, Candida albicans* (Negi and Dave, 2010), *Bacillus subtilis, Staphylococcus aureus, Klebsiella pneumoniae,* and *Shigella* spp. (Joshi et al., 2012). The bark extracts exhibited potent anti-HIV effects, owing to its effect on viral protease and via hampering the interaction of Viral Tat protein to its HIV-1 promoter sequence of LTR (Modi et al., 2013). The antiviral compounds isolated from *A. catechu* can overcome the conventional problem of generation of a drug resistant HIV-1 strain (Márquez et al., 2005) has hence is a promising candidate in drug discovery.

17.3.2 *Acorus calamus* (Distribution: Central Asia, Southern Russia, Siberia, Eastern Europe; Family: Acoraceae; Common Names: Sweet Flag, Calamus)

Acorus calamus, a semiaquatic herb with creeping rhizomes, shows diverse pharmacological properties including antibacterial, insecticidal, antiulcerative, etc. (Pandit et al., 2011). It is a very potent adaptogenic drug. The key bioactive compounds present in *A. calamus* are flavonoid, monoterpene, quinone, sesquiterpene, and phenylpropanoid (Patra and Mitra, 1981). The ethanolic extract has antiproliferative and immunosuppressive properties and is found to inhibit the growth of murine and human cell lines, inhibit mitogen-induced proliferation of peripheral blood mononuclear cells (PBMCs), and the generation of IL-12 and TNF-α (Mehrotra et al., 2003). The volatile oil, petroleum ether, and alcoholic extracts of the leaves of *A. calamus* enhance the phagocytic activity of neutrophils in the concentration range of 5–50 μg/mL (Ravichandiran and Patil Vishal, 2015). A D-galacturonic acid containing pectic polysaccharide isolated from the rhizomes of *A. calamus* at low concentrations can stimulate murine macrophages to produce NO and IL-12 similar to those induced by LPS, thus promoting a Th1 and suppressing the Th2 response. It also lowers serum levels of IgG1 and IgE and induces the secretion of TNF-α secretion by human PBMCs. Thus, the polysaccharide activates the macrophages into M1 type (classically activated macrophages) (Belska et al., 2010). The polysaccharide possibly acts via binding to certain receptors on antigen-presenting cells (APCs) (Schepetkin and Quinn, 2006), releasing immunoregulatory cytokines, and adhesion molecules (Tzianabos, 2000; Retini et al., 2001). Its properties can be used for treating oncological and allergic diseases.

A. calamus restores the hepatic enzymes in acetaminophen-induced liver damage and lowers free radical-induced oxidative stress, hence playing a protective role (Palani et al, 2011). Chronic stress can be detrimental to the immune system. Noise can activate the pituitary–adrenal–cortical axis and the sympathetic adrenal medullary axis thereby increasing the secretion from adrenal glands that directly correlates to stress (Babisch, 2003) and hence can have detrimental effects on the immune status of the body. Studies in rodents have shown that high-intensity noise can reduce the blood leukocytes like eosinophil (Jensen, 1969; Geber et al., 1966) the production of interferons (Chang and Rasmussen, 1965) and hamper the activity of neutrophils (Srikumar et al., 2005). Noise-induced stress reduces the number of $CD4^+$ and $CD8^+$ T-cells, which is reversed by *A. calamus* and its active compound α-asarone. The free radical-induced oxidation of lipids is also prevented by the extract of *A. calamus* and α-asarone (Dharini et al., 2012). In addition to α-asarone, β-asarone, an important chemical constituent of *A. calamus*, is reported to have antifungal activity against filamentous fungi *Microsporum gypseum, Trichophyton rubrum*, and *Penicillium marneffei* (Phongpaichit et al., 2005). The methanolic extract of *A. calamus* rhizomes synthesizes heme peroxidase that inhibits many phytopathogens and thus forms as a part of the plant's defense mechanism. Owing to the presence of this enzyme, *A. calamus* extracts have antifungal properties (Ghosh, 2006). In combination with *Vitex negundo*, the ethanolic extracts of rhizomes of *A. calamus* have anti-helminthic activity (Merekar et al., 2011). Kim et al. (2009a,b) reported that the leaf extracts inhibit inflammatory reactions in HaCaT cells via various mechanisms. In addition, β-asarone also has neuroprotective role. It suppresses neuronal apoptosis by downregulating Bcl2, Bcl-w, and caspase-3 and preventing JNK phosphorylation. It is currently under investigation as a drug in rat models of Alzheimer (Geng et al., 2010). Lectins isolated from the roots of *A. calamus* are potent mitogenic agents for human lymphocytes and murine splenocytes. These lectins significantly inhibited the growth of a J774, a murine macrophage cancer cell line, and of B-cell lymphoma (Bains et al., 2005).

17.3.3 *Allium sativum* [Distribution: Central Asia (Native) to Worldwide; Family: Amaryllidaceae, Common Names: Garlic]

Garlic is an essential medicinal spice and dietary element attributed with immunomodulatory properties owing to the presence of certain proteins in the bulb mainly agglutinins, Alliinase (Gorinstein et al., 2005), antifungal Allivin (Wang and Ng, 2001), and the antimicrobial protein Alliumin (Xia and Ng, 2005). Aged garlic constituents show antiallergic and antitumor properties (Kyo et al., 2001). Alliin a bioactive compound isolated from *A. sativum* is found to increase the expression of proinflammatory cytokine genes like IL-6, MCP-1, and EGR-1 in LPS-stimulated 3T3-L1 adipocytes. It also modulates the cytokine generation, for example, low doses of garlic extract increased IL-10, while decreased IL-12. IL-1α, IFN-γ, TNF-α, IL-6, and IL-8 were found to be reduced on treatment with the extract (Quintero-Fabián et al., 2013). Other *A. sativum*—derived bioactive compounds like allitridin, *S*-allyl-L-cysteine, Caffeic acid, uracil, and diallyl sulfide can inhibit transcription of several proinflammatory cytokine genes like IL-6, MCP-1, TNF-α, IL-1β, and IL-12 by inhibiting the transcription factor NF-κB (Kim et al., 2013; You et al., 2013; Fu et al., 2015a,b; Ho et al., 2014).

Clement et al. found that lectin from garlic non-specifically activates mast cells and basophils. Thus, lectins and agglutinins are potent mitogens and have potential utility in immunomodulation (Clement et al., 2010). Allicin, another compound isolated from garlic, is also reported to have antiparasitic activity against *Plasmodium* (Coppi et al., 2006), *Trypanosoma* (Nok et al., 1996), *Leishmania, Schistosoma, Babesia, Theileria*, and *Entamoeba*. Feng et al. found that Allicin reduced parasitemia when administered in *Plasmodium yoelii*—infected mice due to the generation of proinflammatory cytokines like IFN-γ. Allicin treatment also activated the macrophages and stimulates the expansion of $CD4^+$ T-cells (Feng et al., 2012). Not only the bioactive compounds of *Allium sativum* affect the T-cells but also oil macerated extracts containing active ingredient Z-ajoene affects the B-cells and increased the levels of IgA and interleukines (Washiya et al., 2013). Allicin promotes the maturation of DCs by increasing the expression of CD40: a costimulatory molecule, thus inducing a proinflammatory response in rodent malaria models (Feng et al., 2012). Aged garlic soaked, sliced, and extracted (AGE) in ethanol can inhibit the antigen-specific generation of histamines in rat basophil cell line RBL-2H3. Orally administered AGE significantly decreased the

IgE-mediated skin reactions (Kyo et al., 2001) and induced the proinflammatory cytokines IL-12, INF-γ, and iNOS in *Leishmania*-infected murine macrophages (Gharavi et al., 2011). Contrary to this, AGE upregulated IL-10 and decreased IL-12 production in PBMCs (Hodge et al., 2002), which in effect reduced IFN-γ, IL-2, TNF-α, and IL-6 (Oft, 2014; Gazzinelli et al., 1992). Garlic extracts exerts an anti-inflammatory effect on monocytes and lymphocyte proliferation by upregulating IL-10 and downregulating the production TNF-α on LPS stimulation (Hodge et al., 2002). Diallyl disulfide from *A. sativum* decreased proinflammatory cytokines, NO production in murine macrophages, and leukemic monocyte cell lines (Shin et al., 2013). AGE also affects NK cells and increases its activity against various cancer cell lines (Kyo et al., 1998). Fructooligosaccharides present in AGE show mitogenic potential, activated macrophages, and increased the phagocytic activity, comparable to effects shown by mitogens like zymosan and mannan (Chandrashekar et al., 2011). Immunoproteins isolated from garlic like lectins and agglutinins are also known for their mitogenic properties similar to those of ConA and PHA (Clement et al., 2010). AGE also affects the unique subset of T-cell population by increasing the proliferation of γδ-T-cells that plays crucial role in recognition of pathogen-associated molecular patterns (Nantz et al., 2012).

Bioactive compounds from garlic can also exhibit antiviral activity. For instance, allitridin or diallyl tri-sulfide inhibits T-reg cells in vivo (Li et al., 2013) and thus mounts an antiviral immune response against murine cytomegalovirus (Yi et al., 2005). Fresh garlic extracts also stimulates the proliferation and activation of $CD8^+$ T-cells and causes a delayed-type hypersensitivity (DTH) response (Ebrahimi et al, 2013). *A. sativum* when used in combination with minerals oil enhances serum agglutination, antibody titer and hemolytic activity of serum in *Labeo rohita* against the infections of Gram-negative bacteria *Aeromonas hydrophila* (Dash et al., 2014). Another study in juvenile hybrid *Oreochromis mossambicus* showed that supplementation with garlic increased the lysosomal activity, respiratory burst, and leukocyte count (Ndong and Fall, 2011). Various compounds from *A. sativum* are responsible for allergic reaction like anaphylaxis (Pérez-Pimiento et al., 1999), rhinoconjunctivitis, asthma (Seuri et al., 1993; Cronin, 1987), and urticaria (hives) (Asero et al., 1998) due to the presence of an IgE binding compound called Alliin lyase (Kao et al., 2004).

17.3.4 *Andrographis paniculate* (Distribution: Southeast Asia, China, America, West Indies; Family: Acanthaceae, Common Names: King of Bitters, Kalmegh)

A. paniculata, a well-known medicinal plant in various parts of the world, contains polyphenols, terpenoids (diterpene lactones, entalabdane) xanthones, nocardioides, and flavonoids (flavones) as key bioactive compounds. Aqueous and methanolic extracts of *A. paniculata* have antimicrobial properties that inhibit the growth of fungi and bacteria (Bobbarala et al., 2009). Arabinogalactans, andrographolide, and aqueous extracts are effective against *C. albicans*. Arabinogalactans and aqueous extracts inhibit the growth of *P. aeruginosa, B. subtilis*, and *E. coli*, while andrographolide inhibits the growth of *B. subtilis* (Singha et al., 2003). Xanthones isolated from the roots of *A. paniculata* exhibited antimalarial activity in mice against *Plasmodium berghei* with significantly reduced parasitemia, while 1,2-dihydroxy-6,8-dimethoxy-xantone shows effects against *Plasmodium falciparum* (Xu et al., 2012). Ethanolic (Pongiuluran and Rofaani 2015) and dichloromethane extracts (Chao and Lin, 2010) of *A. paniculata* augment the proliferation of lymphocytes at low concentration.

Studies with aqueous extracts in *O. mossambicus* showed increase in the number of WBCs, RBCs, and thrombocytes (Prasad and Mukthiraj, 2011). It significantly increased the lymphocyte number in type-2 diabetes rat models as well (Radhika et al., 2012). Andrographolide has diverse immunoregulatory effects. On administration of Andrographolide in animals bearing metastatic tumors, antibody dependent cytotoxicity (ADCC) was enhanced when compared to untreated controls. Serum from andrographolide treated mice in the presence of complement showed higher cytotoxicity suggesting that the extract can activate the humoral immune system and generate tumor-specific antibodies that can mediate ADCC (Sheeja and Kuttan, 2010). Andrographolide was tested against HIV-1 infection where it significantly increased the mean $CD4^+$ T-cells and inhibited HIV-induced dysregulation of cell cycle (Calabrese et al., 2000). Andrographolide and its derivatives also inhibited the fusion of HL2/3 cells with TZM-bl cells, which occurs via the interaction of gp120 with CD4 and CCR5, CXCR4, thus inhibiting the HIV virus (Uttekar et al., 2012). Andrographolide also inhibits the expression of Epstein–Barr virus (EBV) (Lin et al., 2008) and HSV-1 (Wiart et al., 2005). It increases secretion of IL-2, IFN-γ by T-cells and the activity of NK cells, thus inhibiting the growth of tumor. It also plays a role in autoimmune disorders such as encephalomyelitis, wherein it interferes with the maturation of DCs. Andrographolide can modulate the innate immune response, regulate the production of antibodies and the generation of antigen-specific splenocytes, and can activate macrophages both via classical and alternative pathways (Wang et al., 2010). The immunomodulating property of the purified andrographolide and neoandrographolide was found to be lower than that of ethanolic extracts indicating that multiple components may bring about such combinatorial immunomodulatory effects (Puri et al., 1993). Andrographolide, 7-*O*-methylwogonin isoandrographolide, and skullcapflavone-1 significantly inhibited inflammatory cytokines IL-6, NO, IL-1β in LPS-stimulated macrophages, and inflammatory mediators like PGE2 and TXB2 in activated promyelocytic leukemia cells. Andrographolide, dehydroandrographolide, and neoandrographolide exert its anti-inflammatory effect by inhibiting the cyclooxygenase (COX) enzyme (Parichatikanond et al., 2010). Another bioactive compound Andrograpanin extracted from the leaves of *A. paniculata* also inhibits proinflammatory cytokines in LPS-stimulated macrophages (Liu et al., 2008). *A. paniculata* is also known for its antioxidant properties. Andrographolide and 14-deoxy-11,12-didehydroandrographolide showed free radical scavenging effect and inhibited lipid peroxidation under DPPH-induced oxidative stress (Akowuah et al., 2009).

17.3.5 *Azadirachta indica* (Distribution: Asia; Family: Meliaceae, Common Names: Neem)

Previous studies have demonstrated that administration of *Azadirachta indica* extract increases the number and the activity of peripheral blood lymphocytes. It also increases the $CD4^+$, $CD8^+$ T-cells, and the markers for T-cell and macrophage activation, namely, CD25 and MAC-3, respectively. *A. indica* extracts also resulted in lesser lung and liver metastases in sarcoma model of Balb/c mice when compared to control mice (Belska et al., 2006). Kumar et al. (2006) showed that Nimbolide isolated from the leaves has antiapoptotic and antiproliferative properties via downregulation of bcl2/bax and upregulation of Apaf-1 and caspase-3. *A. indica* has anticancer properties, and it fortifies the body's immune system. It aids in the proper presentation of tumor-associated antigens to the APCs. Neem leaf preparation (NLP) also acts as an adjuvant to generate antigen (B16MelSAg) specific antiserum in C57BL/6 mice (Baral and Chattopadhyay, 2004). NLP also stimulates the CD40-CD40L interaction that leads to the activation of p38MAPK and generation of proinflammatory cytokines IL-12, thus the activation of NK cells (Bose and Baral, 2007).

Glycoprotein isolated from *A. indica* leaves (neem leaf glycoprotein—NLGP) enhances the expression of IFN-γ that downregulates CXCR3B (responsible for apoptosis in lymphocytes) and thus restored impaired chemotactic movement of PBMCs toward tumor (Chakraborty et al., 2008). NLGP leads to the activation of T-cells and generates Th1 type of cytokine IFN-γ and can effectively activate erythroleukemia and oral cancer cells (Bose et al., 2009).

Administration of aqueous extracts of *A. indica* significantly enhanced the activity of macrophages and facilitates tumor antigen presentation by macrophages and DCs to T and B-lymphocytes and generation of an effector and memory response (Tsang et al., 2011).

17.3.6 *Boerhavia diffusa* (Distribution: Asia, Africa, North America, Caribbean, South America, South Pacific; Family: Nyctaginaceae; Common Names: Tarvine, Punarnava, Red Spiderling)

Major compounds isolated from *Boerhavia diffusa* are Boerhavia acid, isoflavonoids (rotenoids), Punarnavine, sitosterol, Boeravinone, palmitic acid, steroids (ecdysteroid), lignan glycosides, and esters of sitosterol. *B. diffusa* mainly finds its application in the treatment of dropsy, characterized by retention of excessive fluids in tissue and body cavities. The plant extract helps in discharge of these fluids (Bhowmik et al., 2012). Aqueous and ethanolic extracts have antimicrobial properties against *Salmonella typhimurium, Corynebacterium diphtheria, Streptococcus* group, *Shigella dysenteriae, Neisseria gonorrhoeae, Escherichia coli*, and *Bacillus subtilis* (Sangameswaran et al., 2008). Ethyl acetate extracts of *B. diffusa* have antifungal properties and inhibits the growth of *Microsporum gypseum, Microsporum canis, Microsporum fulvum*, etc. (Agrawal et al., 2003). It also has anti-inflammatory and antioxidant properties and scavenges free radicals during oxidative stress (Gacche and Dhole, 2006). Two well-characterized immunostimulants isolated from the roots of *B. diffusa* are Punarnavine and syringaresinol. Aqueous extracts showed significant leukocytosis and lowered the mortality in *E. coli*—induced abdominal sepsis in mice and reduced stress-induced increase in the level of glucose and cholesterol (Mungantiwar et al., 1997). Alkaloid fraction was found to normalize the plasma cortisol levels and reduced DTH reactions in animals (Mungantiwar et al., 1999). In a study, *B. diffusa* extracts enhanced the phagocytic activity of macrophages comparable to the drug levamisole (Sumanth and Mustafa, 2007). *B. diffusa* is also known to have certain adaptogenic effects. The immunomodulatory activity is attributed to the compounds like quercetin, Punarnavine, syringaresinol mono-β-D-glucoside, etc. Ethanolic extracts of *B. diffusa* is found to have immunosuppressive properties and reduces the production of TNF-α and IL-2 in human PBMCs, suppresses human NK cells and the generation of NO in murine macrophages (Mehrotra et al., 2003). Immunosuppressive action is possibly due to the alkaloid/lignin compounds. Pandey et al. found that chloroform and ethanolic extract treatment in RAW 264.7 cell line under LPS stimulation induced NO production, inhibited PHA-induced proliferation, and production of TNF-α and IL-2 from PBMCs. Eupalitin-3-*O*-β-D-galactopyranoside the bioactive compound isolated from ethanolic extracts was found to be most effective (Pandey et al., 2005).

Antiosteoporotic properties (Li et al., 1996) are also attributed to Eupalitin in addition to its anti-inflammatory and immunosuppressive effects. Thus, its anti-inflammatory effects are extrapolated for treatment of rheumatic disorders. Various studies have been performed that show anticancer properties of *B. diffusa* extract. Punarnavine downregulates the expression ERK-1/2, MMP-2, MMP-9, and VEGF (Manu and Kuttan, 2009) and prophylactic as well as simultaneous administration of punarnavine-reduced lung melanoma metastasis. It also enhances the production of IL-2 and IFN-γ; activity of NK cells, and showed enhanced ADCC. The level of proinflammatory cytokines IL-1β, IL-6, and TNF-α was reduced on Punarnavine administration (Manu and Kuttan, 2007). Studies by Srivastava et al. demonstrated the cytotoxic effects of the ethanolic and alkaloid leaf and root extracts of *B. diffusa* in tumor cells line Hela (Srivastava et al., 2011) and U-87 (Srivastava et al., 2005). Methanolic extracts of whole plant reduced cell viability in MCF-7 cell line, and the cells were arrested in G0–G1 phase (Sreeja and Sreeja, 2009). Leyon et al. (2005) found that the methanolic extracts inhibited metastasis in B16F10 melanoma in C57BL/6 mice and reduced the serum parameters of metastasis. Retinoids isolated from root; Boeravinones G and H are efflux inhibitors of cancer-resistance protein (ABCG2) (Ahmed-Belkacem et al., 2007).

17.3.7 *Curcuma longa* [Distribution: Southeast Asia (Native); Family: Zingiberaceae, Common Names: Turmeric]

Curcuma longa a perennial plant is a common culinary ingredient in many parts of Asia and is also known for its therapeutic applications such as an antiseptic, anti-inflammatory blood purifying, and wound-healing agent. Main bioactive compounds isolated from *C. longa* are Curcuminoids, sesquiterpenoids, and turmerones (He et al., 1998). The active component curuminoid, a mixture of curcumin, demethoxycurcumin, and bisdemethoxycurcumin, is a potent anti-inflammatory agent. It has multifaceted role and inhibits COX-2 (Plummer et al, 1999) and activation factor NF-κB (Singh and Aggarwal, 1995). Administration of *C. longa* extracts in CMS model of rats enhanced the activity of NK cells and augmented the levels of IL-6 and TNF-α (Xia et al., 2006). It also modulates the proliferation and responsiveness of macrophages, NK cells (Bhaumik et al., 2000), DCs, and lymphocytes (Jagetia and Aggarwal, 2007). Curcumin shows immunosuppressive effects and inhibits mitogen-induced (PHA, phorbol-12-myristate-13-acetate, Con A) proliferation of human spleen-derived lymphocytes and IL-2, responsible for driving this proliferation (Ranjan et al., 2004). Yadav et al. (2005) found similar immunosuppressive effects of curcumin in human PBMCs, wherein it inhibits IL-2 and expression of NF-κB and PHA-induced proliferation. Curcumin also affects the phosphantigen-induced proliferation of human Vγ9Vδ2 T cell proliferation (Cipriani et al., 2001). Polar extracts mainly containing polysaccharides (that do not contain curcuminoids) have mitogenic potential and increased the number of splenocytes comparable to LPS and ConA. NR-INF-02 (purified polysaccharide) increased the IL-2, IL-6, IL-10, IL-12, IFN-γ, NO, MCP-1, and TNF-α in unstimulated murine macrophages and splenocytes (Chandrasekaran et al., 2013).

17.3.8 *Cynodon dactylon* (Distribution: Tropical and Subtropical Regions; Family: Poaceae, Common Names: Bermuda Grass)

Cynodon dactylon is a perennial grass that is commonly used as a laxative, expectorant, analgesic, etc. It is also used for the treatment of dropsy (Kesari et al., 2006), syphilis (Provin et al., 2008), and diabetes (Singh et al., 2008). Active principles in ethanolic extracts of *C. dactylon*, namely, Cynodin, beta-carotene, hydrocyanic acid, and triticin are powerful antioxidants (Auddy et al., 2003). Santhi et al. found that intraperitoneal administration of *C. dactylon* significantly increased the adhesion of neutrophils to nylon fibers: a model reflecting the process of margination of cells in blood vessels. In addition, the antibody responsiveness in mice to sheep red blood cells (SRBC) was significantly increased owing to heightened responsiveness of macrophages and antibody-synthesizing B-cells

(Mungantiwar et al., 1999). Not only humoral but also cellular immunity was affected on administration of protein fraction of *C. dactylon*. DTH that correlates to cell-mediated immunity increased due to increased responsiveness of T-cells (Santhi and Annapoorani, 2010).

Mangathayaru et al. (2009) showed that oral administration of *C. dactylon* juice increased humoral antibody response against an antigen. *C. dactylon* inhibited the release of TNF-α, IL-6, and IL-1 and promoted anti-inflammatory IL-10 that in turn inhibits proinflammatory cytokines (Moore et al., 2001). Studies done in *Catla catla* show that ethanolic extracts of *C. dactylon* incorporated into the diet of the fish activated non-specific immune mechanism and afforded resistance against *A. hydrophila* infection (Kaleeswaran et al., 2011).

17.3.9 *Ficus benghalensis* [Distribution: Indian Subcontinent (Native); Family: Moraceae; Common Names: Banyan Tree, Banyan Fig]

Ficus benghalensis belonging to the family Moraceae is a very large tree, 20–30 m high, with wide-spreading branches bearing aerial roots. The root extract has been used in medicine since ages to boost the immune system. *Ficus* spp. is used extensively in folk medicines as a vermicide, astringent, hypotensive, and antidysentery drug (Trivedi et al., 1969). The active components isolated from *F. benghalensis* include glucosides (Bhattacharjee, 2008), flavonoids (Mousa et al., 1994), etc. The methanolic and water extracts have immunostimulatory properties and enhances the phagocytic potential of PBMCs. It also induces the proliferation of lymphocytes and hence the generation of cytokines that activate other immune cells (Gabhe et al., 2006). The hydroalcoholic leaf extracts of *F. benghalensis* Linn significantly increased the phagocytic activity of human neutrophils and hence engulfment and clearance of microorganisms by leukocytes, along with free radical scavenging properties and reduction of oxidative stress, thereby showed immunomodulatory and antioxidant activity (Bhanwase and Alagawadi, 2016).

17.3.10 *Murraya koenigii* (Distribution: Asia, South Africa; Family: Rutaceae; Common Names: Curry Leaves)

Murraya koenigii is commonly used in culinary due to its aromatic quality. It is rich in many vitamins and is used for treating stomachaches and as a carminative and analgesic. It has anti-inflammatory, antioxidant, antitumor properties due to the presence of bioactive ingredients like carbazole alkaloid. Many bioactive compounds are present in *M. koenigii* like koenigin, koenine, koenimbine, girinimbin, iso-mahanimbin, koenidine, cyclomahanimbine, tetrahydromahanimbine, carbazole alkaloids, murrayazoline, mahanimbine, murrayastine, etc. Aqueous extracts of *M. koenigii* shows antioxidant activity and protects cardiac tissue against cadmium-induced oxidative stress (Mitra et al., 2012). Leaf extracts also reduced lipid peroxidation in ethanol-induced toxicity in hepatocytes (Sathaye et al., 2011).

Methanolic extracts reduced carrageenan-induced paw edema in albino rats, and this anti-inflammatory effect is comparable to the drug diclofenac (10 mg/kg, p.o.) (Gupta et al., 2010; Bhandari, 2012). Methanolic extracts of *M. koenigii* also significantly increased the NO production from macrophages, which is the principle effector molecule produced by macrophages, hence an indicator of its enhanced cytotoxic activity. It also increased the phagocytic index in carbon-clearance test. Moreover, the humoral antibody response to ovalbumin also increased on treatment with the methanolic extract, thus indicating an overall elevation in humoral response and immunostimulatory effect of the extract on B-cells. There was no significant difference in DTH response on treatment with the extract indicating that it does not have a stimulatory effect on T-cells and hence on cell-mediated immunity (Shah et al., 2008). *M. koenigii* is a potent antimicrobial agent against various antibiotic resistant bacteria such as *P. aeruginosa, S. pneumoniae, S. aureus, K. pneumoniae*, and *E. coli* (Nagappan et al., 2011).

17.3.11 *Ocimum sanctum* (Distribution: Asia, Europe, USA; Family: Lamiaceae; Common Names: Holy Basil)

O. sanctum is anti-inflammatory, analgesic, immunostimulatory and has multidirectional therapeutic uses. Key active compounds are 2-eugenol and its methyl derivatives, 1-stigmast-5-en-3-ol, 2-octadecane, 3-β-caryophyllene, caryophyllene, ursolic acid, methyl carvicol, limatrol, sitosterol, etc. (Wagner et al., 1994). Aqueous extract majorly contains tannins, flavonoids, and alkaloids (Gupta et al., 2002). Studies done in fish showed that administration of *O. sanctum* leaf extract has immunostimulatory effects. It induced an innate immunity, increased the number of phagocytes, RBCs, WBCs, and lymphocytes and thus can be used to afford immunity against various parasitic infections (Nahak and Sahu, 2014). Another study in *L. rohita* demonstrated that *O. sanctum* mounts a nonspecific immune response against *A. hydrophilia* infection and increased the total immunoglobulin level and lysosomal activity, and had a positive effect on hematological and biochemical parameters (Das et al., 2015). The immunostimulatory activity of the leaf extract could be possible due to the presence of bioactive compounds such as eugenol, methyleugenol, caryophyllene (Chopra et al, 2002), ursolic acid, oleanolic acid, and salrigenin (Mukherjee et al., 2005), etc. that are known to have immunomodulatory potential. Ethanolic and aqueous extracts have antimicrobial properties and are effective against different bacterial strains such as *B. pumilus, E coli, S. aureus, S. typhimurium, P. aeruginosa,* etc. (Joshi et al., 2009). *O. sanctum* is also effective against fungi such as *Microsporum canis, Aspergillus fumigatus, Aspergillus niger, Cryptococcus neoformans, Sporotrichum schenckii, Candida albicans* (Ahmad and Beg, 2001; Bano et al., 2017).

Essential oils isolated from *O. sanctum* have antiviral activity against various viruses (Direkbusarakom et al., 1996) such as poliovirus type-3 (Ravi et al., 1997), hepatitis B virus, Infectious hematopoietic Necrosis Virus (IHNV), and Herpes Simplex Virus (HSV). VERO cells when treated with ethanolic extracts (22.5 mg/mL concentration) inhibited the replication of polio type-3 virus. Other *O. sanctum*—derived bioactive compounds like apigenin, linalool, and ursolic acid also exhibit antiviral property and inhibit DNA, RNA, and adenoviruses (Sood et al., 2013). In a study by Mukherjee et al. (2005) the aqueous leaf extracts showed immunomodulatory activity in subclinical trails in bovines and increased the number and activity of neutrophils and lymphocytes and reduced the number of bacteria. *O. sanctum* modulates various cytokines such as IL-2, TNF-α, IFN-γ, etc. during *S. typhimurium* infection that inhibits bacterial growth and is necessary for activation of macrophages and thus aiding in effective clearance of the parasite (Goel et al., 2010).

Oil isolated from the seed of *O. sanctum* regulated both humoral and cell-mediated immune response mediated by GABAergic pathway (Vaghasiya et al., 2010). *O. sanctum* is found to increase the amount of antibody produced, number of WBCs, RBCs, and hemoglobin (Jeba et al., 2011). It also increases the hemagglutination titer in mice (Vaghasiya et al., 2010). *O. sanctum* has free radical scavenging property and limits certain types of cancerous growth. The active component ursolic acid is anticancerous, anti-inflammatory, and inhibits COX (Prashar et al., 1998). Aqueous and alcoholic extracts have antitumor properties and reduced the size of Sarcoma-180 solid tumors in mice (Adhvaryu et al., 2008).

17.3.12 *Panax ginseng* (Distribution: Asia; Family: Araliaceae; Common Names: Ginseng)

Panax ginseng is a well-known oriental medicine and immunomodulator. Main pharmacological components are saponins, polyphenolic compounds, polyacetylenes, triterpenoids, etc. Root, stem, and leaves are commonly made use of in order to boost the immune system. Ginseng extracts containing polysaccharides enhanced the phagocytic activity of macrophages. On treatment with red ginseng acidic polysaccharides (RGAPs), peritoneal macrophages showed increased phagocytic activity (Shin et al., 2002). Ginseng enhances the generation of NO from activated peritoneal macrophages and RAW 264.7 cells that aids in microbe clearance (Friedl et al., 2001). Co-treatment of macrophages with RGAPs and IFN-γ further leads to an increase in the production of IL-1β, TNF-α, and NO (Choi et al., 2008). Polysaccharides from *P. ginseng* increased the secretion of IL-1β and TNF-α from macrophages in vitro (Lim et al., 2002). Treatment of murine macrophage cells J774A.1 with ginseng extract increased the production of IL-12 (Wang et al., 2003). *P. ginseng* also has immunostimulatory effects on DCs (Kim et al., 2009a,b). Administration of ginseng aqueous extracts has immunostimulatory effects on NK cells (Jie et al., 1984) in both immunocompetent and immunosuppressed mice (Kim et al., 1990). Study using blood samples of immunodeficient patients (AIDS and chronic fatigue syndrome) showed that *P. ginseng* enhanced the NK cell activity in PBMCs compared to the placebo group (See et al., 1997). The metabolic end products (M1 and M4) of steroidal ginseng saponin, can drive DCs into maturation and increased the expression of cell-surface markers like MHC class II, CD80, CD83, and CD86. M4 primed mature DCs boosted antigen-presentation ability as evident from production of IFN-γ and 51Cr by the DCs. A reciprocal effect is seen on treatment of DCs with total saponins of *P. ginseng* followed by oxidized-low density lipoprotein. It inhibits the maturation of DCs and downregulates maturation markers like HLA-DR, CD1a, CD40, CD86, etc. It also inhibits TNF-α and IL-12. These reciprocal effects of *P. ginseng* might be due to the activation of different signaling pathways (Su et al., 2010). Ginsan also shows immunostimulatory effects on the maturation of DCs (Kim et al., 2009a,b). Ginsan not only modulates innate immune response but also affects adaptive immune response. Mice orally administered with ginsan prior to infection with *Salmonella* showed higher serum IgG1 and IgG2 and IgA against *Salmonella* (Na et al., 2010). Ginsenoside injected subcutaneously increases serum antibodies specific against *Toxoplasma gondii* and specific IgG, IgG1, IgG2a, and IgG2b responses to H3N2 influenza virus (Yoo et al., 2012; Qu et al., 2011). Both ginsan and ginsenoside stimulated T-cell proliferation and affect cell-mediated immune response. It was also found that splenocytes cultured with ginsan produce proinflammatory IL-2, IFN-γ, IL-1α, and GMSF (Kim et al., 1998). This effect was reversed in mice infected with *S. aureus* (Ahn et al., 2005). Macrophages treated with ginseng radix extract (GRE) produce proinflammatory cytokines such as IFN-γ, IL-1β, TNF-α, and IL-6 in vitro (Liou et al., 2006) possibly through activation of TLR4 signaling (Nakaya et al., 2004) through a non-LPS agent present in GRE. Active compound of ginseng can also downregulate TLR2 and its downstream Myd88 and inhibit proinflammatory cytokine production (Ahn et al., 2005).

In stimulated PBMCs (Larsen et al., 2004) and in mice infected with *P. aeruginosa* (Song et al., 2003), ginseng augmented IL-2 production and mounts a Th1 response. Some studies have shown counteracting function of *P. ginseng* extracts in inhibition of antibody production against ovalbumin (Sumiyoshi et al., 2010) and similar inhibitory effects on long-term administration (Liou et al., 2004).

17.3.13 *Picrorhiza scrophulariiflora* (Distribution: Alpine Himalayas, Tibet; Family: Scrophulariaceae; Common Names: Kutki)

This plant is commonly used in Chinese and Ayurvedic medicine systems to treat various ailments like gastrointestinal, urinary, and skin disorders (Kirtikar, 1935). The rhizomes are rich in iridoid glycosides (Huang et al., 2006), phenolic glycosides, phenylethanoid glycosides (Wang et al., 2004a,b), and terpenoids (Wang et al., 2004a,b). Extracts of *P. scrophulariiflora* inhibits lipid peroxidation and possesses free radical scavenging activity. Scrocaffeside A isolated from *P. scrophulariiflora* is an immunostimulant. In response to Con A and LPS, it significantly increased the number of splenocytes. It also enhanced the activity of macrophages and NK cells even in the absence of stimulation (Coico et al., 2015). The number of mature $CD4^+$ and $CD8^+$ cells and both Th1 and Th2 cytokines also significantly increased. Activated $CD4^+$ T-cells differentiate either into Th1 type cells and secrete TNF-α, IL-12, IL-2, IFN-γ or into Th2 cells and secrete IL-5, IL-10, IL-4, IL-13, and this balance is critical in deciding the immune response and ultimately the disease outcome. Picrogentiosides isolated from *P. scrophulariiflora* enhanced the proliferation of splenocytes comparable to Con A and LPS and enhanced both humoral and cell-mediated immunity (Zou et al., 2010). Picracin and deacetylbaccatin isolated from *P. scrophulariiflora* has antiproliferative effects and inhibited proliferation of T-cells that inhibits the release of IL-2 and subsequent IL-2-IL-2 receptor interaction. Picracin inhibited the release of proinflammatory cytokines IL-1β and TNFα in human monocytes, whereas deacetylbaccatin failed to do so. Moreover, systemic administration of both picracin and deacetylbaccatin decreased paw edema, suggesting its immunosuppressive activity. These compounds have dualistic effects and can induce an inflammatory response on local administration whereas an immunosuppressive effect on systemic administration.

17.3.14 *Terminalia arjuna* (Distribution: Pantropical; Family: Combretaceae; Common Names: Arjuna)

T. arjuna is commonly used in the traditional medicinal system to treat cardiac ailments (Bharani et al., 2002). Treatment of peritoneal macrophages with various doses of *T. arjuna* methanolic bark extracts and gemmo-modified (embryonic tissue) extracts increased its phagocytic potential as evaluated by increased SRBC engulfed by macrophages.

Both the extracts also enhanced the antibody-mediated humoral immune response and cell-mediated phagocytosis. This immunomodulatory property of the *T. arjuna* extract is attributed to the presence of polyphenols and flavonoids (Chiang et al., 2003). Polyphenols attenuate proinflammatory cytokines MMPs, SOCS3 and downregulate TLR2 and NLRP1—an inflammasome component—and upregulate anti-inflammatory cytokines, thus inhibiting the proliferation of lymphocytes and reducing the activity of NK cells (Jung et al., 2012; Ellis et al., 2011). In a study, the *T. arjuna* extracts significantly decreased formalin-induced paw edema due to its anti-inflammatory properties (Halder et al., 2009). Arjunolic acid a triterpene is a strong cardioprotective and antioxidant in rat models (Sumitra et al., 2001). Ethanolic extracts of *T. arjuna* decreases the level of endothelin-1 and inflammatory cytokines like IL-6 and TNF-α (Khaliq et al., 2013) in diabetic rats.

17.3.15 *Tinospora cordifolia* (Distribution: Indian Subcontinent, China; Family: Menispermaceae; Common Names: Guduchi, Moonseed, Giloy)

Tinospora cordifolia belongs to the family Menispermaceae and has a broad spectrum of immunotherapeutic properties ranging from antipyretic, anti-inflammatory, antiallergic antidiabetic, antihepatotoxic, antibacterial properties and has relatively low toxicity. The leaf extract is found to be effective against *E. coli, S. aureus, S. pyrogens, B. subtilis*, and *P. vulgaris* infections. It also helps the cell repair and rejuvenation process. About 10,000 tonnes of the plant is utilized annually for herbal medicine preparation (Singh et al., 2004). The key active components isolated from the plant include phenyl propanoid glycosides such as Cordifolioside A, Cordifolioside B and syringin (Maurya et al., 1996), and immunostimulatory compound—D-glucan (Nair et al., 2004, 2006). The immunomodulatory property of the Guduchi extract was validated by assessing its effect on activating resting macrophages and by estimating the generation of secretory factors like NO and lysosomes (More and Pai, 2011). Lysozymes act as a potent antimicrobial compound against gram-positive species (Shimada et al., 2008). The amount of lysozymes that leads to microbial activation of macrophages was found to be higher on treatment with *T. cordifolia* extract. Water and ethyl acetate extracts of *T. cordifolia* stem increased the phagocytic activity of human neutrophils. The immunostimulatory action may be attributed to the synergistic action of two or more than two active components like Cordifolioside A and syringin. *T. cordifolia* extract is also shown to stimulate the proliferation of stem cells. It increases the total number of WBCs and alpha-esterase-positive cells an indicator of increase in bone marrow cells. The extract also increases the number of antibody producing cells, hence implying its role in fortifying the humoral immune system. The extracts of *T. cordifolia* are also found to be effective against tumors and reduced tumor growth comparable to that of treatment with cyclophosphamide (Mathew and Kuttan, 1999).

17.4 FUTURE PERSPECTIVE

17.4.1 Prebiotics

"Prebiotics" comprise dietary fibers, mainly oligosaccharides, that are not digestible by the host but are acted upon and digested by the gastrointestinal microflora to alter its composition and activity. As prebiotics get digested and fermented by the gut microflora, it selectively stimulate the growth of beneficial bacteria like lactobacilli and bifidobacteria with concomitant reduction of the population of pathogenic bacteria like *Clostridium* in the colon, thereby improving the host immunity (Gibson et al., 2004). Inulin and galacto-oligosaccharides are the main plant-derived prebiotics. The plant fibers also improve the host's immunity by the production of short chain fatty acids like lactate, acetate, propionate, butyrate, etc. (McNeil et al., 1978) that are utilized by the anaerobic bacteria in the gut which boosts a number of immune cells like macrophages, T-helper cells, and NK cells. The fiber derived compound like β-glucan is known to interact directly with immune cells. Large numbers of antigens pass through the gut-associated lymphoid tissues, and there is a cross talk between the immune system and the gut microbiome to enable the host tolerance toward these microorganisms. The gut microbiome can directly affect the mucosal immunity. Germ free mice lacking a microbiota are found to be immune deficient with hampered development of Peyer's patches, spleen, lymph nodes, and T and B-cells (Cebra et al., 1998; Glaister, 1973).

17.4.2 Psychoactive Plants and Immunomodulation

Indigenous people have traditionally used plant-derived psychoactive substances as medicine since time immemorial (Chattopadhyay et al., 2003, 2004; Luna, 1984), but the immunological effects of such substances have been under investigation since the early 1970s. Immune cells express a number of receptors for neurotransmitters, hence a new branch of "neuroimmune communication" has emerged (Quan and Banks, 2007; Szabo and Rajnavolgyi, 2013a,b). Psychedelic ingredients have immunomodulatory potential and interfere with both the innate and adaptive immune system. Tryptamines, the monoamine alkaloids like *N-N*-dimethyltryptamine (DMT), are structurally closely related to neurotransmitter serotonin and hormone melatonin. DMT is mainly found in *Acacia* spp. and is isolated from *Diplopterys cabrerana* and *Psychotria viridis*. Studies have shown that DMT can modulate immune response in human primary cell cultures (Tourino et al., 2013; Szabo et al., 2014). DMT showed cytotoxic, antitumor, and anti-inflammatory activities in human PBMCs co-cultured with glioma cell line (Tourino et al., 2013). Administration of DMT or its methoxy derivative 5-methoxy-*N-N*-dimethytryptamine, in human monocyte-derived DC culture (moDC), enhanced the anti-inflammatory response through sigmar-1 (Szabo and Rajnavolgyi, 2013a,b). DMT treatment in LPS activate human moDC resulted in an enhanced anti-inflammatory response with decreased levels of cytokines IL-1, IL-6, TNF-α, and chemokines CXCL8/IL-8 and elevated level of anti-inflammatory cytokine IL-10. Moreover, the number of NK cells which mounts an antiviral immune response also increased (Dos Santos et al., 2011, 2012). All these studies suggest that plant-derived psychoactive compounds can function as immunomodulators and thus can be promising drug candidates (Szabo, 2015).

17.5 PHYTOMEDICINE AS A NEW TREND

Immune suppression can happen during chronic stress, infections, on exposure to chemical agents during chemotherapy, poor dietary intake, and endogenous autoimmune reactions. Recurrent infections, multi-infections, exposure of the immune system to chemicals and immunosuppressive drugs, resistance of microorganisms to current therapies, and unavailability of effective vaccines against many diseases calls for the use of natural immunostimulants. Plant-based immunostimulants are now used mainly in controlling moderate respiratory and urogenital infections. Extracts of *Baptisia tinctoria, E. purpurea*, and *Thuja occidentalis* are used in the treatment of bacterial infections and for prophylaxis in viral infection in many parts of Europe (Petrunov et al., 2007). The galactose specific lectin isolated from the hemiparasitic plant *V. album* and Glucan Lentinan from the *Lentinus edodes* in combination with 5-fluoruracil is used for the treatment of cancers (Ina et al., 2013) and in curbing the metastasis post—tumor removal (Khan and Mukhtar, 2010). Interestingly, the amino acid L-arginine from various food sources can regulate the growth of malarial parasite *P. falciparum* on one hand and immunomodulation of host cells on the other (Awasthi et al., 2017). A recent study reported that the inositol hexa-phosphoric acid, a nutraceutical from leafy vegetables, can attenuate iron-induced oxidative stress and alleviates iron-overloaded liver injury in mice (Bhowmik et al., 2017). Some plant products currently under various phases of clinical trials include tiotropium from *Datura stramonium*, arteether from *Artemisia annua*, galanthamine (amaryllidaceae type alkaloid) from *Narcissus cultivars*, camptothecin derived alkaloids from stem and bark of *Camptothecin acuminata*, and paclitaxel (alkaloid) from *Taxus canadensis*. According to WHO, out of 225 drugs under the basic and essential category, about 11% are of plant origin (Ansari and Inamdar, 2010). Compared to different synthetic compounds and their diverse pharmacological properties and toxicities, phytochemicals are preferred due to their natural origin, and used for generations without immediate toxic effect. However, it needs to be considered that the plants and animals have evolved through two different pathways following bio-organic evolution. Therefore, the evolution of different biosynthetic and metabolic pathways in plants may not necessarily match with those of animals. As the compatibility between the two systems is a concern, and thus, the tests for toxic effect and efficacy of phytomedicines need to be rigorous.

REFERENCES

Adhvaryu, M.R., Reddy, N.M., Vakharia, B.C., 2008. Prevention of hepatotoxicity due to anti tuberculosis treatment: a novel integrative approach. World J. Gastroenterol. 14 (30), 4753–4762.

Agarwal, S.S., Singh, V.K., 1999. Immunomodulators: a review of studies on Indian medicinal plants and synthetic peptides. Part I: Medicinal plants. Proc. Indian Natl. Sci. Acad. 65, 179–204.

Agrawal, A., Srivastava, S., Srivastava, M.M., 2003. Antifungal activity of *Boerhavia diffusa* against some dermatophytic species of *Microsporum*. Hindustan Antibiot. Bull. 45 (1-4), 1–4.

Ahmad, I., Beg, A.Z., 2001. Antimicrobial and phytochemical studies on 45 Indian medicinal plants against multi-drug resistant human pathogens. J. Ethnopharmacol. 74 (2), 113–123.

Ahmed Abdel-Reheim, M., Messiha, B.A.S., Abo-Saif, A.A., 2017. *Quillaja saponaria* bark saponin protects Wistar rats against ferrous sulphate-induced oxidative and inflammatory liver damage. Pharm. Biol. 55 (1), 1972–1983.

Ahmed-Belkacem, A., Macalou, S., Borrelli, F., Capasso, R., Fattorusso, E., Taglialatela-Scafati, O., et al., 2007. Nonprenylated rotenoids, a new class of potent breast cancer resistance protein inhibitors. J. Med. Chem. 50 (8), 1933–1938.

Ahn, J.Y., Song, J.Y., Yun, Y.S., Jeong, G., Choi, I.S., 2005. Protection of *Staphylococcus aureus*-infected septic mice by suppression of early acute inflammation and enhanced antimicrobial activity by ginsan. FEMS Immunol. Med. Microbiol. 46 (2), 187–197.

Ahn, J., Lee, H., Jang, J., Kim, S., Ha, T., 2013. Anti-obesity effects of glabridin-rich supercritical carbon dioxide extract of licorice in high-fat-fed obese mice. Food Chem. Toxicol. 51, 439–445.

Ahn, J., Chae, H.S., Chin, Y.W., Kim, J., 2017. Dereplication-guided isolation of new phenylpropanoid-substituted diglycosides from *Cistanche salsa* and their inhibitory activity on NO production in macrophage. Molecules 22 (7), 1138.

Akihisa, T., Yokokawa, S., Ogihara, E., Matsumoto, M., Zhang, J., Kikuchi, T., et al., 2017. Melanogenesis-inhibitory and cytotoxic activities of limonoids, alkaloids, and phenolic compounds from *Phellodendron amurense* bark. Chem. Biodivers. 14 (7), e1700105.

Akowuah, G.A., Zhari, I., Mariam, A., Yam, M.F., 2009. Absorption of andrographolides from *Andrographis paniculata* and its effect on CCl 4-induced oxidative stress in rats. Food Chem. Toxicol. 47 (9), 2321–2326.

Alamgir, M., Uddin, S.J., 2010. Recent advances on the ethnomedicinal plants as immunomodulatory agents. In: Chattopadhyay, D. (Ed.), Ethnomedicine: A Source of Complementary Therapeutics. Research Signpost, India, pp. 227–244.

Allison, A.C., Cacabelos, R., Lombardi, V.R., Álvarez, X.A., Vigo, C., 2001. Celastrol, a potent antioxidant and anti-inflammatory drug, as a possible treatment for Alzheimer's disease. Progr. Neuro-Psychopharmacol. Biol. Psychiatry 25 (7), 1341–1357.

Alsuhaibani, S., Khan, M.A., 2017. Immune-stimulatory and therapeutic activity of *Tinospora cordifolia*: double-edged sword against salmonellosis. J. Immunol. Res. 1787803–1787811.

Alzohairy, M.A., 2016. Therapeutics role of *Azadirachta indica* (Neem) and their active constituents in diseases prevention and treatment. Evid. Based Complement. Altern. Med. 2016, 7382506–7382516.

Ansari, J.A., Inamdar, N.N., 2010. The promise of traditional medicines. Int. J. Pharmacol. 6 (6), 808–812.

Arora, D.S., Sood, H., 2017. In vitro antimicrobial potential of extracts and phytoconstituents from *Gymnema sylvestre* R. Br. leaves and their biosafety evaluation. AMB Exp. 7 (1), 115.

Arora, P., Kaushik, D., 2016. Therapeutic potential of *Benincasa cerifera*: a review. Chin. J. Integr. Med. 24, 1–14.

Asero, R., Mistrello, G., Roncarolo, D., Antoniottib, P.L., Falagiani, P., 1998. A case of garlic allergy. J. Allergy Clin. Immunol. 101 (3), 427–428.

Auddy, B., Ferreira, M., Blasina, F., Lafon, L., Arredondo, F., Dajas, F., et al., 2003. Screening of antioxidant activity of three Indian medicinal plants, traditionally used for the management of neurodegenerative diseases. J. Ethnopharmacol. 84 (2), 131–138.

Auttachoat, W., Chitsomboon, B., Peachee, V.L., Guo, T.L., White Jr, K.L., 2004. Immunomodulation by Dok Din Daeng (*Aeginetia indica* Roxb.) extracts in female B6C3F1 mice: (I): stimulation of T cells. Int. Immunopharmacol. 4 (10-11), 1367–1379.

Awasthi, V., Chauhan, R., Chattopadhyay, D., Das, J., 2017. Effect of L-arginine on the growth of *Plasmodium falciparum* and immune modulation of host cells. J. Vector Borne Dis. 54, 139–145.

Azevedo, B.C., Morel, L.J.F., Carmona, F., Cunha, T.M., Contini, S.H.T., Delprete, P.G., et al., 2018. Aqueous extracts from *Uncaria tomentosa* (Willd. ex Schult.) DC reduces bronchial hyper-responsiveness and inflammation in a murine model of asthma. J. Ethnopharmacol. 218, 76–89.

Babisch, W., 2003. Stress hormones in the research on cardiovascular effects of noise. Noise Health 5 (18), 1.

Bag, P., Ojha, D., Mukherjee, H., Halder, U.C., Mondal, S., Nandi, S., et al., 2013. An indole alkaloid from a Tribal folklore inhibits immediate early event in HSV-2 infected cells with therapeutic efficacy in vaginally infected mice. PLoS One 8 (10), e77937.

Bag, P., Ojha, D., Mukherjee, H., Halder, U.C., Mondal, S., Biswas, A., et al., 2014. A dihydro-pyrido-indole potently inhibits HSV-1 infection by interfering with the viral immediate early transcriptional events. Antivir. Res. 105, 126–134.

Bains, J.S., Dhuna, V., Singh, J., Kamboj, S.S., Nijjar, K.K., Agrewala, J.N., 2005. Novel lectins from rhizomes of two *Acorus* species with mitogenic activity and inhibitory potential towards murine cancer cell lines. Int. Immunopharmacol. 5 (9), 1470–1478.

Balasubramani, S.P., Venkatasubramanian, P., Kukkupuni, S.K., Patwardhan, B., 2011. Plant-based rasayana drugs from Ayurveda. Chin. J. Integr. Med. 17 (2), 88–94.

Bano, N., Ahmed, A., Tanveer, M., Khan, G.M., Ansari, M.T., 2017. Pharmacological evaluation of *Ocimum sanctum*. J. Bioequiv. Bioavailab. 9, 387–392.

Baral, R., Chattopadhyay, U., 2004. Neem (*Azadirachta indica*) leaf mediated immune activation causes prophylactic growth inhibition of murine Ehrlich carcinoma and B16 melanoma. Int. Immunopharmacol. 4 (3), 355–366.

Belska, N.V., Guriev, A.M., Danilets, M.G., Trophimova, E.S., Uchasova, E.G., Ligatcheva, A.A., et al., 2006. Enhancement of immune responses to neem leaf extracts (*Azadirachta indica*) correlates with antineoplastic activity in BALB/c-mice. In Vivo 20 (2), 247–251.

Belska, N.V., Guriev, A.M., Danilets, M.G., Trophimova, E.S., Uchasova, E.G., Ligatcheva, A.A., et al., 2010. Water-soluble polysaccharide obtained from *Acorus calamus* L. classically activates macrophages and stimulates Th1 response. Int. Immunopharmacol. 10 (8), 933–942.

Benarba, B., Meddah, B., Aoues, A., 2012. *Bryonia dioica* aqueous extract induces apoptosis through mitochondrial intrinsic pathway in BL41 Burkitt's lymphoma cells. J. Ethnopharmacol. 141 (1), 510–516.

Bhandari, P.R., 2012. Curry leaf (*Murraya koenigii*) or cure leaf: review of its curative properties. J. Med. Nutr. Nutraceuticals 1 (2), 92.

Bhanwase, A.S., Alagawadi, K.R., 2016. Antioxidant and immunomodulatory activity of hydroalcoholic extract and its fractions of leave of *Ficus benghalensis* Linn. Pharmacognosy Res. 8 (1), 50–55.

Bharani, A., Ganguli, A., Mathur, L.K., Jamra, Y., Raman, P.G., 2002. Efficacy of *Terminalia arjuna* in chronic stable angina: a double-blind, placebo-controlled, crossover study comparing *Terminalia arjuna* with isosorbide mononitrate. Indian Heart J. 54 (2), 170–175.

Bhattacharjee, S., 2008. Handbook of Medicinal Plants. Pointer Publishers, Jaipur, Raj, India, p. 118.

Bhattacharya, S.K., Muruganandam, A.V., 2003. Adaptogenic activity of *Withania somnifera*: an experimental study using a rat model of chronic stress. Pharmacol. Biochem. Behav. 75 (3), 547–555.

Bhaumik, S., Jyothi, M.D., Khar, A., 2000. Differential modulation of nitric oxide production by curcumin in host macrophages and NK cells. FEBS Lett. 483 (1), 78–82.

Bhowmik, D., Sampath Kumar, K.P., Paswan, S., Srivatava, S., Yadav, A., Dutta, A., 2012. Traditional Indian herbs *Convolvulus pluricaulis* and its medicinal importance. J. Pharmacognosy Phytochem. 1 (1), 44–51.

Bhowmik, A., Ojha, D., Das, R., Chandra, N.S., Chakraborty, S., Chakraborty, A., et al., 2017. Inositol hexa-phosphoric acid, a nutraceutical, attenuates iron-induced oxidative stress and alleviates liver injury in iron overloaded Mice. Biomed. Pharmacother. 87, 443–450.

Billiau, A., Matthys, P., 2001. Modes of action of Freund's adjuvants in experimental models of autoimmune diseases. J. Leukoc. Biol. 70 (6), 849–860.

Bobbarala, V., Rao, P.K., Rao, G.S., Aryamithra, D., 2009. Bioactivity of *Andrographis paniculata* against selected phytopathogens. J. Pharm. Res. 2 (3), 480–482.

Bos, M., Vennat, B., Meunier, M., Pouget, M., Pourrat, A., Fialip, J., 1996. Procyanidins from tormentil: antioxidant properties towards lipoperoxidation and anti-elastase activity. Biol. Pharm. Bull. 19 (1), 146–148.

Bose, A., Baral, R., 2007. Natural killer cell mediated cytotoxicity of tumor cells initiated by neem leaf preparation is associated with CD40–CD40L–mediated endogenous production of interleukin-12. Hum. Immunol. 68 (10), 823–831.

Bose, A., Chakraborty, K., Sarkar, K., Goswami, S., Chakraborty, T., Pal, S., et al., 2009. Neem leaf glycoprotein induces perforin-mediated tumor cell killing by T and NK cells through differential regulation of IFNγ signaling. J. Immunother. 32 (1), 42–53.

Burnett, B.P., Jia, Q., Zhao, Y., Levy, R.M., 2007. A medicinal extract of *Scutellaria baicalensis* and *Acacia catechu* acts as a dual inhibitor of cyclooxygenase and 5-lipoxygenase to reduce inflammation. J. Med. Food 10 (3), 442–451.

Calabrese, C., Berman, S.H., Babish, J.G., Ma, X., Shinto, L., Dorr, M., et al., 2000. A phase I trial of andrographolide in HIV positive patients and normal volunteers. Phytother. Res. 14 (5), 333–338.

Cebra, J.J., Periwal, S.B., Lee, G., Lee, F., Shroff, K.E., 1998. Development and maintenance of the gut-associated lymphoid tissue (GALT): the roles of enteric bacteria and viruses. J. Immunol. Res. 6 (1-2), 13–18.

Chakraborty, K., Bose, A., Pal, S., Sarkar, K., Goswami, S., Ghosh, D., et al., 2008. Neem leaf glycoprotein restores the impaired chemotactic activity of peripheral blood mononuclear cells from head and neck squamous cell carcinoma patients by maintaining CXCR3/CXCL10 balance. Int. Immunopharmacol. 8 (2), 330–340.

Chandrasekaran, C.V., Kannan Sundarajan, J.R.E., Gururaja, G.M., Mundkinajeddu, D., Agarwal, A., 2013. Immune stimulatory and anti-inflammatory activities of *Curcuma longa* extract and its polysaccharide fraction. Pharmacognosy Res. 5 (2), 71–79.

Chandrashekar, P.M., Prashanth, K.V.H., Venkatesh, Y.P., 2011. Isolation, structural elucidation and immunomodulatory activity of fructans from aged garlic extract. Phytochemistry 72 (2), 255–264.

Chang, S.S., Rasmussen, A.F., 1965. Stress-induced suppression of interferon production in virus-infected mice. Nature 205 (4971), 623–624.

Chang, R., Li, Y., Yang, X., Yue, Y., Dou, L., Wang, Y., et al., 2013. Protective role of deoxyschizandrin and schisantherin A against myocardial ischemia–reperfusion injury in rats. PLoS One 8 (4), e61590.

Chao, W.W., Lin, B.F., 2010. Isolation and identification of bioactive compounds in *Andrographis paniculata* (Chuanxinlian). Chin. Med. 5 (1), 17.

Chattopadhyay, D., Eddouks, M., 2012. Cellular nutrition and nutritional medicine in diabetes and related complications: an overview. In: Eddouks, M., Chattopadhya, D. (Eds.), Phytotherapy in Diabetes and Hypertension. Bentham SciencePublication, USA, 978-1-60805-567-8pp. 03–59, Chapter 1.

Chattopadhyay, D., Arunachalam, G., Mandal, A.B., Bhadra, R., Mandal, S.C., 2003. CNS activity of *Mallotus peltatus* Muell Arg. Leaf extracts: an ethnomedicine of bay islands. J. Ethnopharmacol. 85 (1), 99–105.

Chattopadhyay, D., Arunachalam, G., Bhadra, R., Ghosh, L., Mandal, A.B., 2004. CNS activity of the leaf extracts of *Alstonia macrophylla* Wall ex A. DC: an ethnomedicine of bay islands. Fitoterapia 75 (7-8), 673–682.

Chattopadhyay, D., Mukherjee, H., Bag, P., Ojha, D., Konreddy, A.K., Dutta, S., et al., 2012. Inhibition of NO_2, PGE2, TNF-α and *i*NOS expression by *Shorea robusta* L: an ethnomedicine used for antiinflammatory and analgesic activity. Evid. Based Complement. Altern. Med. 2012, 254849–254862.

Chen, J., Pang, W., Kan, Y., Zhao, L., He, Z., Shi, W., et al., 2018a. Structure of a pectic polysaccharide from *Pseudostellaria heterophylla* and stimulating insulin secretion of ins-1 cell and distributing in rats by oral. Int. J. Biol. Macromol. 106, 456–463.

Chen, X.Y., Zhang, T., Wang, X., Hamann, M.T., Kang, J., Yu, D.Q., et al., 2018b. A Chemical investigation of the leaves of *Morus alba* L. Molecules (Basel, Switzerland) 23 (5), E1018.

Chi, X., Bi, S., Xu, W., Zhang, Y., Liang, S., Hu, S., 2017. Oral administration of tea saponins to relive oxidative stress and immune suppression in chickens. Poult. Sci. 96 (9), 3058–3067.

Chiang, H.C., Lo, Y.J., Lu, F.J., 1994. Xanthine oxidase inhibitors from the leaves of *Alsophila spinulosa* (Hook) Tryon. J. Enzyme Inhib. 8 (1), 61–71.

Chiang, L.C., Ng, L.T., Chiang, W., Chang, M.Y., Lin, C.C., 2003. Immunomodulatory activities of flavonoids, monoterpenoids, triterpenoids, iridoid glycosides and phenolic compounds of *Plantago* species. Planta Med. 69 (07), 600–604.

Choi, H.S., Kim, K.H., Sohn, E., Park, J.D., Kim, B.O., Moon, E.Y., et al., 2008. Red ginseng acidic polysaccharide (RGAP) in combination with IFN-γ results in enhanced macrophage function through activation of the NF-κB pathway. Biosci. Biotechnol. Biochem. 72 (7), 1817–1825.

Choi, E.J., Kwon, H.C., Sohn, Y.C., Nam, C.W., Park, H.B., Kim, C.Y., et al., 2009. Four flavonoids from *Echinosophora koreensis* and their effects on alcohol metabolizing enzymes. Arch. Pharm. Res. 32 (6), 851–855.

Choi, Y.Y., Kim, M.H., Ahn, K.S., Um, J.Y., Lee, S.G., Yang, W.M., 2017. Immunomodulatory effects of *Pseudostellaria heterophylla* (Miquel) Pax on regulation of Th1/Th2 levels in mice with atopic dermatitis. Mol. Med. Rep. 15 (2), 649–656.

Chopra, R., Nayar, S., Chopra, I., 2002. Glossary of Indian Medicinal Plants. National Institute of Science Communication and Information Resources, New Delhi.

Chua, T., Eise, N.T., Simpson, J.S., Ventura, S., 2014. Pharmacological characterization and chemical fractionation of a liposterolic extract of saw palmetto (*Serenoa repens*): effects on rat prostate contractility. J. Ethnopharmacol. 152 (2), 283–291.

Chulet, R., Pradhan, P., 2009. A review on rasayana. Pharmacognosy Rev. 3 (6), 229–234.

Cipriani, B., Borsellino, G., Knowles, H., Tramonti, D., Cavaliere, F., Bernardi, G., et al., 2001. Curcumin inhibits activation of Vγ9Vδ2 T cells by phosphoantigens and induces apoptosis involving apoptosis-inducing factor and large scale DNA fragmentation. J. Immunol. 167 (6), 3454–3462.

Clement, F., Pramod, S.N., Venkatesh, Y.P., 2010. Identity of the immunomodulatory proteins from garlic (*Allium sativum*) with the major garlic lectins or agglutinins. Int. Immunopharmacol. 10 (3), 316–324.

Clifford, L.J., Nair, M.G., Rana, J., Dewitt, D.L., 2002. Bioactivity of alkamides isolated from *Echinacea purpurea* (L.) Moench. Phytomedicine 9 (3), 249–253.

Coico, R., Sunshine, G., Benjamini, E., 2015. Immunology. Wiley-Liss, New Jersey, p. 201.

Čolić, M., Došlov-Kokoruš, Z., Backović, A., Stojanović, D., Antić-Stanković, J., Kovačević, N., 2007. Evaluation of the stimulatory effect of *Epimedium alpinum* L. methanolic extract on the immune response *in vivo*. Die Pharmazie 62 (9), 705–708.

Colombo, M.L., Bosisio, E., 1996. Pharmacological activities of *Chelidonium majus* L. (Papaveraceae). Pharmacol. Res. 33 (2), 127–134.

Coppi, A., Cabinian, M., Mirelman, D., Sinnis, P., 2006. Antimalarial activity of allicin, a biologically active compound from garlic cloves. Antimicrobial. Agents Chemother. 50 (5), 1731–1737.

Cronin, E., 1987. Dermatitis of the hands in caterers. Contact Dermatitis 17 (5), 265–269.

D'Agostino, M., Biagi, C., Feo, V.D., Zollo, F., Pizza, C., 1990. Flavonoids of *Combretum micranthum*. Fitoterapia 61 (5), 477.

Dahanukar, S.A., Thatte, U.M., 1997. Current status of Ayurveda in phytomedicine. Phytomedicine 4 (4), 359–368.

Das, R., Raman, R.P., Saha, H., Singh, R., 2015. Effect of *Ocimum sanctum* Linn (Tulsi) extract on the immunity and survival of *Labeo rohita* (Hamilton) infected with *Aeromonas hydrophila*. Aquacult. Res. 46 (5), 1111–1121.

Dash, P., Yadav, S., Sahoo, P.K., 2014. Immunoadjuvant effect of garlic (*Allium sativum*)–mineral oil suspension on immunity and resistance to *Aeromonas hydrophila* infection in rohu, *Labeo rohita*. Int. Aquat. Res. 6 (3), 167–173.

Davis, L., Kuttan, G., 2000. Immunomodulatory activity of *Withania somnifera*. J. Ethnopharmacol. 71 (1), 193–200.

de Oliveira Melo, M.N., Oliveira, A.P., Wiecikowski, A.F., Carvalho, R.S., de Lima Castro, J., de Oliveira, F.A.G., et al., 2018. Phenolic compounds from *Viscum album* tinctures enhanced antitumor activity in melanoma murine cancer cells. Saudi Pharm. J. 26 (3), 311–322.

De Souza, K.C.B., Bassani, V.L., Schapoval, E.E.S., 2007. Influence of excipients and technological process on anti-inflammatory activity of quercetin and *Achyrocline satureioides* (Lam.) DC extracts by oral route. Phytomedicine 14 (2-3), 102–108.

Dharini, S., Gajalakshmi, G., Devi, R.S., 2012. Effect of *Acorus calamus* Linn and its active principle on noise stress-induced changes on immune system. Int. J. Biomed. Adv. Res. 3, 494–501.

Ding, X., Chi, J., Yang, X., Hao, J., Liu, C., Zhu, C., et al., 2017. Cucurbitacin B synergistically enhances the apoptosis-inducing effect of arsenic trioxide by inhibiting STAT3 phosphorylation in lymphoma Ramos cells. Leuk. Lymphoma 58 (10), 2439–2451.

Direkbusarakom, S., Herunsalee, A., Yoshimizu, M., Ezura, Y., 1996. Antiviral activity of several Thai traditional herbs extracts against fish pathogenic viruses. Fish Pathol. 31 (4), 209–213.

Dobrucka, R., Dlugaszewska, J., Kaczmarek, M., 2018. Cytotoxic and antimicrobial effects of biosynthesized ZnO nanoparticles using of *Chelidonium majus* extract. Biomed. Microdevices 20 (1), 5.

do Rosário, M.M.T., Kangussu-Marcolino, M.M., do Amaral, A.E., Noleto, G.R., de Oliveira Petkowicz, C.L., 2011. Storage xyloglucans: potent macrophages activators. Chem. Biol. Interact. 189 (1), 127–133.

Dorsch, W., Stuppner, H., Wagner, H., Gropp, M., Demoulin, S., Ring, J., 1991. Antiasthmatic effects of *Picrorhiza kurroa*: androsin prevents allergen-and PAF-induced bronchial obstruction in guinea pigs. Int. Arch. Allergy Immunol. 95 (2-3), 128–133.

Dos Santos, R.G., Valle, M., Bouso, J.C., Nomdedéu, J.F., Rodríguez-Espinosa, J., McIlhenny, E.H., et al., 2011. Autonomic, neuroendocrine, and immunological effects of ayahuasca: a comparative study with d-amphetamine. J. Clin. Psychopharmacol. 31 (6), 717–726.

Dos Santos, R.G., Grasa, E., Valle, M., Ballester, M.R., Bouso, J.C., Nomdedéu, J.F., et al., 2012. Pharmacology of ayahuasca administered in two repeated doses. Psychopharmacology 219 (4), 1039–1053.

Doshi, G.M., Une, H.D., Shanbhag, P.P., 2013. Rasayans and non-rasayans herbs: future immunodrug–targets. Pharmacognosy Rev. 7 (14), 92.

Duan, W.J., Yang, J.Y., Chen, L.X., Zhang, L.J., Jiang, Z.H., Cai, X.D., et al., 2009. Monoterpenes from *Paeonia albiflora* and their inhibitory activity on nitric oxide production by lipopolysaccharide-activated microglia. J. Nat. Prod. 72 (9), 1579–1584.

Dwivedi, V.P., Bhattacharya, D., Yadav, V., Singh, D.K., Kumar, S., Singh, M., et al., 2017. The phytochemical Bergenin enhances T helper-1 responses and anti-Mycobacterial immunity by activating the MAP kinase pathway in macrophages. Front. Cell. Infect. Microbiol. 7, 149–158.

Ebrahimi, M., Hassan, Z.M., Mostafaie, A., Mehrjardi, N.Z., Ghazanfari, T., 2013. Purified protein fraction of garlic extract modulates cellular immune response against breast transplanted tumors in BALB/c mice model. Cell J. (Yakhteh) 15 (1), 65–74.

Ellis, L.Z., Liu, W., Luo, Y., Okamoto, M., Qu, D., Dunn, J.H., et al., 2011. Green tea polyphenol epigallocatechin-3-gallate suppresses melanoma growth by inhibiting inflammasome and IL-1β secretion. Biochem. Biophys. Res. Commun. 414 (3), 551–556.

Estko, M., Baumgartner, S., Urech, K., Kunz, M., Regueiro, U., Heusser, P., et al., 2015. Tumour cell derived effects on monocyte/macrophage polarization and function and modulatory potential of *Viscum album* lipophilic extract in vitro. BMC Complement. Altern. Med. 15 (1), 130.

Farias, I., do Carmo Araújo, M., Zimmermann, E.S., Dalmora, S.L., Benedetti, A.L., Alvarez-Silva, M., et al., 2011. *Uncaria tomentosa* stimulates the proliferation of myeloid progenitor cells. J. Ethnopharmacol. 137 (1), 856–863.

Fecka, I., Kucharska, A.Z., Kowalczyk, A., 2015. Quantification of tannins and related polyphenols in commercial products of tormentil (*Potentilla tormentilla*). Phytochem. Anal. 26 (5), 353–366.

Feng, Y., Zhu, X., Wang, Q., Jiang, Y., Shang, H., Cui, L., et al., 2012. Allicin enhances host pro-inflammatory immune responses and protects against acute murine malaria infection. Malar. J. 11 (1), 268.
Friedl, R., Moeslinger, T., Kopp, B., Spieckermann, P.G., 2001. Stimulation of nitric oxide synthesis by the aqueous extract of *Panax ginseng* root in RAW 264.7 cells. Br. J. Pharmacol. 134 (8), 1663–1670.
Fu, R., Zhang, Y.T., Guo, Y.R., Huang, Q.L., Peng, T., Xu, Y., et al., 2013. Antioxidant and anti-inflammatory activities of the phenolic extracts of *Sapium sebiferum* (L.) Roxb. leaves. J. Ehnopharmacol. 147 (2), 517–524.
Fu, E., Tsai, M.C., Chin, Y.T., Tu, H.P., Fu, M.M., Chiang, C.Y., et al., 2015a. The effects of diallyl sulfide upon *Porphyromonas gingivalis* lipopolysaccharide stimulated proinflammatory cytokine expressions and nuclear factor-kappa B activation in human gingival fibroblasts. J. Periodontal Res. 50 (3), 380–388.
Fu, R., Zhang, Y., Peng, T., Guo, Y., Chen, F., 2015b. Phenolic composition and effects on allergic contact dermatitis of phenolic extracts *Sapium sebiferum* (L.) Roxb. Leaves. J. Ethnopharmacol. 162, 176–180.
Gabhe, S.Y., Tatke, P.A., Khan, T.A., 2006. Evaluation of the immunomodulatory activity of the methanol extracts of *Ficus benghalensis* roots in rats. Indian J. Pharmacol. 38 (4), 271–275.
Gacche, R.N., Dhole, N.A., 2006. Antioxidant and possible anti-inflammatory potential of selected medicinal plants prescribed in the Indian traditional system of medicine. Pharm. Biol. 44 (5), 389–395.
Gan, L., Li, X., Zhu, M., Chen, C., Luo, H., Zhou, Q., 2018. Acteoside relieves mesangial cell injury by regulating Th22 cell chemotaxis and proliferation in IgA nephropathy. Ren. Fail. 40 (1), 364–370.
Gazzinelli, R.T., Oswald, I.P., James, S.L., Sher, A., 1992. IL-10 inhibits parasite killing and nitrogen oxide production by IFN-gamma-activated macrophages. J. Immunol. 148 (6), 1792–1796.
Ge, F., Zhu, S., Liu, L., Yan, J., Ji, Y., Sun, Z., 2017. Anti-inflammatory effects of *Fagopyrum cymosum* administered as a potential drug for ulcerative colitis. Exp. Ther. Med. 14 (5), 4745–4754.
Geber, W.F., Anderson, T.A., Van Dyne, B., 1966. Physiologic responses of the albino rat to chronic noise stress. Arch. Environ. Health 12 (6), 751–754.
Geng, Y., Li, C., Liu, J., Xing, G., Zhou, L., Dong, M., et al., 2010. Beta-asarone improves cognitive function by suppressing neuronal apoptosis in the beta-amyloid hippocampus injection rats. Biol. Pharm. Bull. 33 (5), 836–843.
Gharavi, M.J., Nobakht, M., Khademvatan, S.H., Bandani, E., Bakhshayesh, M., Roozbehani, M., 2011. The effect of garlic extract on expression of INFγ and inos genes in macrophages infected with *Leishmania major*. Iranian J. Parasitol. 6 (3), 74.
Ghate, N.B., Hazra, B., Sarkar, R., Mandal, N., 2014. Heartwood extract of *Acacia catechu* induces apoptosis in human breast carcinoma by altering bax/bcl-2 ratio. Pharmacognosy Mag. 10 (37), 27–33.
Ghosh, M., 2006. Antifungal properties of haem peroxidase from *Acorus calamus*. Ann. Bot. 98 (6), 1145–1153.
Gibson, G.R., Probert, H.M., Van Loo, J., Rastall, R.A., Roberfroid, M.B., 2004. Dietary modulation of the human colonic microbiota: updating the concept of prebiotics. Nutr. Res. Rev. 17 (2), 259–275.
Glaister, J.R., 1973. Factors affecting the lymphoid cells in the small intestinal epithelium of the mouse. Int. Arch. Allergy Immunol. 45 (5), 719–730.
Goel, G., Makkar, H.P., Francis, G., Becker, K., 2007. Phorbol esters: structure, biological activity, and toxicity in animals. Int. J. Toxicol. 26 (4), 279–288.

Goel, A., Kumar, S., Bhatia, A.K., 2010. Effect of *Ocimum sanctum* on the development of protective immunity against *Salmonella typhimurium* infection through cytokines. Asia Pac. J. Trop. Med. 3 (9), 682–686.

Gorinstein, S., Drzewiecki, J., Leontowicz, H., Leontowicz, M., Najman, K., Jastrzebski, Z., et al., 2005. Comparison of the bioactive compounds and antioxidant potentials of fresh and cooked Polish, Ukrainian, and Israeli garlic. J. Agric. Food Chem. 53 (7), 2726–2732.

Gracza, L., 1983. Über die Wirkstoffe von Asarum europaeum. Planta Med. 48 (07), 153–157.

Gu, D., Huang, L., Chen, X., Wu, Q., Ding, K., 2018. Structural characterization of a galactan from *Ophiopogon japonicus* and anti-pancreatic cancer activity of its acetylated derivative. Int. J. Biol. Macromol. 113, 907–915.

Guleria, S., Tiku, A.K., Singh, G., Vyas, D., Bhardwaj, A., 2011. Antioxidant activity and protective effect against plasmid DNA strand scission of leaf, bark, and heartwood extracts from *Acacia catechu*. J. Food Sci. 76 (7), C959–C964.

Gupta, S.K., Prakash, J., Srivastava, S., 2002. Validation of claim of Tulsi, *Ocimum sanctum* Linn as a medicinal plant. Indian J. Exp. Biol. 40, 765–773.

Gupta, S., George, M., Singhal, M., Sharma, G.N., Garg, V., 2010. Leaves extract of *Murraya Koenigii* linn for anti-inflammatory and analgesic activity in animal models. J. Adv. Pharm. Technol. Res. 1 (1), 68–77.

Gutierrez, R.M.P., Hoyo-Vadillo, C., 2017. Anti-inflammatory potential of *Petiveria alliacea* on activated RAW264. 7 murine macrophages. Pharmacognosy Mag. 13 (Suppl 2), S174.

Hafeez, B.B., Fischer, J.W., Singh, A., Zhong, W., Mustafa, A., Meske, L., et al., 2015. Plumbagin inhibits prostate carcinogenesis in intact and castrated PTEN knockout mice via targeting PKCε, Stat3, and epithelial-to-mesenchymal transition markers. Cancer Prev. Res. 8 (5), 375–386.

Halder, S., Bharal, N., Mediratta, P.K., Kaur, I., Sharma, K.K., 2009. Anti-inflammatory, immunomodulatory and antinociceptive activity of *Terminalia arjuna* Roxb bark powder in mice and rats. Indian J. Exp. Biol. 47, 577–583.

Han, M.H., Yang, X.W., Zhong, G.Y., Zhang, M., 2007. Bioactive constituents inhibiting TNF-alpha production in fresh rhizome of *Pinellia ternata*. Zhongguo Zhong yao za zhi Zhongguo zhongyao zazhi China Journal of Chinese Materia Medica 32 (17), 1755–1759.

Han, R., Rostami Yazdi, M., Gerdes, S., Mrowietz, U., 2012. Triptolide in the treatment of psoriasis and other immune mediated inflammatory diseases. Br. J. Clin. Pharm. 74 (3), 424–436.

Handayani, S., Susidarti, R.A., Jenie, R.I., Meiyanto, E., 2017. Two active compounds from *Caesalpinia sappan* L. in combination with cisplatin synergistically induce apoptosis and cell cycle arrest on WiDr cells. Adv. Pharm. Bull. 7 (3), 375–380.

He, X.G., Lin, L.Z., Lian, L.Z., Lindenmaier, M., 1998. Liquid chromatography–electrospray mass spectrometric analysis of curcuminoids and sesquiterpenoids in turmeric (*Curcuma longa*). J. Chromatogr. A 818 (1), 127–132.

He, X., Wang, X., Fang, J., Chang, Y., Ning, N., Guo, H., et al., 2017. The genus *Achyranthes*: a review on traditional uses, phytochemistry, and pharmacological activities. J. Ethnopharmacol. 203, 260–278.

Herz, W., 2001. Chemistry of the eupatoriinae. Biochem. Syst. Ecol. 29 (11), 1115–1137.

Ho, C., Choi, E.J., Yoo, G.S., Kim, K.M., Ryu, S.Y., 1998. Desacetylmatricarin, an anti-allergic component from *Taraxacum platycarpum*. Planta Med. 64 (06), 577–578.

Ho, C.Y., Weng, C.J., Jhang, J.J., Cheng, Y.T., Huang, S.M., Yen, G.C., 2014. Diallyl sulfide as a potential dietary agent to reduce TNFα and histamine induced proinflammatory responses in A7r5 cells. Mol. Nutr. Food Res. 58 (5), 1069–1078.

Hoang, M.L., Chen, C.H., Chen, P.C., Roberts, N.J., Dickman, K.G., Yun, B.H., et al., 2016. Aristolochic acid in the etiology of renal cell carcinoma. Cancer Epidemiol. Prev. Biomarkers 25 (12), 1600–1608.

Hodge, G., Hodge, S., Han, P., 2002. *Allium sativum* (garlic) suppresses leukocyte inflammatory cytokine production in vitro: potential therapeutic use in the treatment of inflammatory bowel disease. Cytometry A 48 (4), 209–215.

Hong, H., An, J.C., de La Cruz, J.F., Hwang, S.G., 2017. *Cnidium officinale* Makino extract induces apoptosis through activation of caspase-3 and p53 in human liver cancer HepG2 cells. Exp. Ther. Med. 14 (4), 3191–3197.

Hong, M., Cai, Z., Song, L., Liu, Y., Wang, Q., Feng, X., 2018. *Gynostemma pentaphyllum* attenuates the progression of nonalcoholic fatty liver disease in mice: a biomedical investigation integrated with in silico assay. Evid. Based Complement. Altern. Med. 2018, 8384631–8384643.

Hostettmann, K., Hamburger, M., Hostettmann, M., Marston, A., 1991. New developments in the separation of natural products. Modern Phytochem. Methods 25, 1–32.

Huang, S.X., Zhou, Y., Nie, Q.J., Ding, L.S., Peng, S.L., 2006. Two new iridoid glucosides from *Picrorhiza scrophulariiflora*. J. Asian Nat. Prod. Res. 8 (3), 259–263.

Huang, L., Zhao, H., Huang, B., Zheng, C., Peng, W., Qin, L., 2011. *Acanthopanax senticosus*: review of botany, chemistry and pharmacology. Die Pharmazie 66 (2), 83–97.

Huang, G.C., Lee, C.J., Wang, K.T., Weng, B.C., Chien, T.Y., Tseng, S.H., et al., 2013. Immunomodulatory effects of *Hedysarum polybotrys* extract in mice macrophages, splenocytes and leucopenia. Molecules 18 (12), 14862–14875.

Hwang, H.S., Shim, J.H., 2018. Brazilin and *Caesalpinia sappan* L. extract protect epidermal keratinocytes from oxidative stress by inducing the expression of GPX7. Chin. J. Nat. Med. 16 (3), 203–209.

Ina, K., Kataoka, T., Ando, T., 2013. The use of lentinan for treating gastric cancer. Anticancer Agents Med. Chem. 13 (5), 681–688.

Innocenti, G., Dall'Acqua, S., Sosa, S., Altinier, G., Della Loggia, R., 2005. Topical anti-inflammatory activity of *Solenostemma argel* leaves. J. Ethnopharmacol. 102 (2), 307–310.

Ismail, S., Asad, M., 2009. Immunomodulatory activity of *Acacia catechu*. Indian J. Physiol. Pharmacol. 53, 25–33.

Jafarian, A., Ghannadi, A., Elyasi, A., 2010. The effects of *Allium hirtifolium* Boiss. on cell-mediated immune response in mice. Iranian J. Pharm. Res. 2 (1), 51–55.

Jagetia, G.C., Aggarwal, B.B., 2007. "Spicing up" of the immune system by curcumin. J. Clin. Immunol. 27 (1), 19–35.

Jandú, J.J., Moraes Neto, R.N., Zagmignan, A., de Sousa, E.M., Brelaz-de-Castro, M.C., dos Santos Correia, M.T., et al., 2017. Targeting the immune system with plant lectins to combat microbial infections. Front. Pharmacol. 8, 671.

Jantan, I., Ahmad, W., Bukhari, S.N.A., 2015. Plant-derived immunomodulators: an insight on their preclinical evaluation and clinical trials. Front. Plant Sci. 25 (6), 655.

Jayanti, I., Jalaluddin, M., Avijeeta, A., Ramanna, P.K., Rai, P.M., Nair, R.A., 2018. *In vitro* antimicrobial activity of *Ocimum sanctum* (Tulsi) extracts on *Aggregatibacter actinomycetemcomitans* and *Porphyromonas gingivalis*. J. Contemp. Dent. Pract. 19 (4), 415–419.

Jeba, R.C., Vaidyanathan, R.A.M.A., Rameshkumar, G., 2011. Efficacy of *Ocimum basilicum* for immunomodulatory activity in Wistar albino rat. Int. J. Pharm. Pharm. Sci. 3 (4), 199–203.

Jensen, M.M., 1969. Changes in leukocyte counts associated with vaous stressors. J. Reticuloendothel. Soc. 6 (5), 457.

Jeong, J.Y., Chung, Y.B., Lee, C.C., Park, S.W., Lee, C.K., 1991. Studies on immunopotentiating activities of antitumor polysaccharide from aerial parts of *Taraxacum platycarpum*. Arch. Pharm. Res. 14 (1), 68–72.

Jie, Y.H., Cammisuli, S., Baggiolini, M., 1984. Immunomodulatory effects of Panax ginseng CA Meyer in the mouse. Inflamm. Res. 15 (3), 386–391.

Jo, A.R., Han, H.S., Seo, S., Shin, J.S., Lee, J.Y., Kim, H.J., et al., 2017. Inhibitory effect of moschamine isolated from *Carthamus tinctorius* on LPS-induced inflammatory mediators via AP-1 and STAT1/3 inactivation in RAW 264.7 macrophages. Bioorg. Med. Chem. Lett. 27 (23), 5245–5251.

Joshi, B., Lekhak, S., Sharma, A., 2009. Antibacterial property of different medicinal plants: *Ocimum sanctum*, *Cinnamomum zeylanicum*, *Xanthoxylum armatum* and *Origanum majorana*. Kathmandu Univ. J. Sci. Eng. Technol. 5 (1), 143–150.

Joshi, S., Subedi, Y.P., Paudel, S.K., 2012. Antibacterial and antifungal activity of heartwood of *Acacia catechu* of Nepal. J. Nepal Chem. Soc. 27 (1), 94–99.

Judzentiene, A., Garjonyte, R., Budiene, J., 2016. Variability, toxicity, and antioxidant activity of *Eupatorium cannabinum* (hemp agrimony) essential oils. Pharm. Biol. 54 (6), 945–953.

Jung, M.K., Ha, S., Son, J.A., Song, J.H., Houh, Y., Cho, E., et al., 2012. Polyphenon-60 displays a therapeutic effect on acne by suppression of TLR2 and IL-8 expression via down-regulating the ERK1/2 pathway. Arch. Dermatol. Res. 304 (8), 655–663.

Jung, H.W., Kim, K.H., Park, Y.K., 2017. Inhibitory effect of the extract of *Phellodendron amurense* ruprecht root on collagen-induced arthritis in mice. Chin. J. Integr. Med. 23 (10), 755–762.

Jurkstiene, V., Kondrotas, A.J., Kevelaitis, E., 2007. Immunostimulatory properties of bigroot geranium (*Geranium macrorrhizum* L.) extract. Medicina (Kaunas, Lithuania) 43 (1), 60–64.

Kadri, Y., Nciri, R., Brahmi, N., Saidi, S., Harrath, A.H., Alwasel, S., et al., 2018. Protective effects of *Curcuma longa* against neurobehavioral and neurochemical damage caused by cerium chloride in mice. Environ. Sci. Pollut. Res. 20 (25), 1–11.

Kaleeswaran, B., Ilavenil, S., Ravikumar, S., 2011. Dietary supplementation with *Cynodon dactylon* (L.) enhances innate immunity and disease resistance of Indian major carp, *Catla catla* (Ham.). Fish Shellfish Immunol. 31 (6), 953–962.

Kalia, A., 2011. Textbook of Industrial Pharmacognosy. CBS Publishers & Distributors Pvt, New Delhi.

Kao, S.F., Kuo, H.L., Lee, Y.C., Chiang, H.C., Lin, Y.S., 1994. Immunostimulation by *Alsophila spinulosa* extract fraction VII of both humoral and cellular immune responses. Anticancer Res. 14 (6B), 2439–2443.

Kao, S.H., Hsu, C.H., Su, S.N., Hor, W.T., Chow, L.P., 2004. Identification and immunologic characterization of an allergen, alliin lyase, from garlic (*Allium sativum*). J. Allergy Clin. Immunol. 113 (1), 161–168.

Karkute, S.G., Koley, T.K., Yengkhom, B.K., Tripathi, A., Srivastava, S., Maurya, A., et al., 2018. Anti-diabetic phenolic compounds of black carrot (*Daucus carota* subspecies *sativus* var. *atrorubens* Alef.) inhibit enzymes of glucose metabolism: an *in silico* and *in vitro* validation. Med. Chem. 14 (6), 641–649.

Kartbaeva, E.B., Sakipova, Z.B., Ibragimova, L.N., Kapsalyamova, E.N., Ternynko, I.I., 2015. Compositional study of phenolic compounds of *Cistanche salsa* (CA Mey) G. Beck, growing in the Republic of Kazakhstan. J. Chem. Pharm. Res. 7, 120–122.

Keplinger, K., Laus, G., Wurm, M., Dierich, M.P., Teppner, H., 1998. *Uncaria tomentosa* (Willd.) DC. Ethnomedicinal use and new pharmacological, toxicological and botanical results. J. Ethnopharmacol. 64 (1), 23–34.

Kesari, A.N., Gupta, R.K., Singh, S.K., Diwakar, S., Watal, G., 2006. Hypoglycemic and antihyperglycemic activity of *Aegle marmelos* seed extract in normal and diabetic rats. J. Ethnopharmacol. 107 (3), 374–379.

Khan, N., Mukhtar, H., 2010. Cancer and metastasis: prevention and treatment by green tea. Cancer Metastasis Rev. 29 (3), 435–445.

Khaliq, F., Parveen, A., Singh, S., Gondal, R., Hussain, M.E., Fahim, M., 2013. Improvement in myocardial function by *Terminalia arjuna* in streptozotocin-induced diabetic rats: possible mechanisms. J. Cardiovasc. Pharm. Ther. 18 (5), 481–489.

Khanapur, M., Avadhanula, R.K., Setty, O.H., 2014. *In vitro* antioxidant, antiproliferative, and phytochemical study in different extracts of *Nyctanthes arbortristis* flowers. BioMed Res. Int. 2014, 291271–291279.

Kim, J.Y., Germolec, D.R., Luster, M.I., 1990. Panax ginseng as a potential immunomodulator: studies in mice. Immunopharmacol. Immunotoxicol. 12 (2), 257–276.

Kim, K.H., Lee, Y.S., Jung, I.S., Park, S.Y., Chung, H.Y., Lee, I.R., et al., 1998. Acidic polysaccharide from Panax ginseng, ginsan, induces Th1 cell and macrophage cytokines and generates LAK cells in synergy with rIL-2. Planta Med. 64 (02), 110–115.

Kim, H., Han, T.H., Lee, S.G., 2009a. Anti-inflammatory activity of a water extract of *Acorus calamus* L. leaves on keratinocyte HaCaT cells. J. Ethnopharmacol. 122 (1), 149–156.

Kim, M.H., Byon, Y.Y., Ko, E.J., Song, J.Y., Yun, Y.S., Shin, T., et al., 2009b. Immunomodulatory activity of ginsan, a polysaccharide of *Panax ginseng*, on dendritic cells. Korean J. Physiol. Pharmacol. 13 (3), 169–173.

Kim, S.R., Jung, Y.R., An, H.J., Kim, D.H., Jang, E.J., Choi, Y.J., et al., 2013. Anti-wrinkle and anti-inflammatory effects of active garlic components and the inhibition of MMPs via NF-κB signaling. PLoS One 8 (9), e73877.

Kim, Y.H., Won, Y.S., Yang, X., Kumazoe, M., Yamashita, S., Hara, A., et al., 2016a. Green tea catechin metabolites exert immunoregulatory effects on $CD4^+$ T cell and natural killer cell activities. J. Agric. Food Chem. 64 (18), 3591–3597.

Kim, Y.L., Lee, S.K., Park, K.K., Chung, W.Y., 2016b. The inhibitory effects of *Forsythia koreana* extracts on the metastatic ability of breast cancer cells and bone resorption by osteoclasts. J. Cancer Prev. 21 (2), 88.

Kim, Y.J., Lim, H.S., Kim, B.Y., Seo, C.S., Jeong, S.J., 2017. Quantitative analysis and biological efficacies regarding the neuroprotective and antineuroinflammatory actions of the herbal formula Jodeungsan in HT22 hippocampal cells and BV-2 microglia. Evid. Based Complement. Altern. Med. 2017, 6360836.

Kim, J., Jung, K.H., Yan, H.H., Cheon, M.J., Kang, S., Jin, X., et al., 2018. *Artemisia capillaris* leaves inhibit cell proliferation and induce apoptosis in hepatocellular carcinoma. BMC Complement. Altern. Med. 18 (1), 147.

Kirtikar, K., 1935. Indian Medicinal Plants. Lalit Mohan Basu, Allahabad, pp. 1825–1826.

Kochetkov, N.K., Khorlin, A.J., Vaskovsky, V.E., 1962. The structures of aralosides A and B. Tetrahedron Lett. 3 (16), 713–716.

Kovačević, N., Čolić, M., Backović, A., Došlov-Kokoruš, Z., 2006. Immunomodulatory effects of the methanolic extract of *Epimedium alpinum in vitro*. Fitoterapia 77 (7-8), 561–567.

Kriplani, P., Guarve, K., Baghael, U.S., 2017. *Arnica montana* L: review—a plant of healing. J. Pharm. Pharm. 69 (8), 925–945.

Kumar, S., Suresh, P.K., Vijayababu, M.R., Arunkumar, A., Arunakaran, J., 2006. Anticancer effects of ethanolic neem leaf extract on prostate cancer cell line (PC-3). J. Ethnopharmacol. 105 (1), 246–250.

Kumar, D., Arya, V., Kaur, R., Bhat, Z.A., Gupta, V.K., Kumar, V., 2012. A review of immunomodulators in the Indian traditional health care system. J. Microbiol. Immunol. Infect. 45 (3), 165–184.

Kun-Hua, W., Jian-Hua, M., He-Ping, H., Shan-Lin, G., 2011. Generation of autotetraploid plant of ginger (*Zingiber officinale* Rosc.) and its quality evaluation. Pharmacognosy Mag. 7 (27), 200.

Kyo, E., Uda, N., Suzuki, A., Kakimoto, M., Ushijima, M., Kasuga, S., et al., 1998. Immunomodulation and antitumor activities of aged garlic extract. Phytomedicine 5 (4), 259–267.

Kyo, E., Uda, N., Kasuga, S., Itakura, Y., 2001. Immunomodulatory effects of aged garlic extract. J. Nutr. 131 (3), 1075S–1079S.

Larsen, M.W., Moser, C., HØiby, N., Song, Z., Kharazmi, A., 2004. Ginseng modulates the immune response by induction of interleukin-12 production. Apmis 112 (6), 369–373.

Lasure, A., Van Poel, B., Chimanga, K., Pieters, L., Berge, D.V., Vlientnik, A.J., 1994. Modulation of complement system by flavonoids. Pharm. Pharm. Lett. 4, 32–35.

Latocha, P., 2017. The Nutritional and Health Benefits of Kiwiberry (*Actinidia arguta*)—a review. Plant Foods Hum. Nutr. (Dordrecht, Netherlands) 72 (4), 325.

Lee, J.Y., Kim, C.J., 2010. Arctigenin, a phenylpropanoid dibenzylbutyrolactone lignan, inhibits type I–IV allergic inflammation and pro-inflammatory enzymes. Arch. Pharm. Res. 33 (6), 947–957.

Lee, Y., Jung, J.C., Ali, Z., Khan, I.A., Oh, S., 2012. Anti-inflammatory effect of triterpene saponins isolated from blue cohosh (*Caulophyllum thalictroides*). Evid. Based Complement. Altern. Med. 2012, 798192–798200.

Lee, E., Kim, S.G., Park, N.Y., Park, H.H., Jeong, K.T., Choi, J., et al., 2017a. Anti-inflammatory effects of KOTMIN13: a mixed herbal medicine in LPS-stimulated RAW 264.7 cells and mouse edema models. Pharmacognosy Mag. 13 (50), 216.

Lee, Y.K., Hong, E.Y., Whang, W.K., 2017b. Inhibitory effect of chemical constituents isolated from artemisia iwayomogi on polyol pathway and simultaneous quantification of major bioactive compounds. BioMed Res. Int. 2017, 7375615–7375626.

Leung, K.N., Leung, P.Y., Kong, L.P., Leung, P.K., 2005. Immunomodulatory effects of esculetin (6,7-dihydroxycoumarin) on murine lymphocytes and peritoneal macrophages. Cell. Mol. Immunol. 2 (3), 181–188.

Leyon, P.V., Lini, C.C., Kuttan, G., 2005. Inhibitory effect of *Boerhaavia diffusa* on experimental metastasis by B16F10 melanoma in C57BL/6 mice. Life Sci. 76 (12), 1339–1349.

Li, D.G., Ren, Z.X., 2017. Cordyceps sinensis promotes immune regulation and enhances bacteriostatic activity of PA-824 via IL-10 in *Mycobacterium tuberculosis* disease. Braz. J. Med. Biol. Res. 50 (9), e6188.

Li, J., Li, H., Kadota, S., Namba, T., Miyahara, T., Khan, U.G., 1996. Effects on cultured neonatal mouse calvaria of the flavonoids isolated from *Boerhaavia repen*. J. Nat. Prod. 59 (11), 1015–1018.

Li, X., Wang, H., Liu, C., Chen, R., 2010. Chemical constituents of *Acacia catechu*. Zhongguo Zhong yao za zhi Zhongguo zhongyao zazhi. China Journal of Chinese Materia Medica 35 (11), 1425–1427.

Li, X.C., Yang, L.X., Wang, H.Q., Chen, R.Y., 2011. Phenolic compounds from the aqueous extract of *Acacia catechu*. Chin. Chem. Lett. 22 (11), 1331–1334.

Li, Y.N., Huang, F., Liu, X.L., Shu, S.N., Huang, Y.J., Cheng, H.J., et al., 2013. *Allium sativum* derived allitridin inhibits treg amplification in cytomegalovirus infection. J. Med. Virol. 85 (3), 493–500.

Li, Q., Wang, X., Dai, T., Liu, C., Li, T., McClements, D.J., et al., 2016. Proanthocyanidins, isolated from *Choerospondias axillaris* fruit peels, exhibit potent antioxidant activities *in vitro* and a novel anti-angiogenic property *in vitro* and *in vivo*. J. Agric. Food Chem. 64 (18), 3546–3556.

Liang, H.R., Vuorela, P., Vuorela, H., Hiltunen, R., 1997. Isolation and immunomodulating effect of flavonol glycosides from *Epimedium hunanense*. Planta Med. 63 (04), 316–319.

Lim, D.S., Bae, K.G., Jung, I.S., Kim, C.H., Yun, Y.S., Song, J.Y., 2002. Anti-septicaemic effect of polysaccharide from *Panax ginseng* by macrophage activation. J. Infect. 45 (1), 32–38.

Lima, M.D.R., Lopes, A.P., Martins, C., Brito, G.A., Carneiro, V.C., Goes, P., 2017. The effect of *Calendula officinalis* on oxidative stress and bone loss in experimental periodontitis. Front. Physiol. 8, 440.

Lin, T.P., Chen, S.Y., Duh, P.D., Chang, L.K., Liu, Y.N., 2008. Inhibition of the Epstein–Barr virus lytic cycle by andrographolide. Biol. Pharm. Bull. 31 (11), 2018–2023.

Liou, C.J., Li, M.L., Tseng, J., 2004. Intraperitoneal injection of ginseng extract enhances both immunoglobulin and cytokine production in mice. Am. J. Chin. Med. 32 (01), 75–88.

Liou, C.J., Huang, W.C., Tseng, J., 2006. Short-term oral administration of ginseng extract induces type-1 cytokine production. Immunopharmacol. Immunotoxicol. 28 (2), 227–240.

Liu, Y.L., Frang, Q.N., Zhang, X.C., Feng, X.X., Zhang, L.F., He, X.W., 1983. Chemical constituents of *Fagopyrum cymosum* (Trev) Meisn. Yao xue xue bao Acta Pharmaceutica Sinica 18 (7), 545–547.

Liu, S.Q., Pezzuto, J.M., Kinghorn, A.D., Scheld, H.W., 1988. Additional biologically active constituents of the Chinese tallow tree (*Sapium sebiferum*). J. Nat. Prod. 51 (3), 619–620.

Liu, J., Wang, Z.T., Ge, B.X., 2008. Andrograpanin isolated from *Andrographis paniculata*, exhibits anti-inflammatory property in lipopolysaccharide-induced macrophage cells through down-regulating the p38 MAPKs signaling pathways. Int. Immunopharmacol. 8 (7), 951–958.

Loss-Morais, G., Turchetto-Zolet, A.C., Etges, M., Cagliari, A., Körbes, A.P., Maraschin, F.D.S., et al., 2013. Analysis of castor bean ribosome-inactivating proteins and their gene expression during seed development. Genet. Mol. Biol. 36 (1), 74–86.

Luettig, B., Steinmüller, C., Gifford, G.E., Wagner, H., Lohmann-Matthes, M.L., 1989. Macrophage activation by the polysaccharide arabinogalactan isolated from plant cell cultures of *Echinacea purpurea*. J. Natl. Cancer Inst. 81 (9), 669–675.

Lun, W.U., Zhi-Li, C.H.E.N., Yang, S.U., Qiu-Hong, W.A.N.G., Kuang, H.X., 2015. Cycloartenol triterpenoid saponins from *Cimicifuga simplex* (Ranunculaceae) and their biological effects. Chin. J. Nat. Med. 13 (2), 81–89.

Luna, L.E., 1984. The healing practices of a Peruvian shaman. J. Ethnopharmacol. 11 (2), 123–133.

Lv, J.S., Zhang, L.N., Song, Y.Z., Wang, X.F., Chu, X.Z., 2011. Biological activity exhibited by secondary metabolites of the *Albizia julibrissin* Durazz. pod. Int. Biodeter. Biodegr. 65 (1), 258–264.

Mahajani, S.S., Kulkarni, R.D., 1977. Effect of disodium cromoglycate and *Picrorhiza kurroa* root powder on sensitivity of guinea pigs to histamine and sympathomimetic amines. Int. Arch. Allergy Immunol. 53 (2), 137–144.

Manayi, A., Vazirian, M., Saeidnia, S., 2015. *Echinacea purpurea*: pharmacology, phytochemistry and analysis methods. Pharmacognosy Rev. 9 (17), 63–72.

Mangathayaru, K., Umadevi, M., Reddy, C.U., 2009. Evaluation of the immunomodulatory and DNA protective activities of the shoots of *Cynodon dactylon*. J. Ethnopharmacol. 123 (1), 181–184.

Manna, K., Khan, A., Das, D.K., Kesh, S.B., Das, U., Ghosh, S., et al., 2014. Protective effect of coconut water concentrate and its active component shikimic acid against

hydroperoxide mediated oxidative stress through suppression of NFkB and activation of Nrf2 pathway. J. Ethnopharmacol. 155 (1), 132–146.
Manu, K.A., Kuttan, G., 2007. Effect of punarnavine, an alkaloid from *Boerhaavia diffusa*, on cell-mediated immune responses and TIMP-1 in B16F-10 metastatic melanoma-bearing mice. Immunopharmacol. Immunotoxicol. 29 (3-4), 569–586.
Manu, K.A., Kuttan, G., 2009. Anti-metastatic potential of Punarnavine, an alkaloid from *Boerhaavia diffusa* Linn. Immunobiology 214 (4), 245–255.
Márquez, N., Sancho, R., Bedoya, L.M., Alcamí, J., López-Pérez, J.L., San Feliciano, A., et al., 2005. Mesuol, a natural occurring 4-phenylcoumarin, inhibits HIV-1 replication by targeting the NF-κB pathway. Antivir. Res. 66 (2), 137–145.
Mathew, S., Kuttan, G., 1999. Immunomodulatory and antitumour activities of *Tinospora cordifolia*. Fitoterapia 70 (1), 35–43.
Maurya, R., Wazir, V., Kapil, A., Kapil, R.S., 1996. Cordifoliosides A and B, two new phenylpropene disaccharides from *Tinospora cordifolia* possessing immunostimulant activity. Nat. Prod. Lett. 8 (1), 7–10.
McNeil, N.I., Cummings, J.H., James, W.P., 1978. Short chain fatty acid absorption by the human large intestine. Gut 19 (9), 819–822.
Mehrotra, S., Mishra, K.P., Maurya, R., Srimal, R.C., Yadav, V.S., Pandey, R., et al., 2003. Anticellular and immunosuppressive properties of ethanolic extract of *Acorus calamus* rhizome. Int. Immunopharmacol. 3 (1), 53–61.
Merekar, A.N., Pattan, S.R., Parjane, S.K., Nirmal, S.A., Patel, D.S., Shitre, M.R., 2011. Synergistic anthelmintic activity of rhizomes of *Acorus calamus* and roots of *Vitex negundo*. Pharmacologyonline 3, 209–212.
Mishio, T., Takeda, K., Iwashina, T., 2015. Anthocyanins and other flavonoids as flower pigments from eleven *Centaurea* species. Nat. Prod. Commun. 10 (3), 447–450.
Mitra, E., Ghosh, A.K., Ghosh, D., Mukherjee, D., Chattopadhyay, A., Dutta, S., et al., 2012. Protective effect of aqueous Curry leaf (*Murraya koenigii*) extract against cadmium-induced oxidative stress in rat heart. Food Chem. Toxicol. 50 (5), 1340–1353.
Modi, M., Dezzutti, C.S., Kulshreshtha, S., Rawat, A.K.S., Srivastava, S.K., Malhotra, S., et al., 2013. Extracts from *Acacia catechu* suppress HIV-1 replication by inhibiting the activities of the viral protease and Tat. Virol. J. 10 (1), 309.
Molina-Romo, E.D., Garibay-Escobar, A., Valenzuela-Antelo, O., Ruiz-Bustos, E., Martínez, J.H., Vélazquez, C., et al., 2018. Antiproliferative and apoptotic activities of the medicinal plant *Ziziphus obtusifolia*. Pharmacognosy Res. 10 (1), 55.
Monga, J., Chauhan, C.S., Sharma, M., 2011. Human epithelial carcinoma cytotoxicity and inhibition of DMBA/TPA induced squamous cell carcinoma in Balb/c mice by *Acacia catechu* heartwood. J. Pharm. Pharm. 63 (11), 1470–1482.
Moore, K.W., de Waal Malefyt, R., Coffman, R.L., O'Garra, A., 2001. Interleukin-10 and the interleukin-10 receptor. Annu. Rev. Immunol. 19 (1), 683–765.
More, P., Pai, K., 2011. Immunomodulatory effects of *Tinospora cordifolia* (Guduchi) on macrophage activation. Biol. Med. 3 (2), 134–140.
Mousa, O., Vuorela, P., Kiviranta, J., Törnquist, K., Vuorela, H., Hiltunen, R., 1994. Bioactivity of certain Egyptian Ficus species. Planta Med. 58 (S1), 632–633.
Mukherjee, R., Dash, P.K., Ram, G.C., 2005. Immunotherapeutic potential of *Ocimum sanctum* (L) in bovine subclinical mastitis. Res. Vet. Sci. 79 (1), 37–43.
Mukherjee, P.K., Nema, N.K., Venkatesh, P., Debnath, P.K., 2012. Changing scenario for promotion and development of Ayurveda—way forward. J. Ethnopharmacol. 143 (2), 424–434.
Mukherjee, H., Ojha, D., Bharitkar, Y.P., Ghosh, S., Mondal, S., Kaity, S., et al., 2013. Evaluation of the wound healing activity of *Shorea robusta*, an Indian ethnomedicine, and its isolated constituent(s) in topical formulation. J. Ethnopharmacol. 149 (1), 335–343.

Mungantiwar, A.A., Nair, A.M., Kamal, K.K., Saraf, M.N., 1997. Adaptogenic activity of aqueous extract of the roots of *Boerhaavia diffusa* Linn. Indian Drugs 34 (4), 184–189.

Mungantiwar, A.A., Nair, A.M., Shinde, U.A., Dikshit, V.J., Saraf, M.N., Thakur, V.S., et al., 1999. Studies on the immunomodulatory effects of *Boerhaavia diffusa* alkaloidal fraction. J. Ethnopharmacol. 65 (2), 125–131.

Na, H.S., Lim, Y.J., Yun, Y.S., Kweon, M.N., Lee, H.C., 2010. Ginsan enhances humoral antibody response to orally delivered antigen. Immune Netw. 10 (1), 5–14.

Na-Bangchang, K., Plengsuriyakarn, T., Karbwang, J., 2017. Research and development of *Atractylodes lancea* (Thunb) DC. as a promising candidate for cholangiocarcinoma chemotherapeutics. Evid. Based Complement. Altern. Med. 2017, 5929234–5929249.

Nadkarni, K., Nadkarni, A., 1954. Dr. K.M. Nadkarni's Indian Materia Medica. Bombay Popular Prakashan, Mumbai.

Nadumane, V., Nair, S., 2011. Evaluation of the anticancer and cytotoxic potentials of *Acacia catechu* extracts *in vitro*. J. Nat. Pharm. 2 (4), 190.

Nagappan, T., Ramasamy, P., Wahid, M.E.A., Segaran, T.C., Vairappan, C.S., 2011. Biological activity of carbazole alkaloids and essential oil of *Murraya koenigii* against antibiotic resistant microbes and cancer cell lines. Molecules 16 (11), 9651–9664.

Nagarathna, P.K.M., Reena, K., Reddy, S., Wesley, J., 2013. Review on immunomodulation and immunomodulatory activity of some herbal plants. Int. J. Pharm. Sci. Rev. Res. 22 (1), 223–230.

Nahak, G., Sahu, R., 2014. Immunostimulatory effects of *Ocimum sanctum* Linn. Leaf extracts in *Clarias batrachus* Linn. Asian J. Pharm. Clin. Res. 7, 15.

Naik, G.H., Priyadarsini, K.I., Satav, J.G., Banavalikar, M.M., Sohoni, D.P., Biyani, M.K., et al., 2003. Comparative antioxidant activity of individual herbal components used in Ayurvedic medicine. Phytochemistry 63 (1), 97–104.

Nair, P.R., Rodriguez, S., Ramachandran, R., Alamo, A., Melnick, S.J., Escalon, E., et al., 2004. Immune stimulating properties of a novel polysaccharide from the medicinal plant *Tinospora cordifolia*. Int. Immunopharmacol. 4 (13), 1645–1659.

Nair, P.R., Melnick, S.J., Ramachandran, R., Escalon, E., Ramachandran, C., Bose, C., et al., 2006. Mechanism of macrophage activation by (1, 4)-α-D-glucan isolated from *Tinospora cordifolia*. Int. Immunopharmacol. 6 (12), 1815–1824.

Nair, S.V., Baranwal, G., Chatterjee, M., Sachu, A., Vasudevan, A.K., Bose, C., et al., 2016. Antimicrobial activity of plumbagin, a naturally occurring naphthoquinone from *Plumbago rosea*, against *Staphylococcus aureus* and *Candida albicans*. Int. J. Med. Microbiol. 306 (4), 237–248.

Nakaya, T.A., Kita, M., Kuriyama, H., Iwakura, Y., Imanishi, J., 2004. Panax ginseng induces production of proinflammatory cytokines via toll-like receptor. J. Interferon Cytokine Res. 24 (2), 93–100.

Nantz, M.P., Rowe, C.A., Muller, C.E., Creasy, R.A., Stanilka, J.M., Percival, S.S., 2012. Supplementation with aged garlic extract improves both NK and γδ-T cell function and reduces the severity of cold and flu symptoms: a randomized, double-blind, placebo-controlled nutrition intervention. Clin. Nutr. 31 (3), 337–344.

Navarro-Hoyos, M., Alvarado-Corella, D., Moreira-Gonzalez, I., Arnaez-Serrano, E., Monagas-Juan, M., 2018. Polyphenolic composition and antioxidant activity of aqueous and ethanolic extracts from *Uncaria tomentosa* bark and leaves. Antioxidants 7 (5), 65.

Ndong, D., Fall, J., 2011. The effect of garlic (*Allium sativum*) on growth and immune responses of hybrid tilapia (*Oreochromis niloticus* × *Oreochromis aureus*). J. Clin. Immunol. Immunopathol. Res. 3 (1), 1–9.

Negi, B.S., Dave, B.P., 2010. *In vitro* antimicrobial activity of *Acacia catechu* and its phytochemical analysis. Indian J. Microbiol. 50 (4), 369–374.

Ni, L., Li, L., Qiu, Y., Chen, F.Y., Li, C.J., Ma, J., et al., 2018. Triptergosidols AD, nerolidol-type sesquiterpene glucosides from the leaves of *Tripterygium wilfordii*. Fitoterapia. 128, 187–191.

Nok, A.J., Williams, S., Onyenekwe, P.C., 1996. *Allium sativum*-induced death of African trypanosomes. Parasitol. Res. 82 (7), 634–637.

Oft, M., 2014. IL-10: master switch from tumor-promoting inflammation to antitumor immunity. Cancer Immunol. Res. 2 (3), 194–199.

Ojha, D., Singh, G., Upadhyaya, Y.N., 1969. Clinical evaluation of *Acacia catechu*, Willd. (Khadira) in the treatment of lepromatous leprosy. Int. J. Leprosy Mycobact. Dis. 37 (3), 302–307.

Ojha, D., Mukherjee, H., Jena, A., Mondal, S., Dwivedi, V.P., Mondal, K.C., et al., 2014. *In vivo* and *in vitro* anti-inflammatory activity of *Odina wodier* Roxb, an Indian folk medicine, through inhibition of toll-like receptor 4 signalling pathway. PLoS One 9 (8), e104939.

Ojha, D., Das, R., Sobia, P., Dwivedi, V.P., Ghosh, S., Samanta, A., et al., 2015. *Pedilanthus tithymaloides* inhibits HSV infection through the modulation of NF-kB signalling pathway. PLoS One 10 (9), e0139338.

Olajide, O.A., Makinde, J.M., Okpako, D.T., 2003. Evaluation of the anti-inflammatory property of the extract of *Combretum micranthum* G. Don (Combretaceae). Inflammopharmacology 11 (3), 293–298.

Olennikov, D.N., Kashchenko, N.I., Chirikova, N.K., Akobirshoeva, A., Zilfikarov, I.N., Vennos, C., 2017. Isorhamnetin and quercetin derivatives as anti-acetylcholinesterase principles of marigold (*Calendula officinalis*) flowers and preparations. Int. J. Mol. Sci. 18 (8), 1685.

Palani, S., Raja, S., Kumar, S.N., Kumar, B.S., 2011. Nephroprotective and antioxidant activities of *Salacia oblonga* on acetaminophen-induced toxicity in rats. Nat. Prod. Res. 25 (19), 1876–1880.

Pandey, R., Maurya, R., Singh, G., Sathiamoorthy, B., Naik, S., 2005. Immunosuppressive properties of flavonoids isolated from *Boerhaavia diffusa* Linn. Int. Immunopharmacol. 5 (3), 541–553.

Pandit, S., Mukherjee, P.K., Ponnusankar, S., Venkatesh, M., Srikanth, N., 2011. Metabolism mediated interaction of α-asarone and *Acorus calamus* with CYP3A4 and CYP2D6. Fitoterapia 82 (3), 369–374.

Parichatikanond, W., Suthisisang, C., Dhepakson, P., Herunsalee, A., 2010. Study of anti-inflammatory activities of the pure compounds from *Andrographis paniculata* (burm. f.) Nees and their effects on gene expression. Int. Immunopharmacol. 10 (11), 1361–1373.

Park, S.Y., Park, S.J., Park, T.G., Rajasekar, S., Lee, S.J., Choi, Y.W., 2013. Schizandrin C exerts anti-neuroinflammatory effects by upregulating phase II detoxifying/antioxidant enzymes in microglia. Int. Immunopharmacol. 17 (2), 415–426.

Park, G.H., Song, H.M., Park, S.B., Son, H.J., Um, Y., Kim, H.S., et al., 2018a. Cytotoxic activity of the twigs of *Cinnamomum cassia* through the suppression of cell proliferation and the induction of apoptosis in human colorectal cancer cells. BMC Complement. Altern. Med. 18 (1), 28.

Park, S., Lee, M.S., Jung, S., Lee, S., Kwon, O., Kreuter, M.H., et al., 2018b. *Echinacea purpurea* protects against restraint stress-induced immunosuppression in BALB/c Mice. J. Med. Food 21 (3), 261–268.

Patay, É.B., Bencsik, T., Papp, N., 2016. Phytochemical overview and medicinal importance of Coffea species from the past until now. Asian Pac. J. Trop. Med. 9 (12), 1127–1135.

Patra, A., Mitra, A.K., 1981. Constituents of *Acorus calamus*: structure of acoramone. Carbon-13 NMR spectra of cis-and trans-asarone. J. Nat. Prod. 44 (6), 668–669.

Pattanayak, P., Behera, P., Das, D., Panda, S.K., 2010. *Ocimum sanctum* Linn. A reservoir plant for therapeutic applications: an overview. Pharmacognosy Rev. 4 (7), 95–105.

Peláez, G.L.M., Sierra, J.A., Alzate, F., Holzgrabe, U., Ramirez-Pineda, J.R., 2013. Pentacyclic triterpenes from *Cecropia telenitida* with immunomodulatory activity on dendritic cells. Revista Brasileira de Farmacognosia 23 (5), 754–761.

Pérez-Pimiento, A., Santaolalla, M., De Paz, S., Fernándezparra, B., Domínguez lázaro, A.R., Moneo, I., 1999. Anaphylactic reaction to young garlic. Allergy 54 (6), 626–629.

Petrunov, B., Nenkov, P., Shekerdjiisky, R., 2007. The role of immunostimulants in immunotherapy and immunoprophylaxis. Biotechnol. Biotechnol. Equip. 21 (4), 454–462.

Peumans, W.J., De Ley, M., Broekaert, W.F., 1984. An unusual lectin from stinging nettle (*Urtica dioica*) rhizomes. FEBS Lett. 177 (1), 99–103.

Philip, S., Tom, G., Vasumathi, A.V., 2018. Evaluation of the antiinflammatory activity of *Tinospora cordifolia* (Willd.) Miers chloroform extract—a preclinical study. J. Pharm. Pharmacol. 70 (8), 1113–1125.

Phongpaichit, S., Pujenjob, N., Rukachaisirikul, V., Ongsakul, M., 2005. Antimicrobial activities of the crude methanol extract of *Acorus calamus* Linn. Songklanakarin J. Sci. Technol. 27 (2), 517–523.

Plaza, A., Perrone, A., Balestrieri, M.L., Felice, F., Balestrieri, C., Hamed, A.I., et al., 2005. New unusual pregnane glycosides with antiproliferative activity from *Solenostemma argel*. Steroids 70 (9), 594–603.

Plummer, S.M., Holloway, K.A., Manson, M.M., Munks, R.J., Kaptein, A., Farrow, S., et al., 1999. Inhibition of cyclo-oxygenase 2 expression in colon cells by the chemopreventive agent curcumin involves inhibition of NF-κB activation via the NIK/IKK signalling complex. Oncogene 18 (44), 6013–6020.

Pongituluran, O.B., Rofaani, E., 2015. Antiviral and immunostimulant activities of *Andrographis paniculata*. HAYATI J. Biosci. 22 (2), 67–72.

Prasad, G., Mukthiraj, S., 2011. Effect of methanolic extract of *Andrographis paniculata* (Nees) on growth and haematology of *Oreochromis mossambicus* (Peters). World J. Fish Marine Sci. 3 (6), 473–479.

Prashar, R., Kumar, A., Hewer, A., Cole, K.J., Davis, W., Phillips, D.H., 1998. Inhibition by an extract of *Ocimum sanctum* of DNA-binding activity of 7,12-dimethylbenz [*a*] anthracene in rat hepatocytes *in vitro*. Cancer Lett. 128 (2), 155–160.

Provin, T.L., Wright, A.L., Hons, F.M., Zuberer, D.A., White, R.H., 2008. Seasonal dynamics of soil micronutrients in compost-amended Bermuda grass turf. Bioresour. Technol. 99 (7), 2672–2679.

Puri, A., Saxena, R., Saxena, R.P., Saxena, K.C., Srivastava, V., Tandon, J.S., 1993. Immunostimulant agents from *Andrographis paniculata*. J. Nat. Prod. 56 (7), 995–999.

Qi, X.L., Zhang, Y.Y., Zhao, P., Zhou, L., Wang, X.B., Huang, X.X., et al., 2018. Ent-kaurane diterpenoids with neuroprotective properties from corn silk (*Zea mays*). J. Nat. Prod. 81 (5), 1225–1234.

Qin, S., Huang, K., Fang, Z., Yin, J., Dai, R., 2017. The effect of astragaloside IV on ethanol-induced gastric mucosal injury in rats: involvement of inflammation. Int. Immunopharmacol. 52, 211–217.

Qiu, M., Dong, Y.H., Han, F., Qin, J.M., Zhang, H.N., Du, J.X., et al., 2016. Influence of total flavonoids derived from *Choerospondias axillaris folium* on aconitine-induced antiarrhythmic action and hemodynamics in Wistar rats. J. Toxicol. Environ. Health, A 79 (19), 878–883.

Qu, D.F., Yu, H.J., Liu, Z., Zhang, D.F., Zhou, Q.J., Zhang, H.L., et al., 2011. Ginsenoside Rg1 enhances immune response induced by recombinant *Toxoplasma gondii* SAG1 antigen. Vet. Parasitol. 179 (1), 28–34.

Quan, N., Banks, W.A., 2007. Brain-immune communication pathways. Brain Behav. Immun. 21 (6), 727–735.
Quintans, J., Yokoyama, A., Evavold, B., Hirsch, R., Mayforth, R.D., 1989. Direct activation of murine resting T cells by con A or anti-CD3 Ig. J. Mol. Cell. Immunol. 4 (4), 225–235.
Quintero-Fabián, S., Ortuño-Sahagún, D., Vázquez-Carrera, M., López-Roa, R.I., 2013. Alliin, a garlic (*Allium sativum*) compound, prevents LPS-induced inflammation in 3T3-L1 adipocytes. Mediators Inflamm. 2013, 381815.
Radhika, P., Annapurna, A., Rao, S.N., 2012. Immunostimulant, cerebroprotective & nootropic activities of *Andrographis paniculata* leaves extract in normal and type 2 diabetic rats. Indian J. Med. Res. 135 (5), 636–641.
Radulović, N.S., Stojković, M.B., Mitić, S.S., Randjelović, P.J., Ilić, I.R., Stojanović, N. M., et al., 2012. Exploitation of the antioxidant potential of *Geranium macrorrhizum* (Geraniaceae): hepatoprotective and antimicrobial activities. Nat. Prod. Commun. 7 (12), 1609–1614.
Rahmatullah, M., Hossain, M., Mahmud, A., Sultana, N., Mizanur, S., Mohammad, R., et al., 2013. Antihyperglycemic and antinociceptive activity evaluation of 'khoyer' prepared from boiling the wood of *Acacia catechu* in water. Afr. J. Trad. Complement. Altern. Med. 10 (4), 1–5.
Rajput, Z.I., Hu, S.H., Xiao, C.W., Arijo, A.G., 2007. Adjuvant effects of saponins on animal immune responses. J. Zhejiang Univ. Sci. B 8 (3), 153–161.
Ran, J., Ma, C., Xu, K., Xu, L., He, Y., Moqbel, S.A.A., et al., 2018. Schisandrin B ameliorated chondrocytes inflammation and osteoarthritis via suppression of NF-κB and MAPK signal pathways. Drug Des. Dev. Ther. 12, 1195–1204.
Rani, P., Khullar, N., 2004. Antimicrobial evaluation of some medicinal plants for their antienteric potential against multidrug resistant *Salmonella typhi*. Phytother. Res. 18 (8), 670–673.
Ranjan, D., Chen, C., Johnston, T.D., Jeon, H., Nagabhushan, M., 2004. Curcumin inhibits mitogen stimulated lymphocyte proliferation, NFκB activation, and IL-2 signaling. J. Surg. Res. 121 (2), 171–177.
Rao, P.V., Gan, S.H., 2014. Cinnamon: a multifaceted medicinal plant. Evid. Based Complement. Altern. Med. 2014, 642942–642953.
Ravi, V., Parida, S., Desai, A., Chandramuki, A., GourieDevi, M., Grau, G.E., 1997. Correlation of tumor necrosis factor levels in the serum and cerebrospinal fluid with clinical outcome in Japanese encephalitis patients. J. Med. Virol. 51 (2), 132–136.
Ravichandiran, V., Patil Vishal, S., 2015. *In vitro* evaluation for immunomodulatory activity of *Acorus calamus* on human neutrophils. Int. Res. J. Pharm. 6 (7), 450–452.
Retini, C., Kozel, T.R., Pietrella, D., Monari, C., Bistoni, F., Vecchiarelli, A., 2001. Interdependency of interleukin-10 and interleukin-12 in regulation of T-cell differentiation and effector function of monocytes in response to stimulation with *Cryptococcus neoformans*. Infect. Immun. 69 (10), 6064–6073.
Richardson, S.N., Nsiama, T.K., Walker, A.K., McMullin, D.R., Miller, J.D., 2015. Antimicrobial dihydrobenzofurans and xanthenes from a foliar endophyte of *Pinus strobus*. Phytochemistry 117, 436–443.
Ríos, J.L., 2010. Effects of triterpenes on the immune system. J. Ethnopharmacol. 128 (1), 1–14.
Sahu, A., Rawal, N., Pangburn, M.K., 1999. Inhibition of complement by covalent attachment of rosmarinic acid to activated C3b. Biochem. Pharm. 57 (12), 1439–1446.
Saidi, S., Stavridis, S., Stankov, O., Dohcev, S., Panov, S., 2017. Effects of *Serenoa repens* alcohol extract on benign prostate hyperplasia. Prilozi 38 (2), 123–129.
Saini, P., Gayen, P., Kumar, D., Nayak, A., Mukherjee, N., Mukherjee, S., et al., 2014. Antifilarial effect of ursolic acid from *Nyctanthes arbortristis*: molecular and biochemical evidences. Parasitol. Int. 63 (5), 717–728.

Sairam, K., Rao, C.V., Babu, M.D., Goel, R.K., 2001. Prophylactic and curative effects of *Bacopa monniera* in gastric ulcer models. Phytomedicine 8 (6), 423–430.
Saleem, S., Muhammad, G., Hussain, M.A., Bukhari, S.N.A., 2018. A comprehensive review of phytochemical profile, bioactives for pharmaceuticals, and pharmacological attributes of *Azadirachta indica*. Phytother. Res. 32 (7), 1241–1272.
Sangameswaran, B., Balakrishnan, N., Bhaskar, V.H., Jayakar, B., 2008. Anti-inflammatory and anti-bacterial activity of leaves of *Boerhavia diffusa* L. Pharmacognosy Mag S65–S68.
Santander, S.P., Hernández, J.F., Barreto, C.C., Masayuki, A., Moins-Teisserenc, H., Fiorentino, S., 2012. Immunomodulatory effects of aqueous and organic fractions from *Petiveria alliacea* on human dendritic cells. Am. J. Chin. Med. 40 (04), 833–844.
Santhi, R., Annapoorani, S., 2010. Efficacy of *Cynodon dactylon* for immunomodulatory activity. Drug Invention Today 2 (2), 112–114.
Sathaye, S., Bagul, Y., Gupta, S., Kaur, H., Redkar, R., 2011. Hepatoprotective effects of aqueous leaf extract and crude isolates of *Murraya koenigii* against *in vitro* ethanol-induced hepatotoxicity model. Exp. Toxicol. Pathol. 63 (6), 587–591.
Satoshi, T., Fusao, H., Makoto, I., Axelrod, J., Toshiaki, O., Waxdal, M.J., 1982. Lectin-induced mitosis and phospholipid methylation. Mol. Immunol. 19 (3), 467–475.
Schepetkin, I.A., Quinn, M.T., 2006. Botanical polysaccharides: macrophage immunomodulation and therapeutic potential. Int. Immunopharmacol. 6 (3), 317–333.
Schulze-Koops, H., Burkhardt, H., Kalden, J.R., 1999. What we have learned from trials of immunomodulatory agents in rheumatoid arthritis: Future directions. Drugs Today (Barc) 35, 327–351.
See, D.M., Broumand, N., Sahl, L., Tilles, J.G., 1997. In vitro effects of *Echinacea* and ginseng on natural killer and antibody-dependent cell cytotoxicity in healthy subjects and chronic fatigue syndrome or acquired immunodeficiency syndrome patients. Immunopharmacology 35 (3), 229–235.
Seubwai, W., Vaeteewoottacharn, K., Hiyoshi, M., Suzu, S., Puapairoj, A., Wongkham, C., et al., 2010. Cepharanthine exerts antitumor activity on cholangiocarcinoma by inhibiting NF-κB. Cancer Sci. 101 (7), 1590–1595.
Seuri, M., Taivanen, A., Ruoppi, P., Tukiainen, H., 1993. Three cases of occupational asthma and rhinitis caused by garlic. Clin. Exp. Allergy 23 (12), 1011–1014.
Shah, A.S., Wakade, A.S., Juvekar, A.R., 2008. Immunomodulatory activity of methanolic extract of *Murraya koenigii* (L) Spreng Leaves. Indian J. Exp. Biol. 46, 505–509.
Sharma, S., Sharma, P., Kulurkar, P., Singh, D., Kumar, D., Patial, V., 2017. Iridoid glycosides fraction from *Picrorhiza kurroa* attenuates cyclophosphamide-induced renal toxicity and peripheral neuropathy via PPAR-γ mediated inhibition of inflammation and apoptosis. Phytomedicine 36, 108–117.
Sharrif Moghaddasi, M., 2011. *Aloe vera* their chemicals composition and applications: a review. Int. J. Biol. Med. Res. 2 (1), 466–471.
Shaw, P.C., Lee, K.M., Wong, K.B., 2005. Recent advances in trichosanthin, a ribosome-inactivating protein with multiple pharmacological properties. Toxicon 45 (6), 683–689.
Sheeja, K., Kuttan, G., 2010. *Andrographis paniculata* downregulates proinflammatory cytokine production and augments cell mediated immune response in metastatic tumor-bearing mice. Asian Pac. J. Cancer Prev. 11 (3), 723–729.
Shen, L., Wang, P., Guo, J., Du, G., 2013. Anti-arthritic activity of ethanol extract of *Fagopyrum cymosum* with adjuvant-induced arthritis in rats. Pharm. Biol. 51 (6), 783–789.
Shikov, A.N., Pozharitskaya, O.N., Makarov, V.G., 2016. *Aralia elata* var. *mandshurica* (Rupr. & Maxim.) J. Wen: an overview of pharmacological studies. Phytomedicine 23 (12), 1409–1421.

Shimada, J., Moon, S.K., Lee, H.Y., Takeshita, T., Pan, H., Woo, J.I., et al., 2008. Lysozyme M deficiency leads to an increased susceptibility to *Streptococcus pneumoniae*-induced otitis media. BMC Infect. Dis. 8 (1), 134.

Shin, J.Y., Song, J.Y., Yun, Y.S., Yang, H.O., Rhee, D.K., Pyo, S., 2002. Immunostimulating effects of acidic polysaccharides extract of *Panax ginseng* on macrophage function. Immunopharmacol. Immunotoxicol. 24 (3), 469–482.

Shin, I.S., Hong, J., Jeon, C.M., Shin, N.R., Kwon, O.K., Kim, H.S., et al., 2013. Diallyl-disulfide, an organosulfur compound of garlic, attenuates airway inflammation via activation of the Nrf-2/HO-1 pathway and NF-kappaB suppression. Food Chem. Toxicol. 62, 506–513.

Shin, M.S., Song, J.H., Choi, P., Lee, J.H., Kim, S.Y., Shin, K.S., et al., 2018. Stimulation of innate immune function by *Panax ginseng* after heat processing. J. Agric. Food Chem. 66 (18), 4652.

Shinjyo, N., Fuchino, H., Kawahara, N., Hishida, A., 2018. Determination of (*E*)-ferulic acid content in the root of *Angelica acutiloba*: a simple chemical evaluation method for crude drug quality control. J. Nat. Med. 72 (3), 774–778.

Singh, S., Aggarwal, B.B., 1995. Activation of transcription factor NF-κB is suppressed by curcumin (diferuloylmethane). J. Biol. Chem. 270 (42), 24995–25000.

Singh, S.S., Pandey, S.C., Srivastava, S., Gupta, V.S., Patro, B., Ghosh, A.C., 2003. Chemistry and medicinal properties of *Tinospora cordifolia* (Guduchi). Indian J. Pharmacol. 35 (2), 83–91.

Singh, N., Singh, S.M., Shrivastava, P., 2004. Immunomodulatory and antitumor actions of medicinal plant *Tinospora cordifolia* are mediated through activation of tumor-associated macrophages. Immunopharmacol. Immunotoxicol. 26 (1), 145–162.

Singh, S.K., Rai, P.K., Jaiswal, D., Watal, G., 2008. Evidence-based critical evaluation of glycemic potential of *Cynodon dactylon*. Evid. Based Complement. Altern. Med. 5 (4), 415–420.

Singha, P.K., Roy, S., Dey, S., 2003. Antimicrobial activity of *Andrographis paniculata*. Fitoterapia 74 (7), 692–694.

Soares, G.R., de Moura, C.F.G., Silva, M.J.D., Vilegas, W., Santamarina, A.B., Pisani, L. P., et al., 2018. Protective effects of purple carrot extract (*Daucus carota*) against rat tongue carcinogenesis induced by 4-nitroquinoline 1-oxide. Med. Oncol. 35 (4), 54.

Sohn, H.Y., Son, K.H., Kwon, C.S., Kwon, G.S., Kang, S.S., 2004. Antimicrobial and cytotoxic activity of 18 prenylated flavonoids isolated from medicinal plants: *Morus alba* L., *Morus mongolica* Schneider, *Broussnetia papyrifera* (L.) Vent, *Sophora flavescens* Ait and *Echinosophora koreensis* Nakai. Phytomedicine 11 (7-8), 666–672.

Song, Z., Moser, C., Wu, H., Faber, V., Kharazmi, A., Høiby, N., 2003. Cytokine modulating effect of ginseng treatment in a mouse model of *Pseudomonas aeruginosa* lung infection. J. Cystic Fibros. 2 (3), 112–119.

Song, L., Xiong, D., Hu, M., Kang, X., Pan, Z., Jiao, X., 2016. Immunopotentiation of different adjuvants on humoral and cellular immune responses induced by HA1-2 subunit vaccines of H7N9 influenza in mice. PLoS One 11 (3), e0150678.

Sood, R., Bhatia, S., Bhatnagar, H., Gupta, V., Kumar, M., Dimri, U., et al., 2013. Phytochemical analysis and *in vitro* screening of selected Indian medicinal plants for antiviral activity against highly pathogenic avian influenza virus. Spatula DD 3 (3), 81–88.

Sreeja, S., Sreeja, S., 2009. An *in vitro* study on antiproliferative and antiestrogenic effects of *Boerhaavia diffusa* L. extracts. J. Ethnopharmacol. 126 (2), 221–225.

Srikumar, R., Parthasarathy, N.J., Devi, R.S., 2005. Immunomodulatory activity of triphala on neutrophil functions. Biol. Pharm. Bull. 28 (8), 1398–1403.

Srivastava, R., Saluja, D., Chopra, M., 2005. Isolation and screening of anticancer metabolites from *Boerhavia diffusa*. Indian J. Med. Res. 151 (supplement1), S19.

Srivastava, R., Saluja, D., Dwarakanath, B.S., Chopra, M., 2011. Inhibition of human cervical cancer cell growth by ethanolic extract of *Boerhaavia diffusa* Linn. (Punarnava) root. Evid. Based Complement. Altern. Med. 2011, 427031.

Steinborn, C., Klemd, A.M., Sanchez-Campillo, A.S., Rieger, S., Scheffen, M., Sauer, B., et al., 2017. *Viscum album* neutralizes tumor-induced immunosuppression in a human *in vitro* cell model. PLoS One 12 (7), e0181553.

Stirpe, F., Legg, R.F., Onyon, L.J., Ziska, P., Franz, H., 1980. Inhibition of protein synthesis by a toxic lectin from *Viscum album* L.(mistletoe). Biochem. J. 190 (3), 843–845.

Stohs, S.J., Bagchi, D., 2015. Antioxidant, anti-inflammatory, and chemoprotective properties of *Acacia catechu* heartwood extracts. Phytother. Res. 29 (6), 818–824.

Su, W., Sun, A.J., Xu, D.L., Zhang, H.Q., Yang, L., Yuan, L.Y., et al., 2010. Inhibiting effects of total saponins of panax ginseng on immune maturation of dendritic cells induced by oxidized-low density lipoprotein. Cell. Immunol. 263 (1), 99–104.

Su, Y., Wu, L., Mu, G., Wang, Q., Yang, B., Cheng, G., et al., 2017. 9,19-Cycloartenol glycoside G3 from *Cimicifuga simplex* regulates immune responses by modulating Th17/Treg ratio. Bioorg. Med. Chem. 25 (17), 4917–4923.

Subbotina, M.D., Timchenko, V.N., Vorobyov, M.M., Konunova, Y.S., Aleksandrovih, Y. S., Shushunov, S., 2003. Effect of oral administration of tormentil root extract (*Potentilla tormentilla*) on rotavirus diarrhea in children: a randomized, double blind, controlled trial. Pediatr. Infect. Dis. J. 22 (8), 706–711.

Sumanth, M., Mustafa, S.S., 2007. Antistress, adoptogenic and immunopotentiating activity roots of *Boerhaavia diffusa* in mice. Int. J. Pharmacol. 3 (5), 416–420.

Sumitra, M., Manikandan, P., Kumar, D.A., Arutselvan, N., Balakrishna, K., Manohar, B. M., et al., 2001. Experimental myocardial necrosis in rats: role of arjunolic acid on platelet aggregation, coagulation and antioxidant status. Mol. Cell. Biochem. 224 (1), 135–142.

Sumiyoshi, M., Sakanaka, M., Kimura, Y., 2010. Effects of red ginseng extract on allergic reactions to food in Balb/c mice. J. Ethnopharmacol. 132 (1), 206–212.

Swamy, M.K., Akhtar, M.S., Sinniah, U.R., 2016. Antimicrobial properties of plant essential oils against human pathogens and their mode of action: an updated review. Evid. Based Complement. Altern. Med. 2016, 3012462.

Szabo, A., 2015. Psychedelics and immunomodulation: novel approaches and therapeutic opportunities. Front. Immunol. 6, 358.

Szabo, A., Rajnavolgyi, E., 2013a. Collaboration of Toll-like and RIG-I-like receptors in human dendritic cells: tRIGgering antiviral innate immune responses. Am. J. Clin. Exp. Immunol. 2 (3), 195–207.

Szabo, A., Rajnavolgyi, E., 2013b. The brain–immune–gut triangle: innate immunity in psychiatric and neurological disorders. Curr. Immunol. Rev. 9 (4), 241–248.

Szabo, A., Kovacs, A., Frecska, E., Rajnavolgyi, E., 2014. Psychedelic *N,N*-dimethyltryptamine and 5-methoxy-*N,N*-dimethyltryptamine modulates innate and adaptive inflammatory responses through the sigma-1 receptor of human monocyte-derived dendritic cells. PLoS One 9 (8), e106533.

Tafaghodi, M., Eskandari, M., Khamesipour, A., Jaafari, M.R., 2016. Immunization against cutaneous leishmaniasis by alginate microspheres loaded with autoclaved *Leishmania major* (ALM) and Quillaja Saponins. Iranian J. Pharm. Res. 15 (2), 573.

Tanwar, A., Chawla, R., Ansari, M.M., Thakur, P., Chakotiya, A.S., Goel, R., et al., 2017. In vivo anti-arthritic efficacy of *Camellia sinensis* (L.) in collagen induced arthritis model. Biomed. Pharm. 87, 92–101.

Thatte, U.M., Dahanukar, S.A., 1986. Ayurveda and contemporary scientific thought. Trends Pharm. Sci. 7, 247–251.

Tourino, M.C., Mendes Oliveira, E., Bellé, L.P., Knebel, F.H., Albuquerque, R.C., Dörr, F.A., et al., 2013. Tryptamine and dimethyltryptamine inhibit indoleamine 2,3 dioxygenase and increase the tumor-reactive effect of peripheral blood mononuclear cells. Cell Biochem. Funct. 31 (5), 361–364.

Trivedi, C.P., Shinde, S., Sharma, R.C., 1969. Preliminary phytochemical and pharmacological studies on *Ficus racemosa* (Gular). Indian J. Med. Res. 57 (6), 1070–1074.

Tsang, J.Y.S., Li, D., Ho, D., Peng, J., Xu, A., Lamb, J., et al., 2011. Novel immunomodulatory effects of adiponectin on dendritic cell functions. Int. Immunopharmacol. 11 (5), 604–609.

Tzianabos, A.O., 2000. Polysaccharide immunomodulators as therapeutic agents: structural aspects and biologic function. Clin. Microbiol. Rev. 13 (4), 523–533.

Uttekar, M.M., Das, T., Pawar, R.S., Bhandari, B., Menon, V., Gupta, S.K., et al., 2012. Anti-HIV activity of semisynthetic derivatives of andrographolide and computational study of HIV-1gp120 protein binding. Eur. J. Med. Chem. 56, 368–374.

Vaghasiya, J., Datani, M., Nandkumar, K., Malaviya, S., Jivani, N., 2010. Comparative evaluation of alcoholic and aqueous extracts of *Ocimum sanctum* for immunomodulatory activity. Int. J. Pharm. Biol. Res. 1 (1), 25–29.

Vaghela, M., Iyer, K., Pandita, N., 2018. *In vitro* Inhibitory effect of *Gymnema sylvestre* Extracts and total Gymnemic acids fraction on select Cytochrome P450 activities in Rat liver Microsomes. Eur. J. Drug Metabol. Pharm. 43 (2), 227–237.

Wagner, H., 1999. Immunomodulatory Agents from Plants. Springer Science & Business Media, 13 Chapters with 365 pages. ISBN 978-3-0348-8763-2.

Wagner, H., Kreher, B., Jurcic, K., 1988. *In vitro* stimulation of human granulocytes and lymphocytes by pico-and femtogram quantities of cytostatic agents. Arzneimittel-Forschung 38 (2), 273–275.

Wagner, H., Nörr, H., Winterhoff, H., 1994. Plant adaptogens. Phytomedicine 1 (1), 63–76.

Wang, H.X., Ng, T.B., 2001. Purification of allivin, a novel antifungal protein from bulbs of the round-cloved garlic. Life Sci. 70 (3), 357–365.

Wang, R.J., Yang, B., 2007. Survey of study on the chemical constituents and quality control of *Flos carthami* [J]. Chin. J. Exp. Tradit. Med. Formulae 5, 24.

Wang, H., Actor, J.K., Indrigo, J., Olsen, M., Dasgupta, A., 2003. Asian and Siberian ginseng as a potential modulator of immune function: an *in vitro* cytokine study using mouse macrophages. Clin. Chim. Acta 327 (1), 123–128.

Wang, H., Ye, W.C., Jiang, R.W., Wu, J.J., Mak, T.C., Zhao, S.X., et al., 2004a. Three new cyclopentanoid monoterpenes from *Picrorhiza scrophulariiflora*. Planta Med. 70 (4), 382–384.

Wang, H., Ye, W.C., Xiong, F., Zhao, S.X., 2004b. Phenylethanoid glycosides from root of *Picrorhiza scrophulariiflora*. Zhongguo Zhongyao zazhi China Journal of Chinese Materia Medica 29 (6), 531–534.

Wang, W., Wang, J., Dong, S.F., Liu, C.H., Italiani, P., Sun, S.H., et al., 2010. Immunomodulatory activity of andrographolide on macrophage activation and specific antibody response. Acta Pharm. Sin. 31 (2), 191–201.

Wang, C.C., Yen, J.H., Cheng, Y.C., Lin, C.Y., Hsieh, C.T., Gau, R.J., et al., 2017. *Polygala tenuifolia* extract inhibits lipid accumulation in 3T3-L1 adipocytes and high-fat diet-induced obese mouse model and affects hepatic transcriptome and gut microbiota profiles. Food Nutr. Res. 61 (1), 1379861.

Wang, B., Gupta, G., Singh, M., Veerabhadrappa, K.V.S., Mishra, A., Chinnaboina, G.K., 2018a. Pharmacological evaluation of novel flavone from *Morus alba* in pentylenetetrazole-induced kindling and oxidative stress. J. Environ. Pathol. Toxicol. Oncol. 37 (1), 43–52.

Wang, Y., Kwak, M., Lee, P.C.W., Jin, J.O., 2018b. *Rehmannia glutinosa* polysaccharide promoted activation of human dendritic cells. Int. J. Biol. Macromol. 116 (4), 144.

Wang, Y.X., Liu, Q.Y., Zhang, M., Yang, Z., Pei, X., Wu, X., et al., 2018c. Polysaccharides from *Bupleurum* induce immune reversal in late sepsis. Shock 49 (4), 451–459.

Warashina, T., Umehara, K., Miyase, T., 2012. Constituents from the roots of *Taraxacum platycarpum* and their effect on proliferation of human skin fibroblasts. Chem. Pharm. Bull. 60 (2), 205–212.

Washiya, Y., Nishikawa, T., Fujino, T., 2013. Enhancement of intestinal IgA production by ajoene in mice. Biosci. Biotechnol. Biochem. 77 (11), 2298–2301.

Welch, C., Zhen, J., Bassène, E., Raskin, I., Simon, J.E., Wu, Q., 2017. Bioactive polyphenols in kinkéliba tea (*Combretum micranthum*) and their glucose-lowering activities. J. Food Drug Analysis 26 (5), 9.

Wiart, C., Kumar, K., Yusof, M.Y., Hamimah, H., Fauzi, Z.M., Sulaiman, M., 2005. Antiviral properties of entlabdene diterpenes of *Andrographis paniculata* nees, inhibitors of herpes simplex virus type 1. Phytother. Res. 19 (12), 1069–1070.

Wu, H.P., Lin, Y.K., 2018. Effect of *Eucommia ulmoides* Oliv, *Gynostemma pentaphyllum* (Thunb.) Makino, and *Curcuma longa* L. on Th1-and Th2-cytokine responses and human leukocyte antigen-DR expression in peripheral blood mononuclear cells of septic patients. J. Ehnopharmacol. 217, 195–204.

Wu, Y.C., Hsieh, C.L., 2011. Pharmacological effects of Radix *Angelica Sinensis* (Danggui) on cerebral infarction. Chin. Med. 6 (1), 32.

Wu, Y., Mao, Y., Jin, S., Hou, J., Du, H., Yang, M., et al., 2015. Identification, characterization and structure analysis of a type I ribosome-inactivating protein from *Sapium sebiferum* (Euphorbiaceae). Biochem. Biophys. Res. Commun. 463 (4), 557–562.

Wu, H.C., Horng, C.T., Lee, Y.L., Chen, P.N., Lin, C.Y., Liao, C.Y., et al., 2018a. *Cinnamomum Cassia* extracts suppress human lung cancer cells invasion by reducing u-PA/MMP expression through the FAK to ERK pathways. Int. J. Med. Sci. 15 (2), 115.

Wu, J.J., Guo, Z.Z., Zhu, Y.F., Huang, Z.J., Gong, X., Li, Y.H., et al., 2018b. A systematic review of pharmacokinetic studies on herbal drug Fuzi: Implications for Fuzi as personalized medicine. Phytomedicine 7113 (18), 30036–30039.

Xia, L., Ng, T.B., 2005. Isolation of alliumin, a novel protein with antimicrobial and antiproliferative activities from multiple-cloved garlic bulbs. Peptides 26 (2), 177–183.

Xia, X., Pan, Y., Zhang, W.Y., Cheng, G., Kong, L.D., 2006. Ethanolic extracts from *Curcuma longa* attenuates behavioral, immune, and neuroendocrine alterations in a rat chronic mild stress model. Biol. Pharm. Bull. 29 (5), 938–944.

Xiao, C., Xiao, P., Li, X., Li, X., Li, H., Chen, Y., et al., 2018. Cordyceps sinensis may inhibit Th22 cell chemotaxis to improve kidney function in lgA nephropathy. Am. J. Transl. Res. 10 (3), 857–865.

Xu, C., Chou, G.X., Wang, C.H., Wang, Z.T., 2012. Rare nor-iridoids from the roots of *Andrographis paniculata*. Phytochemistry 77, 275–279.

Xu, K., Jiang, J.S., Feng, Z.M., Yang, Y.N., Li, L., Zang, C.X., et al., 2016. Bioactive sesquiterpenoid and polyacetylene glycosides from *Atractylodes lancea*. J. Nat. Prod. 79 (6), 1567–1575.

Xu, J.Y., Dai, C., Shan, J.J., Xie, T., Xie, H.H., Wang, M.M., et al., 2018. Determination of the effect of *Pinellia ternata* (Thunb.) Breit on nervous system development by proteomics. J. Ehnopharmacol. 213, 221–229.

Yadav, V.S., Mishra, K.P., Singh, D.P., Mehrotra, S., Singh, V.K., 2005. Immunomodulatory effects of curcumin. Immunopharmacol. Immunotoxicol. 27 (3), 485–497.

Yamanoi, T., Aota, K., Momota, Y., Azuma, M., 2017. Treatment with the biscoclaurine alkaloid cepharanthine significantly increases salivary secretion in primary Sjögren's syndrome patients. J. Oral Health Biosci. 29 (2), 39–48.

Yan, X.T., Ding, Y., Li, W., Sun, Y.N., Yang, S.Y., Koh, Y.S., et al., 2014. Two new phenolic compounds from *Artemisia iwayomogi*. Helv. Chim. Acta 97 (2), 283–289.

Yao, H., Cui, P., Xu, D., Liu, Y., Tian, Q., Zhang, F., 2017. A water-soluble polysaccharide from the roots of *Polygala tenuifolia* suppresses ovarian tumor growth and angiogenesis *in vivo*. Int. J. Biol. Macromol. 107 (Pt A), 713–718.

Yavuz Kocaman, A., Güzelkokar, M., 2018. The genotoxic and antigenotoxic potential of the methanolic root extract of *Glycyrrhiza glabra* L. on human peripheral blood lymphocytes. Drug Chem. Toxicol. 2018, 1–8.

Yi, X., Feng, F., Xiang, Z., Ge, L., 2005. The effects of allitridin on the expression of transcription factors T-bet and GATA-3 in mice infected by murine cytomegalovirus. J. Med. Food 8 (3), 332–336.

Yiemwattana, I., Chaisomboon, N., Jamdee, K., 2018. Antibacterial and anti-inflammatory potential of *Morus alba* stem extract. Open Dent. J. 12, 265–274.

Yoo, D.G., Kim, M.C., Park, M.K., Park, K.M., Quan, F.S., Song, J.M., et al., 2012. Protective effect of ginseng polysaccharides on influenza viral infection. PLoS One 7 (3), e33678.

You, S., Nakanishi, E., Kuwata, H., Chen, J., Nakasone, Y., He, X., et al., 2013. Inhibitory effects and molecular mechanisms of garlic organosulfur compounds on the production of inflammatory mediators. Mol. Nutr. Food Res. 57 (11), 2049–2060.

Yu, X., Zhang, N., Lin, W., Wang, C., Gu, W., Ling, C., et al., 2018. Regulatory effects of four ginsenoside monomers in humoral immunity of systemic lupus erythematosus. Exp. Ther. Med. 15 (2), 2097–2103.

Zhang, D., Halaweish, F.T., 2007. Isolation and characterization of ribosome inactivating proteins from cucurbitaceae. Chem. Biodivers. 4 (3), 431–442.

Zhang, M., Su, N., Huang, Q., Zhang, Q., Wang, Y., Li, J., et al., 2017. Phosphorylation and antiaging activity of polysaccharide from *Trichosanthes* peel. J. Food Drug Analysis 25 (4), 976–983.

Zhang, J., Liu, L., Wang, J., Ren, B., Zhang, L., Li, W., 2018a. Formononetin, an isoflavone from *Astragalus membranaceus* inhibits proliferation and metastasis of ovarian cancer cells. J. Ethnopharmacol. 221, 91–99.

Zhang, M.Y., Wu, H.W., Xu, L.P., Yang, H.J., 2018b. Pharmacological effect of *Schisandrae chinensis fructus* and relative active components on cardiovascular and cerebrovascular diseases. Zhongguo Zhong yao za zhi Zhongguo zhongyao zazhi China Journal of Chinese Materia Medica 43 (8), 1536–1546.

Zhou, Y.Y., Xia, X., Peng, W.K., Wang, Q.H., Peng, J.H., Li, Y.L., et al., 2018. The effectiveness and safety of *Tripterygium wilfordii* Hook. F extracts in rheumatoid arthritis: a systematic review and meta-analysis. Front. Pharmacol. 9, 356.

Zhu, N.L., Sun, Z.H., Hu, M.G., Wu, T.Y., Yuan, J.Q., Wu, H.F., et al., 2017. New cassane diterpenoids from *Caesalpinia sappan* and their antiplasmodial activity. Molecules 22 (10), 1751.

Ziauddin, M., Phansalkar, N., Patki, P., Diwanay, S., Patwardhan, B., 1996. Studies on the immunomodulatory effects of Ashwagandha. J. Ethnopharmacol. 50 (2), 69–76.

Zou, L.C., Yan, Z.H., Zhu, T.F., Xiang, H., Deng, X.M., 2010. Two new phenylethanoid glycosides and a new secoiridoid glycoside from the roots of *Picrorhiza scrophulariiflora*. Chin. Chem. Lett. 21 (9), 1103–1106.

FURTHER READING

Agarwal, A., 2013. Immune-stimulatory and anti-inflammatory activities of *Curcuma longa* extract and its polysaccharide fraction. Pharmacognosy Res. 5 (2), 71–79.

SECTION 3

Pharmacokinetics, Interaction, and Toxicity Profile of Phytocompounds

CHAPTER 18

Herb and Modern Drug Interactions: Efficacy, Quality, and Safety Aspects

Zafar Mehmood[1], Mohammad Shavez Khan[2], Faizan Abul Qais[2], Samreen[2] and Iqbal Ahmad[2]
[1]Department of Microbiology, The Himalaya Drug Company, Dehradun, Uttarakhand, India
[2]Department of Agricultural Microbiology, Aligarh Muslim University, Aligarh, Uttar Pradesh, India

18.1 INTRODUCTION

Whole plant or parts of plant (leaves, fruit, roots, bark, flower, etc.) which provide health-promoting or curative properties are generally known as herbal medicine or phytomedicine. Herbalism implies the alternative health care outside conventional medicine. Advances in clinical research and improvement in quality control and analysis have contributed to recent resurgence of the use of alternative and traditional medicinal system and further establishing its importance in treating and preventing diseases (Ekor, 2014). Herbal products are available in both commercial and crude preparations, the latter being more often used in the developing countries and are formulated as mixture (Alissa, 2014). Plant-based medicines have been used long before recorded history; Chinese, Egyptian, and Indian systems of alternative and complimentary describe uses of plants since ages (Pan et al., 2014; Jaiswal et al., 2016). In India only, more than 40% of the total 17,000–18,000 flowering plants find their usage documented in traditional medicine systems. The global market for Indian traditional medicine is estimated to 120 billion USD, reflecting the high demand for Indian natural products (Jaiswal et al., 2016). It is estimated that about 4 billion people (about 80% world population), at least up to some parts of their primary health care, rely on herbal medicinal products (Ekor, 2014). Herbal products are also now becoming mainstream in the developed countries like European countries, North America, and Australia (Braun et al., 2010; Anquez-Traxler, 2011; Blendon et al., 2013). The use of herbal products in health care continues

New Look to Phytomedicine
DOI: https://doi.org/10.1016/B978-0-12-814619-4.00019-7

to grow globally along with continuous introduction of new products; issues related with safety are also increasingly recognized (Ekor, 2014).

Medicinal herbs consist of diverse pharmacologically active compounds, which are responsible for its multi-target effects, in contrast to conventional drugs, that have simple composition and definite mechanism of action. While the drug–drug interaction has been well established in terms of resulting metabolic impact, very little is known about the herb–drug interaction and the available evidence is rather weak (Wang, 2015). The combination of herb and drug in most of medical practices has been proved to be beneficial; nevertheless, reports from some studies have suggested adverse reactions (Tsai et al., 2012). Clinical research reported many cases of adverse effect of herb–drug interaction, although majority of them are devoid of any severe consequences. Most of the reports on adverse effect of herb–drug interaction come from case reports, but with little information and of poor quality. The highest level of evidence regarding the herb–drug interactions comes from case reports coupled with pharmacokinetic trials (Tsai et al., 2012; Izzo et al., 2016).

Coadministration of herbal product with drug may result in cross reactivity of their components with the drug or result into alteration in pharmacokinetics of the drug (Alissa, 2014). Pharmacokinetic studies, necessary to evaluate the overall effect of drug, are very difficult to carry out for medicinal plants because of their complex phytochemical profile (Mazzari and Prieto, 2014). Moreover, the phytopharmacological profile of plants also vary with climatic conditions, postharvest methodologies, part used, etc., making it more difficult to determine the clinical pharmacokinetic and pharmacodynamic effects. Even, standardization for active compound does not rule out the possible variation in other constituent, which possibly results into altered bioavailability and pharmacological activity in human (Alissa, 2014).

Despite the high risk, there is scarcity of literature regarding the herb–drug interaction. This chapter aims to briefly discuss the impact and mechanism of interaction with drug and adulterant-mediated toxicity of some commonly used herbal medicine. In the later sections of this chapter, an attempt has been made to address the clinical implications of herb–drug interaction in patient with specific diseases.

18.2 INTERACTION OF COMMONLY USED HERBS WITH DRUGS

Although herbal drugs or preparation are thought to be safe, the risk of possible herb–drug interaction cannot be completely ruled out (Hussain,

2011). In fact, approximately 30,000 herbal products with more than 1000 unique chemical substances increase the risk of herb–drug interactions (Ismail, 2009). Some of the common herbs with their possible outcome of interaction with drugs are discussed below.

18.2.1 Ginkgo (*Ginkgo biloba*)

Ginkgo (*Ginkgo biloba*) has been used traditionally for the treatment of circulatory disorders and to improve memory loss (Isah, 2015). Other uses of the standardized leaf extracts include management of cardiovascular diseases, cancer, tinnitus, vertigo, age-related muscular degeneration, and psychological disorders like schizophrenia, etc. (Xiong et al., 2014). Pharmacologically active constituents of Ginkgo are flavonoids and terpenoids belonging to different subclasses (Isah, 2015). The suggested therapeutic mechanisms of Ginkgo are its antioxidant effects, inhibition of beta amyloidal peptide, and modulation of different cell signaling receptors and factors (Mahadevan and Park, 2008; Ekor, 2014). Laboratory studies have shown that Ginkgo improves blood circulation and cognition, which is capable of inhibiting platelet activation factor as well as altering the bleeding time. Therefore, it could increase the effect of anticoagulant drugs (Jose Abad et al., 2010). Inductive effect of Ginkgo on CYP450 enzymes has also been thought to be connected with subtherapeutic levels of anticonvulsant drugs in blood serum during fatal seizures (Kupiec and Raj, 2005).

18.2.2 Kava Kava

Kava kava (*Piper methysticum*) is known as central nervous system (CNS) depressant and commonly used as anxiolytic agent. The active principle of the herb is kavalactones, which is reported to be a potent inhibitor of CYP450 enzymes, suggesting high risk of pharmacokinetic interaction with drugs which are metabolized by the same CYP450 (Singh, 2005). Kavalactones are also shown to interact with GABA receptors and sodium and potassium channels (Savage et al., 2017). There are reports of significant interaction of kava with other CNS depressant drugs, and hence, concurrent use with CNS depressant drugs are not advised (Ekor, 2014).

18.2.3 Saw Palmetto (*Serenoa repens*)

Saw palmetto (*Serenoa repens*) is used for treatment of benign prostatic hypertrophy (BPH) in the United States. It is also used as a diuretic and urinary antiseptic (Barnes et al., 2007). In vitro studies have shown that Saw palmetto inhibits alpha adrenergic receptors (Goepel et al., 1999), which could be a possible molecular mechanism of action underling its therapeutic potential (Tachjian et al., 2010). However, clinical investigation did not demonstrate any beneficial effect of Saw palmetto on BPH symptoms or on post void residual bladder volume (Bent et al., 2006). Saw palmetto has been reported to inhibit cyclooxygenase enzyme and increased bleeding when concurrently administered with warfarin (Bressler, 2005).

18.2.4 St. John's Wort

St. John's wort (*Hypericum perforatum*) has been recommended traditionally for various medical conditions but is more commonly used for treatment of depression (Borrelli and Izzo, 2009). Similar to other herbs, the most common for drug interaction in human CYP450 is also shown to be modulated by *H. perforatum* (Gurley et al., 2008; Šemeláková et al., 2016). St. John's wort has also shown to induce the expression of *P*-glycoprotein in vitro and in vivo (Schwarz et al., 2007). The active constituent, hyperforin, has shown to inhibit the reuptake of several neurotransmitters such dopamine, serotonin, glutamate, etc. (Chatterjee et al., 1998). Other constituents like flavonoids, hypericin, and aristoforin are also shown to be interacting molecules (Rahimi and Abdollahi, 2012). St. John's wort has shown to decrease bioavailability of immunosuppressant, antibiotics, and bronchodilator anticoagulant drugs while showing adverse effect when coadministered with oral contraceptives (Velingkar et al., 2017). Although the herb–drug interactions are generally observed with prolong use of *H. perforatum* with a number of prescription drugs used today (Borrelli and Izzo, 2009).

18.2.5 Valerian

Valerian (*Valeriana officinalis*) is commonly used as a sleep aid and is widely available in different commercial preparations (Taibi et al., 2007). Although there has been speculation of possible induction of CYP450 isoenzymes by Valerian which could adversely affect the efficacy of anticancer drugs, however in vitro and preclinical studies have failed to reach on any logical conclusion supporting the adverse interaction of Valerian with anticancer drugs. Conversely, in vivo studies have shown that Valerian and haloperidol cotreatment produce additive affect in experimental rats (Fachinetto et al., 2007). In another study on mice, it was observed that Valerian and alprazolam significantly increased the time spent in the one arm indicating anxiolytic effects (Bhatt et al., 2013).

18.2.6 Echinacea Preparations

Echinacea preparations are obtained from *Echinacea purpurea*, and other *Echinacea* species are most commonly taken by HIV-infected patients, as an immunostimulant and for early treatment of upper respiratory tract infections (Ladenheim et al., 2008; Sharma et al., 2010). The pharmacologically active constituents of Echinacea are caffeic acid derivatives and alkamides which can potentially interact with CYP450 enzymes and increase the bioavailability of certain drug (Gorski et al., 2004). However, clinical studies have shown that coadministration of Echinacea did not affect the pharmacokinetics of antiretroviral drugs (darunavir or ritonavir) and has shown no adverse effect (Moltó et al., 2011). Recently, regulatory effect of *E. purpurea* on efflux transporters has been studied (Awortwe et al., 2017). It was observed that *E. purpurea* modulates the transporters expression via down regulation of related microRNA. The results could be applied to predict possible interaction of Echinacea with related drugs.

18.2.7 Evening Primrose (*Oenothera biennis*)

Evening primrose (*Oenothera biennis*) is a wild plant of medicinal importance, seed oil is traditionally used for treatment of eczema, asthma, rheumatoid arthritis, premenstrual and menopausal syndrome, and other inflammation-related disorders (Dante and Facchinetti, 2011; Ng et al., 2013; Triantafyllidi et al., 2015; Nikfarjam et al., 2016). The main constituent, the seed oil, is gamma linoleic acid along with other phenolic constituents (Munir et al., 2017). The oil may interact with variety of drug such as monoamine oxidase inhibitors, antiepileptic drugs, antipsychotic drugs, analgesics, etc. (Rouhi-Boroujeni et al., 2015). However, no inhibitory effect on CYP3A4 (cytochrome P450 enzyme) was recorded following the treatment with evening primrose oil in human erythrocytes (Loretz and Li, 2018).

18.3 MECHANISM OF HERB–DRUG INTERACTION

The mechanisms for drug interaction can be divided into two general categories, that is, pharmacokinetics and pharmacodynamic.

18.3.1 Pharmacokinetic Interactions

The pharmacokinetic interactions involve the basic knowledge, that is, from absorption till excretion of a drug. It is identified by the changes in drug concentration in serum and controlling the patient's clinical manifestations. Pharmacokinetic interactions involve the entire process from absorption to excretion.

18.3.1.1 Absorption

The interactions that affect the absorption of drug result in either increased or decreased absorption (Boullata and Nace, 2000; Anastasi et al., 2011; Williamson, 2003; Izzo and Ernst, 2001; Scott and Elmer, 2002; Baxter et al., 2013). Absorption of drug is affected by intestinal motility and changes in intestinal pH, complexing mechanisms, and drugs affecting intestinal motility (Williamson, 2003). Numerous herbs such as guar gum (*Cyamopsis tetragonoloba*), aloe leaf (*Aloe vera*), and senna (*Cassia angustifolia*), ingredients of herbal weight-loss products, might reduce the intestinal transit time and subsequently drug absorption. The absorption of *P*-glycoprotein substrates like digoxin has been documented to be deceased by St. John's wort which induces intestinal *P*-glycoprotein (Williamson, 2003; Scott and Elmer, 2002; Baxter et al., 2013). Such type of reduced absorption can be minimized if the drug is consumed 2 h after or 1 h before the herb (Kuhn, 2002). A large number of herbs have water soluble hydrocolloidal carbohydrates (e.g., mucilage and gums) but that are poorly absorbable, for example, rhubarb (*Rheum palmatum*), psyllium (*Plantago ovata*), flaxseed (*Linum usitatissimum*), aloe (*A. vera*), and marshmallow (*Althaea officinalis*), etc. These compounds have tendency to bind other drugs, especially when consumed in powdered forms or as a whole. Absorption of lithium is known to be inhibited by psyllium, an herb containing mucilage. Consumption of aloe and rhubarb can cause diarrhea and reduce the action of drugs particularly that have a narrow therapeutic index (e.g., warfarin and digoxin). Therefore, the binding of herb to the drug can be prevented by alternative consumption, that is, not at the same time.

18.3.1.2 Distribution

The distribution of drug is also affected by the herb. Drugs exhibiting high plasma protein binding affinity (e.g., carbamazepine and warfarin) have a narrow distribution volume and might get displaced by an herb that competes for the same binding sites (Kuhn, 2002; Williamson, 2003;

Scott and Elmer, 2002). Detachment of drug from carrier protein in concurrent drug administration increases the levels of drug in and finally enhancing the therapeutic effect (Boullata and Nace, 2000; Williamson, 2003; Scott and Elmer, 2002). The side effects due to overdose of high protein bound drugs such as carbamazepine and warfarin increase when they are displaced by black willow and meadowsweet (herbs containing pain reducing salicylates). Hence, concurrent administration of such combination of herbs and drugs should be avoided (Kuhn and Winston, 2000; Smith, 2000).

18.3.1.3 Metabolism

18.3.1.3.1 Enzyme Induction

Certain herbs are known to stimulate the production and activity of enzymes that may degrade and eliminate the drug from body, reducing the overall amount of drug (Boullata and Nace, 2000; Williamson, 2003; Scott and Elmer, 2002). For example, cytochrome P450 enzymes are induced by St. John's wort which is responsible for metabolism of numerous drugs (Kuhn, 2002; Boullata and Nace, 2000; Anastasi et al., 2011; Williamson, 2003; Izzo and Ernst, 2001; Scott and Elmer, 2002; Baxter et al., 2013). The efficacy of protease inhibitors (theophylline and carbamazepine) and orally taken contraceptive pill is reduced by this mechanism. Similar is the case for reduction of serum levels of warfarin and digoxin (Anastasi et al., 2011; Williamson, 2003; Izzo and Ernst, 2001; Scott and Elmer, 2002; Baxter et al., 2013). The metabolism of corticosteroids is decreased by liquorice (an herb) leading to toxic effects due to accumulation of corticosteroids. St. John's wort is reported to induce hepatic microsomal enzymes in cytochrome P450 system which increases the metabolism of many drugs including theophylline and digoxin, cyclosporine, and protease inhibitors. In such conditions, the efficacy of dug is hampered; therefore, simultaneous administration of liquorice with these drugs is not recommended (Kuhn and Winston, 2000; Smith, 2000).

18.3.1.3.1.1 Enzyme inhibition The level of drugs or their clearance time may get increased by the production of enzymes required for metabolism or breaking down the drug (Williamson, 2003; Scott and Elmer, 2002; Baxter et al., 2013). In enzyme induction, it may take several days or up to weeks to develop completely, while enzyme inhibition can occur within few days causing a hasty progress of toxicity (Baxter et al., 2013). Licorice has been reported to decrease the metabolism of corticosteroids, causing the toxic and adverse effects due to buildup of corticosteroids (Kuhn, 2002; Baxter et al., 2013; Barnes et al., 2007). There are evidences of in vitro inhibition of cytochrome P450 and isoenzyme CYP3A4 by chamomile and Echinacea. Simultaneous use of drugs such as simvastatin, alprazolam, calcium-channel blockers, and protease inhibitors may increase serum drug levels and cause adverse effects (Izzo and Ernst, 2001; Scott and Elmer, 2002; Baxter et al., 2013; Fetrow and Avila, 2001; Barnes et al., 2007).

18.3.1.3.2 Excretion

The changes in excretion or clearance time may also affect serum drug levels. Herbal diuretics are quite weak and unlikely to cause large problems (Williamson, 2003) Administration of licorice for very long period may cause hypokalemia and water retention and ultimately may interfere with certain medications such as antiarrhythmic and antihypertensive, and agents (Izzo and Ernst, 2001; Scott and Elmer, 2002).

18.3.2 Pharmacodynamic Interactions

18.3.2.1 Additive Interactions

An herb and a drug may produce similar effect and cause amplified drug effect without increasing the amount of either one (Williamson, 2003; Scott and Elmer, 2002; Baxter et al., 2013). Anticoagulants, herbal sedatives, and antihypertensives might increase the effect conventional drug taken of a concurrently for the same purpose. For example, the anticoagulant action of warfarin is enhanced by garlic, gingko, and ginger. Similarly, Valerian increases the hypnotic activity of benzodiazepines (Kuhn, 2002; Williamson, 2003; Scott and Elmer, 2002; Baxter et al., 2013; Fetrow and Avila, 2001; Barnes et al., 2007). Consumption of serotonergic drugs such as selective serotonin reuptake inhibitors (SSRIs) with St John's wort causes serotonin syndrome which can be identified by autonomic dysfunction, altered mental status, and neuromuscular abnormalities (Baxter et al., 2013).

18.3.2.2 Antagonistic Interactions

Sometimes, an herb may produce the contrary effect as that of desired for the drug, thereby reducing the effect of drug (Williamson, 2003; Scott and Elmer, 2002; Baxter et al., 2013). Ephedra or caffeine-containing herbs (cola nut, guarana, mate, and green tea), often used in combination for the additive cardiovascular effects in many herbal weight-loss products, may antagonize the effects of antihypertensive medications (Williamson, 2003; Scott and Elmer, 2002).

18.4 ADVERSE EFFECTS AND INTERACTIONS

Like any other medicinal agents, herbs also have undesirable effects such as toxicity (Kuhn, 2002; Boullata and Nace, 2000; Anastasi et al., 2011; Williamson, 2003; Izzo and Ernst, 2001; Scott and Elmer, 2002; Baxter et al., 2013; Fetrow and Avila, 2001; Barnes et al., 2007). The adverse effect risk of herbs are associated with user's gender, age, genetics, nutrition status, and concurrent disease treatments (Boullata and Nace, 2000; Anastasi et al., 2011; Scott and Elmer, 2002). Clinically, the recognition of side effects of herbs is not on routine basis but such is less frequently reported (Boullata and Nace, 2000; Anastasi et al., 2011; Williamson, 2003; Izzo and Ernst, 2001; Scott and Elmer, 2002; Baxter et al., 2013). There should be knowledge and awareness of toxicological impact of any herbal formulation in presence of prescribed modern drugs. Adverse effects of herb–drug interactions are associated with liver, skin, GI tract, and heart. Administration of *Echinacea* with other drugs known for hepatotoxicity exerts significant hepatotoxic effects and hence should be avoided (Boullata and Nace, 2000). There is very less evidence-based information available about toxicity of herb–drug interactions. Most of the information available are based on speculation or on theoretical interactions, in which many have toxicological profile in vitro only but not in vivo (Williamson, 2003; Izzo and Ernst, 2001; Scott and Elmer, 2002; Baxter et al., 2013). In vivo studies have found weaker/lesser adverse effect compared to those in vitro which warranted a clinical study (Baxter et al., 2013). Herbal component of herb–drug system may cause decrease/increase in the amount of drug available in the blood stream (Williamson, 2003; Baxter et al., 2013). Mostly, herb–drug interactions are based on the same pharmacodynamic and pharmacokinetic principles (Kuhn, 2002; Williamson, 2003; Scott and Elmer, 2002; Baxter et al., 2013).

18.5 INTERACTION RISKS IN SPECIFIC PATIENT POPULATIONS

Herb–drug interactions are known to cause beneficial effects and enhance the health status in patients. However, there are also certain potential risk factors involved in dietary supplements of the patients along with the risk factors involved in modern medication such as cardiovascular medications, anticoagulants, diabetic medications, psychiatric medications, laxatives, and medications for HIV infection. The risk factors involved in individual ailment conditions are discussed below.

18.5.1 Patients Receiving Cardiovascular Medications

St. John's wort is used by patients along with medications of cardiovascular diseases for the management of mood disorders, and it has been observed that it reduces the levels of statins and verapamil in serum (Portoles et al., 2006). There should be close and periodic monitoring of lipid levels and blood pressure if a patient is taking St. John's wort with such drugs. Induction of transport *P*-glycoprotein, cytochrome P450, isoenzymes cytochrome P3A4, cytochrome P1A, and cytochrome P2C9 are the speculated mechanisms of St. John's wort interactions that lead to decrease in levels of medications. A study reported that there was 25% decrease in blood levels of digoxin, when administered with St. John's wort, likely to be due to induction of *P*-glycoprotein and subsequently resulting in decrease in bioavailability of digoxin (Johne et al., 1999; Tian et al., 2005). Ginseng, another common herb, increased the digoxin serum levels in case report of one patient (Mcrae, 1996). Alkaloids, the major constituent of bark or extract of yohimbe (*Pausinystalia yohimbe*), are documented to increase the blood pressure more in hypertensive subjects compared to normotensive patients (De Smet, 1997). This plant is used conventionally for the treatment of erectile dysfunction. The interaction of yohimbine with tricyclic antidepressants causes much more hypertensive effects (Lacomblez et al., 1989). Foxglove (*Digitalis purpurea*), a less used herb, is known to have cardiac glycosides; it exerts an additive effect in presence of digitaloid cardiac glycosides (Brinker, 1997). Therefore, levels of digoxin should be monitored carefully in the patients taking such herbal formulations along with cardiovascular medications.

18.5.2 Patients Receiving Diabetes Medications

The consumption of hypoglycemic herbs affects the glucose balance in both noninsulin-dependent (type 2) and insulin-dependent (type 1) patients. Traditionally, more than 400 plants are used for the management of hyperglycemia. Among these, leaf juice of *Aloe barbadensis* (Yongchaiyudha et al., 1996) and seeds of *Trigonella foenum-graecum*

(Al-Habori and Raman, 1998) are common herbs known for hypoglycemic effects. The fruit of bitter melon/karela (*Momordica charantia*) is reported to improve the glucose tolerance without affecting levels of insulin (Welihinda et al., 1986; Leatherdale et al., 1981). Clinical trials have found that ethanolic fraction of leaves of *Gymnema sylvestre* reduced the requirements of insulin in both type 1 and type 2 diabetes, and the effects were comparable to glibenclamide and *A. barbadensis* juice (Baskaran et al., 1990; Bunyapraphatsara et al., 1996). *Panax ginseng* lowers blood sugar level in patients with diabetes, and this effect might be additive in patients taking insulin or oral hypoglycemics.

There is possibility of additive effects when such herbs with hypoglycemic potential interact with modern antidiabetic drugs. So, there is a need of suitable monitoring of blood sugar level and clear lines of communication between health-care practitioner and patient to avert problems.

18.5.3 Patients Taking Bulk Laxatives

Numerous laxatives such as psyllium are dietary supplements which are not considered as medications by many patients, but such supplements may diminish or slowdown the absorption of certain drugs. The absorption of carbamazepine can be reduced by psyllium, lowering the serum concentration of drug (Etman, 1995). Moreover, it is reported that absorption of lithium was decreased with coadministration of psyllium. Broadly, bulk laxatives should not be taken concurrently with other modern medications, and their dose time should be separated (by hours) for sufficient absorption of drug to take place. Herbal laxatives containing hydrocolloid fiber such as psyllium (*Plantago* spp.), guar gum (*Cyamopsis tetragonolobus*), and konjac (*Amorphophallus rivieri*), when orally taken in adequate amount, may delay gastric emptying and subsequently reducing the absorption rate of carbohydrates and several drugs including glibenclamide, lithium, lovastatin, tricyclics, and digoxin (Stewart, 1992; Brown et al., 1978). Another mechanism involved is the increase in rate of intestinal transit resulting in decrease of intestinally absorbed drugs, for example, species (*Cassia acutifolia, Cassia senna, C. angustifolia, Rhamnus purshiana, Rheum officinale, Rumex crispus, Rhamnus frangula*, and *A. vera*) (Westendorf, 1993).

18.5.4 Patients Receiving Anticoagulants

Certain coumarins and warfarin are known to interact with many drugs, medicinal herbs, and foods. In vitro, ginkgo (*G. biloba*) and garlic (*Allium sativum*), is known to interfere with platelet function and may increase the risk of bleeding in combination with warfarin. Garlic has been reported to inhibit fibrinolytic activity and platelet aggregation in patients with coronary artery disease as well as in healthy subjects and may also be responsible for postoperative bleeding and spontaneous spinal epidural hematoma (Burnham, 1995; Rose et al., 1990). Ginkgolides, an important constituent of ginkgo, antagonizes the platelet-activating factor. Moreover, some anticoagulant coumarins present in herbs may have an additive effect when concurrently administered with pharmaceutical anticoagulants (Fugh-Berman and Ernst, 2001).

A report on 46-year-old woman who was stabilized on warfarin found that administration of *Angelica sinensis*, a Chinese herb, for 4 weeks, doubled prothrombin time (PT) and international normalized ratio. These parameters got normalized within 1 month after withdrawal of the herb (Page and Lawrence, 1999). On the contrary, the extract of this plant did not have any noticeable effect on both warfarin pharmacokinetics and baseline PT. Therefore, caution should be taken in taking warfarin along with *A. sinensis*.

Similarly, extract of *Tanacetum parthenium* and its parthenolide is known to inhibit platelet aggregation by inhibiting the serotonin release. However, no cases of abnormal coagulation tests and bleeding episodes have been reported. Ginger inhibits the platelet aggregation induced by epinephrine, arachidonic acid, and adenosine diphosphate. There are numerous case reports of interactions of warfarin with St. John's wort, garlic, ginkgo, and ginseng (Vaes and Chyka, 2000; Hu et al., 2005). Combination of nonsteroidal anti-inflammatory drugs, mainly aspirin with ginkgo, has found to result in severe bleeding and intracranial bleeding (Meisel et al., 2003; Abebe, 2002; Bebbington et al., 2005). *Panax quinquefolius* has also been reported to decrease serum levels of warfarin in humans, causing the less anticoagulation (Yuan et al., 2004). Clinical trials on *Eleutherococcus senticosus* has not been performed but caution should be taken as it contains a constituent known to inhibit platelet aggregation.

18.5.5 Patients Receiving HIV Medications

It is known that most of the antiretroviral medications are metabolized via *P*-glycoprotein and CYP3A4 systems. As dietary supplements induce these systems, hence such interactions may lead to reduction in serum levels of antiretrovirals. Among dietary supplements, St. John's wort is the most evident to affect these systems (Lee et al., 2006). Previously, it has been found in patients that vitamin C and garlic reduce the serum concentrations, although the clinical study was limited (Gallicano et al., 2003; Mills et al., 2005). In vitro found the inhibitory effect of *Echinacea* species, milk thistle, and goldenseal on CYP450 enzymes, but no relevant effects were obtained in clinical studies (Van den Bout-van den Beukel et al., 2006). Hence, the patients taking HIV medications should be carefully monitored for taking such supplements, especially for St. John's wort as there is risk of a harmful interaction.

18.5.6 Patients Receiving Psychiatric Medications

Serotonin levels could be affected by the interaction of St. John's wort as it has been found in case of serotonin syndrome where patients were receiving a SSRI (Hammerness et al., 2003) and suggested the withdrawal of St. John's wort when an SSRI is being administered and cautioned such patients (Singh, 2005). The levels of psychiatric medications in serum are decreased by St. John's wort as psychiatric drugs are metabolized by CYP450 enzymatic system. Serum concentration of certain benzodiazepines and tricyclic antidepressants has been shown to be affected; however, these alterations might not result in a clinical effect (Izzo, 2004; Markowitz and DeVane, 2001).

18.6 CONCLUSION AND FUTURE DIRECTION

The use of herbal drugs in health care has increased tremendously in the last few decades both in primary health care and management of chronic and life style diseases. In many instances, herbal drugs are used simultaneously with modern drugs. In general, all drugs with a narrow therapeutic index may either have increased adverse effects or be less effective when used in conjunction with herbal products. Concurrent use of herbs may mimic, magnify, or oppose the effect of drugs. The better understanding between of herb–drug interactions has been explored in limited number of cases. In many cases, adverse drug interactions are noticed; therefore, health-care practitioners should make patients cautious against mixing herbs and pharmaceutical drugs. Because physicians are likely to encounter patients who are using herbal remedies, they need to be aware of the purported effects of these products.

Today, our understanding of the interactions between drugs and herbs and between drugs and food is still in its infancy. Herbal medicines still need to be studied scientifically although the experience obtained from their traditional use over the years should not be ignored. Now, it is mandatory to strengthen research in the evaluation of the safety and efficacy of herbal medicines and promote the rational use.

REFERENCES

Abebe, W., 2002. Herbal medication: potential for adverse interactions with analgesic drugs. J. Clin. Pharm. Ther. 27 (6), 391–401.

Al-Habori, M., Raman, A., 1998. Antidiabetic and hypocholesterolaemic effects of fenugreek. Phytother. Res. 12 (4), 233–242.

Alissa, E.M., 2014. Medicinal herbs and therapeutic drugs interactions. Ther. Drug Monit. 36 (4), 413–422.

Anastasi, J.K., Chang, M., Capili, B., 2011. Herbal supplements: talking with your patients. J. Nurse Pract. 7 (1), 29–35.

Anquez-Traxler, C., 2011. The legal and regulatory framework of herbal medicinal products in the European Union: a focus on the traditional herbal medicines category. Drug Inf. J. 45 (1), 15–23.

Awortwe, C., Kaehler, M., Rosenkranz, B., Cascorbi, I., Bruckmueller, H., 2017. MicroRNA-655-3p regulates *Echinacea purpurea* mediated activation of ABCG2. Xenobiotica 1–9.

Barnes, J., Anderson, L.A., Phillipson, J.D., 2007. Herbal Medicines (Electronic version). Pharmaceutical Press, London.

Baskaran, K., Ahamath, B.K., Shanmugasundaram, K.R., Shanmugasundaram, E.R.B., 1990. Antidiabetic effect of a leaf extract from *Gymnema sylvestre* in non-insulin-dependent diabetes mellitus patients. J. Ethnopharmacol. 30 (3), 295–305.

Baxter, K., Driver, S., Williamson, E. (Eds.), 2013. Stockley's Herbal Medicines Interactions. Pharmaceutical Press.

Bebbington, A., Kulkarni, R., Roberts, P., 2005. *Ginkgo biloba*: persistent bleeding after total hip arthroplasty caused by herbal self-medication. J. Arthroplasty 20 (1), 125–126.

Bent, S., Kane, C., Shinohara, K., Neuhaus, J., Hudes, E.S., Goldberg, H., et al., 2006. Saw palmetto for benign prostatic hyperplasia. N. Engl. J. Med. 354 (6), 557–566.

Bhatt, C., Kanaki, N., Nayak, R., Shah, G., 2013. Synergistic potentiation of anti-anxiety activity of valerian and alprazolam by liquorice. Indian J. Pharmacol. 45 (2), 202.

Blendon, R.J., Benson, J.M., Botta, M.D., Weldon, K.J., 2013. Users' views of dietary supplements. JAMA Intern. Med. 173 (1), 74–76.

Borrelli, F., Izzo, A.A., 2009. Herb–drug interactions with St John's wort (*Hypericum perforatum*): an update on clinical observations. AAPS J. 11 (4), 710.

Boullata, J.I., Nace, A.M., 2000. Safety issues with herbal medicine. Pharmacother.: J. Human Pharmacol. Drug Ther. 20 (3), 257–269.

Braun, L.A., Tiralongo, E., Wilkinson, J.M., Spitzer, O., Bailey, M., Poole, S., et al., 2010. Perceptions, use and attitudes of pharmacy customers on complementary medicines and pharmacy practice. BMC Complement. Altern. Med. 10 (1), 38.

Bressler, R., 2005. Herb–drug interactions. Interactions between saw palmetto and prescription medications. Geriatrics (Basel, Switzerland) 60 (11), 32–34.

Brinker, F.J., 1997. Interactions of pharmaceutical and botanical medicines. J. Naturopathic Med. 7 (2), 14–20.

Brown, D.D., Juhl, R.P., Warner, S.L., 1978. Decreased bioavailability of digoxin due to hypocholesterolemic interventions. Circulation 58 (1), 164–172.

Bunyapraphatsara, N., Yongchaiyudha, S., Rungpitarangsi, V., Chokechaijaroenporn, O., 1996. Antidiabetic activity of *Aloe vera* L. juice II. Clinical trial in diabetes mellitus patients in combination with glibenclamide. Phytomedicine 3 (3), 245–248.

Burnham, B.E., 1995. Garlic as a possible risk for postoperative bleeding. Plast. Reconstr. Surg. 95 (1), 213.

Chatterjee, S.S., Bhattacharya, S.K., Wonnemann, M., Singer, A., Müller, W.E., 1998. Hyperforin as a possible antidepressant component of hypericum extracts. Life Sci. 63 (6), 499–510.

Dante, G., Facchinetti, F., 2011. Herbal treatments for alleviating premenstrual symptoms: a systematic review. J. Psychosom. Obstet. Gynaecol. 32 (1), 42–51.

De Smet, P.A.G.M., 1997. Yohimbe alkaloids – general discussion. Adverse Effects of Herbal Drugs. Springer, Berlin, Heidelberg, pp. 181–205.

Ekor, M., 2014. The growing use of herbal medicines: issues relating to adverse reactions and challenges in monitoring safety. Front. Pharmacol. 4, 177.

Etman, M.A., 1995. Effect of a bulk forming laxative on the bioavailability of carbamazepine in man. Drug Dev. Ind. Pharm. 21 (16), 1901–1906.

Fachinetto, R., Villarinho, J.G., Wagner, C., Pereira, R.P., Ávila, D.S., Burger, M.E., et al., 2007. *Valeriana officinalis* does not alter the orofacial dyskinesia induced by haloperidol in rats: role of dopamine transporter. Prog. Neuropsychopharmacol. Biol. Psychiatry 31 (7), 1478–1486.

Fetrow, C.W., Avila, J.R., 2001. Professional's Handbook of Complementary & Alternative Medicines. Springhouse Publishing Company.

Fugh-Berman, A., Ernst, E., 2001. Herb–drug interactions: review and assessment of report reliability. Br. J. Clin. Pharmacol. 52 (5), 587–595.

Gallicano, K., Foster, B., Choudhri, S., 2003. Effect of short-term administration of garlic supplements on single-dose ritonavir pharmacokinetics in healthy volunteers. Br. J. Clin. Pharmacol. 55 (2), 199–202.

Goepel, M., Hecker, U., Krege, S., Rübben, H., Michel, M.C., 1999. Saw palmetto extracts potently and noncompetitively inhibit human α1-adrenoceptors in vitro. Prostate 38 (3), 208–215.

Gorski, J.C., Huang, S.M., Pinto, A., Hamman, M.A., Hilligoss, J.K., et al., 2004. The effect of Echinacea (*Echinacea purpurea* root) on cytochrome P450 activity in vivo. Clin. Pharmacol. Ther. 75 (1), 89–100.

Gurley, B.J., Swain, A., Hubbard, M.A., Williams, D.K., Barone, G., Hartsfield, F., et al., 2008. Clinical assessment of CYP2D6-mediated herb–drug interactions in humans: effects of milk thistle, black cohosh, goldenseal, kava kava, St. John's wort, and Echinacea. Mol. Nutr. Food Res. 52 (7), 755–763.

Hammerness, P., Basch, E., Ulbricht, C., Barrette, E.P., Foppa, I., Basch, S., et al., 2003. St. John's wort: a systematic review of adverse effects and drug interactions for the consultation psychiatrist. Psychosomatics 44 (4), 271–282.

Hu, Z., Yang, X., Ho, P.C.L., Chan, S.Y., Heng, P.W.S., Chan, E., et al., 2005. Herb–drug interactions. Drugs 65 (9), 1239–1282.

Hussain, S., 2011. Patient counseling about herbal-drug interactions. Afr. J. Tradit. Complement. Altern. Med. 8 (5S).

Isah, T., 2015. Rethinking *Ginkgo biloba* L.: medicinal uses and conservation. Pharmacogn. Rev. 9 (18), 140.

Ismail, M.Y.M., 2009. Herb–drug interactions and patient counseling. Int. J. Pharm. Pharm. Sci. 1 (Suppl 1), S151–S161.

Izzo, A.A., 2004. Drug interactions with St. John's wort (*Hypericum perforatum*): a review of the clinical evidence. Int. J. Clin. Pharmacol. Ther. 42 (3), 139–148.

Izzo, A.A., Ernst, E., 2001. Interactions between herbal medicines and prescribed drugs. Drugs 61 (15), 2163–2175.

Izzo, A.A., Hoon-Kim, S., Radhakrishnan, R., Williamson, E.M., 2016. A critical approach to evaluating clinical efficacy, adverse events and drug interactions of herbal remedies. Phytother. Res. 30 (5), 691–700.

Jaiswal, Y., Liang, Z., Zhao, Z., 2016. Botanical drugs in Ayurveda and traditional Chinese medicine. J. Ethnopharmacol. 194, 245–259.

Johne, A., Brockmöller, J., Bauer, S., Maurer, A., Langheinrich, M., Roots, I., 1999. Pharmacokinetic interaction of digoxin with an herbal extract from St John's wort (*Hypericum perforatum*). Clin. Pharmacol. Ther. 66 (4), 338–345.

Jose Abad, M., Miguel Bedoya, L., Bermejo, P., 2010. An update on drug interactions with the herbal medicine *Ginkgo biloba*. Curr. Drug Metab. 11 (2), 171–181.

Kuhn, M.A., 2002. Herbal remedies: drug–herb interactions. Crit. Care Nurse 22 (2), 22–32.

Kuhn, M.A., Winston, D., 2000. Herbal Therapy and Supplements: A Scientific and Traditional Approach. Lippincott Williams & Wilkins.

Kupiec, T., Raj, V., 2005. Fatal seizures due to potential herb–drug interactions with *Ginkgo biloba*. J. Anal. Toxicol. 29 (7), 755–758.

Lacomblez, L., Bensimon, G., Isnard, F., Diquet, B., Lecrubier, Y., Puech, A.J., 1989. Effect of yohimbine on blood pressure in patients with depression and orthostatic hypotension induced by clomipramine. Clin. Pharmacol. Ther. 45 (3), 241–251.

Ladenheim, D., Horn, O., Werneke, U., Phillpot, M., Murungi, A., Theobald, N., et al., 2008. Potential health risks of complementary alternative medicines in HIV patients. HIV Med. 9 (8), 653–659.

Leatherdale, B.A., Panesar, R.K., Singh, G., Atkins, T.W., Bailey, C.J., Bignell, A.H., 1981. Improvement in glucose tolerance due to *Momordica charantia* (Karela). Br. Med. J. (Clin. Res. Ed.). 282 (6279), 1823–1824.

Lee, L.S., Andrade, A.S., Flexner, C., 2006. Interactions between natural health products and antiretroviral drugs: pharmacokinetic and pharmacodynamic effects. Clin. Infect. Dis. 43 (8), 1052–1059.

Loretz, C., Li, A.P., 2018. Evaluation of herb–drug interactions with metmax™ pooled donor human enterocytes: results with twenty eight commonly used herbal supplements. Drug Metab. Pharmacokinet. 33 (1), S62.

Mahadevan, S., Park, Y., 2008. Multifaceted therapeutic benefits of *Ginkgo biloba* L.: chemistry, efficacy, safety, and uses. J. Food Sci. 73 (1), R14–9.

Markowitz, J.S., DeVane, C.L., 2001. The emerging recognition of herb–drug interactions with a focus on St. John's wort (*Hypericum perforatum*). Psychopharmacol. Bull. 35 (1), 53–64.

Mazzari, A.L., Prieto, J.M., 2014. Herbal medicines in Brazil: pharmacokinetic profile and potential herb–drug interactions. Front. Pharmacol. 5, 162.

Mcrae, S., 1996. Elevated serum digoxin levels in a patient taking digoxin and Siberian ginseng. Can. Med. Assoc. J. 155 (3), 293.

Meisel, C., Johne, A., Roots, I., 2003. Fatal intracerebral mass bleeding associated with *Ginkgo biloba* and ibuprofen. Atherosclerosis 167 (2), 367.

Mills, E., Montori, V., Perri, D., Phillips, E., Koren, G., 2005. Natural health product–HIV drug interactions: a systematic review. Int. J. STD AIDS 16 (3), 181–186.

Moltó, J., Valle, M., Miranda, C., Cedeño, S., Negredo, E., Barbanoj, M.J., et al., 2011. Herb–drug interaction between *Echinacea purpurea* and darunavir–ritonavir in HIV-infected patients. Antimicrob. Agents Chemother. 55 (1), 326–330.

Munir, R., Semmar, N., Farman, M., Ahmad, N.S., 2017. An updated review on pharmacological activities and phytochemical constituents of evening primrose (genus *Oenothera*). Asian Pac. J. Trop. Biomed. 7 (11), 1046–1054.

Ng, S.C., Lam, Y.T., Tsoi, K.K.F., Chan, F.K.L., Sung, J.J.Y., Wu, J.C.Y., 2013. Systematic review: the efficacy of herbal therapy in inflammatory bowel disease. Aliment. Pharmacol. Ther. 38 (8), 854–863.

Nikfarjam, M., Bahmani, M., Heidari-Soureshjani, S., 2016. Phytotherapy for depression: a review of the most important medicinal plants of flora of Iran effective on depression. J. Chem. Pharm. Sci. 9 (3), 1242–1247.

Page, R.L., Lawrence, J.D., 1999. Potentiation of warfarin by dong quai. Pharmacother.: J. Human Pharmacol. Drug Ther. 19 (7), 870–876.

Pan, S.Y., Litscher, G., Gao, S.H., Zhou, S.F., Yu, Z.L., Chen, H.Q., et al., 2014. Historical perspective of traditional indigenous medical practices: the current renaissance and conservation of herbal resources. Evid. Based Complement. Alternat. Med. 40, 52–53.

Portoles, A., Terleira, A., Calvo, A., Martinez, I., Resplandy, G., 2006. Effects of *Hypericum perforatum* on ivabradine pharmacokinetics in healthy volunteers: an open-label, pharmacokinetic interaction clinical trial. J. Clin. Pharmacol. 46 (10), 1188–1194.

Rahimi, R., Abdollahi, M., 2012. An update on the ability of St. John's wort to affect the metabolism of other drugs. Expert Opin. Drug Metab. Toxicol. 8 (6), 691–708.

Rose, K.D., Croissant, P.D., Parliament, C.F., Levin, M.B., 1990. Spontaneous spinal epidural hematoma with associated platelet dysfunction from excessive garlic ingestion: a case report. Neurosurgery 26 (5), 880–882.

Rouhi-Boroujeni, H., Rouhi-Boroujeni, H., Gharipour, M., Mohammadizadeh, F., Ahmadi, S., Rafieian-kopaei, M., 2015. Systematic review on safety and drug interaction of herbal therapy in hyperlipidemia: a guide for internist. Acta Biomed. Atenei Parmensis 86 (2), 130–136.

Savage, K., Firth, J., Stough, C., Sarris, J., 2017. GABA-modulating phytomedicines for anxiety: a systematic review of preclinical and clinical evidence. Phytother. Res. 32 (1), 3–18.

Schwarz, U.I., Hanso, H., Oertel, R., Miehlke, S., Kuhlisch, E., Glaeser, H., et al., 2007. Induction of intestinal wP-glycoprotein by St John's wort reduces the oral bioavailability of talinolol. Clin. Pharmacol. Ther. 81 (5), 669–678.

Scott, G.N., Elmer, G.W., 2002. Update on natural product–drug interactions. Am. J. Health Syst. Pharm. 59 (4), 339–347.

Šemeláková, M., Jendželovský, R., Fedoročko, P., 2016. Drug membrane transporters and CYP3A4 are affected by hypericin, hyperforin or aristoforin in colon adenocarcinoma cells. Biomed. Pharmacother. 81, 38–47.

Sharma, S.M., Anderson, M., Schoop, S.R., Hudson, J.B., 2010. Bactericidal and anti-inflammatory properties of a standardized Echinacea extract (Echinaforce®): dual actions against respiratory bacteria. Phytomedicine 17 (8–9), 563–568.

Singh, Y.N., 2005. Potential for interaction of kava and St. John's wort with drugs. J. Ethnopharmacol. 100 (1–2), 108–113.

Smith, M., 2000, March. Drug interactions with natural health products/dietary supplements: a survival guide. In: Complementary and Alternative Medicine: Implications for Clinical Practice and State-of-the-Science Symposia.

Stewart, D.E., 1992. High-fiber diet and serum tricyclic antidepressant levels. J. Clin. Psychopharmacol. 12 (6), 438–440.

Tachjian, A., Maria, V., Jahangir, A., 2010. Use of herbal products and potential interactions in patients with cardiovascular diseases. J. Am. Coll. Cardiol. 55 (6), 515–525.

Taibi, D.M., Landis, C.A., Petry, H., Vitiello, M.V., 2007. A systematic review of valerian as a sleep aid: safe but not effective. Sleep Med. Rev. 11 (3), 209–230.

Tian, R., Koyabu, N., Morimoto, S., Shoyama, Y., Ohtani, H., Sawada, Y., 2005. Functional induction and de-induction of P-glycoprotein by St. John's wort and its

ingredients in a human colon adenocarcinoma cell line. Drug Metab. Dispos. 33 (4), 547–554.
Triantafyllidi, A., Xanthos, T., Papalois, A., Triantafillidis, J.K., 2015. Herbal and plant therapy in patients with inflammatory bowel disease. Ann Gastroenterol.: Quarter. Publicat. Hellenic Soc. Gastroenterol. 28 (2), 210.
Tsai, H.H., Lin, H.W., Simon Pickard, A., Tsai, H.Y., Mahady, G.B., 2012. Evaluation of documented drug interactions and contraindications associated with herbs and dietary supplements: a systematic literature review. Int. J. Clin. Pract. 66 (11), 1056–1078.
Vaes, L.P., Chyka, P.A., 2000. Interactions of warfarin with garlic, ginger, ginkgo, or ginseng: nature of the evidence. Ann. Pharmacother. 34 (12), 1478–1482.
Van den Bout-van den Beukel, C.J., Koopmans, P.P., van der Ven, A.J., De Smet, P.A., Burger, D.M., 2006. Possible drug–metabolism interactions of medicinal herbs with antiretroviral agents. Drug Metab. Rev. 38 (3), 477–514.
Velingkar, V.S., Gupta, G.L., Hegde, N.B., 2017. A current update on phytochemistry, pharmacology and herb–drug interactions of *Hypericum perforatum*. Phytochem. Rev. 16 (4), 725–744.
Wang, X.L., 2015. Potential herb–drug interaction in the prevention of cardiovascular diseases during integrated traditional and western medicine treatment. Chin. J. Integr. Med. 21 (1), 3–9.
Welihinda, J., Karunanayake, E.H., Sheriff, M.H.H., Jayasinghe, K.S.A., 1986. Effect of *Momordica charantia* on the glucose tolerance in maturity onset diabetes. J. Ethnopharmacol. 17 (3), 277–282.
Westendorf, J., 1993. Anthranoid derivatives—general discussion, Adverse Effects of Herbal Drugs, 2. Springer, Berlin, Heidelberg, pp. 105–118.
Williamson, E.M., 2003. Drug interactions between herbal and prescription medicines. Drug Saf. 26 (15), 1075–1092.
Xiong, X.J., Liu, W., Yang, X.C., Feng, B., Zhang, Y.Q., Li, S.J., et al., 2014. *Ginkgo biloba* extract for essential hypertension: a systemic review. Phytomedicine 21 (10), 1131–1136.
Yongchaiyudha, S., Rungpitarangsi, V., Bunyapraphatsara, N., Chokechaijaroenporn, O., 1996. Antidiabetic activity of *Aloe vera* L. juice. I. Clinical trial in new cases of diabetes mellitus. Phytomedicine 3 (3), 241–243.
Yuan, C.S., Wei, G., Dey, L., Karrison, T., Nahlik, L., Maleckar, S., et al., 2004. Brief communication: American ginseng reduces warfarin's effect in healthy patients: a randomized, controlled trial. Ann. Intern. Med. 141 (1), 23–27.

FURTHER READING

Kuhn, M.A., 1998. Pharmacotherapeutics: A Nursing Process Approach, vol. 2. FA Davis Company.
Perlman, B., 1990. Interaction between lithium salts and ispaghula husk. Lancet 335 (8686), 416.
Shanmugasundaram, E.R.B., Rajeswari, G., Baskaran, K., Kumar, B.R., Shanmugasundaram, K.R., Ahmath, B.K., 1990. Use of *Gymnema sylvestre* leaf extract in the control of blood glucose in insulin-dependent diabetes mellitus. J. Ethnopharmacol. 30 (3), 281–294.

SECTION 4

New Dimensions in Phytotherapy Research and Applications

CHAPTER 19

High-Throughput Virtual Screening (HTVS) of Natural Compounds and Exploration of Their Biomolecular Mechanisms: An In Silico Approach

Anupam Dhasmana[1], Sana Raza[2], Roshan Jahan[2], Mohtashim Lohani[3] and Jamal M. Arif[2]
[1]Himalayan School of Biosciences and Cancer Research Institute, Swami Rama Himalayan University, Dehradun, Uttarakhand, India
[2]Department of Biosciences, Integral University, Lucknow, Uttar Pradesh, India
[3]College of Applied Medical Sciences, Jazan University, Jazan, Kingdom of Saudi Arabia

19.1 INTRODUCTION

The pressure to discover and develop new drugs in a shorter time period at a reduced cost has led to a demand for equally effective approaches for drug discovery. Rapid development of these analysis tools has helped to accelerate the drug-discovery process and reduce artifacts. The process of drug discovery involves target identification and validation, discovery of lead compound (identification of relevant compounds that modulate the biological action of the receptor), lead optimization, and clinical trials (Bolten and Gregorio, 2002). Screening refers to the testing of a large number of molecules for their activity in a model system that is representative of a human disease. Drug discovery through high-throughput screening (HTS) employs robotics, data processing and control software, sensitivity detectors, and liquid handling devices, allowing researchers to conduct millions of pharmacological and chemical tests to identify chemical compounds that may modulate a biochemical pathway.

HTS has the potential to ease the bottlenecks associated with the drug-discovery process. It requires minimal compound design or prior

New Look to Phytomedicine
DOI: https://doi.org/10.1016/B978-0-12-814619-4.00020-3

knowledge. The technologies required to screen large libraries have become more efficient.

High-throughput virtual screening (HTVS) or virtual HTS (VHTS) is a popular approach for drug discovery that makes use of computational algorithms to identify novel bioactive molecules (Lionta et al., 2014). Computer-aided drug-discovery and designing methods like structure–activity relationship and others play major role in the development of therapeutically important molecules (Kapetanovic, 2008; Giustiniano et al., 2017). HTVS increases the hit rate of novel drug compounds because it uses a much more targeted search as compared to traditional HTS and combination chemistry. HTVS not only explains the molecular basis of therapeutic activity but also helps to predict the possible derivatives which could improve activity.

HTVS is used for three major purposes: (1) filtering large compound libraries to small sets of predicted active compounds, which may be tested experimentally; (2) lead compound optimization to increase affinity or optimize drug metabolism and pharmacokinetic properties such as potential for absorption, distribution, metabolism, excretion, and toxicity (ADMET); and (3) designing of novel compounds by growing start molecules using functional groups or bringing together fragments into novel chemotypes. Some of the approved drugs that owe their discovery to computer-aided drug designing are captopril, an angiotensin-converting enzyme inhibitor which was approved as antihypertensive drug in 1981 (Talele and Khedkar, 2010); dorzolamide, a carbonic anhydrase inhibitor approved in 1995 (Vijayakrishnan, 2009); saquinavir (approved in 1995), ritonavir, and indinavir (Van Drie, 2007) as therapeutics for treatment of human immunodeficiency virus (HIV); tirofiban, a fibrinogen antagonist approved in 1998 (Hartman et al., 1992).

Natural compounds and their pharmacophore moieties have been extensively used as a source of therapeutic agents (Al-Khodairy et al., 2013). Natural compounds are well known to have high structural and chemical diversity, biochemical specificity, and phyto-molecular potential, which make them potent lead structures for drug discovery (Arif, 2013; Akhtar et al., 2016). Latest advances in structural biology have revealed the diversity of proteins known to be targeted by natural compounds (Arif et al., 2013). The currently used lead generation strategies also focus on the use of natural compounds for drug discovery. Computational biology HTVS methods provide a good platform to screen best suitable natural compounds to target particular biomolecules or diseases,

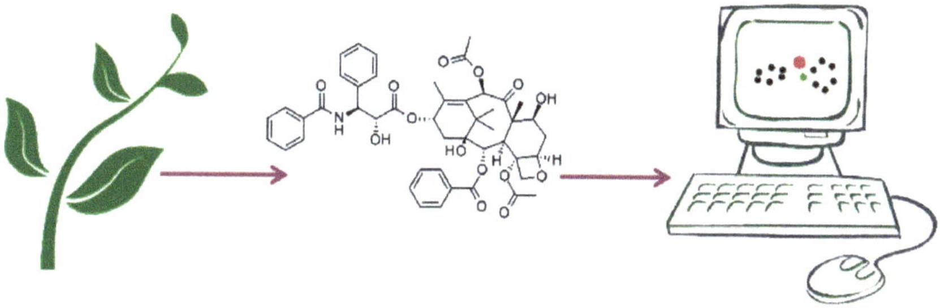

Figure 19.1 Fusion of natural chemistry with computation biology.

where a virtual in silico environment may help in identification of the biochemical potential of natural compounds (Fig. 19.1).

19.2 HIGH-THROUGHPUT VIRTUAL SCREENING

HTVS may be defined as a process of target validation which involves the search for bioactive molecules within a database of interest that match a given query structure. The Protein Data Bank was formed in 1971 at Brookhaven National Laboratory, and Cambridge Crystallographic Data Center, are among the most famous used databases for this purpose. HTVS involves the testing of compounds for their binding activity against target molecules. It may be used to discover ligands for a receptor, ion channels, enzymes, or a pharmacophore target structure. A range of bioactive compounds have been used to study the functions of biomolecules (Khan et al., 2011; Choi et al., 2003; Hoang Nguyen Vo et al., 2016). Identification of the target proteins of a bioactive compound is an essential prerequisite to develop new chemical probes of biological systems and helps to elucidate their mode of action. Among various methods to discover lead compounds and elucidate their biomolecular function, HTVS is notable where small molecular compounds are used as probes to construe the function of the molecule (protein or nucleic acid). A number of virtual HTS tools have proved their aptness in modern-day drug discovery (McInnes, 2007). HTVS can be used to discover high-quality lead molecules to narrow down the range of molecules that need to be tested in vitro and in vivo to reduce the economical investment in chemical synthesis and/or preliminary testing and minimize the cost associated with failure at later stages. HTVS can broadly be divided into two categories—ligand based and structure based (as discussed in Fig. 19.2).

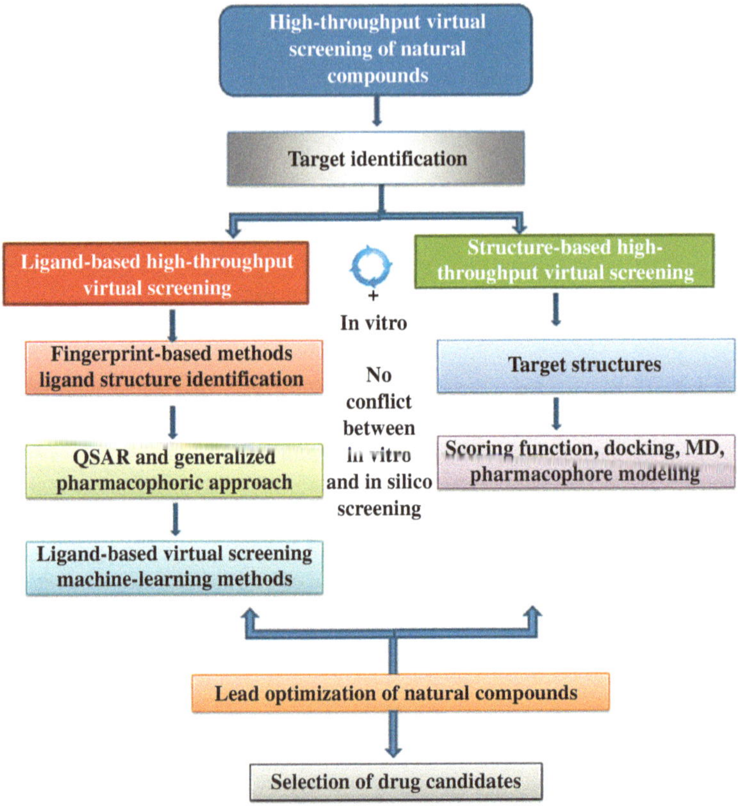

Figure 19.2 Scheme of high-throughput virtual screening.

19.2.1 Ligand-Based High-Throughput Virtual Screening

Ligand-based screening methods harness the chemistry of molecules that are known to bind to a biological target. It utilizes the information of known active and inactive chemical moiety through chemical similarity searches or construction of predictive, quantitative structure–activity relation (QSAR) models (Kalyaanamoorthy and Chen, 2011). Ligand-based methods use ligand information to predict activity, which is based on similarity or dissimilarity to previously known active ligands. The ligands showing similarity to an active ligand are expected to be more active than random ligands.

Ligand-based HTS (LBHTS) is chosen when three-dimensional (3D) structure of the target protein is not available. Having a set of structurally diverse ligands that bind to a receptor, pharmacophoric model of the receptor can be built utilizing the information contained in such a set of ligands. Ligands can then be evaluated to the pharmacophore model to verify their fitness and ability to bind. It uses the information present in known active compounds rather than the structure of the target protein for lead identification.

LBHTS methods can be classified into three major categories.

19.2.1.1 Fingerprint-Based Methods

Structurally related molecules are expected to have similar properties. Databases of structurally similar structures with unknown biological activity may contain compounds with some activity of interest.

19.2.1.2 Generalized Pharmacophoric Approach

A set of structural features in a molecule that is recognized at a receptor site and is responsible for a molecule's biological activity is known as a pharmacophore. This approach is used to identify a pharmacophore pattern, which is common to a set of known active compounds. It subsequently makes use of this pattern for a 3D substructure search. The software helps to find such atoms and functional groups which match the spatial arrangement of this element with the query molecule.

19.2.1.3 Machine-Learning Methods

The machine-learning methods depend on a training set data containing known active and inactive molecules. Logic-based rules are used to describe the features of the substructures and their chemical properties related to activity. These logic-based features provide insights into activity.

19.2.2 Structure-Based High-Throughput Virtual Screening

Structure-based HTS (SBHTS) relies on the knowledge of the structure of target protein to calculate the interaction energy of the compound being tested. It is preferred when a high-resolution structural data of the target protein is available. This type of drug design is based on the information of the 3D structure of the target, as obtained through crystallographic, spectroscopic, or bioinformatics techniques. SBHTS involves the identification of a potential ligand-binding site on the target molecules and docking of ligands to a protein target. This is followed by the application of a *scoring function* to estimate the likeness that a ligand will bind to the protein with high affinity. It is the energy function used to evaluate the binding free energy between protein and ligand. Scoring functions involve ligand database preparation, protein structure preparation, docking calculations, and postprocessing.

In this approach, the 3D structure of target protein is essential to perform in silico HTS. The 3D structure of the target protein may be determined by NMR, X-ray crystallography, or homology modeling. Homology modeling involves the construction of an atomic resolution model of the target protein using its amino acid sequence by in silico searching for a relatively more experimentally resolved 3D structure as template. Structure-based screening requires a computing infrastructure, such as a cluster of Linux systems distributed over networked PCs. Generally, all docking programs predict ligand orientation and conformation into an active site, which is followed by a measure of its fitness into the chosen binding site. Structure based modeling has become a cornerstone of computational biology and medicinal chemistry. Leveraging the knowledge of the biological target and the chemistry of protein–ligand interactions, we can get a better insight into screening of lead molecules for drug discovery.

The key steps in SB-HTVS are as follows:

1. Compound library and target protein preparation.
2. Determination of most favorable binding position.
3. Ranking of docked complexes.

SB-HTVS has been used by several drug-discovery campaigns to identify novel and potent hits (Rehman et al., 2016; Ruiz et al., 2008; Becker et al., 2006; Zhao et al., 2006; Triballeau et al., 2008; Lu et al., 2006; Li et al., 2009; Budzik et al., 2010; Izuhara et al., 2010; Simmons et al., 2010; Roughley et al., 2012).

19.2.2.1 Flexible Protein–Ligand Docking

Flexible protein–ligand docking is a cornerstone for the discovery of new drugs. The integration of different variables allows the prediction of behavior of various chemical compounds and protein molecules and identification of drug leads, through latest programs and algorithms.

This minimizes non-specific interactions of drug molecule and proteins and helps to reduce the cost involved and time consumed.

19.2.2.2 Flexible Protein–Protein Docking

Protein–protein interactions are responsible for a number of biological functions. Predicting these interactions is extremely important to understand human physiology. Association of two biological macromolecules is a fundamental biological phenomenon and remains an unsolved theoretical problem. A variety of tools have been developed to solve protein–protein docking problems for prediction of the geometry of a complex using atom coordinates of its uncomplexed constituents.

19.2.2.3 Hydrophobic Docking

The prevalence of hydrophobic groups at contact sites results in intermolecular atom–atom contacts for correct matches. The representation of molecules according to their hydrophobic groups helps obtain better molecular recognition sites, compared to their full representation.

19.2.3 Docking Algorithms

The docking algorithms may be divided into three main categories.

19.2.3.1 Systematic Methods

All degrees of freedom of a receptor molecule are explored, and ligands are placed into the desired active site. It involves incremental construction, conformational search, Hammerhead algorithm, etc.

19.2.3.2 Random or Stochastic

This includes methods like Monte Carlo and genetic algorithms (GAs). The Monte Carlo method generates several ligand configurations into the protein binding site and then scores the configuration in a multistep procedure. However, the GAs exploit the principles of biological competition and population dynamics to place the ligand in an active site.

19.2.3.3 Molecular Dynamics or Simulation

Based on Newton's equation of motion for an atomic system where atomic forces and masses are used to determine atomic positions over a series of very small time steps, forces between the particles and potential energy are defined by molecular mechanic force fields.

Some of the popular docking programs using the above searching methods are FlexX (incremental construction approaches), GOLD and Autodock (GA), Glide (systematic incremental search techniques), DOCK (shape-based algorithm), and LigandFit (Monte Carlo simulation).

Popular docking software, web servers, and databases are listed in Tables 19.1–19.3.

Table 19.1 List of molecular docking software, their algorithms and websites

Software	Search strategy	Docking	Source/free for academic	Availability
AutoDock	LGA	Rigid/Flexible	Available	http://autodock.scripps.edu
DOCK	Genetic hashing algorithm	Rigid	Available	http://dock.compbio.ucsf.edu
GOLD	GA	Rigid/Flexible	Not available	http://www.ccdc.cam.ac.uk/products/life_sciences/gold/
Glide	Monte Carlo simulated annealing algorithm	Flexible	Not available	http://www.schrodinger.com/productpage/14/5/
FlexX	Fragment-based algorithm	Flexible	Not available	http://www.biosolveit.de/flexx/index.html?ct = 1
FRED	Chemgauss4 scoring function	Rigid/Flexible	Available	http://www.eyesopen.com/oedocking
LigandFit	Monte Carlo conformational search	Rigid/Flexible	Not available	http://accelrys.com
MOE	GAs	Rigid/Flexible	Not available	http://www.chemcomp.com
CDocker	Monte Carlo and GAs	Rigid/Flexible	Not available	http://accelrys.com
VLifeDock	GAs	Rigid/Flexible	Not available	http://www.vlifesciences.com/product/VLifeMDS/VLifeDock.php
Surflex	Using the Hammerhead scoring function	Flexible	Not available	http://www.biopharmics.com/downloads.html
ICM	Monte Carlo minimization procedure	Flexible	Not available	http://www.molsoft.com/docking.html
ParaDocks	Optimization algorithm	Rigid	Available	http://www.paradocks.org
Molegro	MolDock algorithm	Rigid	Not available	http://www.molegro.com/mvd-product.php
Virtual DockerFleksy	Comprises an ensemble-based soft-docking using FlexX-ensemble	Flexible	Not available	http://www.cmbi.ru.nl/software/fleksy/
Hex	Parametric docking and superposition algorithms	Rigid	Available	http://www.hex.lorai.fr
GRAMM	FFT algorithm	Rigid	Available	http://www.bioinformatics.ku.edu/files/vakser/gramm/
HADDOCK	CPORT algorithm	Flexible	Available	http://www.nmr.chem.uu.nl/haddock/

LGA, Lamarckian genetics algorithm; *GA*, Genetic algorithm; *FFT*, Fast Fourier transform.

Table 19.2 Online web server for molecular docking

Resource	Availability
SwissDock	http://www.swissdock.ch
DockingServer	http://www.dockingserver.com/web
1-Click Docking	http://mcule.com/apps/1-click-docking/?utm_source = ccl&utm_medium = maillist&utm_campaign = 1-click-docking
Blaster	http://blaster.docking.org
Mcule Docking	http://mcule.com/subscribe/packages/
Pardock	http://www.scfbio-iitd.res.in/dock/pardock.jsp
FlexPepDock	http://flexpepdock.furmanlab.cs.huji.ac.il/
PatchDock	http://bioinfo3d.cs.tau.ac.il/PatchDock/
BSP-SLIM	http://zhanglab.ccmb.med.umich.edu/BSP-SLIM/
iScreen	http://iscreen.cmu.edu.tw
idTarget	http://idtarget.rcas.sinica.edu.tw

19.2.3.4 Case Study of Docking Approach

The docking methodology may be in a variety of ways. Here, we present a case study where the docking approach was used to identify the mechanistic insight of biological activities of Quinoline, Wild-Type, and Mutant p53. Quinoline is an alkaloid derived from various plant species including Mentha, having antitumor activity.

Ligand file quinoline was downloaded in .mol format from PubChem Chemical Database (PubChem ID 7047). These files could not be directly used by Autodock 4.0 tools (Morris et al., 2009) therefore they were converted into .pdb files using Discovery Studio Visualizer version 3.1. The energy minimization of ligands was done by Chimera (version 1.5.3) using GA Steps 2000 and 0.5 grid units Optimized (Pettersen et al., 2004). RCSB Protein Data Bank was used to obtain the 3D structures of wild-type and mutated p53 proteins. HETATM was removed using Discovery Studio Visualizer. Chimera with the steepest descent steps 1000, steepest descent size 0.02 Å, conjugated gradient steps 1000, and the conjugate gradient step size 0.02 Å was used for conjugate gradient minimization and removal of steric collision (Wang et al., 2004).

To inspect the effect of point mutation on the protein stability change, the difference in folding free energy ($\Delta\Delta G$) was identified by using the I-mutant program. Pymol was used for the visual analysis of the enzymes.

Docking analysis was carried out by using genetic-based algorithm—based tool (Autodock version 4.0). We executed molecular docking

Table 19.3 List of online natural compound databases

Database	Type	No. of compounds	Website
PubChem	Public	>8,000,000 structures	http://pubchem.ncbi.nlm.nih.gov
ChEMBL	Public	1,376,469 structures	http://www.ebi.ac.uk/chembldb/index.php
NCI Set	Public	>140,000 structures	http://dtp.nci.nih.gov/index.html
ChemSpider	Public	>28,000,000 structures	http://www.chemspider.com
TCM	Public	37,170 compounds	http://tcm.cmu.edu.tw
ZINC	Public	>21,000,000 compounds	http://zink.docking.org
ChemBridge	Commercial	>900,000 compounds	http://www.chembridge.com
Specs	Commercial	1,500,000 compounds	http://wwwspecs.net
Asinex	Commercial	>600,000 compounds	http://www.asinex.com
Enamine	Commercial	>1,640,000 compounds	http://www.enamine.net
Maybridge	Commercial	>56,000 molecules	http://www.maybridge.com
WOMBAT	Commercial	331,872 compounds	http://www.sunsetmolecular.com
ChemNevigator	Commercial	>91,500.000 compounds	http://www.chemnavigator.com
ACD	Commercial	>7,000,000 compounds	http://accelrys.com/products/databases/sourcing/available-chemicals-diectory.html
MDDR	Commercial	150,000 compounds	http://accelrys.com/products/databases/bioactivity/mddr.html

simulation methods followed by searching the best conformation of wild-type and mutated p53 tumor suppression proteins and all quinoline complexes on the basis of molecular/interaction binding energy. Water molecules were deleted from the crystal protein structures before docking simulation, and H-atoms were added to both wild-type and mutated p53 proteins. Salvation parameters and Kollman charges were added. Gasteiger charges were added to the compound. Grid box was set to cover the maximum part of protein and chemical compounds. The values were set to $60 \times 60 \times 60$ Å in *X*, *Y*, and *Z* axis of grid point. The default grid points spacing was 0.375 Å. Lamarckian GA (LGA) was used for wild-type and mutated p53 proteins and quinoline for docking calculations. Various LGA parameters such as the population size (ga_pop_size), energy evaluation parameter (ga_num_generation), crossover and mutation rates, and step size were set as 150, 2500,000, 27,000, 0.02, 0.8, and 0.2 Å, respectively. The LGA runs were set to 10 runs. All obtained 10 conformations of both proteins and quinolone complexes were analyzed for the interactions and binding energy of the docked structure using Discovery Studio Visualizer version 3.1.

The I-mutant server was used to investigate the effect of Arg273His mutation on the p53 protein. The overall stability of the Arg273His mutant of p53 was calculated based on atom potentials and torsion angle potentials. The result based on the value of $\Delta\Delta G$ (predicted destabilizing energy) shows that the mutant Arg273His (-1.15 kcal/mol) has significant influence on p53 by the point mutation. A negative $\Delta\Delta G$ is an indication of protein destabilization, which is based on the change in stability of protein upon a point mutation by measuring the free energy difference between wild-type and mutated state. Quinoline showed the highest binding affinity with mutant p53 with -7.02 kcal/mol, whereas wild-type p53 showed -4.98 kcal/mol binding energy with quinoline. These results depict that mutant p53 has better binding affinity than wild-type p53 which may be used for the pharmacological reactivation of mutant p53. Fig. 19.3 shows binding interaction residues.

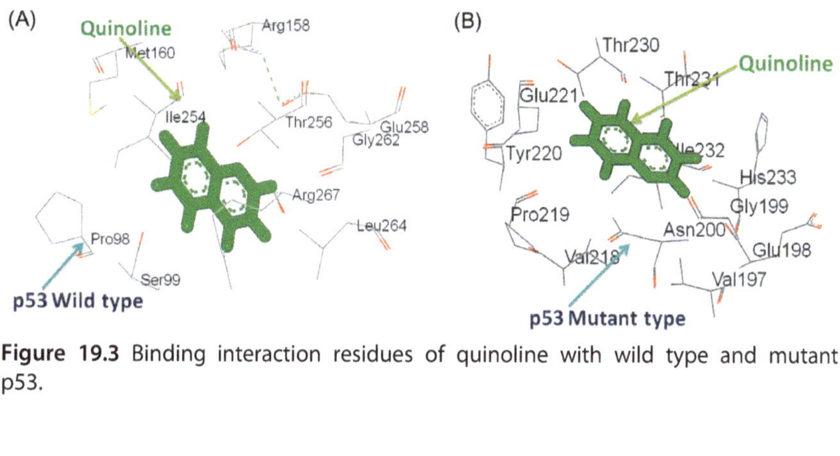

Figure 19.3 Binding interaction residues of quinoline with wild type and mutant p53.

19.3 COMPUTATIONAL METHODS FOR VIRTUAL SCREENING

With the recent advances in computer hardware and improvements in intensive computational algorithms, several docking programs have emerged, which are used for inverse virtual screening (IVS) and HTVS of molecular interactions. HTVS makes use of docking of many ligands against one or more receptors, while IVS docks multiple receptors against one or few ligands. Virtual screening (VS) programs are a useful replacement to the time-consuming, expensive wet bench experimentation. Few open-source, low-cost VS programs are used as nonprofit, institutional budgets to discover new therapeutics.

19.3.1 Amalgamating Virtual High-Throughput Screening, Quantitative Structure–Activity Relationship (QSAR) and Pharmacophore Mapping Methods

19.3.1.1 Quantitative Structure–Activity Relationship

QSAR is a set of tools that are employed to link chemical activities with molecular structure and composition. These techniques are well equipped to deal with large quantities of data, allowing researchers to discern trends in existing dataset. A QSAR model is supported on the hypothesis that there is a connection between the chemical structure and biological function. Based on this hypothesis, QSAR attempts to set up a correlation between numerous properties of molecules and their experimentally known biological functions. QSAR models try to disclose mathematical features between structural features and target response of chemicals compounds (Zhang, 2011).

It was initially used by a scientist from the University of Strasbourg in the 1860s where he surprisingly noticed how the toxicity of alcohols in mammals had improved when a decrease in solubility of water took place. In the 1960s, Corwin Hansch started his pioneer work in pharmaceutical research and began to establish QSAR models using various molecular descriptive parameters to chemical, physical, and biological properties to provide computational estimates for the bioactivity of molecules. The main objective of the QSAR method is to observe the biological responses of a set of molecules, measure it, and statistically relate the measured activity to some molecular structure on their surface. The QSAR product then produces useful equations, images, or models in either two-dimensional (2D) or 3D form that relate these biological responses or physical properties to their molecular structure.

For the main pipeline of QSAR-based approaches for drug-discovery project, the first step engages the sets of active and inactive ligands, and the second constructs a set of mathematical descriptors to explain the attributes like physicochemical and chemical structural properties of those compounds. It is based on machine-learning and statistical approaches that involve the correlation of various electronic, hydrophobic, and steric features with biologic activity to develop quantitative relationships between chemical descriptors and biological activities of compounds to be tested. Classic QSAR is also known as the Hansch–Fujita approach. The predictive QSAR models are helpful in screening large virtual libraries to find out putatively active compounds and to prioritize them for experimental testing. QSAR modeling fulfills the ultimate needs of experimental medicinal chemists to discover novel bioactive compounds. The success of QSAR depends on the quality of the initial set of active or inactive ligands used. QSAR models find application in many disciplines, like risk assessment, toxicity prediction, and making regulatory decisions.

General representation of the QSAR equation:

$$\text{Biological activity} = \text{co} + \text{cd1} + (\text{c2d1})^2 + \text{c3d2} + \text{c4d2}^2 + \cdots.$$

where d is the value of the descriptor for each molecule in the series and c represents a coefficient calculated by fitting variations in the data by regression analysis.

Steps involved in a QSAR study:

Step 1. Molecule Dataset:

A dataset of a series of synthesized molecules tested for its desired biological activity is required for QSAR analysis. For a valid and reliable QSAR model, the activity of all of the chemicals covered must be elicited by a common mechanism.

Biological activity can be of two types:

1. Continuous response: MEC, IC50, ED50, % inhibition
2. Categorical response: Active/inactive

Chemical compounds of interest margin with known biological activity like half-maximum inhibitory concentration (IC50) or reproducible potency (EC50) can be used for model development and validation. Such activity values can be retrieved from various databases such as PubChem's BioAssay. For a confident QSAR analysis, biological data of at least 20 molecules is recommended. The quality of the model totally depends on the quality of the experimental data being used to build the model.

Step 2. Chemical Descriptor Calculation:

Molecular descriptors are numerical representations of the chemical information encoded within a molecular structure by a mathematical procedure. The dimensionality of molecular descriptors used determines the type of QSAR. Numerous software packages are available to calculate a wide variety of descriptors for various chemical structures, for example, Dragon, JOELib, and ADAPT.

Dimensionality of molecular descriptors:

0D: These descriptors are derived from the chemical formula, for example, molecular weight, number and type of atoms, etc.

1D: A list of structural signals, sign that is, a substructural catalog symbol of a molecule can be considered as a one-dimensional (1D) molecular symbol and consists of a list of molecular fragments (e.g., functional groups like −OH, −COOH, −CHO, rings like cyclic, aromatic, and bonds like saturated and unsaturated bonds, substituents, etc.).

2D: 2D or topological information is contained in a molecular graph. It describes how the atoms are bonded in a molecule, both the type of bonding and the interaction of particular atoms (e.g., graph invariants, path count, molecular connectivity indices, etc.).

3D: These are calculated starting from a geometrical or 3D representation of a molecule. These descriptors include molecular surface, molecular volume, and other geometrical properties. For example, 3D-MoRSE descriptors, WHIM descriptors, GETAWAY descriptors, and quantum-chemical descriptors.

4D: In addition to the 3D descriptors, the fourth dimension is generally in terms of different conformations or any other experimental condition such as those derived from GRID or CoMFA methods.

The descriptors may also be classified into four categories viz. geometrical, topological, electronic, and hybrid.

Geometric descriptors: The geometric descriptors include topology of molecules. These descriptors describe features such as path lengths and connectivity. Examples include connectivity indices, distance edge vectors, and eccentricity indices.

Topological descriptors: The topological descriptors include various features related to the electronic environment. Major descriptors of this group include HOMO and LUMO energies, electronegativity, and various atom-centered partial charge descriptors.

Hybrid descriptors: Hybrid descriptors are combinations of electronic or topological descriptors and geometric descriptors. The hydrophobic surface area descriptors, charged partial surface area descriptors, and hydrogen-bonding descriptors are some common types of hybrid descriptors.

Step 3. Generation of QSAR Set:

One of the key steps in QSAR generation is a critical step in modeling. QSAR models are used to screen chemical databases and/or virtual chemical libraries for potentially bioactive molecules. Three mutually exclusive sets of data are generated from the selected dataset of molecules.

The first set generated is known as a training set and is used for model building.

The first set is referred to as a training set. A training set is basically used for model building.

The algorithm used in model building uses the training set to characterize the dataset based on features of the training set. Linear regression models make use of a cross validation set or a test set to examine the predictive ability. It is used to monitor the error rate during training. For a linear model, the training set and cross validation set are combined.

The training set selected to calibrate the model must exhibit a well-balanced distribution and contain representative molecules.

Methods for the division of the dataset into a training and a test set:

1. *Manual selection*: In this scheme, we separate the dataset into test set and training set by manually visualizing the deviation in the chemical and biological features in the given dataset.
2. *Random selection*: The random selection method divides the dataset into a training and a test set by a random distribution process. The points in the given dataset are selected randomly, but it should be uniformly distributed in both the sets. This method creates training and test set by random distribution.
3. *Sphere exclusion method*: A rational method to divide the dataset and create training and test sets. It ensures that the points in both the sets are uniformly distributed w.r.t. chemical and biological space. In this method, we use probe spheres to set a similarity limit.
4. *Others*:
   a. *Experimental design*: Full factorial, fractional factorial, etc. A full factorial design is a type of design in which every setting of every factor appears with every setting of every other factor.

b. *Onion design*: In this kind of plan, the test set or training dataset are divided into a number of subsets that are identified as "shells"/ "layers," and a D-optimal selection is made from each shell.

c. *Cluster analysis*: Cluster analysis is a method of classification of objects into different groups or partitioning of data into subsets or clusters where the members of the subsets or groups share common properties.

d. *Principal component analysis*: Principle component analysis is a statistical method for multidimensional datasets to lower dimension datasets for analysis. It is a mathematical procedure used to transform possibly correlated variables into uncorrelated variables in order to identify the patterns available in data, and to express the data in a form so as to highlight the similarities and differences.

e. *Self-organizing maps (SOM)*: SOMs are data visualization techniques. They reduce the dimensions of data by using self-organizing neural networks. Humans simply cannot visualize high-dimensional data as it is, so these techniques are created to help us understand this high-dimensional data. SOMs help to reduce the dimensions. They produce a map of usually one or two dimensions plotting the similarities in data by grouping similar items together. SOMs accomplish two goals, they help to reduce the dimensions and display similarities.

Step 4. QSAR Model Development:

Analyses of data using a suitable statistical method coupled with a variable selection method help to establish a QSAR model. The statistical methods can be broadly divided into two categories:

Linear and nonlinear methods:

A correlation is established between dependent variable(s), that is, biological activity, and independent variable(s), that is, molecular descriptors. The linear way fits a line between the preferred descriptors and activity; on the other hand, the non-linear way fits a curve between the preferred descriptors and activity. The statistical method used to build the model is decided on the basis of the type of biological activity data. Few commonly used statistical methods are as follows:

Categorical Dependent Variable:

1. Discriminant analysis
2. Logistic regression
3. k-Nearest Neighbor classification

4. Decision trees, and
5. SIMCA

Continuous Dependent Variable:

1. Multiple regressions
2. Principal component regression
3. Continuum regression
4. Partial least squares regression
5. Canonical correlation analysis
6. k-Nearest Neighbor method, and
7. Neural networks

Once a good QSAR set is developed, we need to reduce the original descriptor data to a more convenient size and then choose a number of optimal descriptor subsets so that we can then proceed to build a set of models and select the best one. QSAR relates variations in biological activity to variations in the values of computed or measured properties of a series of molecules and forms of a linear equation:

$$\text{Biological activity} = \text{Const} + (c_1 \times P_1) + (c_2 \times P_2) + \cdots + (c_n \times P_n)$$

The parameters P_1 through P_n are computed for each molecule of the series. The coefficients c_1 through c_n are calculated by lifting variations in the parameters and biological activity.

The quality of any QSAR totally depends on the quality of the data that has been used to derive the model. The quantification of QSAR data is totally dependent upon the correlation coefficient (r). Correlation coefficient measures how strongly the observed data tracks the fitted regression line. Any error in the model or in data leads to a bad fit.

This indicator of fit to the regression line is calculated as

$$r^2 = \frac{\text{Regression variance}}{\text{Original variance}}$$

where,

1. Regression variance = (Original variance − variance observed around regression line).
2. Original variance: sum-of-the-square distances of original data from the mean value.

The value for r^2 falls between 0 and 1.

If r^2 becomes 0, it indicates that there is no relationship between activity and the parameters (s) selected for the study, and if r^2 equals to 1, it indicates that there is a perfect correlation among them.

Multiple regression is a widely used method to build QSAR models. They can be interpreted simply, where the contribution of each descriptor is seen by the magnitude and sign of its regression coefficient. Descriptor coefficient is the relative contribution of the descriptor with other descriptors. The sign depicts direct (+) or inverse (−) relation to activity.

Applications of QSAR:

1. To distinguish drug-like molecules from nondrug-like molecules.
2. To predict boiling points.
3. Drug resistance studies.
4. Toxicity prediction studies.
5. Prediction of physicochemical properties like water solubility, lipophilicity, etc.
6. To study the interactions between the structural domains of proteins. Protein−protein interactions can be analyzed quantitatively for structural variations resulting from site-directed mutagenesis.
7. Prediction of ADME properties like gastrointestinal absorption, blood−brain barrier, drug metabolism, etc.

19.3.1.2 Pharmacophore Mapping

Lemont Kier established the term pharmacophore in 1967. A pharmacophore is an entity having certain steric and electronic features. They demonstrated best possible supramolecular interactions with precise biomolecular targets, having the ability to activate or block their biological process (Wermuth, 2006). It is the spatial arrangement of active functional moiety that a compound or drug must contain to suggest a preferred biologic response. An effective pharmacophore contains information about functional groups that interact with the target as well as the information about the type of non-covalent interactions and interatomic distances between these functional groups. This arrangement can be derived either in a structure-based manner (by mapping the sites of contact between a ligand and binding sites on target) or a ligand-based approach may be used. The structure-based manner involves analyzing one or several co-crystal structures with lead or drug compounds bound. We here focus on the more challenging problem, that is, the ligand-based approach. In order to generate a ligand-based pharmacophore, multiple active compounds are overlaid such that a maximum number of chemical features may overlap geometrically (Wolber et al., 2008). This incorporates molecular flexibility to determine overlapping sites and involves rigid 2D or 3D structural representations. This conformational flexibility is achieved by precomputing the conformational space of each ligand. A general-purpose conformational model is created or the conformations maybe explored by changing molecular coordinates as per the need of the alignment algorithm (Wolber et al., 2008). Catalyst (Accelrys, Inc., San Diego, CA), for example, is one popular pharmacophore-generating software package that uses the "polling" algorithm (Smellie et al., 1995) to generate approximately 250 conformers that are used in its pharmacophore generation algorithm (Acharya et al., 2011). In a study based on HSP90a, 83 known reference molecules were used to generate pharmacophore queries. A total of 25 diverse inhibitors were identified, including three having IC50 values below 10 nM (Al-Sha'er and Taha, 2010).

19.3.1.2.1 Superimposing Active Compounds to Create a Pharmacophore

Molecules are generally aligned using either a point-based or property-based technique. The point-based technique involves superimposing of pairs of points (atoms or chemical features) by minimizing the Euclidean distances. These alignment methods use root-mean-square distance to maximize overlap (Poptodorov et al., 2006). Other kind of technique is molecular property-based alignment tool which is used for the molecular field descriptors to create alignments (Wolber et al., 2008). The fields describe the 3D grids around compounds. Interaction energy for a specific probe at each point is calculated using these fields. Gaussian functions represent the distribution of interaction energies and degree of overlap between Gaussian functions of two aligned compounds. It is used as an objective scoring function to maximize alignment (Poptodorov et al., 2006). One popular field generation method for property-based alignments is GRID (Goodford, 1985). Molecular flexibility is always an important consideration when aligning compounds of interest. Several approaches are used to efficiently sample conformational space such as rigid, flexible, and semiflexible methods. Rigid methods make use of prior knowledge about active conformation of known ligands. They help align only the active conformations. This is however applicable only when the active conformation is well known. Static pre-generated conformations may be used (semiflexible methods), and flexible methods, although computationally expensive, carry out conformational search during the alignment process. This may often use molecular dynamics or random sampling of rotatable bonds. Because the conformational space can increase substantially with an increase in the number of rotatable bonds, strategies are often used to limit the exploration of conformational space through the use of reference geometry (often an active ligand with low flexibility). This method is known as the active analog approach (Marshall et al., 1979).

19.3.1.2.2 Pharmacophore Feature Extraction

A pharmacophore feature map is constructed carefully to balance generalizability along with specificity. A general definition may categorize all functional groups that have similar physiochemical properties (i.e., similar hydrogen-bonding behavior, ionizability) into one group. The specific feature definitions may include specific atom types at specific locations. More general feature definitions lead to an increase in the population of compounds that match the pharmacophore. They allow the identification of novel scaffolds but this leads to an increase in the ratio of false positives. The algorithm used to extract feature maps usually determines the level of feature definition generalizability by user-specified parameters. Hydrogen bond acceptors are commonly used to define pharmacophore maps. Donors, acidic and basic groups, and aromatic and aliphatic hydrophobic moieties are also used (Acharya et al., 2011).

19.3.1.2.3 Pharmacophore Algorithms and Software Packages

Some commonly used software packages for ligand-based pharmacophore generation are DISCO (Martin et al., 1993), GASP (Jones et al., 1995), MOE (Chemical Computing Group, Quebec, Canada), Catalyst (Kurogi and Güner, 2001), Phase (Dixon et al., 2006), and LigandScout (Inte: Ligand, Vienna, Austria). These packages use different approaches to molecular alignment, flexibility, and feature extraction. Catalyst approaches alignment and feature extraction by identifying the common chemical features arranged in 3D space in certain positions. These chemical features focus on those that are expected to be important for interaction between ligand and protein. They include hydrophobic regions, hydrogen-bond donors, hydrogen-bond acceptors, and positive ionizable and negative ionizable regions. Chemical groups that take part in the same type of interaction are considered to be identical. Catalyst contains two algorithms that it uses for the construction of pharmacophores. Hip-Hop is the simpler of the two algorithms. It looks for common 3D arrangements of features for compounds with a threshold activity against the target. It begins with best alignment of only two features (scored by

RMS deviations) and continues to expand the model in order to include more features until no further improvements are possible. This method can only produce a qualitative distinction between the active and inactive predictions. Hypo-Gen, however, makes use of biologic assay data such as IC50 values for active compounds as well as a set of inactive compounds. The pharmacophore construction in Hypo-Gen is initially the same as Hip-Hop; however, it includes additional algorithms, inactive compounds, and experimental values. These algorithms compare the best pharmacophore from the "Hip-Hop" stage with the inactive compounds and the features that are common to the inactive set are eliminated. Ultimately, an optimization is done by Hypo-Gen which tries to refine the predictive capacity of the pharmacophore by making modifications and scoring the precision in predicting the specific experimental activities (Güner, 2000; Kurogi and Güner, 2001).

19.4 IN SILICO ADMET ANALYSIS

QSAR modeling methods have been involved in drug designing processes from several years. Since the last two-and-half decades, another branch of QSAR is in silico ADMET analysis which became very useful and cost effective. In silico ADMET analysis is mainly based on the computational pharmacokinetics and toxicity modeling of investigating drugs. In this section, we discuss some of the available tools which involve in the analysis of computational pharmacokinetics and toxicity.

The in silico ADMET modeling plays a crucial role in the current drug-discovery process because ADMET properties are responsible for the failure of more than 50% of drugs in the clinical phases. Traditionally, ADME tools were functionalized at the end of the drug development pipeline, but in the current scenario, due to the availability of in silico ADMET tools (Table 19.4), it is applied at an initial phase of the drug development

Table 19.4 Online tools for ADMET analysis

S. no.	Name of tool	URL
1.	admetSAR	http://lmmd.ecust.edu.cn/admetsar1/
2.	PreADMET	http://preadmet.bmdrc.org/preadmet/query/query3.php
3.	eADMET	http://www.eadmet.com/en/physprop.php
4.	Tripod	https://tripod.nih.gov/lmcf/
5.	SwissADME	http://www.swissadme.ch/index.php

process, in order to eliminate poor ADME molecules, which leads to significant cost-effective drug development process (Cheng et al., 2013).

19.4.1 Drawbacks of High-Throughput Virtual Screening

Virtual screening is not yet a fully mature technology. Undoubtedly, many virtual screening tools have proved their applicability in the drug-discovery process; however, there are certain issues that continue to remain major drawbacks and hurdles to the accuracy of the virtual screening programs. Limited structural information or missing information, imprecise understanding of the properties of drug-like molecules, inability to map 3D properties onto 2D structures, poor scoring functions, incorrect assessment of existing SAR data, and poor strategies of docking very often create major hurdles in the process of virtual screening.

While programs have achieved a high level of precision and sensitivity in identifying known inhibitors to targets, the reality of the drug screening environment would require scanning a much larger database to obtain even a few promising hits.

Deep insight to chemistry and biology is required to function successfully in a real virtual screening environment, and these algorithms are not plug-and-play things.

The pitfalls of HTVS may be divided into four major categories:

1. Erroneous assumptions and expectations,
2. Data design and content,
3. Choice of software, and
4. Conformational sampling as well as ligand and target flexibility.

These areas must be carefully considered in order to advance the applications of VS programs. The user is always in doubt whether there are some methods that perform better than others, and if "yes," in what particular situations would they be suitable. Of the multitude of settings, parameters, and datasets in the software that the practitioner can choose from, virtual screening becomes much more erroneous if the user is lacking the background knowledge of the VS resource.

19.5 CONCLUSION

Natural products have tremendous potential for therapeutic benefits as traditional and holistic medicines for curing diseases, yielding knowledge to derive bioactive compounds as lead therapeutic agents for drug discovery. The fusion of traditional natural therapeutic agents with modern high-throughput techniques will open the new treasure of medicine to improve the quality of health. This chapter focused on the discussion, identification, optimization, and rapid validation of drug candidates by applying HTVS techniques and methods of drug designing. Here, we have discussed the procurement of natural compounds from many natural compound databases and libraries followed by initial screening of natural compounds. ADMET analysis is an important screening tool which covers the maximum pharmacokinetic properties. Docking tools discussed for ligand–target interaction analysis and explored the biochemical behavior of drugs toward drug targets, for better interaction accuracy, molecular dynamic. Simulation tools help find best the drug binding confirmation modes; and for drug chemotype improvements, QSAR and Pharmacophore were used as potential approaches to identify and design the best drug moieties. These approaches may help open a new area of drug discovery where parents moieties obtained from nature in the form of natural compounds maybe used to develop effective drugs.

REFERENCES

Acharya, C., Coop, A., Polli, J.E., Mackerell Jr., A.D., 2011. Recent advances in ligand-based drug design: relevance and utility of the conformationally sampled pharmacophore approach. Curr. Comput. Aided Drug Des. 7, 10–22.

Akhtar, S., Khan, M.K., Arif, J.M., 2016. Evaluation and elucidation studies of natural aglycones for anticancer potential using apoptosis-related markers: an in silico study. Interdiscip. Sci. 10 (2), 297–310.

Al-Khodairy, F.M., Khan, M.K.A., Kunhi, M., Pulicat, M.S., Arif, J.M., 2013. Erysolin induces apoptosis in human breast cancer cell lines: evidences from *in vitro* and *in silico* studies. Am. J. Bioinform. Res. 3 (3), 62–71.

Al-Sha'er, M.A., Taha, M.O., 2010. Elaborate ligand-based modeling reveals new nanomolar heat shock protein 90α inhibitors. J. Chem. Inform. Model. 50 (9), 1706–1723.

Arif, J.M., 2013. Editorial: In silico exploitation of molecular markers of cancer from wet lab in the mechanism-based screening of potential anticancer compounds. Res. J. Biotechnol. 8 (10), 1–2.

Arif, J.M., Siddiqui, M.H., Akhtar, S., Al-Sagair, O.A., 2013. Exploitation of in silico potential in prediction, validation and elucidation of mechanism of anti-angiogenesis by novel compounds: comparative correlation between wet lab and in silico data. Int. J. Bioinform. Res. Appl. 9 (4), 336–348.

Becker, O.M., Dhanoa, D.S., Marantz, Y., Chen, D., Shacham, S., Cheruku, S., et al., 2006. An integrated in silico 3D model-driven discovery of a novel, potent, and selective amidosulfonamide 5-HT1A agonist (PRX-00023) for the treatment of anxiety and depression. J. Med. Chem. 49 (11), 3116–3135.

Bolten, B.M., Gregorio, T., 2002. Trends in development cycles. Nat. Rev. Drug Discov. 1, 335–336.

Budzik, B., Garzya, V., Walker, G., Woolley-Roberts, M., Pardoe, J., Lucas, A., et al., 2010. Novel N-substituted benzimidazolones as potent, selective, CNS-penetrant, and orally active M(1) mAChR agonists. ACS Med. Chem. Lett. 1, 244–248.

Cheng, F., Li, W., Liu, G., Tang, Y., 2013. In silico ADMET prediction: recent advances, current challenges and future trends. Curr. Top. Med. Chem. 13 (11), 1273–1289.
Choi, Y., Kawazoe, Y., Murakami, K., Misawa, H., Uesugi, M., 2003. Identification of bioactive molecules by adipogenesis profiling of organic compounds. J. Biol. Chem. 278, 7320–7324.
Dixon, S.L., Smondyrev, A.M., Knoll, E.H., Rao, S.N., Shaw, D.E., Friesner, R.A., 2006. PHASE: a new engine for pharmacophore perception, 3D QSAR model development, and 3D database screening: methodology and preliminary results. J. Comput. Aided Mol. Des. 20, 647–671.
Giustiniano, M., Daniele, S., Pelliccia, S., La Pietra, V., Pietrobono, D., Brancaccio, D., et al., 2017. Computer-aided identification and lead optimization of dual murine double minute 2 and 4 binders: structure–activity relationship studies and pharmacological activity. J. Med. Chem. 60 (19), 8115–8130.
Goodford, P.J., 1985. A computational procedure for determining energetically favorable binding sites on biologically important macromolecules. J. Med. Chem. 28, 849–857.
Güner, O.F., 2000. Pharmacophore perception, development, and use in drug design. Molecules 5, 987–989.
Hartman, G.D., Egbertson, M.S., Halczenko, W., Laswell, W.L., Duggan, M.E., Smith, R.L., et al., 1992. Non-peptide fibrinogen receptor antagonists. Discovery and design of exosite inhibitors. J. Med. Chem. 35, 4640–4642.
Hoang Nguyen Vo, T., Tran, N., Nguyen, D., Le, L., 2016. An in silico study on antidiabetic activity of bioactive compounds in *Euphorbia thymifolia* Linn. Springerplus 5 (1), 1359.
Izuhara, Y., Yamaoka, N., Kodama, H., Dan, T., Takizawa, S., Hirayama, N., et al., 2010. A novel inhibitor of plasminogen activator inhibitor-1 provides antithrombotic benefits devoid of bleeding effect in nonhuman primates. J. Cereb. Blood Flow Metab. 30, 904–912.
Jones, G., Willett, P., Glen, R.C., 1995. A genetic algorithm for flexible molecular overlay and pharmacophore elucidation. J. Comput. Aided Mol. Des. 9, 532–549.
Kalyaanamoorthy, S., Chen, Y.P., 2011. Structure-based drug design to augment hit discovery. Drug Discov. Today 16 (17-18), 831–839.
Kapetanovic, I.M., 2008. Computer-aided drug discovery and development (CADDD): *in silico*-chemico-biological approach. Chem. Biol. Interact. 171 (2), 165–176.
Khan, M.S., Akhtar, S., Al-Sagair, O.A., Arif, J.M., 2011. Protective effect of dietary tocotrienols against infection and inflammation-induced hyperlipidemia: an in vivo and in silico study. Phytother. Res. 25, 1586–1595.
Kurogi, Y., Güner, O.F., 2001. Pharmacophore modeling and three-dimensional database searching for drug design using catalyst. Curr. Med. Chem. 8, 1035–1055.
Li, N., Wang, F., Niu, S., Cao, J., Wu, K., Li, Y., et al., 2009. Discovery of novel inhibitors of *Streptococcus pneumoniae* based on the virtual screening with the homology-modeled structure of histidine kinase (VicK). BMC Microbiol. 9, 129.
Lionta, E., George, S., Vassilatis, D.K., Zoe, C., 2014. Structure-based virtual screening for drug discovery: principles, applications and recent advances. Curr. Top. Med. Chem. 14 (16), 1923–1938.
Lu, I.L., Huang, C.F., Peng, Y.H., Lin, Y.T., Hsieh, H.P., Chen, C.T., et al., 2006. Structure-based drug design of a novel family of PPARgamma partial agonists: virtual screening, X-ray crystallography, and in vitro/in vivo biological activities. J. Med. Chem. 49, 2703–2712.
Marshall, G.R., Barry, C.D., Bosshard, H., Dammkoehler, R., Dunn, D., 1979. The conformational parameter in drug design: the active analog approach. Comput. Assist. Drug Des. **112**, 205–226. Chapter 9.

Martin, Y.C., Bures, M.G., Danaher, E.A., DeLazzer, J., Lico, I., Pavlik, P.A., 1993. A fast new approach to pharmacophore mapping and its application to dopaminergic and benzodiazepine agonists. J. Comput. Aided Mol. Des. 7, 83–102.

McInnes, C., 2007. Virtual screening strategies in drug discovery. Curr. Opin. Chem. Biol. 11 (5), 494–502.

Morris, G.M., Huey, R., Lindstrom, W., Sanner, M.F., Belew, R.K., Goodsell, D.S., et al., 2009. Autodock4 and AutoDockTools4: automated docking with selective receptor flexibility. J. Computat. Chem. 16, 2785–2791.

Pettersen, E.F., Goddard, T.D., Huang, C.C., Couch, G.S., Greenblatt, D.M., Meng, E. C., et al., 2004. UCSF Chimera-a visualization system for exploratory research and analysis. J. Comput. Chem. 25, 1605.

Poptodorov, K., Luu, T., Hoffmann, R., 2006. Pharmacophore model generation software tools. Pharm. Pharm. Searches 32, 15–47.

Rehman, A., Akhtar, S., Siddiqui, M.H., Sayeed, U., Ahmad, S.S., Arif, J.M., et al., 2016. Identification of potential leads against 4-hydroxytetrahydrodipicolinate synthase from *Mycobacterium tuberculosis*. Bioinformation 12 (11), 400–406.

Roughley, S., Wright, L., Brough, P., Massey, A., Hubbard, R.E., 2012. Hsp90 inhibitors and drugs from fragment and virtual screening. Top. Curr. Chem. 317, 61–82.

Ruiz, F.M., Gil-Redondo, R., Morreale, A., Ortiz, A.R., Fábrega, C., Bravo, J., 2008. Structure-based discovery of novel non-nucleosidic DNA alkyltransferase inhibitors: virtual screening and *in vitro* and *in vivo* activities. J. Chem. Inf. Model. 48, 844–854.

Simmons, K.J., Chopra, I., Fishwick, C.W., 2010. Structure-based discovery of antibacterial drugs. Nat. Rev. Microbiol. 8, 501–510.

Smellie, A., Teig, S.L., Towbin, P., 1995. Poling—promoting conformational variation. J. Comput. Chem. 16, 171–187.

Talele, T.T., Khedkar, S.A., 2010. Successful applications of computer aided drug discovery: moving drugs from concept to the clinic. Curr. Top. Med. Chem. 10 (1), 127–141.

Triballeau, N., Van Name, E., Laslier, G., Cai, D., Paillard, G., Sorensen, P.W., et al., 2008. High-potency olfactory receptor agonists discovered by virtual high-throughput screening: molecular probes for receptor structure and olfactory function. Neuron 60, 767–774.

Van Drie, J.H., 2007. Computer-aided drug design: the next 20 years. J. Comput. Aided Mol. Des. 21 (10-11), 591–601.

Vijayakrishnan, R., 2009. Structure-based drug design and modern medicine. J. Postgrad. Med. 55 (4), 301.

Wang, J., Wolf, R.M., Caldwell, J.W., Kollman, P.A., Case, D.A., 2004. Development and testing of a general amber force field. J. Comput. Chem. 25, 1157.

Wermuth, C.G., 2006. Pharmacophores: historical perspective and viewpoint from a medicinal chemist. Pharm. Pharm. Searches 32, 1–13.

Wolber, G., Seidel, T., Bendix, F., Langer, T., 2008. Molecule-pharmacophore superpositioning and pattern matching in computational drug design. Drug Discov. Today 13, 23–29.

Zhang, S., 2011. Computer-aided drug discovery and development. Methods Mol. Biol. 716, 23–38.

Zhao, L., Huang, W., Liu, H., Wang, L., Zhong, W., Xiao, J., et al., 2006. FK506-binding protein ligands: structure-based design, synthesis, and neurotrophic/neuroprotective properties of substituted 5,5-dimethyl-2-(4-thiazolidine)carboxylates. J. Med. Chem. 49, 4059–4071.

FURTHER READING

Dhasmana, A., Jamal, Q.M.S., Gupta, R., Siddiqui, M.H., Kesari, K.K., Wadhwa, G., et al., 2016. Titanium dioxide nanoparticles provide protection against polycyclic aromatic hydrocarbon BaP& chrysene induced perturbation of DNA repair machinery: a computational biology approach. Biotechnol. Appl. Biochem. 63 (4), 497−513.

Liu, J., Farmer Jr., J.D., Lane, W.S., Friedman, J., Weissman, I., Schreiber, S.L., 1991. Calcineurin is a common target of cyclophilin-cyclosporin A and FKBP-FK506 complexes. Cell 66, 807−815.

CHAPTER 20

Plant Extracts and Phytocompounds in the Management of Malaria

Meenu Kalkal and Jyoti Das
Immunology Division, National Institute of Malaria Research, New Delhi, India

20.1 INTRODUCTION

Malaria is a life-threatening disease ravaging the tropical and subtropical countries of the world. It is caused by members of the genus *Plasmodium* and is spread via the bite of female *Anopheles* mosquitoes. There are four species of *Plasmodium* which cause malaria to humans: *P. vivax*, *P. ovale*, *P. malariae*, and *P. falciparum*; and the most widespread and severe malaria is caused by *P. falciparum*. However, the parasite *Plasmodium* is transmitted by female *Anopheles* mosquito.

20.1.1 Life Cycle of *Plasmodium* Parasite

The life cycle of the parasite traverses through two hosts. Sexual cycle occurs in mosquito, while asexual cycle is completed in human host (Fig. 20.1). The parasite life cycle starts when infected female mosquito injects sporozoites into the host during its blood meal. Sporozoites then migrate to the liver and invade liver cell or hepatocytes, where they start the asexual *exo*-erythrocytic schizogonic cycle. In human liver, the *Plasmodium* trophozoite initially appears as a mononucleotide round body into the cytoplasm of the host hepatocytes; subsequently, it begins to develop and multiply asexually, to form a mature schizonts (the multinucleated stage of the parasite) and finally release a large number of merozoites. These merozoites after entering the circulation invade the red blood cells to start the erythrocytic cycle. It takes 48 h in *P. falciparum* and *P. vivax*, but in *P. ovale*, it lasts for 50 h, while in *P. malariae*, longer cycle of 72 h is required, which is called development period. The merozoites undergo multiple nuclear divisions along with morphological changes and form schizonts which have appropriate number of

New Look to Phytomedicine
DOI: https://doi.org/10.1016/B978-0-12-814619-4.00021-5

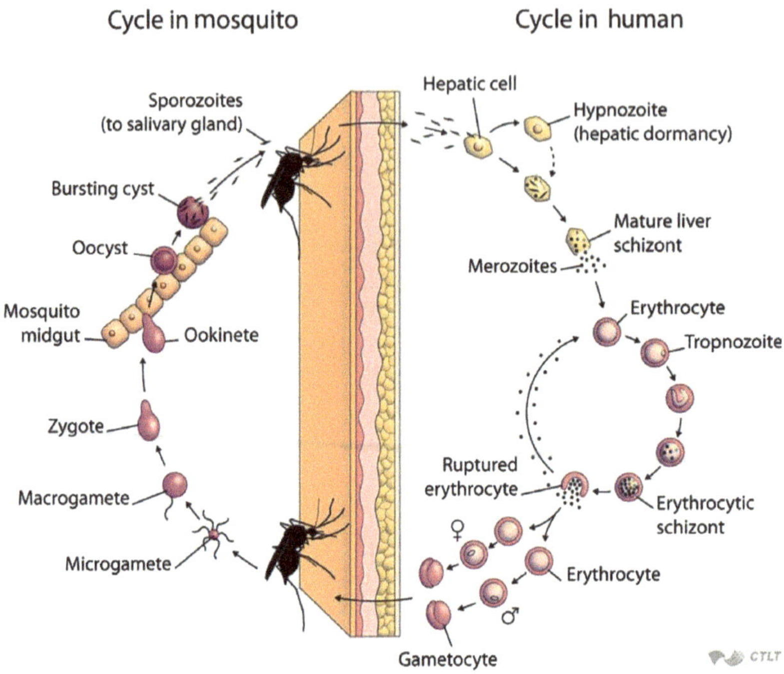

Figure 20.1 Life cycle of malaria parasite. *Johns Hopkins School of Public Health, Open Course Ware, Baltimore, 2018. ⟨http://www.ocw.jhsph.edu/⟩.*

merozoites: approximately 36 for *P. falciparum*, 24 for *P. vivax* and *P. ovale*, and 12 for *P. malariae*. Finally, the erythrocyte ruptures to release the merozoites which invade new erythrocytes to repeat the schizogonic cycle until the process is inhibited by the specific immune response or by chemotherapy. During schizogonic cycle, some of the merozoites differentiate into sexual forms (the gametocytes). These gametocytes enter into the midgut of mosquito during blood meal and initiate the sporogonic or the sexual cycle. The female gametocyte develops into macrogamete after leaving the erythrocyte, while the male gametocyte undergoes nuclear division in which the nucleus divided into eight sperm like flagellated microgamete's, each of which leaves the erythrocyte to fertilize the macrogamete and results in formation of zygote which later develops into elongated and motile ookinete after few hours of blood meal. The ookinete settles beneath the basal lamina and develops into a nonmotile oocyst (Prudêncio et al., 2006). The product of mature oocyst is the sporozoites, which leaves the oocyst and reaches the salivary gland. When the infected mosquito feeds on vertebrate host, these sporozoites along with salivary fluid are injected to start another asexual cycle.

This characteristic feature of the parasite and its ability to multiply rapidly to overcome immune response leads to the disease to a greater extent, globally. In 2016, there were an estimated 216 million cases of malaria, an increase of about 5 million cases over 2015. Therefore, despite a good amount of research efforts to control malaria, morbidity and mortality due to malaria has not changed globally to a great extent.

20.2 MALARIA: CONTROL AND ITS REPERCUSSIONS

Till date, no effective vaccine for malaria prevention strategy is available. Thus, two main strategies being employed are (1) vector control to limit the vector population and malaria transmission and (2) killing of the parasite by treatment with antimalarials. Therefore, treatment with antimalarials is very important in order to kill parasite. Five major groups of drugs have been used as antimalarial agents on the basis of their mechanism of action on *Plasmodium*: quinoline, artemisinin, antifolate, atovaquone, and antibacterial. In the earlier day's quinine, the bark extract of Cinchona tree was the choice of drug for malaria treatment but soon got replaced with various new antimalarial drugs that had less adverse side effects. Since the discovery of chloroquine in 1940s, it became the main choice of antimalarial drug against malaria.

However, during 1973, the emergence of drug resistance in *P. falciparum* against chloroquine (Sehgal et al., 1973) in India has led to the urgency for the development of new antimalarial drugs. Artemisinin is the new effective and widely used antimalarial drug derived from the Chinese herb *Artemisia annua*, being used for the treatment of *P. falciparum* infection (Miller and Su, 2011). Artemisinin derivatives, that is, artesunate, artemether, and arteether, are currently being used in combination with other antimalarial compounds (Rosenthal, 2008). However, resistance to artemisinin has been reported in parts of Southeast Asia like Cambodia, Thailand, and Vietnam, and mutation in the propeller domain of *P. falciparum* Kelch (K13) gene has been associated with the delayed parasite clearance (Cui et al., 2015; Mbengue et al., 2015). Other drugs like anti-folates: sulfa-pyrimethamine and primaquine are also used for malaria treatment in these days.

Besides, therapeutic use of drugs research on development of an effective vaccine was tried globally, but without any success. However, an effective vaccine RTS,S/AS01 targeting circumsporozoite protein of malaria parasite has shown promising results in imparting protection against *P. falciparum* in Phase III Clinical Trials (Agnandji et al., 2011). However, there is still time, and more study is required before it becomes readily available to the malaria exposed populations. Till date, with the reports of insecticide resistance in mosquitoes and drug resistance to the parasite, we are unable to control malaria. It is expected that in coming days, the resistance will spread to other parts of the world, posing a major threat to the efforts being made to control and subsequently eradicate malaria. Thus, there is utmost importance that new drug targets, that disrupt crucial events of the parasite life cycle, need to be identified at the earliest which can help not only in eliminating the pathogen but also complete eradication of malaria. In this context, an alternative to conventional natural and synthetic drugs available, active compounds from antiplasmodial herbs seem to be the best source to find out the new antimalarial agents with higher efficacy, easy availability, and low cost.

20.3 MEDICINAL PLANTS: SOURCE OF NEW ANTIMALARIAL COMPOUNDS

Historically, medicinal plants had huge contribution toward primary health care, particularly for the prevention and treatment of various diseases. Therefore, a good start to combat malaria will be traditional medicinal plants that have been used to treat malaria from many years in local and tribal communities. Numerous research studies have evaluated the potential of medicinal plants, traditionally used in different communities throughout the world, and some studies obtained very promising results so far to kill the *Plasmodium* parasites in vitro and in vivo; however, translation of these results into clinical practice is still a neglected field. In this review, we will focus on selected medicinal plants with good antiplasmodial activity with references and a few specific reviews, as it is out of the scope to describe all such plants in detail.

Till date, more than 1200 medicinal plants have been reported from all over the world for treatment of Malaria (Willcox and Bodeker, 2004). A review estimated that about 752 medicinal plants (Lemma et al., 2017) belonging to 254 genera was tested by various researchers around the globe, but most of those plants are from Africa and Asia region. Traditionally used medicinal plants useful against malaria include administration of infusions (hot teas), decoctions (boiled teas), tinctures (alcohol and water extracts), paste, powder, and macerations or cold-soaking (Builders, 2015) of different plant parts including root, stem, leaf, bark, twigs, flower, fruits, and seed which have been evaluated for their effectiveness against malaria or *Plasmodium* parasite (Builders, 2015). First effective treatment for malaria is quinine, isolated from the bark of Cinchona tree or *Cinchona officinalis*. Later, artemisinin was isolated from *A. annua* followed by febrifugine, a quinazolinone alkaloid isolated from *Dichroa febrifuga* and also found in garden plant *Hydrangea*. However, febrifugine has been clinically tested and found active against *P. vivax* and *P. ovale* but was not recommended to use at commercial level due to liver toxicity (Zhu et al., 2012).

Plants have also been investigated for their antiplasmodial activity in the form of crude extract, active fraction, isolated compounds, or in other forms like oils etc. They usually contain a complex mixture of diverse secondary metabolites. A good amount of studies has been conducted to evaluate the antiplasmodial ability of diverse secondary metabolites, including the isolation of individual compounds from different plant parts and studied their antiplasmodial activity both in vitro and animal model or in vivo model.

20.3.1 Recently Reported Plant Extracts With Good Antiplasmodial Activity

Screening of available literature on various medicinal plants suggests that majority of in vitro and in vivo pharmacological studies have been carried out using crude extracts from whole plants or different parts like leaves, twigs, bark, fruits, fruit pulp, and seeds. These studies validate some of the traditional uses of these plants including treatment of malaria and also report some new applications. However, the efficacy of plants extract may vary with plant parts tested, family of the plant, age and maturity of the plant, harvesting time, genetic characteristics of the plant, geographical location and nature of soil, seasonal variation or climate, method of processing, methods and type of solvent used for extraction, etc.

In vitro studies carried out with extracts from different plants and plant parts revealed antiplasmodial activity of many of those plants against different strains of *P. falciparum*. The current literature revealed that most plant extracts have very less antiplasmodial activity calculated by measuring their ability to inhibit the parasite in vitro. In this review, we tried to compile only those plant extracts that have superior antiplasmodial activity with IC_{50} less than 10 (Table 20.1).

The **in vivo** antiplasmodial activity of different plant extracts has also been evaluated but to a very limited extent. To study the preliminary

Table 20.1 Selected plant extracts with better in vitro antiplasmodial activity

Name of plant	Type of extract	IC_{50} (μg/mL)	References
Nigella sativa, Aristolochia bracteolata	Petroleum ether or chloroform extract	2.35	Ahmed et al. (2010)
Harungana madagascariensis	Ethanolic extract (stem bark)	0.052–0.517	Iwalewa et al. (2008)
Chrozophora oblongifolia	Methanol extract	5.0	Al-Musayeib et al. (2012)
Ficus ingens	Methanol extract	8.4	
Lavandula dentata	Methanol extract	7.1	
Plectranthus barbatus	Methanol extract	6.5	
Quassia africana	Dichloromethane root extract	0.10	Mbatchi et al. (2006)
Lawsonia inermis, Tithonia diversifolia	Dichloromethane: methanol (1:1) combined extract	0.43	Afolayan et al. (2016)
Terminalia ivorensis	Aqueous extract	0.64	Komlaga et al. (2016)
Scutellaria havanensis	Chloroform extract from leaves and stems	7.7	Fernández-Calienes Valdés et al. (2016)
Triclisia gilletii	Leaves, root bark, and stem bark extract	0.02–0.25	Kikueta et al. (2013)
Mezoneuron benthamianum	Ethanolic extract of leaves	6.5	Jansen et al. (2017)
Annona muricata	Root extract	0.09	Yamthe et al. (2015)
Dacryodes edulis	Leaves extract	6.45–8.2	Zofou et al. (2011)
Vernonia amygdalina	Leaves and root extract	8.72	
Staudtia gabonensis	Methanolic extract	0.8	Lekana-Douki et al. (2011)
Adhatoda latibracteata	Dichloromethane extract	07–1.6	
Ocimum gratissimum	Ethyl acetate extract of leaf	1.8	Abiodun et al. (2011)
Trema orientalis	Hexane extract of stem bark	1.93	
Tridax procumbens	Aqueous and ethanol extract	24.8	Appiah-Opong et al. (2011)
Phyllanthus amarus		11.7	

(Continued)

Table 20.1 (Continued)

Name of plant	Type of extract	IC_{50} (μg/mL)	References
Citrullus colocynthis	Methanol extracts against *P. falciparum* 3D7 strain	2.01	Haddad et al. (2017)
	Methanol extracts against *P. falciparum* K1 strain	6.9	
Pavetta corymbosa	Crude methanolic extract	2.042	Koudouvo et al. (2011)
Tamarindus indica	Aqueous extract	4.786	

antimalarial activity of a given extract or compound in animal, *Plasmodium berghei*–infected mouse model is widely used as it is the only available model so far. The in vivo antiplasmodial activity of some promising plants extracts is listed in Table 20.2.

20.4 IN VITRO AND IN VIVO EVALUATION OF PHYTOCOMPOUNDS WITH ANTIMALARIAL ABILITY

A number of reviews have summarized the plants with antimalarial activity from different families (Lemma et al., 2017) as well as various compounds derived from those plants with antimalarial/antiplasmodial activity including the widely distributed phenolics (ellagic acid, epigallocatechin gallate, flavonoids, xanthone, coumarins, and curcumin), naphthopyrones, quinones, widely distributed terpenoids (iridoids, sesquiterpenes, diterpenes, and triterpenes), quassinoids, cucurbitacins (common in *Cucurbitaceae*), alkaloids (indolizidine, indole, and isoquinoline), polyacetylenes, etc. (Wink, 2012). Further literature survey also reviewed some non-alkaloidal compounds which belong to the class of anthraquinones, chromones, flavonoids, limonoids, terpenes, xanthones, miscellaneous, and related compounds (Batista et al., 2009). In addition, substantial research has been done to isolate these individual bioactive compounds from different plant parts (Table 20.3) with antiplasmodial activity. Further, these phytocompounds have been assessed both in vitro and in vivo against *P. falciparum* and *P. berghei*, respectively.

Table 20.2 Selected plant extracts with good in vitro antiplasmodial activity

Name of plant	Type of extract	% Inhibition	References
Rosa damascena	Methanol extract of flowers against *P. berghei*	57.7	Esmaeili et al. (2015)
Vernonia amygdalina	Ethanol extract of leaf against *P. berghei* (ANKA 65)	82.3 (Dose dependent)	Omoregie and Pal (2016)
Croton macrostachyus	Methanolic extract of fresh fruits against *P. berghei*	40–87 (Dose dependent)	Mekonnen (2015)
	Methanolic extract of roots against *P. berghei*	44–89 (Dose dependent)	
Russelia equisetiformis	Ethanolic extract against *P. berghei* (ANKA) infected mice	31.6–86.5 (Dose dependent)	Ojurongbe et al. (2015)
Echinops kebericho	Ethanol extract of root against *P. berghei*-infected mice	16.93–57.29 (Dose dependent)	Toma et al. (2015)
Piper guineense	Ethanolic extracts of leaves and root against *P. berghei*	28.36–62.69 (Dose dependent)	Kabiru et al. (2016)
Maytenus senegalensis	Ethanol extract of root bark against *P. berghei* (ANKA 65)	88.50–98.05 (Dose dependent)	Malebo et al. (2015)
Keetia leucantha	Dichloromethane extract of twig	56.8	Bero et al. (2013)
	Aqueous twigs extract	53.0	
Ajuga remota	Hydro-ethanolic crude extract against *P. berghei* (ANKA 65)	77.34	Nardos and Makonnen (2017)
Triclisia gilletii	Extracts of leaves and bark of root and stem against *P. berghei*	65–75	Kikueta et al. (2013)
Terminalia chebula	Aqueous extract against *P. berghei*	68.89	Pinmai et al. (2011)
Terminalia bellerica		53.40	
Phyllanthus emblica		69.46	

Table 20.3 Selected phytochemicals with antiplasmodial activity

Name of plant	Extract used	Bioactive compounds	IC_{50} (μg/mL)	References
Liriodendron tulipifera	Bark	Aporphine alkaloids: asimilobine, norushinsunine, norglaucine, liriodenine, anonaine, and oxoglaucine	1.2–29.6	Graziose et al. (2011)
	Leaves	Sesquiterpene lactones: peroxyferolide and lipiferolide	6.2 and 1.8	
Cryptolepis sanguinolenta	Ethanol extract of root bark	Cryptolepine	0.031	Kirby et al. (1995)
Cassia fistula L.	Chloroform extract of leaves	Phytol, lutein, di-lineolyl galacto-pyranosylglycerol	5.8–18.9	Grace et al. (2012)
Mezoneuron benthamianum	Ethanolic extract of leaves	Mezobenthamic acid A and B, neocaesalpin H, kaempferol, ethy lgallate, resveratrol, quercetin, 13b-OH-pheophor-bide A, gallic acid, β-sitosterol-glycoside	5.1 ≥ 10	Jansen et al. (2017)

20.5 MODE OF ACTION OF PLANT-DERIVED NATURAL COMPOUNDS

Phytocompounds, used as approved antimalarial agents or drugs, are limited in number, and the effectiveness of any antimalarial therapeutic agent depends upon its dose that must either kill or inhibit the parasite without toxicity to the host. Although a large number of natural plant-derived compounds have been shown to be able to inhibit the growth of the parasite, but in most cases, their mode or mechanism of action is not elucidated. Moreover, the mode of action of these plant-based compounds having antiplasmodial ability is unknown at present. It is possible that the new compounds isolated may have either similar mode of action to the convention drugs or it may be different.

20.6 FUTURE PERSPECTIVE OF DRUG DEVELOPMENT FOR MALARIA FROM PHYTOCOMPOUNDS

High resistance to available antimalarial compounds, either synthetic or naturally derived from plant, there is a strong need for new compounds possessing good antiplasmodial activity with less toxicity to the host. As plant-derived compounds are considered more safe, cheap, and effective candidates for development of phytomedicine, so new compounds from plants having antiplasmodial activity needs to be explored. Around the world, from different countries, literature on traditional plants with antimalarial ability along with quinine, artemisinin, and its semisynthetic derivatives are available, along with the most promising drugs for the treatment of malaria to date are from plants. Crude extracts or bioactive compounds from various plants have been described for their antiplasmodial activity in this review. However, most of them have only been evaluated by in vitro assays and only few of them were evaluated for cytotoxicity and still very limited studies have been carried out in vivo using suitable mouse model. In addition, clinical studies are also required to understand their mode and mechanism of action and bioavailability along with pharmacokinetic and pharmacodynamic studies. However, isolation and purification of individual compounds are difficult and expensive, as they are present in very low amount in different parts of plant. Therefore, there is a strong need to develop better and inexpensive techniques for isolation and purification of phytocompounds with antiplasmodial ability.

Lot of traditional information about the use of plant species having antiplasmodial activity is available, but some parameters like geographical location and abundance of plants, parts of the plant needs to be used, and method of their use should be further explored. Along with this, there is also need to examine duration and dose amount required for treatment. In addition, emphasis should be given for validation and standardization of the most effective medicinal plants or its compound(s) to develop cost-effective phytomedicine in a short time.

20.7 CONCLUSION

The antimalarial ability of various medicinal plants reported in the ancient text has been well validated through various modern scientific investigations. The different parts of the plant, their extracts, and isolated compounds have been shown to possess decent antiplasmodial ability. However, there is a need for further clinical research, particularly involving suitable animal model and human subjects for the development of effective and safe drug formulations. Also, efforts should be made to identify and isolate individual active components from the crude extracts having specific and defined antiplasmodial activity, along with the establishment of their exact mechanism of action.

REFERENCES

Abiodun, O., Gbotosho, G., Ajaiyeoba, E., Happi, T., Falade, M., Wittlin, S., et al., 2011. *In vitro* antiplasmodial activity and toxicity assessment of some plants from Nigerian ethnomedicine. Pharm. Biol. 49 (1), 9–14.

Afolayan, F.I., Adegbolagun, O.M., Irungu, B., Kangethe, L., Orwa, J., Anumudu, C.I., 2016. Antimalarial actions of *Lawsonia inermis, Tithonia diversifolia* and *Chromolaena odorata* in combination. J. Ethnopharmacol. 191, 188–194.

Agnandji, S.T., Lell, B., Soulanoudjingar, S.S., et al., 2011. First results of phase 3 trial of RTS, S/AS01 malaria vaccine in African children. N. Engl. J. Med. 365 (20), 1863–1875.

Ahmed, E.H.M., Nour, B.Y., Mohammed, Y.G., Khalid, H.S., 2010. Antiplasmodial activity of some medicinal plants used in Sudanese folk-medicine. Environ. Health insights 4, 1.

Al-Musayeib, N.M., Mothana, R.A., Matheeussen, A., Cos, P., Maes, L., 2012. *In vitro* antiplasmodial, antileishmanial and antitrypanosomal activities of selected medicinal plants used in the traditional Arabian Peninsula region. BMC Complement. Altern. Med. 12 (1), 49.

Appiah-Opong, R., Nyarko, A.K., Dodoo, D., Gyang, F.N., Koram, K.A., Ayisi, N.K., 2011. Antiplasmodial activity of extracts of *Tridax procumbens* and *Phyllanthus amarus* in *in vitro Plasmodium falciparum* culture systems. Ghana Med. J. 45 (4), 143–150.

Batista, R., De Jesus Silva Júnior, A., De Oliveira, A.B., 2009. Plant-derived antimalarial agents: new leads and efficient Phytomedicine. Part II. Non-alkaloidal natural products. Molecules 14 (8), 3037–3072.

Bero, J., Hérent, M.F., Schmeda-Hirschmann, G., Frédérich, M., Quetin-Leclercq, J., 2013. In vivo antimalarial activity of *Keetia leucantha* twigs extracts and in vitro antiplasmodial effect of their constituents. J. Ethnopharmacol. 149 (1), 176–183.

Builders, M.I., 2015. Plants as antimalarial drugs: a review. World J. Pharm. Pharm. Sci. 8, 1747–1766.

Cui, L., Mharakurwa, S., Ndiaye, D., Rathod, P.K., Rosenthal, P.J., 2015. Antimalarial drug resistance: literature review and activities and findings of the ICEMR network. Am. J. Trop. Med. Hyg. 93, 57–68.

Esmaeili, S., Ghiaee, A., Naghibi, F., Mosaddegh, M., 2015. Antiplasmodial activity and cytotoxicity of plants used in traditional medicine of Iran for the treatment of fever. Iranian J. Pharm. Res. 14 (Suppl.), 103.

Fernández-Calienes Valdés, A., Monzote Fidalgo, L., Sariego Ramos, I., Marrero Delange, D., Morales Rico, C.L., Mendiola Martínez, J., et al., 2016. Antiprotozoal screening of the Cuban native plant *Scutellaria havanensis*. Pharm. Biol. 54 (12), 3197–3202.

Grace, M.H., Lategan, C., Graziose, R., Smith, P.J., Raskin, I., Lila, M.A., 2012. Antiplasmodial activity of the ethnobotanical plant *Cassia fistula*. Nat. Prod. Commun. 7 (10), 1263–1266.

Graziose, R., Rathinasabapathy, T., Lategan, C., Poulev, A., Smith, P.J., Grace, M., et al., 2011. Antiplasmodial activity of aporphine alkaloids and sesquiterpene lactones from *Liriodendron tulipifera* L. J. Ethnopharmacol. 133 (1), 26–30.

Haddad, M.H.F., Mahbodfar, H., Zamani, Z., Ramazani, A., 2017. Antimalarial evaluation of selected medicinal plant extracts used in Iranian traditional medicine. Iranian J. Basic Med. Sci. 20 (4), 415.

Iwalewa, E.O., Omisore, N.O., Adewunmi, C.O., Gbolade, A.A., Ademowo, O.G., Nneji, C., et al., 2008. Anti-protozoan activities of *Harungana madagascariensis* stem bark extract on trichomonads and malaria. J. Ethnopharmacol. 117 (3), 507–511.

Jansen, O., Tchinda, A.T., Loua, J., Esters, V., Cieckiewicz, E., Ledoux, A., et al., 2017. Antiplasmodial activity of *Mezoneuron benthamianum* leaves and identification of its active constituents. J. Ethnopharmacol. 203, 20–26.

Kabiru, A.Y., Ibikunle, G.F., Innalegwu, D.A., Bola, B.M., Madaki, F.M., 2016. *In vivo* antiplasmodial and analgesic effect of crude ethanol extract of *Piper guineense* leaf extract in albino mice. Scientifica 2016, Article ID 8687313, 6 pages.

Kikueta, C.M., Kambu, O.K., Mbenza, A.P., Mavinga, S.T., Mbamu, B.M., Cos, P., et al., 2013. *In vitro* and *in vivo* antimalarial activity and cytotoxicity of extracts and fractions from the leaves, root-bark and stem-bark of *Triclisia gilletii*. J. Ethnopharmacol. 149 (2), 438–442.

Kirby, G.C., Paine, A., Warhurst, D.C., Noamese, B.K., Phillipson, J.D., 1995. *In vitro* and *in vivo* antimalarial activity of cryptolepine, a plant-derived indoloquinoline. Phytother. Res. 9 (5), 359–363.

Komlaga, G., Cojean, S., Dickson, R.A., Beniddir, M.A., Suyyagh-Albouz, S., Mensah, M.L., et al., 2016. Antiplasmodial activity of selected medicinal plants used to treat malaria in Ghana. Parasitol. Res. 115 (8), 3185–3195.

Koudouvo, K., Karou, S.D., Ilboudo, D.P., Kokou, K., Essien, K., Aklikokou, K., et al., 2011. *In vitro* antiplasmodial activity of crude extracts from Togolese medicinal plants. Asian Pac. J. Trop. Med. 4 (2), 129–132.

Lekana-Douki, J.B., Bongui, J.B., Liabagui, S.L.O., Edou, S.E.Z., Zatra, R., Bisvigou, U., et al., 2011. *In vitro* antiplasmodial activity and cytotoxicity of nine plants traditionally used in Gabon. J. Ethnopharmacol. 133 (3), 1103–1108.

Lemma, M.T., Ahmed, A.M., Elhady, M.T., Ngo, H.T., Vu, T.L.H., Sang, T.K., et al., 2017. Medicinal plants for *in vitro* antiplasmodial activities: a systematic review of literature. Parasitol. Int. 66 (6), 713–720.

"Life cycle of the malaria parasite" from Epidemiology of Infectious Diseases. Available at: http://ocw.jhsph.edu. Copyright © Johns Hopkins Bloomberg School of Public Health. Creative Commons BY-NC-SA.

Malebo, H.M., Wiketye, V., Katani, S.J., Kitufe, N.A., Nyigo, V.A., Imeda, C.P., et al., 2015. *In vivo* antiplasmodial and toxicological effect of *Maytenus senegalensis* traditionally used in the treatment of malaria in Tanzania. Malar. J. 14 (1), 79.

Mbatchi, S.F., Mbatchi, B., Banzouzi, J.T., Bansimba, T., Ntandou, G.N., Ouamba, J.M., et al., 2006. *In vitro* antiplasmodial activity of 18 plants used in Congo Brazzaville traditional medicine. J. Ethnopharmacol. 104 (1), 168–174.

Mbengue, A., Bhattacharjee, S., Pandharkar, T., Liu, H., Estiu, G., Stahelin, R.V., et al., 2015. A molecular mechanism of artemisinin resistance in *Plasmodium falciparum* malaria. Nature 520, 683–687.

Mekonnen, L.B., 2015. *In vivo* antimalarial activity of the crude root and fruit extracts of *Croton macrostachyus* (Euphorbiaceae) against *Plasmodium berghei* in mice. J. Tradit. Complement. Med. 5 (3), 168–173.

Miller, L.H., Su, X., 2011. Artemisinin: discovery from the Chinese herbal garden. Cell 146 (6), 855–858.

Nardos, A., Makonnen, E., 2017. *In vivo* antiplasmodial activity and toxicological assessment of hydroethanolic crude extract of *Ajuga remota*. Malar. J. 16 (1), 25.

Ojurongbe, O., Ojo, J.A., Adefokun, D.I., Abiodun, O.O., Odewale, G., Awe, E.O., 2015. *In vivo* antimalarial activities of *Russelia Equisetiformis* in *Plasmodium berghei* infected mice. Indian J. Pharm. Sci. 77 (4), 504.

Omoregie, E.S., Pal, A., 2016. Antiplasmodial, antioxidant and immunomodulatory activities of ethanol extract of *Vernonia amygdalina* del. leaf in Swiss mice. Avicenna J. Phytomed. 6 (2), 236.

Pinmai, K., Hiriote, W., Soonthornchareonnon, N., Jongsakul, K., Sireeratawong, S., Tor-Udom, S., 2011. *In vitro* and *in vivo* antiplasmodial activity and cytotoxicity of water extracts of *Phyllanthus emblica*, *Terminalia chebula*, and *Terminalia bellerica*. J. Med. Assoc. Thai. 93 (12), 120.

Prudêncio, M., Rodriguez, A., Mota, M.M., 2006. The silent path to thousands of merozoites: the *Plasmodium* liver stage. Nat. Rev. Microbiol. 4, 849–856.

Rosenthal, P.J., 2008. Artesunate for the treatment of severe falciparum malaria. N. Engl. J. Med. 358 (17), 1829–1836.

Sehgal, P.N., Sharma, M.I.D., Sharma, S.L., Gogai, S., 1973. Resistance to chloroquine in falciparum malaria in Assam state India. J. Commun. Dis. 5 (4), 175–180.

Toma, A., Deyno, S., Fikru, A., Eyado, A., Beale, A., 2015. *In vivo* antiplasmodial and toxicological effect of crude ethanol extract of *Echinops kebericho* traditionally used in treatment of malaria in Ethiopia. Malar. J. 14 (1), 196.

Willcox, M.L., Bodeker, G., 2004. Traditional herbal medicines for malaria. Br. Med. J. 329 (7475), 1156.

Wink, M., 2012. Medicinal plants: a source of anti-parasitic secondary metabolites. Molecules 17 (11), 12771–12791.

Yamthe, L.R.T., Fokou, P.V.T., Mbouna, C.D.J., Keumoe, R., Ndjakou, B.L., Djouonzo, P.T., et al., 2015. Extracts from *Annona Muricata* L. and *Annona reticulata* L. (Annonaceae) potently and selectively inhibit *Plasmodium falciparum*. Medicines 2 (2), 55–66.

Zhu, S., Chandrashekar, G., Meng, L., Robinson, K., Chatterji, D., 2012. Febrifugine analogue compounds: synthesis and antimalarial evaluation. Bioorg. Med. Chem. 20 (2), 927–932.

Zofou, D., Tene, M., Ngemenya, M.N., Tane, P., Titanji, V.P., 2011. *In vitro* antiplasmodial activity and cytotoxicity of extracts of selected medicinal plants used by traditional healers of Western Cameroon. Malar. Res. Treat. 2011, Article ID 561342.

FURTHER READING

Krettli, A.U., Andrade-Neto, V.F., Brandão, M.D.G.L., Ferrari, W., 2001. The search for new antimalarial drugs from plants used to treat fever and malaria or plants randomly selected: a review. Mem. Inst. Oswaldo Cruz 96 (8), 1033–1042.

CHAPTER 21

Assessment of Antimicrobial Activity of Different Phytochemicals Against Enteric Diseases in Different Animal Models

Hemanta Koley, Debaki Ranjan Howlader and Ushasi Bhaumik
Division of Bacteriology, National Institute of Cholera and Enteric Diseases, Kolkata, West Bengal, India

21.1 INTRODUCTION AND OVERVIEW OF THE INTESTINAL ENVIRONMENT

21.1.1 Evolutionary Concepts

Before the appearance of multicellular life, eukaryotes had started living in close proximity with prokaryotes. During the "early" stages of life, planktonic bacteria exist or coevolved with marine invertebrates. Once the animal body reached a suitable size, exceeding the diffusion radius of the energy-providing molecules, the formation of a gut tube became necessary for capture, concentration, and extraction of soluble environmental materials. After the acquisition of nutrients and minerals, it was not surprising to see that the gut became a favorable place to thrive for the prokaryotes like bacteria (Zhang et al., 2015). The host and microbe thus made a symbiotic dyad where they maintain a constant check and ensure the "sanctity" of the environment. Whenever a breach causes this "pristine" environment get contaminated, the host species suffer from infections with opportunistic as well as obligate pathogens.

21.1.2 Intestinal Microbiota

Like majority of the epithelial surfaces of our body, our gut suits as a home to a staggering number of normal microflora, the commensal one. These groups of flora mainly consist of bacteria, with a certain number of other microorganisms, such as, fungi, viruses, and protozoa (Jandhyala

New Look to Phytomedicine
DOI: https://doi.org/10.1016/B978-0-12-814619-4.00022-7

et al., 2015). Our microbiota contains trillions of bacterial cells, as per popular belief, it is almost 10 times more than our own body cells. Our stomach contains only $10^3 - 10^4$ or less bacteria/g, whereas the jejunum contains $10^5 - 10^6$ bacteria/g and the terminal ileum contains nearly $10^8 - 10^9$ bacteria/g. The most abundant presence of indigenous flora was reported to be found on the colonic epithelial surface, nearly $10^{11} - 10^{14}$ bacteria/g of the intestinal contents. New literature, though, has updated this estimation and came up with a revised proportion for bacteria and human cell number (Sender et al., 2016). Although most of them are harmless, often making symbiosis alliance, they can cause pathogenesis in the right condition, such as immunodeficiency of the host.

21.1.3 Immune Response Toward Commensal Flora With Induction and Immune Tolerance

Questions might arise about the importance of the gut flora to be present in the intestines. After laborious and intensive research for decades, scientists have come up with the possible reasons why the human body need the gut flora to be present. It has been proved that the induction of oral tolerance requires the intestinal tract to be colonized by microbes early in life (Gaboriau-Routhiau et al., 2003). During the developmental stages of life, a huge number of naïve B- and T-lymphocyte receptors are generated to recognize all kinds of different antigens. These receptors can recognize self-antigens as well as foreign antigens. Deleterious reaction with foreign antigens is normally avoided by negative selection or central tolerance. But intestinal or peripheral tolerance cannot be given by this way (Chistiakov et al., 2014). The presence of peripheral tolerance to the food antigens and harmless flora of the gut ensure the host to be healthy instead of developing autoimmune disease such as Crohn's disease and ulcerative colitis. Failure to induce tolerance to food antigens can cause food allergy and celiac diseases too. Thus, the presence of gut bacteria helps in the proper development of the host's immune system.

21.1.4 Mucosal Barrier Functions

The mucosal barrier helps to prevent microbes to come in contact with the healthy host's tissues acting as a mechanical barrier to the microbes (Turner, 2009). More to that, the host's immune system helps in the defense mechanism of the body toward the microbial challenges. Our gut is the main site of our body which comes in contact with most of the antigenic insults in the form of foreign invaders via food and the food particles. This is the reason why our gut alone contains almost 80% of the body's immune cells in the form of mucosal-associated immune system. Hence, anyone can assume that the gut is the safest place of our body that will not be affected by foreign, harmful organisms, but indeed, it is not true. Most of the "exogenous" infectious agents enter their hosts via the mucosal route. The exact reasons which trigger these infections are still to be evaluated in detail. That is why the main treatment till now is the use of antimicrobials whenever a foreign breach is detected. But on the other hand, excessive use of different antimicrobials may induce the occurrence of inflammatory bowel disease in humans (Nitzan et al., 2016).

21.1.5 Intestinal Pathogens and Chemotherapies

To fight against these infections, we need an approach eradicate the causative agent as well as those which helps in the maintenance of the immune homeostasis. Despite the huge growth in the field of the "antimicrobial chemotherapeutic agents" industry, there has been a significant increase in the enteric diseases in the developing countries. Many bacterial species can infect the otherwise healthy human gut, such as the members of Enterobacteriaceae and Vibrionaceae like *Vibrio cholerae*, *Shigella* sp., *Salmonella* sp., etc. These enteric bacteria have been reported to be the causative agents for diarrheal diseases and enteric fever, mainly affecting the developing countries (Petri et al., 2008). Other than the bacteria mentioned above, *Escherichia coli*, *Campylobacter jejuni*, etc. have proven to be the causative agents of diarrheal symptoms in the developed countries too. Since their discovery, different antimicrobials have reduced the disease burden in affected people, but due to their high cost, the usage has been less in poor people. Large-scale production of common antibiotics has caused the price to drop to a certain level where it became available for all. Due to the low-surveillance level, the usage of "over the counter" antibiotics have increased. This has intrigued the emergence of drug-resistant strains worldwide. The Centers for Disease Control and Prevention has classified a number of bacteria as presenting urgent, serious, and concerning threats, many of which are already responsible for placing a substantial clinical and financial burden on the US health-care system, patients, and their families. Thus, the burden of enteric bacterial infections has acquired new heights due to the emergence of MDR strains. In this present scenario, the use of traditional medicines as well as drinks and foods has increased which is showing promising results, causing the infection to be subsided with ease.

21.1.6 Phytochemicals

Plants contain an important source of active ingredients which differ widely in terms of structure and functionality or therapeutic potentials. Phytochemicals are bioactive non-nutrient compounds found in vegetables, fruits, tea, grains, and other plant foods that have certain therapeutic potentials. These phytochemicals include the alkaloids, flavonoids, tannins, terpenoids, glycosides, saponins, and anthraquinones among others. Since antiquity, man has been using various plant products, and some of these products are still in use to treat common maladies. According to a report by World Health Organization (WHO) (2002), 80% people depend on phytochemicals during their illness. The sanitation system and way of living are poor in developing countries. The people from most developing countries come in contact with many microbes that a common man from developed country would not. But due to the presence of various medicinal plants and the knowledge of their usage, people are using these plants instead of antibiotics because of their high cost and unavailability. Some examples are the use of bearberry (*Arctostaphylos uva-ursi*) and cranberry juice (*Vaccinium macrocarpon*) to treat urinary tract infections is reported in different manuals of phytotherapy (Jepson et al., 2000), while species such as lemon balm (*Melissa officinalis*), garlic (*Allium sativum*), and tea tree (*Melaleuca alternifolia*) are described as broad-spectrum antimicrobial agents (Jepson et al., 2000). Different plant parts, such as stems, roots, leaves or their combination, have been employed in the treatment of infectious pathologies in the respiratory system, urinary tract, gastrointestinal and biliary systems, as well as on the skin.

With this knowledge, let us now consider the present developments on the ground of using various phytochemicals in various ailments. In this chapter, we have specifically focused on the use of phytochemicals in the treatment of enteric diseases. Phytochemicals can be used in two different approaches, as follows:

1. Used as a therapeutic agent after the occurrence of the disease or
2. Used as a prophylactic way to prevent the disease from occurring.

21.2 IMPORTANT ANIMAL MODELS OF ENTERIC BACTERIAL INFECTIONS

21.2.1 Animal Models for *Vibrio cholerae*

21.2.1.1 Suckling Mice Model

V. cholerae is able to colonize the small intestines of several suckling mammals, including mice (Klose, 2000). The reason behind the colonization of this bacterium in the suckling mice intestine could be a result of immature host defenses. Toxin-coregulated pilus and its regulator ToxR are important factors for the production of the disease cholera in humans. This model mimics the *V. cholerae* infection in children, which can be treated by using phytochemicals.

21.2.1.2 Mice Ileal Loop Model

Another well-known model for the study of pathogenesis of cholera is the mice ileal loop model (Sawasvirojwong et al., 2013). Here, *V. cholerae* has to be injected in the ligated ileal loops (2–3 cm in length). This model shows the time-course fluid accumulation in the ileal loops. In this model, the level of CT produced can be measured which resembles the human cholera infection. This model might be beneficial for studying pathophysiology of fluid secretion during *V. cholerae* infection and for evaluation of antisecretory therapy of cholera.

21.2.1.3 Rabbit Ileal Loop Model

Among surgical models of *V. cholerae* colonization, adult rabbit ileal loop model is well known (Mosley and Ahmed, 1969). Rabbits were kept on no food diet for 24 h before the surgery. Ileal loops need to be made in rabbit small intestine and inoculated with *V. cholerae* liquid culture via a syringe. After 24 h, rabbits were euthanized and the fluid retention was measured. Blood stained or rice watery fluid was collected and enumerated for the presence of *V. cholerae*. Microscopic examination of the fluid revealed the presence of flecks of mucus, with numerous epithelial cells and vibrios.

21.2.2 Removable Intestinal Tie-Adult Rabbit Diarrhea

Other than rabbit ileal loop model, *Removable Intestinal Tie-Adult Rabbit Diarrhea* is the other surgically manipulated model (Spira et al., 1981). Rabbits fasted for 24 h were used for the colonization study. An aseptic surgical technique was used to make a 4–5-cm midline incision and externalize the intestine at the level of the ileocecal valve. A cecal tie was placed 1–2 cm from the ileocecal junction by using a 3-0 Dexon-S suture (Davis-Geck Inc., American Cyanamid Co., Manati, PR). The exposed intestine was kept moistened with sterile saline. A reversible slip-knot tie was placed on the ileum at a distance of 10 cm proximal to the mesoappendix, using Silastic medical tubing (Dow Corning Corp., Medical Products, Midland, MI), which was externalized through the flank by using a cutting needle. The histologic along with the scanning and transmission electron microscopic evaluations indicated that the mechanism(s) of mucosal damage and diarrhea by *V. cholerae* strains includes bacterial invasion into the luminal epithelial cells, resulting in necrosis and exfoliation of the damaged cells.

The use of different phytochemicals as a prophylactic agent as well as drug has elaborately been mentioned in other part of this chapter. Whether they have anticolonizing activity, is it bactericidal or bacteriostatic, has been assessed by many scientists. Other than these, it is a well-known fact that non-O1 serotype of *V. cholerae* causes dermatitis. The use of phytochemicals as an ointment or a gel-based product in the treatment of this kind of dermatitis can also be done in future.

21.2.3 Animal Models for *Salmonella* sp.

21.2.3.1 Mice Intraperitoneal Mode

Mice can be infected by intraperitoneal route also. Intraperitoneal administration of nucleus tractus solitarius as well as typhoidal salmonellae has been proved to be effective (Simon et al., 2011). However, oral infection allows more accurate control of the administered dose, since orally administered *Salmonella* do not immediately enter a rapid growth phase following inoculation, as can happen with intra peritoneal (IP) infections. In all reports, it has been shown to be the most effective in the case of both vaccine efficacy study and challenge study. Phytochemicals can be used here as a drug where their ability to prevent the infection could be evaluated, for example, polyphenols isolated from edible flower of *Sesbania grandiflora* or Kombucha can be mixed together with the pathogen followed by the administration of the mixture. Then anticolonization activity can be measured by bacterial colonization assays.

21.2.3.2 Rat Model of Infection

Specific pathogen-free male Wistar–Unilever rats were used. The animals were 6 ± 9 weeks of age, were housed individually in makrolon cages,

1 ± 2 weeks prior to inoculation (Naughton et al., 1996). Drinking water and conventional diet were provided ad libitum. In this case, *Salmonella enterica* serovar Enteritidis was used. After 1 ± 2 weeks of rest (i.e., acclimatization), the animals were starved overnight (water ad libitum). After 16 h of starvation, 1 mL of the bacterial suspensions was orally administered by gavages.

21.2.3.3 The Calf Gastroenteritis Model

Calves are regarded as the animal model of choice to study *Salmonella* Typhimurium-induced enterocolitis and diarrhea (Higginson et al., 2016). Cattle are natural hosts for this pathogen and, most importantly, clinical and histological manifestations of *Salmonella* Typhimurium enterocolitis are similar to the human disease. Bovine models have allowed the identification of several *Salmonella*-associated factors required to evoke enterocolitis, including flagella, *aroAD*, lipopolysaccharide, and the SPI1 type III secretion system. This animal model has proven invaluable for correlating in vitro findings with disease parameters in vivo. Unfortunately, the costs and logistics involved in bovine experiments are immense, cattle are outbreed (i.e., genetically diverse), and methods for genetically or immunologically manipulating the bovine host are severely limited when compared with murine models. Same type of treatment can be carried out using the rat and the calf gastrointestinal model.

21.2.3.4 Rabbit Model of Salmonella Infection

New Zealand albino rabbits were used to assess the toxigenicity of *Salmonella* sp. Rabbit intestine was externalized and kept moist in Ringer's solution and washed thoroughly with the same solution (Panda et al., 2014). About 7–9 cm loops were introduced in the rabbit intestine, and each loop was separated by a spacer loop. Bacterial culture was prepared in the conventional way and introduced in the rabbit intestinal loops. Like the rabbit ileal loop for *V. cholerae*, here also, measurement of fluid was evaluated.

21.2.3.5 The Mouse Colitis Model

This model allows the intestinal flora to be diminished by a pretreatment of streptomycin followed by oral inoculation of *Salmonella* Typhimurium (Barthel et al., 2003). Pretreatment of streptomycin kills the resident microflora and helps in the establishment of *Salmonella*. Due to the development of colitis in these streptomycin-pretreated mice, the cecum became small in size. Colitis along with other autoimmune diseases such as the Crohn's disease and ulcerative colitis are growing problems of the developed world. Treating these diseases using various phytochemicals would be a good option for us since they have already proven to be effective against some of these diseases.

21.2.4 Animal Models for *Shigella* sp.

21.2.4.1 Keratoconjunctivitis Model

The ability of shigellae to invade the corneal epithelia of guinea pigs and to spread to contiguous cells, causing keratoconjunctivitis, provides a model system to test the virulence of *Shigella* strains, and the protective efficacy and immunogenicity of *Shigella* vaccines (Hartman et al., 1994). Here, overnight grown culture of *Shigella* was given directly on the guinea pig's eye. *Shigella* invades the corneal cells just like they do in the case of the intestinal epithelium. The keratoconjunctivitis model provides a simple and cost-effective method to test the efficacy of vaccine candidate strains and to monitor the serum immune response to those strains. Phytochemicals can be used here as a drug, as their ability to inhibit *Shigella* sp. was already proved by other groups (Mukherjee et al., 2013). The ability of phytochemicals to treat the conjunctivitis occurred by *Shigella* sp. or any other species can be done after experimentally infecting the eyes of guinea pig with bacteria followed by the use of phytochemicals or their specific active constituents.

21.2.4.2 Intrarectal Model of Guinea Pig

To develop a proper animal model representing human bacillary dysentery, an intrarectal model (Mount and Barron, 1976) in guinea pig was developed by Shim et al. (2007), where bacteria were directly inoculated in rectal path of guinea pig and induced acute inflammation, making this animal model useful for assessing the protective efficacy of *Shigella* vaccine candidates. This animal model reproducing human shigellosis was tested several different strains of *Shigella* by the intrarectal route and was then monitored for signs of clinical dysentery at different time points. In this model, guinea pigs were not starved or antibiotic treated to maintain the natural conditions of the resident microflora in the gut. Within 24 h following challenge, all guinea pigs infected with invasive wild-type *Shigella* strains developed bacillary dysentery characterized by weight loss, tenesmus, and liquid stool mixed with mucus and blood, whereas the control groups infected with phosphate buffer saline or a noninvasive *Shigella* strain did not develop symptoms. Furthermore, robust expression of IL-8, IL-1β, and inducible NO synthase mRNA was detected in the colon from 6 to 24 h following *Shigella* infection. So, it can be concluded from here that the induction of infection was mainly due to the production of high amount of pro-inflammatory cytokines. Literature dictates that the production of pro-inflammatory cytokines in excess is connected to the occurrence of colitis and sepsis. Our assumptions are that the induction of colitis and sepsis can also be prevented by using phytochemicals or their active compounds. After the occurrence of colitis or sepsis in animal models, phytochemicals needs to be administered to see their effects in the animals. Reduction in the level of pro-inflammatory cytokine production can reduce the occurrence of both colitis and sepsis, theoretically. *Shigella* sp. has newborn mice as well as intraperitoneal adult mice models. The possibilities of using phytochemicals or their specific active compounds were already stated above. In the case of *C. jejuni* and *Helicobacter pylori*, the same type of prophylactic approaches as well as use as a potential drug can be carried away.

21.2.4.3 Suckling Mice Model

Neonate mice were inoculated intragastrically with 5×10^9 CFU of virulent *Shigella* strain without any starvation or antibiotic treatment (Mukherjee et al., 2013). Mice were then separated from their mothers and kept aside. Intestines were then pulled out and plated to check the colonization of inoculated bacteria.

21.3 MECHANISM OF ACTION OF VARIOUS PHYTOCHEMICALS ON ENTERIC BACTERIAL INFECTIONS

Following three microflora modulation tools are available so far:

1. The addition of exogenous living microorganisms to foods (i.e., probiotics);
2. The selective stimulation of the growth and activity of beneficial microorganisms indigenous to the gut (i.e., addition of prebiotics); and
3. A combination of both approaches (i.e., synbiotics).

All three approaches attempt to increase the number of bacteria positive for human gastrointestinal health, usually the lactobacilli and bifidobacteria. Another way of defending against enteric pathogens is by using drugs. Our prime focus would be on the use of phytochemicals as prophylactic agents and/or drug. Here, we will discuss about some known traditional phytochemicals and their unknown or less-known effects on enteric pathogens, both in vivo and in vitro.

21.3.1 Kombucha

Kombucha is a popular beverage included among many traditional fermented foods across the world. It has gained its popularity as a traditional medicine due to its various claimed and few proven pharmacological properties, such as detoxification and antioxidant properties, energizing agent, as well as immunomodulatory effects. Fermentation of sugared black tea (SBT) with symbiotic acetic acid bacteria (AAB) and yeasts for 14 days at 28°C results in the formation of Kombucha (Chakravorty et al., 2016).

21.3.1.1 Present Knowledge and Composition

It consists of black tea along with AAB *Acetobacter aceti*, *Acetobacter pasteurianus*, *Gluconobacter oxydans* and yeasts *Saccharomyces* sp., *Zygosaccharomyces kombuchaensis*, *Torulopsis* sp., *Pichia* sp., *Brettanomyces* sp. The interaction between these microorganisms results in the formation of a floating layer on top of the tea. Recently, a group of researcher showed that 30% of this top layer comprises lactic acid bacteria (LAB) along with a good amount of AAB. This optimally mixed culture of LAB and AAB was found to enhance the production of D-saccharic acid 1,4-lactone in Kombucha, which is known for its detoxifying and antioxidant properties (Chakravorty et al., 2016). This compound was found to inhibit hyperglycemia-induced hepatic apoptosis via inhibition of both extrinsic and intrinsic pathways in diabetic rats. Based on past results, scientists are thinking about mixing LAB with Kombucha to increase its health benefits, which will enable the beverage industry to produce higher quality healthy fermented tea.

21.3.1.2 Mechanism of Action

It is believed that most of these beneficial properties may be attributed to the polyphenols, organic acids specifically acetic acid, and a host of other ingredients that are inherent to the raw materials or are produced as a result of the microbial fermentation. Apart from its known roles, Kombucha has also been used as an anticolonization medicine against pathogenic bacterial infection, such as *V. cholerae*, *H. pylori*, *Salmonella* Typhimurium, *Staphylococcus aureus*, *Bacillus cereus*, *Staphylococcus epidermidis*, *Micrococcus luteus*, *E. coli*, *Listeria monocytogenes*, *Pseudomonas aeruginosa*, and other Gram-positive and Gram-negative and pathogenic yeasts, for example, *Candida* sp.

In one study, the progression of the antibacterial activity of Kombucha against enteric bacteria at different fermentation time points was assessed. Unfermented tea samples and acetic acid controls were not able to kill different bacteria efficiently and showed inhibitory activity toward all bacteria, respectively, but the fermented Kombucha showed killing ability of itself. A time-dependant response was also carried out. The 7-day old Kombucha was found to be inhibitory against *E. coli* O157:H7, *V. cholerae* N16961, and *Shigella flexneri* 2a 2457T, except *Salmonella* Typhimurium NCT 572. Optimum inhibitory effects were seen on 14th and 21st day of fermentation. Another negative control, SBT, showed no significant inhibitory effects against these bacteria. The group had also extracted the polyphenols and other polar compounds via ethyl acetate, chloroform, *n*-butanol, and aqueous-extraction methods. The ethyl acetate extract was found to be most potent against enteric bacterial infections. The F1 fraction from TLC was found to be more potent after the chromatography technique was done on the ethyl acetate extract. This F1 fraction was found to be catechin (Rt = 4.481 min, 68.173 ± 7.86 μg/mL) and isorhamnetin (Rt = 8.858 min, 156.384 ± 11.32 μg/mL) via HPLC and ESI–MS analysis (Bhattacharya et al., 2016). Our group is currently working on the active components of Kombucha in more detail.

Morphological alterations were seen in bacterial cells when seen under scanning electron microscope when exposed to the polyphenolic fraction in a concentration-dependent manner. The bacterial membrane integrity was found to be disrupted via a permeabilization assay in both time- and dose-dependent manners, which were proportional to the production of intracellular reactive oxygen species (Bhattacharya et al., 2018).

21.3.1.3 Possible Plan of Work

Kombucha can also be tested in vivo, using mice and rabbit models. In both, mice and rabbits, the well-known *V. cholerae* or *Salmonella* Typhimurium ileal loop model, a fluid accumulation model, is used. As stated above, this is a fluid accumulation model. Specific concentration of Kombucha can be injected in these loops directly followed by the bacterial inoculation. This approach is more of a prophylactic approach rather a medicinal approach, where the bacteria present in the Kombucha mechanically block the binding of the pathogenic bacteria on the walls of the intestine. Also, the byproducts of Kombucha aid in the inhibition in the pathogenesis of the pathogenic bacteria.

Kombucha or its byproducts (specific constituents) can also be used as a medicine in near future. After creating a disease in an animal model, Kombucha or its constituents can be used to treat that infection. Based on the theoretical knowledge, it can be assumed that it will indeed show its medicinal activities in the treatment of enteric and other infections in near future.

21.3.2 Polyphenolic Extracts of Edible Flower of *Sesbania grandiflora*

Flower vegetables are the flowers of the vegetable plants. People are consuming flower vegetables in their diet for ages. Literature dictates that these flower vegetables are useful against diarrhea too. *S. grandiflora* is a small, loosely branching tree of 8–15 m height and 25–30 cm in diameter. The flowers, leaves, barks, etc. are consumed by locals as well as tribes (China et al., 2012).

21.3.2.1 Present Works and Composition

After the collection, sun drying and making a powder of the petals were done, the dried petals were stored at −20°C for further use. Then the dried petals were extracted by using methanol and water to gets the polyphenols. In vitro experiment proved this extract of *S. grandiflora* to be grown promoting toward *Lactobacillus acidophilus*. *Lactobacillus* sp. is known to be present in the human gut in abundance. Promoting this probiotic's growth is the main function of this prebiotic *S. grandiflora* (Mukherjee et al., 2013).

In vitro agar well-diffusion studies proved that the extract was inhibitory toward the growth of both Gram-positive and Gram-negative microorganisms. Of the bacteria tested, *S. aureus* and *S. flexneri* showed the highest susceptibility to the extract compared with *S. Typhi*, *E. coli*, and *V. cholerae*. The extract also showed antagonistic activity toward *S. aureus* in fish (silver pomfret *Pampus argenteus*). As flavonoids are the main constituents of this extract, the observed effect was hypothesized to be the effect of these flavonoids. In other studies, flavonoids isolated from plants were used to assess their activities against different microbes using the same agar diffusion method. An increase in the inhibition of bacterial growth was also observed when quercetin and rutin, quercetin and morin, quercitrin and morin, or morin and rutin were combined. This combination was proved to be effective against *Salmonella* Enteritidis and *B. cereus*. The combination of quercetin and quercitrin, quercetin and morin, morin and rutin, and quercetin and rutin was much more active than either flavonoid alone. *S. grandiflora* was found to contain rutin. Rutin present is a glycosylated flavonoids that retard the growth of *Pseudomonas* and *Staphylococcus* was reported to be the associated microbes of *Aulacophora foveicollis*.

21.3.2.2 Mechanism of Action

Rutin was also type II topoisomerase inhibitor. Rutin also induces topoisomerase IV-mediated DNA cleavage, which leads to the induction of save our ship (SOS) response and growth inhibition in *E. coli*. Thus, it might be interfering with the replication and transcription of the bacteria. Polyphenols also help in the bacterial protein precipitation and enzymatic inhibition. The human intestinal flora was also known to be a producer of flavonoids. So, the increase in the growth of the intestinal bacteria *L. acidophilus* in presence of *S. grandiflora* extract was indeed intriguing. This synergistic effect could be a future treatment regimen for gastrointestinal disorders.

21.3.2.3 Possible Plan of Work

There are several hypothesizes that can be carried out using this extract. The bacteriostatic and bactericidal roles of this extract can extensively be done in the infection of other enteric pathogens. In both, mice and rabbit ileal loop experiments, this extract can be used as a prophylactic agent or a drug like Kombucha described above. Moreover, the flavonoids present in this extract were proved to be effective against *S. aureus*. This organism was found to be the causative agent of various skin disorders. Using this extract as a lotion or ointment could be useful.

21.3.3 *Oxalis corniculata* (Oxalidaceae) Leaf Extract

The extract of a green leafy vegetable *Oxalis corniculata* (Oxalidaceae) was evaluated for the assessment of in vitro antibacterial and in vivo anticolonizing effect against common gut pathogens (Mukherjee et al., 2013).

21.3.3.1 Present Works and Composition

The leaves of a green leafy vegetable of India were separated, sun dried, powdered, and stored in a refrigerator at $-20^{\circ}C$ for further studies. Air-dried leaves were mixed with methanol:water at a ratio of 80:20 and kept in shaking condition for 18 h. Total polyphenol content was then determined. The total polyphenol content of the extract was 910 mg of GAE/g of dry wt. and total flavonoid content was 2.353 g/100 g of extract. It showed the lowest minimum bactericidal concentration (MBC) against *E. coli* (0.1 mg/mL), highest MBC against *S. flexneri* 2a (2457T) (0.14 mg/mL), and intermediate MBC against *S. aureus* (0.12 mg/mL), *Shigella dysenteriae* 1 (NT4907) (0.12 mg/mL), *Shigella sonnei* phase I (IDH00968) (0.12 mg/mL), and *Shigella boydii* 4 (BCH612) (0.12 mg/mL).

21.3.4 Antimicrobial Effects of Different Spices

Antimicrobial activities of spices and herbs and essential oils have been well known for long time. Many studies reported the activities of spices and herbs or essential oils to food-borne pathogenic bacteria. Spices and herbs have been added to foods since ancient times, not only as flavoring agents, but also as folk medicine and food preservative. In addition to imparting characteristic flavors, certain spices and herbs prolong the storage life of foods by preventing rancidity through their antioxidant activity or through bacteriostatic or bactericidal activity (Chawla et al., 2014; Dhiman et al., 2016).

21.3.4.1 Present Knowledge

Basic work regarding spices and their antimicrobial activities indicated that the response of some spices to their antimicrobial properties was somewhat dependent on the temperature. An optimum 30°C was found to be effective for basil, clove, garlic, horseradish, marjoram, oregano, rosemary, and thyme. A decrease of 25°C (i.e., after making the temperature 5°C) makes the antimicrobial property more potent for carom, ginger, Japanese pepper, peppermint, sage, spearmint, and turmeric. The work was carried out on known pathogenic strains such as *Vibrio parahaemolyticus* and *E. coli*. These spices and herbs seem to decrease the survival of the bacterium at low temperature.

21.3.4.2 Mechanism of Action

Marjoram had relatively weak activity against Gram-negative bacteria, but oil extracted from marjoram was proved to be effective against *Vibrio* genera. Other studies concluded that the major antibacterial components of spices and herbs are terpenes such as eugenol and carvacrol. These are generally less dissolved at low temperature, and the antibacterial activity of spices and herbs may be reduced when used in refrigerators. The true activity of spices and herbs is difficult to measure since the low temperature itself could be detrimental for the microbe itself. Garlic is known to have antimicrobial properties since long back. Chemical analyses of garlic cloves have revealed an unusual concentration of sulfur-containing compounds (1%–3%). It contains allicin, which exhibits antibacterial, antifungal, antiparasitic, and antiviral activities. The main antimicrobial effect of allicin is due to its chemical reaction with thiol groups of various enzymes, for example, alcohol dehydrogenase, thioredoxin reductase, and RNA polymerase, which can affect essential metabolism of cysteine proteinase activity involved in the virulence of *Entamoeba histolytica*. Although allicin is the main content that has the proven antimicrobial properties, it is not present in the active form. In presence of pyruvate and alliinase enzyme, allin gets converted into its active form allicin. Allicin has some proven antimicrobial activities. Interestingly, various bacterial strains resistant to antibiotics such as methicillin-resistant *S. aureus* as well as other multidrug-resistant enterotoxicogenic strains of *E. coli*, *Enterococcus*, *S. dysenteriae*, *S. flexneri*, and *S. sonnei* cells were all found to be sensitive to allicin.

Many previous studies have reported the antibacterial activity, phenolic content, or antioxidant activities of spices and herbs. But it was not easy to compare directly the results of different studies and to establish reasonable relationships between antibacterial activity, phenolic content, and antioxidant activity because of the low number of spice and herb samples tested, different determination methods, and different bacterial strains used. The antimicrobial activities of phenolic compounds may involve multiple modes of action. For example, essential oils degrade the cell wall, interact with the composition and disrupt cytoplasmic membrane, damage membrane protein, interfere with membrane integrated enzymes, cause leakage of cellular components, coagulate cytoplasm, deplete the proton motive force, change fatty acid and phospholipid constituents, impair enzymatic mechanisms for energy production and metabolism, alter nutrient uptake and electron transport, influence the synthesis of DNA and RNA, and destroy protein translocation and the function of the mitochondrion in eukaryotes.

Berry fruit polyphenols have also exhibited antiinflammatory activities. We know that inflammation is one of the prime responses to infection. Bacillary dysentery is a result of inflammation. Research has already shown that supplementation with fruit or vegetable extracts high in antioxidants can decrease the enhanced vulnerability to oxidative stress and inflammation that occurs in aging, and that these reductions are expressed as improvements in behavior. Along with these, treatment with berry fruit extracts could provide an alternative view toward infectious disease treatment regime by limiting the inflammation.

21.4 SYNERGISTIC APPROACHES

Synergistic approach means the use of more than one compound when produce better result in combination than the single compound, like polyphenols or phytochemicals as a whole, in the treatment of a single disease. In the past, there has been a wide usage of phytochemicals and their products, both individually and synergistically. In the current time, the use of single phytochemical its extract could provide little or less protection than the synergistic product.

In the literature, there has been a wide usage of this synergistic approach in the treatment of *L. monocytogenes*–induced disease. Thymol and carvacrol, found in oregano extract, have been evaluated and were effective at inhibiting the growth of *L. monocytogenes*. Further, a variety of herbs and spices have been used in suppressing the growth of *L. monocytogenes*. Extracts of oregano and cranberry produced a clear inhibition zone against the same pathogen in agar diffusion assay. Acidification of foods with short-chain organic acids, either by fermentation or by deliberate addition, is an important and widespread mechanism for controlling food-borne pathogens in a variety of foods. Results show that the antimicrobial effect of oregano can be enhanced by addition of cranberry extract at an equivalent phenolic concentration and further that the antimicrobial effect of the extract mixture is significantly enhanced in the presence of lactic acid. Thus, the use of such naturally occurring plant products may provide a potential additional barrier (hurdle technology) to inhibit the growth of food-borne pathogens in food products. Synergistic effects of combined plant extracts provide a wide range of phenolic diversity, significantly increasing or decreasing antimicrobial efficacy. Therefore, a specific phenolic profile can be designed for different foods. Also, the diversity of phenolic types greatly increases functionality for health and wellness (e.g., antioxidants and for oxidation-linked diseases).

Damage to the cell membrane might explain the observed effects, since low-pH exposures can cause sublethal injury to cell membranes, causing disruption of proton motive force, owing to loss of H^+ ATPase. This could make the bacteria more susceptible to phenolic antimicrobial compounds found in oregano–cranberry extract mixtures, which are more enhanced in diphenolics than in cranberry alone. Scientists have postulated that phenolic-containing extracts themselves at higher concentrations may create a low-pH microenvironment owing to proton donation and cell-membrane disruption owing to stacking, which is likely more effective than a low pH alone.

REFERENCES

Barthel, M., Hapfelmeier, S., Quintanilla-Martínez, L., Kremer, M., Rohde, M., Hogardt, M., et al., 2003. Pretreatment of mice with streptomycin provides a *Salmonella enterica* serovar Typhimurium colitis model that allows analysis of both pathogen and host. Infect. Immun. 71 (5), 2839–2858.

Bhattacharya, D., Bhattacharya, S., Patra, M.M., Chakravorty, S., Sarkar, S., Chakraborty, W., et al., 2016. Antibacterial activity of polyphenolic fraction of Kombucha against enteric bacterial pathogens. Curr. Microbiol. 73 (6), 885–896.

Bhattacharya, D., Ghosh, D., Bhattacharya, S., Sarkar, S., Karmakar, P., Koley, H., et al., 2018. Antibacterial activity of polyphenolic fraction of Kombucha against *Vibrio cholerae*: targeting cell membrane. Lett. Appl. Microbiol. 66 (2), 145–152.

Chakravorty, S., Bhattacharya, S., Chatzinotas, A., Chakraborty, W., Bhattacharya, D., Gachhui, R., 2016. Kombucha tea fermentation: microbial and biochemical dynamics. Int. J. Food Microbiol. 220, 63–72.

Chawla, T., Abbasi, N.I., Aishwarya, T., Suneetha, V., 2014. Antimicrobial activity of spices like cloves cardamom and cinnamon on *Bacillus* and *Pseudomonas*. Int. J. Drug Dev. Res. 6 (4).

China, R., Mukherjee, S., Sen, S., Bose, S., Datta, S., Koley, H., et al., 2012. Antimicrobial activity of *Sesbania grandiflora* flower polyphenol extracts on some pathogenic bacteria and growth stimulatory effect on the probiotic organism *Lactobacillus acidophilus*. Microbiol. Res. 167 (8), 500–506.

Chistiakov, D.A., Bobryshev, Y.V., Kozarov, E., Sobenin, I.A., Orekhov, A.N., 2014. Intestinal mucosal tolerance and impact of gut microbiota to mucosal tolerance. Front. Microbiol. 5, 781.

Dhiman, R., Aggarwal, N., Aneja, K.R., Kaur, M., 2016. *In vitro* antimicrobial activity of spices and medicinal herbs against selected microbes associated with juices. Int. J. Microbiol. 2016, 9015802.

Gaboriau-Routhiau, V., Raibaud, P., Dubuquoy, C., Moreau, M.C., 2003. Colonization of gnotobiotic mice with human gut microflora at birth protects against *Escherichia coli* heat-labile enterotoxin-mediated abrogation of oral tolerance. Pediatr. Res. 54 (5), 739–746.

Hartman, A.B., Van de Verg, L.L., Collins Jr., H.H., Tang, D.B., Bendiuk, N.O., Taylor, D.N., et al., 1994. Local immune response and protection in the guinea pig keratoconjunctivitis model following immunization with *Shigella* vaccines. Infect. Immun. 62 (2), 412–420.

Higginson, E.E., Simon, R., Tennant, S.M., 2016. Animal models for salmonellosis: applications in vaccine research. Clin. Vaccine Immunol. 23 (9), 746–756.

Jandhyala, S.M., Talukdar, R., Subramanyam, C., Vuyyuru, H., Sasikala, M., Nageshwar Reddy, D., 2015. Role of the normal gut microbiota. World J. Gastroenterol. 21 (29), 8787–8803.

Jepson, R.G., Mihaljevic, L., Craig, J., 2000. Cranberries for treating urinary tract infections. Cochrane Database Syst. Rev. 2, CD001322. Review. PubMed PMID: 10796775.

Klose, K.E., 2000. The suckling mouse model of cholera. Trends Microbiol. 8 (4), 189–191. Review.

Mosley, W.H., Ahmed, A., 1969. Active and passive immunization in the adult rabbit ileal loop model as an assay for production of antitoxin immunity by cholera vaccines. J. Bacteriol. 100 (1), 547–549.

Mount, D.T., Barron, A.L., 1976. Intrarectal infection of guinea pigs with the agent of guinea pig inclusion conjunctivitis. Proc. Soc. Exp. Biol. Med. 153 (3), 388–391.

Mukherjee, S., Koley, H., Barman, S., Mitra, S., Datta, S., Ghosh, S., et al., 2013. *Oxalis corniculata* (Oxalidaceae) leaf extract exerts *in vitro* antimicrobial and *in vivo* anticolonizing activities against *Shigella dysenteriae* 1 (NT4907) and *Shigella flexneri* 2a (2457T) in induced diarrhea in suckling mice. J. Med. Food 16 (9), 801–809.

Naughton, P.J., Grant, G., Spencer, R.J., Bardocz, S., Pusztai, A., 1996. A rat model of infection by *Salmonella typhimurium* or *S. enteritidis*. J. Appl. Bacteriol. 81 (6), 651–656.

Nitzan, O., Elias, M., Peretz, A., Saliba, W., 2016. Role of antibiotics for treatment of inflammatory bowel disease. World J. Gastroenterol. 22 (3), 1078–1087. Review.

Panda, A., Tatarov, I., Masek, B.J., Hardick, J., Crusan, A., Wakefield, T., et al., 2014. A rabbit model of non-typhoidal *Salmonella* bacteremia. Comp. Immunol. Microbiol. Infect. Dis. 37 (4), 211–220.

Petri Jr., W.A., Miller, M., Binder, H.J., Levine, M.M., Dillingham, R., Guerrant, R.L., 2008. Enteric infections, diarrhea, and their impact on function and development. J. Clin. Invest. 118 (4), 1277–1290. Review.

Sawasvirojwong, S., Srimanote, P., Chatsudthipong, V., Muanprasat, C., 2013. An adult mouse model of *Vibrio cholerae*-induced diarrhea for studying pathogenesis and potential therapy of cholera. PLoS. Negl. Trop. Dis. 7 (6), e2293.

Sender, R., Fuchs, S., Milo, R., 2016. Revised estimates for the number of human and bacteria cells in the body. PLoS Biol. 14 (8), e1002533. 19.

Shim, K.M., Choi, S.H., Jeong, M.J., Kang, S.S., 2007. Effects of aucubin on the healing of oral wounds. In Vivo. Nov–Dec. 21 (6), 1037–1041. PubMed PMID: 18210752.

Simon, R., Tennant, S.M., Galen, J.E., Levine, M.M., 2011. Mouse models to assess the efficacy of non-typhoidal *Salmonella* vaccines: revisiting the role of host innate susceptibility and routes of challenge. Vaccine 29 (32), 5094–5106.

Spira, W.M., Sack, R.B., Froehlich, J.L., 1981. Simple adult rabbit model for *Vibrio cholerae* and enterotoxigenic *Escherichia coli* diarrhea. Infect. Immun. 32 (2), 739–747.

Turner, J.R., 2009. Intestinal mucosal barrier function in health and disease. Nat. Rev. Immunol. 9 (11), 799–809.

WHO, 2002. WHO Monographs on Selected Medicinal Plants, Vol. 2. Geneva. World Health Organization, Switzerland.

Zhang, Y.J., Li, S., Gan, R.Y., Zhou, T., Xu, D.P., Li, H.B., 2015. Impacts of gut bacteria on human health and diseases. Int. J. Mol. Sci. 16 (4), 7493–7519.

CHAPTER 22

Nanoparticles in Ayurvedic Medicine: Potential and Prospects

S. Farooq[1], Zafar Mehmood[2], Faizan Abul Qais[3], Mohammad Shavez Khan[3] and Iqbal Ahmad[3]
[1]The Himalaya Drug Company, Dehradun, Uttarakhand, India
[2]Department of Microbiology, The Himalaya Drug Company, Dehradun, Uttarakhand, India
[3]Department of Agricultural Microbiology, Aligarh Muslim University, Aligarh, Uttar Pradesh, India

22.1 INTRODUCTION

Ayurveda (knowledge of life) is the oldest form of Indian traditional system of medicine. The other traditional system of Indian medicine includes Unani and Siddha. These systems of medicine adopt a holistic approach toward achieving human health by an integrated manner of physical, mental, and spiritual functions of human body (Prakash, 1997). A section of Ayurveda deals with herbo-mineral preparations called *Bhasma* (ash) is known as *Rasa Shastra* (Vedic chemistry) (Kumar et al., 2011). The major therapeutic actions of *Bhasma* are their ability for Immunomodulation and anti-aging property (Rasayana) and ability to target drugs to the site (Yogavahi). Such Ayurvedic preparations are claimed to be nontoxic, absorbed readily, and biocompatible (Sarkar and Chaudhary, 2010).

The term nanotechnology was first defined by in 1974 by Norio Taniguchi in Tokyo Science University. Now this discipline of technology has proved to be useful in physical and chemical sciences but also opens new avenues in medical sciences such as in imaging, sensing, artificial implants, and improving drug delivery (Shi et al., 2010; Boisseau and Loubaton, 2011). Nanotechnology provides opportunity to develop nanoparticles through bottom-up and top-down approaches allowing reengineered materials at nanoscale for developing new improved products (Miyazaki and Islam, 2007). Various types of nanoparticles are available for different applications. The application of nanotechnology in pharmaceutical field includes development of nanomedicine, nanorobots,

New Look to Phytomedicine
DOI: https://doi.org/10.1016/B978-0-12-814619-4.00023-9

biomarkers, biosensors, etc. It is expected that use of nanotechnology may be useful in characterizing and providing scientific evidence for Ayurvedic medicine with special reference to herbo-mineral formulation. In this chapter, an overview of use of metals in Ayurveda, concept, and methods of *Bhasma* preparation and their characteristic along with their nanoparticle nature and significance of herbal constituent are briefly summarized.

22.2 USE OF METALS IN AYURVEDA

The use of metals for therapeutic applications was first evidenced in the Ayurvedic system of medicine. The concept of preparation of *Bhasma* by reduction in particle size of metal is prevailing since beginning of the Ayurveda (Sarkar and Chaudhary, 2010). *Bhasma* is an important Ayurvedic formulation comprising mixture of herbs and metals. In fact, *Rasa Shastra*, an integral part of Ayurveda, deals with numerous formulations based on metal—herb combinations. This branch of Ayurveda describes the characteristics, varieties, processing techniques, therapeutic uses, properties, and development of adverse effects along with their management in a comprehensive manner (Adhikari, 2014). The development of this branch mainly accelerated in the 8th century but their origin can be traced back to ancient texts of Indian civilization. *Charaka Samhita* and *Sushruta Samhita* describe the details of utilization and processing techniques involved in it. Since many centuries, herbo-mineral and metallic preparations become a significant part of Ayurvedic pharmacopoeia and have been practiced throughout India. These formulations are effective even at minute levels when used by following the classical guidelines and, therefore, are considered safe (Sarkar and Chaudhary, 2010). Different types of minerals including essential elements such as sodium, potassium, calcium, manganese, zinc, copper, iron, etc. are integral part of these preparations. Metal-based *Bhasma* (described in the later section of the chapter) includes gold (*Swarna Bhasma*), silver (*Rajata Bhasma*), copper (*Tamra Bhasma*), iron (*Lauha Bhasma*), zinc (*Jasada Bhasma*), alloy of gold, sulfur and mercury (*Makardhwaj Bhasma*), iron oxide (*Mandura Bhasma*), etc.

22.3 PREPARATION OF *BHASMA*: CONCEPT OF SIZE REDUCTION

The process of *Bhasma* preparation can be classified into two groups, namely, (1) metal extraction from its mineral form (*Satpavna*) and (2) conversion of purified metals or alloy into nontoxic *Bhasma*. In *Satpavna* process, raw material (minerals) is processed in series of different steps; every step brings about physiochemical changes in the material along with addition of other constituents. The important step implicated in the procedure for making *Bhasma* is repeated treatment (Wadekar et al., 2005). *Bhasmikaran*, a most important synthesis procedure of *Bhasma*, involves converting of the metallic compound along with organic mixture into the ashes. This process is distinct from modern manufacturing approaches for engineered nanoparticles. During the process, the metals convert specifically into desired chemical compound along with eliminating the toxicity of the metal. Chemically, conversion of zero valent state of metal to their higher oxidation state is pivotal for decreasing the toxicity of metal, which is achieved during the process of *Bhasmikaran* (Palkhiwala and Bakshi, 2014). During these two processes (*Satpavna* and *Bhasmikaran*), metal particle reduces in size, and toxicity of metal oxides is completely destroyed and incorporates the medicinal properties into it. Some of the important steps in preparation of *Bhasma* are listed in Table 22.1 as described by various workers (Prakash, 1997; Sarkar and Chaudhary, 2010; Krishnamachary et al., 2012).

Table 22.1 Different steps involved in *Bhasma* formation and their outcomes

Steps	Description	Physiochemical changes
Shodhana (Purification)	Treatment of metal pieces with herbs, juices, extracts at room temperature or thermal cycling of heating and quenching of metal in water, herbal extracts, oil, etc.	Increased brittleness, reduction in hardness, reduction in particle size Exposed metallic surface reacts with the ingredient and converted into organometallic compound
Marana (conversion to nontoxic powder)	Heating of metal to high temperature like calcination with herbal extract, juices, etc. and rubbed mechanically	Formation of nontoxic compound form. Formation of finer particles
Mardana/ Bhavana (wet grinding)	Metallic powder mixed with more herbal extracts or metals or with inorganic compounds such as borax, lime, etc. with continuous grinding	Formation of new chemical compounds Formation of pellet and briquettes
Jarana (high temperature)	Final heating of the pellets into open air	Oxidation of metal, increase in melting point

There are two commonly used methods for preparation of *Bhasma*, that is, *Kupipakwa* method and *Putapaka* method as described by Sarkar and Chaudhary (2010).

22.3.1 *Putapaka* Method

This is a three-step procedure for the preparation of *Bhasma*. The steps are *Shodhana*, *Bhavana*, and *Marana*. Minerals or metals are first made into coarse powder by hammering, wherein it is melted or heated to red hot followed by quenching in liquid medium for a specific time. *Shodhita* (purified) material is mixed with specific drugs for incineration (*Maraka Dravyas*) and levigated (*Bhavana*) by a specific liquid. In *Bhavana*, materials are grinded for a fixed time in a specific liquid media. *Chakritas* (pellets) prepared from levigated mass and then kept in earthen containers faced together by sealing the junction with mud smeared clothes. In electric muffle furnace or traditional heating grade (*Puta*), the apparatus (*Sarava Samputam*) is heated. After a fixed period of burning, the apparatus is taken out to obtain incinerated powder. The process is repeated for a specified time to obtain *Bhasma* (incinerated metal). The metals with low melting point (e.g., tin, lead, and zinc) are subjected to an extra step called *Jarana* in between *Shodhana* and *Bhavana* process. The molten metal is mixed with plant drug powders and rubbed by iron ladle to make fine powder.

22.3.2 *Kupipakwa* Method

This method involves four steps (*Shodhana*, *Kajjali* preparation, *Bhavana*, and *Kupipakwa*) for the preparation of *Bhasma*. After *Shodhana*, the metals are amalgamated with mercury and mixed with purified sulfur to obtain fine and smooth powder, a step called *Kajjali* preparation. *Kajjali* is levigated for a fixed time in a particular media and allowed to dry. It is then filled in glass bottles (*Kachkupi*), covered with mud cloth, and allowed to *Valuka yantra* (sunbath) for homogenous and indirect heating. Bottles are broken to obtain *Bhasma* (from bottom of bottle) and sublimed product (from neck of bottle). For detailed description, readers are advised to consult specific literature published in review articles/books/Ayurvedic literature (Prakash, 1997; Chaudhary and Singh, 2010; Sarkar and Chaudhary, 2010). A flow chart indicating various steps evolved in the preparation of copper *Bhasma* (Tamra *Bhasma*) is described in Fig. 22.1.

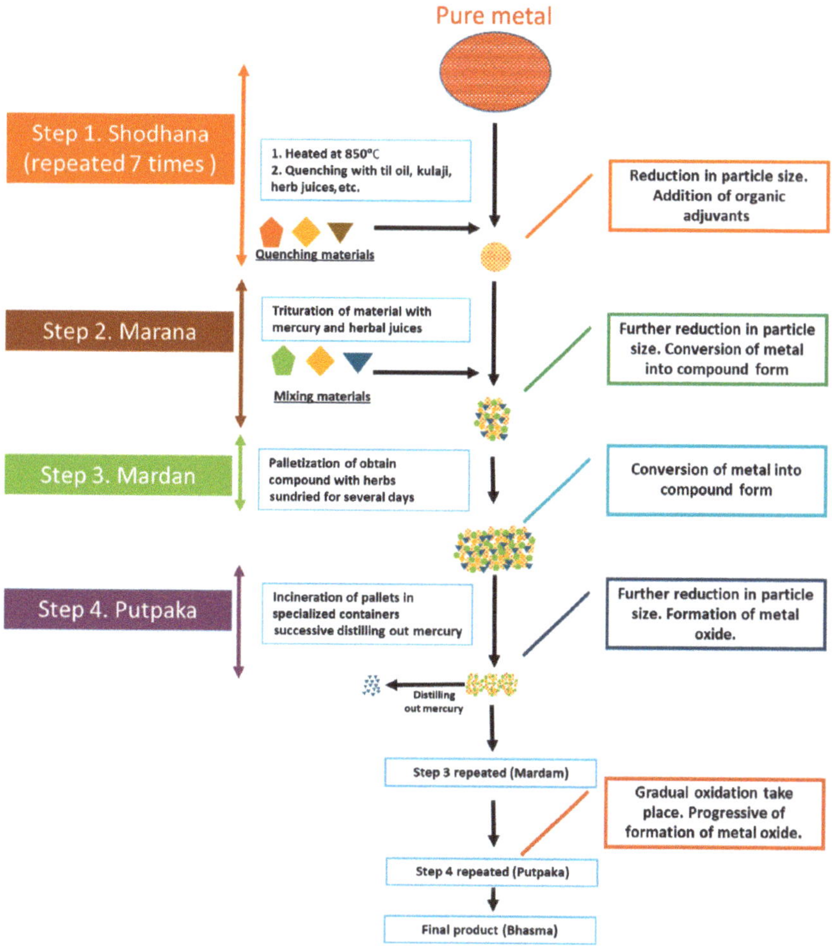

Figure 22.1 Flow chart indicating steps involve in formation of copper *Bhasma* (Tamra *Bhasma*).

22.4 TYPES OF *BHASMA*

The characteristics of some of the common *Bhasma* and their therapeutic uses in different diseases are described briefly, based on Ayurvedic literature such as *Caraka Samhita* (Samhita, 1998), *Rasa Tarangini* (Sadananda, 1998), *Ayurveda Prakasha* (Upadhyaya, 1999), *Rasendra Chudamani* (Mishra, 2004), *Rasaratna samuccaya* (Vagbhatta, 1998), and *Ayurvediya Rasashastra* (Jha, 2000) and reviewed by Galib et al. (2011) are discussed below.

22.4.1 *Rajata* (Silver)

Silver is a noble metal that was used by ancient *Acharyas* for therapeutic purposes since the period of *Charaka* and his contemporize (Galib et al., 2011). As described, *Rajata* or silver is a clear, heavy, and lustrous with metallic sheen, which becomes bright white on cutting or heating. *Rajata*, not having any furrows or ridges, can be accepted for the therapeutic applications. In terms of composition, *Rajata Bhasma* is composed of pure silver metal (52%–59%), ferric oxide (14.33%), free sulfur (0.675%), calcium (10.769%), and silver chloride (0.479%) as well as traces of other metals including potassium, sodium, and aluminum. Silver *Bhasma* has been prescribed for respiratory disorders. The standard prescribed dose range of *Rajata Bhasma* is 30–120 mg (Sarkar and Chaudhary, 2010).

22.4.2 *Swarna* (Gold)

The therapeutic applicability of *Sara Lauha* or *Swarna* is known to Indians since ancient times. The references of application of this *Bhasma* are found in *Charaka Samhita*. The *Swarna Bhasma* comprises metallic gold (96.76%), ferric oxide (0.14%), silica (1.14%), phosphates (0.78%), salt (0.078%), potash (0.16%), and traces of magnesium and copper (Jha, 2000). The elemental form of gold has been used traditionally as an antipruritic agent to relieve itching palms. Some useful formulations of *Swarna* are used to treat chronic diseases such as *Yakshmac* (tuberculosis), Kasa (cough), *Swasa* (respiratory disorders), and *Pandu* (Anemia). The dosage range of *Swarna Bhasma* is 15–30 mg.

22.4.3 *Parada* (Mercury)

There are only few formulations mentioned in *Charaka Samhita* that contains mercury. There is controversy in the reference of therapeutic utility of *Parada* where only a few scholars interpret the term *Rasa* in the verse *chikitsasthana* 7/71 as *Parada*. In *Dwivraniya Chikitsa*, the term *Rasa* is interpreted as *Parada* by the commentator *Chakrapani*. Above two mentioned formulations are only recommended for external applications (Galib et al., 2011).

22.4.4 *Aayasa* or *Loha* (Iron)

Next to *Rajata*, *Swarna* and *Tamra*, *Ayasa* or *Loha* is also known since ancient civilizations, which was used in different dosage during the period of *Charaka*, both for internal and external administrations for many pathological manifestations. Compounds of iron were prescribed in anemia and other related conditions, mainly when blood becomes deficient in iron, or functions of hemopoietic systems are disturbed. It is claimed in *Rasa Shastra* that *Loha* stimulates functional activity of almost all the organs, destroys a number of diseases, promotes life and strength, and also acts as a restorative. According to *Rasa Vagbhata*, among all varieties of *Loha* (Iron), namely, *Munda* (cast iron), *Kanta* (wrought iron), and *Teekshna* (carbon steel), the *Kanta* is the best variety to used. *Chakrapani* has emphasized that there is need of critical care while administration of *Loha*. *Lauha Bhasma* is composed of ferric oxide (96.5%), ferrous oxide (2.5%), calcium oxide (0.3%), and magnesium oxide (0.8%) along with traces of potassium and phosphorus. Formulations of *Loha* were used for the treatment of diseases like *Sula* (chest pain), *Gulma* (abdominal tumor), *Arsha* (piles), *Yakrit Roga* (liver diseases), *Pliha Roga* (spleen disease), *Ksaya* (phthisis), *Pandu* (anemia), *Kamala* (jaundice), etc. The normal dose limit for *Lauha Bhasma* is 30–240 mg (Vagbhatta, 1998).

22.4.5 *Tamra* (Copper)

Copper or *Tamra*, an ancient metal, is known to human civilization since pre-Vedic times that was part of daily livelihood uses. The metals were used in preparation of the alloys mainly brass and bronze. The desired properties of *Tamara* for medicinal uses were metallic sheen, bright reddish in color, heavy, soft with having high tensile strength, and lacking impurities. Apart from the use of *Tamra* in management of diseases like *Krimi* (worms), *Arsha* (piles), *Sthaulya* (obesity), *Ksaya* (phthisis), *Pandu* (anemia), *Swasa* (respiratory disorder), *Kusta* (skin disease), *Kasa* (cough), *Sotha* (edema), *Amlapitta* (acidity), *Sula* (chest pain), *Grahani dosha* (intestinal tract disorder), *Yakrit Roga* (liver disease), etc., the use of *Tamra Patra* (copper vessels) has also been advised in several pharmaceutical procedures. The normal prescription dose for *Tamra Bhasma* is 15–60 mg (Sadananda, 1998; Mishra, 2004).

22.4.6 *Sisaka/Naga* (Lead)

Naga is also an important *Puti Loha* (easy fusing metals) which has been known since ancient times and mentioned as *Sisaka* or *Sisa* in ancient texts. As described by *Charaka*, this metal should be externally used for medicinal purposes especially in the cases of *Mandala Kusta* (dermatological disorders). The desired characteristics for therapeutic purposes in *Naga* are externally black in color and heavy that melt easily and shine with bright black luster when incised. Quantitatively, *Naga Bhasma* contains lead oxide (75.6%) and ferric oxide (7.5%) as major components along with traces of chlorides and carbonates of magnesium and calcium. The normal therapeutic dose of *Naga Bhasma* is 30–120 mg and beneficial for diseases like *Prameha* (spermatorrhoea), *Arsha* (piles), *Gulma* (abdominal tumor), *Sweta Pradara* (leucorrhoea), *Antra sotha* (gastroenteritis), *Grahani roga* (intestinal disorder), etc. (Mishra, 2004; Galib et al., 2011).

22.4.7 Mandura

It is another form of iron that has been used for therapeutic purposes since antiquity in classical Ayurveda. As described in the *Ayurveda Prakasha*, it is the debris or residue collected after heating and beating processes of iron around a blacksmith's anvil. The qualities in *Mandura* for therapeutic purposes as mentioned in ancient literature are smooth to touch, strong, heavy, lack furrows, or fissures. Administration of purified *Mandura* is beneficial in inflammations, jaundice, edematous conditions, etc. It is also a drug for the treatment of anemia (*Pandu*) (Kumar and Garai 2012). Chemical compositions of *Mandura* are ferric oxide (59.14%), chlorides (4.4%), ferrous oxide (26.7%), magnesium (3.9%), sodium (1.7%), and some other elements in trace amounts. There are approximately 30 formulations of *Mandura* mentioned in *Charaka Samhita* having dose range of 30–240 mg (Sadananda, 1998; Upadhyaya, 1999).

22.4.8 *Vanga/Trapu* (Tin)

Vanga, a type of the *Puti Loha* (easy fusing metals) was known to ancient Indian physicians by *Trapu* (tin). The preferable characteristics for therapeutic applications in *Vanga* are bright white in color, smooth, soft, easily melts, shiny, and heavy which are termed as *Khura Vanga*. Chemically, *Vanga Bhasma* is a mixture of stannic oxide (91.4%), potassium (2.9%), ferric oxide (2.9%), calcium oxide (2%), aluminum (2%), and magnesium (0.6%) oxides. *Vanga Bhasma* alone is a drug of choice for *Prameha* (spermatorrhoea) and in combination with other *Puti Loha*, it is beneficial for genitourinary tract. Various formulations of *Vanga Bhasma* are beneficial in the management of *Prameha* (spermatorrhoea), *Swasa* (respiratory disorder), *Kasa* (cough), *Krimi* (worms), *Pandu* (anemia), *Ksaya* (phthisis), *Pradara* (leucorrhea), *Garbhashaya Chyuti* (uterus displacement), etc. in the dose range of 120–240 mg (Galib et al., 2011).

22.4.9 *Kamsya* (Bronze)

Kamsya is an alloy of tin and copper. *Pushpa* variety of *Kamsya* is only acceptable for therapeutic applications which produce sharp sound, are smooth to touch, soft, slightly grayish, turn red on heating, and devoid of impurities. Therapeutic doses of 60–120 mg of *Kamsya Bhasma* are beneficial for diseases like *Kusta* (skin disease), *Krimi* (worms), etc. (Samhita, 1998; Vagbhatta, 1998).

22.4.10 *Pittala* (Brass)

Pittala is another vital *Misra Loha*, known since the period of *Samhita Kala*, which is an alloy of zinc and copper. Several formulations of *Pittala Bhasma* are beneficial in diseases like *Krimi* (worms), *Pandu* (anemia), *Kusta* (skin diseases), etc. (Galib et al., 2011).

22.5 NANOPARTICLE NATURE OF *BHASMA*

Nanoparticles are the new approach to the scientists in the field of biomedical and commercial application. Nano-sized particles can easily enter into the cell and take part in the cellular metabolism, DNA/protein interaction, and could be responsible for altered gene expression (Nel et al., 2009; Huang et al., 2015). Metal nanoparticles have small size and unique chemical properties, which are the important features for the development of various therapeutic uses (Daniel and Astruc, 2004).

Bhasma are nearer to nanocrystallite materials which are solid composed of crystallite with sizes less than 100 nm, at least in one dimension (Meyers et al., 2006). Ayurvedic metallic nanocrystallites or *Bhasma* have unique physicochemical properties such as biocompatibility and ease of surface fictionalization. Steps involve in *Bhasma* preparation such as *Mardana* (trituration) and *Bhavana* (levigation) are responsible for size reduction and formation of metallic nanocrystallites (Sarkar and Chaudhary, 2010). Nanodimensional aspect of *Jasada Bhasma* (zinc) is revealed by physicochemical characterization by X-ray photoelectron spectroscopy, inductively coupled plasma (ICP), energy dispersive spectroscopy, dynamic light scattering, and transmission electron microscopy (TEM). It was observed that the particles are in oxygen-deficient state and many of them are in nanometer size range (Bhowmick et al., 2009). Nanoparticle size of *Lauha Bhasma* was confirmed by surface morphological analysis by scanning electron microscopy (SEM). Spherical nanoparticles with a diameter of about 17 nm were observed (Krishnamachary et al., 2012). A scientific analysis of *Swarna Bhasma* by TEM and atomic force microscopy have demonstrated that the principle ingredient of *Swarna Bhasma* is globular gold nanoparticles of 56–57 nm. Atomic absorption spectroscopy and infra-red (IR) spectroscopy studies reveal that *Swarna Bhasma* is devoid of any other heavy metal or organic material (Brown et al., 2007). However, Mohapatra and Jha (2013) showed the presence of iron, copper, and sulfur in *Swarna Bhasma*. In addition to these elements, the *Bhasma* found to contain potassium, magnesium, aluminum, and silicon in trace amount. *Tarkeshwar Ras*, an amalgamation of different *Bhasma* (such as *Lauha*, *Vanga*, *Abhraka*, and *Rasa Sindhura*) revealed the presence of different elements such as iron, tin, aluminum, etc. The agglomerated particle size was found to be in the range of 0.5–2 μm in diameter (Virupaksha and Kumar, 2012). The TEM study reveals that grain size was significantly reduced in *Swarna Bhasma* to 50–200 nm. Likewise, *Ras Sindoor* (sublimed mercury compound) contains mercury sulfide (crystalline; size, 25–50 nm). This is an organic macromolecule derived from plant extract. Several macro-/trace elements may be present in different amounts, which are bioavailable and accountable for adding to medicinal value of *Ras Sindoor* (Singh et al., 2009). Analysis of *Rajata* (silver) *Bhasma* by SEM revealed particles in the range from 10 to 60 nm and content of silver in *Rajata Bhasma* measured with ICP-AES was found to be 70.56% (Sharma et al., 2016). Physiochemical characterization including X-ray diffraction analysis and SEM results revealed that the crystallite size of *Tamra Bhasma* (copper) powder was less than 100 nm and nanocrystallites of agglomerated size in micrometer (Singh et al., 2017). Some of the *Bhasma* detected with presence of nano-sized particles are described in Table 22.2.

22.6 SIGNIFICANCE OF HERBAL CONSTITUENT IN *BHASMA* CHARACTERISTIC

Manufacturing processes of *Bhasma* are in agreement with present day nanotechnology and are physiochemically similar nanocrystalline materials. One major difference between synthetic nanomaterials and *Bhasma* is the use of different herbs, organic, and mineral material during the preparation. The role of plant and animal products used in the preparation of *Bhasma* is of utmost importance in imparting the overall therapeutic potential. Studies have revealed multidimensional interaction and role of

Table 22.2 Detection of nanoparticles/nanocrystallites in different *Bhasma*

Bhasma	Nanoparticle detected	Size	Reference
Praval Bhasma	Calcium oxide	10–15 μm	Mishra et al. (2014)
Tamra Bhasma	Copper oxide	1–10 μm	Wadekar et al. (2005)
	Different oxides of copper	100 nm	Singh et al. (2017)
Rajata Bhasma	Silver sulfide	1.04 μm	Gokarn (2017)
Mandur Bhasma	Iron oxide	–	Rajurkar (2016)
Shanku Bhasma	Calcium oxide	600 nm	Rasheed and Shivashankar (2017)
Yashada Bhasma	Zinc oxide	5–20 μm	Santhosh et al. (2013)
	Zinc oxide	10–25 nm	Bhowmick et al. (2009)
Vaikrānta Bhasma	Multi-mineral	5–20 μm	Tripathi et al. (2013)
Swarna Bhasma	Gold	9.9 μm	Thakur et al. (2017)
	Gold	56–57 nm	Brown et al. (2007)
Vanga Bhasma	Tin dioxide	10–100 nm	Sumithra et al. (2015)
Muktashukti Bhasma	Calcium carbonate	10–50 μm	Vadnere et al. (2013)
Naga Bhasma	Lead sulfide	~60 nm	Singh et al. (2010)
	Lead oxide	–	Nagarajan et al. (2012)
Trivanga Bhasma	Lead, zinc, and tin oxides	500 nm	Rasheed et al. (2014)
Samagandhak Kajjali	Mercuric sulfide	0.2–10 μm	Thakur et al. (2014)
Mukta Bhasma	Calcite	156 nm	Sawant (2015)

organic constituents during different stages of processing and in functionalizing metal nanoparticle. Krishnamachary et al. (2012) demonstrated the effect of different organic materials on when used during heat treatment in the manufacturing of *Lauha Bhasma*. Sesame oil when used as quenching liquid transforms the raw iron into nanocrystallites, followed by the heat treatment with buttermilk which removes the rust from the raw material, while cow urine and rice gruel successive treatment results in the formation of different complexes of iron and brings about the change in the oxidation state. Further, use of horse gram decoction further reduces the overall particle size to sub 100 nm levels. It has also been demonstrated that successive use of different plant extracts, namely, *Musa × paradisiaca*, *Limonia acidissima*, *Moringa oleifera*, and *Citrus limetta* during manufacturing process of *Kanta centrum* (iron oxide based herbal formulation) bring about physiochemical changes in the metal. It was found that these plant extracts catalyze the reduction of particle size and accelerate the transformation of capping of the iron oxide nanoparticles by constituent minerals. Moreover, during these processes, herbal constituents are found to be associated or adsorbed on the surface of metal nanoparticles. Fourier-transform infrared spectroscopy (FTIR) studies of *Lauha Bhasma* prepared using *Triphala* decoction (Indian gooseberry, chebulic myrobalans, and beleric myrobalans) revealed that the phytoconstituents of the *Triphala* decoction (ellagic acid, chebulagic acid, punicalagin, and corilagin) form coordination compounds with iron oxide. This complexation between metal nanoparticle and different phytocompounds was suggested as functional mechanism of *Bhasma* (nanoparticles) (Krishnamachary et al., 2012).

Recent studies highlighted the bioenhancer potential of nanoparticles prepared using herbal constituents, similar to the preparation *Bhasma*. Cellular permeability studies of *Rajata Bhasma* (silver ash) suggest that *Rajata Bhasma* increases the rate of paracellular transportation of co-incubated molecules without having toxic effect. Authors thus concluded that this increase in cellular uptake could be possible mechanism of action of *Bhasma*, enhancing the delivery of bioactive constituent inside the cells (Mukkavalli et al., 2017). In a similar study, gold *Bhasma* prepared using *putpaka* method which involves heating and quenching of gold with various plant extracts was compared with synthetic citrate capped gold nanoparticles. It was observed via nuclear imaging techniques that the cellular distribution and entry mechanism of Gold *Bhasma* was different from synesthetic nanoparticles. Gold *Bhasma* was found to be nontoxic to HeLa cells, macropinocytosis pathway was involved in transportation and cellular locations includes in vacuole, cytosol, or nucleus (Beaudet et al., 2017).

Availability of scientific evidence for the safety and efficacy of herbal products including *Bhasma* is scanty; it is a common practice that herbal remedies are self-administered by patients without proper guidance (Chandramouli et al., 2010). Lack of significant toxicological data related to use of *Bhasma* still remains one of the major lacunas. Nevertheless, Ayurveda emphasizes the use of specific plant products during the processing of these herbo-metallic preparations (Saper et al., 2008). These herbs are believed to assist the delivery of the drugs to the body and also to contribute to the therapeutic effects. This process of incineration and addition of medicinal herbs is believed to remove impurities and eliminate the harmful effects of the metallic constituents (Kumar et al., 2006). The roles different ingredients involve specially the plant products are of great importance in removing the toxicity of metal or constituent minerals. For example, in the preparation of *Ras Sindoor* (*Bhasma*), a mercury-based herbo-metallic formulation, different herbs such as garlic (antidote for mercury) and extract from the aerial root of *Ficus benghalensis* are extensively used for imparting the medicinal properties as well reducing the toxicity of resultant metallic particles (Kamath et al., 2012). Similarly, *Triphala* decoction (fruits of *Phyllanthus emblica*, *Terminalia bellirica*, and *Terminalia chebula*), *Ziziphus jujuba* decoction, and *Vitex negundo* L. juice are repeatedly used during *Shodhana* (purification) steps to remove the toxic effect of minerals and metals during the process *Abhrak Bhasma* preparations (Wijenayake et al., 2014). Although different factors involve toxicity perspective of *Bhasma* or herbo-metallic nanoparticles, extensive use of plant-based products and critically defined purification steps in *Bhasma* preparations could possibly contribute significantly in reducing the toxicity risks.

22.7 CONCLUSION

Ayurvedic herbo-metallic formulation consisting of *Bhasma* is ancient nanomedicine for treatment of various illness. With the use of modern scientific tools, the nonmaterialistic nature of various types of *Bhasma* has been confirmed. Manufacturing steps of these Ayurvedic herbal nanoparticles (*Bhasma*) are in agreement with modern nanotechnological principles and even provide basis of concept in diminishing the toxic effects of metals. Herbal constituents play an important role in overall therapeutic efficacy of the *Bhasma*; however, focused scientific investigations are needed to explore various aspects of metal—herb interactions during the manufacturing process and in imparting therapeutic potential.

ABBREVIATIONS

SEM scanning electron microscopy
IR infra-red
TEM transmission electron microscopy
ICP inductively coupled plasma
AFM Atomic Force Microscopy
FTIR Fourier-transform infrared spectroscopy

REFERENCES

Adhikari, R., 2014. Ayurvedic *Bhasmas*: overview on nanomaterialistic aspects, applications, and perspectives. In: Adhikari, R., Thapa, S. (Eds.), Infectious Diseases and Nanomedicine I. Springer, New Delhi, pp. 23–32. ISBN: 978-81-322-1777-0.

Beaudet, D., Badilescu, S., Kuruvinashetti, K., Kashani, A.S., Jaunky, D., Ouellette, S., et al., 2017. Comparative study on cellular entry of incinerated ancient gold particles (Swarna *Bhasma*) and chemically synthesized gold particles. Sci. Rep. 7 (1), 10678. Available from: https://doi.org/10.1038/s41598-017-10872-3.

Bhowmick, T.K., Suresh, A.K., Kane, S.G., Joshi, A.C., Bellare, J.R., 2009. Physicochemical characterization of an Indian traditional medicine, Jasada *Bhasma*: detection of nanoparticles containing non-stoichiometric zinc oxide. J. Nanoparticle Res. 11 (3), 655–664.

Boisseau, P., Loubaton, B., 2011. Nanomedicine, nanotechnology in medicine. Comptes Rendus Physique 12 (7), 620–636.

Brown, C.L., Bushell, G., Whitehouse, M.W., Agrawal, D.S., Tupe, S.G., Paknikar, K.M., et al., 2007. Nanogold-pharmaceutics (i) The use of colloidal gold to treat experimentally-induced arthritis in rat models; (ii) Characterization of the gold in Swarna *Bhasma*, a microparticulate used in traditional Indian medicine. Gold Bull. 40 (3), 245–250.

Chandramouli, R., Thirunarayanan, T., Mukeshbabu, K., Sriram, R., 2010. Designing toxicological evaluation of Ayurveda and Siddha products to cater to global compliance—current practical and regulatory perspectives. J. Pharm. Sci. Res. 2, 867–877.

Chaudhary, A., Singh, N., 2010. Herbo mineral formulations (rasaoushadhies) of Ayurveda an amazing inheritance of Ayurvedic pharmaceutics. Ancient Sci. Life 30 (1), 18.

Daniel, M.C., Astruc, D., 2004. Gold nanoparticles: assembly, supramolecular chemistry, quantum-size-related properties, and applications toward biology, catalysis, and nanotechnology. Chem. Rev. 104 (1), 293–346.

Galib, M.B., Mashru, M., Jagtap, C., Patgiri, B.J., Prajapati, P.K., 2011. Therapeutic potentials of metals in ancient India: a review through Charaka Samhita. J. Ayur. Integr. Med. 2 (2), 55–63.

Gokarn, R.A., 2017. Characterization of Rajata *Bhasma* (traditional calcined silver preparation). Intern. J. Green Pharm. 11 (03), 143–148.

Huang, C.L., Hsiao, I.L., Lin, H.C., Wang, C.F., Huang, Y.J., Chuang, C.Y., 2015. Silver nanoparticles affect on gene expression of inflammatory and neurodegenerative responses in mouse brain neural cells. Environ. Res 136, 253–263.

Jha, C.B., 2000. Ayurvediya Rasashastra. Choukhambha Surabharati Prakashan, Varanasi.

Kamath, S.U., Pemiah, B., Sekar, R.K., Krishnaswamy, S., Sethuraman, S., Krishnan, U. M., 2012. Mercury-based traditional herbo-metallic preparations: a toxicological perspective. Arch. Toxicol. 86 (6), 831–838.

Krishnamachary, B., Rajendran, N., Pemiah, B., Krishnaswamy, S., Krishnan, U.M., Sethuraman, S., et al., 2012. Scientific validation of the different purification steps involved in the preparation of an Indian Ayurvedic medicine, Lauha *Bhasma*. J. Ethnopharmacol. 142 (1), 98–104.

Kumar, A., Grai, A.K., 2012. A clinical study on Pandu Roga, iron deficiency anemia, with Trikatrayadi Lauha suspension in children. J. Ayurveda Integr. Med. 3 (4), 215–222. Available from: https://doi.org/10.4103/0975-9476.104446.

Kumar, A., Nair, A.G.C., Reddy, A.V.R., Garg, A.N., 2006. Unique Ayurvedic metallic-herbal preparations, chemical characterization. Biol. Trace Element Res. 109 (3), 231–254.

Kumar, C.S., Moorthi, C., Prabhu, P.C., Jonson, B.B., Venkatnarayan, R., 2011. Standardization of anti-arthritic herbo-mineral preparation. Res. J. Pharma. Biol. Chem. Sci. 2, 679–684.
Meyers, M.A., Mishra, A., Benson, D.J., 2006. Mechanical properties of nanocrystalline materials. Progress Mater. Sci. 51 (4), 427–556.
Mishra, S., 2004. Somadeva's Rasendra Chudamani. Chaukhamba Orientalia, Varanasi.
Mishra, A., Mishra, A.K., Tiwari, O.P., Jha, S., 2014. In-house preparation and characterization of an Ayurvedic *Bhasma*: Praval *Bhasma*. J. Integr. Med. 12 (1), 52–58.
Miyazaki, K., Islam, N., 2007. Nanotechnology systems of innovation—an analysis of industry and academia research activities. Technovation 27 (11), 661–675.
Mohapatra, S., Jha, C.B., 2013. Analytical study of raw Swarna Makshika (Chalcopyrite) and its *Bhasma* through TEM and EDAX. Ayu. 34 (2), 204.
Mukkavalli, S., Chalivendra, V., Singh, B.R., 2017. Physico-chemical analysis of herbally prepared silver nanoparticles and its potential as a drug bioenhancer. Open Nano 2, 19–27.
Nagarajan, S., Pemiah, B., Krishnan, U.M., Rajan, K.S., Krishnaswamy, S., Sethuraman, S., 2012. Physico-chemical characterization of lead based Indian traditional medicine—Naga *Bhasma*. Int. J. Pharm. Pharm. Sci. 4, 69–74.
Nel, A.E., Mädler, L., Velegol, D., Xia, T., Hoek, E.M., Somasundaran, P., et al., 2009. Understanding biophysicochemical interactions at the nano-bio interface. Nat. Mater. 8 (7), 543–557.
Palkhiwala, S., Bakshi, S.R., 2014. Engineered nanoparticles: revisiting safety concerns in light of ethno medicine. Ayu. 35 (3), 237.
Prakash, B., 1997. Use of metals in Ayurvedic medicine. Ind. J. Hist. Sci. 32 (1), 1–7.
Rajurkar, N., 2016. Synthesis and characterization of mandur *Bhasma*. Int. J Pharma. Biol. Arch. 6 (3).
Rasheed, S.P., Shivashankar, M., 2017. Synthesis and characterization of Shanku *Bhasma*—an anti-ulcer herbomineral formulation, IOP Conference Series: Materials Science and Engineering, 263(2). IOP Publishing, p. 022026, ISSN:1757-8981.
Rasheed, A., Naik, M., Haneefa, M., Pillanayil, K., Kumar, A., Pillai, R., et al., 2014. Formulation, characterization and comparative evaluation of Trivanga *Bhasma*: a herbo-mineral Indian traditional medicine. Pak. J. Pharm. Sci. 27 (4), 793–800.
Santhosh, B., Raghuveer, J.P., Rao, V.N., 2013. Analytical study of Yashada *Bhasma* (Zinc based Ayurvedic metallic preparation) with reference to ancient and modern parameters. Open Access Sci. Rep. 2 (1), 582. Available from: https://doi.org/10.4172/scientific/reports. 2013.
Saper, R.B., Phillips, R.S., Sehgal, A., Khouri, N., Davis, R.B., Paquin, J., et al., 2008. Lead, mercury, and arsenic in US- and Indian-manufactured Ayurvedic medicines sold via the Internet. JAMA 300 (8), 915–923.
Sarkar, P.K., Chaudhary, A.K., 2010. Ayurvedic *Bhasma*: the most ancient application of nanomedicine. J. Sci. Res. 69, 901–905.
Sawant, R.S., 2015. Comparative study of mukta *Bhasma* & mukta pishti with reference to their particle size. Inter. J. Ayu. Med. 6 (1), 122–128.
Sharma, R., Bhatt, A., Thakur, M., 2016. Physicochemical characterization and antibacterial activity of Rajata Bhasma and silver nanoparticle. Ayu. 37 (1), 71–75.
Shi, J., Votruba, A.R., Farokhzad, O.C., Langer, R., 2010. Nanotechnology in drug delivery and tissue engineering: from discovery to applications. Nano Lett. 10 (9), 3223–3230.
Singh, S.K., Chaudhary, A., Rai, D.K., Rai, S.B., 2009. Preparation and characterization of a mercury based Indian traditional drug—Ras-Sindoor. Indian Journal of Traditional Knowledge 8 (3), 346–351.

Singh, S.K., Gautam, D.N.S., Kumar, M., Rai, S.B., 2010. Synthesis, characterization and histopathological study of a lead-based Indian traditional drug: Naga *Bhasma*. Ind. J. Pharm. Sci. 72 (1), 24.
Singh, R.K., Kumar, S., Aman, A.K., Karim, S.M., Kumar, S., Kar, M., 2017. Study on physical properties of Ayurvedic nanocrystalline Tamra *Bhasma* by employing modern scientific tools. J. Ayu. Integ. Med. Available from: https://doi.org/10.1016/j.jaim.2017.06.012.
Sumithra, M., Rao, P.R., Nagaratnam, A., Aparna, Y., 2015. Characterization of SnO_2 nanoparticles in the traditionally prepared Ayurvedic medicine. Mater. Today: Proc. 2 (9), 4636–4639.
Thakur, K.S., Vahalia, M.K., Jonnalagadda, V.G., Rashmi, K., Nadkarni, S.D., Gudi, R. V., et al., 2014. Evaluation of structural, chemical characterisation and safety studies of Samagandhak Kajjali, an Indian traditional Ayurvedic drug. J. Pharm. Phytochem. 2 (6), 57–67.
Thakur, K., Gudi, R., Vahalia, M., Shitut, S., Nadkarni, S., 2017. Preparation and characterization of Suvarna *Bhasma* Parada Marit. J. Pharm. 20 (1), 36.
Tripathi, R., Rathore, A.S., Mehra, B.L., Raghubir, R., 2013. Physico-chemical study of Vaikrānta *Bhasma*. Ancient Sci. Life 32 (4), 199.
Upadhyaya, M., 1999. Ayurveda Prakaasha'. 3/151. Choukhambha Bharati Academy, Varanasi.
Vadnere, G.P., Pathan, A.R., Singhai, A.K., 2013. Characterization of indigenous traditional medicine—Muktashukti *Bhasma*. Ind. J. Tradit. Knowl. 12 (3), 483–488.
Vagbhatta, R., 1998. Rasaratnasamuccaya'. Meharchand Lachhmandas Publications., New Delhi.
Virupaksha, G.K., Kumar, N., 2012. Characterization of Tarakeshwara rasa: an Ayurvedic herbomineral formulation. Ayu. 33 (3), 406.
Wadekar, M.P., Rode, C.V., Bendale, Y.N., Patil, K.R., Prabhune, A.A., 2005. Preparation and characterization of a copper based Indian traditional drug: Tamra *Bhasma*. J. Pharm. Biomed. Anal. 39 (5), 951–955.
Wijenayake, A., Pitawala, A., Bandara, R., Abayasekara, C., 2014. The role of herbometallic preparations in traditional medicine—a review on mica drug processing and pharmaceutical applications. J. Ethnophar. 155 (2), 1001–1010.

FURTHER READING

Tucci, P., Porta, G., Agostini, M., Dinsdale, D., Iavicoli, I., Cain, K., et al., 2013. Metabolic effects of TiO_2 nanoparticles, a common component of sunscreens and cosmetics, on human keratinocytes. Cell Death Dis. 4 (3), 549.

CHAPTER 23

Nanoparticle-Based Delivery of Phytomedicines: Challenges and Opportunities

Mohammad Sajid[1], Swaranjit Singh Cameotra[2], Mohd Sajjad Ahmad Khan[3] and Iqbal Ahmad[4]
[1]Cell Biology and Immunology Laboratory, Institute of Microbial Technology, Chandigarh, Punjab, India
[2]1103, Sector 11-C, Chandigarh, Punjab, India
[3]Department Basic Sciences, Biology Unit, Health Track, Imam Abdulrahman Bin Faisal University, Dammam, Kingdom of Saudi Arabia
[4]Department of Agricultural Microbiology, Aligarh Muslim University, Aligarh, Uttar Pradesh, India

23.1 INTRODUCTION

Natural extracts from medicinal plants are used for drug discovery. Extracts from the herbal plants consist of various active chemicals. Phytomedicines produced from natural resources; therefore, they possess several biological activities and functions. The majority of active constituents belong to the family of flavonoids, terpenoids, polyphenols, and tannins. These active chemicals modulate several cellular pathways for the treatment of numerous diseases like cancer, multiple sclerosis, inflammatory bowel disease, metabolic syndrome, diabetes, dengue, etc. (Chaudhary et al., 2015; Farzaei et al., 2017; Graf et al., 2010; Nankar et al., 2017; Singh and Rawat, 2017; Triantafyllidi et al., 2015). Recent studies mainly focused on antimicrobial, antiviral, anticancer, antioxidant, and anti-inflammatory effects of phytodrugs. Despite the enormous use of synthetic drugs, still the majority of people in the developing countries practice traditional phytomedicines for health care. Due to the various side effects of synthetic drugs, a large number of populations are now switching to natural drugs for health benefits. In search of novel drugs, scientific community attracted toward medicinal plants extracts because of their known physiological effects and potential to treat diverse diseases. From 1980–2006, approximately 50% of the approved medicines were plant-derived natural compounds (Ferreira and Pinto, 2010). Commonly, phytomedicines are believed to be nontoxic. However, they exhibit the same physiochemical obstructions as associated with the synthetic

New Look to Phytomedicine
DOI: https://doi.org/10.1016/B978-0-12-814619-4.00024-0

medicines. These problems include toxicity, insolubility, and low bioavailability. Several nanotechnologies provide a solution for these physiochemical and physiological barriers.

The era of nano-sized drug-delivery systems started with the synthesis of the liposome by Bangham et al. (1965). Afterword, a large number of biocompatible molecules were discovered and used for the construction of nanoparticles. Different types of drug-delivery systems are liposomes, niosomes, solid lipid nanoparticles (SLNs), polymeric nanoparticles, polymeric micelles, etc. The nanoparticles have several advantages such as increased targeting and internalization efficiency of a drug. The drug-delivery system increases the therapeutic potential of phytomedicines. Further, nanocarriers protect phytomedicines from thermal and photodegradation. They also decrease the toxicity of phytomedicines. The processes used for the synthesis of nanoparticles are self-assembly, solvent displacement, salting-out, high-pressure homogenization, nanoprecipitation, coprecipitation, supercritical fluid method, and emulsification–diffusion method.

Phytomedicines were fruitfully entrapped into various nanocarriers for their delivery without producing toxic effects (Narayanan et al., 2009; Sanna et al., 2012; Wicki et al., 2015). Nanotechnology provides several advantages for phytomedicines such as it enhances the therapeutic index of plant extracts, regulates the sustained release of active constituents, and decreases required dose for treatment. In this chapter, we summarize the use of different drug-delivery systems like liposomes, niosomes, SLNs, polymeric nanoparticles, and metallic nanoparticles for an effective deliverance of the phytomedicines to their respective targets.

23.2 LIPID-BASED VESICULAR DRUG-DELIVERY SYSTEMS FOR PHYTOMEDICINES

Amphiphilic lipid molecules can assemble into a particular order of one or multiple concentric bilayer structures when dispersed in a polar solvent like water. Vesicular systems provide an excellent solution to the various pharmacological problems associated with the medicines, such as insolubility, short half-life, and toxicity. The insoluble drugs can be made soluble by entrapping them into the vesicular systems. Further, these vesicular systems protect and delay the drug metabolism from the xenobiotic metabolizing enzymes and allow their sustained release inside the body. Another significant advantage of the lipid-based vesicular system is that they protect body tissues for direct contact with toxic drugs, therefore, selectively taken by the body cells (Jain et al., 2014). Hydrophilic as well as hydrophobic drugs can entrap into the lipid-based vesicular systems. In general, the hydrophobic drugs are localized into the internal compartment of the vesicular system, whereas the hydrophobic drugs associated with the layer of lipid membrane of the vesicular system via hydrophobic—hydrophobic interactions. Besides lipids, surfactant molecules were used to synthesize the vesicular nanostructures (Mullertz et al., 2010). The physiochemical properties of the vesicular systems depend on size, elasticity, charge on surface, and the lamellarity of the membrane. These physiochemical and colloidal factors can influence the structure, stability, entrapment efficiency, and drug-delivery potential of the vesicular system. The lipid bilayer can be amalgamated with cholesterol and other charged molecules to increase bilayer plasticity and protection from the aggregation, respectively.

23.2.1 Liposome

Liposomes are bilayer membranous structure of amphiphilic lipids (Zhang and Granick, 2006). In contact with the hydrophobic solvent, these lipids self-assembled into a globular vesicular structure. Liposomes showed an enormous potential to transport drug molecules. The liposome increases the therapeutic potential of the drug by enhancing its half-life, absorption, and bioavailability inside the body. By structure, they can categorize into unilamellar vesicles (ULVs) and multi-lamellar vesicles (MLVs) vesicular system. Further, on their size, ULVs can be classified into small, intermediate, and large ULVs (Vemuri and Rhodes, 1995). The liposome can prepare with synthetic and natural lipids molecules. The dioleoylphosphatidylethanolamine and dioleoylphosphatidylcholine are most commonly used lipid to construct liposomes, as they are an essential component of cell membrane structures. Studies were also done to synthesize liposome from other lipids. Further, to enhance the stability and rigidity of lipid bilayer, cholesterol was also added to a lipid membrane. Cholesterol can attach with other stabilizing molecules such as polyethylene glycol (PEG) for construction of "stealth liposomes" (Immordino et al., 2006). Furthermore, cationic liposomes constructed using charged lipid molecule to increase the delivery of drugs to negatively charged target cells. Based on the interaction between drug and lipid, several methods are developed for the synthesis of liposomes. In contrast to conventional liposomes,

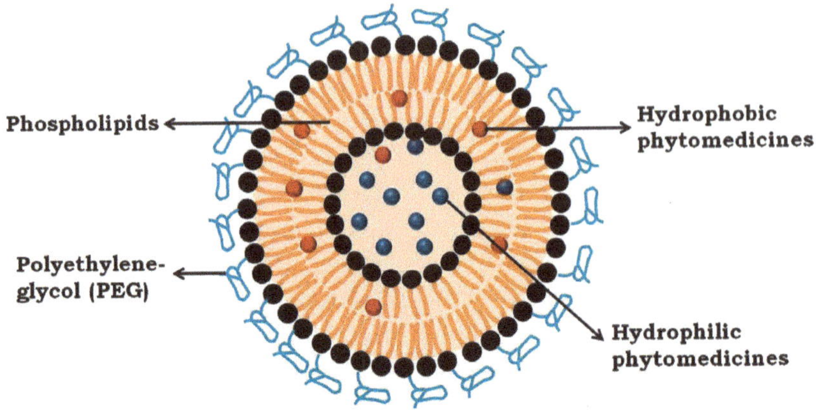

Figure 23.1 ***Liposome-based delivery system for the entrapment of phytomedicines.*** The phospholipid bilayer membrane can cover with PEG molecules. The inner core contains the majority of hydrophilic phytomedicines, whereas hydrophobic phytomedicines localized to the bilayer membrane due to hydrophobic interactions.

attachment of specific ligands for instance immunoglobulins, aptamers, peptides, and small ligands on the exterior of liposomes can selectively deliver the drugs to target cells. The properties mentioned above of liposome make them a prudent delivery molecule for transportation of phytomedicines (Fig. 23.1). The encapsulation of a flavonoid extracted from *Erigeron breviscapus* Vant. breviscapine into liposome considerably increases sustained release, retention time in plasma and delayed the clearance of breviscapine (Xiong et al., 2011).

23.2.1.1 Liposome for Delivery of Antimicrobial Phytomedicines

Antibiotic drug resistance is a great threat for the treatment of several pathogenic microorganisms. Fungal infections are increasing day by day due to antifungal resistance among several pathogenic fungal species. The fungal infections severally affected the immunocompromised individuals (Landlinger et al., 2010). Quite a few antifungal drugs exhibit toxicity and low bioavailability inside the host. Currently, liposomes are the most widely studied and clinically approved lipid-based vesicle systems for the delivery of antimicrobial drugs. Therefore, liposomes provide an alternative itinerary to distribute antifungals to their respective targets. Plant-derived essential oils have antimicrobial activity. However, the majority of these oils are insoluble in water and susceptible to oxidation as well as unstable at high temperature. Liposomal formulations have been constructed to increase the solubility as well as the stability of these oils (Detoni et al., 2012; Gortzi et al., 2006). The essential oils components carvacrol, thymol, *p*-cymene, and γ-terpinene were isolated from *Origanum dictamnus* L. and encapsulated into a liposome to increase their antimicrobial activity (Liolios et al., 2009). The entrapment of oil from *Eucalyptus camaldulensis* into liposome increases its durability and improved antifungal activity (Moghimipour et al., 2012). Eugenol is a natural compound, present in the essential oil of clove plant (*Syzygium aromaticum*). It has a very potent antimicrobial activity (Marchese et al., 2017). However, administration of eugenol causes hepatotoxicity (Soundran et al., 1994). To inhibit its toxic effects, eugenol was successfully encapsulated into the lipid vesicles and can be produced on a large scale (Sebaaly et al., 2016, 2015). Liposome-bearing eugenol has significantly inhibited the growth of foodborne pathogens (Peng et al., 2015). The liposome having various herbal extract with lysozyme significantly suppressed the growth of Gram-positive bacteria *Bacillus subtilis* and *Micrococcus luteus* and Gram-negative bacteria *Escherichia coli* and *Serratia marcescens* (Matouskova et al., 2016). Consequently, liposome having phytomedicines with antimicrobial activity can be a better solution to treat various infectious diseases.

23.2.1.2 Liposome for Delivery of Anticancer Phytomedicines

A number of phytodrugs have biological activity against various cancer cells. Curcumin is a plant-derived polyphenol isolated from turmeric (*Curcuma longa* L.) rhizome. It is one of the most studied phytomedicines because of its excellent biological activities (Zorofchian Moghadamtousi et al., 2014). Curcumin has several cellular signaling targets which play a crucial role in cell survival. Curcumin inhibits cyclooxygenase-2 enzyme activity, whereas it activates caspase pathway and nuclear transcription factor (NF-κB). Curcumin suppresses IL-1, IL-8 secretion and most importantly induces cytochrome-*c* release from mitochondria. Further, curcumin induces tumor suppressor pathway (p53) and ultimately leads to the death of cancer cells. Curcumin successfully showed its bioactivities during in vitro studies. However, in vivo studies monitored that curcumin bioavailability is very low, and high metabolic rate also decreases its distribution in various tissues of the body. Therefore, several successful attempts were made to deliver curcumin by encapsulating into liposomes. Application of liposomal curcumin downregulates proliferation and induces apoptosis by inhibiting the nuclear transcription factor (NF-κB) pathway in cancerous pancreatic cells (Li et al., 2005). Hence, liposomal curcumin can be used as future therapy for pancreatic cancers (Kurzrock and Li, 2005; Ranjan et al., 2013). A 2-hydroxypropyl-γ-cyclodextrin/curcumin liposome considerably showed anticancer ability against osteosarcoma and breast cancer (Dhule et al., 2012). Curcumin liposomes exhibited anti-melanoma activity which can be used for dermal applications (Chen et al., 2012). Mahmud et al. (2016) reported that lower curcumin-to-lipid ratio for PEGylated liposomes could increase its stability in plasma and bioavailability to pancreatic adenocarcinoma cells. Curcumin liposomes significantly inhibited the growth of glioma cells. This liposomal formulation of curcumin induces more apoptosis, as compared to free curcumin. Targeted delivery of curcumin along with quinacrine liposome significantly suppresses the proliferation of glioma cells and glioma stem cells (Wang et al., 2017). An organic solvent-free method was developed for the entrapment of curcumin into liposomes (Cheng et al., 2017).

Paclitaxel (PTX) is a vital phytomedicine which has anticancer activity against carcinoma of breast, ovary, colon, lung, esophagus, blood, urinary tract, head, and neck (Spencer and Faulds, 1994). PTX extracted from the bark of *Taxus brevifolia* Nutt. PTX is an approved drug for the cure of breast and ovarian carcinomas. PTX inhibits proliferation of the cancer cells by interfering in microtubule system of the cell cycle at late G2 or M phase. However, the solubility of PTX is very low (roughly 0.7–30 μg/mL). Further, PTX also induces hypersensitivity and neurotoxicity reactions (Sharma et al., 1997). Therefore, entrapment of PTX into the delivery system increases its anticancer activity and decreases toxicity, as compared to free-form PTX. PEGylated liposome significantly increases solubility (3.39 mg/mL) and stability of PTX but does not increase its anticancer activity when compared with free PTX on breast cancer cell line (Yang et al., 2007). Addition of phosphatidylserine into liposomes can increase the entrapment efficiency of liposome for PTX (Yang et al., 2008). The liposomal PTX formulation was also effective against lung cancer (Zhao et al., 2011).

Celastrol is a plant-derived drug isolated from Chinese herb *Tripterygium wilfordii*. Celastrol possesses a potent anticancer activity, but due to insolubility issue, its use is limited. The entrapment of celastrol into liposome significantly increases its stability in serum, enhances its cellular uptake, and improves anticancer activity against prostate cancer cells (Wolfram et al., 2014). Liposomes having semi-purified Job's tear (*Coix lacryma-jobi* Linn) extract demonstrated increased stability and effective anticancer activity against a colon tumor (Sainakham et al., 2016). It was observed that liposomal entrapment of curcumin in combination with resveratrol induces apoptotic pathway and efficiently suppresses the proliferation of cancer cell (Narayanan et al., 2009). Colchicine, a natural toxin and extracted from *Colchicum autumnale*, has anti-gout and anti-inflammatory activities. Colchicine-loaded liposome enhances its microtubule depolarization potential at a lower concentration when compared with free colchicine (Cauda et al., 2010).

Several types of research also synthesize an advance version of the liposome by covalently attaching the drug molecules to a polar head group of lipid molecules (e.g., phospholipids). These constructs called phytosomes. Phytosome showed better pharmacokinetic and pharmacodynamic profile of herbal extract, as compared to the conventional liposomes. Phytosome-bearing curcumin showed better antiaging activity when compared with another lipid-based vesicular system (Gupta and Dixit, 2011).

23.2.2 Niosome

Niosomes are surfactant molecules based delivery systems that are assembled by the hydration of surfactants, either synthetic or natural (biosurfactant). Niosomes can construct with or without incorporation of cholesterol or other membrane-forming molecules. Like liposomes, niosome can use as a transporter for the hydrophobic and hydrophilic compounds. Niosomes provide improved stability and better release kinetics, as compared to liposomes (Kumar and Rajeshwarrao, 2011). Niosome can be constructed at low cost in combination with phospholipids without much changes in their purity. Niosomes are efficient drug-delivery system since they are nonionic and less toxic. They can be synthesized by chemical surfactants or biosurfactant molecules (Haque et al., 2017). In general, hydrophilic drug mainly resides in the central core of the niosome particle, whereas the hydrophobic drug localized to the membrane of niosome. In addition, variations of niosomes structure can confine their effect to specific targets. Similar to liposomes, niosomes categorize into small ULVs which have size ≤ 0.05 μm, MLV within the size of ≥ 0.05 μm, and large ULVs having size ≥ 0.10 μm. Niosomes were used to deliver various hydrophobic flavanoids molecules. Phytomedicine loaded into niosome has improved stability, lower toxicity, and increased bioavailability (Song et al., 2014). Niosomes were used to transport polyphenolic compound silymarin extracted from seeds of *S. marianum*. Silymarin is applied to treat various liver disorders like hepatitis, jaundice, cirrhosis, and hepatotoxicity caused by xenobiotics. Silymarin and its derivative exhibited antioxidant, antitumor, antiangiogenic, antimetastatic, and anti-inflammatory properties. The niosomal formulations of silymarin were used to prevent the toxicity caused by chemotherapy or radiotherapy to the cancer patients (Ramasamy and Agarwal, 2008). The niosome loaded with *Gymnema sylvestre* extract improved the efficiency of *G. sylvestre* extract and significantly decreases blood–glucose level (Kamble et al., 2013). A niosomal nanocarrier system having *Spermacoce hispida* extract showed extended release kinetics of phytomedicines and increased in vitro antituberculosis activity (Durgadas and Giriraj, 2017). Such studies can lead to the synthesis of environment-friendly and surfactant-based nano-formulations for medicinal use of various insoluble, toxic but efficient antimicrobial drugs obtained from plants.

23.2.2.1 Niosome for Delivery of Antitumor Phytomedicines

Curcumin entrapped in niosomes considerably inhibited the growth of ovarian cancer cells (Xu et al., 2016). Inhalable cationic niosomes having curcumin showed increased apoptosis in lung cancer cells. PEG niosomes ensemble with a tumor-homing and cell-penetrating peptide (CPP) (tLyp-1) increase antitumor property of doxorubicin and curcumin against human glioblastoma (U87) and human mesenchymal stem cells (Ag Seleci et al., 2017). Niosome can efficiently deliver curcumin along with another cytotoxic drug (doxorubicin) to control the growth of cancer cell (Sharma et al., 2015). Another important anticancer phytomedicine, lycopene, is extracted from *Lycopersicum esculentum*. Lycopene is unstable in the presence of light and susceptible to degradation by high temp and oxidation, which makes its use difficult in therapeutic applications. However, niosomes increase lycopene antiproliferative activity against cancer cells (Sharma et al., 2016).

23.2.2.2 Niosomes for Delivery of Anti-inflammatory Phytomedicines

Niosomal formulations were found to be very useful for dermal applications (Muzzalupo et al., 2011). Anti-inflammatory efficacy of drugs can be improved by topical release and confined accumulation of niosomal gels on the skin (Kumbhar et al., 2013; Shahiwala and Misra, 2002). Therefore, these formulations were preferably used in the synthesis of various skin gels to efficiently transport phytomedicines (Junyaprasert et al., 2012;

Yeh et al., 2013). The niosomes showed extended penetration of curcuminoids for transdermal applications (Rungphanichkul et al., 2011). The anti-inflammatory action of four phenylbutanoids (extracted from *Zingiber cassumunar*) was significantly enhanced when entrapped into the niosomal formulation. These niosomal phenylbutanoids inhibited the nitric oxide (NO) production by macrophage, whereas individual phenylbutanoids produce less NO production from macrophages (Kaewchoothong et al., 2012). Curcumin-loaded niosomes to a hydrophilic ointment increase retention time for curcumin inside skin (51.66%), as compared to ointment having only free curcumin (1.64%). Hence, niosome increases the anti-inflammatory properties of curcumin (Agrawal et al., 2015). Further, niosomal formulation increases antioxidant and anti-inflammatory activities of plai oil (Leelarungrayub et al., 2017). Anti-inflammatory effects of herbal extract from *Z. cassumunar* increase when encapsulated into niosomes and these effects are more, in contrast with other steroidal and nonsteroidal drugs (Priprem et al., 2016).

23.2.2.3 Niosomes for Transport of Antiaging/Antioxidant Phytomedicines

Several antiaging phenolic compounds were extracted from *Terminalia chebula*. It includes gallic acid, chebulinic acid, chebulagic acid, isoterchebulin, punicalagin, and 1,3,6-tri-O-galloyl-β-D-glucopyranosehave. These phenolic compounds have antioxidant activities (Saha and Verma, 2016). The niosomes having a semi-purified fraction of gallic acid demonstrated enhanced in vivo antiaging activities, as compared to the unloaded fraction (Manosroi et al., 2011). Liang et al. (2016) reported that encapsulation of epigallocatechin gallate (EGCG) into niosomes considerably increases the antioxidant ability of EGCG, when compared with free EGCG, and increases EGCG bioavailability during intestinal digestion.

In essence, niosomal formulations provide a capable delivery system for controlled transport of phytomedicines. The capabilities of niosomes can be improved by various modification methods. It required further investigation and research efforts to develop a novel niosomes that have appropriate clinical applications. In this regard, biosurfactant molecules can provide an appropriate molecule for the niosome preparation due to their various advantages over chemical surfactant such as low production cost, large-scale manufacture, and similar activity concerning liposomes (Haque et al., 2017).

23.2.3 Solid Lipid Nanoparticles

SLNs are colloidal drug-delivery systems, developed in the late 1980s. It is a combination of liposome and niosome that contain phospholipids and surfactant molecules. SLNs showed advantages over liposome and niosome for delivery of various drugs. SLNs have increased stability, and they provide better protection against degradation of drugs. The production cost of SLNs was also less, as compared to liposome and niosome. SLNs were synthesized for delivery of different traditional phytomedicines (Shi et al., 2012). It also significantly enhances transdermal penetration of phytomedicines inside skin (Sutthanut et al., 2009). SLNs increase the therapeutic potential of eugenol and efficiently inhibited the growth of *Candida* infection during oral candidiasis (Garg and Singh, 2011). Curcumin entangled into SLNs has enhanced antimicrobial activity (Jourghanian et al., 2016). Oxyresveratrol is a potent tyrosinase inhibitor, extracted from the heartwood of *Artocarpus lakoocha*. The SLNs increase the relative bioavailability of oxyresveratrol to 125%, as compared with unformulated oxyresveratrol (Sangsen et al., 2015). A traditional Chinese Phytomedicine triptolide obtained from *T. wilfordii* has immunosuppressive properties and can be used for the autoimmune diseases like rheumatoid arthritis. The entrapment of triptolide in SLNs increases the anti-inflammatory potential of triptolide by enhancing its infiltration into the skin. The SLNs containing curcuminoids from *C. longa* L. showed anti-inflammatory effects (Zamarioli et al., 2015).

Cryptotanshinone (CTS), extracted from the root of *Salvia miltiorrhiza* Bunge, has a prominent anticancer action through inhibiting activation of STAT3 (Lu et al., 2017). The incorporation of CTS inside SLNs considerably increases its bioavailability inside the body (Hu et al., 2010). Another anticancer phytomedicine, frankincense, isolated from the tree of family Bursera ceaes, showed increased antitumor activity when delivering via SLNs in combination with myrrh (Shi et al., 2012). Therefore, SLNs provide a remarkable delivery system for highly hydrophobic phytomedicines. A phytodrug was known as triterpenediol (TPD) from *Boswellia serrata* which suppresses propagation of different cancer cell lines of human origin. In vitro as well as in vivo study showed that TPD–SLNs had more antitumor activity when compared to free TPD (Bhushan et al., 2013). Curcumin in combination with doxorubicin inside SLNs remarkably inhibited proliferation of several multidrug-resistant cancer cells (Naderinezhad et al., 2017).

23.2.4 Polymeric Nanoparticles

Several organic molecules were used to synthesize polymer, and they can further form a nanoparticle with the hydrophilic and hydrophobic region. Therefore, polymeric nanoparticles protect drugs from degradation and enhance its solubility and bioavailability. Different biodegradable polymers have been used to form polymeric nanoparticles. The hydrophobic region is constructed using poly–lactic acid, poly(lactide-*co*-glycolide) (PLGA), poly–glycolic acid, poly(cyanoacrylate), and poly(caprolactone). PEG has been commonly used for the hydrophilic part of the polymeric nanoparticles. The polymeric nanoparticles have a diameter range from 10 to 1000 nm. On the basis of their composition, the polymeric nanoparticles can be termed as "nanocapsules" (NCs) and "nanospheres" (NSs). NC is consists of polymeric molecule membrane which surrounds oily core inside. The phytodrugs can localize to the oily core of the polymeric membrane. NS only contains a polymeric organization in which a phytodrug is present. Polymeric nanoparticles properties like size, charge, structure, and release kinetics can be controlled by choosing appropriate polymer length and solvent during synthesis. Further, the modification can be done on functional groups positioned on the exterior of the polymeric nanoparticles (Fig. 23.2) (Tanihara et al., 1999).

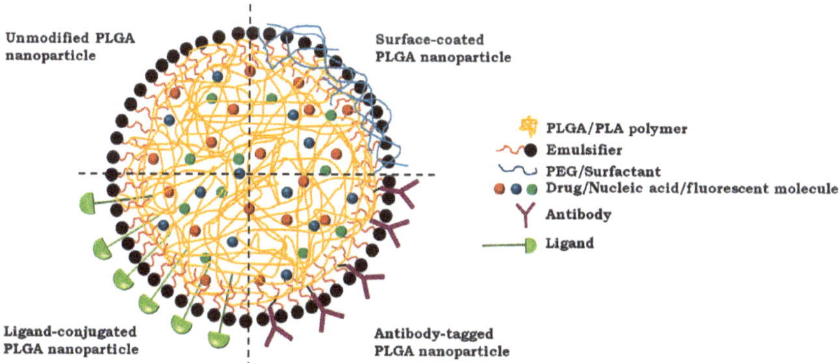

Figure 23.2 ***PLGA polymeric nanoparticles for the protected transport of phytomedicines.*** The PLGA nanoparticles can be unmodified or modified by attaching other molecules to increase stability and secure delivery of phytomedicines. The surface of PLGA nanoparticles can coat with PEG or other surfactant molecules. The polymer of PLGA nanoparticles can tag with antibodies or other ligands for successful targeted delivery of phytomedicines.

23.2.4.1 Poly(Lactide-co-Glycolide) Nanoparticles for Delivery of Phytomedicines

During the last decade, the biocompatible polymers have achieved much attention in the clinical field as an anticancer phytomedicines carrier system against cancer cells and tumor. Several anticancer phytomedicines like curcumin, resveratrol, EGCG, saponins, grape seed extract (GSE), and silymarin were discovered to inhibit the growth of cancer cell. However, their activities were limited by several factors like increased hydrophobicity, low solubility, short half-life, fast metabolism, and high toxicity. Apart from lipid-based nanocarrier, the polymer-based nanoparticles also showed their enormous potential in the delivery of phytomedicines. A variety of anticancer phytomedicines were encapsulated into different biodegradable polymers to enhance their killing potency. An FDA-approved polymer PLGA was used for the entrapment of various phytodrugs. Curcumin-entrapped PLGA nanoparticles significantly inhibited the proliferation of various prostate cancer cell lines such as PC3, LNCaP, and DU145 (Mukerjee and Vishwanatha, 2009). Further, curcumin-PLGA nanoparticles demonstrated enhanced uptake by ovarian (A2780CP) and breast cancer cells (MDA-MB-231), whereas free curcumin showed very less incorporation within these cells. Furthermore, conjugation of PLGA with an antibody against receptor of cancer cell illustrated enhanced localization of curcumin inside the cancerous cells (Yallapu et al., 2010). Dimethylnitrosamine-induced hepatocellular carcinoma was significantly eliminated by curcumin encapsulated in PLGA nanoparticles (Ghosh et al., 2012). The epidermal growth factor receptor (EGFR)−targeting GE11 peptides linked PEGylated PLGA nanoparticles effectively deliver the curcumin to breast cancer cells (MCF-7) (Jin et al., 2017). Therefore, ligand binding to PLGA nanoparticles further increases localized delivery of phytomedicines. The curcumin-loaded hybrid PLGA nanoparticles synthesized using poly(lactic-*co*-glycolic acid)-1,2-distearoyl-glycerol-3-phospho-ethanolamine-*N*-[methoxy (polyethylene glycol)-2000] ammonium salt (PLGA−DSPE−PEG) was effective in decreasing the size of rat glioma-2 (RG2) (Orunoglu et al., 2017). The curcumin-loaded PLGA−PEG−Fe_3O_4 nanoparticles exhibited cytotoxic activity on lung cancer cells (A549) (Sadeghzadeh et al., 2017). Apart from cancer, PLGA-curcumin nanoparticles were also used for the treatment of other diseases (Table 23.1). Curcumin along with an additional anticancer drug inside PLGA nanoparticles significantly inhibited the proliferation of various cancer cells (Table 23.2). Consequently, recent studies demonstrated

Table 23.1 PLGA-curcumin nanoparticles for treatment of various diseases

S. no.	Nanoparticles	Target	Effects	References
1.	PLGA-curcumin nanoparticles	Alzheimer's disease	Destroy amyloid aggregates, exhibit antioxidative property	Mathew et al. (2012)
2.	PLGA-curcumin nanoparticles	Alzheimer's disease	Decreases Aβ aggregates formation	Barbara et al. (2017)
3.	PLGA-curcumin nanoparticles	Malaria	Anti-plasmodial activity	Busari et al. (2017)
4.	PLGA-curcumin nanoparticles	Cerebral malaria	Inhibits the sequestration of parasitized RBCs and $CD8^+$ T cells in the brain, suppresses brain mRNAs synthesis of inflammatory cytokines, chemokine receptor CXCR3, and its ligand CXCL10, increases expression of the anti-inflammatory cytokine IL-10	Dende et al. (2017)
5.	PLGA/CNC/Cur/pDNA-ANG composite nanofibers	Wound healing	Helps in skin regeneration and prevent local infections	Mo et al. (2017)
6.	PLGA-curcumin nanoparticle	Psoriasis	Anti-psoriasis activity	Sun et al. (2017)
7.	PLGA-curcumin nano-formulation	Pain	Antinociceptive effects reduces cytokine release and BDNF in the spinal cord	Pieretti et al. (2017)
8.	PLGA-curcumin nano-formulation	Opioid-induced hyperalgesia	Attenuates hyperalgesia	Hu et al. (2016)

(*Continued*)

Table 23.1 (Continued)

S. no.	Nanoparticles	Target	Effects	References
9.	PLGA-curcumin nanoparticles	Early brain injury after experimental SAH	Upregulation of glutamate transporter-1 and attenuated glutamate concentration of cerebrospinal fluid following SAH inhibits inflammatory response and microglia activation, suppresses SAH-mediated oxidative stress	Zhang et al. (2017)
10.	PLGA-curcumin nanoparticles	Cataract	Regression of diabetic cataract	Grama et al. (2013)
11.	Eudragit S100/PLGA-curcumin microparticles	UC	Decreases colitis in a UC mouse model	Xiao et al. (2015)

SAH, Subarachnoid hemorrhage; *UC*, Ulcerative colitis; *RBCs*, red blood cells; *BDNF*, brain-derived neurotrophic factor.

Table 23.2 Curcumin co-encapsulation with another drug in PLGA nanoparticles

S. no.	Drug combination	Targets	References
1.	Curcumin + doxorubicin	Multidrug-resistant cell lines of chronic myeloid leukemia (CML) and blast-like tumor cells (K-562 cells)	Misra and Sahoo (2011)
2.	Curcumin + chrysin	Colorectal cancer cells (Caco-2)	Lotfi-Attari et al. (2017)
3.	Curcumin + metformin	Breast cancer cells (T47D)	Farajzadeh et al. (2017)
4.	Curcumin + docetaxel	Prostate cancer cells and tumor	Yan et al. (2017)
5.	Curcumin + paclitaxel	Human ovarian cancer cell line (A2780) and glioma	Liu et al. (2016) Cui et al. (2016)
6.	Curcumin + 5-fluorouracil	Squamous cell carcinoma cell line MCF7 and G1 cells	Masloub et al. (2016) Balasubramanian et al. (2014)
7.	Curcumin + SH-aspirin	Human ovarian carcinoma cells (SKOV3)	Zhou et al. (2015)

that curcumin-loaded polymeric nanoparticles are more efficient, as compared to other nano-formulations of curcumin. In essence, PLGA nanoparticles showed immense potential in the delivery of phytomedicines to their targets. Another anticancer herbal drug silymarin was also successfully entrapped into PLGA nanoparticles, and these silymarin-PLGA nanoparticles inhibited prostate cancer cell proliferation through apoptosis (Snima et al., 2014; Xie et al., 2016). A well-known phytodrug GSE was incorporated into PLGA nanoparticles (nano-GSE). Nano-GSE increases availability of the GSE to cancer cells and resulted in a significant increase in the apoptotic index (Narayanan et al., 2010).

23.2.4.2 Chitosan Nanoparticles for Delivery of Phytomedicines

Chitosan is a positively charged polymer of *N*-acetyl-glucosamine and D-glucosamine coupled through β-(1,4)-glycosidic bonds. It is synthesized by the alkaline deacetylation of chitin. Chitosan is famed for the synthesis of polymer-based drug-delivery systems due to its biocompatibility, biodegradability, and nontoxic properties (Ag Seleci et al., 2017). The various bioactive molecules such as antimicrobial or anticancer agents can be localized to the hydrophobic core of the chitosan nanoparticles (Table 23.3). Chitosan is also used as a stabilizing agent for synthesis of a variety of metallic nanoparticles (Ag, Au, TiO_2, and Pd) (Collado-Gonzalez et al., 2017; Goyal et al., 2016; Huang and Yang, 2004). The positive charge on chitosan makes them to actively interact with negatively charged molecules especially antibiotics (Cao and Sun, 2009). Curcumin-encapsulating chitosan-TPP nanoparticles significantly inhibited the growth of *Staphylococcus aureus* and *Pseudomonas aeruginosa*, as compared to free curcumin or sham chitosan-TPP nanoparticles (Mirnejad et al., 2014). Further, curcumin-loaded chitosan phosphate nanoparticles showed proficient antibacterial and antifungal activity without any toxicity on human peripheral blood mononuclear cells (Deka et al., 2016).

An effective antioxidant flavonoid quercetin is isolated from leaves and bark of *Sophora japonica*. It also possesses anticancer and antiviral activities. However, less solubility and low stability of quercetin make it not suitable for treatment. The loading of quercetin into chitosan nanoparticles increases its bioavailability (Zhang et al., 2008). Chitosan nanoparticles in combination with quercetin decrease the cytotoxic effect of various mycotoxins such as aflatoxin-B and ochratoxin-A (Abdel-Wahhab et al., 2015, 2017). The chitosan nanoparticles could be used to deliver phytomedicines and provide protection against toxicity induced by mycotoxins in the high endemic region. Further, lecithin–chitosan nanoparticles increase penetration and accumulation of quercetin into the skin and thus enhance its effect on topical delivery (Tan et al., 2011). Quercetin-loaded lecithin/chitosan nanoparticles demonstrated enhanced antioxidant activity when compared with free quercetin (Souza et al., 2014). Chitosan–caseinophosphopeptide nanoparticles significantly enhance the antioxidant activity of epigallocatechin-3-gallate (Hu et al., 2013). The PLGA nanoparticles coated with chitosan improved the resveratrol loading efficiency and release kinetics, as distinguished from uncoated PLGA nanoparticles (Sanna et al., 2012).

23.2.5 Metallic Nanoparticles for Delivery of Phytomedicines

Metal-based nanoparticles can improve the efficacy of drug molecules due to their biological activity. Hsieh et al. (2011) monitored the anticancer activity of epigallocatechin-3-gallate physically attached to the surface of gold

Table 23.3 Anticancer phytomedicines delivery through chitosan nanoparticles

S. no.	Nanoparticles	Targets	References
1.	Curcumin-loaded dextran sulphated chitosan nanoparticles	Human breast adenocarcinoma (MCF-7), human prostate cancer (PC-3), human osteosarcoma (MG-63)	Anitha et al. (2011a)
2.	Curcumin-loaded O-CMC nanoparticles	Human breast adenocarcinoma (MCF-7), human prostate cancer (PC-3)	Anitha et al. (2011b)
3.	Curcumin-loaded *N,O*-carboxymethyl chitosan nanoparticles	Human breast adenocarcinoma (MCF-7), human prostate cancer (PC-3)	Anitha et al. (2012)
4.	Curcumin-loaded chitosan nanoparticles	Human oral cancer (SCC-9)	Mazzarino et al. (2015)
5.	Curcumin-loaded chitosan–cholesterol micelles	Breast cancer	Muddineti et al. (2017)
6.	EGCG-loaded chitosan–caseinophosphopeptides nanoparticles	Human colon carcinoma (Caco-2)	Hu et al. (2012)
7.	Saponin-loaded chitosan nanoparticles	Human prostate cancer (PC-3), cervical adenocarcinoma (KB)	Rejinold et al. (2011)

nanoparticles (EGCG−pNGs) against bladder tumor cells. EGCG−pNGs complex significantly inhibits tumor size via the mechanism of apoptosis (Hsieh et al., 2011). Further, Chen et al. observed that conjugation of EGCG to gold nanoparticles (EGCG−pNG) enhances the anticancer activity of EGCG against B16F10 murine melanoma cells, as compared to cells treated with EGCG alone. EGCG−pNG also suppresses the growth of tumor in a murine melanoma model (Chen et al., 2014). Similarly, saponin-platinum-II can restore the viability of RAW 264.7 cells after 2,4-dinitrofluorobenzene (DNFB) induced reactive oxygen species (ROS). Therefore, saponin-Pt conjugates demonstrated an effective antioxidant property by decreasing the ROS production and further suppresses the activation of MAP kinase pathway and expression of MIP-2 in response to ROS produce by DNFB (Kim et al., 2009). These results provide insight into the potential usefulness of phytomedicines conjugated to metallic nanoparticles for treatment of various diseases.

23.3 NEW APPROACHES AND CHALLENGES FOR THE DELIVERY OF PHYTOMEDICINES

Drug-delivery systems can provide an apposite platform for delivery of various phytomedicines. They can boost the bioactivities of several herbal drugs. The several new approaches are described into the above headings. Nevertheless, CPPs have proven their role in enhance and targeted delivery of nanoparticles to the cancer cells. Therefore, use of phytomedicines nanoparticles in combination with CPP will be an effective approach to treat the cancer cells. There are some critical challenges associated with the implementation of nano-delivery systems for phytomedicines. One of the significant challenges is to prime the nanoparticles for targeted delivery of phytomedicines to their respective targets without affecting non-disease part of the body. There are several studies done in vitro for phytomedicines evaluation against diseases. Therefore, it is required to perform more research on in vivo animal models to elucidate the effectiveness of phytomedicines. Another big challenge is to scale up production of nanomaterial used for nanoparticle synthesis. In this regard, biosurfactant molecules can provide a solution for large-scale production and development of technologies. Further, there is a need to understand the interactions and effect of nonmaterials on the host.

23.4 FUTURE PROSPECTS

Although the enormous efforts were made by researchers in the field of nanotechnology, still an insufficient number of nanoparticles having phytomedicines is in clinical applications. This is because of several obstacles like less reproducibility of nanoparticles formation, the stability of nanoparticles, the deformed release kinetics of phytomedicines from nanoparticles, low entrapment efficiency of nanoparticles, and ultimately high cost. The advancement in modern scientific nanotechnologies can assist to eliminate pharmacodynamics and pharmacokinetics tribulations associated with phytomedicines entrapment in different delivery systems. Discovery of new molecules is required to increase the stability and longer half-life of nanoparticles. The entrapment and release kinetics of phytomedicines can be improved by the appropriate ratio between drug and nanoparticles. Therefore, improvement in nano-phytomedicine is requisite for construction of a novel drug-delivery system for phytomedicines. Further, merging of nanotechnology with herbal extracts showed significant outcomes in cosmetic industries. Advancement in nanotechnologies can enormously provide help to the cosmetic industries.

23.5 CONCLUSION

In this chapter, we focused on augmentation of various biological activities of phytomedicines entrapped into nanoparticles. Medicinal plants produce a vast array of extremely effective drugs for the treatment of numerous diseases. Further, their use in medicine is restricted due to various physiochemical and pharmacological obstacles. However, many nanoformulations like liposome, niosome, SLNs, and polymeric nanoparticles, metallic nanoparticles, etc. can improve the therapeutic potential of natural medicines. These "nanophyto" delivery systems have proven their potential in cosmetics and medicines.

REFERENCES

Abdel-Wahhab, M.A., Aljawish, A., El-Nekeety, A.A., Abdel-Aiezm, S.H., Abdel-Kader, H.A.M., Rihn, B.H., et al., 2015. Chitosan nanoparticles and quercetin modulate gene expression and prevent the genotoxicity of aflatoxin B(1) in rat liver. Toxicol. Rep. 2, 737–747.

Abdel-Wahhab, M.A., Aljawish, A., El-Nekeety, A.A., Abdel-Aziem, S.H., Hassan, N.S., 2017. Chitosan nanoparticles plus quercetin suppress the oxidative stress, modulate

DNA fragmentation and gene expression in the kidney of rats fed ochratoxin A-contaminated diet. Food Chem. Toxicol. 99 (Suppl. C), 209–221.

Agrawal, R., Sandhu, S.K., Sharma, I., Kaur, I.P., 2015. Development and evaluation of curcumin-loaded elastic vesicles as an effective topical anti-inflammatory formulation. AAPS PharmSciTech 16 (2), 364–374.

Ag Seleci, D., Seleci, M., Stahl, F., Scheper, T., 2017. Tumor homing and penetrating peptide-conjugated niosomes as multi-drug carriers for tumor-targeted drug delivery. RSC Adv. 7 (53), 33378–33384.

Anitha, A., Deepagan, V.G., Divya Rani, V.V., Menon, D., Nair, S.V., Jayakumar, R., 2011a. Preparation, characterization, in vitro drug release and biological studies of curcumin loaded dextran sulphate–chitosan nanoparticles. Carbohydr. Polym. 84 (3), 1158–1164.

Anitha, A., Maya, S., Deepa, N., Chennazhi, K.P., Nair, S.V., Tamura, H., et al., 2011b. Efficient water soluble *O*-carboxymethyl chitosan nanocarrier for the delivery of curcumin to cancer cells. Carbohydr. Polym. 83 (2), 452–461.

Anitha, A., Maya, S., Deepa, N., Chennazhi, K.P., Nair, S.V., Jayakumar, R., 2012. Curcumin-loaded *N,O*-carboxymethyl chitosan nanoparticles for cancer drug delivery. J. Biomater. Sci. Polym. Ed. 23 (11), 1381–1400.

Balasubramanian, S., Girija, A.R., Nagaoka, Y., Iwai, S., Suzuki, M., Kizhikkilot, V., et al., 2014. Curcumin and 5-fluorouracil-loaded, folate- and transferrin-decorated polymeric magnetic nanoformulation: a synergistic cancer therapeutic approach, accelerated by magnetic hyperthermia. Int. J. Nanomed. 9, 437–459.

Bangham, A.D., Standish, M.M., Watkins, J.C., 1965. Diffusion of univalent ions across the lamellae of swollen phospholipids. J. Mol. Biol. 13 (1), 238–252.

Barbara, R., Belletti, D., Pederzoli, F., Masoni, M., Keller, J., Ballestrazzi, A., et al., 2017. Novel curcumin loaded nanoparticles engineered for blood–brain barrier crossing and able to disrupt Abeta aggregates. Int. J. Pharm. 526 (1–2), 413–424.

Bhushan, S., Kakkar, V., Pal, H.C., Guru, S.K., Kumar, A., Mondhe, D.M., et al., 2013. Enhanced anticancer potential of encapsulated solid lipid nanoparticles of TPD: a novel triterpenediol from *Boswellia serrata*. Mol. Pharm. 10 (1), 225–235.

Busari, Z.A., Dauda, K.A., Morenikeji, O.A., Afolayan, F., Oyeyemi, O.T., Meena, J., et al., 2017. Antiplasmodial activity and toxicological assessment of curcumin PLGA-encapsulated nanoparticles. Front. Pharmacol. 8, 622.

Cao, Z., Sun, Y., 2009. Chitosan-based rechargeable long-term antimicrobial and biofilm-controlling systems. J. Biomed. Mater. Res. A 89 (4), 960–967.

Cauda, V., Engelke, H., Sauer, A., Arcizet, D., Bräuchle, C., Rädler, J., et al., 2010. Colchicine-loaded lipid bilayer-coated 50 nm mesoporous nanoparticles efficiently induce microtubule depolymerization upon cell uptake. Nano Lett. 10 (7), 2484–2492.

Chaudhary, T., Chahar, A., Sharma, J.K., Kaur, K., Dang, A., 2015. Phytomedicine in the treatment of cancer: a health technology assessment. J. Clin. Diagn. Res. 9 (12), XC04–XC09.

Chen, C.-C., Hsieh, D.-S., Huang, K.-J., Chan, Y.-L., Hong, P.-D., Yeh, M.-K., et al., 2014. Improving anticancer efficacy of (−)-epigallocatechin-3-gallate gold nanoparticles in murine B16F10 melanoma cells. Drug Des. Devel. Ther. 8, 459–474.

Chen, Y., Wu, Q., Zhang, Z., Yuan, L., Liu, X., Zhou, L., 2012. Preparation of curcumin-loaded liposomes and evaluation of their skin permeation and pharmacodynamics. Molecules 17 (5), 5972–5987.

Cheng, C., Peng, S., Li, Z., Zou, L., Liu, W., Liu, C., 2017. Improved bioavailability of curcumin in liposomes prepared using a pH-driven, organic solvent-free, easily scalable process. RSC Adv. 7 (42), 25978–25986.

Collado-Gonzalez, M., Montalban, M.G., Pena-Garcia, J., Perez-Sanchez, H., Villora, G., Diaz Banos, F.G., 2017. Chitosan as stabilizing agent for negatively charged nanoparticles. Carbohydr. Polym. 161, 63–70.
Cui, Y., Zhang, M., Zeng, F., Jin, H., Xu, Q., Huang, Y., 2016. Dual-targeting magnetic PLGA nanoparticles for codelivery of paclitaxel and curcumin for brain tumor therapy. ACS Appl. Mater. Interfaces 8 (47), 32159–32169.
Deka, C., Aidew, L., Devi, N., Buragohain, A.K., Kakati, D.K., 2016. Synthesis of curcumin-loaded chitosan phosphate nanoparticle and study of its cytotoxicity and antimicrobial activity. J. Biomater. Sci. Polym. Ed. 27 (16), 1659–1673.
Dende, C., Meena, J., Nagarajan, P., Nagaraj, V.A., Panda, A.K., Padmanaban, G., 2017. Nanocurcumin is superior to native curcumin in preventing degenerative changes in experimental cerebral malaria. Sci. Rep. 7 (1), 10062.
Detoni, C.B., de Oliveira, D.M., Santo, I.E., Pedro, A.S., El-Bacha, R., da Silva Velozo, E., et al., 2012. Evaluation of thermal-oxidative stability and antiglioma activity of *Zanthoxylum tingoassuiba* essential oil entrapped into multi- and unilamellar liposomes. J. Liposome Res. 22 (1), 1–7.
Dhule, S.S., Penfornis, P., Frazier, T., Walker, R., Feldman, J., Tan, G., et al., 2012. Curcumin-loaded gamma-cyclodextrin liposomal nanoparticles as delivery vehicles for osteosarcoma. Nanomedicine 8 (4), 440–451.
Durgadas, A., Giriraj, T.K., 2017. Development of novel nano niosomes as drug delivery system of *Spermacoce hispida* extract and in vitro antituberculosis activity. Curr. Nanomater. 2 (1), 17–23.
Farajzadeh, R., Pilehvar-Soltanahmadi, Y., Dadashpour, M., Javidfar, S., Lotfi-Attari, J., Sadeghzadeh, H., et al., 2017. Nano-encapsulated metformin-curcumin in PLGA/PEG inhibits synergistically growth and hTERT gene expression in human breast cancer cells. Artif. Cells Nanomed. Biotechnol 46 (5), 917–925.
Farzaei, M.H., Shahpiri, Z., Bahramsoltani, R., Nia, M.M., Najafi, F., Rahimi, R., 2017. Efficacy and tolerability of phytomedicines in multiple sclerosis patients: a review. CNS Drugs 31 (10), 867–889.
Ferreira, V.F., Pinto, A.C., 2010. A fitoterapia no mundo atual. Quím. Nova 33, 1829.
Garg, A., Singh, S., 2011. Enhancement in antifungal activity of eugenol in immunosuppressed rats through lipid nanocarriers. Colloids Surf., B: Biointerfaces 87 (2), 280–288.
Ghosh, D., Choudhury, S.T., Ghosh, S., Mandal, A.K., Sarkar, S., Ghosh, A., et al., 2012. Nanocapsulated curcumin: oral chemopreventive formulation against diethylnitrosamine induced hepatocellular carcinoma in rat. Chem. Biol. Interact. 195 (3), 206–214.
Gortzi, O., Lalas, S., Chinou, I., Tsaknis, J., 2006. Reevaluation of antimicrobial and antioxidant activity of *Thymus* spp. extracts before and after encapsulation in liposomes. J. Food Prot. 69 (12), 2998–3005.
Goyal, N., Rastogi, D., Jassal, M., Agrawal, A.K., 2016. Chitosan as a potential stabilizing agent for titania nanoparticle dispersions for preparation of multifunctional cotton fabric. Carbohydr. Polym. 154, 167–175.
Graf, B.L., Raskin, I., Cefalu, W.T., Ribnicky, D.M., 2010. Plant-derived therapeutics for the treatment of metabolic syndrome. Curr. Opin. Invest. Drugs (London, England: 2000) 11 (10), 1107–1115.
Grama, C.N., Suryanarayana, P., Patil, M.A., Raghu, G., Balakrishna, N., Kumar, M.N., et al., 2013. Efficacy of biodegradable curcumin nanoparticles in delaying cataract in diabetic rat model. PLoS ONE 8 (10), e78217.
Gupta, N.K., Dixit, V.K., 2011. Development and evaluation of vesicular system for curcumin delivery. Arch. Dermatol. Res. 303 (2), 89–101.

Haque, F., Sajid, M., Cameotra, S.S., Battacharyya, M.S., 2017. Anti-biofilm activity of a sophorolipid–amphotericin B niosomal formulation against *Candida albicans*. Biofouling 33 (9), 768–779.
Hsieh, D.-S., Wang, H., Tan, S.-W., Huang, Y.-H., Tsai, C.-Y., Yeh, M.-K., et al., 2011. The treatment of bladder cancer in a mouse model by epigallocatechin-3-gallate-gold nanoparticles. Biomaterials 32 (30), 7633–7640.
Hu, B., Ting, Y., Zeng, X., Huang, Q., 2012. Cellular uptake and cytotoxicity of chitosan–caseinophosphopeptides nanocomplexes loaded with epigallocatechin gallate. Carbohydr. Polym. 89 (2), 362–370.
Hu, B., Ting, Y., Zeng, X., Huang, Q., 2013. Bioactive peptides/chitosan nanoparticles enhance cellular antioxidant activity of (−)-epigallocatechin-3-gallate. J. Agric. Food Chem. 61 (4), 875–881.
Hu, L., Xing, Q., Meng, J., Shang, C., 2010. Preparation and enhanced oral bioavailability of cryptotanshinone-loaded solid lipid nanoparticles. AAPS PharmSciTech 11 (2), 582–587.
Hu, X., Huang, F., Szymusiak, M., Tian, X., Liu, Y., Wang, Z.J., 2016. PLGA-curcumin attenuates opioid-induced hyperalgesia and inhibits spinal CaMKIIalpha. PLoS ONE 11 (1), e0146393.
Huang, H., Yang, X., 2004. Synthesis of chitosan-stabilized gold nanoparticles in the absence/presence of tripolyphosphate. Biomacromolecules 5 (6), 2340–2346.
Immordino, M.L., Dosio, F., Cattel, L., 2006. Stealth liposomes: review of the basic science, rationale, and clinical applications, existing and potential. Int. J. Nanomed. 1 (3), 297–315.
Jain, S., Jain, V., Mahajan, S.C., 2014. Lipid based vesicular drug delivery systems. Adv. Pharm. 2014, 12.
Jin, H., Pi, J., Zhao, Y., Jiang, J., Li, T., Zeng, X., et al., 2017. EGFR-targeting PLGA–PEG nanoparticles as a curcumin delivery system for breast cancer therapy. Nanoscale 9 (42), 16365–16374.
Jourghanian, P., Ghaffari, S., Ardjmand, M., Haghighat, S., Mohammadnejad, M., 2016. Sustained release curcumin loaded solid lipid nanoparticles. Adv. Pharm. Bull. 6 (1), 17–21.
Junyaprasert, V.B., Singhsa, P., Suksiriworapong, J., Chantasart, D., 2012. Physicochemical properties and skin permeation of Span 60/Tween 60 niosomes of ellagic acid. Int. J. Pharm. 423 (2), 303–311.
Kaewchoothong, A., Tewtrakul, S., Panichayupakaranant, P., 2012. Inhibitory effect of phenylbutanoid-rich *Zingiber cassumunar* extracts on nitric oxide production by murine macrophage-like RAW264.7 cells. Phytother. Res. 26 (12), 1789–1792.
Kamble, B., Talreja, S., Gupta, A., Patil, D., Pathak, D., Moothedath, I., et al., 2013. Development and biological evaluation of *Gymnema sylvestre* extract-loaded nonionic surfactant-based niosomes. Nanomedicine (Lond.) 8 (8), 1295–1305.
Kim, Y.J., Kim, D., Lee, Y., Choi, S.Y., Park, J., Lee, S.Y., et al., 2009. Effects of nanoparticulate saponin-platinum conjugates on 2,4-dinitrofluorobenzene-induced macrophage inflammatory protein-2 gene expression via reactive oxygen species production in RAW 264.7 cells. BMB Rep. 42 (5), 304–309.
Kumar, G.P., Rajeshwarrao, P., 2011. Nonionic surfactant vesicular systems for effective drug delivery—an overview. Acta Pharm. Sin. B 1 (4), 208–219.
Kumbhar, D., Wavikar, P., Vavia, P., 2013. Niosomal gel of lornoxicam for topical delivery: in vitro assessment and pharmacodynamic activity. AAPS PharmSciTech 14 (3), 1072–1082.
Kurzrock, R., Li, L., 2005. Liposome-encapsulated curcumin: in vitro and in vivo effects on proliferation, apoptosis, signaling, and angiogenesis. J. Clin. Oncol. 23 (16), 4091.

Landlinger, C., Preuner, S., Bašková, L., van Grotel, M., Hartwig, N.G., Dworzak, M., et al., 2010. Diagnosis of invasive fungal infections by a real-time panfungal PCR assay in immunocompromised pediatric patients. Leukemia 24, 2032.

Leelarungrayub, J., Manorsoi, J., Manorsoi, A., 2017. Anti-inflammatory activity of niosomes entrapped with plai oil (*Zingiber cassumunar* Roxb.) by therapeutic ultrasound in a rat model. Int. J. Nanomed. 12, 2469–2476.

Li, L., Braiteh, F.S., Kurzrock, R., 2005. Liposome-encapsulated curcumin: in vitro and in vivo effects on proliferation, apoptosis, signaling, and angiogenesis. Cancer 104 (6), 1322–1331.

Liang, R., Chen, L., Yokoyama, W., Williams, P.A., Zhong, F., 2016. Niosomes consisting of Tween-60 and cholesterol improve the chemical stability and antioxidant activity of (−)-epigallocatechin gallate under intestinal tract conditions. J. Agric. Food. Chem. 64 (48), 9180–9188.

Liolios, C.C., Gortzi, O., Lalas, S., Tsaknis, J., Chinou, I., 2009. Liposomal incorporation of carvacrol and thymol isolated from the essential oil of *Origanum dictamnus* L. and in vitro antimicrobial activity. Food Chem. 112 (1), 77–83.

Liu, Z., Zhu, Y.Y., Li, Z.Y., Ning, S.Q., 2016. Evaluation of the efficacy of paclitaxel with curcumin combination in ovarian cancer cells. Oncol. Lett. 12 (5), 3944–3948.

Lotfi-Attari, J., Pilehvar-Soltanahmadi, Y., Dadashpour, M., Alipour, S., Farajzadeh, R., Javidfar, S., et al., 2017. Co-delivery of curcumin and chrysin by polymeric nanoparticles inhibit synergistically growth and hTERT gene expression in human colorectal cancer cells. Nutr. Cancer 69 (8), 1290–1299.

Lu, L., Zhang, S., Li, C., Zhou, C., Li, D., Liu, P., et al., 2017. Cryptotanshinone inhibits human glioma cell proliferation in vitro and in vivo through SHP-2-dependent inhibition of STAT3 activation. Cell Death Dis. 8, e2767.

Mahmud, M., Piwoni, A., Filiczak, N., Janicka, M., Gubernator, J., 2016. Long-circulating curcumin-loaded liposome formulations with high incorporation efficiency, stability and anticancer activity towards pancreatic adenocarcinoma cell lines in vitro. PLoS ONE 11 (12), e0167787.

Manosroi, A., Jantrawut, P., Akihisa, T., Manosroi, W., Manosroi, J., 2011. In vitro and in vivo skin anti-aging evaluation of gel containing niosomes loaded with a semi-purified fraction containing gallic acid from *Terminalia chebula* galls. Pharm. Biol. 49 (11), 1190–1203.

Marchese, A., Barbieri, R., Coppo, E., Orhan, I.E., Daglia, M., Nabavi, S.F., et al., 2017. Antimicrobial activity of eugenol and essential oils containing eugenol: a mechanistic viewpoint. Crit. Rev. Microbiol. 43 (6), 668–689.

Masloub, S.M., Elmalahy, M.H., Sabry, D., Mohamed, W.S., Ahmed, S.H., 2016. Comparative evaluation of PLGA nanoparticle delivery system for 5-fluorouracil and curcumin on squamous cell carcinoma. Arch. Oral Biol. 64, 1–10.

Mathew, A., Fukuda, T., Nagaoka, Y., Hasumura, T., Morimoto, H., Yoshida, Y., et al., 2012. Curcumin loaded-PLGA nanoparticles conjugated with Tet-1 peptide for potential use in Alzheimer's disease. PLoS One 7 (3), e32616.

Matouskova, P., Marova, I., Bokrova, J., Benesova, P., 2016. Effect of encapsulation on antimicrobial activity of herbal extracts with lysozyme. Food Technol. Biotechnol. 54 (3), 304–316.

Mazzarino, L., Loch-Neckel, G., Bubniak Ldos, S., Mazzucco, S., Santos-Silva, M.C., Borsali, R., et al., 2015. Curcumin-loaded chitosan-coated nanoparticles as a new approach for the local treatment of oral cavity cancer. J. Nanosci. Nanotechnol. 15 (1), 781–791.

Mirnejad, R., Mofazzal Jahromi, M.A., Al-Musawi, S., Pirestani, M., Fasihi Ramandi, M., Ahmadi, K., et al., 2014. Curcumin-loaded chitosan tripolyphosphate nanoparticles as a safe, natural and effective antibiotic inhibits the infection of *Staphylococcus aureus* and *Pseudomonas aeruginosa* in vivo. Iran. J. Biotechnol. 12 (3), 1–8.

Misra, R., Sahoo, S.K., 2011. Coformulation of doxorubicin and curcumin in poly(D,L-lactide-*co*-glycolide) nanoparticles suppresses the development of multidrug resistance in K562 cells. Mol. Pharm. 8 (3), 852–866.

Mo, Y., Guo, R., Zhang, Y., Xue, W., Cheng, B., Zhang, Y., 2017. Controlled dual delivery of angiogenin and curcumin by electrospun nanofibers for skin regeneration. Tissue Eng. Part A 23 (13–14), 597–608.

Moghimipour, E., Aghel, N., Zarei Mahmoudabadi, A., Ramezani, Z., Handali, S., 2012. Preparation and characterization of liposomes containing essential oil of *Eucalyptus camaldulensis* Leaf. Jundishapur J. Nat. Pharm. Prod. 7 (3), 117–122.

Mukerjee, A., Vishwanatha, J.K., 2009. Formulation, characterization and evaluation of curcumin-loaded PLGA nanospheres for cancer therapy. Anticancer Res. 29 (10), 3867–3875.

Mullertz, A., Ogbonna, A., Ren, S., Rades, T., 2010. New perspectives on lipid and surfactant based drug delivery systems for oral delivery of poorly soluble drugs. J. Pharm. Pharmacol. 62 (11), 1622–1636.

Muzzalupo, R., Tavano, L., Cassano, R., Trombino, S., Ferrarelli, T., Picci, N., 2011. A new approach for the evaluation of niosomes as effective transdermal drug delivery systems. Eur. J. Pharm. Biopharm. 79 (1), 28–35.

Naderinezhad, S., Amoabediny, G., Haghiralsadat, F., 2017. Co-delivery of hydrophilic and hydrophobic anticancer drugs using biocompatible pH-sensitive lipid-based nanocarriers for multidrug-resistant cancers. RSC Adv. 7 (48), 30008–30019.

Nankar, R., Prabhakar, P.K., Doble, M., 2017. Hybrid drug combination: combination of ferulic acid and metformin as anti-diabetic therapy. Phytomedicine 15 (37), 10–13.

Narayanan, N.K., Nargi, D., Randolph, C., Narayanan, B.A., 2009. Liposome encapsulation of curcumin and resveratrol in combination reduces prostate cancer incidence in PTEN knockout mice. Int. J. Cancer 125 (1), 1–8.

Narayanan, S., Binulal, N.S., Mony, U., Manzoor, K., Nair, S., Menon, D., 2010. Folate targeted polymeric 'green' nanotherapy for cancer. Nanotechnology 21 (28), 285107.

Muddineti, O.S., Kumari, P., Ray, E., Ghosh, B., Biswas, S., 2017. Curcumin-loaded chitosan–cholesterol micelles: evaluation in monolayers and 3D cancer spheroid model. Nanomedicine 12 (12), 1435–1453.

Orunoglu, M., Kaffashi, A., Pehlivan, S.B., Sahin, S., Soylemezoglu, F., Oguz, K.K., et al., 2017. Effects of curcumin-loaded PLGA nanoparticles on the RG2 rat glioma model. Mater. Sci. Eng., C: Mater. Biol. Appl. 78, 32–38.

Peng, S., Zou, L., Liu, W., Gan, L., Liu, W., Liang, R., et al., 2015. Storage stability and antibacterial activity of eugenol nanoliposomes prepared by an ethanol injection-dynamic high-pressure microfluidization method. J. Food Prot. 78 (1), 22–30.

Pieretti, S., Ranjan, A.P., Di Giannuario, A., Mukerjee, A., Marzoli, F., Di Giovannandrea, R., et al., 2017. Curcumin-loaded poly(D,L-lactide-*co*-glycolide) nanovesicles induce antinociceptive effects and reduce pronociceptive cytokine and BDNF release in spinal cord after acute administration in mice. Colloids Surf., B: Biointerfaces 158, 379–386.

Priprem, A., Janpim, K., Nualkaew, S., Mahakunakorn, P., 2016. Topical niosome gel of *Zingiber cassumunar* Roxb. Extract for anti-inflammatory activity enhanced skin permeation and stability of compound D. AAPS PharmSciTech 17 (3), 631–639.

Ramasamy, K., Agarwal, R., 2008. Multitargeted therapy of cancer by silymarin. Cancer Lett. 269 (2), 352–362.

Ranjan, A.P., Mukerjee, A., Helson, L., Gupta, R., Vishwanatha, J.K., 2013. Efficacy of liposomal curcumin in a human pancreatic tumor xenograft model: inhibition of tumor growth and angiogenesis. Anticancer Res. 33 (9), 3603–3609.

Rejinold, N.S., Muthunarayanan, M., Muthuchelian, K., Chennazhi, K.P., Nair, S.V., Jayakumar, R., 2011. Saponin-loaded chitosan nanoparticles and their cytotoxicity to cancer cell lines in vitro. Carbohydr. Polym. 84 (1), 407–416.

Rungphanichkul, N., Nimmannit, U., Muangsiri, W., Rojsitthisak, P., 2011. Preparation of curcuminoid niosomes for enhancement of skin permeation. Pharmazie 66 (8), 570–575.

Sadeghzadeh, H., Pilehvar-Soltanahmadi, Y., Akbarzadeh, A., Dariushnejad, H., Sanjarian, F., Zarghami, N., 2017. The effects of nanoencapsulated curcumin–Fe_3O_4 on proliferation and hTERT gene expression in lung cancer cells. Anticancer Agents Med. Chem. 17 (10), 1363–1373.

Saha, S., Verma, R.J., 2016. Antioxidant activity of polyphenolic extract of *Terminalia chebula* Retzius fruits. J. Taibah Univ. Sci. 10 (6), 805–812.

Sainakham, M., Manosroi, A., Abe, M., Manosroi, W., Manosroi, J., 2016. Potent in vivo anticancer activity and stability of liposomes encapsulated with semi-purified Job's tear (*Coix lacryma-jobi* Linn.) extracts on human colon adenocarcinoma (HT-29) xenografted mice. Drug Delivery 23 (9), 3399–3407.

Sangsen, Y., Wiwattanawongsa, K., Likhitwitayawuid, K., Sritularak, B., Wiwattanapatapee, R., 2015. Modification of oral absorption of oxyresveratrol using lipid based nanoparticles. Colloids Surf., B: Biointerfaces 131, 182–190.

Sanna, V., Roggio, A.M., Siliani, S., Piccinini, M., Marceddu, S., Mariani, A., et al., 2012. Development of novel cationic chitosan- and anionic alginate-coated poly(D,L-lactide-*co*-glycolide) nanoparticles for controlled release and light protection of resveratrol. Int. J. Nanomed. 7, 5501–5516.

Sebaaly, C., Jraij, A., Fessi, H., Charcosset, C., Greige-Gerges, H., 2015. Preparation and characterization of clove essential oil-loaded liposomes. Food Chem. 178 (Suppl. C), 52–62.

Sebaaly, C., Charcosset, C., Stainmesse, S., Fessi, H., Greige-Gerges, H., 2016. Clove essential oil-in-cyclodextrin-in-liposomes in the aqueous and lyophilized states: from laboratory to large scale using a membrane contactor. Carbohydr. Polym. 138, 75–85.

Shahiwala, A., Misra, A., 2002. Studies in topical application of niosomally entrapped Nimesulide. J. Pharm. Pharm. Sci. 5 (3), 220–225.

Sharma, A., Mayhew, E., Bolcsak, L., Cavanaugh, C., Harmon, P., Janoff, A., et al., 1997. Activity of paclitaxel liposome formulations against human ovarian tumor xenografts. Int. J. Cancer 71 (1), 103–107.

Sharma, P.K., Saxena, P., Jaswanth, A., Chalamaiah, M., Tekade, K.R., Balasubramaniam, A., 2016. Novel encapsulation of lycopene in niosomes and assessment of its anticancer activity. J. Bioequivalence Bioavailability 8, 224–232.

Sharma, V., Anandhakumar, S., Sasidharan, M., 2015. Self-degrading niosomes for encapsulation of hydrophilic and hydrophobic drugs: an efficient carrier for cancer multi-drug delivery. Mater. Sci. Eng., C: Mater. Biol. Appl. 56, 393–400.

Shi, F., Zhao, J.H., Liu, Y., Wang, Z., Zhang, Y.T., Feng, N.P., 2012. Preparation and characterization of solid lipid nanoparticles loaded with frankincense and myrrh oil. Int. J. Nanomed. 7, 2033–2043.

Singh, P.K., Rawat, P., 2017. Evolving herbal formulations in management of dengue fever. J. Ayurveda Integr. Med. 8 (3), 207–210.

Snima, K.S., Arunkumar, P., Jayakumar, R., Lakshmanan, V.K., 2014. Silymarin encapsulated poly(D,L-lactic-*co*-glycolic acid) nanoparticles: a prospective candidate for prostate cancer therapy. J. Biomed. Nanotechnol. 10 (4), 559–570.

Song, Q., Li, D., Zhou, Y., Yang, J., Yang, W., Zhou, G., et al., 2014. Enhanced uptake and transport of (+)-catechin and (−)-epigallocatechin gallate in niosomal formulation by human intestinal Caco-2 cells. Int. J. Nanomed. 9, 2157–2165.

Soundran, V., Namagiri, T., Manonayaki, S., Vanithakumari, G., 1994. hepatotoxicity of eugenol. Ancient Sci. Life 13 (3–4), 213–217.
Souza, M.P., Vaz, A.F.M., Correia, M.T.S., Cerqueira, M.A., Vicente, A.A., Carneiro-da-Cunha, M.G., 2014. Quercetin-loaded lecithin/chitosan nanoparticles for functional food applications. Food Bioprocess Technol. 7 (4), 1149–1159.
Spencer, C.M., Faulds, D., 1994. Paclitaxel. A review of its pharmacodynamic and pharmacokinetic properties and therapeutic potential in the treatment of cancer. Drugs 48 (5), 794–847.
Sun, L., Liu, Z., Wang, L., Cun, D., Tong, H.H.Y., Yan, R., et al., 2017. Enhanced topical penetration, system exposure and anti-psoriasis activity of two particle-sized, curcumin-loaded PLGA nanoparticles in hydrogel. J. Control. Release 254, 44–54.
Sutthanut, K., Lu, X., Jay, M., Sripanidkulchai, B., 2009. Solid lipid nanoparticles for topical administration of *Kaempferia parviflora* extracts. J. Biomed. Nanotechnol. 5 (2), 224–232.
Tan, Q., Liu, W., Guo, C., Zhai, G., 2011. Preparation and evaluation of quercetin-loaded lecithin–chitosan nanoparticles for topical delivery. Int. J. Nanomed. 6, 1621–1630.
Tanihara, M., Suzuki, Y., Nishimura, Y., Suzuki, K., Kakimaru, Y., Fukunishi, Y., 1999. A novel microbial infection-responsive drug release system. J. Pharm. Sci. 88 (5), 510–514.
Triantafyllidi, A., Xanthos, T., Papalois, A., Triantafillidis, J.K., 2015. Herbal and plant therapy in patients with inflammatory bowel disease. Ann. Gastroenterol. 28 (2), 210–220.
Vemuri, S., Rhodes, C.T., 1995. Preparation and characterization of liposomes as therapeutic delivery systems: a review. Pharm. Acta Helv. 70 (2), 95–111.
Wang, Y., Ying, X., Xu, H., Yan, H., Li, X., Tang, H., 2017. The functional curcumin liposomes induce apoptosis in C6 glioblastoma cells and C6 glioblastoma stem cells in vitro and in animals. Int. J. Nanomed. 12, 1369–1384.
Wicki, A., Witzigmann, D., Balasubramanian, V., Huwyler, J., 2015. Nanomedicine in cancer therapy: challenges, opportunities, and clinical applications. J. Control. Release 200 (Suppl. C), 138–157.
Wolfram, J., Suri, K., Huang, Y., Molinaro, R., Borsoi, C., Scott, B., et al., 2014. Evaluation of anticancer activity of celastrol liposomes in prostate cancer cells. J. Microencapsulation 31 (5), 501–507.
Xiao, B., Si, X., Zhang, M., Merlin, D., 2015. Oral administration of pH-sensitive curcumin-loaded microparticles for ulcerative colitis therapy. Colloids Surf., B: Biointerfaces 135, 379–385.
Xie, Y., Yi, Y., Hu, X., Shangguan, M., Wang, L., Lu, Y., et al., 2016. Synchronous microencapsulation of multiple components in silymarin into PLGA nanoparticles by an emulsification/solvent evaporation method. Pharm. Dev. Technol. 21 (6), 672–679.
Xiong, F., Wang, H., Chen, Y.J., Geng, K.K., Gu, N., Zhu, J.B., 2011. Characterization, biodistribution and targeting evaluation of breviscapine lipid emulsions following intravenous injection in mice. Drug Delivery 18 (2), 159–165.
Xu, Y.-Q., Chen, W.-R., Tsosie, J.K., Xie, X., Li, P., Wan, J.-B., et al., 2016. Niosome encapsulation of curcumin: characterization and cytotoxic effect on ovarian cancer cells. J. Nanomater. 2016, 9.
Yallapu, M.M., Gupta, B.K., Jaggi, M., Chauhan, S.C., 2010. Fabrication of curcumin encapsulated PLGA nanoparticles for improved therapeutic effects in metastatic cancer cells. J. Colloid Interface Sci. 351 (1), 19–29.

Yan, J., Wang, Y., Jia, Y., Liu, S., Tian, C., Pan, W., et al., 2017. Co-delivery of docetaxel and curcumin prodrug via dual-targeted nanoparticles with synergistic antitumor activity against prostate cancer. Biomed. Pharmacother. 88, 374–383.

Yang, T., Cui, F.D., Choi, M.K., Lin, H., Chung, S.J., Shim, C.K., et al., 2007. Liposome formulation of paclitaxel with enhanced solubility and stability. Drug Delivery 14 (5), 301–308.

Yang, T., Kim, D.-d, Cui, F.-d, 2008. Determination of content and entrapment efficiency of paclitaxel loaded liposome. Chin. J. Pharm. Anal. 28 (2), 231–234.

Yeh, M.I., Huang, H.C., Liaw, J.H., Huang, M.C., Huang, K.F., Hsu, F.L., 2013. Dermal delivery by niosomes of black tea extract as a sunscreen agent. Int. J. Dermatol. 52 (2), 239–245.

Zamarioli, C.M., Martins, R.M., Carvalho, E.C., Freitas, L.A.P., 2015. Nanoparticles containing curcuminoids (*Curcuma longa*): development of topical delivery formulation. Rev. Bras. Farmacogn. 25, 53–60.

Zhang, L., Granick, S., 2006. How to stabilize phospholipid liposomes (using nanoparticles). Nano Lett. 6 (4), 694–698.

Zhang, Y., Yang, Y., Tang, K., Hu, X., Zou, G., 2008. Physicochemical characterization and antioxidant activity of quercetin-loaded chitosan nanoparticles. J. Appl. Polym. Sci. 107 (2), 891–897.

Zhang, Z.Y., Jiang, M., Fang, J., Yang, M.F., Zhang, S., Yin, Y.X., et al., 2017. Enhanced therapeutic potential of nano-curcumin against subarachnoid hemorrhage-induced blood–brain barrier disruption through inhibition of inflammatory response and oxidative stress. Mol. Neurobiol. 54 (1), 1–14.

Zhao, L., Ye, Y., Li, J., Wei, Y.M., 2011. Preparation and the in-vivo evaluation of paclitaxel liposomes for lung targeting delivery in dogs. J. Pharm. Pharmacol. 63 (1), 80–86.

Zhou, L., Duan, X., Zeng, S., Men, K., Zhang, X., Yang, L., et al., 2015. Codelivery of SH-aspirin and curcumin by mPEG–PLGA nanoparticles enhanced antitumor activity by inducing mitochondrial apoptosis. Int. J. Nanomed. 10, 5205–5218.

Zorofchian Moghadamtousi, S., Abdul Kadir, H., Hassandarvish, P., Tajik, H., Abubakar, S., Zandi, K., 2014. A review on antibacterial, antiviral, and antifungal activity of curcumin. BioMed Res. Int. 2014, 12.

CHAPTER 24

Phytomedicine: A Potential Alternative Medicine in Controlling Neurological Disorders

A. Srivastava, P. Srivastava, A. Pandey, V.K. Khanna and A.B. Pant
System Toxicology and Health Risk Assessment Group, CSIR-Indian Institute of Toxicology Research (CSIR-IITR), Lucknow, Uttar Pradesh, India

24.1 INTRODUCTION AND HISTORICAL BACKGROUND

Phytomedicine can be defined as the herbal medicine with therapeutic and healing properties. It came into existence since the advent of human civilization. Sheng Nongs Herbal Book is known as one of the preliminary sources of traditional folk knowledge based on the use of herbs in China and dates back to around 3000 BC. It encompasses the details of almost 365 plants, animals, and minerals that find a place in medication. Our Earth houses approximately 420,000 species of plants; however, there is a lack of appropriate knowledge about them and their varied uses. There are three major areas, namely, food (foodstuffs), medicine (folk and traditional medicines), and research (phytochemical analysis), that predominantly find an immense use of herbal preparations and products and hence can be explored further. Gaining experience from random trials and careful observations from animal studies, people belonging to ancient periods started employing herbs as a therapeutic method against several illnesses. Based on this, the ever so popular Chinese herbal medicine (CHM) as well as Indian herbal medicine, native to and prominently developed in ancient China, Japan, Korea, and India, continue to rule and influence the modern health-care even today. As per the estimate of World Health Organization (WHO), herbal medicines are one of the most sought after primary health-care for around 3.5–4 billion people across the world, and a major portion of traditional medicine involves the plant extract–derived medicines and decoction which may also be termed as the "modern herbal medicine" (Pan et al., 2013).

New Look to Phytomedicine
DOI: https://doi.org/10.1016/B978-0-12-814619-4.00025-2

A herbal medicine or a phytopharmaceutical preparation can be defined as a medicine derived exclusively from a whole plant or parts of plants and manufactured in a crude form or as a purified pharmaceutical formulation. Although with the setting in of the industrial revolution and the advancements in organic chemistry, there was an equivalent increase in the preference for synthetic products as well. However, the WHO emphasizes that between nearly 70% and 95% of the population residing in numerous developing countries still rely more on traditional herbal medicines for their primary medication against diseases (Mohamed et al., 2012). Over the last decade, there has been an enormous rise in the products derived from medicinal plants in terms of interest and use. Earlier what used to be the exclusive domain of health food and specialty stores has now gained considerable popularity into the mainstream market as evident by their impressive sale at some of the major retail outlets, their publicity, and advertising in the media, and due to the entry of various pharmaceutical giants in the area of phytomedicines (Briskin, 2000).

Phytomedicine, in amalgamation with various other health-care fields, has indeed revolutionized and strengthened the foundation of the existing health-care system and occupies a major stake in the industry. Reports gathered from all over the world indicate there are around 35,000 species of plants that are currently being used in herbal therapies/recipes. Although according to research data available only 20% of the total undergoes the stage of phytochemical analysis while 10% reach the biological screening stage. The remaining still need some amount of exploration making use of modern technologies. The future of medicinal plant–derived drugs therefore seems to have tremendous scope for discovering some new and novel therapeutic strategies and products (Khan, 2015).

Our planet is being ruled by plants for over 400 million years now that have successfully survived the test of time even after being challenged time and again by herbivores and microbes. Their defense mechanism is attributed to a range of structurally different secondary metabolites that evolved with time at various stages of development and provide protection against attacks by herbivores, bacteria, fungi, and viruses. Some of these metabolites act as signal compounds that can potentially attract animals that pollinate and disperse the seeds. They additionally act as antioxidants and UV protectants. As far as evolutionary pharmacology is concerned, the secondary metabolites of plant comprise an important collection of bioactive compounds selected and propagated naturally to be used as a remedy against various human infections and health disorders (Wink, 2015).

24.2 PREVALENCE OF HERBAL MEDICINES FOR THERAPY

Pertaining to its safe nature that does not involve any major side effect, approximately 80% of the total world population uses the plant-derived medicine as its first line of primary health-care. These herbal medicines are present in various regulatory formats and models in the form of prescription drugs, traditional medicines, over-the-counter substances, and dietary supplements. However, there is a requirement of harmonization and upgradation in the regulatory processes that is purely a combination of scientific interpretations and the available traditional knowledge. Eventually, the advancements in terms of domestication of wild plants, advent of production biotechnological studies, and genetic enhancement of medicinal herbs, rather than employing the plants growing in wild environment, promise to further offer great advantages. This would additionally ensure superior quality raw material in consistent batches that governs the ultimate efficacy, quality, and safety of the drugs manufactured from them (Pandey et al., 2011). This compelled the regulatory agencies to stringently standardize these herbal medicines so as to evaluate their important safety and efficacy parameters against the set standards and benchmarks. This further does away with the substandard product quality and conquers one of the major challenges in the formulation of phytomedicines, that is, a foolproof safe formulation for human consumption. Phytomedicines must essentially comprise of active plant ingredients or their derivatives. The external addition of any kind of non-plant substance from any source whatsoever takes away the essence of a herbal product (Barbosa et al., 2012).

There has been a recent up swirl among people using herbal remedies against health disorders owing to several important factors. These include the following:

- Their reported efficacy or effectiveness,
- Increasing inclination of the consumers toward natural therapies and a source of alternative medicines,
- The belief that herbal products outdo the chemically manufactured products,
- Poor experience and side effects associated with orthodox pharmaceuticals and conventional therapy against diseases topped up with their

belief that herbal medicines might more effectively treat certain diseases as compared to conventional therapies and medicines,
- The pocket friendly prices of herbal drugs unlike the modern ones,
- Improved quality, safety, and efficacy of herbal medicines with the progress of science and technology,
- If the patients are dissatisfied with the diagnosis and treatment of their physicians and feel that herbal remedies are a much better option, and
- A need for self-medication (Ekor, 2014).

24.3 CANCER

Use of multitargeted phytoconstituents obtained from traditional medicine for chemoprevention of cancer, and its therapy is reportedly more effective than using synthetic agents that target a single molecule (Gullett et al., 2010; Bansal et al., 2013). Evidence shows that phytomedicines or herbal drugs tend to act on multiple therapeutic targets within the multiple distorted signaling cascades and events in cancer (Lin et al., 2015; Tsai et al., 2015). Although previous studies focused on the natural product—mediated direct cytotoxicity, current strategies include multitarget compound identification that can modulate common pathways occurring in cancer or stromal cells. The pharmacological action of curcumin, resveratrol, and lycopene has been consistently highlighted against cancer mechanisms in certain types (Apaya et al., 2016). As cancer is a socioeconomic burden on the patients, several alternatives have been suggested to reduce the cost of cancer treatment and chemotherapy along with improved outcomes. Medicinal herbs and their derivatives are being recognized as vital complementary remedies for cancer. A chunk of clinical reports and studies has reported the superiority of herbal medicines over the immune modulation, survival, and quality of life of cancer patients, on administering these herbal medicines in combination with conventional drugs (Yin et al., 2013). Inclusion of herbal drugs and phytomedicine in the conventional therapy is a much proposed idea now. Phytomedicine involves usage of herb-based traditional medicinal practice employing different plant materials that are both preventive and therapeutic. However, clinical studies estimating the efficacy of these natural drugs in cancer patients sometimes show contradictory results. Their efficacy and safety in cancer therapy is being studied across the world. Phytomedicine when combined with conventional treatment has shown potential positive influence by decreasing the adverse effects and mortality associated with disease and chemotherapy (Chaudhary et al., 2015).

24.4 SICKLE CELL ANEMIA

Primary clinical control of sickle cell anemia utilizes hydroxyurea, folic acid, amino acids supplementation, antimalarial prophylaxis, and penicillin prophylaxis for the management of blood transfusions so as to keep the patient's hemoglobin levels in check. These are expensive with attached risk factors. However, advent of medicinal plants with antisickling properties has been rewarding in the field. This alternative therapy proved to be safer without any reverse sickling (in vitro). Various herbal applications have been identified with ameliorating effects on the disease. The antisickling properties of dried leaves of *Carica papaya* and *Fagara zanthoxyloides* roots were observed in studies to estimate the antioxidant properties of these plant extracts on homozygous sickle cell (SS) erythrocytes under in vitro setup. Results proved that the plants possessed antisickling activity (Imaga, 2013). The Phase IIB (pivotal) trial reported that Niprisan effectively reduced recurring painful sickle cell disease (SCD) crisis when administered for over 6 months without any severe side effects. While this drug appeared to be effective and safe and reduced severe painful crises, further trials are necessary to correctly estimate its role in managing SCD (Oniyangi and Cohall, 2013)."Hémodya," a phytomedicine that plays a role in managing SCD, reduces the concentration of cholesterol in membrane, due to its antioxidant activity. This may further ameliorate the structural and functional integrity of the sickle red cells. Plant extracts possess properties which inhibit the erythrocytes from getting deformed and losing their integrity. Among the important phytochemicals and natural drugs are divanilloylquinic acids and 2-dihyroxymethyl benzoic acid that have been isolated from Fagara. The potential of fermented extracts of garlic and thiocyanate for the treatment of SCD has also been successfully demonstrated (Sahu et al., 2012).

24.5 *HELICOBACTER PYLORI* INFECTIONS

Helicobacter pylori (*H. pylori*) colonizes the human gut in majority of human population. This infection leads to chronic gastritis but may also grow into serious outcomes like peptic ulcer, mucosa-associated lymphoid tissue lymphoma, and gastric carcinoma. Efforts are being put into the investigation of remedial measures and therapeutic alternatives apart from antibiotics. These include probiotics, vaccines, phage therapy, and photodynamic inactivation. The herbs investigated are highly diverse and derived from products with historical background of their use against diseases of *H. pylori* infection. As per several studies, many phytomedicines act against *H. pylori* activity and protect gastrointestinal (GI) tract. Although the mode of action remains elusive, available knowledge finds a correlation between their beneficial action and inhibition of key *H. pylori* enzymes, host immune system modulation, and attenuation of inflammatory responses (Vale and Oleastro, 2014). Recently, numerous studies suggested the complementary function of phytomedicine in *H. pylori* infection that can be reduced and treated with inexpensive, safer, and nontoxic formulations obtained from medicinal herbs. Many plant extracts with anti-*H. pylori* activity have been identified. These include polyphenolic catechins, tannins, carvacrol, cinnamaldehyde, licoisoflavone B, quercetin, berberine, etc. Their reducing power on antibiotic resistance coupled with their antimutagenic properties in *H. pylori* infections was evaluated. The results obtained denoted the significant efficacy of *Teucrium polium* and *Myrtus communis* extracts in prohibition of antibiotic resistance. This treatment may be even more advantageous when clubbed with existing antibiotic regimens for developing more effective protective regimens (Safavi et al., 2016). However, the study of naturally derived medicines against *H. pylori* infection is still in its infancy. Some researchers also reported susceptibility of *H. pylori* to garlic extract in vitro. A high molecular weight component of cranberry juice also inhibits adhesion of *H. pylori* to human gastric mucus, indicating that combining antibiotics and cranberry preparation together might improve *H. pylori* infections (Burger et al., 2000; Shmuely et al., 2004). This may likely occur through inhibition of urease and disrupted energy production at the plasma membrane. This step is the most potential target for eradicating *H. pylori* (Kosikowska and Berlicki, 2011). Ethanolic extracts derived from *Magnolia officinalis Rehd. et Wils.* (Magnoliaceae) and *Cassia obtusifolia* L. (Leguminosae) also inhibit urease (Vítor and Vale, 2011).

24.6 CHRONIC LIVER DISEASE

Chronic liver dysfunctions affect a major section of population worldwide. They involve varied liver pathologies like fatty liver, fibrosis, hepatitis, hepatocellular carcinoma, and cirrhosis. The currently available synthetic drugs have not been found to be very effective agents in treating chronic liver disease and involve unwanted side effects. As a result, phytochemicals and medicinal herbs are being investigated as a means of complementary and alternative therapy for chronic liver diseases. As per earlier studies, medicinal herbs and phytochemicals protect the liver by eliminating virus, inhibiting oxidative injury, suppressing tumorigenesis, and blocking fibrogenesis (Hong et al., 2015).

24.7 MODE OF ACTION

Flavonoids comprise of a native three-ring structure possessing some substitutions and are a low molecular weight group. They are characterized by anti-inflammatory, anti-allergic, antioxidant, antiviral, hepatoprotective, anticarcinogenic, and antithrombotic properties. Falling in the phenolic group of compounds the flavonoids act as strong metal chelators and scavenge free radicals too. They supposedly act by inhibiting the vital enzymes of bacteria (Vale and Oleastro, 2014).

Plants and animals share some of the common basic metabolic pathways. The secondary metabolites present in plants that do not essentially form a part of their primary metabolism render the phytochemicals therapeutically effective. Phytochemicals are diverse in terms of their chemical properties and structure due to the varying secondary plant metabolites in them. There are numerous groups of secondary metabolites that have varied properties. For instance, flavonoids that are antioxidant in nature, act as a protective shield against infection and traumas, incurred due to wounds or insect invasion. The phytochemicals reportedly act through synergy that is one of the key features in the pharmacology of herbal medicines. This means that all the chemical entities present in the herbal formulations act collectively to provide therapeutic relief as compared to one single component bearing the entire load on its own (Larkins and Wynn, 2004).

24.8 ADVANTAGES OF PHYTOMEDICINES OVER CHEMICAL/SYNTHETIC DRUGS

Recently, phytomedicines have gained immense popularity in the pharmaceutical market due to many reasons and factors. The plant extracts are superior in terms of therapeutic efficacy as compared to chemically derived drugs due to the combined action of all its therapeutic constituents in comparison to single isolated components. Also the bio-equivalence of phytomedicines is in sync with synthetic chemotherapeutics. The idea of multidrug and multitarget action has emerged as more promising than the monodrug therapy in conventional methods and that has fairly tilted the balance toward phytotherapeutics. Many health disorders like cancer, diabetes, AIDS, hypertension, malaria, etc. are being treated based on the multidrug therapy, thus shifting the direction from the orthodox conventional monodrug methods. This is based on the fact that several studies and their observations point toward targeting of multiple diseases and their etiologies by this approach being more efficient than the concept of single drug component treating the disease. Herbal therapy is also considered as the more pocket friendly and easily accessible therapeutic approach without any adverse effect associated with them under normal conditions in comparison to the chemically derived medicines (Obodozie, 2012).

Synthetic drugs are known to provide symptomatic relief in most of the disease cases, as observed during various research works. However, the herbal medicines try to improve body's own healing mechanism. Herbal medicines are gentle in action and they try to reestablish the damaged or deficient systems and processes to get rid of the abnormality present in the system. A pharmaceutical drug is developed with the aim to evoke a certain reaction against certain physiological anomaly, and the side effects associated with it are usually considered as a trade-off for the benefits bestowed by these medicines over human health. Herbal medicines, on the other hand, boast of synergistic action mechanism through which they offer therapeutic benefits with hardly any side effect (Karimi et al., 2015).

The affordable nature, popularity among the population, and the chemical free composition of herbal medicines without any side effects has impressed the people all across the globe. Its scope of giving rise to personalized medicine and addressing the issue of side effect associated with chemical drugs has further inclined people towards it. It has been employed to treat diseases other than the life threatening ones and promotion of a sound health. However, its use multiplies further in cases where the conventional drugs fail to impact significantly like in advanced stages of cancer or some unknown infections. Furthermore, herbal drugs are considered far more safe, natural, and nontoxic (Benzie and Wachtel-Galor, 2011).

24.9 NOOTROPICS

Nootropics are also known as smart drugs that are being developed for over three decades and are the predominantly used method for treating cognitive deficits. It has been derived from two words, that is, "noos," pertaining "to mind" and "tropein," signifying "to monitor." In general, it means any given substance that influences the cognitive ability in a positive way (Colucci et al., 2012). They probably act by altering the levels of neurotransmitters, hormones, and enzymes that are available to the brain, through improvement of brain's oxygen supply or stimulation of nerve growth. However, the detailed description of their efficacy seems to be incomplete as yet. This is because of the absence of a scale to quantitatively measure cognition and intelligence. Herbs acting as memory herbs enhance the level of neurotransmitters like acetylcholine and also increase blood flow directed towards the brain, thereby nurturing it with increased supply of oxygen and nutrients, which further refines brain function and memory (Amin and Sharma, 2015).

Nootropics can either be synthetic which are produced in a laboratory like piracetam or can occur naturally as herbal plants like *Ginkgo biloba* and *Panax quinquefolius* (American Ginseng). Natural nootropics aid in promoting the brain function with a simultaneous improvement in brain health. They also act as vasodilators against the small arteries and veins in the brain. When introduced into the system, they tend to increase the blood circulation towards the brain with an upsurge in the vital nutrients, energy, and oxygen flow in brain. They also mitigate the inflammatory responses in the brain. Natural nootropics modulate neurotransmitter concentration in the brain. They reportedly stimulate the release of various neurotransmitters like dopamine as well as uptake of choline, along with cholinergic transmission, turnover of phosphatidylinositol, function of α-amino-3-hydroxy-5-methyl-4-isoxazole propionate receptor, and activity of phosphatase A2. Some of them positively regulate the activity and expression of receptors for acetylcholine or glutamate. These particular characteristics of natural nootropics lead to the long-term potential and enhanced synaptic transmission due to improved levels and activity of neurotransmitters (Suliman et al., 2016).

A number of neurodegenerative or neuropsychiatric diseases can also be treated with certain other potential nootropics. These disorders also include cognitive dysfunctions due to aging. These are the FDA-approved inhibitors of acetylcholinesterase like donepezil, while some are under investigation, like ampakines, nicotinic receptor agonists, glutamate receptor agonists, glycine inhibitors, and phosphodiesterase (PDE) inhibitors (Pieramico et al., 2014). These drugs are more precise in modulating targets (neurotransmitters) like histamine, serotonin, glucocorticoid, neuropeptide receptors, and epigenetic mechanisms. Many laboratory-produced drugs like piracetam and piracetam-like compounds are known to be well tolerated, but their exact effects and mode of action remain somewhat elusive (Gouliaev and Senning, 1994; Gualtieri et al., 2002). Only a few nootropics have made it to clinical trial, while others still remain in the pipeline. They come with their share of drawbacks and loopholes too (Jellen et al., 2015).

24.10 ROLE OF PHYTOMEDICINES IN NEUROPROTECTION

Ayurveda is an age old Indian practice of traditional medicine that involves an extensive use of herbs and herbal preparations that is known to treat various neuropsychiatric disorders. Not just in India, these herbs are being used since decades in various other parts of the world too in folk and traditional medicine for relieving the mind from any kind of stress and anxiety and positively influence the mood. Herbal medicine is continuously rising in its use as alternative therapy against several diseases in both developing as well as developed nations of the world (Rao et al., 2011). Using herbal medicine to treat neuronal anomalies is being practiced since long. Although the actual action of these herbal drugs remains a bit elusive, some of them exhibit anti-inflammatory with/without antioxidant properties in some of the peripheral systems. Since it has been established scientifically that it's the chronic inflammatory responses derived from the neuroglia that are closely associated with the disease pathology in the central nervous system (CNS), herbal drugs and its constituents possessing anti-inflammatory activity are being investigated and proved as potent neuroprotectors in the case of various brain/nervous system disorders. Due to their structural diversity, medicinal herbs are considered as one of the valuable sources for discovering some novel lead compounds that can act against known therapeutic targets making use of various genomics and proteomics techniques followed by high-throughput screening (Kumar and Khanum, 2012).

In the wake of modern pharmacological therapy not coming across as an effective measure against neurodegenerative disorders, the focus is now on the traditional herbal drugs. As per the report generated by WHO, traditional medicine occupies 70%–80% of the total world population's belief as the suitable therapy and the primary health-care measure. These herbal medicines are reportedly safer and well tolerated as compared to the chemically derived medicines in treating chronic illnesses and involve fewer side effects. In its crude form, the traditional medicine is obtained as a standardized herbal extract, formulations derived from it, and also its composite preparations. Moreover, the specific constituents that render a herbal drug its therapeutic efficacy have been identified, isolated and some of them have even been synthesized (Vyawahare et al., 2008).

Currently existing therapeutic regimes options like interventional procedures, synthetic drug, and surgery have so far not been completely successful in improving the normal neural functions due to their inability to repair the neurons that have already been damaged or to regenerate them. They may, at times, lead to sedation and myorelaxation. Due to this lagging feature of conventional therapy in recovery of neural damage and the harmful side effects associated with administration of anti-amnesic drugs, there is an urgent need to look for alternative treatments in the form of herbal or plant-derived medicinal products and drugs to cure such hazardous disorders. The limited or total absence of associated side effects with the herbal formulations further provides them an added advantage. Flavonoids which are naturally anti-oxidative in nature as well as that possess anxiolytic properties are the primary choice to be employed as neuroprotective and anti-amnesic agents (Ferdousy et al., 2016).

Brain is one of the most vulnerable organs of the body as compared to other organs as far as oxidative stress is concerned since the antioxidant defense systems here are not very efficient and active (Rahman et al., 2007). Moreover, some of the neurotransmitters get autoxidized thereby leading to reactive oxygen species (ROS) generation. Oxidative stress being one of the key features of neurodegenerative diseases such as Parkinson's disease (PD), Alzheimer's disease (AD), ischemic diseases, and aging is a well-known fact that contributes majorly toward disease pathogenicity (Esposito et al., 2012). Many polyphenols bear a neuroprotective efficacy since they can easily cross the blood–brain barrier and eat away or scavenge the free radicals of oxygen and nitrogen present at pathological concentrations as well as act as the transition metal ion chelators (Aquilano et al., 2008). The polyphenolic compounds have been observed to possess scavenging activity and can also activate the key enzymes required for antioxidant mechanism in the brain, thus eliminating the ever rising oxidative stress in the diseased brain, thereby recovering the damaged tissue (Esposito et al., 2012). Berries that have the polyphenolic composition are being chosen as a means to improve learning, memory, and cognitive abilities (Chen et al., 2013; de Souza et al., 2014). Based on the findings of preclinical studies flavonoids reportedly manifest their protective actions on the disrupted cognitive abilities of mammals and might reverse the age-dependent damages in the memory and learning capacities. This makes them an attractive choice as a remedy against prevention of brain damage associated with disease conditions like ischemia and other such neurodegenerative diseases leading to a decline in neuronal apoptosis, thus improving learning, memory, and cognitive functions (Subash et al., 2014).

Since excessive generation of oxidative stress is one of the key features of various diseases, administering compounds or products possessing inherent antioxidant and free radical scavenging properties seems to be the best possible solution or therapeutic method against such diseases. Herbal extracts, plant products, and phytoconstituents have strengthened their position as potent free radical scavengers that also inhibit lipid peroxidation. The synthetic compounds with antioxidant characteristics tend to be toxic and/or mutagenic at times, and therefore natural sources of antioxidants are immensely required as alternative methods. Since the chemical composition of herbal medicines and herbal products reflects the physiological functions of a living niche and flora to quite an extent, they are supposedly more compatible with human system (Sen et al., 2010).

24.11 NEUROLOGICAL DISORDERS AND THEIR HERBAL REMEDY

24.11.1 Alzheimer Disease

AD is an old age, progressive, and irreversible neurodegenerative disease that results in a significant memory loss, somewhat abnormal behavior, negative impact on personality, and deterioration in cognitive abilities. There is no cure for Alzheimer's yet, and the drugs that exist for it currently are not promising enough. To this end, Ayurvedic medicinal plants have emerged as one of the most potent sources for lead identification and drug development, and most of the products that have been derived from them are already at the stage of clinical development. In fact, many research studies promote the idea of deriving medicines from Ayurvedic plants or using their constituents to treat AD. Even though the actual mode of action of phytochemicals remains a bit unclear, their wide range boasts of a variety of pharmacological properties like anticholinesterase, anti-inflammatory, anti-amyloidogenic, hypolipidemic, antioxidant, etc. These pharmacologically rich phytochemicals are flavonoids, polyphenols, alkaloids, lignans, sterols, tannins, etc. (Rao et al., 2012). Some medicinal plants with a strong therapeutic efficacy against AD have been investigated. These plants inhibit Acetylcholinesterase (AChE) and oxidative stress—related damages and protect the affected neurons. Besides inhibiting AChE, these plants and their constituents also promote acetylcholine synthesis that improves the ill effects associated with dementia. However, the main underlying mechanism of their neuroprotective aspect remains their antioxidant nature. They also bring about cognitive advantages and have the potential to target the primary and basic pathophysiology associated with the disease. The observations in some studies and the preliminary data obtained indicate that phytomedicines administration can have effects over learning and memory in the cases of mild to moderate AD. Another vital parameter influenced by these plant components is the alterations in the event of Aβ processing, inhibition of cellular apoptosis, and management of oxidative stress and inflammation (Jivad and Rabiei, 2014).

24.11.2 Multiple Sclerosis

Multiple sclerosis (MS) is a disorder of CNS marked by chronic inflammation and demyelination that can impair cognition, mobility, and sensory abilities. It is a leading cause of prevailing disabilities in the world. Different medicinal plants like *Boswellia papyrifera*, *Panax ginseng*, *Aloysia citrodora*, *G. biloba*, *Andrographis paniculata*, etc. have shown positive effects in MS patients. Out of them, *Cannabis sativa* was found to be of highest clinical relevance in alleviating MS symptoms. Besides that, proanthocyanidins, epigallocatechin-3-gallate, ginkgo flavone glycosides, cannabinoids, ginsenosides, boswellic acid, and andrographolide were reported as the primary bioactive components present in medicinal plants that possess therapeutic efficacy against MS. The herbal medicines administered had few mild side effects that were well tolerated (Farzaei et al., 2017). There are two kinds of treatment regime for MS: first those that act to control and regulate the disease process and second those that provide symptomatic relief. The existing pharmacological agents belonging to first category are human recombinant IFN-β and glatiramer acetate that further includes a monoclonal antibody acting against α-4 integrin, drug natalizumab, and mitoxantrone, a chemotherapy agent. However, these therapies have not been much successful where MS is progressive in nature and only partially significant in cases of relapsing MS. They are available exclusively in injectable forms and are associated with several side effects apart from being expensive too. Consequently, the focus is on the development of better and affordable therapies providing symptomatic relief in MS care (Yadav et al., 2010). For this complementary and alternative medicine (CAM), therapies came into picture to mitigate MS symptoms, but they need to be checked in more detail as far as safety and efficacy is concerned. A study based in California and Massachusetts reported that almost 60% of MS patients resorted to CAM for therapy, and every suffering person uses two to three varied forms of CAM. In British Columbia, this percentage is about 67%. Biologically based therapy depends upon the use of herbs, dietary products, and bee venom therapies. Herbs enriched with γ-linoleic acid, like flax seed and rapeseed, are potentially useful for treating MS because of their ability to disrupt metabolism of fatty acids and functions of lymphocytes. Flavonoid rich herbs, berries, and fruits, like blueberry, could also be helpful in the treatment. Herbal energy enhancer products also reduce the feeling of fatigue, a common symptom felt by MS patients. One such group of phytochemical is flavonoids that are colored antioxidants present in plants. They impart color to fruits and vegetables. Based on the epidemiological findings and data, the fruits and vegetables that have high proportion of flavonoids are known to possess antiviral, anti-inflammatory, anti-allergic, and antitumor characteristics. Luteolin, another immune-regulatory plant constituent, is effective against neurodegenerative diseases like MS and acts by inhibiting the inflammation cascades. It is primarily present in leaves of artichoke, rosemary, thyme, and chamomile (Namjooyan et al., 2014).

24.11.3 Parkinson's Disease

PD, a neurodegenerative disorder associated with old age, is characterized by a progressive deterioration in the production of dopamine in the substantia nigra that also forms its key pathological feature. The clinical symptoms observed in PD patients include bradykinesia (troubled initiation in movement), hypokinesia (absence of considerable facial expressions), rest tremor (occurrence of pill-rolling movement in the forearms), rigidity, and non-motor features comprising depression and psychosis autonomic dysfunction. Traditional medicine–based treatment for PD makes use of herbal products across the globe. China is one of the earliest practitioners of herbal prescriptions against PD dating back to approximately 2200 years which continues even now. The health-care practitioners as well as the patients use herbal medicines in conjunction with the conventional methods of treatment in an attempt to reduce the dosage of dopaminergic drugs, the harmful side effects that accompany the continuous and prolonged use of dopaminergic agents, and to finally improve PD symptoms (Kim et al., 2012). Herbal medicine is being used as one of the key and highly popular means of alternative therapy to be imbibed by people suffering from PD. However, there is a lack of strong and rigorous clinical trials to determine how effective it turns out to be in PD. A study focusing on the significance of CHM in Parkinsonism induced by anti-psychosis was conducted in Japan that used a standard formula involving 10 herbs, and it was observed that it significantly reduced the extent of tremors in the patients (Kum et al., 2011). Some more research works based their research on the antiparkinsonian effects of herbal formulations supported with basic and clinical investigations. The tests and experiments were conducted under both in vitro and in vivo setups. A study conducted by Li et al. (2013) enlisted and presented relevant phytochemicals and herbal extracts that possessed antiparkinsonian properties based basically on their pharmacological activities. They acted by regulating or influencing the key signaling cascades involved in PD pathogenesis. These herbal medicines belonged to 24 genera and 18 families, viz., Alpinia, Acanthopanax, Astragalus, etc. These phytomedicines could possibly pave the way for the discovery of valuable sources for drug development that can be employed against PD. These plant species hold a great scope to be explored further and come up as the most promising candidates for PD therapy and hence need further extensive investigations in clinical trials. Further elucidation of active components that are present in some herbal extracts and their compatibility with the other components of the formulations based on the established norms is also a major requirement (Li et al., 2013). The therapeutic benefits of herbs, fruits, vegetables and spices, ornamental and parasitic plants, and fungi are being continuously investigated over the last few decades in cases of PD. Observations and evidence from various studies indicate that phytochemicals are capable of preventing and mitigating even the slightest of signs and symptoms of PD. They act by controlling the generation of ROS, neuroinflammatory responses, production of dopamine, excitotoxicity, mitochondrial function, metal homeostasis, and cellular signaling pathways, all of which are affected in the brain of PD patients. However, further research is required to unearth the exact underlying mechanisms of these compounds that are responsible for attenuation of PD symptoms. Stilbenoids, flavonoids, and alkaloids are the important bioactive derivatives of plants that are known to possess anti-inflammatory and anti-oxidative properties that interest the researchers looking for a possible cure against PD. These natural phytochemicals also aid in maintaining and promoting mitochondrial functions in addition to serving as important cognitive enhancers. Moreover, they also inhibit aggregation of α-synuclein, activation of c-Jun N-terminal kinase, and production of monoamine oxidase (MAO) besides acting as the agonists for dopaminergic neurons. Due to the unwanted side effects and the economic burden of synthetic drugs, natural medicines or herbal drugs present a safer therapeutic option for PD. However, there is still a long way to go to completely prove their worth and efficacy against PD and support it with concrete experimental and clinical evidence (Essa et al., 2014).

24.11.4 Huntington's Disease

Huntington's disease (HD) was first described and identified by George Huntington, a physician from Ohio. It is described as an autosomal-dominant neurodegenerative disorder that is inherited from family and is marked by motor dysfunction that progresses with age and includes chorea and dystonia, fluctuations in emotional quotient, memory, and a drop in weight. The most severely affected areas of brain in HD are the medium spiny neurons in the striatum, and to some extent in cortex. It is further accompanied by loss of γ-amino butyric acid as well as the enkephalin neurons present in basal ganglia in conjunction with the alterations in the amount of *N*-methyl-D-aspartate (NMDA) receptors. Nature presents the best source of medicines and possibly has solutions to every existing health issue of mankind. The innumerable plant species and their products all over the world are a reservoir of pharmacological activities. Some of them are able to directly manifest their pharmacological effects over human body. These naturally occurring herbal products and compounds have inherent antioxidant, and anti-inflammation properties along with calcium antagonization, antiapoptotic effects, and regulation of neuronal functions that collectively exert preventive or therapeutic actions on many of the existing neurodegenerative diseases. This was also particularly proved through experiments in a 3-NP-induced model of neuronal impairment, which also happens to be the accepted animal model for studies on HD, that some plants and their constituent phytochemicals are effective in curbing these neuronal impairments under in vivo conditions (Choudhary et al., 2013).

24.11.5 Cerebral Ischemia

One of the most prevalent and serious disabilities in the world is brain ischemia and reperfusion. Cerebral ischemia and reperfusion is particularly characterized by excessive free radical formation that leads to oxidative stress and generation of ROS that result in oxidative damage of essential biomolecules like proteins, membrane lipids, and nucleic acids. Limited antioxidant capacity of the affected tissue with an elevation in lipid peroxidation level and increased concentration of inhibitors of lipid peroxidation have been linked with pathophysiology of brain ischemia. The cure to many such neurological diseases lies in the medicinal plants obtained from natural resources. Since brain ischemia and reperfusion is characterized by multiple events like cell survival, apoptosis, $Ca2+$ overload, microglia aggregation, ROS accumulation, and inflammation that leads to progressive injury, the exact protective mechanisms exerted by these natural agents over ischemia remains unclear to some extent. But still the need for natural therapies as protective measure against ischemic cerebral injury and associated neurodegenerative disorders is continuously rising (Jivad and Rabiei, 2015). The amelioration of symptoms of several age-related neurological disorders by herbal medicines or phytomedicines is an area that is constantly being focused upon. Some of these phytochemicals include polyphenols like (+)-catechin, quercetin, and resveratrol and exhibit protective action in animal models of various neurological disorders. The mechanisms responsible for this include antioxidation, ischemic preconditioning, inhibition of microglia recruitment, and anti-inflammation. Their efficacy against ischemic brain injury has been proved over time through various independent studies. A few clinical studies regarding the same therapeutic efficacy of phytomedicines in cerebral ischemia have also been successfully conducted. A compound called cinnamophilin, which is extracted from *Cinnamomum philippinense*, exhibited protective effects over ischemic brain by reducing the brain infarction volume and a subsequent improvement in the neurobehavioral abilities when administered 15 min prior to pretreatment or 2 h after the initiation of occlusions in middle cerebral artery occlusion (posttreatment). This suggested that the administration of natural compounds with antioxidant properties is beneficial in clinical treatment of cerebral ischemia and the neuronal diseases associated with it (Wu et al., 2010). Thrombolytic therapy has been so far proven to be the most effective remedy against brain injury and reduced mortality of patients suffering from cerebral infarction. However, when combined with a strong and effective alternate neuronal protection method, remedy may work wonders in treating the disorder. The clinical trials of chemically derived neuroprotective drugs and strategies in stroke have failed to achieve the desired results till now, and therefore herbal approaches toward the same are being developed. People with higher risk for stroke are suggested to take herbal drugs as a means of prophylactic treatment. These herbal drugs belong to ancient Indian Ayurveda system as well as the traditional Chinese medicines that reportedly have all the required therapeutic properties to cure stroke and other such neurological diseases (Gupta et al., 2010).

24.11.6 Dementia

Dementia can also be called as a syndrome involving progressive dysfunction of memory and learning skills, cognitive abilities, behavior, daily life activities, and lifestyle decline. Dementia has so far attacked around 47.5 million people and more worldwide with 7.7 million fresh cases reported and added to the total sufferers each year. Dementia can be of several types wherein vascular dementia (VaD) is claimed as the second most commonly occurring disorder after AD. Its other prevalent forms include PD, frontotemporal dementia, dementia with Lewy bodies, alcohol-related dementia, and HD. VaD has been observed to coexist with other kinds of dementia like AD since many of the postmortem studies have revealed the presence of AD like pathology in around 40% of total patients inflicted with VaD (Chang et al., 2016). Research concerned with ethnobotanicals in memory or cognitive abilities has seen an upsurge over the last few years. The multifaceted properties and medicinal uses of plants or phytochemicals, like traditional uses, strong bioactivities, psychological and clinical efficacy, and their safety in various kinds of dementia have been proven in several studies. Plants like *G. biloba*, sage, etc. show positive influence over cognitive functions by alleviating the symptoms associated with dementia (Perry and Howes, 2011).

24.12 NATURAL PRODUCTS AS THERAPEUTIC AGENTS FOR NEUROLOGICAL DISEASES

Over the last decade, a plethora of literature has emerged explaining how herbal products could possibly influence various diseases to different extents. Herein, we emphasize on how consumption of such products rich in various antioxidants and anti-inflammatory compounds could have a long-term health benefit by possible alterations in the epigenome and moreover prevent the onset of degenerative conditions (Table 24.1). In this section of the chapter, we will be discussing the beneficial effects of phytomedicines and herbal products (phytochemicals) with a special focus on neurodegeneration and neuroprotection.

24.12.1 Genistein

Genistein, an isoflavone, is a compound derived mainly from soy. Various anticancer studies have demonstrated genistein to affect the process of tumorigenesis/carcinogenesis mediated via epigenetic regulations (Zhang and Chen, 2011). Reports are suggestive of possible role of genistein in inhibition of DNA methyltransferases (DNMTs) that could regulate gene expression by preventing DNA methylation at promoter levels. Genistein has also proved to reduce the risk of breast cancer and enhances DNA demethylation of SF1 promoter in endometrial stromal cells present in the endometrium (Khan et al., 2011). Genistein also reduces hypermethylation levels of tumor suppressor genes such as CHD5 and promotes the expression of CHD5 as well as p53 that prevents neuroblastoma (NB) growth and formation of tumor microvessel in vivo (Li et al., 2012). In addition to these, genistein could also act as an inhibitor to reduce the expression of DNMT3b in NB model, thereby paving its way as an adjuvant therapeutic candidate for NB treatment.

24.12.2 Resveratrol

Resveratrol more popular by the name "French Paradox" is capable of preventing various diseases, including neurological disorders such as AD, PD, and stroke. Studies have revealed that such potent effects of resveratrol may not only be attributed to its antioxidant and anti-inflammatory nature but also because of its activation of sirtuin 1 protein (SIRT1) (Baur, 2010).

Table 24.1 Some of the commonly used naturally occurring plants and their products used against neuronal diseases with their sources and targets

S. no.	Herbal drug	Source	Phytochemicals	Disease	Target	Reference
1.	Ginseng	Root of *Panax ginseng*	Polyphenol	Stroke, AD, PD	Proinflammatory markers, free radicals	Ong et al. (2015), Rastogi et al. (2015)
2.	*Bacopa monnieri*	A perennial creeping plant belonging to family Scrophulariaceae	Alkaloid	Epilepsy, mental illnesses, stroke	Signaling pathways specific to disease, cholinesterase	Piyabhan and Wetchateng (2014), Srivastava and Yadav (2016)
3.	*Withania somnifera*	Subtropical undershrub	Glycowithanolides, alkaloids	AD, PD, dementia	Free radicals, lipid peroxidation	Kumar et al. (2017)
4.	Saffron	Stigma of *Crocus sativus* L	Phenols	Depression, cerebral ischemia	Free radicals, cholinesterase	Khazdair et al. (2015)
5.	Curcumin	Rhizome of *Curcuma longa*	Polyphenol	AD, ADHD	Free radicals, histone modification	Wu et al. (2013)
6.	Resveratrol	Red wine, red grape skin	Stilbene	AD, PD, stroke	Free radical, inflammatory markers, SIRT1	Siddiqui et al. (2010)
7.	Genistein	Soy	Isoflavone	Neuroblastoma	DNA methylation	Qian et al. (2012)

AD, Alzheimer's disease; *PD*, Parkinson's disease; *ADHD*, Attention deficit hyperactivity disorder.

24.12.3 Curcumin

Curcumin, the main active component of turmeric, is reported to exhibit antidisease effects in various animal models as well as in humans. Induction of apoptosis in cancer cell lines via curcumin is a well-established fact, and more recently, it has demonstrated to inhibit few epigenetic enzymes (such as HATs, HDAC1, HDAC3, and HDAC8) in vitro (Reuter et al., 2011; Vahid et al., 2015) attributed to its antioxidant, -inflammatory, and -microbial property making it a good candidate for prevention and control of neurodegenerative diseases. Prevalence of AD is less in India with respect to the United States owing to the extensive usage of curcumin in India (Ganguli et al., 2000). Curcumin not only prevents the formation of amyloid oligomers in vitro but also promotes the disaggregation of preformed oligomers (Ono et al., 2004; Yang et al., 2005). In addition to this, it leads to the induction of histone hypoacetylation and triggers poly (ADP-ribose) polymerase- and caspase-3-mediated apoptosis in brain glioma cells. Curcumin also reduces acetylation of H3 and H4 histone proteins, thereby controlling the fate of neural stem cells (Kang et al., 2004).

24.12.4 Omega-3 Fatty Acids

Omega-3 fatty acids constitute an integral part of cell membranes in turn affecting the functional activities of the cell membrane receptors, binding to their substrates, and the downstream signaling pathways, which would ultimately regulate the expression of various proteases or gene expression at the nuclear level. Docosahexaenoic acid (DHA) is required for normal brain and CNS development and learning in infants and proper neurological functioning in adults as well. DHA is preferably taken up by the brain over other fatty acids (Horrocks and Yeo, 1999). Deficits in DHA have been linked with various physiological disorders like rheumatoid arthritis, diabetes mellitus, and cardiovascular disease; few forms of cancer; neurological disorders like depression, AD, attention deficit hyperactivity disorder, and unipolar disorder. DHA has been demonstrated to regulate nerve growth factors to promote the growth of nerves. The multifaceted role of DHA has also been studied and is reported to ameliorate disease symptoms of various disorders. In addition to the neuroprotective function of DHA, it also regulates ion channels like calcium, potassium, and sodium (Vreugdenhil et al., 1996). Such regulation of ion-channels could also provide for control of detrimental effects of over-excitability like in the case of spinal cord injury where excitotoxicity results due to over release of the excitatory neurotransmitter glutamate due to calcium-mediated exocytosis. In addition to its structural contribution, DHA is also known to regulate signal transduction, gene expression, inflammation, as well as ion channel functions (Satkunendrarajah and Fehlings, 2013).

24.12.5 Catechins

Catechins are flavanols which include catechin, (−)-epicatechin, epigallocatechin, (−)-epicatechin gallate, and epigallocatechin gallate and constitute the commonly consumed flavonoids in the United States (Bai et al., 2014). These polyphenolic compounds are constituents of various categories of foods, such as fruits (apples and grapes, especially black grapes), berries (blackberries, cherries, and raspberries), beans (fava beans), and even beverages (tea, cocoa, and red wine). Dark chocolate is also a major source of (−)-epicatechin. Catechins are rich antioxidants also acting as free radical scavengers (Nanjo et al., 1996), metal ion chelators (Mandel et al., 2004; Reznichenko et al., 2006), affecting activities of antioxidant enzyme in the brain (Levites et al., 2001). In various models of neurotoxicity, catechin treatment not only renders neuroprotection but also improves behavioral performance.

24.12.5.1 Crocus sativus L.

The stigma of *Crocus sativus* (Iridaceae), commonly referred to as saffron, is the world's most expensive spice grown all around the world, originally used to treat depression. The attempt to look for cheaper alternatives fueled further research, and the petal of *C. sativus* was found to be as effective as its stigma. Safranal has also exhibited beneficial effects against cerebral ischemia, quinolinic acid–induced oxidative defects in hippocampus of rats. Saffron, crocetin, and safranal owing to their antioxidant functions and anticholinesterase activity have reduced oxidative damage, reduced activation of caspase-3 in neuronal SH-SY5Y cells, and also enhanced memory in aged mice (Papandreou et al., 2006). In a more recent study, saffron has demonstrated protective effects against aluminum-induced neurotoxicity by enhancing antioxidant functions and MAO activity, without a parallel enhancement in cognitive functions in the whole brain and cerebellum (Linardaki et al., 2013).

24.12.5.2 *Bacopa monnieri*

Bacopa monnieri (BM), a perennial creeping plant belonging to the family Scrophulariaceae, is majorly found in wet, damp, and marshy areas throughout India (Aguiar and Borowski, 2013). BM popularly known as Brahmi, a name derived from the Lord Brahma, the creator god of the Hindu pantheon of deities, finds special mention in the Ayurvedic literature for its use in different mental conditions such as anxiety, poor cognitive abilities, and lack of concentration. Studies have also documented its potential as a therapeutic candidate for mental illness and epilepsy. Major active constituents of BM include various alkaloids like brahmine, herpestine, and nicotine; saponins such as D-mannitol and hersaponin, acid A, and monnierin (Aguiar and Borowski, 2013; Le et al., 2015). Bacoside A also possesses enhanced antioxidant defense system (Anbarasi et al., 2006) and memory-enhancement activity as well as could be utilized as nootropics. These herbals work through activation of various pathways to improvise memory and learning abilities that may bring some symptomatic relief to Alzheimer's patients following dementia in the early stages of the disorder (Srivastava and Yadav, 2016).

24.12.5.3 Ginseng

Ginseng, one of the most widely used herbal medicines in the world, has been utilized for centuries to increase immunity and treat disorders; especially in Asian countries. The most commonly used ginseng variant in traditional herbal medicine is ginseng, made from the peeled and dried root of *P. ginseng*. Ginseng has been widely suggested as an effective treatment for a myriad of brain disorders, including stroke as well as acute and chronic neurodegenerative diseases. Ginseng's neuroprotective effects are broadly attributed to its maintenance of homeostasis and mitigation of inflammation via suppression of various proinflammatory markers and oxidative stress by mechanisms involving activation of the cytoprotective transcriptional factor Nrf2, which results in decrease in ROS generation. It could also prevent stroke mediated neuronal demise, thus reducing anatomical and functional stroke damage (Rastogi et al., 2015). *P. ginseng* has been used in traditional Chinese medicine for centuries. Indeed, ginseng extract and its individual ginsenosides have been shown to affect a range of biochemical markers involved in the pathogenesis of PD. Oral administration of this extract could also significantly decrease dopaminergic cell damage, microgliosis, and accumulation of α-synuclein protein aggregates in animal models of PD (Van Kampen et al., 2014). The major ginsenosides such as Rb1, Rg1, Rd, and Re as well as other minor components such as Notoginsenoside R2 and Pseudoginsenoside-F11 have also attracted considerable attention as suitable antiparkinson agents. These compounds exhibit their neuroprotective actions mediated by inhibition of oxidative stress and neuroinflammation, reduction in toxins-induced apoptosis and nigral iron levels, as well as regulation of NMDA receptor channel activities (González-Burgos et al., 2015).

24.12.5.4 *Withania somnifera*

Withania somnifera (WS) (Dunal), popularly known as Ashwagandha, is a subtropical undershrub commonly used in Indian traditional medicines for more than 3000 years and has been categorized as Rasayana in Ayurveda, which is reported to elevate defense system against diseases, arrest aging, revitalize the body, increase resistance against adverse environmental conditions to create a sense of mental well-being. The active component of WS, that is, glycowithanolides possesses the ability to alter the cortical and striatal antioxidant enzyme functions (superoxide dismutase (SOD), catalase (CAT), and glutathione peroxidase) in rats. The other biologically active components include alkaloids (ashwagandha, cuscohygrine, anahygrine, topine, etc.), steroidal compounds, including ergostane type steroidal lactones, withaferin A, withanolides A–Y, withasomniferin A, withasomnidienone, withasomnierose A–C, withanone, etc. Ashwagandha root powder also exerts free radical scavenging activity (Sankar et al., 2007). Clinical trials and animal research also support the usage of WS for a myriad of conditions such as anxiety, cognitive and neurological disorders, senile dementia, AD and PD. *W. somnifera* also demonstrates neuroprotective effects against 6-OHDA-induced Parkinsonism in rats (Ahmad et al., 2005). Perhaps its antioxidant nature and inhibition of lipid peroxidation in vitro and in vivo is reported to underlie its neuroprotective benefits (Bhatnagar et al., 2009).

24.13 CONCLUSION

Based on the innumerable studies and research and clinical data, it becomes clear that designing or developing an effective remedy against neurodegenerative disorders is an urgent requirement. However, the existing synthetic drugs and pharmacological agents being employed against them come with a serious threat of adverse side effects. Consequently, it becomes imperative to find out alternative line of therapy that is a lot safer. To this end, the herbal drugs or the phytomedicines came to the rescue as the naturally extracted drugs that have negligible harmful effects, if at all, and are also pocket friendly. Their inherent antioxidant and anti-inflammatory properties are largely responsible for their neuroprotective effects as proven through various experimental studies. They further regulate the bioenergetics, calcium metabolism, and metal chelation which are also present as some of the key features in neuronal anomalies. These characteristics of phytoconstituents make them an attractive source of therapy against complex neurological disorders. In

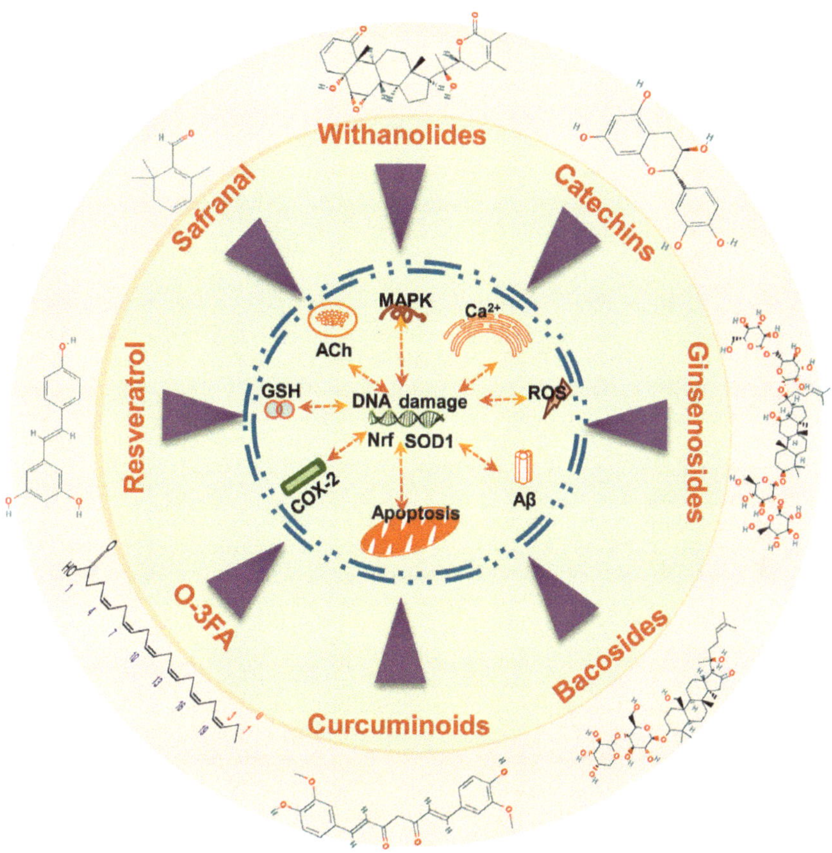

Figure 24.1 Molecular structure and potential activity of selected natural/active constituents or phytochemicals that act on cellular organelle to inhibit apoptosis and DNA damage associated with several major health disorders. These chemical structures can further be modified to fortify the chemical constituents with the help of addition or deletion of a side group.

addition, their multidrug and multitarget potential has been observed to be far superior to the conventional method of single drug with a single therapeutically active component. This could eventually pave the way for developing specific personalized neuroprotective therapies in future. A detailed elucidation of the exact action mechanism of these natural products against some of the severe neurodegenerative disorders would further assist in identification of lead compounds and improve/hasten the drug-development process. The multiple positive effects manifested by phytomedicines over innumerable health hazards have instilled hope in the research and medicine fraternity regarding the therapy for the currently incurable diseases, especially neurodegenerative disorders. Keeping in mind the popularity and effectiveness of herbal drugs, the dream of a healthy world might hopefully materialize soon (Fig. 24.1).

REFERENCES

Aguiar, S., Borowski, T., 2013. Neuropharmacological review of the nootropic herb *Bacopa monnieri*. Rejuvenation Res. 16 (4), 313–326.

Ahmad, M., Saleem, S., Ahmad, A.S., Ansari, M.A., Yousuf, S., Hoda, M.N., et al., 2005. Neuroprotective effects of *Withania somnifera* on 6-hydroxydopamine induced Parkinsonism in rats. Hum. Exp. Toxicol. 24 (3), 137–147.

Amin, H., Sharma, R., 2015. Nootropic efficacy of Satvavajaya Chikitsa and Ayurvedic drug therapy: a comparative clinical exposition. Int. J. Yoga 8 (2), 109.

Anbarasi, K., Vani, G., Balakrishna, K., Devi, C.S., 2006. Effect of bacoside A on brain antioxidant status in cigarette smoke exposed rats. Life Sci. 78 (12), 1378–1384.

Apaya, M.K., Chang, M.-T., Shyur, L.-F., 2016. Phytomedicine polypharmacology: cancer therapy through modulating the tumor microenvironment and oxylipin dynamics. Pharmacol. Ther. 162, 58–68.

Aquilano, K., Baldelli, S., Rotilio, G., Ciriolo, M.R., 2008. Role of nitric oxide synthases in Parkinson's disease: a review on the antioxidant and anti-inflammatory activity of polyphenols. Neurochem. Res. 33 (12), 2416–2426.

Bai, W., Wang, C., Ren, C., 2014. Intakes of total and individual flavonoids by US adults. Int. J. Food Sci. Nutr. 65 (1), 9–20.

Bansal, Y., Sethi, P., Bansal, G., 2013. Coumarin: a potential nucleus for anti-inflammatory molecules. Med. Chem. Res. 22 (7), 3049–3060.

Barbosa, W.L.R., do Nascimento, M.S., do Nascimento Pinto, L., Maia, F.L.C., Sousa, A. J.A.D., Júnior, J.O.C.S., et al., 2012. Selecting medicinal plants for development of phytomedicine and use in primary health care. Bioactive Compounds in Phytomedicine. InTech.

Baur, J.A., 2010. Resveratrol, sirtuins, and the promise of a DR mimetic. Mech. Ageing Dev. 131 (4), 261–269.

Benzie, I.F., Wachtel-Galor, S., 2011. Herbal Medicine: Biomolecular and Clinical Aspects. CRC Press, ISBN-13: 978-1-4398-0713-2.

Bhatnagar, M., Sharma, D., Salvi, M., 2009. Neuroprotective effects of *Withania somnifera* dunal.: a possible mechanism. Neurochem. Res. 34 (11), 1975–1983.

Briskin, D.P., 2000. Medicinal plants and phytomedicines. Linking plant biochemistry and physiology to human health. Plant Physiol. 124 (2), 507–514.

Burger, O., Ofek, I., Tabak, M., Weiss, E.I., Sharon, N., Neeman, I., 2000. A high molecular mass constituent of cranberry juice inhibits *Helicobacter pylori* adhesion to human gastric mucus. FEMS Immunol. Med. Microbiol. 29 (4), 295–301.

Chang, D., Liu, J., Bilinski, K., Xu, L., Steiner, G.Z., Seto, S.W., et al., 2016. Herbal medicine for the treatment of vascular dementia: an overview of scientific evidence. Evid. Based Complement. Alternat. Med. 2016.

Chaudhary, T., Chahar, A., Sharma, J.K., Kaur, K., Dang, A., 2015. Phytomedicine in the treatment of cancer: a health technology assessment. J. Clin. Diagn. Res. 9 (12), XC04.

Chen, L., Xin, X., Zhang, H., Yuan, Q., 2013. Phytochemical properties and antioxidant capacities of commercial raspberry varieties. J. Funct. Foods 5 (1), 508–515.

Choudhary, S., Kumar, P., Malik, J., 2013. Plants and phytochemicals for Huntington's disease. Pharmacogn. Rev. 7 (14), 81.

Colucci, L., Bosco, M., Ziello, A.R., Rea, R., Amenta, F., Fasanaro, A.M., 2012. Effectiveness of nootropic drugs with cholinergic activity in treatment of cognitive deficit: a review. J. Exp. Pharmacol. 4, 163.

de Souza, V.R., Pereira, P.A.P., da Silva, T.L.T., de Oliveira Lima, L.C., Pio, R., Queiroz, F., 2014. Determination of the bioactive compounds, antioxidant activity and chemical composition of Brazilian blackberry, red raspberry, strawberry, blueberry and sweet cherry fruits. Food Chem. 156, 362–368.

Ekor, M., 2014. The growing use of herbal medicines: issues relating to adverse reactions and challenges in monitoring safety. Front. Pharmacol. 4, 177.

Esposito, E., Impellizzeri, D., Mazzon, E., Paterniti, I., Cuzzocrea, S., 2012. Neuroprotective activities of palmitoylethanolamide in an animal model of Parkinson's disease. PLoS ONE 7 (8), e41880.

Essa, M., Braidy, N., Bridge, W., Subash, S., Manivasagam, T., Vijayan, R., et al., 2014. Review of natural products on Parkinson's disease pathology. J. Aging Res. Clin. Pract. 3 (1), 1–8.

Farzaei, M.H., Shahpiri, Z., Bahramsoltani, R., Najafi, F., Rahimi, R., 2017. Efficacy and tolerability of phytomedicines in multiple sclerosis patients: a review. CNS Drugs 1–23.

Ferdousy, S., Rahman, M.A., Al-Amin, M.M., Aklima, J., Chowdhury, J.K.H., 2016. Antioxidative and neuroprotective effects of *Leea macrophylla* methanol root extracts on diazepam-induced memory impairment in amnesic Wistar albino rat. Clin. Phytosci. 2 (1), 17.

Ganguli, M., Chandra, V., Kamboh, M.I., Johnston, J.M., Dodge, H.H., Thelma, B., et al., 2000. Apolipoprotein E polymorphism and Alzheimer disease: the Indo-US cross-national dementia study. Arch. Neurol. 57 (6), 824–830.

González-Burgos, E., Fernandez-Moriano, C., Gómez-Serranillos, M.P., 2015. Potential neuroprotective activity of Ginseng in Parkinson's disease: a review. J. Neuroimmune Pharmacol. 10 (1), 14–29.

Gouliaev, A.H., Senning, A., 1994. Piracetam and other structurally related nootropics. Brain Res. Rev. 19 (2), 180–222.

Gualtieri, F., Manetti, D., Romanelli, M., Ghelardini, C., 2002. Design and study of piracetam-like nootropics, controversial members of the problematic class of cognition-enhancing drugs. Curr. Pharm. Des. 8 (2), 125–138.

Gullett, N.P., Amin, A.R., Bayraktar, S., Pezzuto, J.M., Shin, D.M., Khuri, F.R., et al., 2010. Cancer prevention with natural compounds. Semin. Oncol. 37 (3), 258–281.

Gupta, Y., Briyal, S., Gulati, A., 2010. Therapeutic potential of herbal drugs in cerebral ischemia. Indian J. Physiol. Pharmacol. 54 (2), 99–122.

Hong, M., Li, S., Tan, H.Y., Wang, N., Tsao, S.-W., Feng, Y., 2015. Current status of herbal medicines in chronic liver disease therapy: the biological effects, molecular targets and future prospects. Int. J. Mol. Sci. 16 (12), 28705–28745.

Horrocks, L.A., Yeo, Y.K., 1999. Health benefits of docosahexaenoic acid (DHA). Pharmacol. Res. 40 (3), 211–225.

Imaga, N.A., 2013. Phytomedicines and nutraceuticals: alternative therapeutics for sickle cell anemia. Sci. World J. 2013.

Jellen, L.C., Aliper, A., Buzdin, A., Zhavoronkov, A., 2015. Screening and personalizing nootropic drugs and cognitive modulator regimens in silico. Front. Syst. Neurosci. 9.

Jivad, N., Rabiei, Z., 2014. A review study on medicinal plants used in the treatment of learning and memory impairments. Asian Pac. J. Trop. Biomed. 4 (10), 780–789.

Jivad, N., Rabiei, Z., 2015. Review on herbal medicine on brain ischemia and reperfusion. Asian Pac. J. Trop. Biomed. 5 (10), 789–795.

Kang, J., Liu, Y., Xie, M.-X., Li, S., Jiang, M., Wang, Y.-D., 2004. Interactions of human serum albumin with chlorogenic acid and ferulic acid. Biochim. Biophys. Acta 1674 (2), 205–214.

Karimi, A., Majlesi, M., Rafieian-Kopaei, M., 2015. Herbal versus synthetic drugs; beliefs and facts. J. Nephropharmacol. 4 (1), 27.

Khan, H., 2015. Brilliant future of phytomedicines in the light of latest technological developments. J. Phytopharmacol. 4 (1), 58–60.

Khan, S.I., Aumsuwan, P., Khan, I.A., Walker, L.A., Dasmahapatra, A.K., 2011. Epigenetic events associated with breast cancer and their prevention by dietary components targeting the epigenome. Chem. Res. Toxicol. 25 (1), 61–73.

Khazdair, M.R., Boskabady, M.H., Hosseini, M., Rezaee, R., Tsatsakis, A.M., 2015. The effects of *Crocus sativus* (saffron) and its constituents on nervous system: a review. Avicenna J. Phytomed. 5 (5), 376.

Kim, T.-H., Cho, K.-H., Jung, W.-S., Lee, M.S., 2012. Herbal medicines for Parkinson's disease: a systematic review of randomized controlled trials. PLoS ONE 7 (5), e35695.

Kosikowska, P., Berlicki, Ł., 2011. Urease inhibitors as potential drugs for gastric and urinary tract infections: a patent review. Expert Opin. Ther. Pat. 21 (6), 945–957.

Kum, W.F., Durairajan, S.S.K., Bian, Z.X., Man, S.C., Lam, Y.C., Xie, L.X., et al., 2011. Treatment of idiopathic Parkinson's disease with traditional Chinese herbal medicine: a randomized placebo-controlled pilot clinical study. Evid. Based Complement. Alternat. Med. Available from: https://doi.org/10.1093/ecam/nep116.

Kumar, G., Paliwal, P., Patnaik, R., 2017. *Withania somnifera* phytochemicals confer neuroprotection by inhibition of the catalytic domain of human matrix metalloproteinase-9. Lett. Drug Des. Discovery 14 (6), 718–726.

Kumar, G.P., Khanum, F., 2012. Neuroprotective potential of phytochemicals. Pharmacogn. Rev. 6 (12), 81.

Larkins, N., Wynn, S., 2004. Pharmacognosy: phytomedicines and their mechanisms. Vet. Clin.: Small Anim. Pract. 34 (1), 291–327.

Le, X.T., Pham, H.T.N., Van Nguyen, T., Nguyen, K.M., Tanaka, K., Fujiwara, H., et al., 2015. Protective effects of *Bacopa monnieri* on ischemia-induced cognitive deficits in mice: the possible contribution of bacopaside I and underlying mechanism. J. Ethnopharmacol. 164, 37–45.

Levites, Y., Weinreb, O., Maor, G., Youdim, M.B., Mandel, S., 2001. Green tea polyphenol (−)-epigallocatechin-3-gallate prevents *N*-methyl-4-phenyl-1, 2, 3, 6-tetrahydropyridine-induced dopaminergic neurodegeneration. J. Neurochem. 78 (5), 1073–1082.

Li, H., Xu, W., Huang, Y., Huang, X., Xu, L., Lv, Z., 2012. Genistein demethylates the promoter of CHD5 and inhibits neuroblastoma growth *in vivo*. Int. J. Mol. Med. 30 (5), 1081–1086.

Li, X.-z, Zhang, S.-n, Liu, S.-m, Lu, F., 2013. Recent advances in herbal medicines treating Parkinson's disease. Fitoterapia 84, 273–285.

Lin, F.Y., Hsieh, Y.H., Yang, S.F., Chen, C.T., Tang, C.H., Chou, M.Y., et al., 2015. Resveratrol suppresses TPA-induced matrix metalloproteinase-9 expression through the inhibition of MAPK pathways in oral cancer cells. J. Oral Pathol. Med. 44 (9), 699–706.

Linardaki, Z.I., Orkoula, M.G., Kokkosis, A.G., Lamari, F.N., Margarity, M., 2013. Investigation of the neuroprotective action of saffron (*Crocus sativus* L.) in aluminum-exposed adult mice through behavioral and neurobiochemical assessment. Food Chem. Toxicol. 52, 163–170.

Mandel, S., Maor, G., Youdim, M.B., 2004. Iron and α-synuclein in the substantia nigra of MPTP-treated mice. J. Mol. Neurosci. 24 (3), 401–416.

Mohamed, I., Shuid, A., Borhanuddin, B., Fozi, N., 2012. The application of phytomedicine in modern drug development. Internet J. Herb. Plant Med. 1 (2), 16–22.

Namjooyan, F., Ghanavati, R., Majdinasab, N., Jokari, S., Janbozorgi, M., 2014. Uses of complementary and alternative medicine in multiple sclerosis. J. Tradit. Complement. Med. 4 (3), 145–152.

Nanjo, F., Goto, K., Seto, R., Suzuki, M., Sakai, M., Hara, Y., 1996. Scavenging effects of tea catechins and their derivatives on 1, 1-diphenyl-2-picrylhydrazyl radical. Free Radical Biol. Med. 21 (6), 895–902.

Obodozie, O.O., 2012. Pharmacokinetics and drug interactions of herbal medicines: a missing critical step in the phytomedicine/drug development process. Readings in Advanced Pharmacokinetics—Theory, Methods and Applications. InTech. Available from: https://doi.org/10.5772/33699.

Ong, W.-Y., Farooqui, T., Koh, H.-L., Farooqui, A.A., Ling, E.-A., 2015. Protective effects of ginseng on neurological disorders. Front. Aging Neurosci. 7.

Oniyangi, O., Cohall, D.H., 2013. Phytomedicines (medicines derived from plants) for sickle cell disease. Cochrane Database Syst. Rev. (1).

Ono, K., Hasegawa, K., Naiki, H., Yamada, M., 2004. Curcumin has potent anti-amyloidogenic effects for Alzheimer's β-amyloid fibrils *in vitro*. J. Neurosci. Res. 75 (6), 742–750.

Pan, S.-Y., Zhou, S.-F., Gao, S.-H., Yu, Z.-L., Zhang, S.-F., Tang, M.-K., et al., 2013. New perspectives on how to discover drugs from herbal medicines: CAM's outstanding contribution to modern therapeutics. Evid. Based Complement. Alternat. Med. Available from: https://doi.org/10.1002/14651858.CD004448.pub4.

Pandey, M., Debnath, M., Gupta, S., Chikara, S.K., 2011. Phytomedicine: an ancient approach turning into future potential source of therapeutics. J. Pharmacogn. Phytother. 3 (1), 113–117.

Papandreou, M.A., Kanakis, C.D., Polissiou, M.G., Efthimiopoulos, S., Cordopatis, P., Margarity, M., et al., 2006. Inhibitory activity on amyloid-β aggregation and antioxidant properties of *Crocus sativus* stigmas extract and its crocin constituents. J. Agric. Food Chem. 54 (23), 8762–8768.

Perry, E., Howes, M.J.R., 2011. Medicinal plants and dementia therapy: herbal hopes for brain aging? CNS Neurosci. Ther. 17 (6), 683–698.

Pieramico, V., Esposito, R., Cesinaro, S., Frazzini, V., Sensi, S.L., 2014. Effects of non-pharmacological or pharmacological interventions on cognition and brain plasticity of aging individuals. Front. Syst. Neurosci. 8, 153.

Piyabhan, P., Wetchateng, T., 2014. Neuroprotective effects of *Bacopa monnieri* (Brahmi) on novel object recognition and NMDAR1 immunodensity in the prefrontal cortex, striatum and hippocampus of sub-chronic phencyclidine rat model of schizophrenia. J. Med. Assoc. Thai. 97, S50–S55.

Qian, Y., Guan, T., Huang, M., Cao, L., Li, Y., Cheng, H., et al., 2012. Neuroprotection by the soy isoflavone, genistein, via inhibition of mitochondria-dependent apoptosis pathways and reactive oxygen induced-NF-κB activation in a cerebral ischemia mouse model. Neurochem. Int. 60 (8), 759–767.

Rahman, M., Riaz, M., Desai, U.R., 2007. Synthesis of biologically relevant biflavanoids—a review. Chem. Biodivers. 4 (11), 2495–2527.
Rao, B.V., Srikumar, B.N., Rao, B.S.S., 2011. Herbal remedies to treat anxiety disorders. Different Views of Anxiety Disorders. InTech. Available from: https://doi.org/10.5772/23511.
Rao, R.V., Descamps, O., John, V., Bredesen, D.E., 2012. Ayurvedic medicinal plants for Alzheimer's disease: a review. Alzheimers Res. Ther. 4 (3), 22.
Rastogi, V., Santiago-Moreno, J., Doré, S., 2015. Ginseng: a promising neuroprotective strategy in stroke. Front. Cell. Neurosci. 8, 457.
Reuter, S., Gupta, S.C., Park, B., Goel, A., Aggarwal, B.B., 2011. Epigenetic changes induced by curcumin and other natural compounds. Genes Nutr. 6 (2), 93–108.
Reznichenko, L., Amit, T., Zheng, H., Avramovich-Tirosh, Y., Youdim, M., Mandel, S., 2006. Reduction of iron-regulated amyloid precursor protein and β-amyloid peptide by (−)-epigallocatechin-3-gallate in cell cultures: implications for iron chelation in Alzheimer's disease. J. Neurochem. 97 (2), 527–536.
Safavi, M., Sabourian, R., Foroumadi, A., 2016. Treatment of *Helicobacter pylori* infection: current and future insights. World J. Clin. Cases 4 (1), 5.
Sahu, M., Singh, V., Yadav, S., Harris, K., 2012. Plant extracts with antisickling propensities: a feasible succour towards sickle cell disease management-a mini review. J. Phytol. 4 (3).
Sankar, S.R., Manivasagam, T., Krishnamurti, A., Ramanathan, M., 2007. The neuroprotective effect of *Withania somnifera* root extract in MPTP-intoxicated mice: an analysis of behavioral and biochemical variables. Cell. Mol. Biol. Lett. 12 (4), 473.
Satkunendrarajah, K., Fehlings, M.G., 2013. Do omega-3 polyunsaturated fatty acids ameliorate spinal cord injury?: commentary on: Lim et al., Improved outcome after spinal cord compression injury in mice treated with docosahexaeonic acid. Exp. Neurol. 249, 104–110.
Sen, S., Chakraborty, R., Sridhar, C., Reddy, Y., De, B., 2010. Free radicals, antioxidants, diseases and phytomedicines: current status and future prospect. Int. J. Pharm. Sci. Rev. Res. 3 (1), 91–100.
Shmuely, H., Burger, O., Neeman, I., Yahav, J., Samra, Z., Niv, Y., et al., 2004. Susceptibility of *Helicobacter pylori* isolates to the antiadhesion activity of a high-molecular-weight constituent of cranberry. Diagn. Microbiol. Infect. Dis. 50 (4), 231–235.
Siddiqui, M., Kashyap, M., Kumar, V., Al-Khedhairy, A., Musarrat, J., Pant, A., 2010. Protective potential of trans-resveratrol against 4-hydroxynonenal induced damage in PC12 cells. Toxicol. in vitro 24 (6), 1592–1598.
Srivastava, P., Yadav, R.S., 2016. Efficacy of natural compounds in neurodegenerative disorders. The Benefits of Natural Products for Neurodegenerative Diseases. Springer, pp. 107–123.
Subash, S., Essa, M.M., Al-Adawi, S., Memon, M.A., Manivasagam, T., Akbar, M., 2014. Neuroprotective effects of berry fruits on neurodegenerative diseases. Neural Regen. Res. 9 (16), 1557.
Suliman, N.A., Mat Taib, C.N., Mohd Moklas, M.A., Adenan, M.I., Hidayat Baharuldin, M.T., Basir, R., 2016. Establishing natural nootropics: recent molecular enhancement influenced by natural nootropic. Evid. Based Complement. Alternat. Med. Available from: https://doi.org/10.1155/2016/4391375.
Tsai, C.-F., Kuo, Y.-H., Yeh, W.-L., Wu, C.Y.-J., Lin, H.-Y., Lai, S.-W., et al., 2015. Regulatory effects of caffeic acid phenethyl ester on neuroinflammation in microglial cells. Int. J. Mol. Sci. 16 (3), 5572–5589.
Vahid, F., Zand, H., Nosrat-Mirshekarlou, E., Najafi, R., Hekmatdoost, A., 2015. The role dietary of bioactive compounds on the regulation of histone acetylases and deacetylases: a review. Gene 562 (1), 8–15.

Vale, F.F., Oleastro, M., 2014. Overview of the phytomedicine approaches against *Helicobacter pylori*. World J. Gastroenterol. 20 (19), 5594.

Van Kampen, J.M., Baranowski, D.B., Shaw, C.A., Kay, D.G., 2014. *Panax ginseng* is neuroprotective in a novel progressive model of Parkinson's disease. Exp. Gerontol. 50, 95–105.

Vítor, J., Vale, F.F., 2011. Alternative therapies for *Helicobacter pylori*: probiotics and phytomedicine. Pathog. Dis. 63 (2), 153–164.

Vreugdenhil, M., Bruehl, C., Voskuyl, R., Kang, J., Leaf, A., Wadman, W., 1996. Polyunsaturated fatty acids modulate sodium and calcium currents in CA1 neurons. Proc. Natl. Acad. Sci. U.S.A. 93 (22), 12559–12563.

Vyawahare, N., Ambikar, D., Patil, G., Kamble, P., Chitte, N., 2008. Phytomedicines for Neuroprotection. Electron. J. Pharmacol. Ther. 1, 15.

Wink, M., 2015. Modes of action of herbal medicines and plant secondary metabolites. Medicines 2 (3), 251–286.

Wu, P.-f, Zhang, Z., Wang, F., Chen, J.-g, 2010. Natural compounds from traditional medicinal herbs in the treatment of cerebral ischemia/reperfusion injury. Acta Pharmacol. Sin. 31 (12), 1523–1531.

Wu, J., Li, Q., Wang, X., Yu, S., Li, L., Wu, X., et al., 2013. Neuroprotection by curcumin in ischemic brain injury involves the Akt/Nrf2 pathway. PLoS ONE 8 (3), e59843.

Yadav, V., Shinto, L., Bourdette, D., 2010. Complementary and alternative medicine for the treatment of multiple sclerosis. Expert Rev. Clin. Immunol. 6 (3), 381–395.

Yang, F., Lim, G.P., Begum, A.N., Ubeda, O.J., Simmons, M.R., Ambegaokar, S.S., et al., 2005. Curcumin inhibits formation of amyloid β oligomers and fibrils, binds plaques, and reduces amyloid *in vivo*. J. Biol. Chem. 280 (7), 5892–5901.

Yin, S.-Y., Wei, W.-C., Jian, F.-Y., Yang, N.-S., 2013. Therapeutic applications of herbal medicines for cancer patients. Evid. Based Complement. Alternat. Med. 2013.

Zhang, Y., Chen, H., 2011. Genistein, an epigenome modifier during cancer prevention. Epigenetics 6 (7), 888–891.

INDEX

Note: Page numbers followed by "*f*" and "*t*" refer to figures and tables, respectively.

A

C

D

G

J

K

N

Q

R

T

FSC
www.fsc.org
MIX
Papier | Fördert gute Waldnutzung
FSC® C083411

Kolibri 360 GmbH
Ferdinand-Jühlke-Straße 7
99095 Erfurt, Deutschland
produktsicherheit@kolibri360.de